Environmental Engineering:
Fundamentals, Sustainability, Design

Second Edition

Authors and Editors

James R. Mihelcic
University of South Florida

Julie Beth Zimmerman
Yale University

Contributing Authors

Martin T. Auer
Michigan Technological University

David W. Hand
Michigan Technological University

Richard E. Honrath, Jr.
Michigan Technological University

Mark W. Milke
University of Canterbury

Michael E. Penn
University of Wisconsin-Platteville

Amy L. Stuart
University of South Florida

Noel R. Urban
Michigan Technological University

Brian E. Whitman
Wilkes University

Qiong Zhang
University of South Florida

About the Cover

Richard Buckminster Fuller (1895-1983) was an engineer, architect, poet, and designer. During his life, he pondered the question, "Does humanity have a chance to survive lastingly and successfully on planet Earth, and if so, how?" To begin to answer this question, Fuller ascribed to the "Spaceship Earth" worldview that expresses concern over the use of limited global resources and the behavior of everyone on it to act as a harmonious crew working toward the greater good.

In 1969 Fuller wrote and published a book entitled "Operating Manual for Spaceship Earth." The following quotation from this book reflects his worldview: "Fossil fuels can make all of humanity successful through science's world-engulfing industrial evolution provided that we are not so foolish as to continue to exhaust in a split second of astronomical history the orderly energy savings of billions of years' energy conservation aboard our Spaceship Earth. These energy savings have been put into our Spaceship's life-regeneration-guaranteeing bank account for use only in self-starter functions." To further communicate his ideas, Fuller developed the Dymaxion Map, shown on the cover. This map is a projection of a World map onto the surface of a polyhedron. The projection can be unfolded in many different ways and flattened out to form a two-dimensional map that retains the look and integrity of a globe map. Importantly, the Dymaxion map has no "right way up." Fuller believed that in the universe there was no "up" and "down" or "north" and "south": only "in" and "out." He linked the north-upsuperior/south-down-inferior presentation of most other world maps to cultural bias.

VP & EXECUTIVE PUBLISHER	Don Fowley
EXECUTIVE EDITOR	Linda Ratts
EDITORIAL ASSISTANT	Hope Ellis
MARKETING MANAGER	Christopher Ruel
COVER DESIGN	Kenji Ngieng
PHOTO EDITOR	Mary Ann Price
ASSOCIATE PRODUCTION MANAGER	Joyce Poh

Cover Credit: The Fuller Projection Map design is a trademark of the Buckminster Fuller Institute. © 1938, 1967 & 1992. All rights reserved, www.bfi.org.

This book was set in Palatino by Thomson Digital and printed and bound by Quad/Graphics. The cover was printed by Quad/Graphics.

This book is printed on acid free paper.

Founded in 1807, John Wiley & Sons, Inc. has been a valued source of knowledge and understanding for more than 200 years, helping people around the world meet their needs and fulfill their aspirations. Our company is built on a foundation of principles that include responsibility to the communities we serve and where we live and work. In 2008, we launched a Corporate Citizenship Initiative, a global effort to address the environmental, social, economic, and ethical challenges we face in our business. Among the issues we are addressing are carbon impact, paper specifications and procurement, ethical conduct within our business and among our vendors, and community and charitable support. For more information, please visit our website: www.wiley.com/go/citizenship.

ISBN 978-1-118-74149-8

Printed in the United States of America

V10003266_080518

Preface

Now more than ever, there has been an increasing awareness of the unsustainability trajectory our society is currently following. Occurring simultaneously, there have been numerous proclamations, international meetings, and efforts to assess the current state of affairs and begin to design new technologies, policies, and business models aimed at advancing the goal of a sustainable future. With this in mind, there is an obvious need to continue to manage and remediate legacy environmental challenges from nutrient enrichment of surface waters to groundwater contamination. At the heart of meeting these objectives is training the next generation of engineers, and environmental engineers in particular, to have a deep understanding of the fundamentals of the discipline while also possessing a keen awareness of sustainability. Reorienting the focus environmental engineering is the very motivation for this book – providing both the fundamental training to solve environmental problems as well as the broad understanding of sustainability.

As we move from the stark and egregious environmental problems that gave rise to the field of environmental engineering more than five decades ago to the more complex and global challenges of today, the field of environmental engineering must evolve as well. Looking to the future, there is a clear need for environmental engineers who are able to collaborate across disciplines and communicate broadly to the scientific community, policymakers, and the public. Sustainability presents many opportunities for environmental engineers to evolve from those who characterize, manage, and remediate existing environmental problems to those who are designing and developing new technologies that address sustainability challenges while avoiding unintended consequences. On this journey, it is imperative to honor the great legacy of this discipline – the creativity, passion, and dedication for public good – and continue to serve in the unique role of benefitting people and the planet as we move to face emerging challenges and design a more sustainable future.

This book is motivated by the discussion that is evolving from one centered on describing, characterizing, quantifying, and monitoring current environmental problems to one that is focused on the design and development of innovative new solutions. Innovation requires enhanced skills and tools beyond the fundamental and important environmental engineering curriculum including the ability to think creatively and critically, to work in interdisciplinary teams, and to consider the entire system. As shown in the table below, the very nature of the challenges faced by environmental engineers is changing.

20th Century Environmental Issues	21st Century Environmental Issues
Local	Global
Acute	Chronic
Obvious	Subtle
Immediate	Multigenerational
Discrete	Complex

This shift in focus provides students an opportunity to succeed in engineering practice and actively engage in contributing to a more sustainable future using the knowledge and foundational skills of the environmental engineering discipline. After all, the only reason to study a problem in great detail is to inform its solution and the environmental engineering profession is in a unique and prime position to advance those solutions – and ensure that they are themselves sustainable. That is, having the awareness to ensure that the solutions to sustainability challenges are carefully considered to avoid or minimize the likelihood of legacy problems and unintended consequences. In this way, it is imperative that the idea of sustainability is fully integrated into the fundamental training of environmental engineers, not an afterthought or separate from the very nature of what we do as a profession.

The evolution of the problems themselves and the level of understanding we have about these problems will require engineers to take on new skills, capabilities, and perspective about how we approach our work. It is not that the skills previously learned are antiquated and need to be replaced. Rather, it is that the traditional skills need to be augmented, complemented, and enhanced with new knowledge, new perspectives, and new awareness. The melding of the old and new fundamentals and design skills is the purpose of this text. It is our hope that this text provides engineers with the knowledge and confidence to address 21st century challenges as well as they dealt with the daunting challenges of the 20th century.

Hallmark Features

CHANGES TO THIS SECOND EDITION

In the 2nd edition several key updates were made to the structure and content of this textbook.

- The book is still based on applying foundational principles related to physics, chemistry, biology, risk, mass balances, and sustainability which are applied to the design and operation of technology and strategies used to manage and mitigate environmental problems found in land, water, and air.

- There is continued emphasis on problems important to the United States and the world, with a focus on pollution prevention and resource recovery while still providing information to design treatment processes.

- Chapter 1 was rewritten and is now titled "Sustainable Design, Engineering, and Innovation." It de-emphasizes problems that are driving engineering practice, and instead focuses on the paradigm shift from managing environmental problems with regulations to a framework of sustainability using EPA's Green Book and Path Forward. The Chapter on Air Resources Engineering (Chapter 11) was totally rewritten and now includes discussion and application of Gaussian Plume Models and emphasis of demand management strategies along with traditional air pollution control technologies.

- The text has been reduced from 14 to 11 total chapters which we believe will assist instructors that use the book in a semester course and the text has been aligned with the National Academy of Engineering's focus on Grand Challenges related to managing carbon and nitrogen. With the more pronounced emphasis on innovation and sustainability in the 2nd edition, there are enhancements towards a deeper integration of systems thinking throughout the text and problems. One notable example of this is the recrafting the chapters related to water which now appear as one chapter focused "Water: Quantity and Quality" (Chapter 7) and a second focused on "Wastewater and Stormwater: Collection, Treatment, Resource Recovery" (Chapter 9). In this way, water is considered holistically as a resource including a discussion of water reuse.

- We added several topics brought to the authors' attention by users of the text, e.g., a section on calculating a carbon footprint in Chapter 2 (Environmental Measurements), enhanced section on energy balances in Chapter 4 (Physical Processes), better definition of a watershed and the addition of the Rational Method that is integrated with examples of how land use impacts water quality in Chapter 7 (Water: Quantity and Quality), integration of methods that emphasize resource recovery associated with management of wastewater (Chapter 9), and a section in Chapter 11 (Air Quality Engineering) that emphasizes the use of demand management as solution to air pollution problems. Given the critical need to ensure that sustainability and interdisciplinarity are integral to the training of environmental engineers, the stand-alone chapters "Green Engineering" and "The Built Environment" from the first edition were eliminated, and instead, the relevant content was integrated into other chapters.

- Several educational modules (in powerpoint and video format) to assist an instructor in integration of sustainability and other important environmental engineering topics have been developed and are available as instructor support materials (see below). There is also an increased emphasis on practical field orientated applications of engineering practice and a fifty percent increase in end of chapter problems, for a total of 445. In addition, the solutions manual has been carefully reviewed and updated.

A FOCUS ON SUSTAINABLE DESIGN

Perhaps one of the most important aspects of the textbook is that it will focus the student on the elements of *design*. Design of products, processes, and systems will be essential not only in responding to the environmental issues in ways that our profession has done historically but also in informing the design of new products, processes, and systems to reduce or eliminate problems from occurring in the first place.

To use the tools of green engineering design truly to design for sustainability, students need a command of the framework for this design. The framework perhaps can be summarized in the *four I's*: (1) Inherency, (2) Integration, (3) Interdisciplinary, and (4) International.

Inherency As a reader proceeds through the text; it will become obvious that we are not merely looking at how to change the conditions or circumstances that make a product, process, or system a problem. Readers will understand the *inherent* nature of the material and energy inputs and outputs so that they are able to understand the fundamental basis of the hazard and the root causes of the adverse consequence they seek to address. Only through this inherency approach can we begin to design for sustainability rather than generating elegant technological bandages for flawed conceptions.

Integration Our historical approaches toward many environmental issues have been fragmented—often by media, life cycle, culture, or geographic region. Understanding that energy is inextricably linked to water, water to climate change, climate change to food production, food production to health care, health care to societal development, and so on will be essential in the new paradigm of sustainable design. It is equally necessary to understand that we cannot think about approaching any environmental problem without looking at the problem across all elements of its life cycle. There have been countless attempts to improve environmental circumstances that have resulted in unintended problems that have often been worse than the problem they intended to fix. Attempts to increase drinking water supply in Bangladesh resulted in widespread arsenic poisoning. Attempts to increase crop yields through the production of pesticides in Bhopal, India, resulted in one of the greatest chemical tragedies of our time. Understanding the complex interconnections and ensuring the *integration* of multiple factors in the development of solutions is something that 21st century environmental engineering requires.

Interdisciplinary To achieve the goals of sustainable design, environmental engineers will be working increasingly with a wide array of other disciplines. Technical disciplines of chemistry and biology and other engineering disciplines will be essential but so will the disciplines of economics, systems analysis, health, sociology, and anthropology. This text seeks to introduce the *interdisciplinary* dimensions that will be important to the successful environmental engineer in this century.

International Many well-intentioned engineering solutions fail by not considering the very different context found in the diversity of nations around the world. Although water purification or municipal waste may seem like they can be dealt with through identical processes anywhere in the world, it has been shown repeatedly that the local factors—geographic, climatic, cultural, socioeconomic, political, ethnic, and historical—can all play a role in the success or failure of an environmental engineering solution. The *international* perspective is an important one this textbook emphasizes and incorporates into the fundamentals of the training of environmental engineers.

MATERIAL AND ENERGY BALANCES AND LIFE CYCLE THINKING

The book provides a rigorous development of energy and mass balance concepts with numerous easy-to-follow example problems. It then applies mass and energy balance concepts to a wide range of natural and engineered systems and different environmental media. The book has appropriate coverage of life cycle assessment and provides a life cycle–thinking approach in discussion throughout other chapters.

PEDAGOGY AND ASSESSMENT

Beyond including the elements mentioned previously to prepare engineers for the 21st century, this book also incorporates changes in pedagogy and assessment that provide structure for delivering this new information in a meaningful education experience.

Fink's Taxonomy of Significant Learning One such element is the use of Fink's taxonomy of significant learning in guiding the development of learning objectives for each chapter as well as in example and homework problems. Fink's taxonomy recognizes six domains beyond traditional foundational knowledge, including: foundational knowledge; application of knowledge; integration of knowledge; human dimensions of learning and caring; and learning how to learn. Without much background on the taxonomy, it is clear from these knowledge domain headings alone that these areas recognized by Fink are critical to an engineer tasked with designing solutions to many of today's sustainability challenges.

Important Equations Boxes around important equations indicate for students which are most critical.

Learning Exercises Learning exercises at the end of each chapter include 445 problems that not only ask students to solve traditional numerical problems of assessment and design but also challenge students to research problems and innovate solutions at different levels: campus, apartment, home, city, region, state, or world.

Discussion Topics To further emphasize the importance of the domains of knowledge discussed in the previous paragraph, the book encourages classroom discussions and interaction between students as well as between the students and the instructor. These discussion topics are noted by a symbol in the margin.

Online Resources for Further Learning Online resources for further learning and exploration are listed in margins where appropriate. These resources provide students the opportunity to explore topics in much greater detail and learn of geographical commonalities and uniqueness to specific environmental engineering issues. More important, use of these online resources prepares students better for professional practice by expanding their knowledge of information available at government and nongovernment Web sites.

BOOK WEB SITE

Additional resources for students and instructors are available on the book Web site, located at www.wiley.com/college/mihelcic.

Classroom Materials for Instructors Through an NSF Course, Curriculum, and Laboratory Improvement grant awarded to three of this book's authors (Qiong Zhang, Julie Beth Zimmerman, and James Mihelcic) and to Linda Vanasupa (California Polytechnic State University), we have developed in-depth educational materials (learning objectives, editable slide presentations, assessments, activities) on the following six topics:

1. Systems Thinking
2. (Introduction to) Sustainability
3. Systems Thinking: Population
4. Systems Thinking: Energy
5. Systems Thinking: Material
6. Systems Thinking: Water

All materials are available at the following stable link for download: http://works.bepress.com/lvanasup/

Each set provides an array of classroom materials whose design aligns with educational research on how to foster more significant learning and includes:

- Learning objectives within several critical areas of learning (foundational knowledge, application of knowledge, integration of knowledge, human dimensions of learning and caring, and learning how to learn)

- A set of editable and notated slides for faculty to present lecture material

- Active learning exercises that range from two-minute to three-hour investments; notated guides for faculty using the exercises

- A set of assessment activities that includes learning objectives, criteria for assessment, and standards for judging the criteria

In addition, Linda Vanasupa and Qiong Zhang developed 24 video tutorials related to this material that are published at Open Education Resource (OER) Commons under "The Sustainability Learning Suites." These 24 videos are organized around the themes: systems thinking; sustainable development; energy; water; population; and materials.

http://www.oercommons.org/authoring/1660-the-sustainability-learning-suites/view

These materials have also been submitted for publication at: National Science Digital Library (Nsdl.org).

ADDITIONAL RESOURCES FOR INSTRUCTORS

Additional resources for instructors to support this text include:

- Updated Solutions Manual containing solutions for all 445 end-of-chapter problems in the text.

- Image Gallery with illustrations from the text appropriate for use in lecture slides.

These resources are available only to instructors who adopt the text. Please visit the instructor section of the Web site at www.wiley.com/college/mihelcic to register for a password.

Genesis of the Book

In 1999, we published a book titled *Fundamentals of Environmental Engineering* (John Wiley & Sons). One strength of *Fundamentals of Environmental Engineering* is that it provides in-depth coverage of the basic environmental engineering fundamentals required for design, operation, analysis, and modeling of both natural and engineered systems. The book you are reading now, *Environmental Engineering: Fundamentals, Sustainability, Design*, not only includes updated chapters on those same fundamentals with continued strong emphasis on material and energy balances and inclusion of issues of energy, nutrient management, and carbon, but also includes application of those fundamental skills to design and operate strategies to implement source reduction, resource recovery, and treatment.

Acknowledgements

As we marvel and appreciate all those who have dedicated themselves to leaving the world a better place than they found it—environmental engineers and others—we are grateful for all the talented people who have helped make this book possible and are poised to change the very nature of the field of environmental engineering.

Besides all the individuals who contributed content to the book, the following faculty provided high-quality review and insight through development of the first edition:

Zuhdi Aljobeh, *Valparaiso University*

Robert W. Fuessle, *Bradley University*

Keri Hornbuckle, *University of Iowa*

Benjamin S. Magbanua Jr., *Mississippi State University*

Taha F. Marhaba, *New Jersey Institute of Technology*

William F. McTernan, *Oklahoma State University*

Gbekeloluwa B. Oguntimein, *Morgan State University*

Joseph Reichenberger, *Loyola Marymount University*

Sukalyan Sengupta, *University of Massachusetts*

Thomas Soerens, *University of Arkansas*

Linda Vanasupa (California Polytechnic State University) reviewed the first edition chapters and assisted in developing learning objectives in the context of Fink's taxonomy of significant learning. Linda Phillips (University of South Florida) provided her international perspective, especially regarding integrating service learning with practitioner involvement. The editorial team of Linda Ratts, Hope Ellis, Joyce Poh and Jenny Welter from the first edition have also been a key to success. Their early vision of the book's purpose and attention and contributions to detail, style, and pedagogy have made this a fulfilling and equal partnership.

The following students at the University of South Florida reviewed every chapter of the first edition and provided valuable comments during the editing process: Jonathan Blanchard, Justin Meeks, Colleen Naughton, Kevin Orner, Duncan Peabody, and Steven Worrell. Ezekiel Fugate and Jennifer Ace (Yale University) and Helen E. Muga (University of South Florida) helped us obtain permissions and search for materials in the first edition. We are especially grateful to Colleen Naughton (University of South Florida), Ziad Katirji (Michigan Technological University), and Heather E. Wright Wendel (University of South Florida), who assisted efforts to create, assemble, and proof the *Solutions Manual*. Colleen was responsible for development of the solutions manual for the 2nd Edition.

Finally, thanks to Karen, Paul, Kennedy, Aquinnah, and Mac for embracing the vision of this project over the past several years.

James R. Mihelcic

Julie Beth Zimmerman

About the Authors

James R. Mihelcic is a professor of civil and environmental engineering and a State of Florida 21st Century World Class Scholar at the University of South Florida. He is founder of the Peace Corps Master's International Program in Civil and Environmental Engineering (http://cee.eng.usf.edu/peacecorps) which allows students to combine their graduate studies with international service and research in the Peace Corps as water/sanitation engineers. He is also director of the U.S. EPA National Research Center for Reinventing Aging Infrastructure for Nutrient Management (*RAINmgt*). His teaching and research interests are centered around engineering and sustainability, specifically understanding how global stressors such as climate, land use, and urbanization influence water resources, water quality, water reuse, and selection and provision of water supply and sanitation technologies. Dr. Mihelcic is also an international expert in provision of water, sanitation, and hygiene developing world communities. Dr. Mihelcic is a member of the Environmental Protection Agency's Chartered and Environmental Engineering Science Advisory Boards. He is past president of the Association of Environmental Engineering and Science Professors (AEESP), a Board Certified Environmental Engineering Member, and Board Trustee with the American Academy of Environmental Engineers & Scientists (AAEES). He is lead author for two other textbooks: *Fundamentals of Environmental Engineering* (John Wiley & Sons, 1999) (translated into Spanish) and *Field Guide in Environmental Engineering for Development Workers: Water, Sanitation, Indoor Air* (ASCE Press, 2009).

Dr. Julie Beth Zimmerman is the Donna L. Dubinsky Associate Professor of Environmental Engineering, jointly appointed to the Department of Chemical Engineering, Environmental Engineering Program, and the School of Forestry and Environment. She is also the Sustainability and Innovation Coordinator for the U.S. EPA National Research Center for Reinventing Aging Infrastructure for Nutrient Management (*RAINmgt*). Her research interests broadly focus on green chemistry and engineering with specific emphasis on green downstream processing and life cycle assessment of algal biomass for fuels and value-added chemicals as well as novel biobased sorbents for purification of drinking water and remediation of industrial wastewater. Other ongoing focus areas include the design of safer chemicals from first principles and the implications of nanomaterials on human health and the environment. Further, to enhance the likelihood of successful implementation of these next generation designs, Dr. Zimmerman studies the effectiveness and impediments of current and potential policies developed to advance sustainability. Together, these efforts represent a systematic and holistic approach to addressing the challenges of sustainability to enhance water and resource quality and quantity, to improve environmental protection, and to provide for a higher quality of life. Dr. Zimmerman previously served as an Engineer and program coordinator in the Office of Research and Development at the United States Environmental Protection Agency where she managed sustainability research grants and created EPA's P3 (People, Prosperity, and the Planet) Award program.

Martin T. Auer is a professor of civil and environmental engineering at Michigan Technological University. He teaches introductory courses in environmental engineering and advanced coursework in surface water–quality engineering and mathematical modeling of lakes, reservoirs, and rivers. Dr. Auer's research interests involve field and laboratory studies and mathematical modeling of water quality in lakes and rivers.

David W. Hand is a professor and chair of civil and environmental engineering at Michigan Technological University. He teaches senior-level and graduate courses in drinking water treatment, wastewater treatment, and physical chemical processes in environmental engineering. Dr. Hand's research interests include physical-chemical treatment processes, mass transfer, adsorption, air stripping, homogeneous and heterogeneous advanced oxidation processes, process modeling of water treatment and wastewater treatment processes, and development of engineering software design tools for pollution prevention practice.

Richard E. Honrath was a professor of geological and mining engineering and sciences and of civil and environmental engineering at Michigan Technological University, where he also directed the Atmospheric Sciences graduate program. He taught courses in introductory environmental engineering, advanced air quality engineering and science, and atmospheric chemistry. His research activities involved studies of the large-scale impacts of air pollutant emissions from anthropogenic sources and from wildfires, with a focus on the interaction between transport processes and chemical processing. He also studied photochemistry in ice and snow, including field studies of the interactions among snow, air, and sunlight.

Mark W. Milke is an associate professor/reader at the Department of Civil and Natural Resources Engineering, University of Canterbury, New Zealand, where he has worked since 1991. His research and teaching interests are in solid-waste management, groundwater, and uncertainty analysis. He is a chartered professional engineer in New Zealand.

Michael R. Penn is a professor of civil and environmental engineering at the University of Wisconsin-Platteville. He teaches undergraduate courses in introductory environmental engineering, fluid mechanics, stormwater management, wastewater and drinking water treatment, and solid and hazardous waste management. Dr. Penn's research interests focus on involving undergraduates in studies of agricultural runoff, nutrient cycling in lakes, and urban infrastructure management. Dr. Penn is lead author of the Wiley textbook, *Introduction to Infrastructure: An Introduction to Civil and Environmental Engineering*, intended for first- and second-year undergraduates.

Amy L. Stuart is an associate professor at the University of South Florida, with appointments in the environmental health and environmental engineering programs. She teaches courses on air pollution, numerical methods, environmental modeling, sustainability, and a multi-disciplinary environmental seminar. Dr. Stuart's research is centered on understanding and management of air pollution through development and application of computation models, field measurements, and laboratory chemical analyses. Dr. Stuart is a recipient of a National Science Foundation CAREER grant award in environmental sustainability for work on sustainable urban design to reduce air pollution exposures, resultant health effects, and environmental inequality.

Noel R. Urban is a professor of civil and environmental engineering at Michigan Technological University. His teaching interests focus on environmental chemistry and surface water–quality modeling. His research interests include environmental cycles of major and trace elements, sediment diagenesis and stratigraphy, chemistry of natural organic matter, wetland biogeochemistry, environmental impact and fate of pollutants, influence of organisms on the chemical environment, and the role of the chemical environment in controlling populations.

Brian E. Whitman is an associate professor of environmental engineering at Wilkes University. He teaches courses in water distribution and wastewater collection system design, hydrology, water resources engineering, and water and wastewater treatment process design. Dr. Whitman's research interests include hydraulic modeling of water distribution and wastewater collection systems, environmental microbiology, bioengineering, and reclamation of industrial fly ash. He is the recipient of two Wilkes University Outstanding Faculty Awards and has co-authored three books in the areas of water distribution and wastewater collection system modeling and design.

Qiong Zhang is an assistant professor of civil and environmental engineering at the University of South Florida. She was previously the Operations Manager of the Sustainable Futures Institute at Michigan Technological University and is a research director of the U.S. EPA National Research Center for Reinventing Aging Infrastructure for Nutrient Management (*RAINmgt*). Her teaching interests are in green engineering, water treatment, and environmental assessment for sustainability. Dr. Zhang's research interests lie at the water-energy nexus, process and system modeling, green engineering, and integration of sustainability into engineering curriculum. Her research focuses on exploring and simulating the dynamic interactions between water and energy systems, quantifying the environmental implications of energy systems and energy implications of water and wastewater systems, and seeking technical and nontechnical solutions for integrated water-energy management.

Brief Table of Contents

Detailed Table of Contents

Chapter Six Environmental Risk 246

Chapter Seven Water: Quantity and Quality 296

chapter/One Sustainable Design, Engineering, and Innovation

Julie Beth Zimmerman and
James R. Mihelcic

This chapter discusses the evolution of protecting human health and the environment from regulatory approaches to sustainable development, highlighting critical opportunities for engineers to design appropriate, resilient solutions. Definitions for sustainable development and design are presented. Several emerging topics are presented—green chemistry, biomimicry, green engineering, life cycle thinking, and systems thinking—offering enhancements to engineering fundamentals leading to rigorous and sustainable design solutions.

Learning Objectives

1. Describe the evolution of the protection of human health and the environment from regulatory approaches to sustainability.
2. Relate *The Limits to Growth*, "The Tragedy of the Commons," and the definition of carrying capacity to sustainable development.
3. Define sustainability, sustainable development, and sustainable engineering in your own words and according to others.
4. Redefine engineering problems in a balanced social, economic, and environmental context.
5. Apply life cycle thinking and systems thinking to problem definition and the design and assessment of proposed solutions.
6. Differentiate between traditional indicators and sustainability indicators that measure progress toward achieving the goal of sustainability.
7. Describe several frameworks for sustainable design and understand the importance of design and innovation in advancing sustainability.
8. Discuss the role of regulations and other policy tools, such as voluntary programs, in advancing environmental and human health protection as well as sustainability.

© Ziutograf/iStockphoto

1.1 Background: Evolution from Environmental Protection to Sustainability

In 1962, Rachel Carson (Application 1.1) published *Silent Spring*, establishing the case that there may be reason to be concerned about the impacts of pesticides and environmental pollution on natural systems and human health. Though as early as 1948, there was an industrial air pollution smog release in the milltown of Donora (Pennsylvania) that killed 20 and injured thousands, it was later, in the late 1960s and early 1970s, that numerous clear and startling visual realities of human impacts on the environment took place. This included smog episodes in Los Angeles that obscured visibility, the Cuyahoga River (Ohio) catching on fire in 1969, and the toxic waste and subsequent health effects in neighborhoods such as Love Canal in Niagara Falls, New York.

Through a shared societal value and a growing environmental social movement, the **Environmental Protection Agency (EPA)** was created in 1972. This consolidated in one agency a variety of federal research, monitoring, standard-setting, and enforcement activities with the mission of "protecting human health and the environment." During this same time, Congress passed many of the fundamental and critical environmental regulations, such as the National Environmental Protection Act (NEPA), the Clean Air Act, the Water Pollution Control Act, Wilderness Protection Act, and the Endangered Species Act.

The Environmental Protection Agency (EPA) is an agency of the U.S. federal government that was created for the purpose of protecting human health and the environment by writing and enforcing regulations based on laws passed by Congress (Application 1.2). Its

---

| Application /1.1 | Rachel Carson and the Modern Environmental Movement |

Rachel Carson at Hawk Mountain, Pennsylvania photograph taken ca. 1945 by Shirley Briggs. (Provided courtesy of the Linda Lear Center for Special Collections and Archives, Connecticut College).

Rachel Carson is considered one of the leaders of the modern environmental movement. She was born 15 miles northeast of Pittsburgh in the year 1907. Educated at the undergraduate and graduate levels in science and zoology, she first worked for the government agency that eventually became the U.S. Fish and Wildlife Service. As a scientist, she excelled at communicating complex scientific concepts to the public through clear and accurate writing. She wrote several books, including *The Sea Around Us* (first published in 1951) and *Silent Spring* (first published in 1962).

Silent Spring was a commercial success soon after its publication. It visually captured the fact that songbirds were facing reproductive failure and early death because of manufacturing and prolific use of chemicals such as DDT that had bioaccumulated in their small bodies. Some historians believe that *Silent Spring* was the initial catalyst that led to the creation of the modern environmental movement in the United States along with the U.S. Environmental Protection Agency (EPA).

The EPA has many tools to protect human health and the environment, including partnerships, educational programs, and grants. However, the most significant tool is writing **regulations**, which are mandatory requirements that can be relevant to individuals, businesses, state or local governments, nonprofit organizations, or others.

The **regulatory process** begins with Congress passing a law and then authorizing the EPA to help put that law into effect by creating and enforcing regulations. Of course, there are many checks and balances along the path from law to regulation, including public disclosure of intent to write or modify a regulation, and a public comment period where those potentially affected by the regulation have an opportunity to offer input to the process.

Draft and final federal regulations are published in the **Code of Federal Regulations** (CFR). The number 40 that is associated with environmental regulations (i.e., 40CFR) indicates the section of the CFR related to the environment.

SOURCE: http://www.epa.gov/lawsregs/basics.html

administrator, who is appointed by the president and approved by Congress, leads the agency.

The EPA has its headquarters in Washington, D.C., regional offices for each of the agency's 10 regions (Figure 1.1) and 27 research laboratories. EPA is organized into a number of central program offices as well as regional offices and laboratories, each with its own

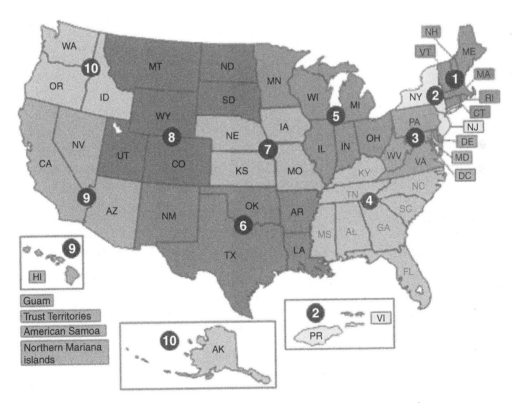

Figure / 1.1   **The EPA's Ten Regions**   Each region has its own regional administrator and other critical functions for carrying out the mission of protecting human health and the environment. EPA headquarters are located in Washington, D.C.

(Adapted from EPA).

regulatory, research, and/or enforcement mandate. The agency conducts environmental assessment, research, and education. It has the responsibility of maintaining and enforcing national standards under a variety of environmental laws, in consultation with state, tribal, and local governments. It delegates some permitting, monitoring, and enforcement responsibility to U.S. states and Native American tribes. EPA enforcement powers include fines, sanctions, and other measures. The agency also works with industries and all levels of government in a wide variety of voluntary pollution prevention programs and energy conservation efforts.

The mission of EPA is to protect human health and the environment. EPA's purpose is to ensure that:

- all Americans are protected from significant risks to human health and the environment where they live, learn, and work;

- national efforts to reduce environmental risk are based on the best available scientific information;

- federal laws protecting human health and the environment are enforced fairly and effectively;

- environmental protection is an integral consideration in U.S. policies concerning natural resources, human health, economic growth, energy, transportation, agriculture, industry, and international trade, and these factors are similarly considered in establishing environmental policy;

- all parts of society—communities, individuals, businesses, and state, local, and tribal governments—have access to accurate information sufficient to effectively participate in managing human health and environmental risks;

- environmental protection contributes to making our communities and ecosystems diverse, sustainable, and economically productive;

- the United States plays a leadership role in working with other nations to protect the global environment.

The Regulatory Process
http://www.epa.gov/lawsregs/regulations/index.html

Access the Code of Federal Regulations
http://www.gpoaccess.gov/cfr/

EPA works closely with the states to implement federal environmental programs. States authorized to manage federal programs must have enforcement authorities that are at least as stringent as federal law. EPA works with officials in state environmental, health, and agricultural agencies on strategic planning, priority-setting, and measurement of results.

While we have made tremendous strides in addressing the most egregious environmental insults and maintained a growing economy, the environmental challenges of today are more complex and subtle than encountered at the start of the modern environmental movement. For example, there are clear connections between emissions to air, land, and water even if the regulations were not written and the EPA was not organized with these considerations.

Furthermore, air and water emissions come from many distributed sources (referred to as **nonpoint source emissions**), so it is much more difficult to identify a specific source that can be regulated and monitored. We also have a much higher level of understanding of the linkages among society, the economy, and the environment. These are recognized as the three **pillars of sustainability** and require that we consider them simultaneously, looking for synergies to achieve mutual benefits. That is, we must create and maintain a prosperous society with high quality of life without the negative impacts that have historically harmed our environment and communities in the name of development. And all of this must be performed while maintaining a sufficient stock of natural resources for current and future generations to maintain an increasing population with an improving quality of life.

Global Environmental Outlook
http://www.unep.org/GEO

Class Discussion
Is it better to live within a determined limit by accepting some restrictions on consumption-fueled growth?

Application/1.3   Tragedy of the Commons

The **Tragedy of the Commons** describes the relationship where individuals or organizations consume shared resources (e.g., air, freshwater; fish from the ocean) and then return their wastes back into the shared resource (e.g., air, land). In this way, the individual or organization receives all of the benefit of the shared resource but distributes the cost across anyone who also uses that resource. The tragedy arises when each individual or organization fails to recognize that every individual and organization is acting in the same way. It is this logic that has led to the current situation in ocean fisheries, the Amazon rain forest, and global climate change. In each case, the consumptive behavior of a few has led to a significant impact on the many and the destruction of the integrity of the shared resource.

Application/1.4   *The Limits to Growth* and Carrying Capacity

The Limits to Growth, published in 1972, warned of the limitations of the world's resources and pointed out there might not be enough resources remaining for the developing world to industrialize. The authors, using mathematical models, argued that "the basic behavior mode of the world system is exponential growth of population and capital, followed by collapse" in a phenomenon known as "carrying capacity." (see Figure 1.2)

Carrying capacity (discussed more in Chapter 5) is a way to think of resource limitations. It refers to the upper limit to population or community size (e.g., biomass) imposed through environmental resistance. In nature, this resistance is related to the availability of renewable resources, such as food, and nonrenewable resources, such as space, as they affect biomass through reproduction, growth, and survival. One solution is to use technological advances to increase the amount of prosperity per unit of resources. Of course, there is a risk that maintaining growth in a limited system by advances in technology can lead to overuse of finite resources—efficiency alone is not an effective indicator of sustainability.

Figure 1.3 provides a timeline of the progression from the start of the domestic environmental movement in the 1960s through the progression to recent major international sustainability activities. Based on the events on the timeline, there is a clear progression from initial regulatory responses to egregious environmental assaults to a more proactive, systematic international dialogue about a broad sustainability agenda

The Story of Stuff
http://www.storyofstuff.com

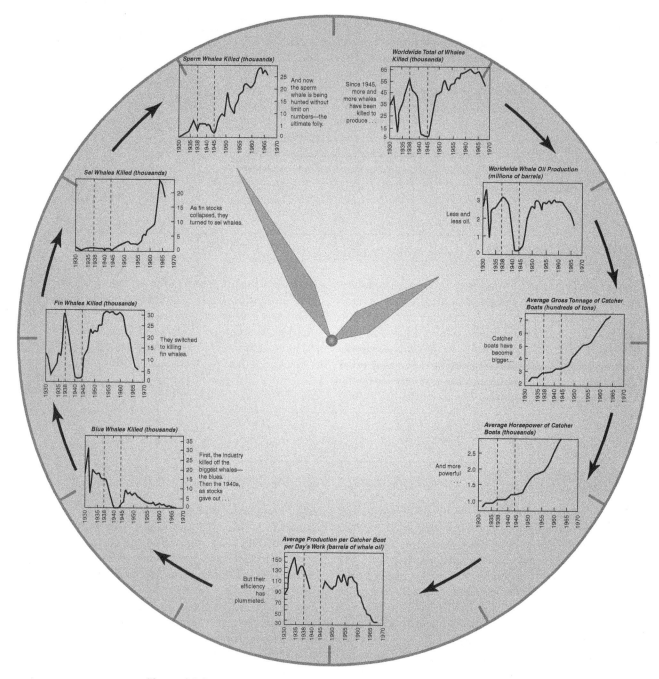

Figure / 1.2 **Limits to Growth and Technology of the Whaling Industry**   Maintaining growth in a limited system by advances in technology will eventually result in extinction for both whales and the whaling industry. As wild pods of whales are destroyed, finding the survivors has become more difficult and has required more effort. As larger whales are killed off, smaller species are exploited to keep the industry alive. Without species limits, large whales are always taken wherever and whenever encountered. Thus, small whales subsidize the extermination of large ones.

(Based on Payne, R. 1968. "Among Wild Whales." *New York Zoological Society Newsletter* (November)).

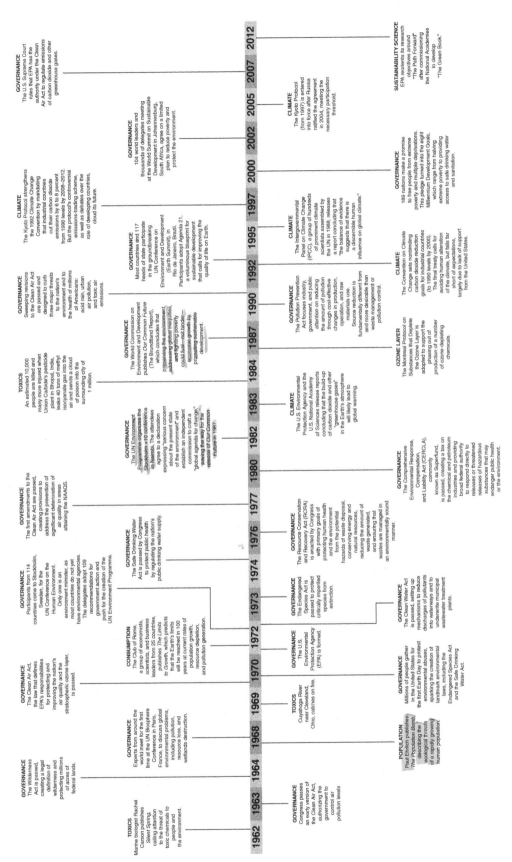

Figure / 1.3 Timeline of critical events leading from a mission of environmental protection to a goal of sustainability.

(Events adopted from www.worldwatch.org).

If you Google the words *sustainability, sustainable development*, and *sustainable engineering*, you will get hundreds of definitions. Try it! The abundance of varying definitions has made it difficult to realize consensus on what sustainability is. However, nearly all of the definitions of sustainability refer to integrating the three elements of the **triple bottom line** (environment, economy, society). Most definitions also extend sustainability criteria to include the aim of meeting the needs of current and future generations.

Sustainability is defined by Merriam-Webster as follows: (1) of, relating to, or being a method of harvesting or using a resource so that the resource is not depleted or permanently damaged and (2) of or relating to a lifestyle involving the use of sustainable methods.

Sustainable development is defined by the Brundtland Commission as "development which meets the needs of the present without compromising the ability of the future to meet its needs."

Sustainable engineering is defined as the design of human and industrial systems to ensure that humankind's use of natural resources and cycles do not lead to diminished quality of life due either to losses in future economic opportunities or to adverse impacts on social conditions, human health, and the environment (Mihelcic et al., 2003).

(Application 1.5). In 1986, the UN World Commission on Environment and Development released *Our Common Future*. This book is also referred to as the **Brundtland Commission** report, because Ms. Gro Brundtland, the former prime minister of Norway, chaired the commission. The Brundtland Commission report defined **sustainable development** as "development which meets the needs of the present without compromising the ability of the future to meet its needs."

This report helped to prompt the 1992 UN Conference on Environment and Development, known as the Earth Summit, held in Rio de Janeiro, Brazil. The conference, the first global conference to specifically address the environment, led to the nonbinding agenda for the 21st century, *Agenda 21*, which set forth goals and recommendations related to environmental, economic, and social issues. In addition, the UN Commission on Sustainable Development was created to oversee the implementation of Agenda 21.

At the 2002 World Summit on Sustainable Development in Johannesburg, South Africa, world leaders reaffirmed the principles of sustainable development adopted at the Earth Summit 10 years earlier. They also adopted the **Millennium Development Goals** (MDGs), listed in Table 1.1. The eight MDGs represent an ambitious agenda for a better world that can guide engineering innovation and practice. This is a good example of the link between policy and engineering: policy can drive engineering innovation, and new engineering advancements can encourage the development of new policies with advanced standards that redefine "best available technologies."

Class Discussion

In which of the MDGs do engineers have a role to play? Are these traditional or emerging roles for engineers to play in society and practice?

Millennium Development Goals

You can go to www.un.org/millenniumgoals/ Go to this URL to learn more about progress toward meeting the MDGs.

1.2   The Path Forward: Operationalizing Sustainability

Given the many definitions of sustainability (refer back to Application 1.5) and the complexity of a systems perspective to include the linkages and feedback between the environment, economy, and society, there are ongoing efforts to move from discussions to operationally

Millennium Development Goals (MDGs)   MDGs are an ambitious agenda embraced by the world community for reducing poverty and improving lives of the global community. Learn more at www.un.org/millenniumgoals/.

Millennium Development Goal	Background	Example Target(s) (of 21 total targets)
1. Eradicate extreme poverty and hunger.	More than a billion people still live on less than $1 a day.	(1a) Halve the proportion of people living on less than $1 a day and those who suffer from hunger.
2. Achieve universal primary education.	As many as 113 million children do not attend school.	(2a) Ensure that all boys and girls complete primary school.
3. Promote gender equality and empower women.	Two-thirds of illiterates are women, and the rate of employment among women is two-thirds that of men.	(3a) Eliminate gender disparities in primary and secondary education, preferably by 2005, and at all levels by 2015.
4. Reduce child mortality.	Every year, nearly 11 million young children die before their fifth birthday, mainly from preventable illnesses.	(4a) Reduce by two-thirds the mortality rate among children under 5 years.
5. Improve maternal health.	In the developing world, the risk of dying in childbirth is one in 48.	(5a) Reduce by three-quarters the ratio of women dying in childbirth.
6. Combat HIV/AIDS, malaria, and other diseases.	40 million people are living with HIV, including 5 million newly infected in 2001.	(6a and 6c) Halt and begin to reverse the spread of HIV/AIDS and the incidence of malaria and other major diseases.
7. Ensure environmental sustainability.	768 million people lack access to safe drinking water and 2.5 billion people lack improved sanitation.	(7a) Integrate the principles of sustainable development into country policies and programs and reverse the loss of environmental resources. (7b) Reduce by half the proportion of people without access to safe drinking water. (7c) Achieve significant improvement in the lives of at least 100 million slum dwellers.
8. Develop a global partnership for development.		(8a) Develop further an open, rule-based, predictable, nondiscriminatory trading and financial system. (8b) Address the special needs of the least-developed countries. (8c) Address the special needs of landlocked countries and small island developing states. (8d) Deal comprehensively with the debt problems of developing countries through national and international measures to make debt sustainable in the long term. (8e) In cooperation with pharmaceutical companies, provide access to affordable, essential drugs in developing countries. (8f) In cooperation with the private sector, make available the benefits of new technologies, especially information and communications.

SOURCE: www.un.org/millenniumgoals/.

Figure / 1.4 Daily Activity in Much of the World of Collecting Water.

(Photo courtesy of James R. Mihelcic).

applying a sustainability framework to organizational and engineering activities. There are often considered to be two broad classes of efforts to operationalize sustainability: top-down and bottom-up. That is, one strategy involves high-level decision-makers initiating activities and establishing organizational structures and incentives to push sustainability into the organization from the top. In the other strategy, people throughout the organization are motivated to pursue their functions in a more sustainable manner and drive sustainability into the organization through grassroots initiatives and self-initiated activities.

There are examples of successful changes from governmental and nongovernmental organizations as well as major corporations that have been realized from both of these approaches, but the most successful examples are when all levels of the organization are working toward sustainability outcomes. A successful example of this evolution to operationalize sustainability can be seen in the **Path Forward** at the Office of Research and Development at the EPA (described in Application 1.6).

Once there is an intention to pursue sustainability, there is a clear need to identify an approach to problem solving that is evolved from previous approaches which had not systematically incorporated triple bottom-line considerations. There are two critical frameworks that can be utilized to support the expanded view necessary to move toward sustainability goals: life cycle thinking and systems thinking. While these two frameworks are related, there are clear differences where life cycle thinking is focused on material and energy flows and the subsequent impacts, while systems thinking can also capture the relationship of political, cultural, social, and economic considerations, and potential feedbacks between these considerations and material and energy flows.

Since 2010, significant changes have been made to EPA's research enterprise. All of EPA's actions and decisions are based on science and research. The EPA has recently embarked on a major effort to realign its research portfolio in order to more effectively address pressing environmental challenges and better serve the Agency's decision-making functions into the future using sustainability as an organizing principle.

In 2010, EPA commissioned a landmark study from the National Academies to provide recommendations on how to systematically operationalize the concept of sustainability into the Agency's entire decision making. The final report entitled "Sustainability and the U.S. EPA" (also known as the "Green Book") outlined several recommendations, including identification of key scientific and analytical tools, indicators, metrics, and benchmarks for sustainability that can be used to track progress toward sustainability goals. EPA scientists have begun to develop the scientific and analytical tools that will be needed in order to respond to and implement sustainability at EPA, including life cycle assessment, ecosystem services valuation, full cost/full benefit accounting, green chemistry, green infrastructure, and more. This effort to develop the tools of sustainability mirrors past EPA efforts to develop the tools for assessing, evaluating, and managing risk.

Access the "Green Book" (Sustainability at the U.S. EPA) at http://www.nap.edu/catalog.php?record_ id=13152#toc

1.2.1   LIFE CYCLE THINKING

Life cycle thinking supports recognizing and understanding how both consuming products and engaging in activities impact the environment from a holistic perspective. That is, **life cycle** considerations take into account the environmental performance of a product, process, or system from acquisition of raw materials to refining those materials, manufacturing, use, and end-of-life management. Figure 1.5a depicts the common **life cycle stages** for a consumer product. In the case of engineering infrastructure, Figure 1.5b depicts the life cycle stages of: (1) site development, (2) materials and product delivery, (3) infrastructure manufacture, (4) infrastructure use, and (5) end-of-life issues associated with infrastructure refurbishment, recycling, and disposal. In some cases, the transportation impacts of moving between these life cycle stages are also considered.

LCA 101
http://www.epa.gov/nrmrl/std/lca/lca.html

There is a need to consider the entire life cycle, because different environmental impacts can occur during different stages. For example, some materials may have an adverse environmental consequence when extracted or processed, but may be relatively benign in use and easy to recycle. Aluminum is such a material. On one hand, smelting of aluminum ore is very energy intensive. This is one reason aluminum is a favored recycled metal. However, an automobile will create the bulk of its environmental impact during the use life stage, not only because of combustion of fossil fuels, but also because of runoff from roads and the use of many fluids during operation. And for buildings, though a vast amount of water, aggregate, chemicals, and energy goes into the production of construction materials, transport of these items to the job site, and construction of a building, the vast amount of water and energy occurs after occupancy, during the operation life stage of the building.

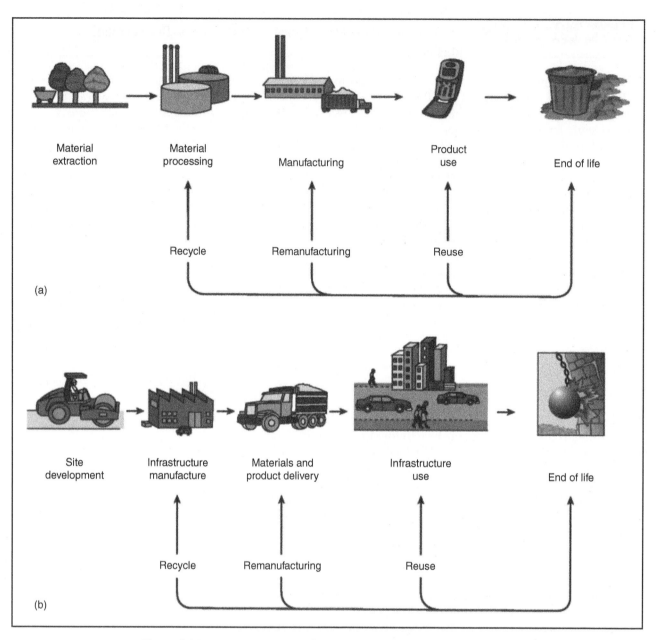

Figure / 1.5 **Common Life Cycle Stages** The most common life cycle stages for (a) a manufactured product and (b) engineered infrastructure.

Figures 1.5a and 1.5b also show, as feedback loops, the potential for recycling, remanufacturing, and reuse. While there are often benefits associated with these various end-of-life handling strategies, they can also carry environmental impacts and should be included when making design or improvement designs and in life cycle considerations.

Further, and potentially most importantly, life cycle thinking will minimize the possibility of shifting impacts from one life cycle stage to another by considering the entire system. For example, efforts to reduce the energy demands of lighting led to the installation of millions of compact fluorescent light bulbs (CFLs) (Application 1.7). However,

The Life Cycle Initiative
http://lcinitiative.unep.fr/

Given the growing concern about the impact of increasing carbon emissions on temperature and climate, there are many strategies proposed to improve energy efficiency, thereby reducing the associated carbon emissions. Electricity production creates about 33 percent of total carbon emissions, while 27 percent of the total carbon emissions result from transportation. Residential electricity is about 33 percent of total electricity (with approximately one-third for industrial and one-third for commercial uses). According to the U.S. Energy Information Administration, the average U.S. household uses 10,000 kWh a year, of which 8.8 percent, or 940 kWh, is lighting.

One effort that has been largely adopted is to reduce the amount of energy, and subsequently carbon emissions, associated with lighting. The United States, and many other countries, are currently phasing out sales of incandescent light bulbs for general lighting. The aim is to force the use and technological development of more energy-efficient lighting alternatives, such as CFLs and light-emitting diode (LED) lamps.

A 100 W incandescent light bulb that runs 3 h a day every day will use around 100 kWh a year. A high-efficiency light uses about one-fourth of the energy of a conventional bulb. Replacing the 100 W bulb with a 25 W CFL would thus save 75 kWh a year. This reduction in electricity use corresponds to a savings of about 150 lb of carbon dioxide (the same as emitted by burning 7.5 gallons of gasoline). Given that 19 percent of global electricity generation is taken for lighting, there is the potential for tremendous savings associated with new alternative lighting technologies.

However, it is important to note that current CFLs contain approximately 4.0 mg of mercury per bulb, raising environmental and human health concerns. Further, initially there were performance considerations associated with CFLs that have led to resistance in the market, including lighting quality and warm-up time. While mercury is not used in the manufacture of LED bulbs, there are still life cycle impacts associated with their production, use, and disposal. However, LED lamps solve many of the performance considerations associated with CFLs. To make the situation even more complex, the cost of CFLs and LEDs is higher than that of incandescent light bulbs. Generally, this extra cost is repaid in the long term, as both lighting technologies use less energy and have longer operating lives than incandescent bulbs.

From this discussion, there are clear opportunities to improve the energy consumption, and subsequent carbon emissions, associated with lighting. However, the technological advances present some trade-offs in terms of mercury use and disposal and performance for CFLs and cost for both CFLs and LEDs. These trade-offs need to be considered and quantified for informed decision making in the present and should be used to guide future design and innovation for improved lighting technologies in the future.

CFLs contain a small amount of mercury. By focusing solely on reducing energy demand and carbon emissions, while not considering the toxicity associated with manufacturing and disposing of CFLs, there is the potential to have a greater environmental and human health impact associated with the mercury, a heavy metal with known neurotoxic effects.

By using life cycle thinking, one can begin to understand and evaluate these potential trade-offs across many environmental and human health endpoints such as energy use, carbon emissions, water use, eutrophication, solid waste production, and toxicity by tracking all of the material and energy inputs associated with not just using energy for lighting but producing and disposing of light bulbs. These trade-offs can be quantified through a tool known as **life cycle assessment (LCA).**

Life cycle thinking supports the goal of improving the overall environmental performance of an engineering design and not simply improving a single stage or endpoint while shifting burdens elsewhere in the life cycle. To effectively capture these impacts across the entire

Class Discussion

How should decisions be made concerning these types of trade-offs (i.e., mercury use for carbon reduction)? Is this a societal decision? Should these decisions be made by companies alone or with public comment? At what scale—local, national, international? How do we systematically weight one potential impact for a potential benefit?

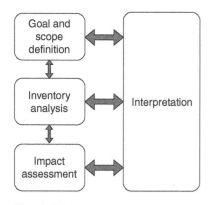

Figure / 1.6 Components of the Life Cycle Assessment (LCA) Framework.

life cycle of the product, process, or system, one must consider the environmental impacts for the entire life cycle through an LCA.

An LCA is a sophisticated way of examining the total environmental impact through every life cycle stage. The LCA framework is depicted in Figure 1.6. LCAs can be used to identify processes, ingredients, and systems that are major contributors to environmental impacts, compare different options within a particular process with the objective of minimizing environmental impacts, and compare two different products or processes that provide the same service.

As shown in Figure 1.6, the first step in performing an LCA is to define the goal and scope. This can be accomplished by answering the following questions:

• What is the purpose of the LCA? Why is the assessment being conducted?

• How will the results be used, and by whom?

• What materials, processes, or products are to be considered?

• Do specific issues need to be addressed?

• How broadly will alternative options be defined?

• What issues or concerns will the study address?

Another item that needs to be addressed at this stage is to define the function and functional unit. The **functional unit** serves as the basis of the LCA, the system boundaries, and the data requirements and assumptions. For example, if you were interested in determining the energy use and associated carbon emissions from reclaiming or desalinating water (over the complete life cycle), the function would be to reclaim treated wastewater or desalinate ocean water. The associated functional unit might therefore be m^3 of reclaimed wastewater or m^3 of desalinated water.

Once the goal, scope, and functional unit have been defined, the next step of an LCA is to develop a flow diagram for the processes being evaluated and conduct an inventory analysis. This involves describing all of the inputs and outputs (including material, energy, and water) in a product's life cycle, beginning with what the product is composed of, where those materials came from, where they go, and the inputs and outputs related to those component materials during their lifetime. It is also necessary to include the inputs and outputs during the product's use, such as whether the product uses electricity or batteries. If the analysis strictly focuses on materials and does not consider energy or other inputs/outputs, it is referred to as a subset of LCA and materials flow analysis.

A **materials flow analysis (MFA)** measures the material flows into a system, the stocks and flows within it, and the outputs from the system. In this case, measurements are based on mass (or volume) loadings instead of concentrations. *Urban materials flow analysis* (sometimes referred to as an **urban metabolism** study) is a method to quantify the flow of materials that enter an urban area (e.g., water, food, and fuel) and the flow of materials that exit an urban area (e.g., manufactured goods, water and air pollutants including greenhouse gases, and solid wastes) (Application 1.8).

example 1

If you are asked to conduct an LCA on two different laundry detergents, what could you use as the functional unit for the analysis?

solution 1

The basis of the LCA could be the weight or volume of each laundry detergent necessary to run 1,000 washing machine cycles. (This says nothing about the performance of the laundry detergents—how clean the clothes are after washing—as that is assumed to be identical for the purpose of the LCA.)

example 2

If you are asked to conduct an LCA on paper versus plastic grocery bags, what could you use as the functional unit for the analysis?

solution 2

The basis of the LCA could be a set volume of groceries to be carried, in which case two plastic bags might be equivalent to one paper bag. Or the functional unit could be related to the weight of groceries carried, in which case you would need to determine whether paper or plastic bags are stronger and how many of each would be needed to carry the specified weight.

Application/1.8   Urban Metabolism and a Case Study on Hong Kong

Urban metabolism studies are important, because planners and engineers can use them for recognizing problems and wasteful growth, setting priorities, and formulating policy. For example, a materials flow analysis performed over 10 years on the quantity of freshwater that enters and exits the Greater Toronto Area found that water inputs had grown 20 percent more than the outputs. Possible explanations for this could be leaking water distribution systems, combined sewer overflow events, and increased use of water for lawn care, all of which would allow inputted water to bypass output monitoring. The analysis also pointed to a need to further develop water conservation because of a fixed availability (or storage capacity) of freshwater.

Figure 1.7 shows the results of a materials flow analysis performed on the city of Hong Kong. Here, 69 percent of the building materials were used for residential purposes, 12 percent for commercial, 18 percent for industrial, and 2 percent for transport

infrastructure. Also, a 3.5 percent measured increase in materials use over the 20-year study period indicated that Hong Kong was still developing into a larger urban system.

During the study period, the city's economy shifted from manufacturing to a service-based center. This resulted in a 10 percent energy shift from the industrial sector to the commercial sector, yet energy consumption rose. The large increase in energy use was attributed to increases in development and residential/occupational comfort and convenience. The rate of use of consumable materials also rose during the study period, with plastics actually increasing 400 percent.

Overall air emissions in Hong Kong decreased; however, air pollutants associated with motor vehicle use and fossil fuel power production (such as NO_x and CO) increased. Land disposal of solid waste rose by 245 percent, creating a dilemma for the space-limited city. Although a large portion of this waste is construction, demolition, and reclamation waste,

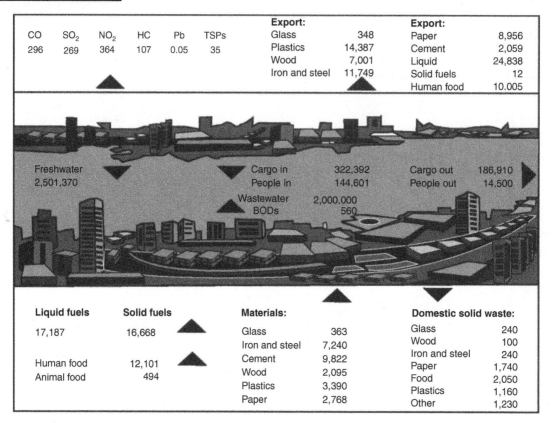

CO	SO$_2$	NO$_2$	HC	Pb	TSPs
296	269	364	107	0.05	35

Export:

Glass	348
Plastics	14,387
Wood	7,001
Iron and steel	11,749

Export:

Paper	8,956
Cement	2,059
Liquid	24,838
Solid fuels	12
Human food	10.005

Freshwater	2,501,370

Cargo in	322,392
People in	144,601

Cargo out	186,910
People out	14,500

Wastewater	2,000,000
BODs	560

Liquid fuels	**Solid fuels**
17,187	16,668

Human food	12,101
Animal food	494

Materials:

Glass	363
Iron and steel	7,240
Cement	9,822
Wood	2,095
Plastics	3,390
Paper	2,768

Domestic solid waste:

Glass	240
Wood	100
Iron and steel	240
Paper	1,740
Food	2,050
Plastics	1,160
Other	1,230

Figure / 1.7 Important Materials Flows into and through the City of Hong Kong All units are in tonnes per day. Arrows are intended to give some indication of the direction of flow of materials.

(Adapted from *AMBIO: A Journal of the Human Environment*, Vol. 30, K. Warren-Rhodes and A. Koenig, "Escalating Trends in the Crash. Urban Metabolism of Hong Kong: 1971–1997," pages 429–438, 2001, with kind permission of Springer Science+Business Media B.V.)

municipal solid waste also rose 80%, with plastics, food scraps, and paper contributing the most to municipal waste.

Though the overall rate of growth for water use declined over the study (10–2 percent) from decreases in agriculture and industrial use, the per capita freshwater consumption rose from 272 to 379 L/day. Water is one of the major waste sinks for the city, due to its large volume of untreated sewage. Biochemical oxygen demand (BOD) loadings increased by 56 percent. Nitrogen discharges also increased substantially. Sewage contamination in Hong Kong waters is now considered a major crisis for the city, having large harmful environmental, economic, and health effects.

One conclusion is that, at its current urban metabolic rate, Hong Kong is exceeding its own natural production and CO$_2$ fixation rates. Materials and energy consumption in the city greatly outweigh the natural assimilation capacity of the local ecosystem. High urban metabolism rates show that, relative to other cities, Hong Kong is more efficient (on a per capita basis) in land, energy, and materials use due to lower material stocks in buildings and transportation infrastructure, has less energy and materials use (domestic consumption), and has higher proportions of space dedicated to parks and open space.

The purpose of an inventory analysis—either a full life cycle or limited to materials—is to quantify what comes in and what goes out, including the energy and material associated with each stage in the life cycle. Inputs include all materials, both renewable and nonrenewable, and energy. It is important to remember that outputs include the desired products as well as by-products and wastes such as emissions to air, water, and land. It is also important to consider the quality of data for inputs and outputs to the system when conducting an inventory analysis.

The third step in an LCA (or MFA) is to conduct an impact assessment. This step involves identifying all the environmental impacts associated with the inputs and outputs detailed in the inventory analysis. In this case, the environmental impacts from across the life cycle are grouped together in broad topics. Environmental impacts can include stressors such as resource depletion, water use, energy use, global warming potential, ozone hole depletion, human toxicity, smog formation, and land use. This step often involves some assumptions about what human health and environmental impact will result from a given emission.

"all": unrealistic

The final step in the impact assessment can be controversial, as it involves weighting these broad environmental impact categories to yield a single score for the overall environmental performance of the product, process, or system being analyzed. This is often a societal consideration that can vary between cultures. For example, Pacific Rim Island nations may give greater weighting to climate change given their vulnerability to sea level rise, while other countries may give greater weighting to human health impacts. This suggests that the total impact score may be distorted by weighting factors. It also means that for an identical life cycle inventory, the resulting decisions from the impact assessment may vary from country to country or organization to organization.

Ultimately, LCA (and MFA) can provide insight into opportunities for improving the environmental impact of given product, process, or system. This can include choosing between two options or identifying areas for improvement for a single option. LCA and MFA are extremely valuable in ensuring that environmental impact is being minimized across the entire life cycle and that impacts are not being shifted from one life cycle stage to another. This leads to a system that is globally optimized to reduce adverse effects of the specified product, process, or system.

Applying Life Cycle Thinking to International Water and Sanitation Development Projects
http://usfmi.weebly.com/thesesreports.html

1.2.2 SYSTEMS THINKING

Beyond tracking the physical inputs and outputs to a system, **systems thinking** considers component parts of a system as having added characteristics or features when functioning within a system rather than in isolation. This suggests that systems should be viewed in a holistic manner. Systems as a whole can be better understood when the linkages and interactions between components are considered in addition to understanding the individual components. An example of the benefit of using life cycle thinking and systems thinking for the issue of assessing the potential environmental impact of biofuels is presented in Application 1.9.

Class Discussion
Are biofuels sustainable? From a life cycle and systems perspective, biofuels may make sense from a carbon perspective, but what other endpoints may be critically important to their successful implementation? Can these tools help to evaluate the impacts of biofuels on food availability and pricing?

The nature of systems thinking makes it extremely effective for solving the most difficult types of problems. For example, sustainability challenges are quite complex, depend on interactions and interdependencies, and are currently managed or mitigated through disparate mechanisms. In this way, policies or technologies may be implemented with well-articulated goals, but can lead to unintended consequences because all of the potential system feedbacks were not considered.

One way to begin a systems analysis is through a **causal loop diagram (CLD)**. CLDs provide a means to articulate the dynamic, interconnected nature of complex systems. These diagrams consist of arrows connecting variables (things that change over time) in a way that shows how one variable affects another. Each arrow in a CLD is labeled with an *s* or *o*. An *s* means that when the first variable changes, the second one changes in the same direction. (For example, increased profits lead to increased investments in research and development.) An *o* means that the first variable causes a change in the opposite direction of the second variable. (For example, more green engineering innovations can lead to reduced environmental and human health liabilities.)

In CLDs, the arrows come together to form loops, and each loop is labeled with an *R* or *B* (Figure 1.9). *R* means *reinforcing*—that is, the causal relationships within the loop create exponential growth or collapse. For instance, Figure 1.9 shows that the more fossil fuel–based energy consumed, the more carbon dioxide that is emitted, as the global temperatures increase, and the more energy that needs to be consumed. *B* means *balancing*—that is, the causal influences in the loop keep the variables in equilibrium. For example, in Figure 1.9, the more profits generated by a company, the more research and development investments that can be made, which will lead to

---

Application / 1.9   Life Cycle and Systems Thinking Applied to Biofuels

A recent example where the relevance of life cycle thinking and systems thinking was made clear was the proposal to use biobased fuels to replace a portion of the U.S. transportation fuel portfolio. There has been significant emphasis placed on alleviating dependence on fossil fuel by producing fuel energy from agricultural products. One of the clearest examples of this is the emphasis in the United States on producing ethanol from corn. Whether the economics of producing ethanol from corn is considered by monetizing life cycle emissions or direct environmental impacts (including water, fertilizer, and pesticide application), corn-based ethanol may require (per unit of fuel produced) more fossil fuel and fertilizer inputs that emit large amounts of greenhouse gases, particulate matter, and nutrients than the current petroleum-based production.

This is not to suggest that producing energy from biobased resources is not an appropriate or ultimately sustainable strategy. It is rather to suggest that pursing renewable energy in a way that only addresses the singular goal of reducing use of finite resources can lead to increased environmental and human health impacts and even greater stress on the earth's systems without using life cycle and systems thinking frameworks. Figure 1.8 shows the environmental impact of biofuels created from different crops sources. Note how this supposed "greener" fuel can have significant and varied environmental impacts across the life cycle. These impacts are also highly dependent on the feedstock choice and production location.

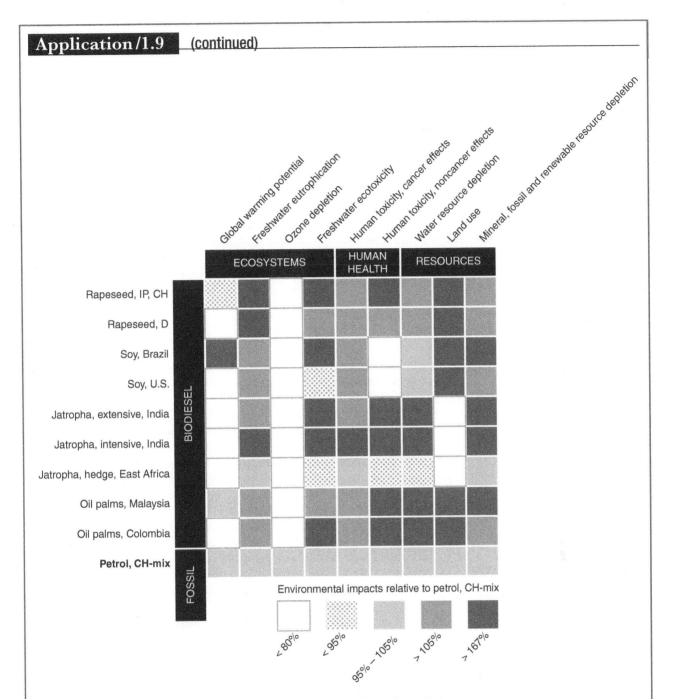

Figure / 1.8 Overview of the Diversity of Environmental Effects from Various Renewable Feedstocks for the Production of Biodiesel Environmental impacts are reported relative to the production of petroleum-based petrol mix with lighter shading indicating less impact and darker shading indicating greater impact than the conventional system. Based on material from the *Seattle Post-Intelligencer* (2008).

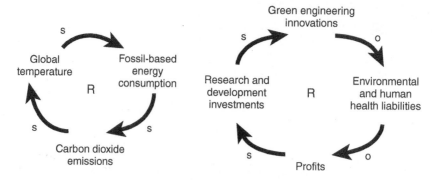

Figure / 1.9 **Examples of Reinforcing and Balancing CLDs** Each arrow in a CLD is labeled with an "*s*" or an "*o*". An *s* means that when the first variable changes, the second one changes in the same direction. An *o* means that the first variable causes a change in the opposite direction of the second variables. *R* means reinforcing—that is, the causal relationship within the loop to create exponential growth or collapse. *B* means balancing—that is, the casual influences in the loop keep variables in equilibrium.

more green engineering innovations, reducing the number of environmental and human health liabilities, which leads to greater potential profits.

CLDs can contain many different *R* and *B* loops, all connected with arrows. Drawing these diagrams can develop a deep understanding of the system dynamics. Through this process, opportunities for improvements will be highlighted. For example, the links between finite resource consumption for energy production, carbon emissions, and global temperatures may lead us to find new sources of renewable energy.

Further, it is through systems thinking that we can also begin to understand the **resilience** of a system. Resiliency is a very important concept for sustainable systems because it is the capacity of system to survive, adapt, and grow in the face of unforeseen changes, even catastrophic incidents (Fiksel, 2003). Resilience is a common feature of complex systems, such as companies, cities, or ecosystems. Given the uncertainty and vulnerability around sustainability challenges such as climate change, water scarcity, and energy demands, sustainable designs likely will need to incorporate resilience as a fundamental concept.

The idea of designing engineered systems for resilience would be to introduce more distributed and/or smaller systems that can continue to effectively function in uncertain situations with greater resilience. Examples include power generation and rainwater harvesting at the household or community level, and decentralized wastewater treatment. Again, it is necessary to consider the life cycle impacts of the entire system when designing a new, distributed system with more redundancy to replace a more centralized system. This is in order to understand the potential trade-offs between environmental and human health impacts for resiliency gains. This is where the lifetime of a given system becomes a crucial factor in LCA.

example/ 1.2   Distributed Systems That May Improve Functionality and Resilience

Provide an example of a distributed system composed of independent yet interactive elements that may deliver improved functionality and greater resilience. What are the potential benefits in terms of sustainability?

solution

A collection of distributed electric generators (for instance, fuel cells) connected to a power grid may be more reliable and fault-tolerant than centralized power generation (Fiksel, 2003). The sustainability benefits may include the following:

- Reduced resources necessary for transmission and distribution
- Reduced losses due to long-distance transmission and distribution, so less total energy needs to be generated to provide the same amount to the end user
- Possible credit given to owner for net reductions in area emissions
- Lower overall emissions if distributed energy source is cleaner than alternative (e.g., fuel cells, landfill gas recovery, biomass)
- Potential for reduced emissions by producing energy only to meet current demand (much more flexibility in production levels with distributed systems)

1.3   Engineering for Sustainability

Engineers, in particular, have a unique role to play in the Path Forward to a sustainable future. This is because they have a direct effect on the design and development of products, processes, and systems, as well as on natural systems through material selection, project siting, and the end-of-life management of chemicals, materials, and products. Engineers play a significant and vital role in nearly all aspects of our lives. They provide basic services such as water, sanitation, mobility, energy, food, health care, and shelter, in addition to advances such as real-time communications and space exploration. The implementation of all of these engineering achievements can lead to benefits as well as problems in terms of the environment, economy, and society. The adverse impacts of traditional engineering design, often implemented without a sustainability perspective, can be found all around us in the form of water use inefficiencies, depletion of finite material and energy resources, chemicals with unintentional toxicity impacts, and degradation of natural systems.

Engineers must develop and implement solutions with an understanding of the potential benefits and impacts over the lifetime of the design. In this way, the traditions of innovation, creativity, and brilliance that engineers use to find new solutions to any challenge can be applied to designing sustainable solutions—that is, solutions that not only address grand societal challenges but also are in, and of themselves, sustainable by not creating legacy adverse impacts on the

Green Chemistry
http://www.epa.gov/greenchemistry

Biomimicry
http://www.biomimicry.net

iStockphoto.

© Chanyut Sribua-rawd/iStockphoto.

environment and society. Mutual benefits resulting from this green engineering view of design include a competitive and growing economy in the global marketplace, improved quality of life for people, and enhanced protection and restoration of natural systems.

1.3.1 FRAMEWORKS FOR SUSTAINABLE DESIGN

To support the design of these sustainable solutions, the **Principles of Green Engineering** (Application 1.10) were developed to provide a framework for thinking in terms of sustainable design criteria that, if followed, can lead to useful advances for a wide range of engineering problems.

Green chemistry is a field devoted to the design of chemical products and processes that reduce or eliminate the use and generation of hazardous materials (Anastas and Warner, 1998). Green chemistry focuses on addressing hazard through molecular design and the processes used to synthesize those molecules.

The fields of green chemistry and green engineering also use the lessons and processes of nature to inspire design through biomimicry (Benyus, 2002). **Biomimicry** (from *bios*, meaning life, and *mimesis*, meaning to imitate) is a design discipline that studies nature's best ideas and then imitates these designs and processes to solve human problems. Studying a leaf to invent a better solar cell is an example of this "innovation inspired by nature" (Benyus, 2002).

Application /1.10   The Principles of Green Engineering (from Anastas and Zimmerman, 2003)

Green engineering *is the design, discovery, and implementation of engineering solutions with an awareness of these potential benefits and impacts throughout the lifetime of the design. The goal of green engineering is to minimize adverse impacts while simultaneously maximizing benefits to the economy, society, and the environment.*

The 12 Principles of Green Engineering

1. Designers need to strive to ensure that all material and energy inputs and outputs are as inherently nonhazardous as possible.

2. It is better to prevent waste than to treat or clean up waste after it is formed.

3. Separation and purification operations should be a component of the design framework.

4. System components should be designed to maximize mass, energy, and temporal efficiency.

5. System components should be output pulled rather than input pushed through the use of energy and materials.

6. Embedded entropy and complexity must be viewed as an investment when making design choices on recycle, reuse, or beneficial disposition.

7. Targeted durability, not immortality, should be a design goal.

8. Design for unnecessary capacity or capability should be considered a design flaw. This includes engineering "one size fits all" solutions.

9. Multi-component products should strive for material unification to promote disassembly and value retention (minimize material diversity).

10. Design of processes and systems must include integration of interconnectivity with available energy and materials flows.

11. Performance metrics include designing for performance in commercial "afterlife."

12. Design should be based on renewable and readily available inputs throughout the life cycle.

Application/1.11   Examples of Green Chemistry

The fundamental research of green chemistry has been brought to bear on a diverse set of challenges, including energy, agriculture, pharmaceuticals and health care, biotechnology, nanotechnology, consumer products, and materials. In each case, green chemistry has been successfully demonstrated to reduce intrinsic hazard, to improve material and energy efficiency, and to ingrain a life cycle perspective.

Some examples of green chemistry that illustrate the breadth of applicability include:

- a dramatically more effective fire extinguishing agent that eliminates halon and utilizes water in combination with an advanced surfactant;

- production of large-scale pharmaceutical active ingredients without the typical generation of thousands of pounds of toxic waste per pound of product;

- elimination of arsenic from wood preservatives that are used in lumber applied to household decks and playground equipment;

- introduction of the first commodity bio-based plastic that has the performance qualities needed for a multimillion pound application, as a food packaging;

- a new solvent system that eliminates large-scale ultrapure water usage in computer chip manufacture, replacing it with liquid carbon dioxide, which allows for the production of the next generation of nano-based chips.

Application/1.12   Examples of Biomimicry

Three levels in biology can be distinguished from which innovative and sustainable technology can be modeled:

- Mimicking natural methods of manufacture of chemical compounds to create new ones

- Imitating mechanisms found in nature (e.g., velcro)

- Studying organizational principles from social behavior of organisms, such as the flocking behavior of birds or the emergent behavior of bees and ants

Pigment-Free Color: There can significant environmental impacts associated with dyes, inks, coatings, and paints. Looking to natural systems for ideas of how to create color, one quickly finds that nature uses structure rather than pigment to offer the brilliant hues seen on butterflies, peacocks, and hummingbirds. The colors seen result from light scattering off regularly spaced melanin rods and interference effects through thin layers of keratin. Qualcomm is mimicking this strategy to create screens for electronic devices.

Preservatives: One of the emerging chemical classes of concern are anti-microbials used in a range of applications from personal care products to industrial systems. Using biomimicry as a tool, one would look for organisms that inherently demonstrate this desirable trait. For example, red and green algae produce halogenated metabolites, primarily utilizing bromide, that have demonstrated anti-microbial activity. Based on this approach, Nalco developed a product, Stabrex^{TM}, a chlorine alternative to maintaining industrial cooling systems.

Clean without chemicals: There are many environmental and human health concerns associated with certain classes of detergents and soaps. So how does nature provide the service of cleanliness without potentially toxic chemicals? One example to consider how the lotus plant that prevents dirt from interfering with photosynthesis. Lotus leaves have rough hydrophobic surfaces that allow dirt to be carried away by drops of water that "ball up" and roll off the surface. A number of new products have emerged based on this "lotus-effect," including Lotusan paint that provides a similar molecular-structure to the lotus leaf such that dirt is carried away by the rain providing "self-cleaning" building exteriors.

Examples are based on *Biomimicry: Innovation Inspired by Nature*, Janine M. Benyus, with permission of Harper-Collins Publishers.

1.3.2 THE IMPORTANCE OF DESIGN AND INNOVATION IN ADVANCING SUSTAINABILITY

Embedded in the discussion of sustainability and engineering is the word *design*. **Design** is the engineering stage where the greatest influence can be achieved in terms of sustainable outcomes. At the design stage, engineers are able to select and evaluate the characteristics of the final outcome. This can include material, chemical, and energy inputs; effectiveness and efficiency; aesthetics and form; and intended specifications such as quality, safety, and performance.

The design state also represents the time for innovation, brainstorming, and creativity, offering an occasion to integrate sustainability goals into the specifications of the product, process, or system. *Sustainability should not be viewed as a design constraint. It should be utilized as an opportunity to leapfrog existing ideas or designs and drive innovative solutions that consider systematic benefits and impacts over the lifetime of the design.*

This potential is shown in Figure 1.10. This figure demonstrates that allowing an increased number of degrees of freedom to solve a challenge, address a need, or provide a service creates more design space to generate sustainable solutions.

For a given investment (time, energy, resources, capital), potential benefits can be realized. These benefits include increased market share, reduced environmental impact, minimized harm to human health, and improved quality of life. In the case in which constraints require merely optimizing the existing solution or making incremental improvements, some modest gains can be achieved. However, if the degrees of freedom within the design space can be increased, more benefits can be realized. This is because the engineer has an

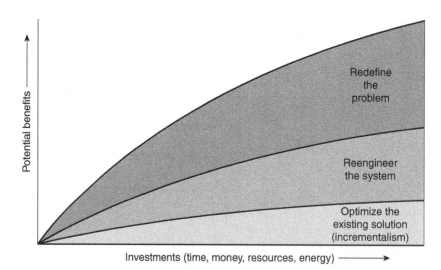

Figure / 1.10 **Increasing Potential Benefits with Increasing Degrees of Design Freedom for a Given Investment** Note that allowing an increased number of degrees of freedom to solve a problem frees up more design space to innovate and generate sustainable solutions.

opportunity to design a new solution that may appear very different in form but provides the same service. This may pose challenges if the new design is too embedded into an existing and constrained system. Ultimately, the most benefits can be achieved when the engineer designs with the most degrees of freedom—at the highest system scale—to ensure that each component within the system is sustainable, performs with the other system components, and meets the overall intended purpose.

example/ 1.3   Degrees of Freedom and Sustainable Design

In 2004, the average miles per gallon for a car on the road in the United States was 22. In response to concerns about global climate change, policy makers and engineers are working toward a more innovative technical and management strategies to improve gas mileage and lower carbon dioxide emissions. What are the design opportunities for improvement scaled with increasing degrees of freedom and what are the potential benefits?

solution

Table 1.2 gives three design solutions. As the degrees of freedom in the design increase, engineers in this example have more flexibility to innovate a solution to the problem.

Table / 1.2

Three Design Solutions Investigated in Example 1.3

		Increasing degrees of freedom	
		$\longrightarrow$	
	Incremental Improvement	Reengineer the System	Redefine the System Boundary
Design solution	Improve the efficiency of the Carnot engine; use lighter-weight materials (composites instead of metals)	Use a hybrid electric or fuel cell system for energy; change the shape of the car for improved aerodynamics; capture waste, heat, and energy for reuse	Meet mobility needs without individual car; implement a public transit system; design communities so commercial districts and employment are within walking and cycling distance; provide access to desired goods and services without vehicular transportation
Potential realized benefits	Moderate fuel savings; moderate reductions in CO_2 emissions	Improved fuel savings; improved reductions in CO_2 emissions; improved material and energy efficiency	Elimination of the environmental impacts associated with the entire automobile life cycle; maximized fuel savings and CO_2 reductions; improved infrastructure; denser development (smart growth); improved health of society from walking and less air pollution

Figure 1.11 **Percent Costs Incurred versus Design Timeline** The costs can be thought of as economic or environmental. During the design phase, approximately 70 percent of the cost becomes fixed for development, manufacture, and use.

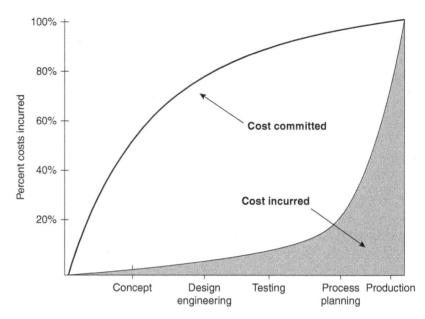

The design phase also offers unique opportunities in the life cycle of an engineered product, process, or system. As shown in Figure 1.11, it is at the design phase of a typical product that 70–75 percent of the cost is set, even though these costs will not be realized until much later in the product life cycle. The environmental costs are analogous to economic ones. For example, it is also at the design phase that materials are specified. This often dictates the production process as well as operation and maintenance procedures (i.e., painting, coating, rust inhibiting, cleaning, and lubricating).

As soon as a material is specified as a design decision, the entire life cycle of that material from acquisition through processing as well as the end of life is now included as a part of the environmental impacts of the designed product, process, or system. Therefore, it is at the design phase that the engineer has the greatest ability to affect the environmental impacts associated with the final outcome.

As an example, think of all the materials and products that go into construction and furnishing a building. At this point, the engineer needs to vision the future in regard to how these materials will be maintained, what cleaning agents will be used, what the water and energy demands of the building will be, what will happen to the building after its useful life is over, and what the fate of these materials at the end of the building's life will be. In terms of transportation systems, an engineer can think beyond the design of a new highway intended to relieve urban congestion, because data clearly shows that these new transportation corridors will become congested in just a few years after the highway is completed.

It is also important to note that it is at the design phase that the engineer has the opportunity to incorporate increased efficiency, reduced waste of water, materials, and energy, reduce costs, and most importantly, impart new performance and capabilities. While many of the other attributes listed can be achieved through "end of the pipe" control technologies, it is only by working at the design phase

EPA's Design for Environment Program
http://www.epa.gov/oppt/dfe

that the actual product, process, or system characteristics can be changed. Choosing a material that is inherently nontoxic has tremendous benefits in terms of human health and environmental impacts as well as eliminating the need to control the circumstances in which this chemical is used and how it is handled. Adding new performance and/ or capabilities often brings improved environmental characteristics as well as offering the opportunity for improved competitiveness and market share making this a better design for many reasons. Simply controlling or minimizing waste through manufacturing or even end of life cannot alter or improve the fundamental nature of the design, which adds value while offering an improved environmental profile. There is even a new recognition that wastewater should be viewed as a source of water, energy, and nutrients, and not just something to be remediated to the minimum standard as cheaply and quickly as possible. This can have a tremendous impact on the design of next generation wastewater treatment and resource recovery systems.

1.4   Measuring Sustainability

An **indicator**, in general, is something that points to an issue or condition. Its purpose is to show you how well a system is working. If there is a problem, an indicator can help you determine what direction to take to address the issue. Indicators are as varied as the types of systems they monitor. However, there are certain characteristics that effective indicators have in common (Sustainable Measures, 2007) as given in Table 1.3.

An example of an indicator is the gas gauge in your car. The gas gauge shows you how much gasoline is left in your car. If the gauge shows that the tank is almost empty, you know it is time to fill up. Another example of an indicator is a midterm report card. It shows a student and instructor whether they are doing well enough to go to the next grade or whether extra help is needed. Both of these indicators provide information to help prevent or solve problems, hopefully before they become too severe. Another example of a common one-dimensional indicator of economic progress is gross domestic product (GDP). Note, however, that many argue that GDP is insufficient to be used as a sustainability indicator, because it measures economic productivity in areas that would not be considered in a vision of a more sustainable world (e.g., economics of prisons, pollution control, and cancer treatment).

While the Principles of Green Engineering provide a framework for designers, many engaged in sustainability efforts also develop metrics or indicators to monitor their progress in meeting sustainability goals. A **sustainability indicator** measures the progress toward achieving a goal of sustainability. Sustainability indicators should be a collection of indictors that represent the multidimensional nature of sustainability, considering environmental, social, and economic facets. In terms of campus sustainability indicators, the University Leaders for a Sustainable Future (ULSF, 2008) states that "Sustainability implies that the critical activities of a higher education institution are (at a minimum) ecologically sound, socially just and economically viable, and that they will continue to be so for future generations." Table 1.4 provides a

Table / 1.3
Characteristics and Intentions of Effective Indicators
Relevant
Easy to understand by all stakeholders
Reliable
Quantifiable
Based on accessible data

SOURCE: From the Community Indicators Consortium (www.communityindicators.net).

National Transportation Statistics: Transportation, Energy, Environment
http://www.bts.gov

Sustainable Seattle
http://www.sustainableseattle.org

University Leaders for a Sustainable Future
http://www.ulsf.org

Table / 1.4

Traditional Indicators versus Sustainability Indicators for a Community and What They Say about Sustainability

Economic indicators	Traditional	Median income Per capita income relative to the U.S. average size of the economy as measured by gross national product (GNP) and GDP
	Sustainable	Number of hours of paid employment at the average wage required to support basic needs Wages paid in the local economy that are spent in the local economy Dollars spent in the local economy that pay for local labor and local natural resources Percent of local economy based on renewable local resources
	Emphasis of sustainability indicator	What wage can buy Defines basic needs in terms of sustainable consumption Local financial resilience
Environmental indicators	Traditional	Ambient levels of pollution in air and water Tons of solid waste generated Cost of fuel
	Sustainable	Use and generation of toxic materials (both in production and by end user) Vehicle miles traveled Percent of products produced that are durable, repairable, or readily recyclable or compostable Total energy used from all sources Ratio of renewable energy used at renewable rate to nonrenewable energy
	Emphasis of sustainability indicator	Measuring activities causing pollution Conservative and cyclical use of materials Use of resources at sustainable rate
Social indicators	Traditional	Number of registered voters SAT and other standardized-test scores
	Sustainable	Number of voters who vote in elections Number of voters who attend town meetings Number of students trained for jobs that are available in the local economy Number of students who go to college and come back to the community
	Emphasis of sustainability indicator	Participation in democratic process Ability to participate in the democratic process Matching job skills and training to needs of the local economy

SOURCE: Hart (2007).

comparison of traditional versus sustainability indicators for a community, and what new information they provide about progress toward sustainability that is not captured by more traditional indicators (Hart, 2007).

Several quantitative sustainability metrics are heavily utilized by engineers. One of these metrics is the **efficiency factor (or E factor)**, which is a measure of material efficiencies, that is the waste generation for materials. While efficiencies of all types have always been a component of good design, the generation of waste, particularly hazardous waste, can be considered a design flaw. As given in Equation 1.1 (and demonstrated in Example 1.4), the E factor measures the efficiency of various chemical industries in terms of the kilograms of material inputs relative to the kilograms of final product (Sheldon, 2007). It does not consider chemicals or materials that are not directly involved in the synthesis, such as solvents and rinse water. A higher value for the E factor means more waste is produced and thus there is a greater potential for adverse impact on human health and the environment. Manufacturers would thus strive to develop processes where the E factor approaches one:

$$E\ factor = \frac{\sum kg\ inputs}{\sum kg\ product} \qquad (1.1)$$

example/1.4   Determining the E Factor

Calculate the E factor for the desired product, given the following chemical production process:

$$CH_3CH_2CH_2CH_2OH + NaBr + H_2SO_4 \rightarrow CH_3CH_2CH_2CH_2Br + NaHSO_4 + H_2O$$

Table 1.5 provides details about the molecules involved.

Table / 1.5

Information Needed for Example 1.4

Type	Molecular Formula	Molecular Weight	Weight (g)	Moles
Reactant	$CH_3CH_2CH_2CH_2OH$	74.12	0.8 (added)	0.80 (added)
Reactant	NaBr	102.91	1.33 (added)	1.33 (added)
Reactant	H_2SO_4	98.08	2.0 (added)	2.0 (added)
Desired product	$CH_3CH_2CH_2CH_2Br$	137.03	1.48	0.011
Auxiliary	$NaHSO_4$			
Auxiliary	H_2O			

solution

$$E\ factor = \frac{\sum kg\ inputs}{\sum kg\ product}$$

$$E\ factor = \frac{0.0008 + 0.00133 + 0.002}{0.00148} = 2.8$$

In this example, 2.8 times more mass of material inputs are required than are obtained in the final product. This is not close to the value of zero we would want to set as a goal if the company had zero waste as a sustainability goal.

According to Sheldon, the current bulk chemical industries have E factors of less than 1–5, compared with 5 to greater than 50 for fine chemicals, and 25 to more than 100 for pharmaceuticals. This shows that today there is great opportunity to reduce waste production during chemical manufacturing.

Be aware also that this type of calculation is only a measure of mass efficiency and does not consider the toxicity of the materials used or generated (see Chapter 6 for more information on toxicity and hazard).

1.5 Policies Driving Green Engineering and Sustainability

There is a close, albeit often unrecognized, link between policy and engineering design. **Policies** are plans or courses of action, as of a government or other organization, intended to influence and determine decisions, actions, and other matters. Governmental policies are often aimed at protecting the public good in much the same way that green chemistry and green engineering are aimed at protecting human health and the environment. Policy can be a powerful driver influencing engineering design in terms of which material and energy sources are used through subsidies and/or strict regulations on emissions. In this way, policy can play a significant role in supporting engineering design for sustainability. There are two main types of policies that can affect design at this scale: regulations and voluntary programs.

1.5.1 REGULATIONS

Product Policy Institute

http://www.productpolicy.org/content/about-epr

Extended Producer Responsibility in California

http://www.calrecycle.ca.gov/epr/

A **regulation** is a legal restriction promulgated by government administrative agencies through rulemaking supported by a threat of sanction or a fine. While there a traditional environmental regulations focused on end of pipe releases, there is an emerging policy area focused on sustainable design. Two of the most established examples include **extended product responsibility (EPR)** initiatives and banning specific substances.

Extended product responsibilities, such as the European Union's (EU) Waste Electrical and Electronic Equipment directive, hold the original manufacturer responsible for their products throughout the life cycle. This directive aims to minimize the impact of electrical and electronic goods on the environment by increasing reuse and recycling and reducing the amount of electrical and electronic equipment going to landfills. It seeks to achieve this by making producers responsible for financing the collection, treatment, and recovery of waste electrical equipment and by obliging distributors (sellers) to allow consumers to return their waste equipment free of charge. This drives engineers to design electrical and electronic equipment with the Principles of Green Engineering. For example, these designs consider end-of-life management and aim for ease of disassembly, recovery of complex components, and minimized material diversity. One positive impact of this approach from a company's perspective is that it reconnects the consumer with the manufacturer at the end-of-the-life, life stage.

Another policy approach to driving engineering design toward sustainability goals is banning specific substances of concern. An example closely tied to the Electrical and Electronic Equipment directive is the EU's Restriction of Hazardous Substances (RoHS). RoHS is focused on "the restriction of the use of certain hazardous substances in electrical and electronic equipment." This Directive bans the placing on the EU market of new electrical and electronic equipment containing more than agreed levels of lead, cadmium, mercury, hexavalent chromium, polybrominated biphenyl (PBB), and polybrominated diphenyl ether (PBDE) flame retardants. By banning these chemicals of concern in significant levels, this directive is driving the implementation of green chemistry and green engineering principles in terms of designing alternative chemicals and materials that reduce or eliminate the use and generation of hazardous substances and preventing pollution.

European Commission Environmental Initiatives
http://ec.europa.eu/environment/index_en.htm

1.5.2   VOLUNTARY PROGRAMS

Another policy strategy for encouraging green engineering design is through **voluntary programs**. Voluntary programs are not mandated by law or enforceable, but are meant to encourage and motivate desirable behaviors. The government, industry, or third-party nongovernmental organizations can sponsor these programs. While there are many different varieties of voluntary programs, two types that have been established with success are eco-labeling and preferential purchasing.

Environmental standards allow for an environmental assessment of a product's impact on factors such as air pollution, wildlife habitat, energy, natural resources, ozone depletion and global warming, and toxic contamination. Companies that meet environmental standards for their specific product or service can apply an **eco-label**. Eco-labels attempt to provide an indicator to consumers of the product's environmental performance (e.g., "recycled packaging" or "no toxic emissions"). Independent third parties, such as Green Seal, United States Green Buildings Council, and EnergyStar, provide nonbiased verification of environmental labels and certifications and are the most reliable originators of eco-labels.

First-party eco-labels are self-awarded, and so are not independently verified. In the United States, these sorts of labels are governed by the Federal Trade Commission's (FTC) guide for the use of environmental marketing claims and must be accurate. The FTC has brought action against several manufacturers for violating truth-in-advertising laws.

To further support these programs, many organizations are implementing **environmentally preferable or preferential purchasing** policies. These policies can be implemented by any organization (even your college and university!) and mandate a preference to purchase products from office supplies to computers to industrial chemicals with improved environmental and human health profiles. By specifying purchases of this type, organizations are creating a demand in the marketplace for products and services with reduced impacts on human health and the environment, a very powerful tool to drive innovation in this area and to reduce costs of these products through economies of scale. Companies highlighted in the EPA report, "Private Sector Pioneers: How Companies Are Incorporating Environmentally Preferable Purchasing" (EPA, 1999), have achieved a variety of savings as given in Table 1.6.

EnergyStar
http://www.energystar.gov

Green Seal
http://www.greenseal.org

Green Buildings
http://www.usgbc.org

Table / 1.6
Savings Realized from Environmentally Preferential Purchasing Programs
Reduced material costs for manufacturers
Reduced repair and replacement costs when using more durable and repairable equipment
Reduced disposal costs by generating less waste
Improved product design and performance of the product(s)
Increased employee safety and health at the facility

As noted in the EPA report on environmental preferential purchasing, many companies adopted environmental purchasing policies for traditional business reasons as listed in Table 1.7. Although these reasons result in intangible benefits, there are specific examples of measurable reduced costs associated with environmentally preferable products. These include a lower purchase price (e.g., remanufactured products), reduced operational costs (e.g., energy efficiency), reduced disposal costs (e.g., more durable products), and reduced hazardous management costs (e.g., less toxic products). In addition, purchasing environmentally preferable products may reduce an organization's potential future liability, improve the work environment, and minimize risks to workers.

1.6 Designing Tomorrow

By considering the fundamental concepts of sustainability, engineers can contribute to addressing the challenges traditionally associated with economic growth and development. This new awareness provides the potential to design a better tomorrow—one in which our products, processes, and systems are more sustainable including being inherently benign to human health and the environment, minimizing material and energy use, and considering the entire life cycle.

Key Terms

- biomimicry
- Brundtland Commission
- carrying capacity
- Carson, Rachel
- causal loop diagram (CLD)
- Code of Federal Regulations (CFR)
- design
- eco-label
- efficiency factor (E factor)
- environmentally preferable or preferential purchasing
- Environmental Protection Agency (EPA)
- extended product responsibility (EPR)

- functional unit
- green chemistry
- green engineering
- indicator
- life cycle
- life cycle assessment (LCA)
- life cycle stages
- life cycle thinking
- *The Limits to Growth*
- materials flow analysis (MFA)
- Millennium Development Goals (MDGs)
- nonpoint source emissions
- Path Forward
- pillars of sustainability

- policies
- Principles of Green Engineering
- regulations
- regulatory process
- resilience
- sustainable development
- sustainable engineering
- sustainability
- sustainability indicator
- systems thinking
- Tragedy of the Commons
- triple bottom line
- urban metabolism
- voluntary programs

chapter/One Problems

1.1 Write an official one-page office memo to your instructor that provides definitions for: (a) sustainable development (by the Bruntland Commission), (b) sustainability (according to the American Academy of Environmental Engineers and Scientists (AAEES) Body of Knowledge), (c) sustainability (according to the American Society of Civil Engineers (ASCE) Body of Knowledge), and (d) sustainable development (according to the National Society of Professional Engineers (NSPE) Code of Ethics).

1.2 Write your own definition of sustainable development as it applies to your engineering profession. Explain its appropriateness and applicability in two to three sentences.

1.3 Identify three definitions of sustainability from three sources (e.g., local, state, or federal government; industry; environmental organization; international organization; financial or investment organization). Compare and contrast those definitions with the Brundtland Commission definition. How do the definitions reflect their sources?

1.4 Relate the "Tragedy of the Commons" to a local environmental issue. Be specific about what you mean in terms of the "commons" for this particular example, and carefully explain how these "commons" are being damaged for current and future generations.

1.5 Research the progress that two countries of your choice (or your instructor's choice) have made in meeting each of the eight MDGs. Summarize the results in a table. Among other sources, you might consult the UN's MDG web site, www.un.org/millenniumgoals/.

1.6 Go to the U.S. Department of Energy's web site (www.doe.gov) and research energy consumption in the household, commercial, industrial, and transportation sectors. Develop a table on how this specific energy consumption relates to the percent of U.S. and global CO_2 emissions. Identify a sustainable solution for each sector that would reduce energy use and CO_2 emissions.

1.7 As a consumer interested in reducing your carbon emissions, (a) which should you do: (1) install more efficient lighting for your home, or (2) buy a car that gets more miles per gallon? To answer this, consider that a 100 W light bulb that is run 3 h a day every day will use around 100 kWh a year. A high-efficiency light uses about 25 percent of a conventional light bulb. Replacing it with a 25 W compact fluorescent bulb would save 75 kWh a year. This would equal 150 lb of carbon dioxide or the same amount of carbon dioxide emissions associated with burning 7.5 gallons of gasoline. (b) Given that the average U.S. household uses 10,000 kWh a year of which 8.8 percent is lighting, how many gallons of gas and pounds of CO_2 could be saved by switching all of the bulbs in a home? (c) For comparison, if you drove 12,000 miles a year and upgraded from a car that gets the national average of 20 miles per gallon (mpg) to one that got 30 mpg, how much would you reduce your gas consumption and CO_2 emissions on an annual basis? (d) What if you upgraded to a car that gets 30–37 mpg? (Combustion of 100 gallons of gasoline releases 2,000 lb of carbon dioxide.)

1.8 Visit EPA's Presidential Green Chemistry Challenge Award web site at www.epa.gov/greenchemistry/pubs/pgcc/past.html. Select a past award-winning project. Based on the description of this project, what are the environmental, economic, and social benefits of this green chemistry advance?

1.9 Discuss whether shoe A (leather) or shoe B (synthetic) is better for the environment based on the data given in Table 1.8. Is it possible to weight one aspect (air, water, land pollution, or solid waste) as being more important than another? How? Why? Who makes these decisions for our society?

1.10 To compare plastic and paper bags in terms of acquisition of raw materials, manufacturing and processing, use, and disposal, we will use data provided by Franklin Associates, a nationally known consulting firm whose clients include the U.S. EPA as well as many companies and industry groups. In 1990, Franklin Associates compared plastic bags to paper bags in terms of their energy and air/water emissions in manufacture, use, and disposal. Table 1.9 presents the results of their study.

(a) Which bag would you choose if you were most concerned about air pollution? (Note that the information does not tell you whether these are toxic air emissions or greenhouse gas emissions.) (b) If you assume that two plastic bags equal one paper bag,

Table / 1.8

Hypothetical Life Cycle Environmental Impacts of Shoes on a Basis of per 100 Pairs of Shoes Produced

Product	Energy Use (BTU)	Raw Material Consumption	Water Use (gallons)	Air Pollution (lb)	Water Pollution	Hazardous and Solid Waste
Shoe A (leather)	1	Limited supply; some renewable	2	4	2 lb organic chemicals	2 lb hazardous sludge
Shoe B (synthetic)	2	Large supply; not renewable	4	1	8 lb inert inorganic chemicals	1 lb hazardous sludge; 3 lb nonhazardous sludge

Table / 1.9

Results of Study Comparing Plastic and Paper Bags

Life Cycle Stages	Air Emissions (pollutants) (oz/bag)		Energy Required (BTU/bag)	
	Paper	Plastic	Paper	Plastic
Materials manufacture, product manufacture, product use	0.0516	0.0146	905	464
Raw materials acquisition, product disposal	0.0510	0.0045	724	185

does the choice change? (c) Compare the energy required to produce each bag. Which bag takes less energy to produce?

1.11 You are preparing a life cycle analysis of three different electrification options for powering your 1,200 ft² home in rural Connecticut. The options you are considering include: (1) just using your local grid, (2) putting a solar installation on your roof, or (3) building a transmission extension to join up with your neighbor's already-built wind turbine. Write a possible goal, scope, function, and functional unit for this LCA. Explain your reasoning.

1.12 Consider the full life cycle of each of the three electrification options (possibly beyond whatever you have selected for the scope of your LCA) in Problem 1.11. Discuss which of the life cycle stages

is most impactful for each electrification type. You will need to take into account the life cycle impacts of primary through final energy in each case. As a reminder, life cycle stages typically include resource extraction, manufacture, transportation, use, and end of life.

1.13 Draw CLD for corn-based ethanol production using the following variables: climate change, corn-based ethanol use, fertilizer demand, CO_2 emissions, fuel demand, fossil fuel use, and corn demand.

1.14 (a) Is centralized drinking water treatment and distribution more resilient than point of use water treatment technologies? Why, or why not? (b) Does it matter whether these water treatment systems are implemented in the developing or developed world?

1.15 The design team for a building project was formed at your company last week, and they have already held two meetings. Why is it so important for you to get involved immediately in the design process?

1.16 Provide an example of a product either commercially available or currently under development that uses biomimicry as the basis for its design. Explain how the design is mimicking a product, process, or system found in nature.

1.17 Two reactants, benzyl alcohol and tosyl chloride, react in the presence of an auxiliary, triethylamine, and the solvent toluene to produce the product sulfonate ester (Table 1.10). (a) Calculate the E factor for the reaction. (b) What would happen to the E factor if the solvents and auxiliary chemicals were included in the calculation? (c) Should these types of materials and chemicals be included in an efficiency measure? Why, or why not?

Table / 1.10

Useful Information Needed to Solve Problem 1.17

Reactant	Benzyl alcohol	10.81 g	0.10 mole	MW 108.1 g/mole
Reactant	Tosyl chloride	21.9 g	0.115 mole	MW 190.65 g/mole
Solvent	Toluene	500 g		
Auxiliary	Triethylamine	15 g		MW 101 g/mole
Product	Sulfonate ester	23.6 g	0.09 mole	MW 262.29 g/mole

1.18 Choose three of the Principles of Green Engineering. For each one, (a) explain the principle in your own words; (b) find an example (commercially available or under development), and explain how it demonstrates the principle; and (c) describe the associated environmental, economic, and societal benefits, identifying which ones are tangible and which ones are intangible.

1.19 (a) Develop five sustainability metrics or indicators for a corporation or an industrial sector analogous to those presented for communities in Table 1.4. (b) Compare them with traditional business metrics or indicators. (c) Describe what new information can be determined from the new sustainability metrics or indicators.

1.20 A car company has developed a new car, ecoCar, that gets 100 mpg, but the cost is slightly higher than cars currently on the market. What type of incentives could the manufacturer offer or ask Congress to implement to encourage customers to buy the new ecoCar?

1.21 Do you agree or disagree with the following statement? Explain why, or why not, in three to five sentences. "Technology-forcing pollution regulations are preferable to standards- or outcome-based regulations."

1.22 You are about to buy a car that will last 7 years before you have to buy a new one, and Congress has just passed a new tax on greenhouse gases. Assume a 5 percent annual interest rate. You have two options: (a) Purchase a used car for $12,000, upgrade the catalytic converter at a cost of $1,000, and pay a $500 annual carbon tax. This car has a salvage value of $2,000. (b) Purchase a new car for $16,500 and pay only $100 annually in carbon tax. This car has a salvage value of $4,500. Based on the annualized cost of these two options, which car would you buy?

References

Anastas, P. T., 2012. Fundamental changes to EPA's research enterprise: The Path Forward. *Environmental Science and Technology*, 46: 580–586.

Anastas, P.T., and J. C. Warner, 1998. *Green Chemistry: Theory and Practice*. Oxford: Oxford University Press.

Anastas, P.T., and J. B. Zimmerman, 2003. Design through the twelve principles of green engineering. *Environmental Science and Technology*, 37(5): 94A–101A.

Benyus, J. M. 2002. *Biomimicry: Innovation Inspired Design*. New York: Harper Perennial.

Environmental Protection Agency (EPA). 1999. *Private Sector Pioneers: How Companies Are Incorporating Environmentally Preferential Purchasing*. Report No. EPA742-R-99-01.

Fiksel, J. 2003. Designing resilient, sustainable systems. *Environmental Science and Technology*, 37: 5330–5339.

Hart, M. 2007. *Sustainable Measures* web site, www.sustainablemeasures.com

Meadows, D.H., D. L. Meadows, J. Randers, and W. W. Behrens III., 1972. *The Limits to Growth*. London: Earth Island Limited.

Mihelcic, J. R., J. C. Crittenden, M. J. Small, D. R. Shonnard, D. R. Hokanson, Q. Zhang, H. Chen, S. A. Sorby, V. U. James, J. W. Sutherland, and J. L. Schnoor. 2003. Sustainability science and engineering: Emergence of a new metadiscipline. *Environmental Science and Technology*, 37(23): 5314–5324.

Payne, R. 1968. "Among Wild Whales." *New York Zoological Society Newsletter* (November).

Seattle Post-Intelligencer, 2008. Bio-debatable: food versus fuel, May 3.

Sheldon, R.A. 2007. The E factor: Fifteen years on. *Green Chemistry* 9: 1273–1283.

University Leaders for a Sustainable Future (ULSF). 2008. *Sustainability Assessment Questionnaire*. ULSF web site, www.ulsf.org/programs_saq.html.

Warren-Rhodes, K., and A. Koenig, 2001. Escalating trends in the urban metabolism of Hong Kong: 1971–1997. *AMBIO: A Journal of the Human Environment*, 30: 429–438.

chapter/Two Environmental Measurements

James R. Mihelcic, Richard E. Honrath Jr., Noel R. Urban, Julie Beth Zimmerman

In this chapter, readers become familiar with the different units used to measure pollutant levels in aqueous (water), soil/sediment, atmospheric, and global systems. Coverage is provided on sources and atmospheric concentrations of carbon dioxide and other greenhouse gases, and methods to report their emissions that includes determining a carbon footprint.

Chapter Contents

Learning Objectives

1. Calculate chemical concentration in mass/mass, mass/volume, volume/volume, mole/mole, mole/volume, and equivalent/volume units.
2. Convert chemical concentration from mass per volume or mass per mass units to a parts per million or parts per billion basis.
3. Calculate chemical concentration in units of partial pressure.
4. Calculate chemical concentration in common constituent units such as water hardness, carbon equivalents, and carbon dioxide equivalents.
5. Convert concentrations of individual chemical species for nitrogen and phosphorus species to common constituent units for these nutrients.
6. Use the ideal gas law to convert between units of ppm_v and $\mu g/m^3$.
7. Describe historic and current atmospheric concentrations of the major greenhouse gases: carbon dioxide, methane, and nitrous oxide.
8. Describe the primary sources of the major greenhouse gases—carbon dioxide, methane, and nitrous oxide—that are associated with operation of environmental and civil engineering infrastructures.
9. Understand regulations and reporting requirements associated with emissions of greenhouse gases.
10. Utilize global warming potentials to determine the mass of greenhouse gas emissions in carbon dioxide equivalents.
11. Use eGRID to calculate the greenhouse gas emissions associated with electricity generation and the carbon footprint of different infrastructures that make up the built environment.
12. Calculate particle concentrations in air and water.
13. Represent specific chemical concentration in mixtures to a direct effect such as oxygen depletion to express units of biochemical oxygen demand and chemical oxygen demand.

© Tony Freeman/PhotoEdit

37

2.1 Mass Concentration Units

Chemical concentration is one of the most important determinants in almost all aspects of chemical fate, transport, and treatment in both natural and engineered systems. This is because concentration is the driving force that controls the movement of chemicals within and between environmental media, as well as the rate of many chemical reactions. In addition, concentration often determines the severity of adverse effects, such as toxicity, bioconcentration, and climate change.

Concentrations of chemicals are routinely expressed in a variety of units. The choice of units to use in a given situation depends on the chemical, where it is located (air, water, or soil/sediments) and how the measurement will be used. It is therefore necessary to become familiar with the units used and methods for converting between different sets of units. Representation of concentration usually falls into one of the categories listed in Table 2.1.

Important prefixes to know include pico (10^{-12}, abbreviated as p), nano (10^{-9}, abbreviated as n), micro (10^{-6}, abbreviated as μ), milli (10^{-3}, abbreviated as m), and kilo (10^{+3}, abbreviated as k). Other important units are the tonne (which is also called the metric ton by some in the United States), which equals 1,000 kg (or 2,204 lb), and the common ton, which equals 2,000 lb. In addition, 1 teragram (Tg) = 10^{12} g = 1 million metric tons.

Concentration units based on chemical mass include mass chemical per total mass and mass chemical per total volume. In these descriptions, m_i is used to represent the mass of the chemical referred to as chemical i.

2.1.1 MASS/MASS UNITS

Mass/mass concentrations are commonly expressed as parts per million, parts per billion, parts per trillion, and so on. For example, 1 mg of a solute placed in 1 kg of solvent equals 1 ppm_m. **Parts per million by mass** (referred to as **ppm** or $\textbf{ppm}_\textbf{m}$) is defined as the number of units of mass of chemical per million units of total mass. Thus, we can express the previous example mathematically:

$$\text{ppm}_\text{m} = \text{g of } i \text{ in } 10^6 \text{ g total} \qquad (2.1)$$

Clear Water Act Analytical Methods

http://www.epa.gov/waterscience/methods

Air Pollution Monitoring Techniques

http://www.epa.gov/ttn/amtic/

Table / 2.1

Common Units of Concentration Used in Environmental Measurements

Representation	Example	Typical Units
Mass chemical/total mass	mg/kg in soil	mg/kg, ppm_m
Mass chemical/total volume	mg/L in water or air	mg/L, $\mu\text{g/m}^3$
Volume chemical/total volume	volume fraction in air	ppm_v
Moles chemical/total volume	moles/L in water	M

SOURCE: Mihelcic (1999); reprinted with permission of John Wiley & Sons, Inc.

This definition is equivalent to the following general formula, which is used to calculate ppm_m concentration from measurements of chemical mass in a sample of total mass m_{total}:

$$ppm_m = \frac{m_i}{m_{total}} \times 10^6 \qquad (2.2)$$

Note that the factor 10^6 in Equation 2.2 is really a conversion factor. It has the implicit units of ppm_m/mass fraction (mass fraction = m_i/m_{total}), as given in Equation 2.3:

$$ppm_m = \frac{m_i}{m_{total}} \times 10^6 \frac{ppm_m}{\text{mass fraction}} \qquad (2.3)$$

In Equation 2.3, m_i/m_{total} is defined as the mass fraction, and the conversion factor of 10^6 is similar to the conversion factor of 10^2 used to convert fractions to percentages. For example, the expression $0.25 = 25\%$ can be thought of as:

$$0.25 = 0.25 \times 100\% = 25\% \qquad (2.4)$$

Similar definitions are used for the units ppb_m, ppt_m, and percent by mass. That is, 1 **ppb_m** equals 1 **part per billion** or 1 g of a chemical per billion (10^9) g total, so that the number of ppb_m in a sample is equal to $m_i/m_{total} \times 10^9$. And 1 **$ppt_m$** usually means 1 **part per trillion** (10^{12}). However, be cautious about interpreting ppt values, because they may refer to either parts per thousand or parts per trillion.

Mass/mass concentrations can also be reported with the units explicitly shown (e.g., mg/kg or µg/kg). In soils and sediments, 1 ppm_m equals 1 mg of pollutant per kg of solid (mg/kg), and 1 ppb_m equals 1 µg/kg. **Percent by mass** is analogously equal to the number of grams of pollutant per 100 g total.

© Anthony Rosenberg/iStockphoto.

---

example/2.1   Concentration in Soil

A 1 kg sample of soil is analyzed for the chemical solvent trichloroethylene (TCE). The analysis indicates that the sample contains 5.0 mg of TCE. What is the TCE concentration in ppm_m and ppb_m?

solution

$$[TCE] = \frac{5.0 \text{ mg TCE}}{1.0 \text{ kg soil}} = \frac{0.005 \text{ g TCE}}{10^3 \text{ g soil}}$$

$$= \frac{5 \times 10^{-6} \text{ g TCE}}{\text{g soil}} \times 10^6 = 5 \text{ ppm}_m = 5{,}000 \text{ ppb}_m$$

Note that in soil and sediments, mg/kg equals ppm_m, and µg/kg equals ppb_m.

2.1.2 MASS/VOLUME UNITS: mg/L AND $\mu g/m^3$

In the atmosphere, it is common to use concentration units of mass per volume of air, such as mg/m^3 and $\mu g/m^3$. In water, mass/volume concentration units of mg/L and $\mu g/L$ are common. In most aqueous systems, ppm_m is equivalent to mg/L. This is because the density of pure water is approximately 1,000 g/L (demonstrated in Example 2.2).

The density of pure water is actually 1,000 g/L at 5°C. At 20°C, the density has decreased slightly to 998.2 g/L. This equality is strictly true only for *dilute* solutions, in which any dissolved material does not contribute significantly to the mass of the water, and the total density remains approximately 1,000 g/L. Most wastewaters, reclaimed waters, and natural waters can be considered dilute, except perhaps seawaters, brines, and some recycled streams.

---

example/2.2 Concentration in Water

One liter of water is analyzed and found to contain 5.0 mg of TCE. What is the TCE concentration in mg/L and ppm_m?

solution

$$[TCE] = \frac{5.0 \text{ mg TCE}}{1.0 \text{ L H}_2\text{O}} = \frac{5.0 \text{ mg}}{L}$$

To convert to ppm_m, a mass/mass unit, it is necessary to convert the volume of water to mass of water. To do this, divide by the density of water, which is approximately 1,000 g/L:

$$TCE = \frac{5.0 \text{ mg TCE}}{1.0 \text{ L H}_2\text{O}} \times \frac{1.0 \text{ L H}_2\text{O}}{1,000 \text{ g H}_2\text{O}}$$

$$= \frac{5.0 \text{ mg TCE}}{1,000 \text{ g total}} = \frac{5.0 \times 10^{-6} \text{ g TCE}}{\text{g total}} \times \frac{10^6 \text{ ppm}_m}{\text{mass fraction}}$$

$$= 5.0 \text{ ppm}_m$$

In most dilute aqueous systems, mg/L is equivalent to ppm_m.

In this example, the TCE concentration is well above the allowable U.S. drinking water standard for TCE, 5 $\mu g/L$ (or 5 ppb), which was set to protect human health. Five ppb is a small value. Think of it this way: Earth's population exceeds 6 billion people, meaning that 30 individuals in one of your classes constitute a human concentration of approximately 5 ppb!

2.2 Volume/Volume and Mole/Mole Units

Units of volume fraction or mole fraction are frequently used for gas concentrations. The most common volume fraction units are **parts per million by volume** (referred to as **ppm** or ppm_v), defined as:

$$ppm_v = \frac{V_i}{V_{total}} \times 10^6 \qquad (2.5)$$

example/ 2.3   Concentration in Air

What is the carbon monoxide (CO) concentration expressed in $\mu g/m^3$ of a 10 L gas mixture that contains 10^{-6} mole of CO?

solution

In this case, the measured quantities are presented in units of moles of the chemical per total volume. To convert to mass of the chemical per total volume, convert the moles of chemical to mass of chemical by multiplying moles by CO's molecular weight. The molecular weight of CO (28 g/mole) is equal to 12 (atomic weight of C) plus 16 (atomic weight of O).

$$[CO] = \frac{1.0 \times 10^{-6} \text{ mole CO}}{10 \text{ L total}} \times \frac{28 \text{ g CO}}{\text{mole CO}}$$

$$= \frac{28 \times 10^{-6} \text{ g CO}}{10 \text{ L total}} \times \frac{10^6 \ \mu g}{g} \times \frac{10^3 \text{ L}}{m^3} = \frac{2,800 \ \mu g}{m^3}$$

where V_i/V_{total} is the volume fraction and 10^6 is a conversion factor, with units of 10^6 ppm$_v$ per volume fraction.

Other common units for gaseous pollutants are **parts per billion (10^9) by volume (ppb$_v$)**. Table 2.2 provides examples of the change in the atmospheric concentration of three major **greenhouse gases (GHGs)** since preindustrial times, around the year 1750.

The advantage of volume/volume units is that gaseous concentrations reported in these units do not change as a gas is compressed or expanded. Atmospheric concentrations expressed as mass per volume (e.g., $\mu g/m^3$) decrease as the gas expands, since the pollutant mass remains constant but the volume increases. Both mass/volume units, such as $\mu g/m^3$, and ppm$_v$ units are frequently used to express gaseous concentrations. (See Equation 2.9 for conversion between $\mu g/m^3$ and ppm$_v$.)

Table / 2.2

Change in Atmospheric Concentration of Major GHGs Since Preindustrial Times

	2011 Atmospheric Concentration	Preindustrial Atmospheric Concentration	Percent Change Since Preindustrial Times
Carbon dioxide (CO_2)	391 ppm	280 ppm	+140%
Methane (CH_4)	1,813 ppb	700 ppb	+259%
Nitrous oxide (N_2O)	324 ppb	270 ppb	+120%

SOURCE: Data from World Meteorological Organization (2012).

2.2.1 USING THE IDEAL GAS LAW TO CONVERT ppm$_v$ TO µg/m^3

The ideal gas law can be used to convert gaseous concentrations between mass/volume and volume/volume units. The **ideal gas law** states that *pressure (P)* times *volume occupied (V)* equals *the number of moles (n)* times the *gas constant (R)* times the *absolute temperature (T)* in degrees Kelvin or Rankine. This is written in the familiar form of

$$PV = nRT \tag{2.6}$$

In Equation 2.6, the **universal gas constant**, R, may be expressed in many different sets of units. Some of the most common values for R are:

0.08205 L-atm/mole-K

8.205×10^{-5} m^3-atm/mole-K

82.05 cm^3-atm/mole-K

1.99×10^{-3} kcal/mole-K

8.314 J/mole-K

1.987 cal/mole-K

62,358 cm^3-torr/mole-K

62,358 cm^3-mm Hg/mole-K

Because the gas constant may be expressed in different units, always be careful of its units and cancel them out to ensure you are using the correct value of R.

The ideal gas law also states that the volume occupied by a given number of molecules of any gas is the same, no matter what the molecular weight or composition of the gas, as long as the pressure and temperature are constant. The ideal gas law can be rearranged to show the volume occupied by n moles of gas:

$$V = n\frac{RT}{P} \tag{2.7}$$

At standard conditions ($P = 1$ atm and $T = 273.15$ K), 1 mole of any pure gas will occupy a volume of 22.4 L. This result can be derived by using the corresponding value of R (0.08205 L-atm/mole-K) and the form of the ideal gas law provided in Equation 2.7. At other temperatures and pressures, this volume varies as determined by Equation 2.7.

In Example 2.4, the terms RT/P cancel out. This demonstrates an important point that is useful in calculating volume fraction or mole fraction concentrations: *For gases, volume ratios and mole ratios are equivalent.* This is clear from the ideal gas law, because at constant temperature

A gas mixture contains 0.001 mole of sulfur dioxide (SO_2) and 0.999 mole of air. What is the SO_2 concentration, expressed in units of ppm_v?

solution

The concentration in ppm_v is determined using Equation 2.5.

$$[SO_2] = \frac{V_{SO_2}}{V_{total}} \times 10^6$$

To solve, convert the number of moles of SO_2 to volume using the ideal gas law (Equation 2.6) and the total number of moles to volume. Then divide the two expressions:

$$V_{SO_2} = 0.001 \text{ mole } SO_2 \times \frac{RT}{P}$$

$$V_{total} = (0.999 + 0.001) \text{ mole total} \times \frac{RT}{P}$$

$$= (1.000) \text{ mole total} \times \frac{RT}{P}$$

Substitute these volume terms for ppm_v:

$$ppm_v = \frac{0.001 \text{ mole } SO_2 \times \frac{RT}{P}}{1.000 \text{ mole total} \times \frac{RT}{P}} \times 10^6$$

$$ppm_v = \frac{0.001 \text{ L } SO_2}{1.000 \text{ L total}} \times 10^6 = 1{,}000 \text{ } ppm_v$$

Note also that the **mole ratio** (moles i/moles total) is sometimes referred to as the **mole fraction**, X.

and pressure, the volume occupied by a gas is proportional to the number of moles. Therefore, Equation 2.5 is equivalent to Equation 2.8:

$$ppm_v = \frac{\text{moles } i}{\text{moles total}} \times 10^6 \qquad (2.8)$$

The solution to Example 2.4 could have been found simply by using Equation 2.8 and determining the mole ratio. Therefore, in any given problem, you can use either units of volume or units of moles to calculate ppm_v. Being aware of this will save unnecessary conversions between moles and volume.

Example 2.5 and Equation 2.9 show how to use the ideal gas law to convert concentrations between $\mu g/m^3$ and ppm_v. Example 2.5 demonstrates a useful way to write the conversion for air concentrations

example/2.5   Conversion of Gas Concentration between ppb$_v$ and μg/m^3

The concentration of SO_2 is measured in air to be 100 ppb$_v$. What is this concentration in units of μg/m^3? Assume the temperature is 28°C and pressure is 1 atm. Remember that T expressed in K is equal to T expressed in °C plus 273.15.

solution

To accomplish this conversion, use the ideal gas law to convert the volume of SO_2 to moles of SO_2, resulting in units of moles/L. This can be converted to μg/m^3 using the molecular weight of SO_2 (which equals 64). This method will be used to develop a general formula for converting between ppm$_v$ and μg/m^3.

First, use the definition of ppb$_v$ to obtain a volume ratio for SO_2:

$$100\ \text{ppb}_v = \frac{100\ \text{m}^3\ SO_2}{10^9\ \text{m}^3\ \text{air solution}}$$

Now convert the volume of SO_2 in the numerator to units of mass. This is done in two steps. First, convert the volume to a number of moles, using a rearranged format of the ideal gas law (Equation 2.6), $n/V = P/RT$, and the given temperature and pressure:

$$\frac{100\ \text{m}^3\ SO_2}{10^9\ \text{m}^3\ \text{air solution}} \times \frac{P}{RT} = \frac{100\ \text{m}^3\ SO_2}{10^9\ \text{m}^3\ \text{air solution}} \times \frac{1\ \text{atm}}{8.205 \times 10^{-5}\ \frac{\text{m}^3\text{-atm}}{\text{mole-K}}(301\text{K})} = \frac{4.05 \times 10^{-6}\ \text{mole}\ SO_2}{\text{m}^3\ \text{air}}$$

In the second step, convert the moles of SO_2 to mass of SO_2 by using the molecular weight of SO_2:

$$\frac{4.05 \times 10^{-6}\ \text{mole}\ SO_2}{\text{m}^3\ \text{air}} \times \frac{64\ \text{g}\ SO_2}{\text{mole}\ SO_2} \times \frac{10^6\ \mu\text{g}}{\text{g}} = \frac{260\ \mu\text{g}}{\text{m}^3}$$

between units of μg/m^3 and ppm$_v$:

$$\frac{\mu\text{g}}{\text{m}^3} = \text{ppm}_v \times \text{MW} \times \frac{1{,}000P}{RT} \qquad (2.9)$$

where MW is the molecular weight of the chemical species, R equals 0.08205 L-atm/mole-K, T is the temperature in K, and 1,000 is a conversion factor (1,000 L = m^3). Note that for 0°C, RT has a value of 22.4 L-atm/mole, while at 20°C, RT has a value of 24.1 L-atm/mole.

Nitrogen Dioxide Air Pollution

www.epa.gov/air/nitrogenoxides

Atmospheric NO$_2$ Air Quality Over Time

www.epa.gov/airtrends/nitrogen.html

2.3   Partial-Pressure Units

In the atmosphere, concentrations of chemicals in the gas and particulate phases may be determined separately. A substance will exist in the gas phase if the atmospheric temperature is above the substance's boiling (or sublimation) point or if its concentration is below the saturated vapor pressure of the chemical at a specified

Table / 2.3

Composition of the Atmosphere

Compound	Concentration (% volume or moles)	Concentration (ppm$_v$)
Nitrogen (N$_2$)	78.1	781,000
Oxygen (O$_2$)	20.9	209,000
Argon (Ar)	0.93	9,300
Carbon dioxide (CO$_2$)	0.039	391
Neon (Ne)	0.0018	18
Helium (He)	0.0005	5
Methane (CH$_4$)	0.00018	1.813
Krypton (Kr)	0.00011	1.1
Hydrogen (H$_2$)	0.00005	0.50
Nitrous oxide (N$_2$O)	0.000032	0.324
Ozone (O$_3$)	0.000002	0.020

SOURCE: 2011 Values updated from Mihelcic (1999); with permission of John Wiley & Sons, Inc.

temperature (vapor pressure is defined in Chapter 3). The major and minor gaseous constituents of the atmosphere all have boiling points well below atmospheric temperatures. Concentrations of these species typically are expressed as either volume fractions (for example, percent, ppm$_v$, or ppb$_v$) or partial pressures (units of atmospheres).

Table 2.3 summarizes the concentrations of the most abundant atmospheric gaseous constituents, including carbon dioxide and methane. Carbon dioxide is the largest human contributor to GHGs in the atmosphere. The global atmospheric concentration of carbon dioxide has increased to 391 ppm$_v$ in 2011 from preindustrial revolution levels of 280 ppm$_v$. Global atmospheric concentrations of methane recorded in 2011 have reached 1,813 ppb. This recorded methane concentration greatly exceeds the natural range of 320–790 ppb$_v$, measured in ice cores, that dates over the past 650,000 years. According to the **Intergovernmental Panel on Climate Change** (IPCC, see www.ipcc.ch), it is very likely that this increase in methane concentration is due to agricultural land use, population growth, and energy use associated with burning fossil fuels.

The total pressure exerted by a gas mixture may be considered as the sum of the partial pressures exerted by each component of the mixture. The **partial pressure** of each component is equal to the pressure that would be exerted if all of the other components of the mixture were suddenly removed. Partial pressure is commonly written as P_i, where i refers to a particular gas. For example, the partial pressure of oxygen in the atmosphere P_{O_2} is 0.21 atm.

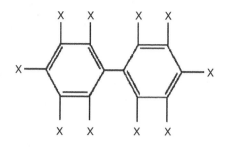

X = chlorine or hydrogen

Figure / 2.1 **Chemical Structure of Polychlorinated Biphenyls (PCBs)**
PCBs are a family of compounds produced commercially by chlorinating biphenyl. Chlorine atoms can be placed at any or all of 10 available sites, with 209 possible PCB congeners. The great stability of PCBs caused them to have a wide range of uses, including serving as coolants in transformers and as hydraulic fluids and solvents. However, the chemical properties that resulted in this stability also resulted in a chemical that did not degrade easily, bioaccumulated in the food chain, and was hazardous to humans and wildlife. The 1976 Toxic Substances Control Act (TSCA) banned the manufacture of PCBs and PCB-containing products. TSCA also established strict regulations regarding the future use and sale of PCBs. PCBs typically were sold as mixtures commonly referred to as Arochlors. For example, the Arochlor 1260 mixture consists of 60 percent chlorine by weight, meaning the individual PCBs in the mixture primarily are substituted with 6–9 chlorines per biphenyl molecule. In contrast, Arochlor 1242 consists of 42 percent chlorine by weight; thus, it primarily consists of PCBs with 1–6 substituted chlorines per biphenyl molecule.

(From Mihelcic (1999). Reprinted with permission of John Wiley & Sons, Inc.).

Remember, the ideal gas law states that, at a given temperature and volume, pressure is directly proportional to the number of moles of gas present; therefore, pressure fractions are identical to mole fractions (and volume fractions). For this reason, partial pressure can be calculated as the product of the mole or volume fraction and the total pressure. For example:

$$P_i = [\text{volume fraction}_i \text{ or mole fraction}_i \times P_{\text{total}}]$$
$$= \left[(\text{ppm}_v)_i \times 10^{-6} \times P_{\text{total}} \right] \tag{2.10}$$

Rearranging Equation 2.10 shows that ppm_v values can be calculated from partial pressures as follows:

$$\text{ppm}_v = \frac{P_i}{P_{\text{total}}} \times 10^6 \tag{2.11}$$

Partial pressure can thus be added to the list of unit types that can be used to calculate ppm_v. That is, either volume (Equation 2.5), moles (Equation 2.8), or partial pressures (Equation 2.11) can be used in ppm_v calculations.

Example 2.6 applies these principles to the partial pressure of a formerly popular family of chemical compounds known as PCBs, as illustrated in Figure 2.1.

2.4   Mole/Volume Units

Units of **moles per liter** (molarity, M) are used to report concentrations of compounds dissolved in water. **Molarity** is defined as the number of moles of compound per liter of solution. Concentrations expressed in these units are read as **molar**.

Molarity, M, should not be confused with molality, m. Molarity is usually used in equilibrium calculations and throughout the remainder of this book. **Molality** is the number of moles of a solute added to exactly 1 L of solvent. Thus, the actual volume of a molal solution is slightly larger than 1 L. Molality is more likely to be used when properties of the solvent, such as boiling and freezing points, are a concern. Therefore, it is rarely used in environmental situations.

example/2.6 Concentration as Partial Pressure

The concentration of gas-phase PCBs in the air above Lake Superior was measured to be 450 picograms per cubic meter (pg/m^3). What is the partial pressure (in atm) of PCBs? Assume the temperature is $0°C$, the atmospheric pressure is 1 atm, and the average molecular weight of PCBs is 325.

solution

The partial pressure is defined as the mole or volume fraction times the total gas pressure. First, find the number of moles of PCBs in a liter of air. Then use the ideal gas law (Equation 2.7) to calculate that 1 mole of gas at $0°C$ and 1 atm occupies 22.4 L. Substitute this value into the first expression to determine the mole fraction of PCBs:

$$450 \frac{pg}{m^3 \text{ air}} \times \frac{\text{mole}}{325 \text{ g}} \times 10^{-12} \frac{g}{pg} \times 10^{-3} \frac{m^3}{L} = 1.38 \times 10^{-15} \frac{\text{mole PCB}}{\text{L air}}$$

$$1.38 \times 10^{-15} \frac{\text{mole PCB}}{\text{L air}} \times \frac{22.4 \text{ L}}{\text{mole air}} = 3.1 \times 10^{-14} \frac{\text{mole PCB}}{\text{mole air}}$$

Multiplying the mole fraction by the total pressure (1 atm) (see Equation 2.10) yields the PCB partial pressure of 3.1×10^{-14} atm.

example/2.7 Concentration as Partial Pressure Corrected for Moisture

What would be the partial pressure (in atm) of carbon dioxide (CO_2) when the barometer reads 29.0 inches of Hg, the relative humidity is 80 percent, and the temperature is $70°F$? Use Table 2.3 to obtain the concentration of CO_2 in dry air.

solution

The partial-pressure concentration units in Table 2.3 are for dry air, so the partial pressure must first be corrected for the moisture present in the air. In dry air, the CO_2 concentration is 391 ppm_v. The partial pressure will be this volume fraction times the total pressure of dry air. The total pressure of dry air is the total atmospheric pressure (29.0 in. Hg) minus the contribution of water vapor. The vapor pressure of water at $70°F$ is 0.36 lb/in^2. Thus, the total pressure of dry air is

$$P_{total} - P_{water} = 29.0 \text{ in. Hg} - \left[0.36 \frac{lb}{in.^2} \times \frac{29.9 \text{ in. Hg}}{14.7 \text{ lb/in.}^2} \times 0.8\right]$$

$$= 28.4 \text{ in. Hg}$$

The partial pressure of CO_2 would be:

$$\text{vol. fraction} \times P_{total} = 391 \text{ ppm}_v \times \frac{10^{-6} \text{ vol. fraction}}{ppm_v} \times \left[28.4 \text{ in. Hg} \times \frac{1 \text{ atm}}{29.9 \text{ in. Hg}}\right] = 3.7 \times 10^{-4} \text{ atm}$$

example/ 2.8 Concentration as Molarity

The concentration of TCE is 5 ppm. Convert this to units of molarity. The molecular weight of TCE is 131.5 g/mole.

solution

Remember, in water, ppm_m is equivalent to mg/L, so the concentration of TCE is 5.0 mg/L. Conversion to molarity units requires only the molecular weight:

$$\frac{5.0 \text{ mg TCE}}{L} \times \frac{1 \text{ g}}{10^3 \text{ mg}} \times \frac{1 \text{ mole}}{131.5 \text{ g}} = \frac{3.8 \times 10^{-5} \text{ moles}}{L}$$

$$= 3.8 \times 10^{-5} \text{ M}$$

Often, concentrations below 1 M are expressed in units of millimoles per liter, or millimolar ($1 \text{ mM} = 10^{-3}$ moles/L), or in micromoles per liter, or micromolar ($1 \mu M = 10^{-6}$ moles/L). Thus, the concentration of TCE could be expressed as 0.038 mM or 38 μM.

example/ 2.9 Concentration as Molarity

The concentration of alachlor, a common herbicide, in the Mississippi River was found to range from 0.04 to 0.1 μg/L. What is the concentration range in nmoles/L? The molecular formula for alachlor is $C_{14}H_{20}O_2NCl$, and its molecular weight is 270.

solution

The lowest concentration range in nmoles/L can be found as follows:

$$\frac{0.04 \ \mu g}{L} \times \frac{mole}{270 \text{ g}} \times \frac{10^{-6} \text{ g}}{\mu g} \times \frac{10^9 \text{ nmole}}{mole} = \frac{0.15 \text{ nmole}}{L}$$

Similarly, the upper limit (0.1 μg/L) can be calculated as 0.37 nmoles/L.

2.5 Other Types of Units

Concentrations can also be expressed as normality, expressed as a common constituent, or represented by effect.

2.5.1 NORMALITY

Normality (equivalents/L) typically is used in defining the chemistry of water, especially in instances where acid–base and oxidation–reduction reactions are taking place. Normality is also used frequently in the laboratory during the analytical measurement of water constituents. For example, "Standard Methods for the Examination of Water and Wastewater" has many examples where concentrations of chemical reagents are prepared and reported in units of normality and not molarity.

Reporting concentration on an **equivalent basis** is useful because if two chemical species react and the two species reacting have the same strength on an equivalent basis, a 1 mL volume of reactant number 1 will react with a 1 mL volume of reactant number 2. In acid–base chemistry, the number

example/2.10 Calculations of Equivalent Weight

What are the equivalent weights of HCl, H_2SO_4, NaOH, $CaCO_3$, and aqueous CO_2?

solution

To find the equivalent weight of each compound, divide the molecular weight by the number of equivalents:

$$\text{eqv wt of HCl} = \frac{(1+35.5)\text{g/mole}}{1\text{ eqv/mole}} = \frac{36.5\text{ g}}{\text{eqv}}$$

$$\text{eqv wt of } H_2SO_4 = \frac{(2\times1)+32+(4\times16)\text{g/mole}}{2\text{ eqv/mole}} = \frac{49\text{ g}}{\text{eqv}}$$

$$\text{eqv wt of NaOH} = \frac{(23+16+1)\text{g/mole}}{1\text{ eqv/mole}} = \frac{40\text{ g}}{\text{eqv}}$$

$$\text{eqv wt of } CaCO_3 = \frac{40+12+(3\times16)\text{g/mole}}{2\text{ eqv/mole}} = \frac{50\text{ g}}{\text{eqv}}$$

Determining the equivalent weight of aqueous CO_2 requires additional information. Aqueous carbon dioxide is not an acid until it hydrates in water and forms carbonic acid ($CO_2 + H_2O \rightarrow H_2CO_3$). So aqueous CO_2 really has 2 eqv/mole. Thus, one can see that the equivalent weight of aqueous carbon dioxide is

$$\frac{12+(2\times16)\text{g/mole}}{2\text{ eqv/mole}} = \frac{22\text{ g}}{\text{eqv}}$$

of equivalents per mole of acid equals the number of moles of H^+ the acid can potentially donate. For example, HCl has 1 equivalent/mole, H_2SO_4 has 2 equivalents/mole, and H_3PO_4 has 3 equivalents/mole. Likewise, the number of equivalents per mole of a base equals the number of moles of H^+ that will react with 1 mole of the base. Thus, NaOH has 1 equivalent/ mole, $CaCO_3$ has 2 equivalents/mole, and PO_4^{3-} has 3 equivalents/mole.

In oxidation–reduction reactions, the number of equivalents is related to how many electrons a species donates or accepts. For example, the number of equivalents of Na^+ is 1 (where e^- equals an electron) because $Na \rightarrow Na^+ + e^-$. Likewise, the number of equivalents for Ca^{2+} is 2 because $Ca \rightarrow Ca^{2+} + 2e^-$. The **equivalent weight** (in grams (g) per equivalent (eqv)) of a species is defined as the molecular weight of the species divided by the number of equivalents in the species (g/mole divided by eqv/mole equals g/eqv).

All aqueous solutions must maintain charge neutrality. Another way to state this is that the sum of all cations on an equivalent basis must equal the sum of all anions on an equivalent basis. Thus, water samples can be checked to determine whether something is incorrect in the analyses or a constituent is missing. Example 2.12 showed how this is done.

example/2.11 Calculation of Normality

What is the normality (N) of 1 M solutions of HCl and H_2SO_4?

solution

$$1\,M\,HCl = \frac{1\,mole\,HCl}{L} \times \frac{1\,eqv}{mole} = \frac{1\,eqv}{L} = 1\,N$$

$$1\,M\,H_2SO_4 = \frac{1\,mole\,H_2SO_4}{L} \times \frac{2\,eqv}{mole} = \frac{2\,eqv}{L} = 2\,N$$

Note that on an equivalent basis, a 1 M solution of sulfuric acid is twice as strong as a 1 M solution of HCl.

example/2.12 Use of Equivalents in Determining the Accuracy of a Water Analysis

Prof. Mihelcic was in the city of Dunedin in New Zealand to view yellow-eyed penguins and albatrosses. The label on a bottle of New Zealand mineral water he purchased there states that a chemical analysis of the mineral water resulted in the following cations and anions being identified with corresponding concentrations (in mg/L):

$$[Ca^{2+}] = 2.9 \quad [Mg^{2+}] = 2.0 \quad [Na^+] = 11.5 \quad [K^+] = 3.3$$
$$[SO_4^{2-}] = 4.7 \quad [Fl^-] = 0.09 \quad [Cl^-] = 7.7$$

Is the analysis correct?

solution

First, convert all concentrations of major ions to an equivalent basis. To do this, multiply the concentration in mg/L by a unit conversion (g/1,000 mg) and then divide by the equivalent weight of each substance (g/eqv). Then sum the concentrations of all cations and anions on an equivalent basis. A solution with less than 5 percent error generally is considered acceptable.

Cations

$$[Ca^{2+}] = \frac{1.45 \times 10^{-4}\,eqv}{L}$$

$$[Mg^{2+}] = \frac{1.67 \times 10^{-4}\,eqv}{L}$$

$$[Na^+] = \frac{5 \times 10^{-4}\,eqv}{L}$$

$$[K^+] = \frac{8.5 \times 10^{-5}\,eqv}{L}$$

Anions

$$[SO_4^{2-}] = \frac{9.75 \times 10^{-5}\,eqv}{L}$$

$$[Fl^-] = \frac{4.73 \times 10^{-6}\,eqv}{L}$$

$$[Cl^-] = \frac{2.17 \times 10^{-4}\,eqv}{L}$$

The total amount of cations equals 9.87×10^{-4} eqv/L, and the total amount of anions equals 3.2×10^{-4} eqv/L.

The analysis is not within 5 percent. The analysis resulted in more than three times more cations than anions on an equivalent basis. Therefore, either of two conclusions is possible: (1) One or more of the reported concentrations are incorrect, assuming all major cations and anions are accounted for. (2) One or more important anions were not accounted for by the chemical analysis. (Bicarbonate, HCO_3^-, would be a good guess for the missing anion, as it is a common anion in most natural waters.)

2.5.2 CONCENTRATION AS A COMMON CONSTITUENT

Concentrations can be reported as a **common constituent** and can therefore include contributions from a number of different chemical compounds. GHGs, nitrogen, and phosphorus are chemicals that have their concentration typically reported as a common constituent.

For example, the phosphorus in a lake, estuary, untreated, or reclaimed wastewater may be present in inorganic forms called orthophosphates (H_3PO_4, $H_2PO_4^-$, HPO_4^{2-}, PO_4^{3-} HPO_4^{2-} complexes), polyphosphates (e.g., $H_4P_2O_7$ and $H_3P_3O_{10}^{2-}$), metaphosphates (e.g., $HP_3O_9^{2-}$), and/or organic phosphates. Because phosphorus can be chemically converted between these forms and can thus be found in several of these forms, it makes sense at some times to report the total P concentration, without specifying which form(s) are present. Thus, each concentration for every individual form of phosphorus is converted to mg P/L using the molecular weight of the individual species, the molecular weight of P (which is 32), and simple stoichiometry. These converted concentrations of each individual species can then be added to determine the total phosphorus concentration. The concentration is then reported in units of mg/L as phosphorus (written as mg P/L, mg/L as P, or mg/L P).

The alkalinity and hardness of a water typically are reported by determining all of the individual species that contribute to either alkalinity or hardness, then converting each of these species to **units of mg CaCO$_3$/L**, and finally summing up the contribution of each species. Hardness is thus typically expressed as mg/L as $CaCO_3$.

The **hardness** of a water is caused by the presence of divalent cations in water. Ca^{2+} and Mg^{2+} are by far the most abundant divalent cations in natural waters, though Fe^{2+}, Mn^{2+}, and Sr^{2+} may contribute as well.

Nitrogen and Phosphorus Pollution Policy and Data
www.epa.gov/nandppolicy/index.html

Nutrient-Caused Hypoxia in the Gulf of Mexico
toxics.usgs.gov/hypoxia

The Nitrogen Cycle
www.esrl.noaa.gov/gmd/outreach/lesson_plans/The Nitrogen Cycle.pdf

---

example/ 2.13   Nitrogen Concentrations as a Common Constituent

A water contains two nitrogen species. The concentration of NH_3 is 30 mg/L NH_3, and the concentration of NO_3^- is 5 mg/L NO_3^-. What is the total nitrogen concentration in units of mg N/L?

solution

Use the appropriate molecular weight and stoichiometry to convert each individual species to the requested units of mg N/L and then add the contribution of each species:

$$\frac{30 \text{ mg } NH_3}{L} \times \frac{\text{mole } NH_3}{17 \text{ g}} \times \frac{\text{mole N}}{\text{mole } NH_3} \times \frac{14 \text{ g}}{\text{mole N}}$$

$$= \frac{24.7 \text{ mg } NH_3 - N}{L}$$

$$\frac{5 \text{ mg } NO_3^-}{L} \times \frac{\text{mole } NO_3^-}{62 \text{ g}} \times \frac{\text{mole N}}{\text{mole } NO_3} \times \frac{14 \text{ g}}{\text{mole N}}$$

$$= \frac{1.1 \text{ mg } NO_3^- - N}{L}$$

$$\text{total nitrogen concentration} = 24.7 + 1.1 = \frac{25.8 \text{ mg N}}{L}$$

In Michigan, Wisconsin, and Minnesota, untreated waters usually have a hardness of 121–180 mg/L as $CaCO_3$. In Illinois, Iowa, and Florida water is harder, with many values greater than 180 mg/L as $CaCO_3$.

To find the total hardness of a water sample, sum the contributions of all divalent cations after converting their concentrations to a common constituent. To convert the concentration of specific cations (from mg/L) to hardness (as mg/L $CaCO_3$), use the following expression, where M^{2+} represents a divalent cation:

$$\frac{M^{2+} \text{ in mg}}{L} \times \frac{50}{\text{eqv wt of } M^{2+} \text{ in g/eqv}} = \frac{mg}{L} \text{ as } CaCO_3 \qquad (2.13)$$

The 50 in Equation 2.13 represents the equivalent weight of calcium carbonate (100 g $CaCO_3$/2 equivalents). The equivalent weights (in units of g/eqv) of other divalent cations are Mg, 24/2; Ca, 40/2; Mn, 55/2; Fe, 56/2; and Sr, 88/2.

2.5.3 CONCENTRATIONS OF CARBON DIOXIDE AND OTHER GHGs

The **Kyoto Protocol** is a global agreement to regulate six major GHGs. It was adopted in Kyoto, Japan, in 1997 and entered into force in 2005. It sets binding targets for 37 industrialized countries and the European Union to reduce GHG emissions. Each gas has a different ability to

---

example/2.14 Determination of a Water's Hardness

Water has the following chemical composition: $[Ca^+] = 15$ mg/L; $[Mg^{2+}] = 10$ mg/L; $[SO_4^{2-}] = 30$ mg/L. What is the total hardness in units of mg/L as $CaCO_3$?

solution

Find the contribution of hardness from each divalent cation. Anions and all nondivalent cations are not included in the calculation.

$$\frac{15 \text{ mg } Ca^{2+}}{L} \times \left(\frac{\frac{50 \text{ g } CaCO_3}{eqv}}{\frac{40 \text{ g } Ca^{2+}}{2 \text{ eqv}}} \right) = \frac{38 \text{ mg}}{L} \text{ as } CaCO_3$$

$$\frac{10 \text{ mg } Mg^{2+}}{L} \times \left(\frac{\frac{50 \text{ g } CaCO_3}{eqv}}{\frac{24 \text{ g } Mg^{2+}}{2 \text{ eqv}}} \right) = \frac{42 \text{ mg}}{L} \text{ as } CaCO_3$$

Therefore, the total hardness is $38 + 42 = 80$ mg/L as $CaCO_3$. This water is moderately hard.

Note that if reduced iron (Fe^{2+}) or manganese (Mn^{2+}) were present, they would be included in the hardness calculation.

absorb heat in the atmosphere (the radiative forcing), so each differs in its global warming potential (GWP). The Kyoto Protocol has been ratified by 191 states (i.e., countries). However, it was not adopted by many large emitters of GHGs, including the United States. Furthermore, in 2011, Canada renounced their earlier support.

Though the U.S. government has not ratified the Kyoto Protocol, in 2007, the U.S. Supreme Court ruled that EPA has the authority under the Clean Air Act to regulate emissions of carbon dioxide and other GHGs. In October 30, 2009, EPA published a rule in the Federal Register (40 CFR Part 98) that required mandatory reporting of GHGs from large sources. The implementation of this rule is referred to as the **Greenhouse Gas Reporting Program**. It applies to a wide range of GHG emitters that includes fossil fuel suppliers, industrial gas suppliers, and facilities that inject CO_2 underground for sequestration. This movement to regulate GHGs as air pollutants was further confirmed in 2012 when the U.S. Court of Appeals for the District of Columbia unanimously upheld the first ever proposed regulations to regulate emissions of GHGs.

The **global warming potential (GWP)** is a multiplier used to compare the emissions of different greenhouse gases to a common constituent, in this case carbon dioxide. The GWP is determined over a set time period, typically 100 years, over which the radiative forcing of the specific gas would result. GWPs allow policy makers to compare emissions and reductions of specific gases.

Carbon dioxide equivalents are a metric measure used to compare the mass emissions of greenhouse gases to a common constituent, based on the specific gas's global warming potential. Units are mass based and typically a million metric tons of carbon dioxide equivalents. Table 2.4 provides global warming potentials for the six major greenhouse gases. Note that an equivalent mass release of two greenhouse gases does not have the same impact on global warming. For example, from Table 2.4, we can see that 1 ton of methane emissions equates to 25 tons of carbon dioxide emissions.

Greenhouse Gas Reporting Program
www.epa.gov/ghgreporting

Regulating Greenhouse Gases
www.epa.gov/climatechange/endangerment

United Nations Framework Convention on Climate Change
http://unfccc.int

Intergovernmental Panel on Climate Change
www.ipcc.ch

Table / 2.4

100-Year Global Warming Potentials (GWPs) Used to Convert Mass Greenhouse Gas Emissions to Carbon Dioxide Equivalents (CO_2e)

Type of Emission	Multiplier for CO_2 Equivalents (CO_2e)
Carbon dioxide	1
Methane	25
Nitrous oxide	298
Hydrofluorocarbons (HFCs)	124–14,800 (depends on specific HFC)
Perfluorocarbons (PFCs)	7,390–12,200 (depends on specific PFC)
Sulfur hexafluoride (SF_6)	22,800

SOURCE: Values from *Climate Change 2007: A Physical Science Basis*, Intergovernmental Panel on Climate Change. Note that EPA reports that they use 100-year GWPs listed in the IPCC's Second Assessment Report to be consistent with the international standards under the United Nations Framework Convention on Climate Change.

Table / 2.5

U.S. Greenhouse Gas Emissions from Sources Relevant to Environmental and Civil Engineering   Total GHG emissions in 2010 were 6,821.8 Tg CO_2 equivalents. $1 \text{ Tg} = 10^{12}$ g or 1 million metric tons.

Source (Gas)	CO_2 Equivalents (Tg)	Source (Gas)	CO_2 Equivalents (Tg)
Fossil fuel combustion (CO_2)	5,387.8	Agricultural soil management (N_2O)	207.8
Iron and steel production (CO_2)	54.3	Manure management (N_2O)	18.3
Cement production (CO_2)	30.5	Wastewater treatment (N_2O)	5.0
Transportation (CO_2)	1,745.5	Composting (N_2O)	1.7
Soda ash manufacture and consumption (CO_2)	3.7	Mobile combustion (N_2O)	20.6
Landfills (CH_4)	107.8	Stationary combustion (N_2O)	22.6
Manure management (CH_4)	52.0	Substitution of ozone-depleting substances (HFCs)	114.6
Wastewater treatment (CH_4)	16.3	Electrical transmission and distribution (SF_6)	11.8
Rice cultivation (CH_4)	8.6	Semiconductor manufacture (HFCs, PFCs)	4.4

SOURCE: Data from EPA (2012).

Learn the Impact of Climate Change Where You Live

www.epa.gov/climatechange

Greenhouse gas emissions are sometimes also reported as **carbon equivalents**. In this case, the mass of carbon dioxide equivalents is multiplied by 12/44 to obtain carbon equivalents. The multiplier 12/44 is the molecular weight of carbon (C) divided by the molecular weight of carbon dioxide (CO_2). Table 2.5 gives some relevant U.S. greenhouse gas emissions in units of CO_2 equivalents (abbreviated CO_2e). Note the expected large contribution associated with energy use from burning fossil fuels. But also note the amount of greenhouse gas emissions associated with other human activities. It is clear from this table that sustainable development will require that every engineering assignment consider how to reduce the overall emissions of greenhouse gases.

The largest amount of greenhouse gas emissions are from carbon dioxide. In 2010, total U.S. greenhouse gas emissions were 6,821.8 Tg CO_2e (a Tg equals 1 million metric tons) of which 5,706.4 CO_2e were carbon dioxide emissions.

Table 2.6 provides the breakdown of the largest sources of carbon dioxide emissions in the United States over time. Note how the sources of emissions listed in this table are associated with decisions engineers make that impact the design and operation of infrastructure associated with electricity production, transportation, and buildings. There has been a 10.5 percent increase in greenhouse gas emissions since 1990 in the United States, with an average increase in emissions

Table / 2.6

Largest Sources of Carbon Dioxide Emissions in the United States Over 20 Years

Source	1990	2007	2010
Fossil fuel combustion	4,738.3	6,118.6	5,387.8
Electricity generation	1,820.8	2,412.8	2,258.4
Transportation	1,485.9	1,893.9	1,745.5
Industrial	846.4	844.4	777.8
Residential	338.3	341.6	340.2
Commercial	219.0	218.9	224.2
Total CO_2 emissions	5,100.5	6,107.6	5,706.4

of about 0.5%. There was a slight decrease in emissions from 2007 to 2009 because of the economic downturn. Emissions increased again from 2009 to 2010, primarily because of an increase in economic output that caused an increase in energy consumption and much warmer summer conditions that resulted in an increase in electricity demand for cooling buildings that is currently provided primarily by burning coal and natural gas.

In terms of waste management, the primary source of greenhouse gas emissions comes from solid waste management (versus management of drinking water, wastewater, and reclaimed water). EPA reports that in 2010 landfills accounted for approximately 16.2 percent of total U.S. anthropogenic methane emissions. This source is the third largest contribution of methane in the United States with only natural gas systems and enteric fermentation associated with domesticated animals being larger. Of the 264 Tg of CO_2e produced by U.S. landfills, only 107.8 Tg CO_2e were emitted because of recovery, flaring, and oxidizing produced methane. In comparison, wastewater treatment and reclamation accounted for approximately 2.5 percent of methane emissions and composting of organic waste accounted for less than 1 percent of total methane emissions. Note there are also biogenic emissions of CO_2, N_2O, and CH_4 associated with the treatment of wastewater as the complex organic matter that makes up wastewater (measured as biochemical or chemical oxygen demand) decomposes to more simpler chemical forms such as CO_2.

A **carbon footprint** is defined as the total greenhouse gas emissions (reported in carbon equivalents) that are associated with a product, service, company, or other entity such as a household or water treatment plant. It consists of direct and indirect greenhouse gas emissions. **Direct emissions** are from sources owned or controlled by the reporting entity. **Indirect emissions** are a consequence of activities of the reporting entity, but they occur at other sources

Household Carbon Footprint Calculator

http://www.epa.gov/climatechange/ghgemissions/ind-calculator.html

that are owned or controlled by another entity (Greenhouse Gas Protocol, 2012). Table 2.7 gives a further categorization of direct and indirect greenhouse gas emissions.

Categorization of Direct and Indirect Greenhouse Gas Emissions

Emission Type	Explanation
Scope 1 emissions	All direct emissions (i.e., sources owned or controlled by the reporting entity)
Scope 2 emissions	Indirect emissions from consumption of purchased electricity, heat, or steam
Scope 3 emissions	Other indirect emissions (e.g., extraction and production of purchased materials and fuels, transport-related activities in vehicles not owned or controlled by a reporting entity, outsourced activities, waste disposal)

SOURCE: Extracted from Greenhouse Gas Protocol (2012).

example / 2.15 Carbon Equivalents as a Common Constituent

The U.S. greenhouse gas emissions reported in the year 2010 were 5,706.4 teragrams (Tg) CO_2e of carbon dioxide (CO_2), 666.5 Tg CO_2e of methane (CH_4), and 306.2 Tg CO_2e of N_2O. How many gigagrams (Gg) of CH_4 and N_2O were emitted in 2010? There are 1,000 gigagrams in one teragram.

solution

The solution requires a unit conversion:

$$TgCO_2e = (Gg \text{ of gas}) \times GWP \times \frac{Tg}{1,000\,Gg}$$

For methane:

$$666.5\,Tg\,CO_2e = (Gg \text{ of methane gas}) \times 25 \times \frac{Tg}{1,000\,Gg}$$

2.67×10^4 Gg of methane were emitted in 2010.

For N_2O:

$$306.2\,Tg\,CO_2e = (Gg \text{ of } N_2O \text{ gas}) \times 298 \times \frac{Tg}{1,000\,Gg}$$

1.03×10^3 Gg of nitrogen oxide were emitted in 2010.

   If you go to the U.S. Environmental Protection Agency web site (www.epa.gov), you can learn more about emissions and sinks of different greenhouse gases in the United States. The Intergovernmental Panel on Climate Change web site (www.ipcc.ch) has updated information on the status of global climate change.

The **Emissions and Generation Resource Integrated Database (eGRID)** allows a user to develop greenhouse gas inventories and carbon footprints. eGRID determines GHG emissions associated with electricity generation (e.g., MWh, GWh) by converting electrical usage into lb of CO_2, CH_4, and N_2O emissions and lb of CO_2e.

What is unique about eGRID is it makes this conversion using the energy mix portfolio that is unique to a particular region of the United States. This is because the greenhouse gas emissions associated with electricity generation from consuming a specific amount of electricity differs around the country. The reason for this is because a region's energy mix used to produce electricity can consist of coal, natural gas, nuclear, hydro, biomass, wind, and solar. eGRID thus provides conversion factors that allow a user to convert electricity usage (reported as MWh or GWh) to lb of CO_2, CH_4, N_2O, and CO_2e.

Table 2.8 provides a few examples of these conversion factors for different regions of the United States. Note how eGRID CO_2e output emissions are 0.2–1.4% greater than CO_2 output emissions because of the addition of CH_4 and N_2O emissions.

eGRID is based on generation of electricity and does not account for line losses from the point of generation to the point of consumption. That is, 100 kWh of electricity consumption requires slightly more than 100 kWh of electricity generation. In terms of the magnitude, line losses differ around the country: 2.795% in Alaska, 3.691% in Hawaii, 5.333% out west, 6.177% in Texas, and 6.409% in the east (with a U.S. average of 6.179%). Thus, if a user wants to account for line losses in the estimation of greenhouse gas emissions, they would have to divide the eGRID generated greenhouse gas emissions by (1 percent line losses/100) to determine the total greenhouse gas emissions that result from consumption of electricity.

Table / 2.8

Comparison of Greenhouse Gas Emission Rates

eGRID subregion name	CO_2 (lb/MWh)	CH_4 (lb/GWh)	N_2O (lb/GWh)	CO_2e (lb/MWh)
WECC California	724.12	30.24	8.08	727.26
SERC Virginia/Carolina	1,134.88	23.77	19.79	1,141.51
SERC Midwest	1,830.51	21.15	30.50	1,840.41
FRCC all (Florida)	1,318.57	45.92	16.94	1,324.79
United States	1,329.35	27.27	20.60	1,336.31

Data from eGRID2007 version 1.1, year 2005 data. See http://www.epa.gov/egrid for data for all 26 U.S. subregions.

example / 2.16   Determine Carbon Footprint from Electricity Consumption Data

Assume you own a building in Virginia or the Carolinas and you consume 11,000 kWh of electricity per year for heating, cooling, lighting, and operation of electronics and appliances. What is the amount of direct greenhouse gas emissions associated with CO_2, CH_4, and N_2O (and the overall carbon footprint) for operating the building? Ignore line losses in your calculations.

solution

Using the conversion factors provided by eGRID (and listed in Table 2.8 for the subregion of Virginia and the Carolinas), you can determine that the emissions of specific greenhouse gas emissions associated with operating this building as 12,484 lb CO_2, 261 lb CH_4, and 218 lb N_2O. There are 1,000 kW in 1 MW and 1,000,000 kW in 1 GW. These emissions do not account for line losses, which are 6.409 percent in the eastern United States. To account for line losses, divide these eGRID-generated emission values by $(1-6.409/100)$.

You can determine the carbon footprint by one of two methods. The easiest is to multiply the electricity consumption of 11,000 kWh by the CO_2e conversion factor of 1,141.51 lb CO_2e/MWh provided by eGRID (and listed in Table 2.8).

$$11,000 \text{ kW} \times 1,141.51 \text{ lb } CO_2e/\text{MWh} \times \text{MW}/1,000 \text{ kW} = 12,556 \text{ lb } CO_2 = 12,556 \text{ lb } CO_2e$$

This results in a value of 12,556 lb CO_2e. You can find the solution in a longer manner, summing the contribution from each of the three greenhouse gases accounted for by eGRID, using the GWPs listed in Table 2.4.

$$11,000 \text{ kW} \times 1,134.88 \text{ lb } CO_2/\text{MWh} \times \text{MW}/1,000 \text{ kW} = 12,484 \text{ lb } CO_2 = 12,484 \text{ lb } CO_2e$$

$$11,000 \text{ kW} \times 23.77 \text{ lb } CH_4/\text{GWh} \times \text{GW}/10^6 \text{ kW} = 0.26 \text{ lb } CH_4 \times 25 \text{ lb } CO_2e/\text{lb } CH_4 = 6.5 \text{ lb } CO_2e$$

$$11,000 \text{ kW} \times 19.79 \text{ lb } N_2O/\text{GWh} \times \text{GW}/10^6 \text{ kW} = 0.22 \text{ lb } CH_4 \times 298 \text{ lb } CO_2e/\text{lb } N_2O = 65.5 \text{ lb } CO_2e$$

The total GHG emissions in CO_2e are the sum of these three values and equals 12,556 lb CO_2e. Note the large amount of CO_2 emissions from electricity generation here compared to the contribution of CH_4 and N_2O (even with their higher GWPs). This value could also be referred to as the carbon footprint of the building for 1 year when only considering direct emissions. And again, these emissions do not account for line losses, which are 6.409% in the eastern United States. To account for line losses, divide these eGRID-generated emission values by $(1-6.409/100)$.

If the building installs solar panels on site to reduce the grid-supplied electricity use by 2,500 kWh/year, the carbon footprint associated with grid-supplied electricity would decrease to 9,703 lb CO_2e, a reduction of 2,854 lb CO_2e.

(Example adapted from Rothschild et al., 2009.)

2.5.4   REPORTING PARTICLE CONCENTRATIONS IN AIR AND WATER

The concentration of particles in an air sample is determined by pulling a known volume (for instance, several thousand m^3) of air through a filter. The increase in weight of the filter due to collection of particles on it can be determined. Dividing this value by the volume of air passed through the filter gives the **total suspended particulate (TSP)** concentration in units of g/m^3 or $\mu g/m^3$.

In aquatic systems and in the analytical determination of metals, the solid phase is distinguished by filtration using a filter opening of 0.45 μm. This size typically determines the cutoff between the *dissolved* and *particulate* phases. In water quality, solids are divided into a *dissolved* or *suspended* fraction. This is done by a combination of filtration and evaporation procedures. Each of these two types of solids can be further broken down into a *fixed* and *volatile* fraction. Figure 2.2 shows the

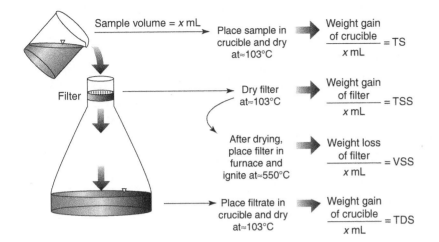

Figure / 2.2 Analytical Differences between Total Solids (TS), Total Suspended Solids (TSS), Volatile Suspended Solids (VSS), and Total Dissolved Solids (TDS).

(From Mihelcic (1999). Reprinted with permission of John Wiley & Sons, Inc.).

analytical differences between total solids (TS), total suspended solids (TSS), total dissolved solids (TDS), and volatile suspended solids (VSS).

Total solids (TS) are determined by placing a well-mixed water sample of known volume in a drying dish and evaporating the water at 103°C to 105°C. The increase in the weight of the drying dish is due to the total solids, so to determine total solids, divide the increase in weight gain of the drying dish by the sample volume. Concentrations typically are reported in mg/L.

To determine **total dissolved solids (TDS)** and **total suspended solids (TSS)**, first filter a well-mixed sample of known volume through a glass-fiber filter with a 2 μm opening. The suspended solids are the particles caught on the filter. To determine the concentration of TSS, dry the filter at 103−105°C, determine the weight increase in the filter, and then divide this weight gain by the sample volume. Results are given in mg/L. Suspended solids collected on the filter may harm aquatic ecosystems by impairing light penetration or acting as a source of nutrients or oxygen-depleting organic matter. Also, a water high in suspended solids may be unsuited for human consumption or swimming.

The TDS are determined by collecting the sample that passes through the filter, drying this filtrate at 103−105°C, and then determining the weight gain of the drying dish. This weight gain divided by the sample volume is the concentration of TDS, stated in mg/L. Dissolved solids tend to be less organic in composition and consist of dissolved cations and anions. For example, hard waters are also high in dissolved solids.

TS, TDS, and TSS can be further broken down into a fixed and volatile fraction. For example, the volatile portion of the TSS is termed the **volatile suspended solids (VSS)**, and the fixed portion is termed the **fixed suspended solids (FSS)**. The way to determine the volatile fraction of a sample is to take each sample just discussed and ignite it in a furnace at 500°C (±50°C). The weight loss due to this high-temperature ignition provides the volatile fraction, and the fixed fraction is what sample remains after ignition.

In wastewater treatment and resource recovery plants, the suspended solids or volatile fraction of suspended solids are used as a measure of the number of microorganisms in the biological treatment process. Figure 2.3 shows how to relate the various solid determinations.

TS	=	TDS	+	TSS
		=		=
TVS	=	VDS	+	VSS
		+		+
TFS	=	FDS	+	FSS

Figure / 2.3 Relationships among the Various Measurements of Solids in Aqueous Samples   For example, if the TSS and VSS are measured, the FSS can be determined by difference.

(From Mihelcic (1999). Reprinted with permission of John Wiley & Sons, Inc.).

2.5.5 REPRESENTATION BY EFFECT

In some cases, the actual concentration of a specific substance is not used at all, especially in instances where mixtures of ill-defined chemicals are present (e.g., in untreated sewage). Instead, **representation by effect** is used. With this approach, the strength of the solution or mixture is defined by some common factor on which all

---

example/2.17   Determining Concentrations of Solids in a Water Sample

A laboratory provides the following analysis obtained from a 50 mL sample of wastewater: total solids $= 200$ mg/L, total suspended solids $= 160$ mg/L, fixed suspended solids $= 40$ mg/L, and volatile suspended solids $= 120$ mg/L.

1. What is the concentration of total dissolved solids of this sample?

2. Suppose this sample was filtered through a glass-fiber filter, and the filter was then placed in a muffle furnace at 550°C overnight. What would be the weight of the solids (in mg) remaining on the filter after the night in the furnace?

3. Is this sample turbid? Estimate the percent of the solids that are organic matter.

solution

1. Refer to Figure 2.3 to see the relationship between the various forms of solids. TDS equals TS minus TSS; thus,

$$TDS = \frac{200\ mg}{L} - \frac{160\ mg}{L} = \frac{40\ mg}{L}$$

2. The solids remaining on the filter are suspended solids. (Dissolved solids would pass through the filter.) Because the filter was subjected to a temperature of 550°C, the measurement was made for the volatile and fixed fraction of the suspended solids, i.e., the VSS and FSS. However, during the ignition phase, the volatile fraction was burned off, while what remained on the filter was the inert or fixed fraction of the suspended solids. Thus, this problem is requesting the fixed fraction of the suspended solids. Accordingly, the 50 mL sample had FSS of 40 mg/L. Therefore,

$$FSS = \frac{40\ mg}{L} = \frac{\text{wt of suspended solids remaining on filter after ignition}}{\text{mL sample}}$$
$$= \frac{x}{50\ mL}$$

The unknown, x, can be solved for and equals 2 mg.

3. The sample is turbid because of the suspended particles, measured as TSS. If the sample was allowed to sit for some time period, the suspended solids would settle, and the overlaying water might not appear turbid. The solids found in this sample contain at least 60-percent organic matter. The total solids concentration is 200 mg/L, and of this, 120 mg/L are volatile suspended solids. Because volatile solids consist primarily of organic matter, we can conclude that approximately 60 percent (120/200) of the solids are organic.

the chemicals within the mixture depend. An example is oxygen depletion from biological and chemical decomposition of the chemical mixture. For many organic-bearing wastes, instead of identifying the hundreds of individual compounds that may be present, it is more convenient to report the effect, in units of the milligrams of oxygen that can be consumed per liter of water. This unit is referred to as either **biochemical oxygen demand (BOD)** or **chemical oxygen demand (COD)**.

Key Terms

- biochemical oxygen demand (BOD)
- carbon dioxide equivalents
- carbon equivalents
- carbon footprint
- chemical oxygen demand (COD)
- common constituent
- direct emissions
- Emissions and Generation Resource Integrated Database (eGRID)
- equivalent basis
- equivalent weight
- fixed suspended solids (FSS)
- global warming potential (GWP)
- Greenhouse Gas Reporting Program

- greenhouse gases (GHGs)
- hardness
- ideal gas law
- indirect emissions
- Intergovernmental Panel on Climate Change
- Kyoto Protocol
- mass/mass concentrations
- molality
- molar
- molarity
- mole fraction
- mole ratio
- moles per liter
- normality
- partial pressure
- parts per billion by mass (ppb_m)

- parts per billion by volume (ppb_v)
- parts per million by mass (ppm or ppm_m)
- parts per million by volume (ppm_v)
- parts per trillion by mass (ppt_m)
- percent by mass
- representation by effect
- total dissolved solids (TDS)
- total solids (TS)
- total suspended particulates (TSP)
- total suspended solids (TSS)
- units of mg $CaCO_3/L$
- universal gas constant
- volatile suspended solids (VSS)

chapter/Two **Problems**

2.1 (a) During drinking water treatment, 17 lb of chlorine are added daily to disinfect 5 million gallons of water. What is the aqueous concentration of chlorine in mg/L? (b) The *chlorine demand* is the concentration of chlorine used during disinfection. The *chlorine residual* is the concentration of chlorine that remains after treatment so the water maintains its disinfecting power in the distribution system. If the chlorine residual is 0.20 mg/L, what is the chlorine demand in mg/L?

2.2 A water sample contains 10 mg NO_3^-/L. What is the concentration in (a) ppm_m, (b) moles/L, (c) mg NO_3^--N, and (d) ppb_m?

2.3 A liquid sample has a concentration of iron (Fe) of 5.6 mg/L. The density of the liquid is 2,000 g/L. What is the Fe concentration in ppm_m?

2.4 Coliform bacteria (e.g., *Escherichia coli*) are excreted in large numbers in human and animal feces. Water that meets a standard of less than one coliform per 100 mL is considered safe for human consumption. Is a 1 L water sample that contains nine coliforms safe for human consumption?

2.5 The treated effluent from a domestic wastewater treatment plant contains ammonia at 9 mg N/L and nitrite at 0.5 mg N/L. Convert these concentrations to mg NH_3/L and mg NO_2^-/L.

2.6 Nitrate concentrations exceeding 44.3 mg NO_3^-/L are a concern in drinking water due to the infant disease known as methemoglobinemia. Nitrate concentrations near three rural wells were reported as 0.01 mg/L NO_3^--N, 1.3 mg/L NO_3^--N, and 20 mg/L NO_3^--N. Do any of these wells exceed the 44.3 ppm_m level?

2.7 Sulfate (SO_4^{2-}) concentration is 10 mg SO_4^{2-}/L and monohydrogen sulfide (HS^-) concentration is 2 mg HS^-/L. What is the total inorganic sulfur concentration in mg S/L?

2.8 Suppose you must determine the amount of hydrogen halides (HCl, HBr, and HF) in the flue gas leaving a chemical reactor. The emission sampling train for hydrogen halide determination calls for a total of 200 mL of 0.1 N H_2SO_4 as an absorbing solution. The absorbing solution will be located on the impingers of the sampling train. (a) How many grams of H_2SO_4 should be added to water to create 200 mL of a 0.1 N H_2SO_4 solution? (b) Calculate the molarity of the 0.1 N H_2SO_4 solution (problem from EPA Air Pollution Training Institute).

2.9 The concentration of cadmium (Cd) in a liquid is known to be 130 ppm at 20°C. Calculate the total quantity of cadmium present in a 1 gallon sample. The sample has a density of 62.4 lb/ft^3 (problem from EPA Air Pollution Training Institute).

2.10 As a quality control check, a sample of acetone is taken from a process to determine the concentration of suspended particulate matter. An 850 mL sample was placed in a beaker and evaporated. The remaining suspended solids were determined to have a mass of 0.001 g. The specific gravity of acetone is 0.79 g/cm^3. (a) Determine the concentration of the sample as mg/L. (b) Determine the concentration of the sample as ppm (problem from EPA Air Pollution Training Institute).

2.11 A paper mill produces paper from wood pulp. Pulp production (at the pulp plant) begins with digesting the wood chips in a solution of sodium hydroxide and sodium sulfide. The sodium hydroxide is diluted with water (shown in the reaction below) prior to being sent to the digester:

$$NaOH + H_2O \rightarrow Na^+ + OH^- + H_2O$$

If 4 kg of sodium hydroxide is added for each 1,000 L of water, determine the following: (a) the molarity of the resulting solution, (b) the normality of the resulting solution (problem from EPA Air Pollution Training Institute).

2.12 In Florida, advanced wastewater treatment standards require that treated effluent have no more than 5 ppm BOD_5, 5 ppm TSS, 3 ppm total nitrogen (TN), and 1 ppm total phosphorus (TP). (a) What is the wastewater standard for TN and TP in mg/L? (b) If all of the nitrogen is transformed to nitrate during the advanced treatment, what is the effluent limit of nitrate in mg/L? (c) If your laboratory had obtained and processed 200 mL sample of treated wastewater for the TSS test, how many mg of suspended solids were captured on the filter for this sample?

2.13 Mirex (MW = 540) is a fully chlorinated organic pesticide manufactured to control fire ants. Due to its structure, mirex is very unreactive, so it persists in the environment. Water samples from Lake Erie have had mirex measured as high as 0.002 µg/L, and lake trout samples have had 0.002 µg/g. (a) In the water samples, what is the aqueous concentration of mirex in units of (i) ppb_m, (ii) ppt_m, and (iii) µM? (b) In the fish samples, what is the concentration of mirex in (i) ppm_m and (ii) ppb_m?

2.14 Total mercury concentrations in the San Francisco Bay Area are reported to be 1.25 ng/L in water, 8 mg/L in rain, 2.1 mg/m³ in air, and 250 ng in 1 g of dry sediment. Report all these concentrations in ppt. Assume the air temperature is 20°C.

2.15 Leachate is produced when precipitation infiltrates a sanitary landfill, contacts the waste material, and appears at the bottom of the stored waste. Assume 6 kg of benzene (molecular formula of C_6H_6) were placed in the landfill and it is all dissolved in the 100,000 gallons of leachate produced during 1 year. What is the benzene concentration in the leachate during this 1 year in (a) mg/L, (b) ppb_m, and (c) moles/L?

2.16 Chlorophenols impart unpleasant taste and odor to drinking water at concentrations as low as 5 mg/m³. They are formed when the chlorine disinfection process is applied to phenol-containing waters. What is the threshold for unpleasant taste and odor in units of (a) mg/L, (b) µg/L, (c) ppm_m, and (d) ppb_m?

2.17 The concentration of monochloroacetic acid in rainwater collected in Zurich was 7.8 nanomoles/L. Given that the formula for monochloroacetic acid is $CH_2ClCOOH$, calculate the concentration in µg/L.

2.18 Assume that concentrations of Pb, Cu, and Mn in rainwater collected in Minneapolis were found to be 9.5, 2.0, and 8.6 µg/L, respectively. Express these concentrations as nmoles/L, given that the atomic weights are 207, 63.5, and 55, respectively.

2.19 The dissolved oxygen (DO) concentration is measured as 0.5 mg/L in the anoxic zone and 8 mg/L near the end of a 108-ft-long aerated biological reactor. What are these two DO concentrations in units of (a) ppm_m and (b) moles/L?

2.20 Assume that the average concentration of chlordane—a chlorinated pesticide now banned in the United States—in the atmosphere above the

Arctic Circle in Norway is 0.6 pg/m³. In this measurement, approximately 90 percent of this compound is present in the gas phase; the remainder is adsorbed to particles. For this problem, assume that all the compound occurs in the gas phase, the humidity is negligibly low, and the average barometric pressure is 1 atm. Calculate the partial pressure of chlordane. The molecular formula for chlordane is $C_{10}Cl_8H_6$. The average air temperature through the period of measurement was −5°C.

2.21 What is the concentration in (a) ppm_v and (b) percent by volume of carbon monoxide (CO) with a concentration of 103 µg/m³? Assume a temperature of 25°C and pressure of 1 atm.

2.22 Ice-resurfacing machines use internal combustion engines that give off exhaust containing CO and NO_x. Average CO concentrations measured in local ice rinks have been reported as high as 107 ppm_v and as low as 36 ppm_v. How do these concentrations compare with an outdoor air quality 1 h standard of 35 mg/m³? Assume the temperature equals 20°C.

2.23 Formaldehyde is commonly found in the indoor air of improperly designed and constructed buildings. If the concentration of formaldehyde in a home is 0.7 ppm_v and the inside volume is 800 m³, what mass (in grams) of formaldehyde vapor is inside the home? Assume $T = 298$ K and $P = 1$ atm. The molecular weight of formaldehyde is 30.

2.24 The concentration of ozone (O_3) in Beijing on a summer day ($T = 30$°C, $P = 1$ atm) is 125 ppb_v. What is the O_3 concentration in units of (a) µg/m³ and (b) moles of O_3 per 10^6 moles of air?

2.25 The National Ambient Air Quality Standard (NAAQS) for sulfur dioxide (SO_2) is 0.14 ppm_v (24 h average). (a) What is the concentration in µg/m³ assuming an air temperature of 25°C? (b) What is the concentration in moles SO_2 per 10^6 moles of air?

2.26 A balloon is filled with exactly 10 g of nitrogen (N_2) and 2 g of oxygen (O_2). The pressure in the room is 1.0 atm and the temperature is 25°C. (a) What is the oxygen concentration in the balloon, expressed as percent by volume? (b) What is the volume (in liters) of the balloon after it has been blown up?

2.27 A gas mixture contains 1.5×10^{-5} mole CO and has a total of 1 mole. What is the CO concentration in ppm_v?

2.28 "Clean" air might have a sulfur dioxide (SO_2) concentration of 0.01 ppm_v, while "polluted" air might have a concentration of 2 ppm$_v$. Convert these two concentrations to $\mu g/m^3$. Assume a temperature of 298 K.

2.29 Carbon monoxide (CO) affects the oxygen-carrying capacity of your lungs. Exposure to 50 ppm$_v$ CO for 90 min has been found to impair one's ability to discriminate stopping distance; therefore, motorists in heavily polluted areas may be more prone to accidents. Are motorists at a greater risk of accidents if the CO concentration is 65 mg/m^3? Assume a temperature of 298 K.

2.30 Humans produce 0.8–1.6 L of urine per day. The annual mass of phosphorus in this urine on a per capita basis ranges from 0.2 to 0.4 kg P. (a) What is the maximum concentration of phosphorus in human urine in mg P/L? (b) What is the concentration in moles P/L? (c) Most of this phosphorus is present as HPO_4^{2-}. What is the concentration of phosphorus in mg HPO_4^{2-}/L?

2.31 Assume 66% of phosphorus in human excrement in found in urine (the remaining 34 percent is found in feces). Assume humans produce 1 L of urine per day and the annual mass of phosphorus in this urine is 0.3 kg P. If indoor water usage is 80 gallons per capita per day in a single individual apartment, what is concentration (in mg P/L) in the wastewater that is discharged from the apartment unit? Account for phosphorus in urine and feces.

2.32 A dry cleaning facility owned by JMA Inc. has been observed to have impacted 6,000 gallons of groundwater with 0.70 lb of tetrachloroethylene (PCE). Assuming all the PCE is present in the dissolved phase and the chemical pollutant is evenly distributed throughout the impacted volume of the groundwater, what is the concentration of PCE in groundwater in ppm?

2.33 A dry cleaning facility has been observed to have impacted 20 m^3 of a saturated groundwater aquifer (porosity of 0.30) with 0.70 lb of tetrachloroethylene (PCE) (molecular formula of C_2Cl_4). A bioremediation system is utilized that degrades all of the PCE present to ethene through the process of reductive dechlorination (molecular formula of ethene is C_2H_4). How many moles/L of chlorine are present in the impacted volume of aquifer after all the PCE is dechlorinated? Porosity is defined as the number of voids (that can fill with air or water) divided by the volume (that includes voids and solids).

2.34 Copper was used as a fungicide in citrus orchards that are being considered for a stormwater retention project as part of the Everglades restoration effort. Copper accumulates in apple snails, a primary food source of the federally endangered bird called the Everglades Snail Kite. The professional engineer at BTA Inc. is considering two former citrus production areas for construction of a stormwater treatment area—a 1,500-acre site with soils having a copper content of 220 ppm and a 2,000-acre site with 160 lb of copper uniformly distributed in the top 6–8 in of soil. Which site would our female engineer recommend for the project if the ecological threshold to support the Everglades Snail Kite is 85 mg Cu/kg? Ecological threshold means that the soil concentration of copper cannot exceed this value. Assume the plow layer of soil (upper 6–8 in.) for the 2,000-acre site weighs 2,000,000 lb.

2.35 The Department of Environmental Quality determined that toxaphene concentrations in soil exceeding 60 $\mu g/kg$ (regulatory action level) can pose a threat to underlying groundwater. (a) If a 100 g sample of soil contains 10^{-5} g of toxaphene, what are the (a) toxaphene soil and (b) regulatory action level concentrations reported in units of ppb$_m$?

2.36 Polycyclic aromatic hydrocarbons (PAHs) are a class of organic chemicals associated with the combustion of fossil fuels. Undeveloped areas may have total PAH soil concentrations of 5 $\mu g/kg$, while urban areas may have soil concentrations that range from 600 to 3,000 $\mu g/kg$. What is the concentration of PAHs in undeveloped areas in units of ppm$_m$?

2.37 The concentration of toluene (C_7H_8) in subsurface soil samples collected after an underground storage tank was removed indicated the toluene concentration was 5 mg/kg. What is the toluene concentration in ppm$_m$?

2.38 While visiting Zagreb, Croatia, Arthur Van de Lay visits the Mimara Art Museum and then takes in the great architecture of the city. He stops at a café in the old town and orders a bottle of mineral water. The reported chemical concentration of this water is $[Na^+] = 0.65$ mg/L, $[K^+] = 0.4$ mg/L, $[Mg^{2+}] = 19$ mg/L, $[Ca^{2+}] = 35$ mg/L, $[Cl^-] = 0.8$ mg/L, $[SO_4^{2-}] = 14.3$ mg/L, $[HCO_3^-] = 189$ mg/L, $[NO_3^-] = 3.8$ mg/L

The pH of the water is 7.3. (a) What is the hardness of the water in mg/L CaCO$_3$? (b) Is the chemical analysis correct?

2.39 The city of Melbourne, Florida, has a surface water treatment plant that produces 20 MGD of potable drinking water. The water source has hardness measured as 94 mg/L as CaCO$_3$ and after treatment, the hardness is reduced to 85 mg/L as CaCO$_3$. (a) Is the treated water soft, moderately hard, or hard? (b) Assuming all the hardness is derived from calcium ion, what would the concentration of calcium be in the treated water (mg Ca^{2+}/L). (c) Assuming all the hardness is derived from magnesium ion, what would the concentration of magnesium be in the treated water (mg Mg^{2+}/L)?

2.40 A laboratory provides the following analysis obtained from a 50 mL sample of water. [Ca^{2+}] = 60 mg/L, [Mg^{2+}] = 10 mg/L, [Fe^{2+}] = 5 mg/L, [Fe^{3+}] = 10 mg/L, Total solids = 200 mg/L, suspended solids = 160 mg/L, fixed suspended solids = 40 mg/L, and volatile suspended solids = 120 mg/L. (a) What is the hardness of this water sample in units of mg/L as CaCO$_3$? (b) What is the concentration of total dissolved solids of this sample? (c) If this sample was filtered through a glass-fiber filter, and then the filter was placed in a muffle furnace at 550°C overnight, what would be the weight of the solids (in mg) remaining on the filter after the night in the furnace?

2.41 In 2010 landfills in the United States produced approximately 107.8 Tg CO$_2$e of methane emissions. Wastewater treatment plants emitted 16.3 Tg CO$_2$e of methane. (a) How many pounds and metric tons of methane (reported as CO$_2$ equivalents) did landfills and wastewater plants emit in 2010? (b) What percent of the total 2010 methane emissions (and greenhouse gas emissions) do these two sources contribute (total methane emissions in 2010 were 666.5 Tg CO$_2$e, and total greenhouse gas emissions in 2010 were 6,821.8 CO$_2$e).

2.42 Mobile combustion of N$_2$O in 2010 emitted 20.6 Tg CO$_2$e. How many Gg of N$_2$O was this?

2.43 Reverse osmosis is used to treat brackish groundwater water and requires 1 kWh of energy per 1 m^3 of treated water. In comparison, reverse osmosis of seawater requires 4 kWh of energy per 1 m^3 of treated water (this difference is because of the higher TDS concentration of seawater). According to eGRID, the carbon dioxide equivalent

emission rate is 1,324.79 lb CO$_2$e/MWh in Florida and 727.26 lb CO$_2$e/MWh in California. Estimate the carbon footprint of using reverse osmosis to desalinate 1 m^3 brackish groundwater and 1 m^3 seawater in Florida and California. Ignore line losses in your estimate.

2.44 Your home in Texas averages 24 kWh/day of electricity use. (a) What is your annual estimate of individual greenhouse gas emissions of CO$_2$, CH$_4$, and N$_2$O for operating the home? (b) What is the carbon footprint (in lb CO$_2$e) for living in the home for 1 year with and without line losses included in the estimate?

2.45 You are considering installing a 10 kW solar system that will provide 14,000 kW h of electricity per year (assume you live in the eGRID subregion SERC Midwest). Assuming your electricity consumption remains the same, how much is your carbon footprint reduced every year (in pounds of CO$_2$) if you install the solar panels?

2.46 You are considering purchasing a new television set and wish to factor in the energy consequences of your purchase. You are considering a 55 in. screen model (screen is 49.75 in. high and 29.75 in. wide) and a 32 in. screen model (screen is 29.1 in. wide and 17.5 in. high). Research shows that for these particular models, the 55 in. screen consumes 0.10 W/in.^{2} and the 32 in. screen consumes 0.17 W/in.^{2} (a) Compare the two televisions by determining the power rating (number of watts) associated with each television size. Report your answer in watts, N-m/s, and J/s. (b) How many kWh of energy are consumed by each screen if you operate the television for 3 h per day? (c) Assuming that operating the television for 1 kWh produces 0.5453 kg of CO$_2$, compare the two screen sizes in terms of their carbon footprint calculated over a 365-day year (assume you operate the television 3 h per day).

2.47 A laboratory provides the following solids analysis for a wastewater sample: TS = 200 mg/L; TDS = 30 mg/L; FSS = 30 mg/L. (a) What is the total suspended solids concentration of this sample? (b) Does this sample have appreciable organic matter? Why or why not?

2.48 A 100 mL water sample is collected from the activated sludge process of municipal wastewater treatment. The sample is placed in a drying dish (weight = 0.5000 g before the sample is added) and then placed in an oven at 104°C until all moisture is evaporated. The weight of the dried dish is recorded as 0.5625 g. A similar 100 mL sample is filtered and the 100 mL liquid sample that passes through the filter is collected and placed in another drying dish (weight $= 0.5000$ g before the sample is added). This sample is dried at 104°C, and the dried dish's weight is recorded as 0.5325 g. Determine the concentration (in mg/L) of (a) total solids, (b) total suspended solids, (c) total dissolved solids, and (d) volatile suspended solids (assume VSS $= 0.7 \times$ TSS).

References

Environmental Protection Agency (EPA), 2012. Inventory of U.S. Greenhouse Gas Emissions and Sinks: 1990–2010, April, 430-R-12-001.

Greenhouse Gas Protocol, http://www.ghgprotocol.org, retrieved June 21, 2012.

Mihelcic, J. R., 1999. *Fundamentals of Environmental Engineering*. New York: John Wiley & Sons.

Rothschild, S. S, D. Quiroz, M. Salhotra, and A. Diem, 2009. The value of eGRID and eGRIDweb to GHG inventories, 13 pages, retrieved from http://www.epa.gov/egrid June 21, 2012.

World Meteorological Organization (WMO), 2012. *Greenhouse Gas Bulletin*, No. 8, November 19, 2012, Geneva, Switzerland.

chapter/Three Chemistry

James R. Mihelcic and
Noel R. Urban

This chapter presents several important chemical processes that describe the behavior of chemicals in both engineered and natural systems. The chapter begins with a discussion of the difference between activity and concentration. It then covers reaction stoichiometry and thermodynamic laws, followed by application of these principles to a variety of equilibrium processes. The basis of chemical kinetics is then explained, as are the rate laws commonly encountered in environmental problems.

Learning Objectives

1. Use ionic strength to calculate activity coefficients for electrolytes and nonelectrolytes.
2. Write balanced chemical reactions.
3. Relate the first and second laws of thermodynamics to engineering practice.
4. Write and apply equilibrium expressions for volatilization, air–water, acid–base, oxidation–reduction, precipitation–dissolution, and sorption reactions.
5. Apply different forms of Henry's law constant to specific environmental engineering situations.
6. Apply mass balance principles to predict the partitioning of chemicals among different environmental media.
7. Estimate how concentrations will change during the course of reactions using kinetic rate expressions for zero-order, first-order, and pseudo first-order reactions.
8. Determine how temperature affects the reaction rate.

PhotoDisc, Inc./Getty Images

3.1 Approaches in Environmental Chemistry

Chemistry is the study of the composition, reactions, and characteristics of matter. It is important because the ultimate fate of many chemicals discharged to air, water, soil, and treatment facilities is controlled by their reactivity and chemical speciation. Design, construction, and operation of treatment processes thus depend on fundamental chemical processes. Furthermore, individuals who predict (model) how chemicals move through indoor environments, groundwater, surface water, soil, the atmosphere, or a reactor are interested in whether a chemical degrades over time and how to mathematically describe the rate of chemical disappearance or equilibrium conditions.

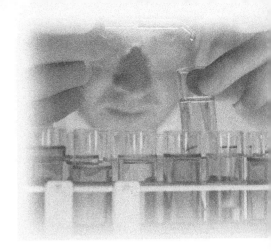

Two very different approaches are used in evaluating a chemical's fate and treatment: kinetics and equilibrium. **Kinetics** deals with the rates of reactions, and **equilibrium** deals with the final result or stopping place of reactions. The kinetic approach is appropriate when the reaction is slow relative to our time frame or when we are interested in the rate of change of concentration. The equilibrium approach is useful whenever reactions are very fast, whenever we want to know in which direction a reaction will go, or whenever we want to know the final, stable conditions that will exist at equilibrium. If reactions happen very rapidly relative to the time frame of our interest, the final conditions that result from the reaction are likely to be of more interest than the rates at which the reaction occurs. In this case, an equilibrium approach is used. Examples of rapid reactions in the aqueous phase include acid–base reactions, complexation reactions, and some phase-transfer reactions, such as volatilization.

3.2 Activity and Concentration

For a substance dissolved in a solvent, the **activity** can be thought of as the effective or apparent concentration, or that portion of the true mole-based concentration of a species that participates in a chemical reaction, normalized to the standard state concentration. In many environmental situations, *activity* and *concentration* are used interchangeably. Places where they begin to greatly differ include seawater, briny groundwater, recycled streams, and highly concentrated waste streams. Activity typically is designated by { } brackets, and concentration by [ ] brackets.

In an **ideal system**, the molar free energy of a solute in water depends on the mole fraction. However, this fraction does not reflect the effect of other dissolved species or the composition of the water, both of which also affect a solute's molar free energy. Chemical species interact by covalent bonding, van der Waals interactions, volume exclusion effects, and long-range electrostatic forces (repulsion and attraction between ions). In dilute aqueous systems, most interactions are caused by long-range electrostatic forces. On a molecular scale, these interactions can lead to local variations in the electron potential of the solution, resulting in a decrease in the total free energy of the system.

The use of activity instead of concentration accounts for these non-ideal effects. Activity is related to concentration by use of activity coefficients. **Activity coefficients** depend on the solution's ionic strength. Several equations (not described here in detail), developed specifically for either electrolytes (ions) or nonelectrolytes (uncharged

UNEP Chemicals Branch
http://www.chem.unep.ch/

species), express the activity coefficient of an individual species as a function of the ionic strength.

The **ionic strength** of a solution (referred to as I or μ) has units of moles/liter and is a measure of the long-range electrostatic interactions in that solution. Ionic strength can be calculated as follows:

$$\mu = 1/2\sum_i C_i z_i^2 \tag{3.1}$$

where C is the molar concentration of an ionic species i in solution and z_i is the charge of the ion. In most natural waters, the ionic strength is derived primarily from the major background cations and anions. Freshwaters typically have an ionic strength of 0.001–0.01 M, and the ocean has an ionic strength of approximately 0.7 M. The ionic strength of aqueous systems rarely exceeds 0.7 M. Fortunately, it can be correlated to easily measured water-quality parameters such as total dissolved solids (TDS) or specific conductance:

$$\mu = 2.5 \times 10^{-5}\,(\text{TDS}) \tag{3.2}$$

where TDS is in mg/L, or

$$\mu = 1.6 \times 10^{-5}\,(\text{specific conductance}) \tag{3.3}$$

where specific conductance is in micromhos per centimeter (μmho/ cm) and is measured with a conductivity meter.

The methods for calculating activity coefficients for electrolytes and nonelectrolytes are summarized in Figure 3.1. *Electrolytes* (for example, Pb^{2+}, SO_4^{2-}, HCO_3^{2-}) have a charge associated with them; *nonelectrolytes* (for example, O_2, H_2SO_4, C_6H_6) do not.

---

example/3.1 Calculating Ionic Strength and Activity Coefficients for Electrolytes

Calculate the ionic strength and all the individual activity coefficients for a 1 L solution of water at 15°C in which 0.01 mole of $FeCl_3$ and 0.005 mole of H_2SO_4 are dissolved.

solution

After the two compounds are placed in water, they will completely dissociate to form 0.01 M Fe^{3+}, 0.01 M H^+, 0.03 M Cl^-, and 0.005 M SO_4^{2-}. The ionic strength is calculated by Equation 3.1:

$$\mu = 1/2[0.01(3+)^2 + 0.01(1+)^2 + 0.03(1-)^2 + 0.005(2-)^2] = 0.075\,\text{M}$$

This ionic strength is relatively high but still much less than that of seawater. The Güntelberg approximation (see Figure 3.1) is useful for calculating activity coefficients for electrolytes when $\mu < 0.1$ M. The value of "A" depends on temperature and equals 0.49 at 0°C, 0.50 at 15°C, and 0.51 at 25°C:

$$\gamma(H^+) = 0.78,\, \gamma(Cl^-) = 0.78,\, \gamma(SO_4^{2-}) = 0.38,\, \gamma(Fe^{3+}) = 0.11$$

The activity coefficients of ions with higher valence deviate much more from 1.0 for a given ionic strength; that is, for electrolytes, use of activity coefficients is much more important for ions with a higher valence, because they are strongly influenced by the presence of other ions. Thus, while at a particular ionic strength, it may not be important to calculate activity coefficients for monovalent ions, it may be very important for di-, tri-, and tetravalent ions.

An air stripper is used to remove benzene (C_6H_6) from seawater and freshwater. Assume the ionic strength of seawater is 0.7 M and that of freshwater is 0.001 M. What is the activity coefficient for benzene in seawater and freshwater?

solution

Because benzene is a nonelectrolyte, use the expression in Figure 3.1 to determine the activity coefficients. The value for k_s (the salting-out coefficient) for benzene is 0.195.

$$\log \gamma = k_s \times \mu$$
$$\log \gamma = 0.195 \times (0.001 \, M): \text{results in } \gamma \text{ (freshwater)} = 1$$
$$\log \gamma = 0.195 \times (0.7 \, M): \text{results in } \gamma \text{ (seawater)} = 1.4$$

For freshwater, the activity coefficient does not deviate much from 1. It turns out there is little deviation for *nonelectrolytes* when $\mu < 0.1$ M. Therefore, determining activity coefficients for non-electrolytes becomes important for solutions with high ionic strengths. For most dilute environmental systems, activity coefficients for electrolytes and nonelectrolytes usually are assumed to be equal to 1. Places where they can gain importance are in the ocean, estuaries, briny groundwater, and some recycled or reused waste streams.

STEP 1

After deciding whether ionic strength effects are important in a particular situation, calculate ionic strength from

$$\mu = \frac{1}{2} \Sigma_i C_i z_i^2 \quad \text{(Equation 3.1)}$$

or

estimate ionic strength after measuring the solution's total dissolved solids (Equation 3.2) or conductivity (Equation 3.3)

STEP 2

If species is an electrolyte, γ will *always be* ≤ 1

If species is a nonelectrolyte, γ will *always be* ≥ 1

For low ionic strengths, $\mu < 0.1$ M,

For high ionic strengths, $\mu < 0.5$ M,

For all ionic strengths, use

use the Güntelberg (or similar) approximation:

use the Davies (or similar) approximation:

$$\log \gamma_i = ks \times \mu$$

$$\log \gamma_i = \frac{-A \, z_i^2 \sqrt{\mu}}{1 + \sqrt{\mu}}$$

$$\log \gamma_i = -A \, z_i^2 \left(\frac{\sqrt{\mu}}{1 + \sqrt{\mu}} - 0.3\mu \right)$$

Figure / 3.1   **Two-Step Process to Determine Activity Coefficients for Electrolytes and Nonelectrolytes** ($A \approx 0.5$ for temperatures of 0–25°C).

(From Mihelcic (1999). Reprinted with permission of John Wiley & Sons, Inc.).

Original Artwork, **"Periodic Circles-2"** by Princess Simpson Rashid. (Courtesy of the Artist). www.princessrashid.com.

© Elena Korenbaum/iStockphoto.

3.3   Reaction Stoichiometry

The **law of conservation of mass** states that in a closed system, the mass of material present remains constant; the material may change form, but the total mass remains the same.

When this law is combined with our understanding that elements may combine with one another in numerous ways but are not converted from one to another (except for nuclear reactions), we arrive at the basis for reaction **stoichiometry**: in a closed system, the number of atoms of each element present remains constant. Therefore, in any single chemical reaction, the number of atoms of each element must be the same on both sides of the reaction equation.

A corollary of the law of conservation of mass is that electrical charges are also conserved; that is, the sum of charges on each side of a reaction equation must be equal. Electrical charges result from the balance between the numbers of protons and electrons present. Protons and electrons both have mass, and neither is converted into other subatomic particles during chemical reactions. Therefore, the total number of protons and electrons must remain constant in a closed system. It follows that the balance between the number of protons and electrons also must remain constant in a closed system. This means that reactions must be balanced in terms of mass and charge, and stoichiometry can be used not only for converting units of concentration but also for calculating chemical inputs and outputs.

3.4   Thermodynamic Laws

As the roots of the word imply (*thermo* equals heat; *dynamo* equals change), **thermodynamics** deals with conversions of energy from one form to another. Table 3.1 provides an overview of the **first law of thermodynamics** and **second law of thermodynamics**. Figure 3.2 illustrates the change in free energy (G) during a reaction. In Figure 3.2, a process could proceed if it reduced the free energy from its value at point A in the direction of point C, but it could not proceed if it raised the energy in the direction of point B. The process could proceed from A as far as point C, but it could not go further toward point D. A reaction could also proceed from point D toward point C or point E. This is because moving in either direction results in a decrease in free energy.

Point E is called a **local equilibrium**. It is not the minimum possible energy point of the system (point C is), but to leave point E requires an input of energy. Hence, if the free energy of a system under all conditions could be quantified, we could then determine the changes that could occur spontaneously in that system (that is, any changes that would cause a decrease in the free energy).

ΔG is the free-energy change under *ambient conditions* (the prevailing environmental conditions). The value of ΔG is calculated according to the following relationship:

$$\Delta G = \Delta G^0 + RT \ln (Q) \tag{3.4}$$

Table / 3.1

Overview of the First and Second Laws of Thermodynamics

Law	What It Tells Us	Mathematical Expression	What It Means to Us
First law of thermodynamics	Energy is conserved; it may be converted from one form to another, but the total amount in a closed system is constant. In an open system, one must account for fluxes across the system boundaries.	For an open system: $$dU = dQ - dW + dG$$ where U = internal energy content, Q = heat content, W = work done, and G = energy of chemical inputs	This relationship demonstrates that the chemical potential (the energy within the chemical bonds of a molecule) constitutes a part of the total energy of the system. In a closed system (in which case, the third term on the right would be absent), reactions that change the chemical potential without changing the internal energy content must result in equivalent changes in heat content and in the pressure–volume work performed.
Second law of thermodynamics	All systems tend to lose useful energy and approach a state of minimum free energy or an equilibrium state. Thus, a process will proceed spontaneously (without energy put into the system from the outside) only if the process leads to a decrease in the free energy of the system (that is, $\Delta G < 0$).	Formal definition of *Gibbs free energy*: $$G = \sum_i \mu_i \times N_i = H - T \times S$$	The Gibbs free energy is related to the system's enthalpy (H), entropy (S), and temperature (T). The energy of inter- and intra-molecular bonds that bind various atoms and molecules together is termed *enthalpy*, while *entropy* refers to the disorder of the system. The chemical potential of all substances present, μ_i, multiplied by the abundance of those substances, N_i, is equal to the combination of enthalpy and entropy present.

where ΔG^0 is the change in free energy determined under *standard conditions*, R is the gas constant, T is the ambient temperature in K, and Q is the reaction quotient. ΔG^0 is determined from reaction stoichiometry and tabulated values as described in most chemistry books. Other references provide detail on determination and application of this term (see, for example, Mihelcic, 1999).

The reaction quotient Q is defined as the product of the activities (apparent concentrations of the reaction products) raised to the power of their stoichiometric coefficients, divided by the product of the activities (or concentrations) of the reactants raised to the

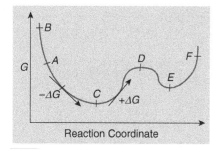

Figure / 3.2 **Change in Free Energy (G) during a Reaction** If the change leads to a decrease in free energy (that is, for the forward reaction, if the slope of a tangent to the curve is negative), then the reaction can proceed spontaneously. Points C and E represent possible equilibrium points because the slopes of tangents at these points would be zero.

(From Mihelcic (1999). Reprinted with permission of John Wiley & Sons, Inc.).

power of their stoichiometric coefficients. Thus, for the generalized reaction,

$$aA + bB \leftrightarrow cC + dD \qquad (3.5)$$

in which a moles of compound A react with b moles of compound B to form c moles of compound C and d moles of compound D, Q is given by

$$Q = \frac{\{C\}^c\{D\}^d}{\{A\}^a\{B\}^b} \qquad (3.6)$$

As noted in Example 3.2, activity coefficients (γ) are usually assumed to equal 1; thus, Q can be calculated based on concentrations. Table 3.2 describes the four rules used to determine what value to use for the activity (concentration) $[i]$ in Equation 3.6. Following these rules is essential to make activities and reaction quotients dimensionless.

Table / 3.2

Rules for Determining Value of $[i]$ These rules determine what value to use for the activity (that is concentration) termed $[i]$ of a chemical species i. Following these rules is essential to make activities and reaction quotients dimensionless.

Rule 1	For liquids (for example, water): $[i]$ is equal to the mole fraction of the solvent. In aqueous solutions, the mole fraction of water can be assumed to equal 1. Thus, $[H_2O]$ always equals 1.
Rule 2	For pure solids in equilibrium with a solution (for example, $CaCO_{3(s)}$, $Fe(OH)_{3(s)}$): $[i]$ always equals 1.
Rule 3	For gases in equilibrium with a solution (for example, $CO_{2(g)}$, $O_{2(g)}$): $[i]$ equals the partial pressure of the gas (units of atm).
Rule 4	For compounds dissolved in water: $[i]$ is always reported in units of moles/L (not mg/L or ppm_m).

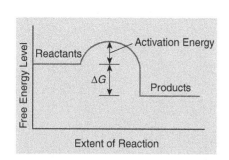

Figure / 3.3 **Energetic Relationships Required for a Reaction to Occur** ΔG is called the Gibbs free-energy change.

(From Mihelcic (1999). Reprinted with permission of John Wiley & Sons, Inc.).

Only reactions that result in thermodynamically favorable changes in their energy state can occur. This change in energy state is called **Gibbs free-energy** change and is denoted ΔG. It is this change in energy state that defines the equilibrium condition. However, not all reactions that occur would result in a favorable change in Gibbs free energy, and the magnitude of this energy change seldom is related to the rate of the reaction. For a reaction to occur, it generally is necessary that atoms collide and that this collision have the right orientation and enough energy to overcome the **activation energy** required for the reaction. These energetic relationships are shown in Figure 3.3.

Equilibrium is defined as the state (or position) with the minimum possible free energy. This occurred at point C in Figure 3.2. The value of the slope at the point of equilibrium (point C) is 0. In other words, the change in free energy is zero at equilibrium. If the change in free energy (ΔG) is equal to 0 at equilibrium, the reaction quotient at equilibrium (see Equation 3.6) is usually written with a special symbol, K, and is provided a special name, the **equilibrium constant**.

The equilibrium constant for the reaction written as Equation 3.5 is given by the equilibrium reaction quotient, Q_{eqn}:

$$Q_{eqn} = \frac{[C]^c [D]^d}{[A]^a [B]^b} = K \qquad (3.7)$$

The equilibrium constant is useful because it provides the ratio of the concentration (or activity) of individual reactants and products for any reaction at equilibrium. Remember, activity coefficients must be included if conditions are not ideal and these coefficients are raised to appropriate stoichiometric values.

Do not confuse the equilibrium constant, K, with the reaction rate constant, k, which we will discuss later in this chapter. K is constant for a specific reaction (as long as temperature is constant). As reviewed in Figure 3.4, equilibrium constants and partition coefficients are defined for reactions that describe volatilization (saturation vapor pressure), air–water exchange (Henry's law constant, K_H), acid–base chemistry (K_a and K_b), oxidation–reduction reactions (K), precipitation–dissolution reactions (K_{sp}), and sorptive partitioning (K_d, K_p, K_{oc}, K).

Application / 3.1   Effect of Temperature on the Equilibrium Constant

Most tabulated equilibrium constants are recorded at 25°C. The **van't Hoff relationship** (Equation 3.8) is used to convert equilibrium constants to temperatures other than those for which the tabulated values are provided. Van't Hoff discovered that the equilibrium constant (K) varied with absolute temperature and the enthalpy of a reaction (ΔH^0). Van't Hoff proposed the following expression to describe this:

$$\frac{d \ln K}{dT} = \frac{\Delta H^0}{RT^2} \qquad (3.8)$$

Here ΔH^0 is found from the heat of formation (ΔH_f^0) for the reaction of interest determined at

standard conditions. Most temperatures encountered in environmental problems are relatively small (for example, 0–40°C). Therefore, the temperature differences are not that large. If ΔH^0 is assumed not to change over the temperature range investigated, Equation 3.8 can be integrated to yield

$$\ln \left[\frac{K_2}{K_1} \right] = \frac{\Delta H^0}{R} \times \left(\frac{1}{T_1} - \frac{1}{T_2} \right) \qquad (3.9)$$

Equation 3.9 can be used to calculate an equilibrium constant for any temperature (that is, temperature 2, T_2) if the equilibrium constant is known at another absolute temperature (T_1, which is usually 20°C or 25°C).

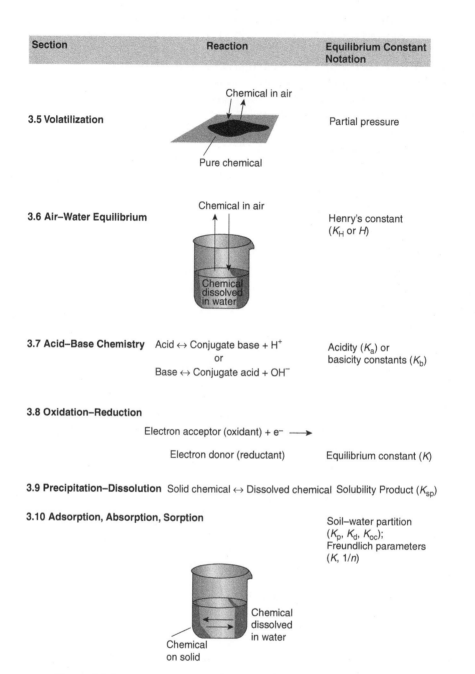

Section	Reaction	Equilibrium Constant Notation
3.5 Volatilization		Partial pressure
3.6 Air–Water Equilibrium		Henry's constant (K_H or H)
3.7 Acid–Base Chemistry	Acid $\leftrightarrow$ Conjugate base + H^+ or Base $\leftrightarrow$ Conjugate acid + OH^-	Acidity (K_a) or basicity constants (K_b)
3.8 Oxidation–Reduction	Electron acceptor (oxidant) + $e^- \longrightarrow$ Electron donor (reductant)	Equilibrium constant (K)
3.9 Precipitation–Dissolution	Solid chemical $\leftrightarrow$ Dissolved chemical	Solubility Product (K_{sp})
3.10 Adsorption, Absorption, Sorption		Soil–water partition (K_p, K_d, K_{oc}); Freundlich parameters (K, $1/n$)

Figure / 3.4 Important Equilibrium Processes for Environmental Engineering.

(Adapted from Mihelcic (1999). Reprinted with permission of John Wiley & Sons, Inc.).

3.5 Volatilization

A key step in the transfer of pollutants between different environmental media is volatilization. All liquids and solids exist in equilibrium with a gas or vapor phase. **Volatilization** (synonymous with *evaporation* for the case of water) is the transformation of a compound from its liquid state to its gaseous state. *Sublimation* is the word used for transformation from the solid to gaseous state. The reverse reaction is termed *condensation*.

Most people have first-hand experience with the phenomenon of sublimation of water. The water vapor in the atmosphere (the humidity) is a function of temperature. Modern refrigerators prevent frost buildup by maintaining a low humidity inside the refrigerator; any ice that forms is sublimed or vaporized. Similarly, the amount of snow on the ground decreases in periods between snowfalls, partially due to the sublimation of the snow. Many organic pollutants volatilize more readily than water. The fumes from gasoline, paint thinners, waxes, and glue attest to the volatility of organic chemicals contained in these commonly used products. Volatilization of chemicals can result in regional and long-range transport of the chemicals to places far away, where adverse environmental effects can be detected (see Figure 3.5).

The equilibrium between a gas and a pure liquid or solid phase is determined by the saturated vapor pressure of a compound. **Saturated vapor pressure** is defined as that partial pressure of the gas phase of a

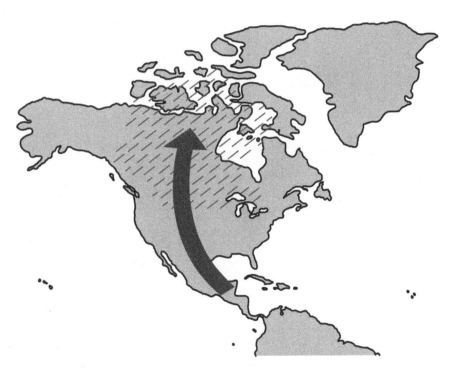

Figure / 3.5 **Spread of Chemicals Through Volatilization** The process of exporting toxic chemicals to other countries that then return by atmospheric transport has been termed "the circle of poison." Persistent organic pollutants (POPs) become concentrated in the food chain, where they can cause toxic effects on animal reproduction, development, and immunological function. The U.S. State Department has termed POPs "one of the great environmental challenges the world faces." POPs include polychlorinated biphenyls (PCBs), polychlorinated dibenzo-*p*-dioxins, and furans, and pesticides such as DDT, toxaphene, chlordane, and heptachlor.

(From Mihelcic (1999). Reprinted with permission of John Wiley & Sons, Inc.).

Though banned for use in many developed countries, POPs are still manufactured for export and/or remain widely used and unregulated in many parts of the world. These chemicals volatilize more easily in the warm surface temperatures found in the southern United States and subtropical and tropical regions of the world. They then condense and deposit in high latitudes where temperatures are cooler. Wildlife such as seals, killer whales, and polar bears along with human populations, such as the Inuit, that reside in the Arctic are unfairly burdened with the environmental risk associated with production and use of these chemicals. Given this information, what equitable and sustainable solutions that consider current and future generations can you think of?

substance that exists in equilibrium with the liquid or solid phase of the substance at a given temperature.

The more volatile a compound, the higher its saturated vapor pressure. For example, the saturated vapor pressure of the solvent tetrachloroethylene (PCE) is 0.025 atm at 25°C, while the saturated vapor pressure of the pesticide lindane is 10^{-6} atm at the same temperature. Clearly, lindane is much less volatile than PCE. For the sake of comparison, water at 25°C has a slightly higher saturated vapor pressure (0.031 atm) than PCE. In other words, if containers or spills of PCE were left exposed to the air in the presence of containers of water, there would be about as much PCE as water in the atmosphere of an indoor-air environment.

The equilibrium between gas and liquid phases can be expressed in the usual form of a chemical reaction with an equilibrium constant:

$$H_2O_{(l)} \leftrightarrow H_2O_{(g)} \tag{3.10}$$

Equation 3.10 indicates that liquid water is in equilibrium with gaseous water (water vapor). The equilibrium constant (called the *saturated vapor pressure*) for this reaction is

$$K = \frac{[H_2O_{(g)}]}{[H_2O_{(l)}]} = P_{H_2O} \tag{3.11}$$

where P_{H_2O} is the partial pressure of water. Because the concentration (assumed to equal the activity, that is, $\gamma = 1.0$) of a pure liquid is defined as 1.0 (remember Rule 1 in Table 3.2), the equilibrium constant is simply equal to the concentration in the vapor phase (called the saturated vapor pressure). One way of expressing gas-phase concentrations is as partial pressures; hence, the equilibrium constant for volatilization often is expressed in units of atmospheres.

If a mixture of miscible (mutually soluble) liquids—rather than a pure liquid—was present, the denominator in Equation 3.11 would be the concentration of the individual liquid (A) in mole fractions, X_A:

$$K = \frac{[A_{(g)}]}{[A_{(l)}]} = \frac{P_A}{X_A} \tag{3.12}$$

Equation 3.12 is known as **Raoult's law**. The constant, K, equals the saturated vapor pressure. Raoult's law is useful whenever a mixture of chemicals (for example, gasoline, diesel fuel, or kerosene) is spilled.

The vapor pressure for all compounds increases with temperature, and at the boiling point of the compound, the vapor pressure equals atmospheric pressure. This statement has practical consequences. First, atmospheric concentrations of volatile substances tend to be higher in summer than in winter, in the day versus at night, and in warmer locations. Second, for any structurally similar group of liquid chemicals exposed to the air, the equilibrium gas-phase concentrations will decrease in order of increasing boiling points.

One feature of environmental problems is that they seldom are confined to just one medium. For example, a lot of the mercury discharged into the environment is first emitted as an air pollutant, but its most damaging effects occur in lakes after it moves through the atmosphere, is deposited into a lake, and then undergoes a biological transformation process called **methylation**. This allows mercury to bioaccumulate in fish, a process that has resulted in thousands of fishing advisories in the U.S. lakes.

Most mercury is released into the environment from combustion of coal associated with electricity production. Even with concerns over use of fossil fuels, U.S. electricity production in the year 2030 is still expected to originate primarily from combustion of coal, not sources of renewable energy. In addition, China is expected to double its coal consumption by 2020, and the migration of its population from rural to urban areas is resulting in increased energy use per capita. China is also expected to surpass the United States in greenhouse gas emissions by the year 2009, mainly because of its plan to consume its vast stores of coal.

The burning of fossil fuels such as coal releases carbon dioxide and other greenhouse gases into the atmosphere. It also incurs other future economic, social, and environmental costs that will be assumed by current and future generations—all from the environmental release of the neurotoxin mercury. Deploying engineering systems and public policies that consider systems and life cycle thinking, conserve energy, use renewable energy sources, and right size buildings can thus have several mutually beneficial impacts to the economy, society, and the environment.

example/3.3   Calculation of Gaseous Concentration in a Confined Area

On a Friday afternoon, a worker spills 1 L of tetrachloroethylene (PCE) on a laboratory floor. The worker immediately closes all the windows and doors and turns off the ventilation in order to avoid contaminating the rest of the building. The worker notifies the appropriate safety authority, but it is Monday morning before the safety official stops by with a crew to clean up the laboratory. Should the cleanup crew bring a mop or an air pump to clean up the room? The volume of the laboratory is 340 m^3, and the temperature in the room is 25°C. For PCE, the vapor pressure is 0.025 atm, the liquid density at 25°C is 1.62 g/cm^3, and the molecular weight is 166 g/mole.

solution

PCE is a volatile chemical. The problem asks how much of the 1 L of spilled PCE remained on the floor versus how much volatilized into the air. If any PCE remained on the floor, the partial pressure of PCE in the air would be 0.025 atm. The ideal gas law can be used to solve for the number of moles present in the air (the term n/V would provide the concentration):

$$n = \frac{PV}{RT} = \frac{(0.025 \text{ atm}) \times (340 \text{ m}^3) \times \left(\frac{1{,}000 \text{ L}}{\text{m}^3}\right)}{\frac{0.08205 \text{ L-atm}}{\text{mole-K}} \times (298 \text{ K})} = 348 \text{ moles}$$

The density of PCE can be used to determine that the 1 L spill weighs 1,620 g. Using the molecular weight of PCE, the 1 L spill would contain 9.8 moles of PCE. This is much less than the amount that could potentially volatilize into the air in the room (348 moles), assuming equilibrium has been attained. Thus, it can be concluded that no PCE would remain on the floor, and it would be entirely in the air. The cleanup crew should arrive at work equipped with air pumps and filters.

example/3.3 (continued)

This problem demonstrates another important point: chemistry and engineering need to become "green." If a green chemical (with zero hazard) were substituted for the solvent PCE in the required use, there would be no risk and, thus, no concern related to the spill. Better yet, perhaps the process that the PCE was used for could be changed so that no chemical is required at all. This type of thinking would result in reduced health costs, because workers would not be exposed to toxic chemicals. Other savings would result because there would be no requirement to pay the cleanup crew for remediation, no energy needed for the remediation phase, less paperwork associated with regulations that govern the handling and storage of the PCE, and no future liability associated with storage and use of PCE. The company might also be able to increase its market share by promoting that its facility is more socially and environmentally responsible.

Green Chemistry
http://www.epa.gov/greenchemistry

3.6 Air–Water Equilibrium

The **Henry's law constant**, K_H, is used to describe a chemical's equilibrium between the air and water (often termed the dissolved or aqueous) phases. This situation is referred to as air–water equilibrium. **Henry's law** is just a special case of Raoult's law (Equation 3.12) applied to dilute systems (most environmental situations are dilute). Because the mole fraction of a dissolved substance in a dilute system is a very small number, concentrations such as moles/L typically are used rather than mole fractions. Equation 3.12 can also be used to estimate Henry's law constants in the absence of reliable experimental data. To determine a Henry's law constant for a particular chemical, divide the saturated vapor pressure of the chemical by its aqueous solubility.

The units of Henry's law constant vary depending on whether the air–water exchange reaction is written in the forward direction for transfer from the gas phase into aqueous phase or from the aqueous phase into the gas phase. In addition, Henry's law constants may also be unitless. Thus, it is important to use the proper units, understand why particular units are used, and be able to convert between different units.

3.6.1 HENRY'S LAW CONSTANT WITH UNITS FOR A GAS DISSOLVING IN A LIQUID

The air–water exchange of a gas (in this case, oxygen) from the atmosphere into water in the forward direction (depicted in Figure 3.4) can be written as

$$O_{2(g)} \leftrightarrow O_{2(aq)} \qquad (3.13)$$

The equilibrium expression for this reaction is

$$K_H = \frac{[O_{2(aq)}]}{[O_{2(g)}]} = \frac{[O_{2(aq)}]}{P_{O_2}} \qquad (3.14)$$

The value of the Henry's law constant, K_H, at 25°C for oxygen is 1.29×10^{-3} moles/L-atm. In this case, the units of K_H are moles/L-atm.

example/3.4 Using Henry's Law Constant to Determine the Aqueous Solubility of Oxygen

Calculate the concentration of dissolved oxygen (units of moles/L and mg/L) in a water equilibrated with the atmosphere at 25°C. The Henry's law constant for oxygen at 25°C is 1.29×10^{-3} mole/L-atm.

solution

The partial pressure of oxygen in the atmosphere is 0.21 atm. Equation 3.14 can be rearranged to yield

$$K_H \times P_{O_2} = [O_{2(aq)}] = \left(1.29 \times 10^{-3} \frac{\text{mole}}{\text{L-atm}}\right) \times 0.21 \text{ atm}$$

$$= 2.7 \times 10^{-4} \frac{\text{mole}}{\text{L}}$$

Thus, the solubility of oxygen at this temperature is 2.7×10^{-4} moles/L. If this value is multiplied by the molecular weight of oxygen (32 g/mole), the solubility can be reported as 8.7 mg/L.

The reaction was written as oxygen gas transferring into the aqueous phase in the forward direction because in this case we are concerned with how the composition of the gas affects the composition of the aqueous solution. Thus, the equilibrated dissolved oxygen saturation concentration in surface waters is a function of the partial pressure of oxygen in the atmosphere and the Henry's law constant.

The concentration of **dissolved oxygen** in water equilibrated with the atmosphere is 14.4 mg/L at 0°C and 9.2 mg/L at 20°C. This value demonstrates that oxygen solubility in water depends on water temperature (one reason trout like colder waters). For the reaction described in Equation 3.13, the change in heat of formation (ΔH^0) at standard conditions is −3.9 kcal. Because ΔH^0 is negative, Equation 3.13 could be written as

$$O_{2(g)} \leftrightarrow O_{2(aq)} + \text{heat} \qquad \textbf{(3.15)}$$

An increase in the temperature (or adding heat to the system) will, according to Le Châtelier's principle, favor the reaction that tends to diminish the increase in temperature. The effect is to drive the reaction in Equation 3.15 to the left, which consumes heat, diminishing the temperature increase in the process. Therefore, at equilibrium, more oxygen will be present in the gas phase at an increased temperature; thus, the solubility of dissolved oxygen will be lower at the increased temperature.

3.6.2 DIMENSIONLESS HENRY'S LAW CONSTANT FOR A SPECIES TRANSFERRING FROM THE LIQUID PHASE INTO THE GAS PHASE

In the case for the transfer of a chemical dissolved in the aqueous phase into the atmosphere, the chemical equilibrium between the gas and liquid phase chemical is described by a reaction written in reverse of Equation 3.13. For example, for the chemical trichloro-ethylene (TCE) transferring from the aqueous phase to the gaseous phase (as would be done if you were air stripping the chemical out of water):

$$TCE_{(aq)} \leftrightarrow TCE_{(g)} \qquad \text{(3.16)}$$

In this case, the equilibrium expression for this reaction is written as

$$K_H = \frac{[TCE_{(g)}]}{[TCE_{(aq)}]} \qquad \text{(3.17)}$$

where the gas-phase TCE is described by units of moles/liter of gas, not as partial pressure. Accordingly, the Henry's law constant, K_H, has units of moles per liter of gas divided by moles/liter of water, which cancel out. Therefore, the Henry's law constant in this case is termed *dimensionless* by some. In fact, it really has units of liters of water per liters of air. Other units of Henry's law constant include atm and L-atm/mole.

Henry's law constants that have units and those without units can be related using the ideal gas law. Several unit conversions for Henry's law constant are provided in Table 3.3.

Table / 3.3

Unit Conversion of Henry's Law Constants

$$K_H\left(\frac{L_{H_2O}}{L_{Air}}\right) = \frac{K_H\left(\frac{L\text{-atm}}{mole}\right)}{RT}$$

$$K_H\left(\frac{L\text{-atm}}{mole}\right) = K_H\left(\frac{L_{H_2O}}{L_{Air}}\right) \times RT$$

$$K_H\left(\frac{L_{H_2O}}{L_{Air}}\right) = \frac{K_H(atm)}{RT \times 55.6 \frac{mole\ H_2O}{L_{H_2O}}}$$

$$K_H\left(\frac{L\text{-atm}}{mole}\right) = \frac{K_H(atm)}{55.6 \frac{mole\ H_2O}{L_{H_2O}}}$$

$$K_H(atm) = K_H\left(\frac{L\text{-atm}}{mole}\right) \times 55.6 \frac{mole\ H_2O}{L_{H_2O}}$$

$$K_H(atm) = K_H\left(\frac{L_{H_2O}}{L_{Air}}\right) \times RT \times 55.6 \frac{mole\ H_2O}{L_{H_2O}}$$

$$R = 0.08205 \frac{atm\text{-}L}{mole\text{-}K}$$

SOURCE: From Mihelcic (1999); reprinted with permission of John Wiley & Sons, Inc.

example/3.5 **Conversion between Dimensionless and Nondimensionless Henry's Law Constants**

The Henry's law constant for the reaction transferring oxygen from air into water is 1.29×10^{-3} moles/ L-atm at 25°C. What is the dimensionless K_H for the transfer of oxygen from water into air at 25°C?

solution

The problem is requesting a Henry's law constant for the reverse reaction. Therefore, the Henry's law constant provided equals the inverse of 1.29×10^{-3} moles/L-atm, or 775 L-atm/mole for the transfer of aqueous oxygen into the gas phase. Solve using the ideal gas law:

$$K_H(\text{dimensionless}) = \frac{\dfrac{775 \text{ L-atm}}{\text{mole}}}{\left(\dfrac{0.08205 \text{ L-atm}}{\text{mole-K}}\right) \times (298 \text{ K})} = 32$$

3.7 Acid–Base Chemistry

Acid–base chemistry is important in treatment of pollution and in understanding the fate and toxicity of chemicals discharged to the environment.

3.7.1 pH

By definition, the **pH** of a solution is

$$pH = -\log[H^+] \qquad (3.18)$$

where $[H^+]$ is the concentration of the hydrogen ion. The pH scale in aqueous systems ranges from 0 to 14, with acidic solutions having a pH below 7, basic solutions having a pH above 7, and neutral solutions having a pH near 7. Ninety-five percent of all natural waters have a pH between 6 and 9. Rainwater not affected by anthropogenic acid-rain emissions has a pH of approximately 5.6 due to the presence of dissolved carbon dioxide that originates in the atmosphere.

The concentrations of OH^- and H^+ are related to one another through the equilibrium reaction for the dissociation of water:

$$H_2O \leftrightarrow H^+ + OH^- \qquad (3.19)$$

The equilibrium constant for the dissociation of water (K_w) for Equation 3.19 equals 10^{-14} at 25°C. Thus,

$$K_w = 10^{-14} = [H^+] \times [OH^-] \qquad (3.20)$$

Equation 3.20 allows the determination of the concentration of H^+ or OH^- if the other is known. Table 3.4 gives the range of K_w at

Table / 3.4

Dissociation Constant for Water at Various Temperatures and Resulting pH of a Neutral Solution

Temperature (°C)	K_w	pH of Neutral Solution
0	0.12×10^{-14}	7.47
15	0.45×10^{-14}	7.18
20	0.68×10^{-14}	7.08
25	1.01×10^{-14}	7.00
30	1.47×10^{-14}	6.92

SOURCE: From Mihelcic (1999); reprinted with permission of John Wiley & Sons, Inc.

temperatures of environmental significance. At 25°C in pure water, $[H^+]$ equals $[OH^-]$; thus $[H^+] = 10^{-7}$, and the pH of pure water is equal to 7.00. However, at 15°C, $[H^+]$ equals $10^{-7.18}$, so the pH of a neutral solution at this temperature is equal to 7.18.

3.7.2 DEFINITION OF ACIDS AND BASES AND THEIR EQUILIBRIUM CONSTANTS

Acids and bases are substances that react with hydrogen ions (H^+). An **acid** is defined as a species that can release or donate a hydrogen ion (also called a proton). A **base** is defined as a chemical species that can accept or combine with a proton. Equation 3.21 shows an example of an acid (HA) associated with a conjugate base (A^-):

$$HA \leftrightarrow H^+ + A^- \tag{3.21}$$

Acids that have a strong tendency to dissociate (this means that the reaction in Equation 3.21 goes far to the right) are called *strong acids*, while acids that have less of a tendency to dissociate (this means that the reaction in Equation 3.21 goes just a little to the right) are called *weak acids*.

The strength of an acid is indicated by the magnitude of the equilibrium constant for the dissociation reaction. The equilibrium constant for the reaction depicted in Equation 3.21 is

$$K_a = \frac{[H^+][A^-]}{[HA]} \tag{3.22}$$

where K_a = is the equilibrium constant for the reaction when an acid is added to water. At equilibrium, a strong acid will dissociate and show high concentrations of H^+ and A^- and a smaller concentration of HA. This means that when a strong acid is added to water, the result is a

Table / 3.5

Common Acids and Bases and Their Equilibrium Constants When Added to Water at 25°C

Acids			Bases		
	Name	$pK_a = -\log K_a$		Name	$pK_b = -\log K_b$
HCl	Hydrochloric	−3	Cl^-	Chloride ion	17
H_2SO_4	Sulfuric	−3	HSO_4^-	Bisulfate ion	17
HNO_3	Nitric	−1	NO_3^-	Nitrate ion	15
HSO_4^-	Bisulfate	1.9	SO_4^{2-}	Sulfate ion	12.1
H_3PO_4	Phosphoric	2.1	$H_2PO_4^-$	Dihydrogen phosphate	11.9
CH_3COOH	Acetic	4.7	CH_3COO^-	Acetate ion	9.3
$H_2CO_3^*$	Carbon dioxide and carbonic acid	6.3	HCO_3^-	Bicarbonate	7.7
H_2S	Hydrogen sulfide	7.1	HS^-	Bisulfide	6.9
$H_2PO_4^-$	Dihydrogen phosphate	7.2	HPO_4^{2-}	Monohydrogen phosphate	6.8
HCN	Hydrocyanic	9.2	CN^-	Cyanide ion	4.8
NH_4^+	Ammonium ion	9.3	NH_3	Ammonia	4.7
HCO_3^-	Bicarbonate	10.3	CO_3^{2-}	Carbonate	3.7
HPO_4^{2-}	Monohydrogen phosphate	12.3	PO_4^{3-}	Phosphate	1.7
NH_3	Ammonia	23	NH_2^-	Amide	−9

SOURCE: From Mihelcic (1999); reprinted with permission of John Wiley & Sons, Inc.

much larger negative free-energy change than when adding a weaker acid. Thus, for strong acids, the equilibrium constant K_a will be large (and ΔG will be very negative). Similarly, the K_a for a weak acid will be small (and ΔG will be less negative).

Just as pH equals $-\log[H^+]$, **pK_a** is the negative logarithm of the acid dissociation constant (that is, $pK_a = -\log(K_a)$). Table 3.5 provides values of equilibrium constants for some acids and bases of environmental importance. The table shows that the pK_a of a weak acid is larger than the pK_a of a strong acid.

The pK_a of an acid is related to the pH at which the acid will dissociate. Strong acids are those that have a pK_a below 2. They can be assumed to dissociate almost completely in water in the pH range 3.5–14. HCl, HNO_3, H_2SO_4, and $HClO_4$ are four very strong acids commonly encountered in environmental situations. Likewise, their conjugate bases (Cl^-, NO_3^-, SO_4^{2-}, and ClO_4^-) are so weak that in the pH range of 3.5–14, they are assumed to never exist with protons.

Class Discussion

What is better for protection of human health and the environment for current and future generations: (1) meeting regulatory requirements to transform nitrogen to less toxic forms before discharge, (2) transforming nitrogen from aqueous to gaseous species so it is removed from the water before discharge, or (3) reclaiming treated wastewater and reusing the water and dissolved nitrogen for irrigation?

example/3.6 Acid–Base Equilibrium

What percentage of total ammonia (that is, $NH_3 + NH_4^+$) is present as NH_3 at a pH of 7? The pK_a for NH_4^+ is 9.3; therefore,

$$K_a = 10^{-9.3} = \frac{[NH_3][H^+]}{[NH_4^+]}$$

solution

The problem is requesting

$$\frac{[NH_3]}{([NH_4^+] + [NH_3])} \times 100\%$$

Solving this problem requires another independent equation because the preceding expression has two unknowns. The equilibrium expression for the NH_4^+/NH_3 system provides the second required equation:

$$10^{-9.3} = \frac{[NH_3] \times [H^+]}{[NH_4^+]} = \frac{[NH_3] \times [10^{-7}]}{[NH_4^+]}$$

Thus, at $pH = 7$, $[NH_4^+] = 200 \times [NH_3]$. This expression can be substituted into the first expression, yielding

$$\frac{[NH_3]}{(200 \times [NH_3] + [NH_3])} \times 100\% = 0.5\%$$

At this neutral pH, almost all of the total ammonia of a system exists as ammonium ion (NH_4^+). In fact, only 0.5 percent exists as NH_3!

The form of total ammonia most toxic to aquatic life is NH_3. It is toxic to several fish species at concentrations above 0.2 mg/L. Thus, wastewater discharges with a pH less than 9 will have most of the total ammonia in the less toxic NH_4^+ form. This is one reason why some wastewater discharge permits for ammonia specify that the pH of the discharge must also be less than 9.

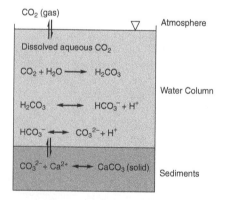

Figure / 3.6 **Important Components of the Carbonate System.**

(From Mihelcic (1999). Reprinted with permission of John Wiley & Sons, Inc.).

3.7.3 CARBONATE SYSTEM, ALKALINITY, AND BUFFERING CAPACITY

Figure 3.6 shows the important components of the **carbonate system**. The concentration of **dissolved carbon dioxide** in water equilibrated with the atmosphere (partial pressure of CO_2 is $10^{-3.5}$ atm) is 10^{-5} moles/L. This is a significant amount of carbon dioxide dissolved in water. This reaction can be written as follows:

$$CO_{2(g)} \leftrightarrow CO_{2(aq)} \tag{3.23}$$

where $K_H = 10^{-1.5}$ moles/L-atm.

Upon dissolving in water, dissolved CO_2 undergoes a hydration reaction by reacting with water to form carbonic acid:

$$CO_{2(aq)} + H_2O \leftrightarrow H_2CO_3 \tag{3.24}$$

where $K = 10^{-2.8}$. This reaction has important implications for the chemistry of water in contact with the atmosphere. First, water in contact with the atmosphere (for example, rain) has the relatively strong acid, carbonic acid, dissolved in it. Thus, the pH of rainwater not impacted by anthropogenic emissions will be below 7. The pH of "unpolluted" rainwater is approximately 5.6. Thus, acid rain, which typically has measured pH values of 3.5–4.5, is approximately 10–100 times more acidic than natural rainwater, but not 10,000 times more acidic, because natural rainwater is not neutral with a pH of 7.0.

In addition, because natural rainwater is slightly acidic and the partial pressure of carbon dioxide in soil may also be high from biological activity, water that contacts rocks and minerals can dissolve ions into solution. Inorganic constituents dissolved in freshwater and the dissolved salts in the oceans have their origin in minerals and the atmosphere. Carbon dioxide from the atmosphere provides an acid that can react with the bases of rocks, releasing the rock constituents into water, where they can either remain dissolved or precipitate into a solid phase.

It is difficult to distinguish analytically the difference between $CO_{2(aq)}$ and true H_2CO_3. Therefore, the term $H_2CO_3{}^*$ has been defined to equal the concentration of $CO_{2(aq)}$ plus the concentration of true H_2CO_3. However, $H_2CO_3{}^*$ can be approximated by $CO_{2(aq)}$ because true H_2CO_3 makes up only about 0.16 percent of $H_2CO_3{}^*$. Thus, the concentration of $H_2CO_3{}^*$ in waters equilibrated with the atmosphere is approximately 10^{-5} M.

$H_2CO_3{}^*$ is in equilibrium with bicarbonate ion as follows:

$$H_2CO_3{}^* \leftrightarrow HCO_3{}^- + H^+ \qquad \textbf{(3.25)}$$

where $K_{a1} = 10^{-6.3}$. Also, bicarbonate is in equilibrium with carbonate ion as follows:

$$HCO_3{}^- \leftrightarrow CO_3{}^{2-} + H^+ \qquad \textbf{(3.26)}$$

where $K_{a2} = 10^{-10.3}$.

According to our definition of an acid and base, bicarbonate can act as either an acid or a base. Bicarbonate and carbonate are also common bases in water. The *total inorganic carbon* content of a water sample is defined as follows:

$$\text{Total inorganic carbon} = [H_2CO_3{}^*] + [HCO_3{}^-] + [CO_3{}^{2-}] \qquad \textbf{(3.27)}$$

In the pH range of most natural waters (pH 6–9), $H_2CO_3{}^*$ and $CO_3{}^{2-}$ are small relative to $HCO_3{}^-$. Therefore, $HCO_3{}^-$ is the predominant component in Equation 3.27.

Table 3.6 provides definitions and descriptions of two important terms related to the carbonate system: **alkalinity** and **buffering capacity**. In the majority of natural freshwaters, alkalinity is caused primarily by $HCO_3{}^-$, $CO_3{}^{2-}$, and OH^-. In some natural waters and industrial waters, other salts of weak acids that may be important in determining a solution's alkalinity are borates, phosphates, ammonia, and organic acids. For example, anaerobic digester supernatant and municipal wastewaters contain large amounts of bases such as ammonia (NH_3), phosphates ($HPO_4{}^{2-}$ and $PO_4{}^{3-}$), and bases of various organic acids. The

Class Discussion
Some scientists have suggested that we add large quantities of iron to the world's oceans to precipitate out carbonate, thus shifting carbonate system chemistry so the oceans take up more carbon dioxide from the atmosphere. Do you consider this "geoengineering" of the environment a sustainable solution for the current problem of global emissions of carbon dioxide to the atmosphere?

Table / 3.6

Explanation of Alkalinity and Buffering Capacity

Term	Definition
Alkalinity	Measure of a water's capacity to neutralize acids
	Alkalinity (moles/L) $= [HCO_3^-] + 2[CO_3^{2-}] + [OH^-] - [H^+]$
	In most natural waters near pH $= 6$–8, the concentration of bicarbonate (HCO_3^-) is significantly greater than that of carbonate (CO_3^{2-}) or hydroxide (OH^-); therefore, the total alkalinity can be approximated by the bicarbonate concentration.
Buffering capacity	Ability of a water to resist changes in pH when either acidic or alkaline material is added.
	In most freshwater systems, the buffering capacity is due primarily to the bases (OH^-, CO_3^{2-}, HCO_3^-) and acids ($H^+, H_2CO_3^*, HCO_3^-$).

bases of silica ($H_3SiO_4^-$) and boric acid ($B(OH)_4^-$) can contribute to alkalinity in the oceans.

In most natural waters, the buffering capacity is due primarily to the bases (OH^-, CO_3^{2-}, HCO_3^-) and acids ($H^+, H_2CO_3^*, HCO_3^-$). Many lakes in the United States (for instance, in New England and the upper Midwest) have a low buffering capacity and consequently have been strongly influenced by acidic deposition (acid rain). This is because the geology of the basins that underlie these lakes is such that the slow dissolution of the underlying rocks and minerals does not result in the release of much alkalinity.

Learn more about ocean acidification

http://pmel.noaa.gov/co2/story/
What+is+Ocean+Acidication%3F

---

example / 3.7 Do Changes in Atmospheric CO_2 Concentration Impact the Chemistry of the World's Oceans

The National Oceanic and Atmospheric Administration (NOAA) has measured atmospheric CO_2 concentrations at Mauna Loa (Hawaii) for over 50 years. Their monthly mean atmospheric CO_2 concentrations are shown to be increasing (see Figure 4.14). Monthly mean atmospheric CO_2 concentrations were approximately 315 ppm in 1960 and have increased to 392 ppm in August 2012. Levels were only 275 ppm prior to the Industrial Revolution time period. Scientists predict that without serious mitigation efforts, CO_2 levels may increase to 556 ppm by 2050. Determine the concentration of CO_2 (in moles/L) in water equilibrated with these four atmospheric concentrations. The Henry's law constant for CO_2 is $10^{-1.5}$ moles/L-atm.

solution

Preindustrial Time Period: $10^{-1.5}$ moles/L-atm $\times$ 0.275 atm $= 8.70 \times 10^{-3}$ moles/L

1960: $10^{-1.5}$ moles/L-atm $\times$ 0.315 atm $= 9.96 \times 10^{-3}$ moles/L

2012: $10^{-1.5}$ moles/L-atm $\times$ 0.392 atm $= 1.24 \times 10^{-2}$ moles/L

2050: $10^{-1.5}$ moles/L-atm $\times$ 0.556 atm $= 1.75 \times 10^{-2}$ moles/L

In actuality, the oceans absorb approximately 25 percent of all the CO_2 released into the atmosphere. However, these very simple calculations demonstrate how increases in the concentration of atmospheric CO_2 can increase concentrations of dissolved CO_2 in Earth's oceans. Remember from the discussion of Equation 3.24 that dissolved aqueous CO_2 undergoes hydration to form carbonic acid (H_2CO_3). Thus, increasing levels of CO_2 in the atmosphere are leading to increases in carbonic acid concentrations in the world's oceans. In fact, since the Industrial Revolution, the pH of the oceans has decreased by approximately 0.1 pH unit.

This might not seem like a lot, but remember that pH is a logarithmic scale (Equation 3.18); therefore, this 0.1 pH unit decrease is equivalent to about a 25 percent increase in acidity. In addition, the amount of carbonate (CO_3^{2-}) dissolved in water is impacted by change in pH.

As shown in Equation 3.26, a decrease in pH (which by definition is associated with an increase in H^+ concentration) will decrease the amount of carbonate present in water as the reaction shifts to the left. This change in carbonate chemistry may negatively impact some food chains in parts of the ocean because carbonate is the building block of coral and skeletons and shells of other marine organisms. Decreasing pH is also shown to cause dissolution (Section 3.9) of carbonate solids that make up the skeleton and shell of marine organisms.

What impact do you expect this to have on the many food resources provided by Earth's oceans? You may want to research the percentage of the global population that is dependent on obtaining food from Earth's oceans. You can also research and discuss in class what impact this might have on communities that are dependent on the ocean's fisheries for their economic and social well-being.

3.8   Oxidation–Reduction

Some chemical reactions occur because electrons are transferred between different chemical species. These reactions are called **oxidation–reduction** or **redox reactions**. Oxidation–reduction reactions control the fate and speciation of many metals and organic pollutants in natural environments, and numerous treatment processes employ redox chemistry. Also, many biological processes are just redox reactions mediated by microorganisms. The most commonly used wastewater treatment processes involve redox reactions that oxidize organic carbon to CO_2 (while reducing oxygen to water) and oxidize and reduce various forms of nitrogen.

For molecules composed of single, charged atoms, the **oxidation state** is simply the charge on the atom; for example, the oxidation state of Cu^{2+} is $+2$. In molecules containing multiple atoms, each atom is assigned an oxidation state according to the conventions provided in Table 3.7.

In a redox reaction, a molecule's oxidation state either goes up (in which case the molecule is *oxidized*) or down (in which case the molecule is *reduced*). Oxidized species can be depicted as reacting with free electrons (e^-) in half-reactions such as the following:

electron acceptor (oxidant) $+ e^- \rightarrow$ electron donor (reductant)   **(3.28)**

In this reaction, the species gaining the electron (the **electron acceptor** or *oxidant*) is reduced to form the corresponding reduced species;

Table / 3.7

Conventions for Assigning Oxidation State to Common Atoms (H, O, N, S) in Molecules

1. The overall charge on a molecule = Σ charges (oxidation state) of its individual atoms.

2. The atoms in the molecule of concern have the following oxidation state; however, these numbers should be set equal to other numbers in reverse order (apply a different number to S before applying N, and so on) such that Convention 1 is always satisfied.

Atom	Oxidation State
H^+	+1
O	-2
N	-3
S	-2

reduced molecules can donate electrons (the **electron donor**) and serve as *reductants*.

Consider two examples. In the first, ammonia nitrogen (oxidation state of -3) can be converted through nitrification and denitrification to N_2 gas (oxidation state of 0). In addition, important atmospheric pollutants include NO (oxidation state of +2) and NO_2 (oxidation state of +4). This conversion of nitrogen to different compounds occurs through many redox reactions. In the second example, acid rain is caused by emissions of SO_2 (sulfur oxidation state of +4), which is oxidized in the atmosphere to sulfate ion, SO_4^{2-} (sulfur oxidation state of +6). Sulfate ion returns to Earth's surface in dry or wet deposition as sulfuric acid.

example/ 3.8   Determining Oxidation States

Determine the oxidation states of sulfur in sulfate (SO_4^{2-}) and bisulfide (HS^-).

solution

We expect the sulfur in sulfate to be more highly oxidized (due to the presence of oxygen in the molecule) than in bisulfide (due to the presence of hydrogen). The overall charge of -2 on sulfate must be maintained, and since the charge on each oxygen atom is -2 (see Table 3.7), the charge on the sulfur in sulfate must be $-2 - 4(-2) = +6$. To maintain the overall charge of -1 on bisulfide, the charge on sulfur must be $-1 - (+1) = -2$. Here the charge on H^+ was +1 (see Table 3.7). As expected, the sulfur found in sulfate is more oxidized than sulfur in bisulfide.

3.9 Precipitation–Dissolution

Precipitation–dissolution reactions involve the dissolution of a solid to form soluble species (or the reverse process whereby soluble species react to precipitate out of solution as a solid). Common precipitates include hydroxide, carbonate, and sulfide minerals.

A reaction that sometimes occurs in homes is the precipitation of $CaCO_3$. If waters are hard, this compound forms a scale in tea kettles, hot-water heaters, and pipes. Much effort is devoted to preventing excessive precipitation of $CaCO_3$ in municipal and industrial settings, and the process of removing divalent cations from water is referred to as water softening.

The reaction common to all of these situations is the conversion of a solid salt into dissolved components. In this example, the solid is calcium carbonate:

$$CaCO_{3(s)} \leftrightarrow Ca^{2+} + CO_3^{2-} \tag{3.29}$$

Here, the subscript (s) denotes that the species is a solid. The equilibrium constant for such a reaction is referred to as the solubility product, K_{sp}. At equilibrium for the reaction in Equation 3.29, the K_{sp} is equal to Q:

$$K_{sp} = \frac{[Ca^{2+}][CO_3^{2-}]}{[CaCO_{3(s)}]} = [Ca^{2+}][CO_3^{2-}] \tag{3.30}$$

Solubility is defined as the maximum quantity (generally expressed as mass) of a substance (the solute) that can dissolve in a unit volume of solvent under specified conditions. Because the activity (which we assume equals concentration) of a solid is defined as equal to 1.0 (Rule 2 of Table 3.2), the equilibrium constant, K_{sp}, is equal to the solubility product. Thus, if we know the equilibrium constant and the concentration of one of the species, we can determine the concentration of the other species.

No precipitate will form if the product of the concentrations of the ions is less than K_{sp} (in Equation 3.30, Ca^{2+} and CO_3^{2-} are the species). This solution is described as *undersaturated*. Likewise, if the product of the concentrations of the ions exceeds K_{sp}, the solution is described as *supersaturated*, and the solid species will precipitate until the product of the ion concentrations equals K_{sp}. Table 3.8 provides some important solubility products and the associated reactions.

Karst formations are created from dissolution of limestone ($CaCO_3$) and dolomite ($CaMg(CO_3)_2$). You can recognize Karst terrain because of the presence of springs, caves, and sinkholes. Figure 3.7 shows the location of the 13 major Karst groundwater aquifers in the United States. These formations are important because they are a major source of water supply. The hydrology of a Karst formation is a challenge to manage because the majority of water flow occurs through fissures, fractures, and conduits that are interconnected, thus making it easy to contaminate and exploit.

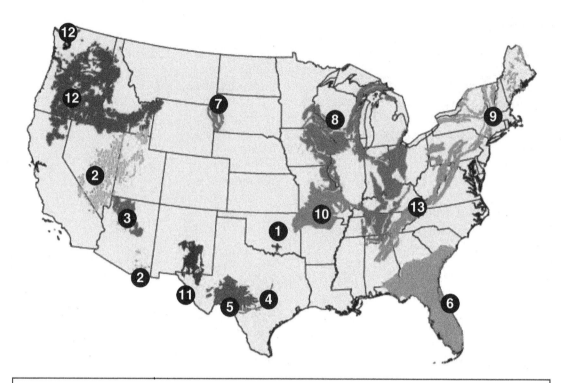

Principal Karst Aquifers

1   Arbuckle-Simpson aquifer – Underlies more than 500 square miles in south central Oklahoma and is the principal water source for about 40,000 people. Many springs and small karst features, but only a few air-filled caves.

2   Basin and Range and Bear River range carbonate aquifers – Some fractured carbonate rocks beneath alluvial basin fill. Includes areas near Cedar Break National Monument, Great Basin National Park, and the Bear River Range.

3   Colorado Plateau Karst.

4   Edwards Balcones Fault Zone aquifer – Highly faulted and fractured carbonate rocks of Cretaceous age in an area of about 4,000 square miles in south-central Texas. Primary drinking-water supply for San Antonio, Texas.

5   Edwards-Trinity Plateau aquifer – Consists of rocks of Cretaceous age that are present in an area of about 35,500 square miles in west-central Texas.

6   Upper Floridan and Biscayne aquifers

7   Madison aquifer – An important water resource in the northern plains states, where surface-water supplies are limited and population is increasing. It is one of the largest confined aquifer systems in the United States.

8   Midwest Paleozoic Carbonate aquifers – Karst developed in several Paleozoic aquifers that span the Midwest from Michigan to Tennessee. Contains some of the longest mapped caves in the world, including Kentucky's Mammoth Cave National Park.

9   New England karst aquifers – Solution terrain in crystalline limestones and marbles mainly in northeastern Maine, western Vermont, and western Massachusetts.

10   Ozark Plateau karst aquifers – Paleozoic carbonate rocks underlying several mid-continent states. Comprises two aquifers (Springfield and Ozark) and an intervening confining unit, and yields modest amounts of water.

11   Roswell Basin aquifer – An eastward-dipping carbonate aquifer overlain by a leaky evaporitic confining unit and an unconfined alluvial aquifer. Decades of intensive pumping have caused substantial declines in hydraulic head.

12   Pacific Northwest basalt aquifers – Late Cenozoic basalt lava fields that contain lava tubes, fissures, open sinkholes, and caves formed by extrusion of the still-liquid portion of the lava.

13   Valley and Ridge, Piedmont, and Blue Ridge aquifers – Extensive areas of karst within complex geologic structures, resulting in highly variable karst aquifer characteristics. Includes the Great Valley aquifer, an important water resource for many cities.

Figure / 3.7   Principal Karst Aquifers of the United States.

(Adapted from U.S. Geological Survey).

Table / 3.8

Common Precipitation–Dissolution Reactions, the Associated Solubility Product K_{sp}, and Significance

Equilibrium Equation	K_{sp} at 25°C	Significance
$CaCO_{3(s)} \leftrightarrow Ca^{2+} + CO_3^{2-}$	3.3×10^{-9}	Hardness removal, scaling, ocean sequestration of carbon dioxide
$MgCO_{3(s)} \leftrightarrow Mg^{2+} + CO_3^{2-}$	3.5×10^{-5}	Hardness removal, scaling
$Ca(OH)_{2(s)} \leftrightarrow Ca^{2+} + 2OH^-$	6.3×10^{-6}	Hardness removal
$Mg(OH)_{2(s)} \leftrightarrow Mg^{2+} + 2OH^-$	6.9×10^{-12}	Hardness removal
$Cu(OH)_{2(s)} \leftrightarrow Cu^{2+} + 2OH^-$	7.8×10^{-20}	Heavy-metal removal
$Zn(OH)_{2(s)} \leftrightarrow Zn^{2+} + 2OH^-$	3.2×10^{-16}	Heavy-metal removal
$Al(OH)_{3(s)} \leftrightarrow Al^{3+} + 3OH^-$	6.3×10^{-32}	Coagulation
$Fe(OH)_{3(s)} \leftrightarrow Fe^{3+} + 3OH^-$	6×10^{-38}	Coagulation, iron removal
$CaSO_{4(s)} \leftrightarrow Ca^{2+} + SO_4^{2-}$	4.4×10^{-5}	Flue gas desulfurization
$MgNH_4PO_4 \cdot 6H_2O_{(s)} \leftrightarrow$ $Mg^{2+} + NH_4^+ + PO_4^{3-} + 6H_2O$	$5.5 \times 10^{-14} - 2.5 \times 10^{-13}$	Struvite precipitation for recovery of phosphorus from urine at wastewater treatment plant, struvite precipitation in waterless urinals

SOURCE: Adapted from Mihelcic (1999); reprinted with permission of John Wiley & Sons, Inc.

example / 3.9   Precipitation–Dissolution Equilibrium

What pH is required to reduce a high concentration of dissolved Mg^{2+} to 43 mg/L? K_{sp} for the following reaction is $10^{-11.16}$.

$$Mg(OH)_{2(s)} \leftrightarrow Mg^{2+} + 2OH^-$$

solution

In this situation, the dissolved magnesium is removed from solution as a hydroxide precipitate. First, the concentration of Mg^{2+} is converted from mg/L to moles/L:

$$[Mg^{2+}] = \frac{43 \text{ mg}}{L} \times \frac{g}{1{,}000 \text{ mg}} \times \frac{1 \text{ mole}}{24 \text{ g}} = 0.0018 \text{ M}$$

Then, the equilibrium relationship is written as

$$10^{-11.16} = \frac{[Mg^{2+}] \times [OH^-]^2}{[Mg(OH)_{2(s)}]}$$

Substituting values for all the known parameters,

$$10^{-11.16} = \frac{[0.0018] \times [OH^-]^2}{1}$$

Solve for $[OH^-] = 6.2 \times 10^{-5}$ M. This results in $[H^+] = 10^{-9.79}$ M, so pH = 9.79. At this pH, any magnesium in excess of 0.0018 M will precipitate as $Mg(OH)_{2(s)}$ because the solubility of Mg^{2+} will be exceeded.

3.10 Adsorption, Absorption, and Sorption

Sorption is a nonspecific term that can refer to either or both process(es) of **adsorption** of a chemical at the solid surface and/or **absorption** (partitioning) of the chemical into the volume of the solid. In the case of organic pollutants, sorption is a key process determining fate, and the chemical is commonly absorbed into the organic fraction of the particle due to favorable energetics of this process. The *sorbate* (adsorbate or absorbate) is the substance transferred from the gas or liquid phase to the solid phase. The *sorbent* (adsorbent or absorbent) is the solid material onto or into which the sorbate accumulates. Solids that sorb chemicals may be either natural (for example, surface soil, harbor or river sediment, aquifer material) or anthropogenic (for example, activated carbon) materials.

Figure 3.8 shows a schematic of sorption processes for naphthalene sorbing to a natural solid such as a soil particle from the water phase. Why does this sorption occur? From a thermodynamic viewpoint, molecules always prefer to be in a lower energy state. A molecule adsorbed onto a surface has a lower energy state on a surface than in the aqueous phase. Therefore, during the process of equilibration, the molecule is attracted to the surface and a lower energy state. Attraction of a molecule to a surface can be caused by physical and/or chemical forces. Electrostatic forces govern the interactions between most adsorbates and adsorbents. These forces include dipole–dipole interactions, dispersion interactions or London–van der Waals force, and hydrogen bonding. During sorption to soils and sediments, **hydrophobic partitioning**—a phenomenon driven by entropy changes—can also account for the interaction of a hydrophobic (water-fearing) organic chemical with a surface.

Table 3.9 provides examples of some common sorption isotherms and related partitioning phenomena. A *sorption isotherm* is a relationship that describes the affinity of a compound for a solid in water or gas at constant temperature (*iso* means constant and *therm* refers to temperature). The two sorption isotherms covered in Table 3.9 are the **Freundlich isotherm** and the **linear isotherm**. Figure 3.9 shows the relationship between the Freundlich and linear isotherms for various

Figure 3.8 Sorption of an Organic Chemical (Naphthalene) onto a Natural Material such as a Soil or Sediment Particle This typically occurs when the sorbate either sorbs onto reactive surface sites (adsorption) or absorbs or partitions into organic matter that coats the particle (the sorbent). The sorption process influences the mobility, natural degradation, and engineered remediation of pollutants.

(From Mihelcic (2009). Reprinted with permission of John Wiley & Sons, Inc.).

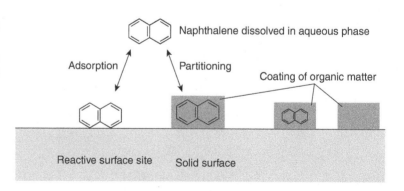

Table / 3.9

Common Terms Used to Describe Sorption Isotherms and Other Partitioning Phenomena

Isotherm or Other Partitioning Term	Usually Presented as	Symbols and Units	Common Application
Freundlich isotherm	$q = KC^{1/n}$ (Equation 3.31)	q = mass of adsorbate adsorbed per unit mass of adsorbent after equilibrium (mg/g). C = mass of adsorbate in the aqueous phase after equilibrium (mg/L). K = Freundlich isotherm capacity parameter $((mg/g)(L/mg)^{1/n})$. $1/n$ = Freundlich isotherm intensity parameter (unitless).	Drinking water and air treatment where adsorbents such as activated carbon are used
Linear isotherm Special case of Freundlich isotherm where $1/n$ (that is, dilute systems)	$K = \dfrac{q}{C}$ (Equation 3.32)	q and C are same as Freundlich isotherm. K = soil- or sediment–water partition (or distribution) coefficient, also written as K_P or K_d (units of cm^3/g or L/kg).	Dilute systems, especially soil, sediment, and groundwater
Normalizing K for organic carbon[*]	$K_{oc} = \dfrac{K}{f_{oc}}$ (Equation 3.33)	K is the same as the linear isotherm (also referred to as K_p, K_d). f_{oc} is the fraction of organic carbon for a specific soil. K_{oc} has units of cm^3/g organic carbon (or L/kg organic carbon) and sediment (1% organic carbon equals an f_{oc} of 0.01).	Soil, sediment, and groundwater
Octanol–water partition coefficient	$K_{ow} = \dfrac{[A]_{octanol}}{[A]_{water}}$ (Equation 3.34)	$[A]_{octanol}$ is concentration of chemical dissolved in octanol ($C_8H_{17}OH$), and $[A]_{water}$ is concentration of same chemical dissolved in same volume of water. K_{ow} is unitless and usually reported as $\log K_{ow}$.	Helps determine the hydrophobicity of a chemical. Can be related to other environmental properties such as K_{oc} and bioconcentration factors

[*]It has been shown that for soils and sediments with a fraction of organic carbon (f_{oc}) greater than 0.001 (0.1%) and low equilibrium solute concentrations ($<10^{-5}$ molar or 1/2 the aqueous solubility), the soil–water partition coefficient (K_p) can be normalized to the soil's organic carbon content.

ranges of $1/n$. Here, $1/n$ is the Freundlich isotherm intensity parameter (unitless).

A problem with the value of the **soil–water partition coefficient,** K (Equation 3.32, given in Table 3.9), is that it is chemical- and sorbent-specific. Thus, although K could be measured for every relevant system, this would be time-consuming and costly. Fortunately, when the solute is a neutral, nonpolar organic chemical, the soil–water partition coefficient can be normalized for organic carbon, in which case it remains chemical-specific but no longer sorbent-specific. K_{oc} is called the **soil–water partition coefficient normalized to organic carbon.** K_{oc} has units of cm^3/g organic carbon or L/kg organic carbon (see Equation 3.33 in Table 3.9).

Persistent Organic Pollutants
http://www.chem.unep.ch/pops

Class Discussion

Should a chemical like DDT be banned globally or considered a viable solution to the unfair burden of malaria that inflicts many parts of the developing world, especially Africa. What equitable solutions that consider future generations of humans and wildlife can you think of?

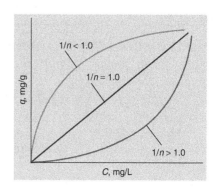

Figure / 3.9 **Freundlich Isotherm Plotted for Different Values of** $1/n$ For values of $1/n$ less than 1, the isotherm is considered favorable for sorption because low values of the sorbate liquid-phase concentration yield large values of the solid-phase concentration. This means that it is energetically favorable for the sorbate to be sorbed. At higher aqueous concentrations, the ability of the solid to sorb the chemical decreases as the active sorption sites become saturated with sorbate molecules. For $1/n$ values greater than 1, the isotherm is considered unfavorable for sorption because high values of the liquid-phase sorbate concentration are required to get sorption to occur on the sorbent. However, as sorption occurs, the surface is modified by the sorbing chemical and made more favorable for additional sorption. If the $1/n$ value equals 1, the isotherm is termed a linear isotherm.

(From Mihelcic (1999). Reprinted with permission of John Wiley & Sons, Inc.).

example/3.10 Adsorption Isotherm Data Analysis

A methyl tertiary-butyl ether (MTBE) adsorption isotherm was performed on a sample of activated carbon. The isotherm was performed at 15°C using 0.250 L amber bottles with an initial MTBE concentration, C_0, of 150 mg/L. The three left columns of Table 3.10 provide the isotherm data. Determine the Freundlich isotherm parameters (K and $1/n$).

Table / 3.10

Isotherm Data and Results Used in Example 3.10

Isotherm Data			Results		
Initial MTBE Concentration, C_0 (mg/L)	Mass of GAC, M (g)	MTBE Equilibrium Liquid-Phase Concentration, C, mg/L	$q = (V/M)$ $\times (C_0 - C)$ (mg/g)	log q	log C
150	0.155	79.76	113.290	2.0542	1.9018
150	0.339	42.06	79.602	1.9009	1.6239
150	0.589	24.78	53.149	1.7255	1.3941
150	0.956	12.98	35.832	1.5543	1.1133
150	1.71	6.03	21.048	1.3232	0.7803
150	2.4	4.64	15.142	1.1802	0.6665
150	2.9	3.49	12.630	1.1014	0.5428
150	4.2	1.69	8.828	0.9459	0.2279

example/3.10 (continued)

solution

The values of the MTBE adsorbed for each isotherm point (q) and the logarithm values of C and q can be determined and inputted into Table 3.10 (three left columns). To determine the Freundlich isotherm parameters, fit the logs of the isotherm data, log q versus log C, using the linear form of Equation 3.31 (Table 3.9), expressed as

$$\log q = \log K + \left(\frac{1}{n}\right)\log C$$

Graph log q versus log C, as shown in Figure 3.10, and use a linear regression to fit the data to determine K and $1/n$.

From Figure 3.10, the linear form of Equation 3.31 with values for K and $1/n$ added is expressed as

$$\log q = 0.761 + (0.6906)\log C$$

Here, $\log K = 0.761$, so $K = 10^{0.761} = 5.77 (\text{mg/g})(\text{L/mg})^{1/n}$. Thus, $K = 5.77(\text{mg/g})(\text{L/mg})^{1/n}$, and $1/n = 0.6906$.

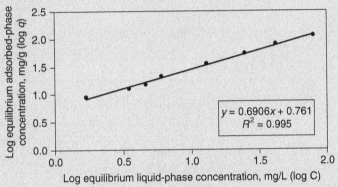

Figure / 3.10   Freundlich Isotherm Data Graphed for Example 3.10 to Determine K and $1/n$.

For systems with a relatively high amount of organic carbon (greater than 0.1 percent), K_{oc} can be directly correlated to a parameter called the **octanol–water partition coefficient, K_{ow}**, of a chemical. Values of K_{ow} range over many orders of magnitude, so K_{ow} usually is reported as log K_{ow}. Table 3.11 lists some typical values of log K_{ow} for a wide variety of chemicals. Values of K_{ow} for environmentally significant chemicals range from approximately 10^1 to 10^7 (log K_{ow} range of 1–7). The higher the value, the greater the tendency of the compound to partition from the water into an organic phase. Chemicals with high values of K_{ow} are hydrophobic (water-fearing).

The magnitude of an organic chemical's K_{ow} can tell a lot about the chemical's ultimate fate in the environment. For example, the values in Table 3.11 indicate that very hydrophobic chemicals such as 2,3,7,8-tetrachlorodibenzo-p-dioxin are more likely to bioaccumulate in the lipid portions of humans and animals. Conversely, chemicals such as benzene, trichloroethylene (TCE), tetrachloroethylene (PCE), and toluene are frequently identified as groundwater contaminants because they are relatively soluble and easily dissolve in groundwater recharge that is

Table / 3.11

Examples of log K_{ow} for Some Environmentally Significant Chemicals

Chemical	log K_{ow}
Phthalic acid	0.73
Benzene	2.17
Trichloroethylene	2.42
Tetrachloroethylene	2.88
Toluene	2.69
2,4,-Dichlorophenoxyacetic acid	2.81
Naphthalene	3.33
1,2,4,5-Tetrachlorobenzene	4.05
Phenanthrene	4.57
Pyrene	5.13
Hexabromobiphenyl	6.39
2,3,7,8-Tetrachlorodibenzo-p-dioxin	6.64
Decabromobiphenyl	8.58

infiltrating vertically toward an underlying aquifer. This is in contrast to pyrene or 2,3,7,8-tetrachlorodibenzo-p-dioxin, which are both likely to be confined near the soil's surface in the location of the spill.

Figure 3.11 shows how K_{oc} and K_{ow} are linearly correlated for a set of 72 chemicals that span many ranges of hydrophobicity. K_{ow} has also

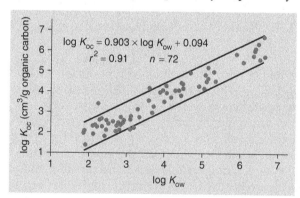

Figure / 3.11 Scatter Plot of log K_{oc}(cm^3/g Organic Carbon) versus log K_{ow} for 72 Chemicals The relationship is given by the equation $\log(K_{oc}[\text{cm}^3/\text{g}]) = 0.903 \log (K_{ow}) + 0.094$ ($n = 72$, $r^2 = 0.91$). The heavy lines represent the 90 percent confidence intervals for the correlation. Individuals seeking values of K_{oc} should consult a data set that has undergone a quality check or use an appropriate, statistically validated correlation to estimate the value of K_{oc}.

(From Baker, J. R., J. R. Mihelcic, D. C. Luehrs, and J. P. Hickey. "Evaluation of Estimation Methods for Organic Carbon Normalized Sorption Coefficients." *Water Environment Research*, 69:136–145, 1997. Copyright WEF, reprinted with permission).

been correlated to other environmental properties such as bioconcentration factors and aquatic toxicity. K_{oc} can then be related to the site-specific soil–water partition coefficient (K_p) by knowledge of the system's organic carbon content, using Equation 3.33 (Table 3.9).

---

example / 3.11   Determination of K_{oc} from K_{ow}

The log K_{ow} for anthracene is 4.68. What is anthracene's soil–water partition coefficient normalized to organic carbon?

solution

Use an appropriate correlation between log K_{oc} and log K_{ow} (such as provided in Figure 3.11). Note that this correlation requests log K_{oc}, not K_{ow}:

$$\log K_{oc} = 0.903(4.68) + 0.094 = 4.32$$

Therefore, $K_{oc} = 10^{4.32}$ cm^3/g organic carbon.

---

example / 3.12   Use of K_{oc} to Predict Aqueous Concentration

Anthracene has contaminated harbor sediments, and the solid portion of sediments is in equilibrium with the pore water. If the organic carbon content of sediments is 5 percent and the solid sediment anthracene concentration is 50 µg/kg sediment, what is the pore water concentration of anthracene at equilibrium? In Example 3.11, log K_{oc} for anthracene was estimated to be 4.32.

solution

An organic carbon (OC) content of 5 percent means that the fraction of organic carbon, f_{oc}, is 0.05. Use Equation 3.33 (from Table 3.9) to find the sediment-specific partition coefficient, K:

$$K \frac{\text{cm}^3}{\text{g sediment}} = \frac{10^{4.32}\ \text{cm}^3}{\text{g OC}} \times \frac{0.05\ \text{g OC}}{\text{g sediment}} = 1{,}045\ \text{cm}^3/\text{g sediment}$$

The equilibrium aqueous-phase concentration, C, is then derived from the equilibrium expression given in Equation 3.32 (Table 3.9):

$$C = \frac{q}{K} = \frac{\dfrac{50\ \mu g}{\text{kg sediment}} \times \dfrac{\text{kg}}{1{,}000\ \text{g}}}{\dfrac{1{,}045\ \text{cm}^3}{\text{g sediment}}} \times \frac{\text{cm}^3}{\text{mL}} \times \frac{1{,}000\ \text{mL}}{\text{L}} = \frac{0.048\ \mu g}{\text{L}}$$

Note that the aqueous-phase concentration of anthracene is relatively low compared with the sediment-phase concentration (50 ppb$_m$ in the sediments and 0.048 ppb$_m$ in the pore water). This is because anthracene is hydrophobic. Its aqueous solubility is low (and K_{ow} is high), so it prefers to partition into the solid phase.

Also, the solid phase is high in organic carbon content. A sand-gravel aquifer would be much lower in organic carbon (f_{oc} very low); therefore, less of the anthracene would partition from the aqueous into the solid phase.

example/3.13 Partitioning of Chemical between Air, Water, and Soil Phases

A student uses a reactor to mimic the environment for a class demonstration. The sealed 1 L reactor contains 500 mL water, 200 mL soil (1 percent organic carbon and density of 2.1 g/cm^3), and 300 mL air. The temperature of the reactor is 25°C. After adding 100 µg TCE to the reactor, the student incubates the reactor until equilibrium is achieved between all three phases. The Henry's law constant for TCE is 10.7 L-atm/mole at 25°C, and TCE has a log K_{ow} of 2.42. Assuming that no chemical or biological degradation of TCE occurs during the incubation, what is the aqueous-phase concentration of TCE at equilibrium? What is the mass of TCE in the aqueous, air, and sorbed phases after equilibrium is attained?

solution

Set up a simplified mass balance that equates the total mass of TCE added to the mass of TCE in each phase at equilibrium:

$$\text{Total mass of TCE added} = [\text{mass of aqueous TCE}]$$

$$+ [\text{mass of gaseous TCE}] + [\text{mass of sorbed TCE}]$$

$$100\,µg = [V_{aq} \times C_{aq}] + [V_{air} \times C_{air}] + [M_{soil} \times C_{sorbed}]$$

The problem is requesting C_{aq}. The three known parameters are $V_{aq} = 500\,mL$, $V_{air} = 300\,mL$, and mass of soil $= M_{soil} = V_{soil} \times$ density of soil $= 200\,mL \times cm^3/mL \times 2.1\,g/cm^3 = 420\,g$. The three unknowns are C_{aq}, C_{air}, and C_{sorbed}; however, C_{air} can be related to C_{aq} by a Henry's law constant, and C_{sorbed} can be related to C_{aq} by a soil–water partition coefficient.

Convert the Henry's law constant to dimensionless form. $K_H = 10.7$ L-atm/mole (by the units, we can tell this Henry's law constant is for the reaction written in the following direction: $C_{aq} \leftrightarrow C_{air}$). Convert to dimensionless form using the ideal gas law (see Table 3.3):

$$\frac{\dfrac{10.7\,\text{L-atm}}{\text{mole}}}{\dfrac{0.0825\,\text{L-atm}}{\text{mole-K}}(298\,K)} = 0.44$$

The Henry's law constant of 0.44 is equal to C_{air}/C_{aq}, so $C_{air} = 0.44\,C_{aq}$.

Determine the soil–water partition coefficient. Remember, $K = K_{oc} \times f_{oc}$, and 1 percent organic carbon means $f_{oc} = 0.01$. Because K_{oc} and K are not provided, estimate K_{oc} from K_{ow}: $\log K_{oc} = 0.903 \times 2.42 + 0.094 = 2.28$. Therefore, $K_{oc} = 10^{2.28}$, and

$$K = 10^{2.28} \times 0.01 = \frac{1.9\,cm^3}{g} \quad \text{and} \quad C_{sorbed} = \frac{1.9\,cm^3}{g} \times C_{aq}$$

Accordingly, substitute into the mass balance so all concentrations are in terms of C_{aq}:

$$100\,µg = [500\,mL \times C_{aq}] + [300\,mL \times 0.44\,C_{aq}] + \left[420\,g \times \frac{1.9\,cm^3}{g} \times \frac{mL}{cm^3} \times C_{aq}\right]$$

$$100\,µg = C_{aq}\left\{500\,mL + [300\,mL \times 0.44] + \left[420\,g \times \frac{1.9\,cm^3}{g} \times \frac{mL}{cm^3}\right]\right\}$$

$$100\,µg = C_{aq}[500\,mL + 132\,mL + 798\,mL]$$

$$C_{aq} = 0.070\,µg/mL = 0.070\,mg/mL = 70\,ppb_m$$

The total mass of TCE in the aqueous phase is 35 µg; in the air phase, it is 9.2 µg; sorbed to soil, it is 55.8 µg.

The mass of chemical found in each of the three phases is a function of the combined effects of partitioning between each phase. The amount of chemical that partitions to each phase is based on the physical/chemical properties of the chemical (for example, Henry's law constant, log K_{ow}) and soil/sediment properties (f_{oc}). This is very important when determining where a chemical migrates in the environment or an engineered system, as well as in determining what method of treatment should be selected.

3.11 Kinetics

The kinetic approach to environmental chemistry addresses the rate of reactions. Concepts include the rate law, zero-order and first-order reactions, half-life, and factors that affect the rate of reaction.

3.11.1 THE RATE LAW

The **rate law** expresses the dependence of the reaction rate on measurable, environmental parameters. Of particular interest is the dependence of the rate on the concentrations of the reactants. Other parameters that may influence the reaction rate include temperature and the presence of catalysts (including microorganisms).

The rate of an irreversible reaction and the exact form of the rate law depend on the mechanism of the reaction. Consider the hydrolysis of dichloromethane (DCM). In this reaction, one molecule of DCM reacts with a hydroxide ion (OH^-) to produce chloromethanol (CM) and chloride ion:

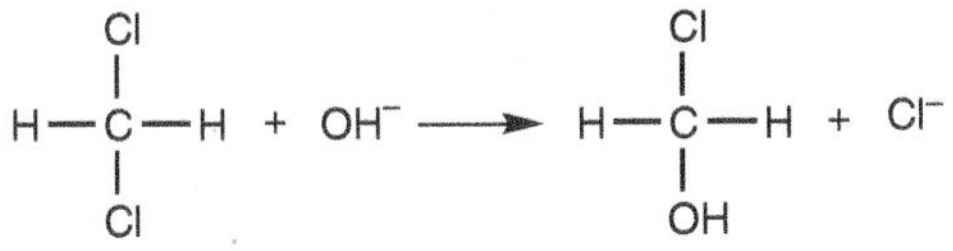

For the reaction depicted here to occur, one molecule of DCM must collide and react with one molecule of OH^-. The rate of an irreversible binary reaction is proportional to the concentration of each chemical species. For the hydrolysis of DCM, it can be written as

$$R = k[DCM][OH^-] = -d[DCM]/dt$$
$$= -d[OH^-]/dt = d[CM]/dt = d[Cl^-]/dt \qquad (3.35)$$

where R is the rate of reaction, k is the rate constant for this particular reaction, [DCM] is the concentration of DCM, [OH^-] is the concentration of hydroxide ion, [CM] is the concentration of CM, [Cl^-] is the concentration of chloride ion, and t is time. The negative signs in Equation 3.35 indicate that the products' concentrations are decreasing over time.

The bold portion on the left side of Equation 3.35 is referred to as the reaction's *rate law*, which expresses the dependence of the reaction rate on the concentrations of the reactants. The rate law in this case would be called first order with respect to DCM and first order with respect to OH^-. The term **first order** indicates that each species is raised to the first power. The rate law is second order overall because it involves the product of two species, each raised to the first power. Because the reaction was depicted as irreversible, it was assumed that the concentration of products did not influence the rate of the forward reaction.

To generalize these terms, a hypothetical rate law can be constructed for a generic irreversible reaction of a moles of species A reacting with b moles of species B to yield products, P. The rate law is written as

$$R = k[A]^a[B]^b$$ (3.36)

This reaction would be termed ath order with respect to A and bth order with respect to B. The **overall order** of the reaction would be $(a + b)$. This reaction is termed an **elementary reaction** because the reaction order is controlled by the stoichiometry of the reaction. That is, a equals the molar stoichiometric coefficient of species A and b equals the molar stoichiometric coefficient for B.

The order of a reaction should be determined experimentally, because it often does not correspond to the reaction stoichiometry. This is because the mechanism or steps of the reaction do not always correspond to that shown in the reaction equation.

The collision-based reaction of the hydrolysis of dichloromethane can be contrasted with some biological transformations of organic chemicals that occur in treatment plants or natural environments where soils and sediments are present. In some of these situations, zero-order transformations are observed. A reaction is termed zero-order when it does not depend on the concentration of the compound involved in the reaction. Zero-order kinetics can be due to several items, including the rate-limiting diffusion of oxygen from the air into the aqueous phase, which may be slower than the demand for oxygen by the microorganism biodegrading the chemical. Another explanation for an observation of zero-order kinetics is the slow, rate-limiting movement of a chemical (required by the microorganisms for energy and growth) that has a low water solubility (ppb_m and ppm_m range) from an oil or soil/sediment phase into the aqueous phase, where the chemical is then available for the organism to utilize.

One chemical that has been observed to have zero-order kinetics of biodegradation is 2,4-D, an herbicide commonly used by farmers and households. 2,4-D can be transported into a river or lake by horizontal runoff or vertical migration to groundwater that is hydraulically connected to a lake or river. It has been found to disappear in lake water according to zero-order kinetics. The rate law for this type of reaction can be written as

$$R = -d[2, 4\text{-D}]/dt = k$$ (3.37)

3.11.2 ZERO-ORDER AND FIRST-ORDER REACTIONS

Many environmental situations can be described by zero-order or first-order kinetics. Figure 3.12 compares the major differences between these two types of kinetics. In this section, we discuss these kinetic expressions in depth by first constructing a generic chemical reaction whereby a chemical, C, is converted to some unknown products:

$$C \rightarrow \text{products} \qquad (3.38)$$

The rate law that describes the decrease in concentration of chemical C with time can be written as

$$d[C]/dt = -k[C]^n \qquad (3.39)$$

Here, $[C]$ is the concentration of C, t is time, k is a rate constant that has units dependent on the order of the reaction, and the reaction order, n, typically is an integer (0, 1, 2).

ZERO-ORDER REACTION   If n is 0, Equation 3.39 becomes

$$d[C]/dt = -k \qquad (3.40)$$

This is the rate law describing a zero-order reaction. Here, the rate of disappearance of C with time is zero-order with respect to C, and the overall order of the reaction is zero-order.

Equation 3.40 can be rearranged and integrated for the following conditions; at time 0, the concentration of C equals C_0, and at some future time t, the concentration equals C:

$$\int_{C_0}^{C} d[C] = -k \int_{0}^{t} dt \qquad (3.41)$$

Reaction Order	Rate Law	Integrated Form of Rate Law	Plot of Concentration versus Time	Linearized Plot of Concentration versus Time	Half-Life, t	Example Units of Rate Constant, k
Zero	$\dfrac{d[C]}{dt} = -k$	$[C] = [C_0] - kt$		Same as $[C]$ vs. time	$\dfrac{0.5[C_0]}{k}$	moles/L-s mg/L-s
First	$\dfrac{d[C]}{dt} = -k[C]$	$[C] = [C_0]e^{-kt}$			$\dfrac{0.693}{k}$	s^{-1}, min^{-1}, h^{-1}, day^{-1}

Figure / 3.12   **Summary of Zero- and First-Order Rate Expressions**   Note the differences between each of these expressions.

(From Mihelcic (1999). Reprinted with permission of John Wiley & Sons, Inc.).

Integration of Equation 3.41 yields

$$[C] = [C]_0 - kt \tag{3.42}$$

A reaction is zero-order if concentration data plotted versus time result in a straight line (illustrated in Figure 3.12). The slope of the resulting line is the zero-order rate constant k, which has units of concentration/time (for example, moles/liter-day).

FIRST-ORDER REACTION   If $n = 1$, Equation 3.39 becomes

$$d[C]/dt = -k[C] \tag{3.43}$$

This is the rate law for a first-order reaction. Here, the rate of disappearance of C with time is first-order with respect to [C], and the overall order of the reaction is first-order.

Equation 3.43 can be rearranged and integrated for the same two conditions used in Equation 3.40 to obtain an expression that describes the concentration of C with time:

$$[C] = [C]_0 e^{-kt} \tag{3.44}$$

Here, k is the first-order reaction rate constant and has units of time^{-1} (for example, h^{-1}, day^{-1}).

A reaction is first-order when the natural logarithm of concentration data plotted versus time results in a straight line. The slope of this straight line is the first-order rate constant, k, as illustrated in Figure 3.12.

There are some important things to note about first- and zero-order chemical reactions. First, when comparing the concentration over time in the two reactions (as shown in the figure), the rate of the first-order reaction (slope of concentration data versus time) decreases over time, while in the zero-order reaction, the slope remains constant over time. This suggests that the rate of a zero-order reaction is independent of chemical concentration (see Equation 3.42), while the rate of a first-order reaction is dependent on the concentration of the chemical (see Equation 3.44). Thus, a chemical whose disappearance follows concentration-dependent kinetics, like first-order, will disappear more slowly as its concentration decreases.

3.11.3   PSEUDO FIRST-ORDER REACTIONS

There are many circumstances in which the concentration of one participant in a reaction remains constant during the reaction. For example, if the concentration of one reactant initially is much higher than the concentration of another, it is impossible for the reaction to cause a significant change in the concentration of the substance with the high initial concentration. Alternatively, if the concentration of one substance is buffered at a constant value (for example, pH in a lake does not change because it is buffered by the dissolution and precipitation of alkalinity-containing solid $CaCO_3$), then the concentration of the buffered species will not change, even if the substance participates in a reaction. A **pseudo first-order** reaction is used in these situations. It can be modeled as if it were a first-order reaction. Consider the following *irreversible elementary reaction*:

$$aA + bB \rightarrow cC + dD \qquad \textbf{(3.45)}$$

---

example/3.14   Use of Rate Law

How long will it take the carbon monoxide (CO) concentration in a room to decrease by 99 percent after the source of carbon monoxide is removed and the windows are opened? Assume the first-order rate constant for CO removal (due to dilution by incoming clean air) is 1.2/h. No chemical reaction is occurring.

solution

This is a first-order reaction, so use Equation 3.44. Let $[CO]_0$ equal the initial CO concentration. When 99 percent of the CO goes away, $[CO] = 0.01 \times [CO]_0$. Therefore,

$$0.01 = [CO]_0 = [CO]_0 \, e^{-kt}$$

where $k = 1.2/h$. Solve for t, which equals 3.8 h.

The rate law for this reaction is

$$\boxed{R = k[A]^a[B]^b} \qquad \textbf{(3.46)}$$

If the concentration of A does not change significantly during the reaction for one of the reasons previously discussed (that is, $[A_0] \gg [B_0]$ or $[A] \cong [A_0]$), the concentration of A may be assumed to remain constant and can be incorporated into the rate constant, k. The rate law then becomes

$$R = k'[B]^b \qquad \textbf{(3.47)}$$

where k' is the pseudo first-order rate constant and equals $k[A_0]^a$. This manipulation greatly simplifies the rate law for the disappearance of substance B:

$$d[B]/dt = -k'[B]^b \qquad \textbf{(3.48)}$$

If b is equal to 1, then the solution of Equation 3.48 is identical to that for Equation 3.44. In this case, the pseudo first-order expression can be written as follows:

$$\boxed{[B] = [B_0]e^{-k't}} \qquad \textbf{(3.49)}$$

example/3.15 Pseudo First-Order Reaction

Lake Silbersee is located in the German city of Nuremberg. The lake's water quality has been diminished because of high hydrogen sulfide concentrations (which have a rotten-egg smell) that originate from a nearby leaking landfill. To combat the problem, the city decided to aerate the lake in an attempt to oxidize the odorous H_2S to nonodorous sulfate ion according to the following oxidation reaction:

$$H_2S + 2O_2 \rightarrow SO_4^{2-} + 2H^+$$

It has been determined experimentally that the reaction follows first-order kinetics with respect to both oxygen and hydrogen sulfide concentrations:

$$d[H_2S]/dt = -k[H_2S][O_2]$$

The present rate of aeration maintains the oxygen concentration in the lake at 2 mg/L. The rate constant k for the reaction was determined experimentally to be 1,000 L/mole-day. If the aeration completely inhibited anaerobic respiration and thus stopped the production of sulfide, how long would it take to reduce the H_2S concentration in the lake from 500 to 1 μM?

solution

The dissolved oxygen of the lake is maintained at a constant value and therefore is a constant. It can be combined with the rate constant to make a pseudo first-order rate constant. Thus,

$$[H_2S] = [H_2S]_0 \, e^{-k't}$$

where $k' = k[O_2]$

$$1 \, \mu M = 500 \, \mu M \times e^{\left\{- \frac{1,000\,L}{\text{mole-day}} \times \frac{2\,mg}{L} \times \frac{g}{1,000\,mg} \times \frac{\text{mole}}{32\,mg} \times t\right\}}$$

Solve for the time: $t = 100$ days.

3.11.4 HALF-LIFE AND ITS RELATIONSHIP TO THE RATE CONSTANT

It often is useful to express a reaction in terms of the time required to react one-half of the concentration initially present. The **half-life**, $t_{1/2}$, is defined as the time required for the concentration of a chemical to decrease by one-half (for example, $[C] = 0.5 \times [C]_0$). The relationship between half-life and the reaction rate constant depends on the order of the reaction, as shown in Figure 3.12.

For zero-order reactions, the half-life can be related to the zero-order rate constant, k. To do this, substitute $[C] = 0.5 \times [C]_0$ into Equation 3.42:

$$0.5[C]_0 = [C]_0 - kt_{1/2} \tag{3.50}$$

Equation 3.50 can be solved for the half-life:

$$\boxed{t_{1/2} = \frac{0.5 \times [C]_0}{k}} \tag{3.51}$$

Likewise, for a first-order reaction, the half-life can be related to the first-order rate constant, k. In this case, substitute $[C] = 0.5 \times [C]_0$ into Equation 3.44:

$$0.5[C]_0 = [C]_0 e^{-kt} \tag{3.52}$$

Is Nuclear Power Safe?
www.ucsusa.org/nuclear_power

Radon: Number One Source of Natural Radiation
www.epa.gov/radon

The half-life for a first-order relationship then is given by

$$t_{1/2} = \frac{0.693}{k}$$

(3.53)

---

example/3.16   Converting a Rate Constant to Half-Life

Subsurface half-lives for benzene, TCE, and toluene are listed as 69, 231, and 12 days, respectively. What are the first-order rate constants for all three chemicals?

solution

The model only accepts concentration-dependent, first-order rate constants. Thus, to solve the problem, convert half-life to a first-order rate constant with the use of Equation 3.53:
    For benzene,

$$t_{1/2} = \frac{0.693}{k} = \frac{0.693}{69 \, \text{days}} = 0.01/\text{day}$$

Similarly, $k_{TCE} = 0.058/\text{day}$ and $k_{toluene} = 0.058/\text{day}$.

---

example/3.17   Use of Half-Life in Determining First-Order Decay

The 2011 Fukushima nuclear disaster was the largest since the Chernobyl disaster of 1986. It occurred after the Tohoku earthquake and tsunami and consisted of a several nuclear meltdowns and releases of radioactive materials. One and a half years after the disaster, Japan still bans the sale of 36 fish species caught off the coast of Fukushima, destroying the social and economic livelihood of the local region. A 2012 scientific paper reported two greenling fish close to shore had greater than 25,000 becquerels (Bq) per kg of fish (wet weight) from the presence of radioactive cesium. This is 250 times greater than the government's safety limit (in comparison, the U.S. threshold is 1,200 Bq per kg of fish (wet weight)). Assume that the only reaction by which cesium is lost from the fish is through radioactive decay and the half-life for this isotope is 3 years. Calculate the concentration of radioactive cesium in a Fukushima fish after 5 years. (Note: A becquerel is a measure of radioactivity; 1 becquerel equals 1 radioactive disintegration per second.) Would regulators allow this fish to be consumed in Japan or the United States?

solution

Because the half-life equals 3 years, the rate constant k can be determined from Equation 3.53:

$$k = \frac{0.693}{t_{1/2}} = \frac{0.693}{3 \, \text{year}} = 0.23/\text{year}$$

Therefore:

$$[^{137}Cs]_{t=5} = [^{137}Cs]_{t=0} \exp(-kt) = 25{,}000 \, \text{Bq/kg} \times \exp\left(\frac{-0.23}{\text{year}} \times 5 \, \text{year}\right) = 7{,}916 \, \text{Bq/kg}$$

This fish is not safe to eat according to regulations in Japan and the United States. The value greatly exceeds the threshold values of 100 and 1,200 Bq per kg of fish (wet weight) set by each country's government. A problem to work on outside of the classroom is, how many years would it take this fish to reach safe levels set by each country?

3.11.5 EFFECT OF TEMPERATURE ON RATE CONSTANTS

Rate constants typically are determined and compiled for temperatures at 20°C or 25°C. However, groundwaters usually have temperatures around 8–12°C, and surface waters, wastewaters, and soils generally have temperatures ranging from 0°C to 30°C. Thus, when a different temperature is encountered, you must first determine if the effect of temperature is important, and second, if important, determine how to convert the rate constant for the new temperature.

The **Arrhenius equation** is used to adjust rate constants for changes in temperature. It is written as

$$k = A\, e^{-(Ea/RT)} \tag{3.54}$$

where k is the rate constant of a particular order, A is termed the pre-exponential factor (same units as k), Ea is the **activation energy** (kcal/mole), R is the gas constant, and T is temperature (K). The activation energy, Ea, is the energy required for the collision to result in a reaction. The pre-exponential factor is related to the number of collisions per time, so it is different for gas- and liquid-phase reactions. The pre-exponential factor, A, has a small dependence on temperature for many reactions; however, most environmental situations span a relatively small temperature range. Its value depends to a great extent on the number of molecules that collide in a reaction. For example, unimolecular reactions exhibit values of A that can be several orders of magnitude greater than bimolecular reactions.

A plot of $\ln(k)$ versus $1/T$ can be used to determine Ea and A. After Ea and A are known for a particular reaction, Equation 3.54 can be used to adjust a rate constant for changes in temperature.

The Arrhenius equation is the basis for another commonly used relationship between rate constants and temperature used for biological processes over narrow temperature ranges. The *carbonaceous biochemical oxygen demand (CBOD)* rate constant, k, known at a particular temperature, typically is converted to other temperatures using the following expression:

$$k_{T_2} = k_{T_1} \times \Theta^{(T_2 - T_1)} \tag{3.55}$$

where Θ is a **dimensionless temperature coefficient**. In fact, Θ equals $\exp\{Ea \div [R \times T_1 \times T_2]\}$, as can be seen from the Arrhenius equation. Θ is temperature dependent and has been found to range from 1.056 to 1.13 for biological decay of municipal sewage.

The rate constant for carbonaceous biochemical oxygen demand (CBOD) at 20°C is 0.1/day. What is the rate constant at 30°C? Assume $\Theta = 1.072$.

solution

Using Equation 3.55,

$$k_{30} = 0.1/\text{day}\left[1.072^{(30°C-20°C)}\right] = 0.2/\text{day}$$

This example demonstrates that, for biological systems used in wastewater treatment and resource recovery, we would often observe a doubling in the biological reaction with every 10°C increase in the temperature.

Key Terms

- absorption
- acid
- activation energy
- activity
- activity coefficients
- adsorption
- alkalinity
- Arrhenius equation
- base
- buffering capacity
- carbonate system
- dimensionless temperature coefficient
- dissolved carbon dioxide
- dissolved oxygen
- electron acceptor
- electron donor
- elementary reaction
- equilibrium
- equilibrium constant (K)
- first law of thermodynamics
- first order

- Freundlich isotherm
- Gibbs free-energy
- half-life
- Henry's law
- Henry's law constant (K_H)
- hydrophobic partitioning
- ideal system
- ionic strength
- kinetics
- K_a
- K_b
- K_p
- K_w
- law of conservation of mass
- linear isotherm
- local equilibrium
- methylation
- octanol–water partition coefficient (K_{ow})
- overall order
- oxidation–reduction

- oxidation state
- pH
- photosynthesis
- pK_a
- precipitation–dissolution
- pseudo first-order
- Raoult's law
- rate law
- redox reactions
- saturated vapor pressure
- second law of thermodynamics
- soil–water partition coefficient
- soil–water partition coefficient normalized to organic carbon (K_{oc})
- sorption
- stoichiometry
- thermodynamics
- van't Hoff relationship
- volatilization
- zero order

3.1 How many grams of NaCl would you need to add to a 1 L water sample (pH = 7) so the ionic strength equaled 0.1 M?

3.2 You are studying the feasibility of using a reverse osmosis membrane system to desalinate seawater (TDS = 35,000 mg/L) and inland brackish groundwater (TDS typically ranges from 1,000 to 10,000 mg/L). (a) Estimate the ionic strength of the seawater and brackish water. (b) A conductivity meter provides a reading of 7,800 μmho/cm when placed in one of these water samples. Which water source is the sample from?

3.3 Calculate the ionic strength and individual activity coefficients for a 1 L solution in which 0.02 moles of $Mg(OH)_2$, 0.01 moles of $FeCl_3$, and 0.01 moles of HCl are dissolved.

3.4 Hydrogen sulfide is an odor-causing chemical found at many wastewater collection and treatment facilities. The following expression describes hydrogen sulfide gas reacting with aqueous-phase hydrogen sulfide (a diprotic acid).

$$H_2S_{(gas)} = H_2S_{(aqueous)}$$

Use your understanding of chemical equilibrium and thermodynamics to determine the Henry's constant (moles/L-atm) for this reaction at 25°C. The change in free energy of formation at standard conditions (units of kcal/mole) is as follows: $H_2S_{(gas)} = -7.892$, $H_2S_{(aqueous)} = -6.54$, $HS^-_{(aqueous)} = +3.01$, $SO_4^{2-} = -177.34$.

3.5 The reaction of divalent manganese with oxygen in aqueous solution is given as follows:

$$Mn^{2+} + \tfrac{1}{2}O_{2(aqueous)} + H_2O = MnO_{2(solid)} + 2H^+$$

The equilibrium constant (K) for this reaction is 23.7. It has been found that a lake water sample that contains no oxygen at 25°C, pH = 8.5, originally contained 0.6 mg/L of Mn^{2+}. The sample was aerated (atmospheric conditions of the dissolved oxygen concentration is 9.2 mg/L) and after 10 days of contact with atmospheric oxygen, the Mn^{2+} concentration was 0.4 mg/L. The molecular weight of Mn is 55, O is 16, and H is 1. The change in free energy of formation at standard conditions (units of kcal/mole)

are as follows: $Mn^{2+} = -54.4$, $O_{2 \text{ (aqueous)}} = +3.93$, $H_2O = -56.69$, $MnO_{2 \text{ (solid)}} = -111.1$, $H^+ = 0$. (a) Assuming that the pH remains constant during aeration, will the precipitate continue to form after the measurement on the tenth day? Assume ideal conditions. (b) What should the Mn^{2+} concentration be (in moles/L) at equilibrium, assuming that pH and presence of dissolved oxygen are the same as in part "A"? Assume ideal conditions. (c) What should the Mn^{2+} concentration (in moles/L) be at equilibrium if 2×10^{-3} moles/liter of NaCl are added to the solution and the pH is adjusted to 2? (problem based on Snoeyink and Jenkins, 1980).

3.6 Phosphate ion reacts in water to form monohydrogen phosphate according to the following reaction:

$$PO_4{}^{3-} + H_2O = HPO_4{}^{2-} + OH^-$$

The equilibrium constant for this reaction is $10^{-1.97}$. (a) Given that this is a dilute system (so you can assume ideal conditions), temperature is 298 K and the total combined phosphate/monohydrogen phosphate is 10^{-4} M, what percentage of the total concentration is in the phosphate ion form at pH = 11? (b) Will the reaction proceed as written at pH = 9 when $[PO_4{}^{3-}] = 10^{-6.8}$ and $[HPO_4{}^{2-}] = 10^{-4}$ M? And if not, which direction will the reaction proceed?

3.7 The chemical 1,4-dichlorobenzene (1,4-DCB) is sometimes used as a disinfectant in public lavatories. At 20°C (68°F), the vapor pressure is 5.3×10^{-4} atm. (a) What would be the concentration in the air in units of g/m^3? The molecular weight of 1,4-DCB is 147 g/mole. (b) An alternative disinfectant is 1-bromo-4-chlorobenzene (1,4-CB). The boiling point of 1,4-CB is 196°C, whereas the boiling point of 1,4-DCB is 180°C. Which compound would cause the highest concentrations in the air in lavatories? (Explain your answer.)

3.8 The boiling temperatures of chloroform (an anesthetic), carbon tetrachloride (commonly used in the past for dry cleaning), and tetrachloroethylene (previously used as a degreasing agent) are 61.7°C, 76.5°C, and 121°C. The vapor pressure of a chemical is directly proportional to the inverse of the chemical's boiling point. If a large quantity of these compounds were spilled in the environment, which compound would you predict to have higher concentrations in the air above the site? (Explain your answer.)

3.9 What would be the saturation concentration (mole/L) of oxygen (O_2) in a river in winter when the air temperature is 0°C if the Henry's law constant at this temperature is 2.28×10^{-3} mole/L-atm? What would the answer be in units of mg/L?

3.10 The log Henry's law constant (units of L-atm/mole and measured at 25°C) for trichloroethylene is 1.03; for tetrachloroethylene, 1.44; for 1,2-dimethylbenzene, 0.71; and for parathion, −3.42. (a) What is the dimensionless Henry's law constant for each of these chemicals? (b) Rank the chemicals in order of ease of stripping from water to air.

3.11 The dimensionless Henry's law constant for trichloroethylene (TCE) at 25°C is 0.4. A sealed glass vial is prepared that has an air volume of 4 mL overlying an aqueous volume of 36 mL. TCE is added to the aqueous phase so that initially it has an aqueous-phase concentration of 100 ppb. After the system equilibrates, what will be the concentration (in units of μg/L) of TCE in the aqueous phase?

3.12 The Henry's law constant for H_2S is 0.1 mole/L-atm, and

$$H_2S_{(aq)} \rightleftharpoons HS^- + H^+$$

where $K_a = 10^{-7}$. If you bubble pure H_2S gas into a beaker of water, what is the concentration of HS^- at a pH of 5 in (a) moles/L, (b) mg/L, and (c) ppm_m?

3.13 Determine the equilibrium pH of aqueous solutions of the following strong acids or bases: (a) 15 mg/L of HSO_4^-, (b) 10 mM NaOH, and (c) 2,500 μg/L of HNO_3.

3.14 What would be the pH if 10^{-2} moles of hydrofluoric acid (HF) were added to 1 L pure water? The pK_a of HF is 3.2.

3.15 When Cl_2 gas is added to water during the disinfection of drinking water, it hydrolyzes with the water to form HOCl. The disinfection power of the acid HOCl is 88 times better than its conjugate base, OCl^-. The pK_a for HOCl is 7.5. (a) What percentage of the total disinfection power ($HOCl + OCl^-$) exists in the acid form at pH = 6? (b) At pH = 7?

3.16 A 1 L aqueous solution is prepared at 25°C with 10^{-4} moles of hydrocyanic acid (HCN) and 10^{-3} moles of disodium carbonate (Na_2CO_3) and reaches equilibrium. (a) List the eight unknown chemical species here (water is not unknown). (b) List (do not solve) all four equilibrium expressions that describe this system, making sure to include the value for the equilibrium constants.

3.17 For the endothermic reaction,

$$SO_{2(g)} = S_{(s)} + O_2$$

will an increase in temperature increase, decrease, or have no effect on the reaction's equilibrium constant?

3.18 What pH is required to reduce a high concentration of a dissolved Mg^{2+} to 25 mg/L? The solubility product for the following reaction is $10^{-11.16}$.

$$Mg(OH)_{2(s)} = Mg^{2+} + 2OH^-$$

3.19 (a) What is the solubility (in moles/L) of CaF_2 in pure water at 25°C? (b) What is the solubility of CaF_2 if the temperature is raised 10°C? (c) Does the solubility of CaF_2 increase, decrease, or remain the same if the ionic strength is raised? (Explain your answer.)

3.20 At a wastewater treatment plant, $FeCl_{3(s)}$ is added to remove excess phosphate from the effluent. Assume the following reactions occur:

$$FeCl_{3(s)} \rightleftharpoons Fe^{3+} + 3Cl^-$$
$$FePO_{4(s)} \rightleftharpoons Fe^{3+} + PO_4^{3-}$$

The equilibrium constant for the second reaction is $K_{sp} = 10^{-26.4}$. What concentration of Fe^{3+} is needed to maintain the phosphate concentration below the limit of 1 mg P/L?

3.21 One method to remove metals from water is to raise the pH and cause them to precipitate as their metal hydroxides. (a) For the following reaction, compute the standard free energy of reaction:

$$Cd^{2+} + 2OH^- \rightleftharpoons Cd(OH)_{2(s)}$$

(b) The pH of water initially was 6.8 and then was raised to 8.0. Is the dissolved cadmium concentration reduced to below 100 mg/L at the final pH? Assume the temperature of the water is 25°C.

3.22 Naphthalene has a log K_{ow} of 3.33. Estimate its soil–water partition coefficient normalized to organic carbon and the 95 percent confidence interval of your estimate.

3.23 Atrazine, an herbicide widely used for corn, is a common groundwater pollutant in the corn-producing regions of the United States. The log K_{ow} for atrazine is 2.65. Calculate the fraction of total atrazine that will be adsorbed to the soil given that the soil has an organic carbon content of 2.5 percent. The bulk density of the soil is 1.25 g/cm^3; this means that each cubic centimeter of soil (soil plus water) contains 1.25 g soil particles. The porosity of the soil is 0.4.

3.24 Mercury concentrations in San Francisco Bay were measured to be 8 ng/L in rain water, 1.25 ng/L dissolved in the Bay water, and 250 ng/gm dry weight of sediment. Using the information provided and assuming equilibrium, what is the sediment-water partition coefficient for mercury in the sediments (units of cm^3 per gram dry weight of sediment)?

3.25 Given the following general reaction:
$$A + 2B + 3C \rightarrow P + 4Q$$
Show how the change in concentration of C with time is related to the change in concentration of A, B, P, and Q with time.

3.26 Which of the following statements about the study of chemical kinetics is true? (a) temperature has no effect on the rate of a reaction, (b) changes in reactant concentration do not affect the rate at which a reaction occurs, (c) the addition of a catalyst to a reaction will speed up the reaction but it will not ultimately result in a larger mass of product, (d) for the same reactants, the larger the surface area, the slower a reaction will occur (problem from EPA Air Pollution Training Institute, http://www.epa.gov/apti/bces/).

3.27 Peridisulfate ($S_2O_8^{2-}$) reacts with thiosulfate ($S_2O_3^{2-}$) according to the following reaction:
$$S_2O_8^{2-} + 2S_2O_3^{2-} \rightarrow 2SO_4^{2-} + S_4O_6^{2-}$$
(a) Show how the change in peridisulfate concentration with time is related to the change in concentration with time of the other three species. (b) If the reaction is elementary and irreversible, what is the overall order of the reaction?

3.28 A first-order reaction that results in the destruction of a pollutant has a rate constant of 0.1/day. (a) How many days will it take for 90 percent of the chemical to be destroyed? (b) How long will it take for 99 percent of the chemical to be destroyed? (c) How long will it take for 99.9 percent of the chemical to be destroyed?

3.29 A bacteria strain has been isolated that can cometabolize tetrachloroethane (TCA). This strain can be used for the bioremediation of hazardous-waste sites contaminated with TCA. Assume that the biodegradation rate is independent of TCA concentration (that is, the reaction is zero-order). In a bioreactor, the rate for TCA removal was 1 μg/L-min. What water retention time would be required to reduce the concentration from 1 mg/L in the influent to 1 μg/L in the effluent of a reactor? Assume the reactor is completely mixed.

3.30 Assume PO_4^{3-} is removed from municipal wastewater through precipitation with Fe^{3+} according

to the following reaction: $PO_4^{3-} + Fe^{3+} \rightarrow FePO_{4(s)}$. The rate law for this reaction is
$$\frac{d[PO_4^{3-}]}{dt} = -k[Fe^{3+}][PO_4^{3-}]$$
(a) What is the reaction order with respect to PO_4^{3-}?
(b) What order is this reaction overall?

3.31 Obtain the World Health Organization (WHO) report on "Urine diversion: Hygienic risks and microbial guidelines for reuse." Review Figure 1.2. (a) How many grams of N, P, and K are excreted every day in a Swedish person's urine?

3.32 Obtain the World Health Organization (WHO) report on "Urine diversion: Hygienic risks and microbial guidelines for reuse." Read Chapter 4 (Pathogenic microorganisms in urine). Answer the following questions. (a) Is the urine in a healthy individual's bladder sterile or nonsterile? (b) What concentration of dermal bacteria is picked up during urination (bacteria/mL)? (c) What percent of urinary tract infections are caused by *Escherichia coli*?

3.33 Ammonia (NH_3) is a common constituent of many natural waters and wastewaters. When water-containing ammonia is treated at a water treatment plant, the ammonia reacts with the disinfectant hypochlorous acid (HOCl) in solution to form monochloroamine (NH_2Cl) as follows:
$$NH_3 + HOCl \rightarrow NH_2Cl + H_2O$$
The rate law for this reaction is
$$\frac{d[NH_3]}{dt} = -k[HOCl][NH_3]$$
(a) What is the reaction order with respect to NH_3?
(b) What order is this reaction overall? (c) If the HOCl concentration is held constant and equals 10^{-4} M, and the rate constant equals 5.1×10^6 L/mole-s, calculate the time required to reduce the concentration of NH_3 to one-half its original value.

3.34 Nitrogen dioxide (NO_2) concentrations are measured in an air-quality study and decrease from 5 to 2 ppm_v in 4 min with a particular light intensity. (a) What is the first-order rate constant for this reaction? (b) What is the half-life of NO_2 during this study? (c) What would the rate constant need to be changed to in order to decrease the time required to lower the NO_2 concentration from 5 to 2 ppm_v in 1.5 min?

3.35 Assume that municipal solid waste is 30 percent organic carbon by wet weight. The organic carbon in the solid waste decays by first-order kinetics after placed in a landfill with reported rate constants for a dry climate (0.02/year), moderate climate (0.038/year), and wet climate (0.057/year). Dry climate is defined as precipitation plus recirculated leachate being less than 20 in./year; moderate climate as precipitation plus recirculated leachate ranges from 20 to 40 in./year; and a wet climate having precipitation plus recirculated leachate greater than 40 in/year. Estimate the time it takes for 20 and 90 percent of the organic carbon contained in a municipal solid-waste landfill to decay in the three different climates. In practice, this will be the period when greenhouse gases should be captured from the landfill.

3.36 On March 11, 2011, a massive earthquake and tsunami triggered a major disaster at Japan's Fukushima nuclear plant. A plume extending to the northwest of the site deposited significant amounts of iodine-131, cesium-134, and cesium-137 up to 30 miles away. Iodine-131 has an 8-day half-life and cesium-137 has a 3-year half-life. Determine how long it will take 99 percent of the iodine-131 and 99 percent of the cesium-137 to naturally decay (you can learn about "U.S. Nuclear Power Safety One Year after Fukushima" by reading the report written by D. Lochbaum and E. Lyman, located on the web site of the Union of Concerned Scientists, http://www.ucsusa.org/publications/publications-nuclear-power.html).

3.37 After the Chernobyl nuclear accident, the concentration of ^{137}Cs in milk was proportional to the concentration of ^{137}Cs in the grass that cows consumed. The concentration in the grass was, in turn, proportional to the concentration in the soil. Assume that the only reaction by which ^{137}Cs was lost from the soil was through radioactive decay and the half-life for this isotope is 3 years. Calculate the concentration of ^{137}Cs in cow's milk after 5 years (units of Bq/L) if the concentration in milk shortly after the accident was 12,000 becquerels (Bq) per liter (a becquerel is a measure of radioactivity; 1 becquerel equals 1 radioactive disintegration per second).

3.38 Table 3.12 shows the annual mean growth rate (units of ppm CO_2/year) measured at Mauna Loa (Hawaii). The annual mean rate of growth of CO_2 in a given year is the difference in concentration between the end of December and the start of January of that year. The National Oceanic and Atmospheric Administration (NOAA) reports that the annual growth rate is similar to the global growth rate of CO_2

in the atmosphere (Dr. Pieter Tans, NOAA/ESRL, http://www.esrl.noaa.gov/gmd/ccgg/trends/, and Dr. Ralph Keeling, Scripps Institution of Oceanography, scrippsco2.ucsd.edu/). (a) What is the average growth rate of CO_2 in the atmosphere over this 20-year period (ppm CO_2/year)? (b) Review the shape of the figure showing the atmospheric CO_2 measurements made at Mauna Loa over the past 50 years (Figure 4.14 or the web site referred to above). Does the data follow a first-order or zero-order reactor order? Explain your answer. (c) Assume that the 1959 monthly mean concentration of CO_2 measured at Mauna Loa was 315 ppm. Using the average growth rate you determined in part (a) over the 20-year period and the appropriate reactor order, what atmospheric CO_2 concentration would you estimate for the year 1980, 2012, and 2050?

3.39 If the rate constant for the degradation of biochemical oxygen demand (BOD) at 20°C is 0.23/day, what is the value of the BOD rate constant at 5°C and 25°C? Assume that Θ equals 1.1.

3.40 Excess nitrogen inputs to estuaries have been scientifically linked to poor water quality and degradation of ecosystem habitat. The nitrogen loading to Narragansett Bay was estimated to be 8,444,631 kg N/year and to Chesapeake Bay is 147,839,494 kg N/year. The watershed area for Narragansett Bay is 310,464 ha and for Chesapeake Bay is 10,951,074 ha. The nitrogen loading rates are estimated for Galveston Bay to be 16.5 kg N per ha per year, 26.9 kg N per ha per year for Tampa Bay, 49.0 kg N per ha per year for Massachusetts Bay, and 20.2 kg N per ha per year for Delaware Bay. Rank the loading rates from lowest to highest for these six estuaries.

3.41 Excess nitrogen inputs to estuaries have been scientifically linked to poor water quality and degradation of ecosystem habitat. Perform a library search for the paper title "Nitrogen inputs to seventy-four southern New England estuaries: Application of a watershed nitrogen model" (Latimer and Charpentier, 2010). Based on this article, what is the percent contribution of the following four sources of nitrogen to the watershed of the New England estuaries? (a) Direct atmospheric deposition to the estuaries, (b) wastewater, (c) indirect atmospheric deposition to the watershed of the estuary, and (d) fertilizer runoff from lawns, golf courses, and agriculture.

Table / 3.12

Year	1959	1960	1961	1962	1963	1964	1965	1966	1967	1968	1969	1970	1971	1972	1973	1974	1975	1976	1977	1978	1979	1980
ppm/year	0.94	0.54	0.95	0.64	0.71	0.28	1.02	1.24	0.74	1.03	1.31	1.06	0.85	1.69	1.22	0.78	1.13	0.84	2.10	1.30	1.75	1.73

References

Baker, J. R., J. R. Mihelcic, D. C. Luehrs, and J. P. Hickey, 1997. Evaluation of estimation methods for organic carbon normalized sorption coefficients. *Water Environment Research, 69*: 136–145.

Latimer, J. S., and M. A. Charpentier, 2010. "Nitrogen inputs to seventy-four southern New England estuaries: Application of a watershed nitrogen model." *Estuarine, Coastal and Shelf Science, 89*: 125–136.

Mihelcic, J. R., 1999. *Fundamentals of Environmental Engineering.* New York: John Wiley & Sons.

Snoeyink, V. L., and D. Jenkins, 1980. *Water Chemistry.* New York: John Wiley & Sons.

chapter/Four Physical Processes

Richard E. Honrath Jr.,
James R. Mihelcic, Julie Beth
Zimmerman, Qiong Zhang

In this chapter, readers will learn about the physical processes that are important in the movement of pollutants through the environment and processes used to control and treat pollutant emissions. The chapter begins with a study of the use of material and energy balances and the processes of advection and dispersion. Energy balances are applied to a wide range of topics: the greenhouse effect and climate change, household energy losses, energy efficiency, and the urban heat island effect. The final section of this chapter extends previous descriptions of transport processes with a look at movement of fluids and particles in fluids; specifically turbulent and mechanical dispersion and gravitational settling that follows Stokes' Law.

Learning Objectives

1. Use the law of conservation of mass to write a mass balance that includes rate of chemical production or disappearance.
2. Determine whether a situation is at steady or nonsteady state, and apply this information to the mass balance.
3. Differentiate batch reactors, completed mixed flow reactors, and plug-flow reactors.
4. Relate a reactor's retention time to reactor volume and flow.
5. Differentiate forms of energy, and write an energy balance.
6. Relate an energy balance to the greenhouse effect, household energy losses and application of energy efficiency, and the urban heat island effect.
7. Relate temperature change to sea level rise under different population, economic growth, and energy management scenarios.
8. Describe magnitude and specific types of materials flows associated with the built environment and the implications of these flows for design, planning, and management.
9. Calculate heat loss from buildings through the building skin and from infiltration.
10. Relate heat loss in buildings to degree-days and the R factor of building materials.
11. Determine heat input from passive solar and storage of heat using thermal walls.
12. Relate features of the built environment, such as street and building geometry, location and number of trees and water, building materials, and nonpervious surfaces, to the urban heat island effect.
13. Differentiate and employ the transport processes of advection, dispersion, and diffusion.
14. Apply Fick's law and Stokes' law to environmental engineering problems.

4.1 Mass Balances

The **law of conservation of mass** states that mass can neither be produced nor destroyed. Conservation of mass and conservation of energy provide the basis for two commonly used tools: the **mass balance** and the energy balance. This section discusses mass balances, and energy balances are the topic of Section 4.2.

The principle of conservation of mass means that if the amount of a chemical increases somewhere (for example, in a lake), then that increase cannot be the result of some "magical" formation. The chemical must have been either carried into the lake from elsewhere or produced via chemical or biological reaction from other compounds that were already in the lake. Similarly, if reactions produced the mass increase of this chemical, they must also have caused a corresponding decrease in the mass of some other compound(s).

In terms of sustainability, this same principle of mass balance can be thought of in terms of the use of finite material and energy sources. For example, the consumption of fossil-based energy sources—oil, gas, and coal—must maintain a mass balance. As a result, as these resources are combusted for energy, the original source is depleted, and wastes are generated in the form of emissions to the air, land, and water. While the mass of carbon remains constant, much of it is removed from the energy-intensive form of oil, gas, or coal and is converted to carbon dioxide, a greenhouse gas.

Conservation of mass provides a basis for compiling a budget of the mass of any chemical. In the case of a lake, this budget keeps track of the amounts of chemical entering and leaving the lake and the amounts formed or destroyed by chemical reaction. This budget can be balanced over a given time period, much as a checkbook is balanced. Equation 4.1 describes the mass balance:

$$\text{mass at time } t + \Delta t = \text{mass at time } t$$

$$+ \left(\begin{array}{c} \text{mass entering} \\ \text{from } t \text{ to } t + \Delta t \end{array} \right) - \left(\begin{array}{c} \text{mass exiting} \\ \text{from } t \text{ to } t + \Delta t \end{array} \right)$$

$$+ \left(\begin{array}{c} \text{net mass of chemical produced} \\ \text{from other compounds by} \\ \text{reactions between } t \text{ and } t + \Delta t \end{array} \right) \quad \textbf{(4.1)}$$

Each term of Equation 4.1 has units of mass. This form of balance is most useful when there is a clear beginning and end to the balance period (Δt), so that the change in mass over the balance period can be determined. Continuing our earlier analogy, when balancing a checkbook, a balance period of 1 month is often used.

In environmental problems, however, it is usually more convenient to work with values of **mass flux**—the rate at which mass enters or leaves a system. To develop an equation in terms of mass flux, the mass balance equation is divided by Δt to produce an equation with units of mass per unit time. Dividing Equation 4.1 by Δt and moving the first term on the right (mass at time t) to the left-hand side yields:

$$\frac{(\text{mass at time } t + \Delta t) - (\text{mass at time } t)}{\Delta t} = \frac{\begin{pmatrix} \text{mass entering from} \\ t \text{ to } t + \Delta t \end{pmatrix}}{\Delta t}$$

$$-\frac{\begin{pmatrix} \text{mass exiting from} \\ t \text{ to } t + \Delta t \end{pmatrix}}{\Delta t} + \frac{\begin{pmatrix} \text{net chemical} \\ \text{production from} \\ t \text{ to } t + \Delta t \end{pmatrix}}{\Delta t} \qquad \textbf{(4.2)}$$

Note that each term in Equation 4.2 has units of mass/time. The left side of Equation 4.2 is equal to $\Delta m / \Delta t$.

In the limit as $\Delta t \to 0$, the left side becomes dm/dt, the rate of change of chemical mass in the lake. As $\Delta t \to 0$, the first term on the right side of Equation 4.2 becomes the rate at which mass enters the lake (the mass flux into the lake), and the second term becomes the rate at which mass exits the lake (the mass flux out of the lake). The last term of Equation 4.2 is the *net rate* of chemical production or loss.

The symbol $\dot{m}$ refers to a mass flux with units of mass/time. Substituting mass flux, the equation for mass balances can be written as follows:

$$\begin{pmatrix} \text{mass} \\ \text{accumulation} \\ \text{rate} \end{pmatrix} = (\text{mass flux in}) - (\text{mass flux out}) + \begin{pmatrix} \text{net rate of} \\ \text{chemical} \\ \text{production} \end{pmatrix}$$

or

$$\boxed{\frac{dm}{dt} = \dot{m}_{\text{in}} - \dot{m}_{\text{out}} + \dot{m}_{\text{reaction}}} \qquad \textbf{(4.3)}$$

Equation 4.3 is the governing equation for mass balances used throughout environmental engineering and science.

4.1.1 CONTROL VOLUME

A mass balance is meaningful only in terms of a specific region of space, which has boundaries across which the terms $\dot{m}_{\text{in}}$ and $\dot{m}_{\text{out}}$ are determined. This region is called the **control volume**.

In the previous example, we used a lake as our control volume and included mass fluxes into and out of the lake. Theoretically, any volume of any shape and location can be used as a control volume. Realistically, however, certain control volumes are more useful than others. The most important attribute of a control volume is that it has boundaries over which $\dot{m}_{\text{in}}$ and $\dot{m}_{\text{out}}$ can be calculated.

4.1.2 TERMS OF THE MASS BALANCE EQUATION FOR A CMFR

A well-mixed tank is an analogue for many control volumes used in environmental situations. For example, in the lake example, it might be reasonable to assume that the chemicals discharged into the lake are mixed throughout the entire lake. Such a system is called a **completely mixed flow reactor (CMFR)**. Other terms, most commonly *continuously stirred tank reactor (CSTR)*, are also used for such systems. A schematic diagram of a CMFR is shown in Figure 4.1.

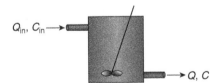

Figure / 4.1 Schematic Diagram of a CMFR The stir bar is used as a symbol to indicate that the CMFR is well mixed.

(From Mihelcic (1999). Reprinted with permission of John Wiley & Sons, Inc.).

The following discussion describes each term in a mass balance of a hypothetical compound within the CMFR.

MASS ACCUMULATION RATE (dm/dt)

The rate of change of mass within the control volume, dm/dt, is referred to as the **mass accumulation rate**. To directly measure the mass accumulation rate would require determining the total mass within the control volume of the compound for which the mass balance is being conducted. This is usually difficult, but it is seldom necessary. If the control volume is well mixed, then the concentration of the compound is the same throughout the control volume, and the mass in the control volume is equal to the product of that concentration, C, and the volume, V. (To ensure that $C \times V$ has units of mass/time, express C in units of mass/volume.) Expressing mass as $C \times V$, the mass accumulation rate is equal to

$$\frac{dm}{dt} = \frac{d(VC)}{dt} \tag{4.4}$$

In most cases (and in all cases in this text), the volume is constant and can be moved outside the derivative, resulting in

$$\boxed{\frac{dm}{dt} = V\frac{dC}{dt}} \tag{4.5}$$

In any mass balance situation, once a sufficient amount of time has passed, conditions will approach **steady state**, meaning that conditions no longer change with time. In steady-state conditions, the concentration—and hence the mass—within the control volume remains constant. In this case, $dm/dt = 0$. If, however, insufficient time has passed since a flow, inlet concentration, reaction term, or other problem condition has changed, the mass in the control volume will vary with time, and the mass balance will be **nonsteady state**.

The amount of time that must pass before steady state is reached depends on the conditions of the problem. To see why, consider the approach to steady state of the amount of water in two large, initially empty sinks. In the first sink, the faucet is opened halfway and the drain is opened slightly. Initially, the mass of water in the sink increases over time, since the faucet flow exceeds the flow rate out of the drain. Conditions are changing, so this is a nonsteady-state situation. However, as the water level in the sink rises, the flow rate out of the drain will increase, and eventually the drain flow will equal the faucet flow. At this point, the water level will cease rising, and the situation will have reached steady state.

If this experiment is repeated with a second sink, but this time with the drain opened fully, the drain flow will increase more rapidly and will equal the faucet flow while the water level in the sink is still low. In this case, steady state will be reached more rapidly. In general, the speed at which steady state is approached depends on the magnitude of the mass flux terms, relative to the total mass in the control volume.

Determining whether or not a mass balance problem is steady state is something of an art. However, if conditions of the problem have changed recently, then the problem is probably a nonsteady state. Conversely, if conditions have remained constant for a very long time, it is probably a steady-state problem. Treating a steady-state problem as nonsteady state will always result in the correct answer, while treating a nonsteady-state problem as steady state will not. This does not mean that all problems should be treated as nonsteady state, however. Nonsteady-state solutions generally are more difficult, so it is advantageous to identify steady state whenever present.

In terms of emissions to the environment, steady state is often equated with nature's ability to assimilate wastes at the rate at which they are released. For example, in the case of **carbon dioxide emissions** released from burning fossil fuels, at steady state the rate of emissions would equal the total of all removal rates from the atmosphere. These include uptake by the oceans and the small fraction of uptake by plants for photosynthesis that is not balanced by respiration, which releases carbon dioxide. Eventually, as the carbon dioxide concentration in the atmosphere rises, the rate of uptake by the oceans will balance the rate of emissions from fossil-fuel burning. However, for that to happen, the concentration in the atmosphere would have to increase significantly, and the dissolved carbon dioxide would have to become well mixed throughout the ocean. Since these processes take centuries to millennia, carbon dioxide emissions accumulate in the atmosphere, where they contribute to the greenhouse effect.

A similar situation can occur for the release of industrial chemicals to the environment. Currently, the ease of assimilation by the environment is often ignored when chemicals are selected or designed and manufactured for uses that result in release to the environment. In many cases, the result is accumulation in the environment in a system that is not at steady state. This is of particular concern with chemicals that bioaccumulate (build up in organisms), becoming more concentrated in organisms further up the food chain.

Indoor Air in Large Buildings
http://www.epa.gov/iaq/largebldgs/

Greenhouse Gas Emissions from Transportation
http://www.epa.gov/otaq/climate

---

example/4.1   Determining Whether a Problem is Steady State

For each of the following mass balance problems, determine whether a steady-state or nonsteady-state mass balance would be appropriate.

1. Vision a mass balance on chloride (Cl^-) dissolved in a lake. Two rivers bring chloride into the lake, and one river removes chloride. No significant chemical reactions occur, as chloride is soluble and nonreactive. What is the annual average concentration of chloride in the lake?

2. A degradation reaction within a well-mixed tank is used to destroy a pollutant. Inlet concentration and flow are held constant, and the system has been operating for several days. What is the pollutant concentration in the effluent, given the inlet flow and concentration and the first-order decay rate constant?

3. The source of pollutant in problem 2 is removed, resulting in an instantaneous decline of the inlet concentration to zero. How long would it take until the outlet concentration reaches 10 percent of its initial value?

solution

1. Over an annual period, river flows and concentrations can be assumed to be relatively constant. Since conditions are not changing, and since a single value independent of time is requested for chloride concentration, the problem is steady state.

2. Again, conditions in the problem are constant and have remained so for a long time, so the problem is steady state. Note that the presence or absence of a chemical reaction does not provide any information on whether the problem is steady state.

3. Two clues reveal that this problem is nonsteady state. First, conditions have changed recently: the inlet concentration dropped to zero. Second, the solution requires calculation of a time period, which means conditions must be varying with time.

MASS FLUX IN ($\dot{m}_{in}$) Often, the volumetric flow rate, Q, of each input stream entering the control volume is known. In Figure 4.1, the pipe has a flow rate of Q_{in}, with corresponding chemical concentration of C_{in}. The *mass flux into the CMFR* is then given by the following equation:

$$\dot{m}_{in} = Q_{in} \times C_{in} \qquad (4.6)$$

If it is not immediately clear how $Q \times C$ results in a mass flux, consider the units of each term:

$$\dot{m} = Q \times C$$

$$\frac{\text{mass}}{\text{time}} = \frac{\text{volume}}{\text{time}} \times \frac{\text{mass}}{\text{volume}}$$

Note that the concentration must be expressed in units of mass/volume.

If the volumetric flow rate is not known, it may be calculated from other parameters. For example, if the fluid velocity v and the cross-sectional area A of the pipe are known, then $Q = v \times A$.

In some situations, mass may enter the control volume through direct emission into the volume. In this case, the emissions are frequently specified in mass flux units mass/time, which can be used in a mass balance directly. For example, if a mass balance is performed on the air pollutant carbon monoxide over a city, we would use estimates of the total carbon monoxide emissions (in units of tons/day) from automobiles and power plants in the city.

Another way to describe the flux is in terms of a flux density, J, times the area through which the flux occurs. J has units of mass/area-time and is discussed further under the topic of diffusion. This type of flux notation is most useful at interfaces where there is no fluid flow, such as the interface between the air and water at the surface of a lake.

Often, the mass flux is composed of several terms. For example, a tank may have more than one inlet, or the air over a city may receive carbon monoxide blowing from an upwind urban area in addition to its own emissions. In such cases, $\dot{m}_{in}$ is the sum of all individual contributions to mass input fluxes.

MASS FLUX OUT ($\dot{m}_{out}$) In most cases, there is only one effluent flow from a CMFR. Then the mass flux out may be calculated as $\dot{m}_{in}$, which was calculated in Equation 4.6:

$$\dot{m}_{out} = Q_{out} \times C_{out} \qquad (4.7)$$

In the case of a well-mixed control volume, the concentration is constant throughout. Therefore, the concentration in flow exiting the control volume is referred to simply as C, the concentration in the control volume, and

$$\dot{m}_{out} = Q_{out} \times C \qquad (4.8)$$

NET RATE OF CHEMICAL REACTION ($\dot{m}_{reaction}$) The term $\dot{m}_{reaction}$ or $\dot{m}_{rxn}$ refers to the net rate of production of a compound from chemical or biological reactions. It has units of mass/time. Thus, if other compounds react to form the compound, $\dot{m}_{rxn}$ will be greater than zero; if the compound reacts to form some other compound(s), resulting in a loss, $\dot{m}_{rxn}$ will be negative.

Although the chemical-reaction term in a mass balance has units of mass/time, chemical-reaction rates are usually expressed in terms of concentration, not mass. Thus, to calculate $\dot{m}_{rxn}$, we multiply the rate of change of concentration by the CMFR volume to obtain the rate of change of mass within the control volume:

$$\dot{m}_{rxn} = V \times \left(\frac{dC}{dt}\right)_{\text{reaction only}} \qquad (4.9)$$

where $(dC/dt)_{\text{reaction only}}$ is obtained from the rate law for the reaction and is equal to the rate of change in concentration that would occur if the reaction took place in isolation, with no influent or effluent flows.

Mass flux due to reaction may take various forms. The following are the most common:

- **Conservative compound.** Compounds with no chemical formation or loss within the control volume are termed **conservative compounds**. Conservative compounds are not affected by chemical or biological reactions, so $(dC/dt)_{\text{reaction only}} = \dot{m}_{reaction} = 0$. The term *conservative* is used for these compounds because their mass is truly conserved: what goes in equals what goes out.

- **Zero-order decay.** The rate of loss of the compound is constant. For a compound with **zero-order decay**, $(dC/dt)_{\text{reaction only}}$ equals $-k$, and $\dot{m}_{rxn}$ equals $-Vk$. Zero-order reactions are discussed in Chapter 3.

- **First-order decay.** For a compound with **first-order decay**, the rate of loss of the compound is directly proportional to its concentration: $(dC/dt)_{\text{reaction only}}$ equals $-kC$. For such a compound, $\dot{m}_{rxn}$ equals $-VkC$. First-order reactions are discussed in Chapter 3.

- **Production at a rate dependent on the concentrations of other compounds in the CMFR.** In this situation, the chemical is produced by reactions involving other compounds in the CMFR, and $(dC/dt)_{\text{reaction only}}$ is greater than zero.

Learn about the Chesapeake Bay
http://www.chesapeakebay.net/

Tampa Bay Estuary Program
http://www.tbep.org

STEPS IN MASS BALANCE PROBLEMS   Solution of mass balance problems involving CMFRs generally will be straightforward if the problem is done carefully. Most difficulties in solving mass balance problems arise from uncertainty regarding the location of control volume boundaries or values of the individual terms in the mass balance. Therefore, the following steps will assist in solving each mass balance problem:

1. Draw a schematic diagram of the situation and identify the control volume and all influent and effluent flows. All mass flows that are known or to be calculated must cross the control volume boundaries, and it should be reasonable to assume that the control volume is well mixed.

2. Write the mass balance equation in general form:

$$\frac{dm}{dt} = \dot{m}_{in} - \dot{m}_{out} + \dot{m}_{rxn}$$

3. Determine whether the problem is steady state $(dm/dt = 0)$ or nonsteady state $(dm/dt = V \times dC/dt)$.

4. Determine whether the compound being balanced is conservative $(\dot{m}_{rxn} = 0)$ or nonconservative $(\dot{m}_{rxn}$ must be determined based on the reaction kinetics and Equation 4.9).

5. Replace $\dot{m}_{in}$ and $\dot{m}_{out}$ with known or required values, as just described.

6. Finally, solve the problem. This will require solution of a differential equation in nonsteady-state problems and solution of an algebraic equation in steady-state problems.

4.1.3   REACTOR ANALYSIS: THE CMFR

Reactor analysis refers to the use of mass balances to analyze pollutant concentrations in a control volume that is either a chemical reactor or a natural system modeled as a chemical reactor. Ideal reactors can be divided into two types: completely mixed flow reactors (CMFRs) and plug-flow reactors (PFRs). CMFRs are used to model well-mixed environmental reservoirs. PFRs, described in Section 4.1.5, behave essentially like pipes and are used to model situations such as downstream transport in a river in which fluid is not mixed in the upstream–downstream direction.

This section presents several examples involving CMFRs in different combinations of steady-state or nonsteady-state conditions and conservative or nonconservative compounds, as summarized in Table 4.1. Example 4.2 demonstrates the use of CMFR analysis to determine the concentration of a substance resulting from the mixing of two or more influent flows.

Examples 4.3 through 4.5 refer to the tank depicted in Figure 4.1 and demonstrate steady-state and nonsteady-state situations with and

Table / 4.1

Summary of CMFR Examples

Example Number	Form of dm/dt	Form of $\dot{m}_{reaction}$
Example 4.2	Steady state	Conservative
Example 4.3	Steady state	First-order decay
Example 4.4	Nonsteady state	First-order decay
Example 4.5	Nonsteady state	Conservative

SOURCE: Mihelcic (1999). Reprinted with permission of John Wiley & Sons, Inc.

without first-order chemical decay. Calculations analogous to those in Examples 4.3 through 4.5 can be used to determine the concentration of pollutants exiting a treatment reactor, the rate of increase of pollutant concentrations within a lake resulting from a new pollutant source, or the period required for pollutant levels to decay from a lake or reactor once a source is removed.

example/ 4.2   **Steady-State CMFR with Conservative Chemical: The Mixing Problem**

A pipe from a municipal wastewater treatment plant discharges 1.0 m³/s of poorly treated effluent containing 5.0 mg/L of phosphorus compounds (reported as mg P/L) into a river with an upstream flow rate of 25 m³/s and a background phosphorus concentration of 0.010 mg P/L (see Figure 4.2). What is the resulting concentration of phosphorus (in mg/L) in the river just downstream of the plant outflow?

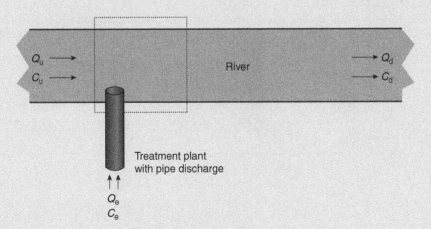

Figure / 4.2   **Mixing Problem Used in Example 4.2**   The control volume is indicated by the area inside the dotted lines.

(From Mihelcic (1999). Reprinted with permission of John Wiley & Sons, Inc.).

solution

To solve this problem, apply two mass balances: one to determine the downstream volumetric flow rate (Q_d) and a second to determine the downstream phosphorus concentration (C_d). First, a control volume must be selected. To ensure that the input and output fluxes cross the control volume boundaries, the control volume must cross the river upstream and downstream of the plant's outlet and must also cross the discharge pipe. The selected control volume is shown in Figure 4.2 within dotted lines. It is assumed to extend downriver far enough that the discharged wastewater and the river water become well mixed before leaving the control volume. As long as that assumption is met, it makes no difference to the analysis how far downstream the control volume extends.

Before beginning the analysis, determine whether this is a steady-state or nonsteady-state problem and whether the chemical-reaction term will be nonzero. Because the problem statement does not refer to time, and it seems reasonable to assume that both the river and waste stream discharge have been flowing for some time and will continue to flow, this is a steady-state problem. In addition, this problem concerns the concentration resulting from rapid mixing of the river and effluent flows.

Therefore, we can define our control volume to be small and can safely assume that chemical or biological degradation is insignificant during the time spent in the control volume, so we treat this as a steady-state problem.

1. Determine the downstream flow rate, Q_d. To find Q_d, conduct a mass balance on the total river water mass. In this case, the "concentration" of river water in (mass/volume) units is simply the density of the water, ρ:

$$\frac{dm}{dt} = \dot{m}_{in} - \dot{m}_{out} + \dot{m}_{rxn}$$
$$= \rho Q_{in} - \rho Q_{out} + 0$$

where the term $\dot{m}_{rxn}$ has been set to zero because the mass of water is conserved. Since this is a steady-state problem, $dm/dt = 0$. Therefore, as long as the density ρ is constant, $Q_{in} = Q_{out}$, and $(Q_u + Q_e) = 26 \, \text{m}^3/\text{s} = Q_d$.

2. Determine the phosphorus concentration downstream of the discharge pipe, C_d. To find C_d, use the standard mass balance equation with steady-state conditions and with no chemical formation or decay:

$$\frac{dm}{dt} = \dot{m}_{in} - \dot{m}_{out} + \dot{m}_{rxn}$$

$$0 = (C_u Q_u + C_e Q_e) - C_d Q_d + 0$$

Solve for C_d:

$$C_d = \frac{C_u Q_u + C_e Q_e}{Q_d}$$

$$= \frac{(0.010 \, \text{mg/L}) \, (25 \, \text{m}^3/\text{s}) + (5.0 \, \text{mg/L}) \, (1.0 \, \text{m}^3/\text{s})}{26 \, \text{m}^3/\text{s}}$$

$$= 0.20 \, \text{mg/L}$$

example/4.3 Steady-State CMFR with First-Order Decay

The CMFR shown in Figure 4.1 is used to treat an industrial waste, using a reaction that destroys the pollutant according to first-order kinetics, with $k = 0.216/$day. The reactor volume is 500 m³, the volumetric flow rate of the single inlet and exit is 50 m³/day, and the inlet pollutant concentration is 100 mg/L. What is the outlet concentration after treatment?

solution

An obvious control volume is the tank itself. The problem requests a single, constant outlet concentration, and all problem conditions are constant. Therefore, this is a steady-state problem ($dm/dt = 0$).

The mass balance equation with a first-order decay term $\left([dC/dt]_{\text{reaction only}} = -kC \text{ and } \dot{m}_{\text{rxn}} = -VkC \right)$ is:

$$\frac{dm}{dt} = \dot{m}_{\text{in}} - \dot{m}_{\text{out}} + \dot{m}_{\text{rxn}}$$

$$0 = QC_{\text{in}} - QC - VkC$$

Solve for C:

$$C = C_{\text{in}} \times \frac{Q}{Q + kV}$$

$$= C_{\text{in}} \times \frac{1}{1 + \left(k \times \dfrac{V}{Q} \right)}$$

Substituting the given values, the numerical solution is:

$$C = 100 \,\text{mg/L} \times \frac{50 \,\text{m}^3/\text{day}}{50 \,\text{m}^3/\text{d} + (0.216/\text{day})(500 \,\text{m}^3)}$$

$$= 32 \,\text{mg/L}$$

example/4.4 Nonsteady-State CMFR with First-Order Decay

The manufacturing process that generates the waste in Example 4.3 has to be shut down, and starting at $t = 0$, the concentration C_{in} entering the CMFR is set to 0. What is the outlet concentration as a function of time after the concentration is set to 0? How long does it take the tank concentration to reach 10 percent of its initial, steady-state value?

solution

The tank is again the control volume. In this case, the problem is clearly nonsteady-state, because conditions change as a function of time. The mass balance equation is:

$$\frac{dm}{dt} = \dot{m}_{\text{in}} - \dot{m}_{\text{out}} + \dot{m}_{\text{rxn}}$$

$$V\frac{dC}{dt} = 0 - QC - kCV$$

Solve for dC/dt:

$$\frac{dC}{dt} = -\left(\frac{Q}{V} + k\right)C$$

To determine C as a function of time, the preceding differential equation must be solved. Rearrange and integrate:

$$\int_{C_0}^{C_t} \frac{dC}{dt} = \int_0^t -\left(\frac{Q}{V} + k\right)dt$$

Integration yields

$$\ln C - \ln C_0 = -\left(\frac{Q}{V} + k\right)t$$

Because $\ln x - \ln y$ is equal to $\ln(x/y)$, we can rewrite this equation as

$$\ln\left(\frac{C}{C_0}\right) = -\left(\frac{Q}{V} + k\right)t$$

which yields

$$\frac{C_t}{C_0} = e^{-(Q/V + k)t}$$

We can verify that this solution is reasonable by considering what happens at $t = 0$ and $t = \infty$. At $t = 0$, the exponential term is equal to 1, and $C = C_0$, as expected. As $t \to \infty$, the exponential term approaches 0, and concentration declines to 0—again as expected—since C_{in} is equal to 0.

We can now plug in values to determine the dependence of C on time. Example 4.3 provides Q and V. The initial concentration is equal to the concentration before C_{in} was set to 0, which was found to be 32 mg/L in Example 4.3. Plugging in these values yields the outlet concentration as a function of time:

$$C_t = 32 \text{ mg/L} \times \exp\left[-\left(\frac{50 \text{ m}^3/\text{day}}{500 \text{ m}^3} + \frac{0.216}{\text{day}}\right)t\right]$$

$$= 32 \text{ mg/L} \times \exp\left(-\frac{0.316}{\text{day}}t\right)$$

This solution is plotted in Figure 4.3a.

How long will it take the concentration to reach 10 percent of its initial, steady-state value? That is, at what value of t is $C_t/C_0 = 0.10$? At the time when $C_t/C_0 = 0.10$,

$$\frac{C}{C_0} = 0.10 = \exp\left(-\frac{0.316}{\text{day}}t\right)$$

example/4.4 (continued)

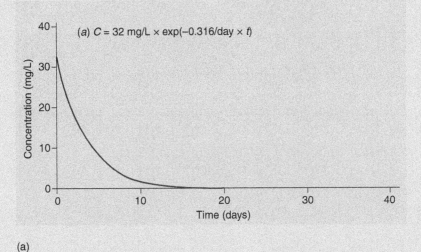

(a)

(b)

Figure / 4.3 **Concentration versus Time Profiles for the Solutions to Examples 4.4 and 4.5** (a) First-order decay in concentration resulting from the removal of $\dot{m}_{in}$ at time zero. The decay in concentration results from the sum of chemical-reaction loss $\dot{m}_{rxn}$ and the mass flux out term $\dot{m}_{out}$. (b) Exponential approach to steady-state conditions when a reactor is started with initial concentration equal to zero. In the absence of a chemical-reaction loss term, concentration in the reactor exponentially approaches the inlet concentration.

From Mihelcic (1999). Reprinted with permission of John Wiley & Sons, Inc.

Taking the natural logarithm of both sides,

$$\ln 0.10 = -2.303 = -\frac{0.316}{\text{day}}t$$

Therefore, $t = 7.3$ days.

example/4.5 Nonsteady-State CMFR, Conservative Substance

The CMFR reactor depicted in Figure 4.1 is filled with clean water prior to being started. After start-up, a waste stream containing 100 mg/L of a conservative pollutant is added to the reactor at a flow rate of 50 m^3/day. The volume of the reactor is 500 m^3. What is the concentration exiting the reactor as a function of time after it is started?

solution

Again, the tank will serve as a control volume. We are told that the pollutant is conservative, so $\dot{m}_{rxn} = 0$. The problem asks for concentration as a function of time, so the mass balance must be nonsteady state. The mass balance equation is

$$\frac{dm}{dt} = \dot{m}_{in} - \dot{m}_{out} + \dot{m}_{rxn}$$

$$V\frac{dC}{dt} = QC_{in} - QC + 0$$

Solve for dC/dt:

$$\frac{dC}{dt} = -\left(\frac{Q}{V}\right)(C - C_{in})$$

Because of the extra term on the right (C_{in}), this equation cannot be immediately solved. However, with a change of variables, we can transform the mass balance equation into a simpler form that can be integrated directly, using the same method as in Example 4.4. Let $y = (C - C_{in})$. Then $dy/dt = (dC/dt) - d(C_{in}/dt)$. Since C_{in} is constant, $dC_{in}/dt = 0$, so $dy/dt = dC/dt$. Therefore, the last of the preceding equations is equivalent to

$$\frac{dy}{dt} = -\frac{Q}{V}y$$

Rearrange and integrate:

$$\int_{y(0)}^{y(t)} \frac{dy}{y} = \int_0^t -\frac{Q}{V}dt$$

Integration yields:

$$\ln\left(\frac{y(t)}{y(0)}\right) = -\frac{Q}{V}t$$

or

$$\frac{y(t)}{y(0)} = e^{-(Q/V)t}$$

Replacing y with $(C - C_{in})$ results in the following equation:

$$\frac{C - C_{in}}{C_0 - C_{in}} = e^{-(Q/V)t}$$

Since clean water is present in the tank at start-up, $C_0 = 0$:

$$\frac{C - C_{in}}{-C_{in}} = e^{-(Q/V)t}$$

Rearrange to solve for C:

$$C - C_{in} = -C_{in} e^{-(Q/V)t}$$
$$C = C_{in} \times \left(1 - e^{-(Q/V)t}\right)$$

This is the solution to the question posed in the problem statement.

Note what happens as $t \to \infty : e^{-(Q/V)t} \to 0$, and $C \to C_{in}$. This is not surprising, since the substance is conservative. If the reactor is run long enough, the concentration in the reactor will eventually reach the inlet concentration. This final equation (plotted in Figure 4.3b) provides C as a function of time. This can be used to determine how long it would take for the concentration to reach, say, 90 percent of the inlet value.

4.1.4 BATCH REACTOR

A reactor that has no inlet or outlet flows is termed a **batch reactor**. It is essentially a tank in which a reaction is allowed to occur. After one batch is treated, the reactor is emptied, and a second batch can be treated. Because there are no flows, $\dot{m}_{in} = 0$, and $\dot{m}_{out} = 0$. Therefore, the mass balance equation reduces to

$$\frac{dm}{dt} = \dot{m}_{rxn} \tag{4.10}$$

or

$$V\frac{dC}{dt} = V\left(\frac{dC}{dt}\right)_{reaction\ only} \tag{4.11}$$

Simplifying:

$$\frac{dC}{dt} = \left(\frac{dC}{dt}\right)_{reaction\ only} \tag{4.12}$$

Thus, in a batch reactor, the change in concentration with time is simply that which results from the chemical reaction. For example, for a first-order decay reaction, $r = -kC$. Thus,

$$\frac{dC}{dt} = -kC \tag{4.13}$$

or

$$\boxed{\frac{C_t}{C_0} = e^{-kt}} \tag{4.14}$$

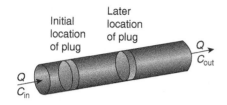

Figure / 4.4 Schematic Diagram of a Plug-Flow Reactor.

(From Mihelcic (1999). Reprinted with permission of John Wiley & Sons, Inc.).

4.1.5 PLUG-FLOW REACTOR

The **plug-flow reactor (PFR)** is used to model the chemical transformation of compounds as they are transported in systems resembling pipes. A schematic diagram of a PFR is shown in Figure 4.4. PFR pipes may represent a river, a region between two mountain ranges through which air flows, or a variety of other engineered or natural conduits through which liquids or gases flow. Of course, a pipe in this model can even represent a pipe. Figure 4.5 illustrates examples of a PFR in an engineered system (Figure 4.5a) and a PFR in a natural system (Figure 4.5b).

As fluid flows down the PFR, the fluid is mixed in the radial direction, but mixing does not occur in the axial direction. That is, each plug of fluid is considered a separate entity as it flows down the pipe. However, time passes as the plug of fluid moves downstream (or downwind). Thus, there is an implicit time dependence, even in steady-state PFR problems. However, because the velocity of the fluid (v) in the PFR is constant, time and downstream distance (x) are interchangeable, and $t = x/v$. That is, a fluid plug always takes an amount of time equal to x/v to travel a distance x down the reactor. This observation can be used with the mass balance formulations just given to determine how chemical concentrations vary during flow through a PFR.

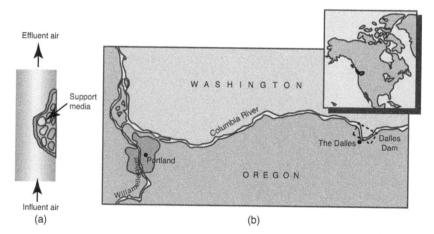

Figure / 4.5 Examples of Plug-Flow Reactors in Engineered and Natural Systems (a) Packed-tower biofilters are used to remove odorous air emissions, such as hydrogen sulfide (H_2S), from gas-phase emissions. Biofilters consist of a column packed with a support medium, such as rocks, plastic rings, or activated carbon, on which a biofilm is grown. Contaminated water or air is passed through the filter, and bacterial degradation results in the desired reduction of pollutant emissions. (b) The Columbia River flows 1,200 mi from its source in Canada to the Pacific Ocean. Before reaching the ocean, the Columbia River flows southward into the United States and forms the border between Oregon and Washington. Shown is a stretch of the river near The Dalles, Washington, where the river once narrowed and spilled over a series of rapids, christened *les Dalles*, or the trough, by early French explorers. A large dam has since been constructed near The Dalles. The section of the river downstream of the dam could be modeled as a PFR.

(From Mihelcic (1999). Reprinted with permission of John Wiley & Sons, Inc.).

To develop the equation governing concentration as a function of distance down a PFR, we will analyze the evolution of concentration with time within a single fluid plug. The plug is assumed to be well mixed in the radial direction but does not mix at all with the fluid ahead or behind it. As the plug flows downstream, chemical decay occurs, and concentration decreases. The mass balance for mass within this moving plug is the same as that for a batch reactor:

$$\frac{dm}{dt} = \dot{m}_{in} - \dot{m}_{out} + \dot{m}_{rxn} \tag{4.15}$$

$$V\frac{dC}{dt} = 0 - 0 + V\left(\frac{dC}{dt}\right)_{\text{reaction only}} \tag{4.16}$$

where $\dot{m}_{in}$ and $\dot{m}_{out}$ are set equal to zero because there is no mass exchange across the plug boundaries.

Equation 4.16 can be used to determine concentration as a function of flow time within the PFR for any reaction kinetics. In the case of first-order decay,

$$V(dC/dt)_{\text{reaction only}} = -VkC$$

and

$$V\frac{dC}{dt} = -VkC \tag{4.17}$$

which results in

$$\frac{C_t}{C_0} = \exp(-kt) \tag{4.18}$$

It is generally desirable to express the concentration at the outlet of PFR in terms of the inlet concentration and PFR length or volume, rather than time spent in the PFR. In a PFR of length L, each plug travels for a period $\theta = L/v = L \times A/Q$, where A is the cross-sectional area of the PFR and Q is the flow rate. The product of length and cross-sectional area is simply the PFR volume, so Equation 4.18 is equivalent to

$$\boxed{\frac{C_{out}}{C_{in}} = \exp\left(-\frac{kV}{Q}\right)} \tag{4.19}$$

Equation 4.19 has no time dependence. Although concentration within a given plug changes over time as that plug flows downstream, the concentration at a given fixed location within the PFR is constant with respect to time, since all plugs reaching that location have spent an identical period in the PFR.

COMPARISON OF THE PFR TO THE CMFR   The ideal CMFR and the PFR are fundamentally different and thus behave differently. When a parcel of fluid enters the CMFR, it is immediately mixed throughout the entire volume of the CMFR. In contrast, each parcel of fluid entering the PFR remains separate during its passage through the reactor.

Mississippi River Basin
http://www.epa.gov/msbasin/

Gulf of Mexico Program
http://www.epa.gov/gmpo/

Figure / 4.6 Comparison of (a) Completely Mixed Flow Reactor and (b) Plug-Flow Reactor.

(From Mihelcic (1999). Reprinted with permission of John Wiley & Sons, Inc.)

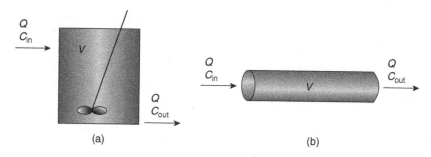

(a)

(b)

To highlight these differences, consider an example involving the continuous addition of a pollutant to each reactor, with destruction of the pollutant within the reactor according to first-order kinetics. The two reactors are depicted in Figure 4.6. This example assumes the incoming concentration (C_{in}), the flow rate (Q), and the first-order reaction rate constant (k) are known and are the same for both reactors. Consider two common problems:

1. If the volume V is known (the same for both reactors), what is the resulting outlet concentration (C_{out}) exiting the CMFR and PFR?

2. If an outlet concentration is specified, what volume of reactor is required for the CMFR and for the PFR? Table 4.2 summarizes the results of this comparison and lists the input variables.

The results given in Table 4.2 indicate that, for equal reactor volumes, the PFR is more efficient than the CMFR and, for equal outlet concentrations, a smaller PFR is required. Why is this? The answer has to do with the fundamental difference between the two reactors—fluid parcels entering the PFR travel downstream without mixing, while fluid parcels entering the CMFR are immediately mixed with the low-concentration fluid within the reactor. Since the rate of chemical reaction is proportional to concentration, the rate of chemical reaction

Table / 4.2

Comparison of CMFR and PFR Performance*

Example 1. Determine C_{out}, given $V = 100$ L, $Q = 5.0$ L/s, $k = 0.05$/s.	
CMFR	**PFR**
$C_{out} = C_{in}/(1 + kV/Q)$ $C_{out}/C_{in} = 0.50$	$C_{out} = C_{in}\exp(-kV/Q)$ $C_{out}/C_{in} = 0.37$
Example 2. Determine V, given $C_{out}/C_{in} = 0.5$, $Q = 5.0$ L/s, $k = 0.05$/s.	
CMFR	**PFR**
$V = (C_{in}/C_{out} - 1) \times (Q/k)$ $V = 100$ L	$V = -(Q/k) \ln (C_{out}/C_{in})$ $V = 69$ L

*Example 1 compares the effluent concentration (C_{out}) for a PFR and CMFR of the same volume; Example 2 compares the volume required for each reactor type if a given percent removal is required. SOURCE: Mihelcic (1999). Reprinted with permission of John Wiley & Sons, Inc.

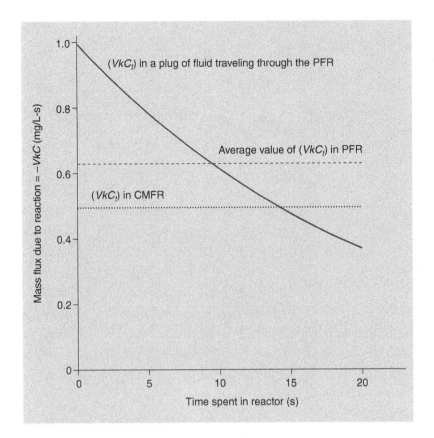

Figure / 4.7 Origin of the Higher Destruction Efficiency of a PFR Under Conditions of First-Order Decay   The rate of chemical destruction $(\dot{m}_{rxn} = -VkC$ is shown as a function of time spent in the reactor for a PFR (solid line) and CMFR (dotted line) for the conditions given in Example 1 of Table 4.2. Since concentration changes as each plug passes through the PFR $(C = C_{in}\exp[-k\Delta t])$, the value of $\dot{m}_{rxn}$ changes. The average rate of destruction in the PFR is shown by a dashed line. The rate of chemical destruction is constant throughout the well-mixed CMFR and is equal to $-VkC$. Since the high inlet concentration is diluted immediately on entering the CMFR, the rate of reaction is lower than that throughout most of the PFR and is lower than the average rate of reaction in the PFR.

(From Mihelcic (1999). Reprinted with permission of John Wiley & Sons, Inc.).

within the CMFR is reduced relative to that within the PFR. This effect is illustrated in Figure 4.7. The mass flux due to reaction is equal to $-VkC$ in both reactors. However, in the PFR, concentration decreases exponentially as each plug passes through the PFR, as shown by the solid curve in Figure 4.7. The average mass flux due to reaction in the PFR is simply the average value of this curve—the value indicated by the dashed line in Figure 4.7. In contrast, dilution as the incoming fluid is mixed into the CMFR immediately reduces the influent concentration to that within the CMFR, resulting in a reduced rate of destruction, indicated by the dotted line in Figure 4.7.

Response to Inlet Spikes.   CMFRs and PFRs also differ in their response to spikes in the inlet concentration. In many pollution-control systems, inlet concentrations or flows are not constant. For example, flow into municipal wastewater treatment plants varies dramatically over the course of each day. It is often necessary to ensure that a temporary increase in inlet concentration does not result in excessive outlet concentrations. And as will be seen in Chapter 9, low-impact development technology such as bioretention cells are designed to first store storm-water from a developed urban area and then release it back into the environment at a slow rate, reducing spikes in inlet concentrations and reducing the chance of overburdening wastewater treatment facilities.

Source reduction is always the preferred alternative to treatment. However, when source reduction techniques are not in place, reducing

Figure / 4.8 **Response of a CMFR and PFR to a Temporary Increase in Inlet Concentration** The influent concentration, shown in the lower, inset figure, increases to 2.0 during the period $t = 0$–15 s. The resulting concentrations exiting the CMFR and PFR of Example 2 in Table 4.2 are shown as a function of time before, during, and after the temporary doubling of inlet concentration. The concentration exiting the CMFR is shown with a dashed line; the concentration exiting the PFR is shown with a solid line. The maximum concentration reached in the CMFR effluent is less than that reached in the PFR effluent because the increased inlet concentration is diluted by the volume of lower-concentration fluid within the CMFR.

From Mihelcic (1999). Reprinted with permission of John Wiley & Sons, Inc.

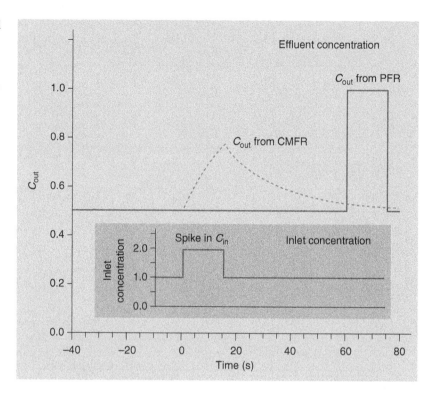

or eliminating spikes in outlet concentration requires the use of CMFRs as a result of the mixing that occurs within CMFRs but not within PFRs. Consider the effect of a temporary doubling of the concentration entering a PFR and CMFR: each is designed to reduce the influent concentration by the same amount with the flow, first-order decay rate constant, and required degree of destruction equal to the values given in Example 2 of Table 4.2.

The resulting changes in outlet concentration for the PFR and CMFR are shown in Figure 4.8. The concentration in fluid exiting the CMFR begins to rise immediately after the inlet concentration increases, as the more concentrated flow is mixed throughout the CMFR. The outlet concentration does not immediately double in response to the doubled inlet concentration, however, because the higher-concentration influent flow is diluted by the volume of low-concentration fluid within the CMFR. The CMFR outlet concentration rises exponentially and would eventually double, but the inlet spike does not last long enough for this to occur. In contrast, the outlet concentration exiting the PFR does not change until enough time has passed for the first plug of higher-concentration fluid to traverse the length of the PFR. At that time, the outlet concentration doubles, and it remains elevated for a period equal to the duration of the inlet spike.

Selection of CMFR or PFR. Selection of a CMFR or PFR in an engineered system is based on the considerations just described: control efficiency as a function of reactor size and response to changing inlet conditions. In many cases, the optimal choice is to use a CMFR to reduce

example/ 4.6 Required Volume for a PFR

Determine the volume required for a PFR to obtain the same degree of pollutant reduction as the CMFR in Example 4.3. Assume that the flow rate and first-order decay rate constant are unchanged ($Q = 50 \, m^3/day$ and $k = 0.216/day$).

solution

The CMFR in Example 4.3 achieved a pollutant decrease of $C_{out}/C_{in} = 32/100 = 0.32$. From Equation 4.19,

$$\frac{C_{out}}{C_{in}} = e^{-(kV/Q)}$$

or

$$0.32 = \exp - \left(\frac{0.216/day \times V}{50 \, m^3/day} \right)$$

Solve for V:

$$V = \ln 0.32 \times \frac{50 \, m^3/day}{-0.216/day}$$

$$= 264 \, m^3$$

As expected, this volume is smaller than the 500 m^3 required for the CMFR in Example 4.3.

sensitivity to spikes, followed by a PFR for efficient use of resources. Deciding between a CMFR and PFR has other environmental implications. If one reactor design is found to be more efficient than the other for a given set of operating conditions, using the more efficient design can cut energy requirements, waste production, and use of operating materials.

In natural systems, the choice is based on whether or not the system is mixed (in which case a CMFR would be used to model the system) or flows downstream without mixing (requiring use of a PFR). In some cases, it is necessary to use both the CMFR and PFR models. A common example of this involves effluent flow into a river. A CMFR is used to define a mixing problem, as was done in Example 4.2. This sets the inlet concentration for a PFR, which is used to model degradation of the pollutant as it flows further downstream. (This type of problem is investigated in Chapter 7 for dissolved oxygen in rivers.)

4.1.6 RETENTION TIME AND OTHER EXPRESSIONS FOR V/Q

A number of terms—including **retention time**, *detention time*, and *residence time*—refer to the average period spent in a given control volume, θ. The retention time is given by

$$\theta = \frac{V}{Q} \tag{4.20}$$

where V is the volume of the reactor and Q is the total volumetric flow rate exiting the reactor. Examples 4.7 and 4.8 illustrate the calculation and application of retention time.

example / 4.7   Retention Time in a CMFR and a PFR

Calculate the retention times in the CMFR of Example 4.3 and the PFR of Example 4.6.

solution

For the CMFR,

$$\theta = \frac{V}{Q} = \frac{500 \text{ m}^3}{50 \text{ m}^3/\text{day}} = 10 \text{ days}$$

For the PFR,

$$\theta = \frac{V}{Q} = \frac{264 \text{ m}^3}{50 \text{ m}^3/\text{day}} = 5.3 \text{ days}$$

example / 4.8   Retention Times for the Great Lakes

The Great Lakes region is shown in Figure 4.9. Calculate the retention times for Lake Michigan and Lake Ontario, using the data provided in Table 4.3.

solution

For Lake Michigan,

$$\theta = \frac{4,900 \times 10^3 \text{ m}^3}{36 \times 10^9 \text{ m}^3/\text{year}} = 136 \text{ years}$$

For Lake Ontario,

$$\theta = \frac{1,634 \times 10^9 \text{ m}^3}{212 \times 10^9 \text{ m}^3/\text{year}} = 8 \text{ years}$$

Table / 4.3

Volume and Flows for the Great Lakes

Lake	Volume 10^9 m^3	Outflow 10^9 m^3/year
Superior	12,000	67
Michigan	4,900	36
Huron	3,500	161
Erie	468	182
Ontario	1,634	211

SOURCE: Mihelcic (1999). Reprinted with permission of John Wiley & Sons, Inc.

These values mean that Lake Michigan changes its water volume completely once every 136 years and Lake Ontario once every 8 years. The higher flow and smaller volume of Lake Ontario result in a significantly shorter retention time. This means pollutant concentrations can increase in Lake Ontario much more quickly than they can in Lake Michigan and will drop much more quickly in Lake Ontario if a pollutant source is eliminated—provided that flow out of the lakes is the dominant pollutant sink.

These values of θ can be used to determine whether it would be appropriate to model the lakes as CMFRs in a mass balance problem. Temperate lakes generally are mixed twice per year. Therefore, over the period required for water to flush through Lakes Michigan and Ontario, the lakes would be mixed many times. It would therefore be appropriate to model the lakes as CMFRs in mass balances involving pollutants that do not decay significantly in less than approximately 1 year.

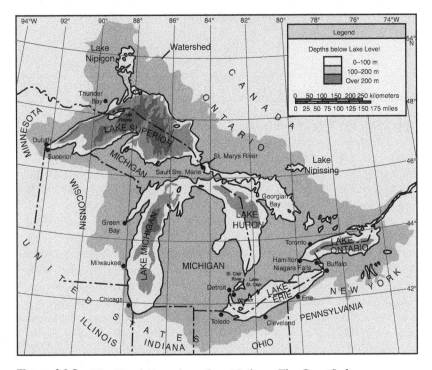

Figure / 4.9 **The North American Great Lakes** The Great Lakes are an important part of the physical and cultural heritage of North America. The Great Lakes contain approximately 18 percent of the world's supply of freshwater, making them the largest system of available surface freshwater (only the polar ice caps contain more freshwater). The first humans arrived in the area approximately 10,000 years ago. Around 6,000 years ago, copper mining began along the south shore of Lake Superior, and hunting/fishing communities were established throughout the area. Population in the region in the 16th century is estimated between 60,000 and 117,000—a level that resulted in few human disturbances. Today, the combined Canadian and U.S. population in the region exceeds 33 million. Increases in human settlement and exploitation over the past 200 years have caused many disturbances to the ecosystem. Today, the outflow from the Great Lakes is less than 1 percent per year. Therefore, pollutants that enter the lakes by air, direct discharge, or from nonpoint pollution sources may remain in the system for a long period of time.

(From Mihelcic (1999). Reprinted with permission of John Wiley & Sons, Inc.).

Nutrient Pollution

http://www.epa.gov/nutrientpollution/

Gulf of Mexico Integrated Science

gulfsci.usgs.gov

Lake of the Sky: Lake Tahoe Basin Research

gallery.usgs.gov/videos/431#.
ULjLOGfkvQu

4.1.7 MATERIALS FLOW ANALYSIS AND URBAN METABOLISM

As discussed in Chapter 1, if the inventory analysis step of a life cycle assessment (LCA) only focuses on materials, it is referred to as a **materials flow analysis (MFA)**. Chapter 1 explained that an MFA measures the material flows into a system, the stocks and flows within it, and the outputs from the system. In this case, measurements are based on mass (or volume) loadings instead of concentrations. An urban MFA (sometimes referred to as an **urban metabolism** study) is a method to quantify the flow of materials that enter an urban area (for example, water, food, and fuel) and the flow of materials that exit an urban area (for example, manufactured goods, water and air pollutants including greenhouse gases, and solid wastes). Chapter 1 also presented the results from an urban metabolism study performed on the city of Hong Kong (Application 1.8).

Application / 4.1 **Flow of Nutrients into the Gulf of Mexico**

Excessive nutrients (that is, nitrogen and phosphorus) that make their way to surface waters (for example, lakes, estuaries, near shore coastal zones) can fuel the growth of large amounts of algae. When algae dies and decays, it consumes oxygen. This process can result in a zone of low dissolved oxygen (also referred to as a "hypoxic zone") that can threaten the ecological health of the water body and the economic and social well-being of communities that depend on water quality for fishing and tourism (water-quality issues like this are covered in Chapter 7).

A major example of this hypoxic zone (also referred to as a dead zone) is in the Northern Gulf of Mexico, a location of one of the most productive fisheries in the United States. Two issues in managing this enormous environmental problem are: (1) there is an enormous land area (that encompasses 31 states) that makes up the Mississippi River watershed and (2) there are many different types of land use in the watershed that result in wide variety of sources that discharge nutrients. A joint federal-state Gulf of Mexico Hypoxia Task Force is evaluating recommendations by EPA's Science Advisory Board to set reduction targets of at least 45 percent for both nitrogen and phosphorus in an effort to shrink the size of the hypoxic zone.

Figure 4.10 shows that the delivery of phosphorus to the Gulf of Mexico is highest from watersheds in the central and eastern portions of the Mississippi River Basin. The same holds true for nitrogen. Nine states contribute more than 75 percent of the nitrogen and phosphorus that reaches the Gulf of Mexico. These states include Illinois, Iowa, Indiana, Missouri, Arkansas, Kentucky, Tennessee, Ohio, and Mississippi. However, these nine states only make up one-third of the land area that drains to the Mississippi River (which encompasses a total of 31 states).

Figure 4.11 shows the sources of phosphorus and nitrogen delivered to the Gulf of Mexico. This figure indicates that 66 percent of nitrogen originates from cultivation of crops (primarily corn and soybeans), with animal grazing and manure contributing about 5 percent. Atmospheric contributions of nitrogen also are important, accounting for 16 percent of the total nitrogen input to the Gulf of Mexico. In contrast, there are no major atmospheric emissions of phosphorus and animal manure on pasture and range lands contribute nearly as much phosphorus as cultivated crops (37 versus 43 percent).

Figure 4.11 clearly shows that agricultural sources contribute more than 70 percent of the nitrogen and phosphorus delivered to the Gulf of Mexico, versus only 9–12 percent originating from urban sources. These urban sources include nonpoint fertilizer runoff from residential and commercial landscaping and point source discharges from wastewater treatment facilities. Such findings suggest the dominance of agricultural nonpoint sources; however, urban discharges tend to be concentrated, especially in coastal areas.

Information and much of the text obtained by the U.S. Geological Survey.

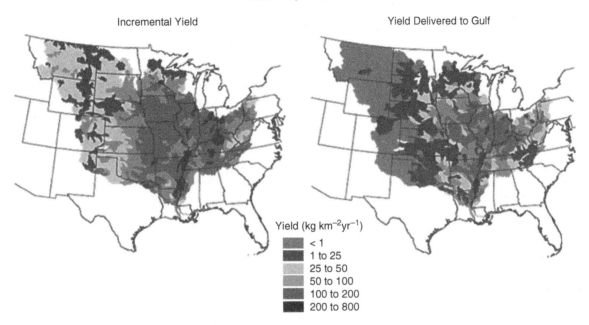

Total Phosphorus

Incremental Yield   Yield Delivered to Gulf

Yield (kg km^{-2}yr^{-1})

■ < 1
■ 1 to 25
■ 25 to 50
■ 50 to 100
■ 100 to 200
■ 200 to 800

Figure / 4.10   Total Phosphorus Yield (kg/km²-year) Delivered to the Gulf of Mexico from the Mississippi River Watershed   The delivery of phosphorus and nitrogen to the Gulf of Mexico is highest from watersheds in the central and eastern portions of the basin.

(Redrawn from U.S. Department of Interior, U.S. Geological Survey).

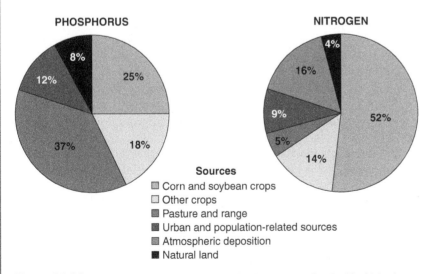

PHOSPHORUS

8%
12%
25%
37%
18%

NITROGEN

4%
16%
9%
5%
52%
14%

Sources
☐ Corn and soybean crops
☐ Other crops
▨ Pasture and range
■ Urban and population-related sources
▨ Atmospheric deposition
■ Natural land

Figure / 4.11   Sources of Phosphorus and Nitrogen to the Gulf of Mexico.

(Redrawn from U.S. Department of Interior, U.S. Geological Survey).

4.2 Energy Balances

Modern society is dependent on the use of energy. Such use requires transformations in the form of energy and control of energy flows. For example, when coal is burned at a power plant, the chemical energy present in the coal is converted to heat, which is then converted in the plant's generators to electrical energy. Eventually, the electrical energy is converted back into heat for warmth or used to do work. However, energy flows and transformation can also cause environmental problems. For example, thermal heat energy from electrical power plants can result in increased temperatures in rivers used for cooling water, greenhouse pollutants in the atmosphere alter Earth's energy balance and may cause significant increases in global temperatures, and burning of fossil fuels to produce energy is associated with emissions of pollutants.

The movement of energy and changes in its form can be tracked using **energy balances**, which are analogous to mass balances. The **first law of thermodynamics** states that energy can neither be produced nor destroyed. Conservation of energy provides a basis for energy balances, just as the law of conservation of mass provides a basis for mass balances. However, all energy balances are treated as conservative; as long as all possible forms of energy are considered (and in the absence of nuclear reactions), there is no term in energy balances that is analogous to the chemical-reaction term in mass balances.

4.2.1 FORMS OF ENERGY

Energy and the Environment
http://www.epa.gov/energy/

The forms of energy can be divided into two types: *internal* and *external*. Energy that is part of the molecular structure or organization of a given substance is internal. Energy resulting from the location or motion of a substance is external. Examples of external energy include *gravitational potential energy* and *kinetic energy*. Gravitational potential energy is the energy gained when a mass is moved to a higher location above the Earth. Kinetic energy is the energy that results from the movement of objects. When a rock thrown off a cliff accelerates toward the ground, the sum of kinetic and potential energy is conserved (neglecting friction); as the rock falls, it loses potential energy but increases in speed, gaining kinetic energy. Table 4.4 gives the mathematical representations of common forms of energy encountered in environmental engineering.

Heat is a form of internal energy—it results from the random motions of atoms. Heat is thus really a form of kinetic energy, although it is considered separately because the motion of the atoms cannot be seen. When a pot of water is heated, energy is added to the water. That energy is stored in the form of internal energy, and the change in internal energy of the water is expressed as follows:

$$\text{change in internal energy} = (\text{mass of } H_2O) \times c \times \Delta T \qquad \textbf{(4.21)}$$

where c is the heat capacity or specific heat of the water, with units of energy/mass-temperature. Heat capacity is a property of a given material. For water, the heat capacity is 4,184 J/kg-°C (1 Btu/lb-°F).

Table / 4.4

Some Common Forms of Energy

	Representation for Energy or Change in Energy
Heat internal energy	$\Delta E = \text{mass} \times c \times \Delta T$
Chemical internal energy	$\Delta E = \Delta H_{rxn}$ at constant volume
Gravitational potential	$\Delta E = \text{mass} \times \Delta \text{height}$
Kinetic energy	$E = \dfrac{\text{mass} \times (\text{velocity})^2}{2}$
Electromagnetic energy	$E = \text{Planck's constant} \times \text{photon frequency}$

SOURCE: Mihelcic (1999). Reprinted with permission of John Wiley & Sons, Inc.

Application / 4.2 Another Energy Classification System: Renewable and Nonrenewable

Energy resources can be described as renewable and nonrenewable. **Renewable energy** sources can be replaced at a rate equal to or faster than the rate at which they are used (the source is continuously available). The sun, wind, and tides are examples of renewable energy feedstocks. **Nonrenewable energy** sources, on the other hand, are consumed faster than they can be replenished. Fossil-based feedstocks are considered nonrenewable because they cannot be replenished as fast as they are consumed (the source is finite). Each type of energy source, renewable and nonrenewable, falls within the categories listed in Table 4.4. For example, the energy contained in fossil fuels is present in the form of chemical internal energy, wind power comes from kinetic energy, and solar power uses electromagnetic energy.

Chemical internal energy reflects the energy in the chemical bonds of a substance. This form of energy is composed of two parts:

1. **The strengths of the atomic bonds in the substance.** When chemical reactions occur, if the sum of the internal energies of the products is less than that for the reactants, a reduction in chemical internal energy has occurred. As a result of the conservation of energy, this leftover energy must show up in a different form. Usually, the energy is released as heat. The most common example of this is the combustion of fuel, in which hydrocarbons and oxygen react to form carbon dioxide and water. The chemical bonds in carbon dioxide and water are much lower in energy than those in hydrocarbons, so combustion releases a significant amount of heat.

2. **The energy in the interactions between molecules.** Solids and liquids form as a result of interactions between adjacent molecules. These bonds are much weaker than the chemical bonds between atoms in molecules, but are still important in many energy balances. The energy required to break these bonds is referred to as *latent heat*. Values of latent heat are tabulated for various substances

The Debate over Nuclear Power

Con: http://www.nrdc.org/nuclear/
plants/contents.asp
Pro: http://energy.gov/ne/nuclear-
reactor-technologies

Class Discussion

Some advocate nuclear power as a source of energy to replace fossil fuels. Others see it as a security risk and having generational problems related to storing waste. Using a definition of sustainable development, does nuclear power have a role in our transformation toward a sustainable future?

for the phase changes from solid to liquid and from liquid to gas. The latent heat of condensation for a given substance is equal to the heat released when a unit of mass of the substance condenses to form a liquid. (An equal amount of energy is required for evaporation.) The latent heat of fusion is equal to the heat released when a unit of mass solidifies. (Again, an equal amount of energy is required to melt the substance.)

4.2.2 CONDUCTING AN ENERGY BALANCE

In analogy with the mass balance equation (Equation 4.3), the following equation can be used to conduct energy balances:

$$\begin{pmatrix} \text{change in internal} \\ \text{plus external energy} \\ \text{per unit time} \end{pmatrix} = (\text{energy flux in}) - (\text{energy flux out})$$

or

$$\frac{dE}{dt} = \dot{E}_{in} - \dot{E}_{out} \qquad (4.22)$$

The use of this relationship is illustrated in Examples 4.9 and 4.10. This same approach for calculating heat balances can be used to investigate the energy efficiency of different products, processes, and systems. Later in this chapter, a heat balance will be used for decision making about the energy efficiency of a building.

---

example/ 4.9   Heating Water: Scenario 1

A 40-gallon electric water heater heats water entering the house, which has a temperature of 10°C as it enters the heater. The heating level is set to the maximum while several people take consecutive showers. If, at the maximum heating level, the heater uses 5 kW of electricity and the water use rate is a continuous 2 gallons/min, what is the temperature of the water exiting the heater? Assume that the system is at steady state and the heater is 100 percent efficient; that is, it is perfectly insulated, and all of the energy used heats the water.

solution

The control volume is the water heater. Because the system is at steady state, dE/dt is equal to zero. The energy flux added by the electric heater heats water entering the water heater to the temperature at the outlet. The energy balance is thus

$$\frac{dE}{dt} = 0 = \dot{E}_{in} - \dot{E}_{out}$$

The energy flux into the water heater comes from two sources: the heat content of the water entering the heater and the electrical heating element. The heat content of the water entering the heater is the product of the water-mass flux, the heat capacity, and the inlet temperature. The energy added by the heater is given as 5 kW.

The energy flux out of the water heater is just the internal energy of the water leaving the system ($\dot{m}_{H_2O} \times c \times T_{out}$). There is no net conversion of other forms of energy. Therefore, the energy balance may be rewritten as follows:

$$0 = (\dot{m}_{H_2O}cT_{in} + 5\,kW) - \dot{m}_{H_2O}cT_{out}$$

example/ 4.9   (continued)

Each term of this equation is an *energy flux* and has the units of energy/time. To solve, place each term in the same units—in this case, watts (1 W equals 1 J/s, and 1,000 W = 1 kW). In addition, the water flow rate (gallons/min) needs to be converted to units of mass of water per unit time using the density of water. Combining the first and third terms,

$$0 = \dot{m}_{H_2O}c(T_{out} - T_{in}) + 5\,kW$$

$$0 = \frac{2\,gal\,H_2O}{min} \times \frac{3.785\,L}{gal} \times \frac{1.0\,kg}{L} \times \frac{4{,}184\,J}{kg \times °C} \times (T_{in} - T_{out}) + \frac{5{,}000\,J}{s} \times \frac{60\,s}{min}$$

$$= 3.16 \times 10^4 \frac{J}{min \times °C} \times (T_{in} - T_{out}) + 3.00 \times 10^5 \frac{J}{min}$$

Solve for T_{out}:

$$T_{out} = T_{in} + 9.5°C = (10 + 9.5) = 19.5°C$$

This is a cold shower! But it makes sense; many people have taken such a cold shower after the hot water in the tank was used up by previous showers.

example/ 4.10   Heating Water: Scenario 2

Example 4.9 showed that it is necessary to wait until the water in the tank is reheated (hopefully, by passive solar energy!) before taking a hot shower. How long would it take the temperature to reach 54°C if no hot water were used during the heating period and the water temperature entered the heater at 20°C?

solution

In this case, assuming the homeowner is not taking advantage of solar energy, the only energy input is the electrical heat, and no energy is leaving the tank. Therefore, the rate of increase in internal energy is equal to the rate at which electrical energy is used:

$$\frac{dE}{dt} = \dot{E}_{in} - \dot{E}_{out} = \dot{E}_{in} - 0$$

From Table 4.4, $\Delta E = mass \times c \times \Delta T$, so we can express the relationship as follows:

$$\frac{dE}{dt} = \frac{(mass\ of\ H_2O) \times c \times \Delta T}{\Delta t}$$

and

$$\frac{(mass\ of\ H_2O) \times c \times \Delta T}{\Delta t} = \dot{E}_{in} = 5{,}000\,J/s$$

This expression can be solved for the change in time, Δt, given that ΔT is equal to $54°C - 20°C = 34°C$:

$$\Delta t = \frac{(mass\ of\ H_2O) \times c \times \Delta T}{5{,}000\,J/s}$$

$$= \frac{\left(40\,gal\,H_2O \times \frac{3.785\,L}{gal} \times \frac{1.0\,kg}{L}\right)\left(4{,}184\frac{J}{kg \times °C}\right)(54°C - 20°C)}{5{,}000\,J/s}$$

$$= 4.3 \times 10^3\,s = 1.2\,h$$

© M. Eric Honeycutt/iStockphoto.

Although we have neglected heat loss in Examples 4.9 and 4.10, in the real world heat loss often significantly affects the energy efficiency of processes, systems, and products. For example, heat loss through a poorly insulated hot-water heater tank requires additional energy to maintain the water at the desired temperature and can be a significant fraction of the total energy required. Similarly, windows and poor insulation in a house often act as conduits for heat loss, increasing the amount of energy required to maintain a comfortable temperature.

The additional energy use due to largely avoidable heat loss is significant: the energy used to offset unwanted heat losses and gains through windows in residential and commercial buildings costs the United States tens of billions of dollars every year. However, when properly selected and installed, windows can help minimize a home's heating, cooling, and lighting costs. This notion is further explored in this chapter through a discussion of resistance values (R values), heat balance, and energy efficiency.

example/4.11 Thermal Pollution from Power Plants

The second law of thermodynamics states that heat energy cannot be converted to work with 100 percent efficiency. As a result, a significant fraction of the heat released in electrical power plants is lost as waste heat; in modern large power plants, this loss accounts for 65–70 percent of the total heat released from combustion.

A typical coal-fired electric power plant produces 1,000 MW of electricity by burning fuel with an energy content of 2,800 MW; 340 MW are lost as heat up the smokestack, leaving 2,460 MW to power turbines that drive a generator to produce electricity. However, the thermal efficiency of the turbines is only 42 percent. That means 42 percent of this power goes to drive the generator, but the rest (58 percent of 2,460 = 1,430 MW) is waste heat that must be removed by cooling water. Assume that cooling water from an adjacent river, which has a total flow rate of 100 m^3/s, is used to remove the waste heat. How much will the temperature of the river rise as a result of the addition of this heat?

solution

This problem is similar to Example 4.9, because a specified amount of heat is added to a flow of water, and the resulting temperature rise must be determined. An energy balance can be written over the region of the river to which the heat is added. Here, T_{in} represents the temperature of the water upstream, and T_{out} represents the temperature after heating:

$$\frac{dE}{dt} = \dot{E}_{in} - \dot{E}_{out}$$

$$0 = (1{,}430 \text{ MW of heat from power plant}) + (\dot{m}_{H_2O} \times c \times T_{H_2O_{in}}) - (\dot{m}_{H_2O} \times c \times T_{H_2O_{out}})$$

Rearranging,

$$\dot{m}_{H_2O} \times c \times (T_{out} - T_{in}) = 1{,}430 \text{ MW}$$

The remainder of this problem is essentially unit conversion. To obtain $\dot{m}_{H_2O}$ requires multiplication of the given river volumetric flow rate by the density of water (approximately $1{,}000 \text{ kg/m}^3$). The heat capacity of water, $c = 4{,}184 \text{ J/kg-°C}$, also is required. Thus,

$$\left(100 \frac{\text{m}^3}{\text{s}} \times 1{,}000 \frac{\text{kg}}{\text{m}^3}\right) \times \left(4{,}184 \frac{\text{J}}{\text{kg} \times °\text{C}}\right) \times \Delta T = 1{,}430 \times 10^6 \text{ J/s}$$

Solving for ΔT:

$$\Delta T = 3.4°\text{C}$$

Consideration of this temperature increase is important, as the Henry's law constant for oxygen changes with temperature. This results in a reduced dissolved-oxygen concentration in the river in warmer water, which may be harmful to aquatic life.

4.2.3 IMPACT OF GREENHOUSE GAS EMISSIONS ON EARTH'S ENERGY BALANCE

Earth's average surface temperature is determined by a balance between the energy provided by the sun and the energy radiated away by Earth to space. The energy radiated to space is emitted in the form of infrared radiation. As illustrated in Figure 4.12, some of this infrared radiation is absorbed in the atmosphere. The gases responsible for this absorption are called **greenhouse gases**; without them, Earth would not be habitable, as demonstrated in Application 4.3.

Changes in atmospheric concentrations of carbon dioxide—the most important greenhouse gas—over time are shown in Figure 4.14. Increasing atmospheric concentrations of carbon dioxide—as well as those of methane, nitrous oxide, chlorofluorocarbons, and tropospheric ozone, which have occurred as a result of human activities—increase the value of $E_{greenhouse}$. This enhanced greenhouse effect, termed the **anthropogenic greenhouse effect**, is currently equivalent to an increase in the energy flux to Earth of approximately 2 W/m^2. Projections indicate that the increase could be as high as 5 W/m^2 over the next 50 years.

© Mehmet Salih Guler/iStockphoto.

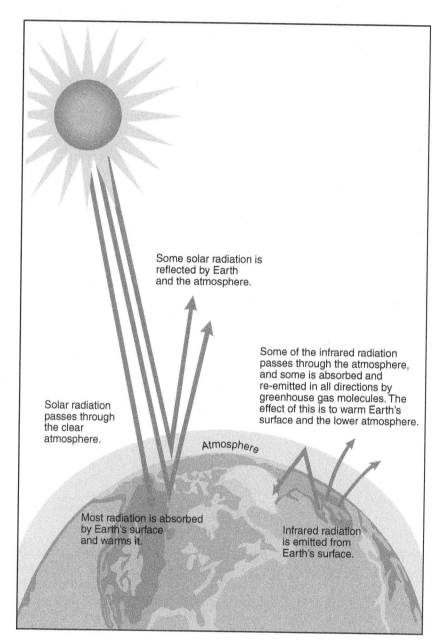

Some solar radiation is
reflected by Earth
and the atmosphere.

Some of the infrared radiation
passes through the atmosphere,
and some is absorbed and
re-emitted in all directions by
greenhouse gas molecules. The
effect of this is to warm Earth's
surface and the lower atmosphere.

Solar radiation
passes through
the clear
atmosphere.

Atmosphere

Most radiation is absorbed
by Earth's surface
and warms it.

Infrared radiation
is emitted from
Earth's surface.

Figure / 4.12 The Greenhouse Effect

(Redrawn from *Our Changing Planet: The FY 1996 U.S. Global Change Research Program*, Report
by the Subcommittee on Global Change Research, Committee on Environment and Natural Resources
Research of the National Science and Technology Council (supplement to the President's fiscal year
1996 Budget)).

As the energy absorbed by greenhouse gases increases, some
other term in the energy balance must respond to maintain steady
state. If the solar radiation absorbed by Earth remains constant, then
Earth's average temperature must increase. The magnitude of the
resulting temperature increase depends on the response of the complex global climate system, including changes in cloudiness and
ocean circulation.

The energy balance of the Earth is being increasingly altered by human activities, mainly through the addition to the atmosphere of carbon dioxide from fossil-fuel combustion. Calculate the global average temperature of Earth without greenhouse gases and show the effect greenhouse gases have on Earth's energy balance.

solution

An energy balance can be written with the entire Earth as the control volume. For this system, the goal is to calculate Earth's annual average temperature. Over time periods of at least 1 year, it is reasonable to assume that the system is at steady state. The energy balance is

$$\frac{dE}{dt} = 0 = \dot{E}_{in} - \dot{E}_{out}$$

The energy flux in is equal to the solar energy intercepted by Earth. At Earth's distance from the sun, the sun's radiation is 1,368 W/m^2, referred to as S. Earth intercepts an amount of energy equal to S times the cross-sectional area of the Earth: $S \times \pi r_e^2$. However, because Earth reflects approximately 30 percent of this energy back to space, $\dot{E}_{in}$ equals only 70 percent of this value:

$$\dot{E}_{in} = 0.7 S \pi R_e^2$$

The second term, $\dot{E}_{out}$, is equal to the energy radiated to space by Earth. The energy emitted per unit surface area of Earth is given by Boltzmann's law:

$$(\text{Energy flux per unit area}) = \sigma T^4$$

where σ is Boltzmann's constant, equal to 5.67×10^{-8} W/m^2-K^4. To obtain $\dot{E}_{out}$, this value is multiplied by Earth's total surface area, $4\pi R_e^2$. (The total surface area of the sphere is used here because

energy is radiated away from Earth during both day and night.)

$$\dot{E}_{out} = 4\pi R_e^2 \sigma T^4$$

To solve the energy balance, set $\dot{E}_{in}$ equal to $\dot{E}_{out}$:

$$4\pi R_e^2 \sigma T^4 = 0.7 S \pi R_e^2$$

Simplify:

$$T^4 = \frac{0.7S}{4\sigma}$$

Plugging in the values for S and σ yields Earth's average annual temperature: $T = 255$ K or $-18°C$.

This is too cold! In fact, the globally averaged temperature at Earth's surface is much warmer: 287 K. The reason for the difference is the presence of gases in the atmosphere that absorb the infrared radiation emitted by Earth and prevent it from reaching space. These gases, which include water vapor, CO_2, CH_4, and N_2O, were neglected in the initial energy balance. To include their influence, we can add a new term in the energy balance: the energy flux absorbed and retained by these gases. If the impact of greenhouse gas absorption is given by $E_{greenhouse}$, then the corrected $\dot{E}_{out}$ term is

$$\dot{E}_{out} = 4\pi R_e^2 \sigma T^4 - E_{greenhouse}$$

The reduction in $\dot{E}_{out}$ that results from greenhouse gas absorption is sufficient to cause the higher observed surface temperature. Clearly, this is largely a natural phenomenon, since surface temperatures were well above 255 K long before humans began burning fossil fuels. However, human activities—primarily the burning of fossil fuels—are changing the atmospheric composition to a significant extent and are increasing the magnitude of the greenhouse effect.

Glazing is a very important component of windows, solar hot-water heaters, greenhouses, and other technologies that incorporate passive solar heating to trap the heat associated with incoming solar radiation. Materials used for glazing include glass, acrylics, polycarbonates, and polyethylene. The glazing allows shortwave radiation from the sun to pass through. After passing through the glazing, the shortwave radiation is then absorbed by surfaces and materials such as water and masonry that make excellent collectors of this solar energy. Some longwave radiation is emitted from these surfaces. The longwave radiation cannot easily pass through the glazing material, so the collector heats up. As shown in Figure 4.13, glazing materials function similarly to the greenhouse gases that trap solar radiation and lead to climate change.

Figure / 4.13   **Function of Glazing Materials**   Glazing materials allow shortwave radiation from the sun to pass through the glazing, but reflected longwave radiation cannot pass through the glazing. This is similar to the greenhouse gases that trap solar radiation and lead to climate change.

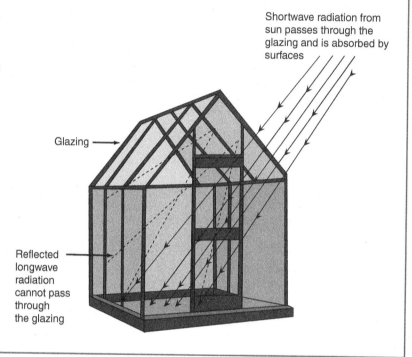

Shortwave radiation from sun passes through the glazing and is absorbed by surfaces

Glazing

Reflected longwave radiation cannot pass through the glazing

Figure / 4.14   **Global Average Carbon Dioxide Concentration Trend**   These measurements of CO_2 were made at the Mauna Loa, Hawaii, observatory by the National Oceanic and Atmospheric Administration. The annual increase of approximately 0.5 percent per year is attributed to fossil-fuel combustion and deforestation. The annual cycle is the result of photosynthesis and respiration, which result in a drawdown of CO_2 during the summer growing season and an increase during winter. The weekly average concentration reached 399.50 ppm on May 5, 2013.

(Redrawn from data provided by NOAA; http://www.esrl.noaa.gov/gmd/ccgg/trends/.).

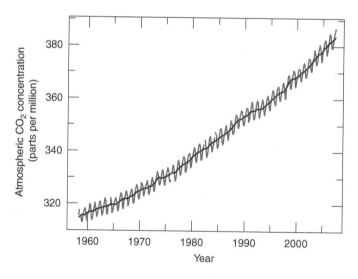

The **Intergovernmental Panel on Climate Change (IPCC)** (co-winners of the 2007 Nobel Peace Prize) was established by the World Meteorological Organization and the United Nations Environment Programme (UNEP) to assess scientific, technical, and socioeconomic information related to better understanding of climate change. (For more information, see http://www.ipcc.ch/.) More than 2,000 notable scientists make up the IPCC.

Current global climate models predict that the anthropogenic greenhouse effect will cause a global average temperature increase (relative to 1990) of 1.1–6.4°C by 2099 (IPCC, 2007b). Resulting alterations to global and regional climate are predicted to include increased rainfall and increased frequency of severe storms, although some regions of the planet may experience increased frequency of drought or even regional cooling as a result of changes in atmospheric and oceanic circulation patterns. Several different scenarios for economic and population growth, material and energy efficiency technologies, and consumption patterns and the resulting predicted temperature changes and sea level rise are provided in Table 4.5.

Class Discussion

Using the scenarios and outcomes listed in Table 4.5, how do population and continued use of fossil fuels affect the warming of the Earth and the rise in sea level?

Table / 4.5

Temperature Change and Sea Level Rise Resulting from Various Future Scenarios Scenarios include economic and population growth, material and energy efficiency technology development, and consumption patterns for 2090–2099.

Scenario	Temperature Change (°C at 2090–2099 relative to 1980–1999)		Sea Level Rise (m at 2090–2099 relative to 1980–1999)
	Best Estimate	*Likely Range*	*Model-Based Range**
B1: rapid economic growth toward a service and information economy; population peaks in midcentury and then declines; reductions in material intensity; clean/resource-efficient technologies; global solutions to sustainability, including improved equity	1.8	1.1–2.9	0.18–0.38
A1T: rapid economic growth; population peaks in midcentury and then declines; rapid introduction of new and efficient technologies; convergence among regions; nonfossil-fuel energy sources	2.4	1.4–3.8	0.20–0.45
B2: local solutions to sustainability; continuously increasing population; intermediate levels of economic development; less rapid and more diverse technological change	2.4	1.4–3.8	0.20–0.43
A1B: same as A1T except balance between fossil and nonfossil-fuel energy sources	2.8	1.7–4.4	0.21–0.48
A2: self-reliance and preservation of local identities; continuously increasing population; economic development that is primarily regionally oriented; slow and fragmented per capita economic growth and technological change	3.4	2.0–5.4	0.23–0.51
A1Fl: same as A1T except fossil-intensive energy sources	4.0	2.4–6.4	0.26–0.59

*Excluding future rapid dynamic changes in ice flow in the large glacial regions of Greenland and Antarctica. Based on IPCC (2007a).

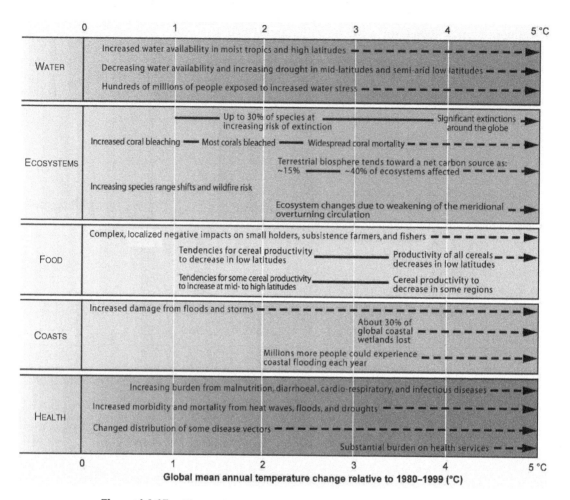

Figure / 4.15 Illustrative examples of global impacts projected for climate changes (and sea level and atmospheric carbon dioxide where relevant) associated with different amounts of increase in global average surface temperature in the 21st century. The black lines link impacts and dotted arrows indicate impacts continuing with increasing temperature. Entries are placed so that the left-hand side of the text indicates the approximate onset of a given impact. Quantitative entries for water stress and flooding represent the additional impacts of climate change relative to the conditions projected across the range of Special Report on Emissions Scenarios (SRES) scenarios A1FI, A2, B1, and B2 (see Endbox 3).

(Adaptation to climate change is not included in these estimations. Confidence levels for all statements are high with permission of Intergovernmental Panel on Climate Change: Impacts, Adaptation, and Vulnerability, Summary for Policymakers, Table SPM.2, 2007).

The global consequences of warming will be significant. Figure 4.15 provides some expected impacts on water, ecosystems, food, coastal areas, and health as they relate to the specific increase in global mean temperature. Not only are ecosystems and wildlife heavily dependent on climate, but human health and the economy are as well.

The impact of climate change will of course differ by location. For example, small island nations, parts of the developing world, and particular U.S. geographical regions will be affected to a greater extent. Some industries will be affected more than others. Economic sectors that depend on agriculture will struggle with more variability in weather patterns, and the insurance industry will have a difficult time responding to more catastrophic weather events.

4.2.4 ENERGY EFFICIENCY IN BUILDINGS: INSULATION, INFILTRATION, AND THERMAL WALLS

Previously in this chapter, we developed an energy mass balance expression (Equation 4.22) and then applied it to heating water and thermal pollution.

Similarly, an **energy balance** can be used to describe a **heat balance** in a building to demonstrate methods to design and construct buildings that are more energy efficient. In a building, the heat balance can be written as follows:

$$\begin{bmatrix} \text{change in internal} \\ \text{plus external energy} \\ \text{per unit time} \end{bmatrix} = \begin{bmatrix} \text{heat into} \\ \text{building} \end{bmatrix} - \begin{bmatrix} \text{heat loss} \\ \text{from building} \end{bmatrix} \quad (4.23)$$

In many scenarios with buildings, it is assumed that the building temperature is maintained at a constant value. Thus, the change in internal plus external energy per unit time in Equation 4.23 equals zero. In this case, after the heat loss is determined, a heating system (passive solar and/or mechanical) can be sized to counter the heat loss.

The *heat loss* from the building is related to losses through the building skin (walls, ceilings, windows, doors) and through airflow that occurs through any cracks or holes in the building (infiltration). The heat added into a conventional building is typically from conversion of nonrenewable fuels such as natural gas, oil, or electricity. Sustainable heating requires that the building be oriented toward the sun, be insulated, and have a heating system designed to take advantage of the input of the sun's energy through passive solar design or use of renewable energy.

HEAT LOSS IN A BUILDING   For demonstration purposes, let us develop a heat balance on the heat loss associated with a 3,000 ft^2 home. There are several ways to perform this analysis. For our analysis, we will introduce and use a term called *degree-day*. We will also use Btu as the measure of energy $(1\,J = 9.4787 \times 10^{-4}\,\text{Btu})$. A Btu is defined as the amount of heat that must be added to 1 lb water to raise its temperature by 1°F.

To simplify the calculation, we assume that the 3,000 ft^2 home is a simple cube; thus, the four exposed walls are approximately 14.4 ft (width) by 14.4 ft (length) by 14.4 ft (height). The area of each wall is then approximately 207 ft^2. The roof area of the cube would also be 207 ft^2. This building is assumed to have insulation specifications of R-19 walls and an R-30 ceiling, and the air infiltration rate is stated to be 0.50 air changes per hour for heating.

HEAT LOSS THROUGH BUILDING SKIN   The heat loss through the skin of the building (Btu/°F-day) is determined as follows:

$$\text{Heat loss} = \frac{1}{R} \times A \times t \quad (4.24)$$

The *R value* is a measure of resistance to heat flow. The inverse of R ($1/R$) is defined as the flow of Btu through a 1 ft^2 section of building skin for 1 h, during which the temperature difference between the inside and outside of the building skin is 1°F. In Equation 4.24, A is the area of a particular section of the skin (wall, window, door, and ceiling) and t is time (usually 24 h).

The daily total heat loss through the four walls and the ceiling can be determined as follows:

$$\text{heat loss} = \left[\left(\frac{1}{19}\frac{\text{Btu}}{\text{ft}^2\text{-}°\text{F-h}}\right)\right] \times 4 \text{ walls} \times 207 \text{ ft}^2 \times \frac{24 \text{ h}}{\text{day}}$$

$$+ \left[\left(\frac{1}{30}\frac{\text{Btu}}{\text{ft}^2\text{-}°\text{F-h}}\right) \times 1 \text{ ceiling} \times 207 \text{ ft}^2 \times \frac{24 \text{ h}}{\text{day}}\right] \quad \textbf{(4.25)}$$

Solving Equation 4.25 results in

$$\text{heat loss} = 1,046\frac{\text{Btu}}{°\text{F-day}} + 6.9\frac{\text{Btu}}{°\text{F-day}} = 1,053\frac{\text{Btu}}{°\text{F-day}} \quad \textbf{(4.26)}$$

The unit of "°F-day" in Equation 4.26 is defined as a **degree-day**. Defined for heating, a degree-day is the number of degrees Fahrenheit below 65°F for 24 h. Application 14.5 discusses degree-days in more detail. In our example, the value determined in Equation 4.26 can be written as 1,053 Btu/degree-day.

Once the total heat loss (in units of Btu/°F-day) is determined, that value can be multiplied by the total number of degree-days for heating in a particular location for the period of time of interest (day, month, or year). The resulting value will be the total energy requirements for heating the structure over that time period.

In our calculation, the heat loss through the actual building skin would be different if we broke the building skin down in greater detail to the area associated with the specific components of the building skin (siding, doors, and windows) and the specific R values associated with these components. In this case, we would determine the heat loss through each component of the building skin and then add up those amounts to find the total heat loss.

Application / 4.5   Degree-Days

You may have seen the term *degree-day* used on your gas or electric bill. A **degree-day** is an index that reflects demand for energy that is used to heat or cool a building. The NOAA Climate Prediction Center provides degree-day data for almost 200 major weather stations in the United States (www.cpc.ncep.noaa .gov/). The baseline used for computations is 65°F.

A degree-day defined for heating is the number of degrees Fahrenheit below 65°F for a particular time period. Thus, if the mean daily temperature on a particular winter day was 32°F, this would equate to 33 degree-days for heating over that 24 h period.

A degree-day defined for cooling is the number of degrees Fahrenheit above 65°F for a particular time period. Likewise, if a mean daily temperature for a summer temperature was reported as 85°F, this would equate to 20 degree-days for cooling over that 24 h period.

Degree-days can be summed up for a week, month, or year to determine energy demand associated with heating and cooling.

example/4.12   Determining the Importance of Insulation in Minimizing Heat Loss through a Building Skin

Determine the heat loss through an insulated and uninsulated wall.* Each wall contains the following materials, which have the R factors given in the following table:

Component of Wall	R Factor
1 in. stucco on outside of wall	0.20
1/2 in. sheathing under stucco	1.32
1/2 in. drywall on inside of wall	0.45
Inside air film along inside of wall	0.68
Outside air film along outside of wall	0.17

The 3.5 in air space in the uninsulated wall has an R factor of 1.01. If 3.5 in fiberglass insulation is placed in this space, it will have an R factor of 11.0.

solution

Remember that Equation 4.24 allowed us to determine the heat loss through the skin of the building (Btu/°F-day) as follows:

$$\text{heat loss} = \frac{1}{R} \times A \times t$$

For the uninsulated wall, the combined R value equals

$$0.17 + 0.20 + 1.32 + 0.45 + 0.68 + 1.01 = 3.73$$

For the insulated wall, the combined R value equals

$$0.17 + 0.20 + 1.32 + 0.45 + 0.68 + 11.0 = 13.72$$

The heat loss through the uninsulated wall thus equals

$$\frac{1}{3.73} \times 100 \text{ ft}^2 \times \frac{24 \text{ h}}{\text{day}} = 643 \frac{\text{Btu}}{\text{°F-day}} = 643 \frac{\text{Btu}}{\text{degree-day}}$$

And the heat loss through the insulated wall equals

$$\frac{1}{13.72} \times 100 \text{ ft}^2 \times \frac{24 \text{ h}}{\text{day}} = 175 \frac{\text{Btu}}{\text{°F-day}} = 175 \frac{\text{Btu}}{\text{degree-day}}$$

Note that the wall with only 3.5 in fiberglass insulation added has much less heat loss through the building skin. By knowing the number of degree-days for a particular date of the year that requires heating, we can also determine the days heating requirement.

*This example is based on Wilson (1979).

HEAT LOSS FROM INFILTRATION   To determine the **heat loss due to infiltration**, we must know the room size. For our simplified calculation, we will assume that the 3,000 ft² home is one giant room and the air infiltration rate is 0.50 air change per hour for heating. The heat loss associated with infiltration is the amount of energy required to heat the air lost from the room every day through cracks and holes in the building envelope. For a particular volume of room or building, this can be determined as follows:

$$\begin{bmatrix} \text{heat loss from} \\ \text{infiltration} \end{bmatrix} = \text{volume} \times \begin{bmatrix} \text{air} \\ \text{infiltration} \\ \text{rate} \end{bmatrix} \times \begin{bmatrix} \text{heat to raise} \\ \text{temperature of} \\ \text{the air 1°F} \end{bmatrix} \quad \textbf{(4.27)}$$

Heat capacity is the term used to describe the heat required to raise the temperature of air. At sea level, 0.018 Btu energy is needed to increase the temperature of 1 ft³ air by 1°F. (At 2,000 ft elevation, this value is 0.017; at 5,000 ft elevation, this value is 0.015.)

Note in Equation 4.27, the importance of rightsizing a building because heat lost due to infiltration is directly related to the volume of the particular space being analyzed. (The same is true for energy requirements related to cooling.) One particular popular design feature in U.S. homes today is not only to oversize a residential home, but also design an entry space with a large, high cathedral-type ceiling. After reading the remainder of this section, you will be able to estimate the energy required to heat such unsustainable design features.

Assuming the building in our example is located at sea level, the heat loss by infiltration is written as follows:

$$3{,}000 \text{ ft}^3 \times \left(\frac{0.5 \text{ air change}}{\text{h}} \right) \times 0.018 \frac{\text{Btu}}{\text{ft}^3\text{-}°\text{F}} \times \frac{24 \text{ h}}{\text{day}} = 648 \frac{\text{Btu}}{°\text{F-day}} \quad \textbf{(4.28)}$$

Again, using our method of degree-days, the value of 648 Btu/°F-day can be written as 648 Btu/degree-day.

Note the magnitude of this value compared with the value we determined for heat loss through the building's skin (Equation 4.26). The magnitude of this value is why it is important to make a building airtight by specification and proper installation of weather stripping, caulk, gasketing, and so on.

TOTAL HEAT LOSS   To determine the building's total heating load in our example, we can sum the heat loss through the building skin and heat loss from infiltration:

$$1{,}053 \frac{\text{Btu}}{°\text{F-day}} + 648 \frac{\text{Btu}}{°\text{F-day}} = 1{,}701 \frac{\text{Btu}}{°\text{F-day}} = 1{,}701 \frac{\text{Btu}}{\text{degree-day}} \quad \textbf{(4.29)}$$

The total energy demand to meet the heat lost is found from the following expression:

$$\text{Total energy demand} = \text{total heat lost} \times \begin{bmatrix} \text{degree-days for} \\ \text{heating for} \\ \text{time period} \end{bmatrix} \quad \textbf{(4.30)}$$

Assume again that the average temperature on a particular winter day is 33°F. Remember from the earlier definition of a degree-day that the 33°F temperature would result in 32 degree-days (65°F − 33°F = 32 degree-days) for that particular day. Thus, for our example, where the average temperature was 33°F, this would mean that the building would require the following amount of energy input for daily heating:

$$1,071 \frac{\text{Btu}}{\text{degree-day}} \times 32 \text{ degree-days} = 5.44 \times 10^4 \text{ Btu} \qquad \textbf{(4.31)}$$

PASSIVE SOLAR GAIN AND THERMAL WALLS   In the example used in this section to determine the energy required to make up for heat loss, we determined that for a particular winter day where the average temperature is 33°F, 5.44×10^4 Btus of energy are required to heat the house. Equation 4.23 included a term called *heat into building.* This added heat into the building can be derived from nonrenewable or renewable energy. Fortunately, all or part of this heat input can be derived by taking advantage of the energy provided by the sun. This heat input is called **passive solar gain**.

Thermal walls take advantage of passive solar energy and thermal conduction to transfer heat from warmer to cooler areas. They typically employ a large concrete or masonry wall to collect and store solar energy and then distribute this energy as heat into a building space. A masonry floor or fireplace also can accomplish this to a lesser extent. Figure 4.16 shows examples of how thermal walls can be incorporated into more sustainable building design that takes advantage of natural ventilation and overhangs. The Anasazi cliff dwellings of the American Southwest incorporate many of these design features.

Thermal walls can be sized to account for a particular fraction of the total heating load. The calculation first requires the determination of the heating losses, as was just performed in this section. Then, for a particular location and some assumptions related to the thermal conductivity and volumetric heat capacity of the wall material as well as the type of glazing that is placed between the wall and the sun, the percentage of the heating load that can be accounted for based on a particular area of thermal wall can be calculated. Due to space constraints, we will not go into these calculations. Readers are directed elsewhere (for example, Wilson, 1979).

As we have written in the past (Mihelcic et al., 2007), an ideal material for constructing a thermal wall would be readily available, inexpensive, nontoxic, and have optimal thermal properties (for example, heat capacity and conductivity). Water has a higher volumetric heat capacity (62 Btu/cu ft-°F) than wood, adobe, and concrete. These materials have heat capacity values that range in the 20s. Water is also an ideal material to release stored thermal energy as heat into a building space, because fluids can use convection to distribute heat. Also, the thermal conductivity of water (0.35 Btu-ft/ft^2-h-°F) is much higher than for wood and dry adobe. Thus, a thermal wall constructed of water will provide a larger fraction of the required heating load than a similarly sized wall built of concrete or stone. Simple water-filled thermal walls can be constructed of 55-gallon drums painted black and placed on the southerly side of a building behind some type of glazing material.

© Chris Williams/iStockphoto.

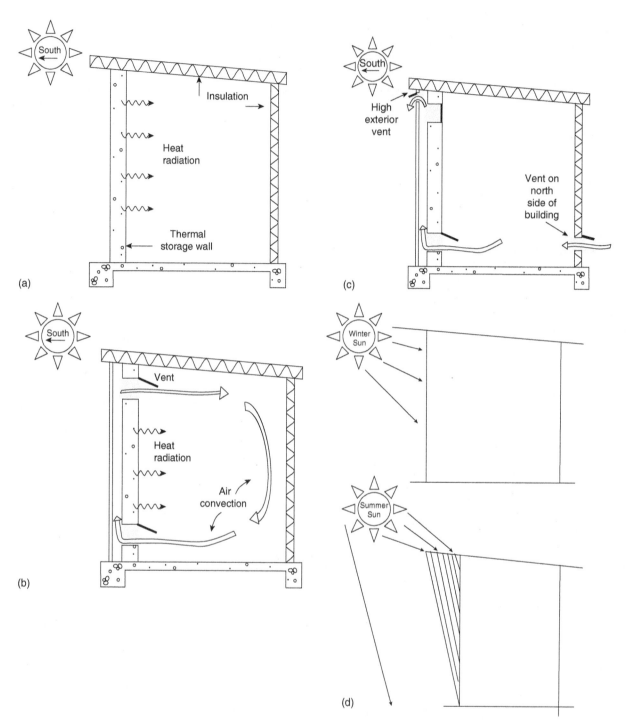

Figure / 4.16 **Examples of Passive Solar Design and Ventilation Applicable to Northern Hemisphere** These methods can be used to eliminate or minimize the need for mechanical heating and cooling. (a) Thermal walls use heat transfer to collect and dissipate heat. (b) Ventilation systems can use convection to provide natural heating. (c) Ventilation systems can provide natural cooling. (d) Overhangs take advantage of thermal properties of the sun during winter months while minimizing the sun's impact during warmer summer months.

(Adapted from Wilson (1979) with permission of the New Mexico Solar Energy Association (www.nmsea.org)).

Because of its high effective conductivity, water is an especially attractive material in instances where heat is required early in the day. Examples of places where heat is required early in the day are schools and offices. In residential situations where a family may be gone during much of the day and heat is needed in the evening, a conventional mass wall may be a better choice because it releases its stored energy more slowly.

4.2.5   URBAN HEAT ISLAND

The term **heat island** refers to urban air and surface temperatures that are higher than in nearby rural areas. Many cities and suburbs have air temperatures up to 10°F (5.6°C) warmer than the surrounding natural land cover. Figure 4.17 shows a typical city's heat island profile. Urban temperatures are typically lower at the urban–rural border than in dense downtown areas. The sketch also shows how parks and open land create cooler areas. This is one reason that greening the built environment provides social and environmental benefits.

   Heat islands form as cities replace natural land cover with pavement, buildings, and other infrastructure (referred to as the built environment). Displacing trees and vegetation minimizes the natural cooling effects of shading and evaporation of water from soil and leaves (evapotranspiration). **Nonpervious materials** have significantly different thermal bulk properties (including heat capacity and thermal conductivity) and surface radiative properties (albedo and emissivity) than the surrounding rural areas. This initiates a change in the energy balance of the urban area, often causing it to reach higher temperatures—measured both on the surface and in the air—than its surroundings (Oke, 1982). Tall buildings and narrow streets can heat air that is trapped between them, thus reducing airflow. This is referred to as the *canyon effect*. Waste heat from vehicles, factories, and air conditioners may add warmth to their surroundings, further exacerbating the heat island effect.

   Urban heat islands can impair a city's public health, air quality, energy demand, and infrastructure costs in several ways (Rosenfeld

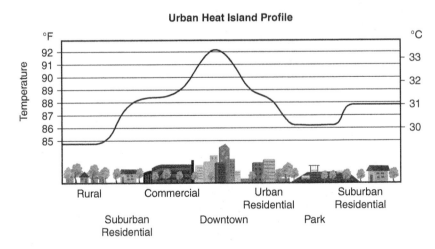

Urban Heat Island Profile

Figure / 4.17   **Urban Heat Island Profile**   This profile shows that increased temperatures of up to 10°F (5.6°C) can be found in dense downtown areas, as compared with surrounding rural, suburban, and open areas. The geometry of streets and buildings, along with the built environment's dependence on masonry, concrete, and asphalt structures that have high thermal bulk properties that store the sun's energy, have helped create this problem.

(Adapted from United States Environmental Protection Agency (2007)).

et al., 1997). Heat islands prolong and intensify heat waves in cities, making residents and workers uncomfortable and putting them at increased risk for heat exhaustion and heatstroke. In addition, high concentrations of ground-level ozone aggravate respiratory problems such as asthma, putting children and the elderly at particular risk. Hotter temperatures and reduced airflow increase demand for air conditioning, increasing energy use when demand is already high. This in turn contributes to power shortages and raises energy expenditures at a time when energy costs are at their highest. Urban heat islands contribute to global warming by increasing the demand for electricity to cool our buildings.

The study of urban heat islands is complicated, though. For example, in cooler climates during the winter, the urban heat island effect can cause nighttime temperatures to be less severe, which would require less heating. Also, fewer snowfall and frost events may occur, and changes in melting patterns of snowfall may change the urban hydrology of snowmelt.

To further investigate the causes of urban heat islands, an energy balance can be written on a shallow layer (that is the control volume) at the urban land surface containing air and surface elements, as shown in Figure 4.18:

$$Q^* + Q_A - Q_H - Q_E - Q_G = \Delta Q_S \qquad \textbf{(4.32)}$$

The terms in Equation 4.32 are defined in the caption for Figure 4.18. Here Q^* is the net radiation, the sum of the incoming and outgoing

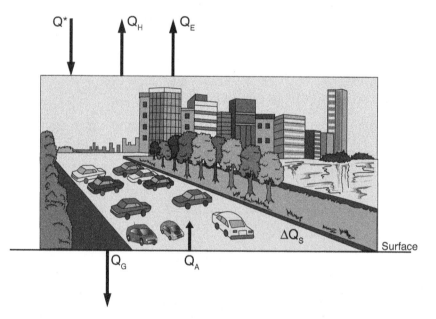

Figure / 4.18 **Energy Balance Written for a Shallow Layer at the Urban Land Surface** This layer contains air and surface elements that make up the built environment. Q^* is the net radiation, Q_H is the sensible heat flux, Q_E is the latent heat flux, Q_G is the ground heat flux, Q_A is the anthropogenic heat discharge, and ΔQ_S is the energy stored or withdrawn from the layer.

shortwave and longwave radiation. Incoming solar shortwave radiation is a function of solar zenith angle, and a fraction of it is then reflected as outgoing shortwave radiation, which depends on the solar **albedo** of the surface. The higher the albedo of the surface, the more solar energy is reflected back into the atmosphere and leaves the shallow layer shown in Figure 4.18. Incoming longwave radiation is emitted from the sky and surrounding environment. Outgoing longwave radiation includes both that emitted from the surface and the reflected incoming longwave radiation.

Q_A is the total anthropogenic heat discharge in the box. The first two terms $(Q^* + Q_A)$ are balanced by the sensible heat flux (Q_H), latent heat flux (Q_E), and ground heat flux (Q_G). Sensible heat is heat energy transferred between the surface and air. When the surface is warmer than the air above, heat will be transferred upward into the air and leave the box via conduction followed by convection. The latent heat flux is produced by transpiration of vegetation and evaporation of land surface water, which removes heat from the surface in the form of water vapor. The ground heat flux is the flux of heat transferred from the surface downward to subsurface via conduction. Finally, ΔQ_S is a term to account for energy that is stored or withdrawn from the layer. The ambient temperature within the layer will be influenced by ΔQ_S. Later, in Table 4.6, we will investigate how these energy balance terms are related to the layout of, and materials incorporated into, the built environment.

The magnitude of the urban heat island can be described as the difference in temperature between urban (u) and rural (r) monitoring stations (ΔT_{u-r}). ΔT_{u-r} will be greatest on clear, cool nights, but it also has been found to depend on street geometry. In the most dense section of the urban environment, the magnitude of this loss term (part of Q^*) is controlled by how well the sky is viewed at ground level. This sky view factor has been found to be approximated by the ratio of building height to street width (H/W). The maximum ΔT_{u-r} (in °C) can be related to the street geometry by the following expression (Oke, 1981):

$$\text{maximum } \Delta T_{u-r} = 7.45 + 3.97 \ln\left(\frac{H}{W}\right) \quad \textbf{(4.33)}$$

While climate considerations related to street geometry can be designed into new urban development, in existing cities little can be done to modify the effect of the street canyon on climate. In such cases, climate can be modified by selection of surfaces, coatings, and vegetation while also reducing the amount of mechanical waste heat that cities produce.

Table 4.6 relates many of the terms in the energy balance (Equation 4.32) with engineered features of the urban environment. Some features are related to the physical geometry of the street layout. Others include modification of surfaces, materials choices, use of impervious pavements, preservation of wetlands, and incorporation of green roofs and low-impact development technologies for storm water control.

Urban Heat Island Mitigation
http://www.epa.gov/heatisland/mitigation/index.htm

Class Discussion

Urban landscapes laid out in a more spread-out, horizontal direction (versus the densely populated vertical direction common in places like Manhattan) will have a less extreme urban heat island. Is this a more, or less, sustainable approach to populating an urban area? Obviously the answer is not easy and will require more thinking and analysis. The question does demonstrate why sustainable solutions require engineers to think beyond their individual project and take a systems approach to solving problems, which incorporates a regional and global outlook.

Features of Urban Environment Related to Terms in the Heat Island Energy Balance Designing and modifying an urban environment to modify climate processes requires an understanding of this balance.

Energy Balance Term	Feature of an Urban Environment That Alters the Energy Balance Term	Engineering Modifications That Reduce Intensity of Urban Heat Island
Net shortwave and longwave radiation, Q^*	Canyon geometry of the street and building	Canyon geometry influences the way shortwave radiation enters and is absorbed by the built environment and the way longwave radiation is reflected out of the urban canopy.
Heat added by humans (Q_{human})	Emission of waste heat from buildings, factories, and vehicles	Though this is a small term in the overall energy balance, buildings can be designed to reduce the need for mechanical cooling. Cities can be planned so they are dependent on mechanical engines to move people and goods.
Sensible heat flux, Q_H	Types of engineering materials	Increasing the surface albedo of paints and roofing materials will limit the surface–air sensible heat flux. Albedo is a measure of the amount of solar energy reflected by the surface. Narrow canyon geometry can result in reduced air flow, which decreases the effect of Q_H.
Latent heat flux, Q_E	Types of engineering materials and storm water management	The latent heat flux out of the system is the result of water evaporation. The energy is carried out in the form of water vapor (in the form of the higher energy in the water molecules in the vapor form). The heat is taken from the vegetation or water. This is the same process as sweat, where one's body is cooled with the heat going away in the form of latent heat. Impervious and nonvegetated surfaces hinder evaporative cooling (unless water is sprinkled on them). Low-impact development recognizes that leaving some standing water on the surface is not bad and vegetation such as green roofs and trees is an important feature of the urban built environment.
Increased storage of heat	Different abilities to store heat in different types of construction materials	The thermal conductivity of asphalt and concrete are similar (1.94 versus 2.11 $J/m^3 - K$, respectively). The thermal admittance of asphalt and concrete results in increased storage of heat. Urban surfaces heat up faster than natural and impervious surfaces that retain water. Built-environment materials have a high ability to store and release heat. Paved surfaces are thick and in contact with an underlying ground surface. Buildings, though, have a thinner skin that separates indoor and outdoor air. Surfaces with higher albedo will reduce the stored heat.

SOURCE: Based on Mills (2004).

4.3 Buildings: Right Sizing and Energy

The average American now spends over 85 percent of his or her time indoors. This fact, along with the large material flows required to construct, operate, and maintain a building, has important consequences for engineers. In the United States, buildings use approximately

example/4.13 Urban Heat Island and Street Geometry

Assume a downtown area has two 12 ft. travel lanes for vehicles, two 12 ft. bus lanes, two 12 ft. metered parking lanes, and a 12 ft. sidewalk on each side. This is all surrounded by 10-story buildings that are 125 ft. tall. What is the maximum urban heat island impact that can be expected?

solution

The maximum urban heat island in the downtown core can be estimated using Equation 4.33. The street width includes the roadway and the sidewalk areas.

$$\text{maximum } \Delta T_{u-r} = 7.45 + 3.97 \ln (125 \text{ ft.}/96 \text{ ft.}) = 8.5°C$$

Note how this example shows the importance of street and building geometry (referred to as the *street canyon*) on the urban heat island. Try doing this example again for the same street size but shorter buildings. A neighborhood with the same street profile but 40 ft tall buildings will have a maximum heat island impact of 4.0°C (Cambridge Systematics, 2005). Try doing the example again for an old historic city with narrow streets but shorter building heights. What do you discover about urban heat island intensity in the urban core as it relates to street and building geometry and population density?

one-third of the total energy, two-thirds of the electricity, and one-eighth of the water and transform land that provides valuable ecological services. Buildings also account for 40 percent of global raw-materials use (3 billion tons annually).

The six components of a building are: (1) foundation, (2) super-structure, (3) exterior envelope, (4) interior partitions, (5) mechanical systems, and (6) furnishings. Each of these components, during each stage of a building's life cycle, has a potential adverse impact on human health, as well as issues of energy use, water use, biodiversity, and use and release of hazardous chemicals. **Energy efficiency** of the building enve-lope is a function of the building's size, how well insulated the structure is, how airtight the structure is, and how the building's glazed area (for example, its windows) is oriented to take advantage of solar heating gain.

Table 4.7 gives that while the average household size in the United States decreased from 3.67 members in 1940 to 2.62 in 2002,

Table / 4.7

Then and Now: Increasing Size of the American Home

	Then	Now
Average number of occupants	3.67 in 1940	2.62 in 2002
Average size of home	100 m² (1,100 ft²) in the 1940s and 1950s	217 m² (2,340 ft²) in 2002
Garage	48% of single-family homes had a garage for 2 or more cars in 1967	82% of homes had a garage for 2 or more cars in 2002
Air-conditioning	46% of new homes had central air-conditioning in 1975	87% of new homes had air-conditioning in 2002

Table / 4.8

Materials Used to Construct a 2,082 ft² (193 m²) House in the United States Larger homes are thought to consume more materials on a square-foot basis because they tend to have larger ceilings and more features.

Component	Quantity
Framing lumber	32.7 m²
Sheathing	1,073 m²
Concrete	15.35 tonnes
Exterior siding	280 m²
Roofing	264 m²
Insulation	284 m²
Interior wall materials	516 m²
Flooring (tile, wood, carpeting, resilient flooring)	193 m²
Ductwork	69 m
Windows	18
Cabinets	18
Interior doors	12
Closet doors	6
Exterior doors	3
Patio door	1
Garage doors	2
Fireplace	1
Toilets	3
Bathtubs	2
Shower stall	1
Bathroom sinks	3
Kitchen sinks	1
Range	1
Refrigerator	1
Dishwasher	1
Disposal unit	1
Range hood	1
Clothes washer	1
Clothes drier	1

SOURCE: From Wilson and Boehland, *Journal of Industrial Ecology*, MIT Press Journals, copyright (2005).

the average home size increased from 1,100 to 2,340 ft^2. The increased size of residential dwellings has large implications for regional and global materials flows, along with materials usage and pollution production during the home's life. In terms of residential construction, Table 4.8 lists the materials used to construct a 2,082 ft^2 U.S. home. Even appliances that are touted as being more energy-efficient are consuming more and more energy, because of their larger size (think of television size).

Rightsizing residential, commercial, and institutional buildings is a major design tool to save materials and produce less pollution during *all* stages of the building's life cycle. As an example, in a recent study, different energy insulation scenarios were applied to 1,500 ft^2 and 3,000 ft^2 homes located in two North American cities with different climates (Boston and St. Louis). Table 4.9 compares the heating and cooling energy requirements associated with each building. Also compared is the past home of this book's lead author, located in the northern United States.

The data in Table 4.9 show that when floor area is halved, heating costs are reduced slightly more than half, and cooling costs are reduced by about one-third. The smaller, but less energy efficient, older house still uses less energy than the new and better-insulated larger house. Besides the energy required to heat and cool larger spaces, larger homes also require longer runs for ducting and hot-water pipes, which causes energy losses associated with the conveyance of warm air, chilled air, and hot water (Wilson and Boehland, 2005).

The highly insulated home located in the northern Midwest (remodeled by the lead author of this book) has zero cooling costs. It has no mechanical air-conditioning system, which negates the need for the materials associated with a cooling and delivery system along with energy associated to chill air. Besides being in a relatively cool geographic location, the building is designed so that insulation stores cool air obtained by opening windows during the night. The strategic placement of windows that capture prevailing breezes and the use of tree shading and a shaded porch also replace the need for mechanical cooling.

The highly insulated home is also designed to take advantage of passive solar heating in the winter, which requires no other heating source on sunny winter days. The house also incorporates use of extensive water-efficient appliances and a solar hot-water heating system. Hanging clothes outside to dry (even in the winter) is preferred over mechanical drying. Energy gains are made not only from not having to pump and treat water, but also in the energy savings associated with heating water.

Some home designers now espouse this different approach to house design—one focused not on size, but on quality and functionality, where space is designed to be used with what is termed *space efficiency*. This type of house design can use much less materials, water, and energy throughout the various life stages of building construction, occupancy, and end of life.

Ways to Save Energy
http://www.energysavers.gov/

Comparative Annual Energy Use for Small versus Large Houses R factor is a measure of resistance to heat flow. R-19 is comparable to RSI-3.3 in the metric system.

House	Location	Relative Energy Standard[a]	Heating (million Btu)	Cooling (million Btu)	Heating Cost ($)[b]	Cooling Cost ($)[c]
3,000 ft²	Boston, MA	Good	73	19	445	190
3,000 sq. ft²	St. Louis, MO	Good	61	29	378	294
1,500 ft²	Boston, MA	Good	35	13	217	131
1,500 ft²	St. Louis, MO	Good	29	20	181	198
1,500 ft²	Boston, MA	Poor	48	12	297	124
1,500 ft²	St. Louis, MO	Poor	40	21	247	206
1,500 ft²	Northern U.S.	High	27[d]	0[e]	240	0

[a]"Good" means a moderately insulated home with R-19 walls, R-30 ceilings, double-low-e vinyl windows, R-4.4 doors, R-6 insulation in air ducts, and infiltration of 0.50 air change per hour for heating and 0.25 air change per hour for cooling.

"Poor" means a poorly insulated home with R-13 walls, R-19 attic, insulated glass vinyl windows, R-2.1 doors, and infiltration of 0.50 air change per hour for heating and 0.25 air change per hour for cooling. Air ducts are not insulated.

"High" means the home is carefully designed and constructed to be airtight. It has R-25 walls, R-50 in the attic, double-low-e vinyl windows, R-14 doors, and infiltration of 0.20 air change per hour for heating.

[b]Heating costs assume natural gas costs $0.50 cents per 100,000 Btu.

[c]Cooling costs assumed to be $0.10 per kWh.

[d]Heating consumes two cords of hardwood and assumes 17 million usable Btu per cord.

[e]No air-conditioning is installed. Building insulation stores cool air obtained during the night, and strategic placement of windows, tree shading, and use of porch contribute to no need for mechanical cooling.

SOURCE: Adapted from Wilson and Boehland (2005). With permission of Wiley-Blackwell.

Class Discussion

Investigate the minimum insulation requirements for new construction in your area and compare those requirements with the data in Table 4.9 and this example. Why haven't more consumers taken advantage of cost and energy saving strategies such as installing insulation, energy-efficient windows and doors, or tankless hot-water heaters?

4.4 Mass Transport Processes

Transport processes move chemicals from where they are generated, resulting in impacts that can be distant from the pollution source. In addition, transport processes are used in the design of treatment systems. Here, our discussion has two purposes: to provide an understanding of the processes that cause pollutant transport and to present and apply the mathematical formulas used to calculate the resulting pollutant fluxes.

4.4.1 ADVECTION AND DISPERSION

Transport processes in the environment can be divided into two categories: advection and dispersion. **Advection** refers to transport with the mean fluid flow. For example, if the wind is blowing toward the east, advection will carry any pollutants present in the atmosphere toward the east. Similarly, if a bag of dye is emptied into the center of a river, advection will carry the resulting spot of dye downstream. In contrast, **dispersion** refers to the transport of compounds through the action of random motions. Dispersion works to eliminate sharp discontinuities in concentration and results in smoother, flatter concentration profiles. Advective and dispersive processes usually can be considered independently. For the spot of dye in a river, while advection moves the center of mass of the dye downstream, dispersion spreads out the concentrated spot of dye to a larger, less concentrated region.

DEFINITION OF THE MASS FLUX DENSITY Mass flux ($\dot{m}$, with units of mass/time), discussed earlier in this chapter, calculates the rates at which mass is transported into and out of a control volume in mass balances. Because mass balance calculations are always made with reference to a specific control volume, this value clearly refers to the rate at which mass is transported *across the boundary of the control volume*. However, in calculations of advective and dispersive fluxes, a specific, well-defined control volume will not be created. Instead, we determine the **flux density** across an imaginary plane oriented perpendicular to the direction of mass transfer.

The resulting mass flux density is defined as the rate of mass transferred across the plane per unit time per unit area. The symbol J will be used to represent the flux density, expressed as the rate per unit area at which mass is transported across an imaginary plane. J has units of (mass/time-length squared).

The total mass flux across a boundary ($\dot{m}$) can be calculated from the flux density. To do this, simply multiply J by the area of the boundary:

$$\boxed{\dot{m} = J \times A} \qquad \textbf{(4.34)}$$

The mass transfer process that J describes can result from advection, dispersion, or a combination of both processes.

CALCULATION OF THE ADVECTIVE FLUX The **advective flux** refers to the movement of a compound along with flowing air or water. The advective-flux density depends simply on concentration and flow velocity:

$$J = C \times v \qquad \textbf{(4.35)}$$

The fluid velocity, v, is a vector quantity. It has both magnitude and direction, and the flux J refers to the movement of pollutant mass in the same direction as the fluid flow. The coordinate system is generally defined so that the x-axis is oriented in the direction of fluid flow. In this case, the flux J will reflect a flux in the x-direction, and the fact that J is really a vector quantity will be ignored.

DISPERSION Dispersion results from random motions of two types: the random motion of molecules and the random eddies that arise in turbulent flow. Dispersion from the random molecular motion is termed *molecular diffusion*; dispersion that results from turbulent eddies is called *turbulent dispersion* or *eddy dispersion*.

example/4.14 **Calculation of the Advective-Flux Density**

Calculate the average flux density J of phosphorus downstream of the wastewater discharge of Example 4.2. The cross-sectional area of the river is 30 m².

solution

In Example 4.2, the following conditions downstream of the spot where a pipe discharged to a river were determined: volumetric flow rate $Q = 26$ m³/s and downstream concentration $C_d = 0.20$ mg/L. The average river velocity is $v = Q/A = (26\,\text{m}^3/\text{s})/(30\,\text{m}^2) = 0.87$ m/s. Using the definition of flux density (Equation 4.35), we can solve for J:

$$J = \left[(0.20\,\text{mg/L}) \times \frac{10^3\,\text{L}}{\text{m}^3} \right] \times (0.87\,\text{m/s})$$

$$= 174\,\text{mg/m}^2\text{-s or } 0.17\,\text{g/m}^2\text{-s}$$

Fick's Law   **Fick's law** is used to calculate the dispersive-flux density. It can be derived by analyzing the mass transfer that results from the random motion of gas molecules.[1] The purpose of this derivation is to provide a qualitative and intuitive understanding of why diffusion occurs, and the derivation is useful only for that purpose. In problems where it is necessary to calculate the diffusive flux, we will use Fick's law (Equation 4.45, derived later in this section).

Consider a box that is initially divided into two parts, as shown in Figure 4.19. Each side of the box has a height and depth of one unit, and a width of length Δx. Initially, the left portion of the box is filled with 10 molecules of gas x, and the right side is filled with 20 molecules of gas y, as shown in the top half of Figure 4.19. What happens if the divider is removed?

Molecules are never stationary. All of the molecules in the box are constantly moving around, and at any moment, they have some probability of crossing the imaginary line at the center of the box. Assume that the molecules on each side are counted every Δt seconds. The probability that a molecule crosses the central line during the period between observations can be defined as k, which is assumed to equal 20 percent (any value would do for the present purpose). The first time the box is checked, after a period Δt, 20 percent of the molecules originally on the left will have moved to the right, and 20 percent of the molecules originally on the right will have moved to the left. Counting the molecules on each side gives the situation shown in the bottom of Figure 4.19. Eight molecules of x remain on the left, and two have moved to the right, while 16 molecules of y remain on the right, four having moved to the left.

Because the boxes are equal in size, the concentration within each box is proportional to the number of molecules within it. Therefore, the random motion of molecules in the boxes has reduced the

---

[1] This derivation is based closely on one presented by Fischer et al. (1979).

166 Chapter 4 Physical Processes

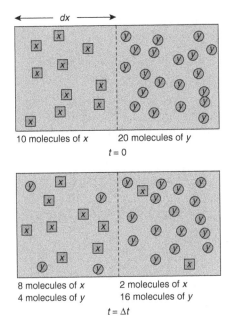

10 molecules of x   20 molecules of y

t = 0

8 molecules of x   2 molecules of x
4 molecules of y   16 molecules of y

t = Δt

Figure / 4.19 **Diffusion of Gas Molecules in a Box**   A box is divided into two regions of equal size. Ten gas molecules of one type (x) are added to the left side, while 20 gas molecules of another type (y) are added to the right side. Although they are distinguishable, the two types of molecules are identical in every physical respect. At time, the divider separating the two regions is removed. As a result of random motion, each molecule within the box has a 20 percent probability of moving to the opposite side of the box during each time interval of duration Δt. The result after one time interval is shown in the bottom figure.

(From Fischer et al., *Mixing in Inland and Coastal Waters*, Copyright Elsevier (1979)).

concentration difference between the boxes, with the difference falling from $(20 - 0)$ to $(16 - 4)$ for the molecules of y, and from $(10 - 0)$ to $(8 - 2)$ for the molecules of x. This result leads to a fundamental property of dispersive processes: dispersion moves mass from regions of high concentration to regions of low concentration and reduces concentration gradients.

The flux density J can also be derived for the two-box experiment. For this calculation, the situation shown in Figure 4.19 is used again, with the probability of any molecule crossing the central boundary during a period Δt equal to k. Since each molecule can be considered independently, the movement of a single molecule type—say, molecule y—can be analyzed.

Let m_L be the total mass of molecule y in the left half of the box, and m_R equal the mass in the right half. Since our box has unit height and depth, the area perpendicular to the direction of diffusion is one square unit. Thus, the flux density (the flux per unit area) is just equal to the rate of mass transfer across the boundary. The amount of mass transferred from left to right in a single time step is equal to km_L, since each molecule has a probability k of crossing the boundary, while the amount transferred from right to left during the same period is km_R. Thus, the net rate of mass flux from left to right across the boundary is equal to $(km_L - km_R)$ divided by Δt:

$$J = \frac{k}{\Delta t}(m_L - m_R) \tag{4.36}$$

Since it is more convenient to work with concentrations than with total mass values, Equation 4.36 needs to be converted to concentration units. The concentration in each half of the box is given by

$$C_L = \frac{m_L}{\Delta x \times (\text{height}) \times (\text{depth})} \tag{4.37}$$

Because the height and depth both equal 1, we can simplify:

$$= \frac{m_L}{\Delta x} \tag{4.38}$$

For the right side of the box,

$$C_R = \frac{m_R}{\Delta x} \tag{4.39}$$

Substituting $C\Delta x$ for the mass in each half of the box, we can solve for the flux density:

$$J = \frac{k}{\Delta t}(C_L \Delta x - C_R \Delta x) \tag{4.40}$$

$$= \frac{k}{\Delta t}(\Delta x)(C_L - C_R) \tag{4.41}$$

Finally, note that as $\Delta x \to 0, (C_R - C_L)/\Delta x \to dC/dx$. Therefore, if we multiply Equation 4.41 by $(\Delta x/\Delta x)$,

$$J = \frac{k}{\Delta t}(\Delta x)(C_L - C_R)\frac{\Delta x}{\Delta x} \tag{4.42}$$

$$= \frac{k}{\Delta t}(\Delta x)^2 \frac{(C_L - C_R)}{\Delta x} \tag{4.43}$$

we obtain

$$J = -\frac{k}{\Delta t}(\Delta x)^2 \frac{dC}{dx} \tag{4.44}$$

The negative sign in this equation is simply a result of the convention that flux is positive when it flows from left to right, while the derivative is positive when concentration increases toward the right.

Equation 4.44 states that the flux of mass across an imaginary boundary is proportional to the concentration gradient at the boundary. Since the resulting flux cannot depend on arbitrary values of Δt or Δx, the factor $k(\Delta x)^2/\Delta t$ must be constant. This product is the value called the **diffusion coefficient**, D. Rewriting Equation 4.44 results in Fick's law:

$$J = -D\frac{dC}{dx}$$

(4.45)

The units of the diffusion coefficient are clear from an analysis of the units of Equation 4.45 or from the units of the parameters in Equation 4.44; the diffusion coefficient has the same units as $k(\Delta x)^2/\Delta t$. Since k is a probability and thus has no units, the units of D are (length^2/time). Diffusion coefficients are commonly reported in units of cm^2/s.

Note the form of Equation 4.45:

$$\text{flux density} = (\text{constant}) \times (\text{concentration gradient})$$

(4.46)

This form of equation will also appear in Chapter 7 when Darcy's law is covered. Darcy's law governs the rate at which water flows through porous media, as in groundwater flow. The same equation also governs heat transfer, replacing the concentration gradient with a temperature gradient.

Molecular Diffusion The molecules-in-a-box analysis used earlier is essentially an analysis of molecular diffusion. Purely molecular diffusion is relatively slow. Table 4.10 lists typical values for the diffusion coefficient. These values are approximately 10^{-2} to 10^{-1} cm^2/s for gases and much lower, around 10^{-5} cm^2/s, for liquids. The difference in diffusion coefficient between gases and liquids is understandable because gas molecules are free to move much greater distances before being stopped by bumping into another molecule.

The diffusion coefficient also varies with temperature and the molecular weight of the diffusing molecule. This is because the average

Table / 4.10

Selected Molecular-Diffusion Coefficients in Water and Air

Compound	Temperature (°C)	Diffusion Coefficient (cm^2/s)
Methanol in H_2O	15	1.26×10^{-5}
Ethanol* in H_2O	15	1.00×10^{-5}
Acetic acid in H_2O	20	1.19×10^{-5}
Ethylbenzene in H_2O	20	8.1×10^{-6}
CO_2 in air	20	0.151

*Of the two similar compounds methanol and ethanol, the less massive compound, methanol, has the higher diffusion coefficient.
SOURCE: Mihelcic (1999). Reprinted with permission of John Wiley & Sons, Inc.

speed of the random molecular motions depends on the kinetic energy of the molecules. As heat is added to a material and temperature increases, the thermal energy is converted to random kinetic energy of the molecules, and the molecules move faster. This results in an increase in the diffusion coefficient with increasing temperature. However, if the molecules have differing molecular weights, a heavier molecule moves more slowly at a given temperature, so the diffusion coefficient decreases with increasing molecular weight.

example/4.15 Molecular Diffusion

The transport of polychlorinated biphenyls (PCBs) from the atmosphere into the Great Lakes is of concern because of health impacts on aquatic life and on people and wildlife that eat fish from the lakes. PCB transport is limited by molecular diffusion across a thin stagnant film at the surface of the lake, as shown in Figure 4.20. Calculate the flux density J and the total annual amount of PCBs deposited into Lake Superior if the transport is by molecular diffusion, the PCB concentration in the air just above the lake's surface is 100×10^{-12} g/m^3, and the concentration at a height of 2.0 cm above the water surface is $450 \times 10^{-12} =$ g/m^3. The diffusion coefficient for PCBs is equal to 0.044 cm^2/s, and the surface area of Lake Superior is 8.2×10^{10} m^2. (The PCB concentration in the air at the air–water interface is determined by Henry's law equilibrium with dissolved PCBs.)

Figure / 4.20 **Variation of PCB Concentration with Height above Lake Superior** C_{air} is the PCB concentration in the atmosphere above the lake, and C^* is the concentration at the air–water interface, which is determined by Henry's law equilibrium with the dissolved PCB concentration. The flux of PCBs into the lake is determined by the rate of diffusion across a stagnant film above the lake.

(From Mihelcic (1999). Reprinted with permission of John Wiley & Sons, Inc.).

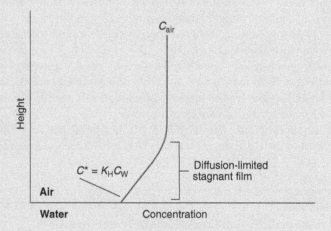

solution

To calculate the flux density, first determine the concentration gradient. Assume that concentration changes linearly with height between the surface and 2.0 cm, as no concentration information was provided between those two heights. The gradient is then

$$\frac{dC}{dz} = \frac{450 \times 10^{-12} \text{ g/m}^3 - 100 \times 10^{-12} \text{ g/m}^3}{2.0 \text{ cm} - 0 \text{ cm}} \times \frac{10^2 \text{ cm}}{\text{m}}$$

$$= 1.8 \times 10^{-8} \text{ g/m}^4$$

Fick's law (Equation 4.45) can be used to calculate the flux density:

$$J = -D\frac{dC}{dz}$$
$$= -(0.044 \text{ cm}^2/\text{s}) \times 1.8 \times 10^{-8} \text{ g/m}^4 \times \frac{\text{m}^2}{10^4 \text{ cm}^2} \times \frac{3.15 \times 10^7 \text{ s}}{\text{yr}}$$
$$= -2.4 \times 10^{-6} \text{ g/m}^2\text{-yr}$$

Here, the negative sign indicates the flux is downward, but it is not necessary to pay attention to the sign to determine that. Remember, diffusion always transports mass from higher-concentration to lower-concentration regions.

The total depositional flux is given by $\dot{m} = J \times A$:

$$\dot{m} = -2.4 \times 10^{-6} \text{ g/m}^2\text{-yr} \times 8.2 \times 10^{10} \text{ m}^2$$
$$= -2.0 \times 10^5 \text{ g/yr}$$

Thus, the total annual input of PCBs into Lake Superior from the atmosphere is approximately 200 kg. Although this is a small annual flux for such a large lake, PCBs do not readily degrade in the environment, and they bioaccumulate in the fish, resulting in unhealthy levels.

Turbulent Dispersion In **turbulent dispersion**, mass is transferred through the mixing of *turbulent eddies* within the fluid. This is fundamentally different from the processes that determine molecular diffusion. In turbulent dispersion, the random motion of the *fluid* does the mixing, while in molecular diffusion, the random motion of the pollutant molecules is important.

Random motions of the fluid are generally present in the form of whorls, or eddies. These are familiar in the form of eddies or whirlpools in rivers but occur in all forms of fluid flow. The size of turbulent eddies is several orders of magnitude larger than the mean free path of individual molecules, so turbulence moves mass much faster than does molecular diffusion. As a result, turbulent, or eddy dispersion coefficients used in Fick's law are generally several orders of magnitude larger than molecular-diffusion coefficients.

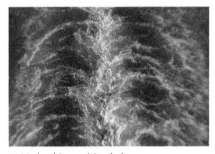
© Michael Braun/iStockphoto.

The value of the turbulent-dispersion coefficient depends on properties of the fluid flow. It does not depend on molecular properties of the compound being dispersed (as did the molecular-diffusion coefficient), because in turbulence, the molecules are simply carried along with the macroscale flow. For flow in pipes or streams, the most important flow property determining the turbulent-dispersion coefficient is the flow velocity. Turbulence is present only at flow velocities above a critical level, and the degree of turbulence is correlated with velocity. More precisely, the presence or absence of turbulence depends on the

Reynolds number, a dimensionless number that depends on velocity, width of the river or pipe, and viscosity of the fluid. In addition, the degree of turbulence depends on the material over which the flow occurs, so that flow over bumpy surfaces will be more turbulent than flow over smooth surfaces, and the increased turbulence will cause more rapid mixing. In lakes and in the atmosphere, buoyant mixing that results from temperature-induced density gradients can also cause turbulent mixing, even in the absence of currents.

Except in the case of transport across a boundary, such as at the air–water interface considered in Example 4.15, turbulent dispersion almost always entirely dominates molecular diffusion. This is because even an occasional amount of weak turbulence will cause more mixing than several days of molecular diffusion.

Fick's law applies to turbulent dispersion just as it does to molecular diffusion. Thus, flux density calculations are the same for both processes; only the magnitude of the dispersion coefficient is different.

Mechanical Dispersion   The final dispersion process considered in this chapter is similar to turbulence in that it is a result of variations in the movement of the fluid that carries a chemical. In **mechanical dispersion**, these variations are the result of (1) variations in the flow pathways taken by different fluid parcels that originate in nearby locations or (2) variations in the speed at which fluid travels in different regions.

Dispersion in groundwater flow provides a good example of the first process. Figure 4.21 shows a magnified depiction of the pores through which groundwater flows within a subsurface sample. (Note that, as shown in Figure 4.21, groundwater movement is not a result of underground rivers or creeks, but rather is caused by the flow of water through the pores of the soil, sand, or other material underground.) Because transport through the soil is limited to the pores between soil particles, each fluid particle takes a convoluted path through the soil, and as it is transported horizontally with the mean flow, it is displaced vertically a distance that depends on the exact flow path it took. The great variety of possible flow paths results in a random displacement in the directions perpendicular to the mean flow path. Thus, a spot of dye introduced into the groundwater flow between points B and C in the figure would be spread out, or dispersed, into the region between points B' and C' as it flows through the soil.

Figure / 4.21   Process of Mechanical Dispersion in Groundwater Flow
Two fluid parcels starting near each other at locations B and C are dispersed to locations farther apart (B' and C') during transport through the soil pores, while parcels from A and B are brought closer together, resulting in mixing of water from the two regions.

(From Hemond and Fechner (1994). Copyright Elsevier).

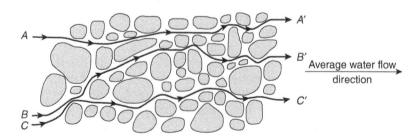

The second type of mechanical dispersion results from differences in flow speed. Anywhere that a flowing fluid contacts a stationary object, the speed at which the fluid moves will be slower near the object. For example, the speed of water flowing down a river is fastest in the center of a river and can be very slow near the edges. Thus, if a line of dye were somehow laid across the river at one point, it would be stretched out in the upstream/downstream direction as it flowed down the river, with the center part of the line moving faster than the edges. This type of dispersion spreads things out in the longitudinal direction in the direction of flow. This is in contrast to mechanical dispersion in groundwater, which spreads things out in the direction perpendicular to the direction of mean flow.

4.4.2 MOVEMENT OF A PARTICLE IN A FLUID: STOKES' LAW

The movement of a particle in a fluid is determined by a balance of the viscous drag forces resisting the particle movement with gravitational or other forces that cause the movement. In this section, a force balance on a particle is used to derive the relationship between particle size and settling velocity known as Stokes' law, and Stokes' law is used in examples involving particle-settling chambers.

GRAVITATIONAL SETTLING   Consider the settling particle shown in Figure 4.22. To determine the velocity at which it falls (the settling velocity), a force balance will be conducted. Three forces act on the particle: the downward gravitational force, an upward buoyancy force, and an upward drag force.

The gravitational force F_g is equal to the gravitational constant g times the mass of the particle, m_p. In terms of particle density ρ_P and diameter D_p, m_p is equal to $(\rho_P \pi / 6 D_P^3)$. Therefore,

$$F_g = \rho_P \frac{\pi}{6} D_P^3 g \qquad (4.47)$$

The buoyancy force F_B is a net upward force that results from the increase of pressure with depth within the fluid. The buoyancy force is equal to the gravitational constant times the mass of the fluid displaced by the particle:

$$F_B = \rho_f \frac{\pi}{6} D_P^3 g \qquad (4.48)$$

where ρ_f is equal to the fluid density.

The only remaining force to determine is the drag force, F_D. The drag force is the result of frictional resistance to the flow of fluid past the surface of the particle. This resistance depends on the speed at which the particle is falling through the fluid, the size of the particle, and the *viscosity*, or resistance to shear, of the fluid. Viscosity is essentially what one would qualitatively call the "thickness" of the fluid. Honey has a high viscosity, water has a relatively low viscosity, and the viscosity of air is much lower yet.

Over a wide range of conditions, the friction force can be correlated with the Reynolds number. Most particle-settling situations involve

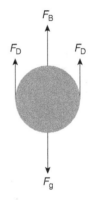

Figure / 4.22   Forces Acting on a Particle Settling through Air or Water   The gravitational force F_g is in the downward direction and is counteracted by the buoyancy force F_B and the drag force F_D.

(From Mihelcic (1999). Reprinted with permission of John Wiley & Sons, Inc.).

creeping flow conditions (Reynolds number less than 1). In this case, the Stokes' drag force can be used:

$$F_D = 3\pi\mu D_P v_r \tag{4.49}$$

where μ is the fluid viscosity (units of g/cm-s) and v_r is the downward velocity of the particle relative to the fluid.

The net downward force acting on the particle is equal to the vector sum of all forces acting on the particle:

$$F_{down} = F_g - F_B - F_D \tag{4.50}$$

$$= \rho_P \frac{\pi}{6} D_P^3 g - \rho_f \frac{\pi}{6} D_P^3 g - 3\pi\mu D_P v_r \tag{4.51}$$

$$= (\rho_P - \rho_f) \frac{\pi}{6} D_P^3 g - 3\pi\mu D_P v_r \tag{4.52}$$

The particle will respond to this force according to Newton's second law (force equals mass times acceleration). Thus,

$$F_{down} = m_p \times \text{acceleration} \tag{4.53}$$

$$= m_P \times \frac{dv_r}{dt} \tag{4.54}$$

This differential equation can be solved to determine the time-varying velocity of a particle that is initially at rest. The solution indicates that, in almost all cases of environmental interest, the period of time required before the particle reaches its final settling velocity is very short (much less than 1 s). For this reason, in this text, only the final (terminal) settling velocity is considered.

When the particle has reached terminal velocity, it is no longer accelerating, so $dv/dt = 0$. Thus, from Equation 4.54, $F_{down} = 0$. Setting F_{down} equal to zero and noting that v_r is equal to the settling velocity v_s at terminal velocity, Equation 4.52 can be rearranged to yield

$$(\rho_P - \rho_f) \frac{\pi}{6} D_P^3 g = 3\pi\mu D_P v_s \tag{4.55}$$

Solving for v_s:

$$v_s = \frac{g(\rho_P - \rho_f)}{18\,\mu} D_P^2 \tag{4.56}$$

Equation 4.56 is referred to as **Stokes' law**. The resulting settling velocity is often called the Stokes' velocity. Stokes' law, so called because it is based on the Stokes' drag force, is the fundamental equation used to calculate terminal settling velocities of particles in both air and water. It is used in the design of treatment systems to remove particles from exhaust gases and from drinking water and wastewater as well as in analyses of settling particles in lakes and in the atmosphere. Several examples of the use of Stokes' law are provided in Chapter 9.

An important implication of Stokes' law is that the settling velocity increases as the *square* of the particle diameter, so larger particles settle much faster than smaller particles. This result is used in drinking-water treatment: coagulation and flocculation are used to get small particles to come together and form larger particles, which can then be removed by gravitational settling in a reasonable amount of time. This process results in reduction in the turbidity (increase in the clarity) of the water. In contrast, particles with very small diameters settle extremely slowly. As a result, atmospheric particles with diameters less than 1–10 μm generally fall more slowly than the speed of turbulent eddies of air, with the result that they are not removed by gravitational settling.

Note that particle–particle interactions have been ignored in this derivation. Thus, Stokes' law is valid for *discrete particle settling*. In situations where particle concentration is extremely high, particles form agglomerations or mats, and Stokes' law may no longer be valid.

Key Terms

- advection
- advective flux
- albedo
- anthropogenic greenhouse effect
- batch reactor
- carbon dioxide emissions
- completely mixed flow reactor (CMFR)
- conservative compound
- control volume
- degree-day
- diffusion coefficient
- dispersion
- energy balance
- energy efficiency
- Fick's law
- first law of thermodynamics

- first-order decay
- flux density
- glazing
- greenhouse gases
- heat
- heat balance
- heat island
- heat loss due to infiltration
- Intergovernmental Panel on Climate Change (IPCC)
- law of conservation of mass
- mass accumulation rate
- mass balance
- mass flux
- materials flows analysis (MFA)
- mechanical dispersion

- nonpervious materials
- nonrenewable energy
- nonsteady state
- passive solar gain
- plug-flow reactor (PFR)
- reactor analysis
- renewable energy
- retention time
- rightsizing
- steady state
- Stokes' law
- thermal walls
- turbulent dispersion
- urban metabolism
- zero-order decay

4.1 A waste stabilization pond is used to treat a dilute municipal wastewater before the liquid is discharged into a river. The inflow to the pond has a flow rate of $Q = 4,000$ m^3/day and a BOD concentration of $C_{in} = 25$ mg/L. The volume of the pond is 20,000 m^3. The purpose of the pond is to allow time for the decay of BOD to occur before discharge into the environment. BOD decays in the pond with a first-order rate constant equal to 0.25/day. What is the BOD concentration at the outflow of the pond, in units of mg/L?

4.2 A mixture of two gas flows is used to calibrate an air pollution measurement instrument. The calibration system is shown in Figure 4.23. If the calibration gas concentration C_{cal} is 4.90 ppm$_v$, the calibration gas flow rate Q_{cal} is 0.010 L/min, and the total gas flow rate Q_{total} is 1.000 L/min, what is the concentration of calibration gas after mixing (C_d)? Assume the concentration upstream of the mixing point is zero.

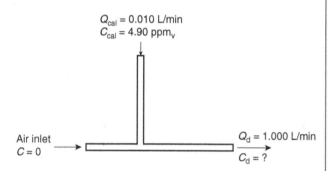

$Q_{cal} = 0.010$ L/min
$C_{cal} = 4.90$ ppm$_v$

Air inlet
$C = 0$

$Q_d = 1.000$ L/min
$C_d = ?$

Figure / 4.23 **Gas Calibration System.**

(From Mihelcic (1999). Reprinted with permission of John Wiley & Sons, Inc.).

4.3 Consider a house into which radon is emitted through cracks in the basement. The total volume of the house is 650 m^3 (assume the volume is well mixed throughout). The radon source emits 250 pCi/s. (A picoCurie (pCi) is a unit proportional to the amount of radon gas and indicates the amount of radioactivity of the gas.) Air inflow and outflow can be modeled as a flow of clean air into the house of 722 m^3/h and an equal air flow out. Radon can be considered conservative in this problem. (a) What is the retention time of the house? (b) What is the steady-state concentration of radon in the house (units of pCi/L)?

4.4 You are in an old spy movie and have been locked into a small room (volume 1,000 ft^3). You suddenly realize a poison gas has just started entering the room through a ventilation duct. You are safe as long as the concentration is less than 100 mg/m^3. If the ventilation air flow rate in the room is 100 ft^3/min and the incoming gas concentration is 200 mg/m^3, how long do you have to escape?

4.5 In the simplified depiction of an ice rink with an ice-resurfacing machine operating (shown in Figure 4.24), points 1 and 3 represent the ventilation air intake and exhaust for the entire ice rink, and point 2 is the resurfacing machine's exhaust. Given that C indicates the concentration of carbon monoxide (CO), conditions at each point are as follows: point 1: $Q_1 = 3.0$ m^3/s, $C_1 = 10$ mg/m^3; point 2: emission rate$= 8$ mg/s of nonreactive CO; point 3: Q_3, C_3 unknown. The ice rink's volume (V) is 5.0×10^4 m^3. (a) Define a control volume as the interior of the ice rink. What is the mass flux of CO into the control volume, in units of mg/s? (b) Assume that the resurfacing machine has been operating for a very long time and that the air within the ice rink is well mixed. What is the concentration of CO within the ice rink, in units of mg/m^3?

Figure / 4.24 **Schematic Diagram of an Ice-Resurfacing Machine in an Ice Rink.**

(From Mihelcic (1999). Reprinted with permission of John Wiley & Sons, Inc.).

4.6 Poorly treated municipal wastewater is discharged to a stream. The river flow rate upstream of the discharge point is $Q_u = 8.7$ m^3/s. The discharge occurs at a flow of $Q_d = 0.9$ m^3/s and has a BOD concentration of 50.0 mg/L. Assume the upstream BOD concentration is negligible. (a) What is the BOD concentration just downstream of the discharge point? (b) If the stream has a cross-sectional area of 10 m², what is the BOD concentration 50 km downstream? (BOD is removed with a first-order decay rate constant equal to 0.20/day.)

4.7 As shown in Figure 4.25, during an air emission test, the inlet gas stream to a fabric filter is 100,000 actual ft³/min (ACFM) and the particulate loading is 2 grains/actual cubic feet (ACF). The outlet gas stream from the fabric filter is 109,000 actual ft³/min and the particulate loading is 0.025 grains/actual ft³/min. What is the maximum quantity of ash that will have to be removed per hour from the fabric filter hopper based on these test results? Assume that 7,000 grains of particles equals 1 lb (problem from EPA Air Pollution Training Institute).

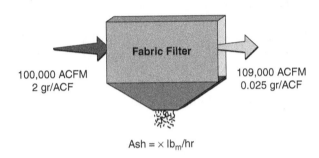

Figure / 4.25 Schematic Diagram of Fabric Filter Used to Remove Particulates from Air.

(Adapted from EPA).

4.8 Two towns, located directly across from each other, operate municipal wastewater treatment plants situated along a river. The river flow is 50 million gallons per day (50 MGD). Coliform counts are used as a measure to determine a water's ability to transmit disease to humans. The coliform count in the river upstream of the two treatment plants is 3 coliforms/ 100 mL. Town 1 discharges 3 MGD of wastewater with a coliform count of 50 coliforms/100 mL, and town 2 discharges 10 MGD of wastewater with a coliform count of 20 coliforms/100 mL. Assume the state requires the downstream coliform count not exceeding 5 coliforms/100 mL. (a) Is the state water-quality standard being met downstream? (Assume coliforms do not die by the time they are measured downstream.) (b) If the state standard downstream is not met, the state has informed town 1 that it must treat its sewage further so the downstream standard is met. Use a mass balance approach to show that the state's request is unfeasible.

4.9 How much water must be continually added to the wet scrubber shown in Figure 4.26 in order to

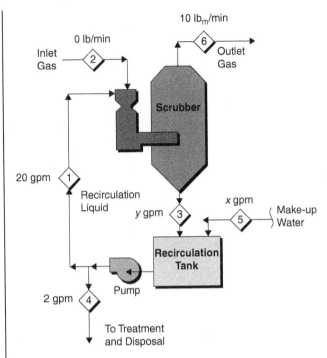

Figure / 4.26 Schematic Diagram of Web Scrubber Used to Remove Particulates from Air.

(Adapted from EPA).

keep the unit running? Each of the streams is identified by a number located in a diamond symbol. Stream 1 is the recirculation liquid flow stream back to the scrubber and it is 20 gallons per minute (gpm). The liquid being withdrawn for treatment and disposal (stream 4) is 2 gpm. Assume that inlet gas stream (number 2) is completely dry and that the outlet stream (number 6) has 10 lb_m/min of moisture evaporated in the scrubber. The water being added to the scrubber is stream number 5. One gallon of water weighs 8.34 lb (problem from EPA Air Pollution Training Institute).

4.10 In the winter, a stream flows at 10 m³/s and receives discharge from a pipe that contains road runoff. The pipe has a flow of 5 m³/s. The stream's chloride concentration just upstream of the pipe's discharge is 12 mg/L, and the runoff pipe's discharge has a chloride concentration of 40 mg/L. Chloride is a conservative substance. (a) Does wintertime salt usage on the road elevate the downstream chloride concentration above 20 mg/L? (b) What is the maximum daily mass of chloride (metric tons/day) that can be discharged through the road runoff pipe without exceeding the water-quality standard?

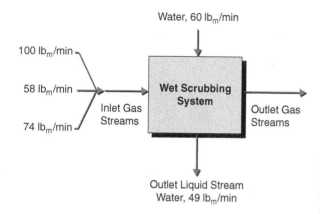

Water, 60 lb$_m$/min

100 lb$_m$/min

58 lb$_m$/min → Inlet Gas Streams → **Wet Scrubbing System** → Outlet Gas Streams

74 lb$_m$/min

Outlet Liquid Stream
Water, 49 lb$_m$/min

Figure / 4.27 Schematic Diagram of Web Scrubber Used to Remove Particulates from Air.

(Adapted from EPA).

4.11 A wet scrubbing system has three separate inlet streams (Figure 4.27). The mass flow rates in these inlet streams are 100, 58, and 74 lb$_m$/min. The water spray into the scrubber is 60 lb$_m$/min, and some of this spray evaporates and leaves with the gas stream. The water stream leaving the scrubber is 49 lb$_m$/min. What is the mass of the gas stream leaving the scrubber? (problem from EPA Air Pollution Training Institute)

4.12 Calculate the hydraulic residence times (the retention time) for Lake Superior and for Lake Erie using data in Table 4.3.

4.13 The total flow at a wastewater treatment plant is 600 m^3/day. Two biological aeration basins are used to remove BOD from the wastewater and are operated in parallel. They each have a volume of 25,000 L. In hours, what is the aeration period of each tank?

4.14 You are designing a reactor that uses chlorine in a PFR or CMFR to destroy pathogens in water. A minimum contact time of 30 min is required to reduce the pathogen concentration from 100 pathogens/L to below 1 pathogen/L through a first-order decay process. You plan on treating water at a rate of 1,000 gal/min. (a) What is the first-order decay rate constant? (b) What is the minimum size (in gallons) of the reactor required for a PFR? (c) What size (in gallons) of CMFR would be required to reach the same outlet concentration? (d) Which type of reactor would you select if your treatment objective stated that "no discharge can ever be greater than 1 pathogen/L"? Explain your reasoning. (e) If the desired chlorine residual in the treated water after it leaves the reactor is 0.20 mg/L and the chlorine demand used during treatment is 0.15 mg/L,

what must be the daily mass of chlorine added to the reactor (in grams)?

4.15 The concentration of BOD in a river just downstream of a wastewater treatment plant's effluent pipe is 75 mg/L. If the BOD is destroyed through a first-order reaction with a rate constant equal to 0.05/day, what is the BOD concentration 50 km downstream? The velocity of the river is 15 km/day.

4.16 A 1.0×10^6 gallon reactor is used in a water reclamation plant. The influent concentration is 100 mg/L, the effluent concentration is 25 mg/L, and the flow rate through the reactor is 500 gallons/min. (a) What is the first-order rate constant for decay of BOD in the reactor? Assume the reactor can be modeled as a CMFR. Report your answer in units per hour. (b) Assume the reactor should be modeled as a PFR with first-order decay, *not* as a CMFR. In that case, what must be the first-order decay rate constant within the PFR reactor? (c) It has been determined that the outlet concentration is too high, so the residence time in the reactor must be doubled. Assuming all other variables remain constant, what must be the volume of the new CMFR?

4.17 You are to design a reactor for removal of reduced iron (Fe^{2+}) from water. The influent water has an iron concentration of 10 mg/L, and this must be reduced to 0.1 mg/L. The water has a pH of 6.5 and the plan is to oxidize the iron to Fe^{3+} using pure oxygen gas, then remove the resulting particulate matter in a sedimentation basin. It has been found that the reduction in Fe^{2+} concentration over time equals $K_{apparent} \times$ [Fe^{2+}], where $K_{apparent}$ equals: $8 \times 10^{13} \times$ [partial pressure of O$_2$] $\times K_w^2/[H^+]^2$. The units of $K_{apparent}$ determined from this expression are min^{-1} and the partial pressure of oxygen is 0.21 atm and the dissociation constant for water, K_w, equals 10^{-14}. Determine the volume (m^3) of a plug flow reactor to treat 1 MGD of water.

4.18 How many watts of power would it take to heat 1 L of water (weighing 1.0 kg) by 10°C in 1.0 h? Assume no heat losses occur, so all of the energy expended goes into heating the water.

4.19 Your house has a 40-gallon electric water heater that heats water to a temperature of 110°F. Several friends are visiting you over the weekend and they are taking consecutive showers. Assume that at the maximum heating level, the heater uses 5 kW of electricity. The water use rate is a continuous 2 gpm with the new water-saving showerhead you recently installed. Your very old showerhead had used 5 gpm!

You replaced the showerhead because you learned that heating water was the second highest energy use in your home. What is the temperature of the water exiting the heater (a) using the old showerhead and (b) the new efficient showerhead? Assume the system is at steady state so all of the energy used heats the water.

4.20 (a) Determine the heat loss (in Btu/°F-day and Btu/degree-day) through a 120 ft² insulated wall described in the following table. (b) Determine the heat loss through the same wall when a 3 ft. by 7 ft. door (R factor $= 4.4$) is inserted into the wall.

Component of Wall	R Factor
2 in. Styrofoam board insulation on outside of wall under siding	10
Old cedar log wall	20
Fiberglass insulation on inside of wall	11
1/2 in. drywall on inside of wall	0.45
Inside air film along inside of wall	0.68
Outside air film along outside of wall	0.17

4.21 Look up (a) the total degree-days for heating and (b) the total degree-days for cooling for your university town or city (or hometown).

4.22 In Section 4.2.4, we worked out a problem where the combined heat loss from a hypothetical 3,000 ft² building was 1,053 Btu/degree-days. Determine the total energy requirements (in Btu) to heat that hypothetical building for the locations in the following table.

Location	Heating Degree-Days
Anchorage, AK	541
Winslow, AZ	70
Yuma, AZ	0
Rochester, NY	237
Pittsburgh, PA	106
Rapid City, SD	193

4.23 Go to the Weather Channel web site (www.weather.com) and look up the monthly average temperature for a major metropolitan area and nearby rural area anywhere in the world over a 12-month period. Use the data you looked up to estimate the magnitude of the urban heat island effect for that city. Graph your data in two figures and determine the temperature differences in each month.

4.24 Identify an urban core of a major metropolitan area that you are familiar with or that is close to your college or university. Calculate the magnitude of the maximum urban heat island impact in the urban core. Provide some detailed alternatives for reducing the urban heat island in this core area and relate them to specific items in the energy balance performed on the urban canopy.

4.25 Assume a small downtown area has two 12 ft. travel lanes with 6 ft. sidewalks on each side. This is all surrounded by buildings that are 25 ft. tall. What is the maximum urban heat island impact that can be expected?

4.26 Using the systems thinking approach, draw a systems diagram for urban heat islands, including feedback mechanisms for increased energy demands for cooling and refrigeration, increased air pollution from these increased energy demands, and other effects such as global warming and public health.

4.27 The concentration of a pollutant along a quiescent water-containing tube is shown in Figure 4.28. The diffusion coefficient for this pollutant in water is equal to 10^{-5} cm²/s. (a) What is the initial pollutant flux density in the x-direction at the following locations: $x = 0.5, 1.5, 2.5, 3.5$, and 4.5? (b) If the diameter of the tube is 3 cm, what is the initial flux of pollutant mass in the x-direction at the same locations? (c) As time passes, this diffusive flux will change the shape of the concentration profile. Draw a sketch of

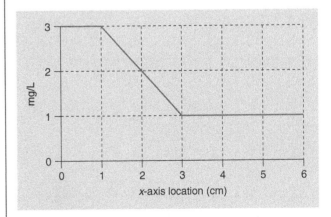

Figure / 4.28 Hypothetical Concentration Profile in a Closed Pipe.

(From Mihelcic (1999). Reprinted with permission of John Wiley & Sons, Inc.).

concentration in the tube versus x-axis—location showing what the shape at a later time might look like. (It is not necessary to do any calculations to draw this sketch.) Assume that the concentration at $x = 0$ is held at 3 mg/L and the concentration at $x = 6$ is held at 1 mg/L. (d) Describe, in one paragraph, why the concentration profile changed in the way that you sketched in your solution to part (c).

4.28 The tube in problem 4.27 is connected to a source of flowing water, and water is passed through the tube at a rate of 100 cm^3/s. If the pollutant concentration in the water is constant at 2 mg/L, find: (a) the mass flux density of the pollutant through the tube due to advection and (b) the total mass flux through the tube due to advection.

4.29 The following conditions exist downstream of the point where treated effluent from an advanced wastewater treatment facility has removed the phosphorus concentration to 1 mg P/L. The river characteristics just downstream of the discharge point are cross-sectional area equal to 20 m^2 and a volumetric flow rate of 17 m^3/s. Determine the average flux density of phosphorus downstream of the discharge point.

4.30 Calculate the settling velocity of a particle with 100 μm diameter and a specific gravity of 2.4 in 10°C water ($\mu = 1.308 \times 10^{-3}$ N-s/m^2 and the density of water equals 999.7 kg/m^3).

4.31 Calculate the settling velocity of a particle with 10 μm diameter and a specific gravity of 1.05 in 15°C water ($\mu = 1.140 \times 10^{-3}$ N-s/m^2 and the density of water equals 999.1 kg/m^3).

4.32 One type of pathogen commonly found in the developing world are helminths (that is, parasitic intestinal worms). These worm species are generally transmitted in a soil (or biosolids) environment, directly from one human host to another. The eggs of helminths develop into their infective state in a soil environment. (a) Determine the settling velocity for *Ascaris lumbricoides*, in a wastewater stabilization lagoon which have a diameter of 50 μm, density of 1.11 g/cm^3, and assumed spherical shape. (b) Determine the settling velocity for hookworm eggs which have a diameter of 60 μm, density of 1.055 g/cm^3, and assumed spherical shape. Assume the wastewater is 15°C ($\mu = 1.140 \times 10^{-3}$ N-s/m^2 and the density of water equals 999.1 kg/m^3).

References

Cambridge Systematics, 2005. Cool Pavement Report. Draft report prepared for Heat Island Reduction Initiative, U.S. Environmental Protection Agency.

Fischer, H. B., E. J. List, J. Imberger, and N. H. Brooks, 1979. *Mixing in Inland and Coastal Waters*. New York: Academic Press.

Hemond, H. F., and E. J. Fechner, 1994. *Chemical Fate and Transport in the Environment*. San Diego: Academic Press.

Intergovernmental Panel on Climate Change (IPCC), 2007a. Summary for policymakers. In *Climate Change 2007: The Physical Science Basis*. Contribution of Working Group I to the Fourth Assessment Report of the Intergovernmental Panel on Climate Change, S. Solomon, D. Qin, M. Manning, Z. Chen, M. Marquis, K. B. Averyt, M. Tignor, and H. L. Miller, Eds. New York: Cambridge University Press.

Intergovernmental Panel on Climate Change (IPCC), 2007b. Summary for policymakers. In: *Climate Change 2007: Impacts, Adaptation and Vulnerability. Contribution of Working Group II to the Fourth Assessment Report of the Intergovernmental Panel on Climate Change*, M. L. Parry, O. F. Canziani, J. P. Palutikof, P. J. van der Linden and C. E. Hanson, Eds. Cambridge, UK: Cambridge University Press, 7–22.

Mihelcic, J. R., 1999. *Fundamentals of Environmental Engineering*. New York: John Wiley & Sons, Inc.

Mihelcic, J. R., J. B. Zimmerman, and A. Ramaswami, 2007. Integrating developed and developing world knowledge into global discussions and strategies for sustainability, part 1: Science and technology. *Environmental Science and Technology*, *41*(10): 3415–3421.

Mills, G., 2004. The Urban Canopy Layer Heat Island. IAUC Teaching Resources, compiled for the International Association for Urban Climate Teaching Resource Committee, www.urban-climate.org/UHI_Canopy.pdf, accessed October 30, 2007.

Oke, T. R., 1981. Canyon geometry and the nocturnal urban heat island: comparison of scale model and field observations. *International Journal of Climatology*, *1*: 237–254.

Oke, T. R., 1982. The energetic basis of the urban heat island. *Quarterly Journal of the Royal Meteorological Society*, *108*(455): 1–24.

Rosenfeld, A., J. Romm, H. Akbari, and A. Lloyd, 1997. Painting the town white—and green. *Technology Review*, February/March.

United States Environmental Protection Agency, 2007. Urban Heat Island Basic information. www.epa.gov/heatisland, last accessed August 31, 2007.

Wilson, A., 1979. *Thermal Storage Wall Design Manual*. Santa Fe: New Mexico Solar Energy Association.

Wilson, A., and J. Boehland, 2005. Small is beautiful: U.S. house size, resource use, and the environment. *Journal of Industrial Ecology*, *9*(1–2): 277–287.

chapter/Five Biology

Martin T. Auer, James R. Mihelcic,
Michael R. Penn,
and Julie Beth Zimmerman

*In this chapter, readers are intro-
duced to the fundamental biological
principles governing ecosystems,
with special attention to processes
that mediate the fate of chemical
substances in natural and engineered environments. The chapter
begins with a discussion of ecosystem structure and function, including a description of population
dynamics, that is, organism growth
and attendant demand on resources.
Ecological footprint and the IPAT
equation are discussed to explain
the relationship between resource
limitations and population and consumption. Production and consumption are then examined, leading to
consideration of ecosystem trophic
structure and energy flow. The chapter also introduces material flow in
ecosystems, focusing on key biogeochemical cycles (for example, oxygen,
carbon, nitrogen, sulfur, and phosphorus) and effects of human activity on
these flows. Finally, concepts relating
to human and ecosystem health are
explored, including biomagnification,
biodiversity, and ecosystem health.*

© Derek Dammann/iStockphoto

Chapter Contents

Learning Objectives

1. Describe the relationships among individual organisms, species,
and populations in ecosystem functions and structure.
2. Distinguish the exponential, logistic, and Monod models
for population growth over time.
3. Identify and use the appropriate model to calculate changes
in population over time.
4. Determine the carrying capacity of a population, and articulate how
carrying capacity is affected by environmental conditions.
5. Use the Monod growth limitation model to calculate yield, substrate
utilization, or biomass growth, and relate these terms to carrying
capacity.
6. Discuss how human population growth, consumption, technology, and
carrying capacity are related to the IPAT equation and ecological footprint.
7. Describe interconnections and energy/material transfer within a food
web or ecosystem.
8. Define the following terms: biochemical oxygen demand (BOD), 5-day
biochemical oxygen demand (BOD_5), ultimate biochemical oxygen
demand (BOD_U), carbonaceous biochemical oxygen demand (CBOD),
nitrogenous biochemical oxygen demand (NBOD), and theoretical
oxygen demand (ThOD).
9. Describe the approach for calculating ThOD and the laboratory
procedures for determining BOD.
10. Summarize the roles of photosynthesis and respiration in capturing
and efficiently transferring energy in ecosystems.
11. Describe the flow of oxygen, carbon, nitrogen, sulfur, and phosphorus
through ecosystems and the impact of human activities on these flows.
12. Demonstrate how biological processes are related to issues of energy
production and global carbon cycling.
13. Discuss the significance and application of bioaccumulation factors
(BAFs) and bioconcentration factors (BCFs).
14. Describe the benefits, threats, and indicators of biodiversity in relation
to society, the economy, and ecosystem health and function.

B iology is defined as the scientific study of life and living things, often taken to include their origin, diversity, structure, activities, and distribution.

Biology includes the study of biotic effects. **Biotic** effects—those produced by or involving organisms—are important in many phases of environmental engineering. This chapter's exploration of environmental biology will focus on those activities—the ways organisms are affected by and have an effect on the environment. These include: (1) effects on humans (for example, infectious disease); (2) impacts on the environment (for example, species introductions); (3) impacts by humans (for example, endangered species); (4) mediation of environmental transformation (for example, breakdown of toxic chemicals); and (5) utilization in the treatment of contaminated air, water, and soil.

5.1 Ecosystem Structure and Function

In Figure 5.1, the Earth is conceptualized as comprising "great spheres" of living and nonliving material. The *atmosphere* (air), *hydrosphere* (water), and *lithosphere* (soil) constitute the **abiotic**, or nonliving, component. The **biosphere** contains all of the living things on Earth. Any intersection of the biosphere with the nonliving spheres—living things and their attendant abiotic environment—constitutes an **ecosystem**. Examples include natural (lake, grassland, forest, and desert) and engineered (biological waste treatment) ecosystems (Figure 5.2). Taken together, all of the ecosystems of the world make up the *ecosphere*. **Ecology** is the

Figure / 5.1 Earth's Great Spheres of Living and Nonliving Material   The atmosphere, hydrosphere, and lithosphere are the nonliving components, and the biosphere contains all the living components. The ecosphere is the intersection of the abiotic spheres and the biotic component.

(Kupchella and Hyland, *Environmental Science*, 1st edn, © 1986. Reprinted with permission of Pearson Education, Inc., Upper Saddle River, NJ).

Atmosphere

Biosphere

Hydrosphere   Lithosphere

Lake	Grassland	Biological waste treatment

Physical–chemical environment: water, coupled to the atmosphere and lake sediments and influenced by the meteorology characteristic of a specific latitude and altitude.	Physical–chemical environment: soil, coupled to the atmosphere and soil-water reserves and influenced by the meteorology characteristic of a specific latitude and altitude.	Physical–chemical environment: wastewater, largely uninfluenced by the meteorology characteristic of a specific latitude and altitude.
Energy source: the sun	Energy source: the sun	Energy source: organic wastes (originally from the sun)
Primary production: algae, aquatic plants, and certain bacteria	Primary producers: grasses and flowers	Primary producers: none
Energy transfer: zooplankton, fish	Energy transfer: grasshoppers, ground squirrel, coyote	Energy transfer: bacteria, protozoans

Figure / 5.2 An Ecosystem: Plants, Animals, Microorganisms, and Their Physical–Chemical Environment.

study of structure and function of the ecosphere and its ecosystems: interactions between living things and their abiotic environment.

Although the field of taxonomy (classification of organisms) is highly dynamic and home to vigorous debate, biologists today place living things within one of three domains: (1) the Archaea, (2) the Bacteria, and (3) the Eukarya. The Archaea and the Bacteria are **prokaryotes**, meaning that their cellular contents, such as pigments and nuclear material, are not segregated within cellular structures (for example, chloroplasts and the nucleus). While members of the Archaea and the Bacteria are similar in physical appearance, they differ in several important ways, including cellular composition and genetic structure. In our functional treatment of organisms, we consider the term *bacteria* to include members of both the domain Archaea and the domain Bacteria. The third domain, Eukarya, consists of organisms with segregated or compartmentalized organization—**eukaryotes**—possessing a nucleus and organelles such as chloroplasts. The domain Eukarya may be further divided into four kingdoms: (1) Protista (protists), (2) Fungi, (3) Plantae (plants), and (4) Animalia (animals).

Feeding strategies, of importance in many environmental engineering applications, also differ among the various kingdoms and domains. Some organisms obtain their food by **absorption** (uptake of dissolved nutrients, as in kingdom Fungi), some through **photosynthesis** (fixation of light energy into simple organic molecules, as in kingdom Plantae), and some by *ingestion* (intake of particulate nutrients, as in kingdom Animalia). Some members of the kingdoms Plantae and Protista combine phototrophy and heterotrophy in mixotrophy, a practice where nutrition comes from photosynthesis and the uptake of dissolved and/or particulate organic carbon.

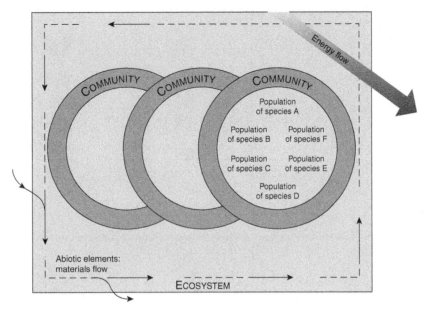

Figure / 5.3 **Biotic Component of an Ecosystem, Organized According to Species, Populations, and Communities**   In this schematic, energy flows through and chemicals cycle largely within the ecosystem. Natural, engineered, and industrial environments may be considered as ecosystems. For example, various biological processes employed for wastewater treatment and resource recovery (activated sludge, wetland, lagoon) have communities composed of a variety of microorganism populations. The nature of the ecosystem is determined by the physical design of the unit processes and by the chemical and biological character of the wastewater entering the system.

(From Mihelcic (1999). Reprinted with permission of John Wiley & Sons, Inc.).

Domains may be subdivided into kingdoms, phyla, classes, orders, families, genera, and species. A **species** is a group of individuals that possesses a common gene pool and that can successfully interbreed. Each species is assigned a scientific name (genus plus species) in Latin, to avoid confusion associated with common names. Under this system of binomial nomenclature, *Sander vitreus* is the scientific name for the fish species commonly referred to as walleye, walleye pike, pike, pike perch, pickerel, yellow pike, yellow pickerel, yellow pike perch, or yellow walleye.

All of the members of a species in a given area make up a **population**— for example, the walleye population of a lake. All of the populations (of different species) that interact in a given system make up the **community**—for example, the fish community of a lake. Finally, as shown in Figure 5.3, all of the communities plus the abiotic factors make up the ecosystem (here, a lake) and the ecosystems, the ecosphere.

5.1.1   MAJOR ORGANISM GROUPS

A wide variety of organisms are encountered in **natural systems** (for example, lakes and rivers, wetlands, and soil) and **engineered systems** (for example, wastewater treatment and resource recovery plants, landfills, constructed wetlands, and bioretention cells). Features of major organism groups especially important in environmental engineering are illustrated in Figure 5.4. More than half of the endangered species

Learn More on Ecosystems

http://www.epa.gov/research/ecoscience/

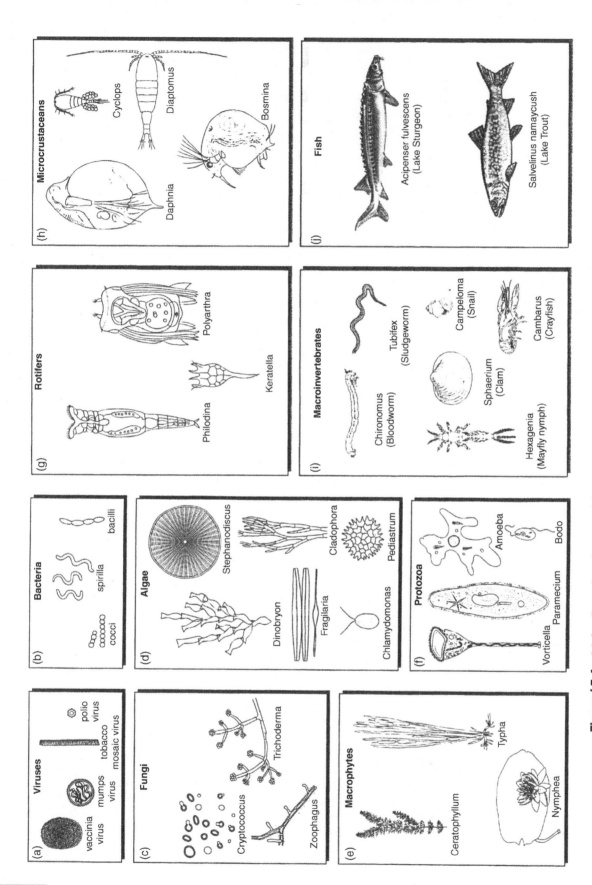

Figure / 5.4 Major Organism Groups with Representative Members The groups are: (a) viruses, (b) bacteria, (c) fungi, (d) algae, (e) macrophytes, (f) protozoa, (g) rotifers, (h) microcrustaceans, (i) macroinvertebrates, and (j) fish.

(From Mihelcic [1999]. Reprinted with permission of John Wiley & Sons, Inc.).

in the United States are higher plants. The latest report from the **International Panel on Climate Change (IPCC)** indicates, with very high confidence, that recent warming is strongly affecting terrestrial biological systems, including poleward and upward shifts in species ranges, earlier spring greening, and greater potential for disturbances from pests and fire (IPCC, 2007a). At the same time, higher plants significantly contribute to the removal of greenhouse gases from the atmosphere, removing as much 300×10^{12} kg of carbon dioxide from the atmosphere every year—the equivalent of CO_2 emissions from approximately 40 billion automobiles.

As we move from consideration of individual organisms to populations and communities of organisms, we should not lose sight of the attributes of the species that make their roles in the ecosystem special. Furthermore, interaction among organism groups results in highly dynamic communities in both natural and engineered systems. Seasonal cycles and natural and human perturbations of the environment

Climate Change and Ecosystems

http://www.epa.gov/climatechange/effects/eco.html

Application / 5.1   Producing Biodiesel from Algae

Biodiesel is a renewable fuel that can be manufactured from vegetable oils, animal fats, waste restaurant grease, and algae. While a number of feedstocks are currently being explored for biodiesel production, algae have emerged as one of the most promising sources. Algae are fast growing (with doubling times for many species on the order of hours) and can thrive in virtually any climate, overcoming a limitation of producing biodiesel from crops. The yield of oil from algae (some species have a 50 percent oil content) is much higher than those from traditional oilseeds—

potentially producing 250 times the amount of oil per acre as soybeans. Algae grown in photo-bioreactors like those shown in Figure 5.5 are harvested and pressed. Oil collected by pressing can then be converted to biodiesel through traditional transesterification reactions—the same ones used for vegetable oil. Overall, algae are remarkable and efficient biological factories capable of taking waste carbon dioxide, such as from nutrient laden domestic wastewater or power plant emissions, and converting it into a high-density liquid form of energy: natural oil.

Figure / 5.5 Conceptual Rendering of Large-Scale Photo-Bioreactors Used to Grow Algae   Provided with sunlight, water, carbon dioxide, and some added nutrients, algae can produce about 10,000 gallons of biodiesel per acre per year.

often lead to dramatic shifts in population size and community structure. For example, the transparency or clarity of lakes varies with the quantity of soil particles and fertilizers delivered from terrestrial sources by tributary streams. The abundance of algae may, in turn, fluctuate with the size of the microcrustacean populations that graze on them and with the availability of nutrients introduced from the watershed. In some lakes, water clarity can go from "crystal clear" to "pea soup" to "crystal clear" over a matter of days as algal and microcrustacean populations wax and wane.

5.2 Population Dynamics

Population dynamics play a role in the fate of fecal bacteria discharged to surface waters, the efficiency of microorganisms in biological treatment, and substrate–organism interactions in the cleanup of contaminated soils. Other applications include control of nuisance algae growth in lakes, biomanipulation as a management approach for surface water quality, and the transfer of toxic chemicals through the food chain. Our ability to manage and protect the environment can be enhanced through an understanding of **population dynamics**, for example, by simulating or modeling the response of populations to environmental stimuli. In studying population dynamics, it is important to remember that, like other organisms, humans represent a population, one that may grow exponentially and experience the stress of approaching its carrying capacity.

5.2.1 UNITS OF EXPRESSION FOR POPULATION SIZE

Although it is the individual that is born, the individual that reproduces, and the individual that dies, in an environmental context these events are best appreciated by examining entire populations. And while it is possible to characterize individual populations through direct enumeration (for example, the number of alligators), the populations making up natural or engineered ecosystems include organisms of widely differing sizes. Thus, a "head count" provides a poor representation of population size and function where an estimate of all living material or biomass is desired.

An alternative approach is to use a common constituent such as dry weight (g DW), organic carbon content (g C), or, for plants, chlorophyll content (g Chl). For example, we might report plant biomass as $g\,DW/m^2$ for grasslands, metric tons C/hectare for forests, and mg chl/m^3 for lakes. In the wastewater treatment and resource recovery process, microorganisms exist in a mixture with waste solids. Here, biomass is typically expressed as total suspended solids (TSS) or volatile suspended solids (VSS).

5.2.2 MODELS OF POPULATION GROWTH

A mass balance can be applied to the study of population dynamics in living organisms. Consider the case of the algal or bacterial community of a lake or river or the community of microorganisms in a reactor used for waste treatment and resource recovery. The mass balance on biomass in a batch reactor may be written as follows:

$$V\frac{dX}{dt} = QX_{in} - QX \pm \text{reaction} \tag{5.1}$$

V is volume (L), X is biomass (mg/L), t is time (days), Q is flow (L/day), and *reaction* refers to all the kinetic processes mediating the growth or death of the organisms. Each term in Equation 5.1 has units of mass per time (mg/day).

To simplify the conceptual development of the models that follow, the flow terms will be ignored here (thus, Q is equal to 0 in a batch reactor). Assuming that first-order kinetics adequately describe the reaction term (in this case, population growth), Equation 5.1 can be rewritten as follows:

$$V\frac{dX}{dt} = VkX \tag{5.2}$$

To simplify, dividing both sides of Equation 5.2 by V:

$$\frac{dX}{dt} = kX \tag{5.3}$$

where k is the first-order rate coefficient (time^{-1}). Because the reaction term is describing growth, the right side of Equation 5.3 is positive.

We will use this equation to develop realistic, but not overly complex, models to simulate the rates of organism growth in a batch reactor. Three models are introduced here, describing unlimited (exponential), space-limited (logistic), and resource-limited (Monod) growth.

EXPONENTIAL OR UNLIMITED GROWTH   The population dynamics of many organisms, from bacteria to humans, can be described using a simple expression, the **exponential-growth model**:

$$\frac{dX}{dt} = \mu_{max}X \tag{5.4}$$

Equation 5.4 is identical to Equation 5.3, with μ_{max}, the maximum specific growth rate coefficient (day^{-1}) being a special case of the first-order rate constant k. The coefficient μ_{max} describes the condition where a full complement of energy reserves may be directed to growth, unaffected by feedback from crowding or resource competition or limitations.

In addition to directing energy reserves toward growth, organisms must pay a "cost of doing business." Here, energy reserves mobilized through respiration are used to support cell maintenance and reproduction. In the terminology of wastewater engineering, this is *endogenous* (derived within) decay. The organism's respiratory demand may be represented as in Equation 5.4, using a first-order respiration or decay coefficient:

$$\frac{dX}{dt} = -k_d X \tag{5.5}$$

where k_d is the respiration rate coefficient (day^{-1}). Here, the right-side term is negative because it represents a loss of biomass. In some situations, the definition of k_d is expanded to include other losses, such as settling and predation.

Equations 5.4 and 5.5 may be combined for a batch reactor:

$$\frac{dX}{dt} = (\mu_{max} - k_d)X \tag{5.6}$$

and integrated to yield

$$X_t = X_0 e^{(\mu_{max} - k_d)t} \tag{5.7}$$

where X_t is the biomass at some time t and X_0 is the initial biomass, reported as numbers or as a surrogate concentration, such as mg DW/L. The term $(\mu_{max} - k_d)$ may also be thought of as the net effect of energy applied to growth minus the energy applied to respiration and termed μ_{net}:

$$X_t = X_0 e^{(\mu_{net}t)} \tag{5.8}$$

The expression utilized in Equation 5.7 will be retained here for clarity.

---

example/5.1   Exponential Growth and the Effect of the Specific Growth Rate on the Rate of Growth

Consider a population or community in a batch reactor with an initial biomass (X_0) of 2 mg DW/L, a maximum specific growth rate (μ_{max}) of 1.1/day, and a respiration rate coefficient of 0.1/day. Determine the biomass concentration (mg DW/L) over a time period of 10 days.

solution

Assume exponential growth. The biomass at any time is given by Equation 5.7:

$$X_t = X_0 e^{(\mu_{max} - k_d)t}$$

and

$$X_t = 2 \times e^{(1.1/day - 0.1/day)t}$$

Table 5.1 and Figure 5.6a present the results.

The J-shaped form of Figure 5.6a is typical of exponential growth. The steepness of the curve is determined by the value of the net specific growth rate coefficient ($\mu_{net} = \mu_{max} - k_d$). The influence of the value of μ_{net} on the shape of the growth curve is shown in Figure 5.6b.

Note the similarity between the exponential-growth models (Equation 5.4) applied here for living organisms in a batch reactor,

$$\frac{dX}{dt} = \mu_{max}X$$

Table / 5.1

Results of Calculations in Example 5.1 for Biomass as a Function of Time Using the Exponential-Growth Model

Time (days)	Biomass (mg DW/L)	Time (days)	Biomass (mg DW/L)
1	5	6	807
2	15	7	2,193
3	40	8	5,962
4	109	9	16,206
5	297	10	44,053

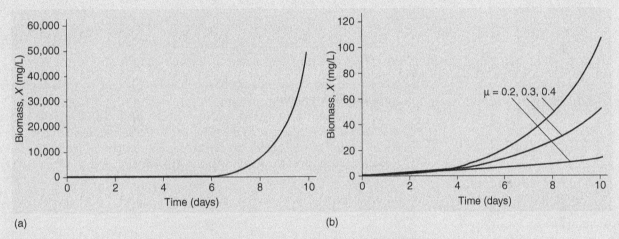

(a)   (b)

Figure / 5.6   **Effect of Specific Growth Rate on Exponential Growth**   (a) Exponential population growth as determined in Example 5.1. (b) Population growth according to the exponential model for three values of the specific growth rate coefficient. As μ increases, the rate of population growth (dX/dt) also increases.

(From Mihelcic (1999). Reprinted with permission of John Wiley & Sons, Inc.).

with the expression for first-order decay introduced in Chapters 3 and 4 for application to chemical losses in a batch reactor:

$$\frac{dC}{dt} = -kC$$

Both of these are first-order expressions; that is, both rates are a direct function of concentration (organism or chemical). However, organism concentrations typically increase exponentially (growth), while chemical concentrations decrease exponentially (decay).

LOGISTIC GROWTH: THE EFFECT OF CARRYING CAPACITY If we examine the predictions generated by the exponential-growth model a bit further along in time, we observe some interesting biomass levels. For example, in 100 days, the biomass simulated in Example 5.1 would reach 5.4×10^{43} mg DW/L! Does this make any sense? No wonder the exponential-growth model is sometimes called *unlimited growth*: there are no constraints or upper bounds on biomass.

The exponential-growth model has some appropriate applications, and we can learn much from this simple approach. However, the **logistic-growth model** provides a framework more in tune with our concept of how populations and communities behave. Here, we invoke a **carrying capacity**, or upper limit, to population or community size (biomass) imposed by environmental conditions. Figure 5.7 illustrates the concept of carrying capacity and identifies space limitation and population-dependent losses such as disease and predation as components of environmental conditions. Food limitation is not addressed, as the concept of carrying capacity is limited here to space-related or nonrenewable resources.

The logistic-growth model is developed for a batch reactor by modifying the exponential-growth model (Equation 5.6) to account for carrying-capacity effects:

$$\frac{dX}{dt} = (\mu_{max} - k_d)\left(1 - \frac{X}{K}\right)X \qquad (5.9)$$

where K is the carrying capacity (mg DW/L), that is, the maximum sustainable population biomass.

To appreciate the way in which carrying capacity mediates the rate of population growth, examine the behavior of the second term in parentheses in Equation 5.9. Note that when population size is small ($X \ll K$), Equation 5.9 reduces to the exponential-growth model, and as the carrying

Figure / 5.7 **Effect on Biomass of Limitation by Nonrenewable Resources as Manifested through Carrying Capacity** According to the logistic-growth model, environmental resistance (represented by the downward pressure of the hand) reduces the growth rate. At some time, the population reaches a carrying capacity that the population cannot exceed.

(Based on Enger et al., 2010; figure from Mihelcic (1999). Reprinted with permission of McGraw-Hill and John Wiley & Sons, Inc.).

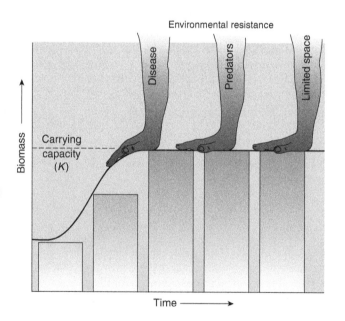

capacity is approached ($X \rightarrow K$), the population growth rate approaches zero. Equation 5.9 can be integrated for a batch reactor to yield:

$$X_t = \frac{K}{1 + \left[\left(\dfrac{K - X_0}{X_0}\right)e^{-(\mu_{max} - k_d)t}\right]} \qquad (5.10)$$

Equation 5.10 permits calculation of biomass as a function of time according to the logistic-growth model.

---

example/5.2 Logistic Growth

Consider the population from Example 5.1 ($X_0 = 2$ mg DW/L; $\mu_{max} = 1.1$/day; $k_d = 0.1$/day), but with a carrying capacity (K) of 5,000 mg DW/L. Determine the population biomass over a time period of 10 days.

solution

Use the carrying-capacity term, K, to apply the logistic-growth model. Use Equation 5.10 to solve for the biomass concentration over the 10-day period. Table 5.2 and Figure 5.8a show the population biomass over time. In this example, the specific growth rate begins to decrease after several days and approaches zero. Also, the biomass concentration levels off over time as the carrying capacity (in this case, 5,000 mg DW/L) is approached.

Figure 5.8b compares the exponential and logistic-growth models. Note that both models predict the same population behavior at low population numbers. This suggests that the exponential model may be appropriately applied under certain conditions.

Table / 5.2

Population over Time Determined for Logistic Growth Examined in Example 5.2

Time (days)	Biomass (mg DW/L)	$\mu_{max} - k_d$ (day^{-1})
0	2	1.000
1	7	0.999
2	22	0.996
3	72	0.986
4	232	0.954
5	695	0.861
6	1,745	0.651
7	3,201	0.360
8	4,276	0.145
9	4,757	0.049
10	4,924	0.015
11	4,977	0.005
12	4,993	0.001
13	4,998	<0.001
14	4,999	<0.001
15	5,000	0.000

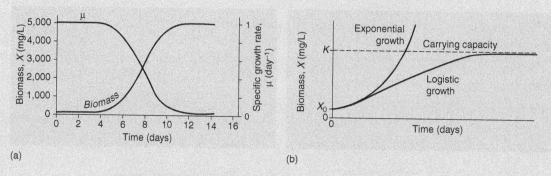

Figure / 5.8   **Application of the Logistic-Growth Model**   (a) Population biomass and specific growth rate according to the logistic model as determined in Example 5.2. (b) A comparison of the exponential and logistic-growth models. Both predict a similar population response in the early stages when population is small. However, the exponential model predicts that unlimited growth will continue, while the logistic model predicts an approach to the carrying capacity.

(From Mihelcic (1999). Reprinted with permission of John Wiley & Sons, Inc.).

RESOURCE-LIMITED GROWTH: THE MONOD MODEL   In nature, it is more common for organisms to reach the limits imposed by reserves of **renewable resources**—for example, food—than to approach the limits established by carrying capacity. The relationship between nutrients and the population or community growth rate can be described using the **Monod model**. In this model, the maximum specific growth rate is modified to account for the effects of limitation, in this case by renewable resources:

$$\mu = \mu_{max} \frac{S}{K_s + S}$$ (5.11)

where μ is the specific growth rate (day^{-1}), S is the nutrient or substrate concentration (mg S/L), and K_s is the half-saturation constant (mg S/L).

The **substrate** or "food" in Equation 5.11 may be either a macronutrient (for instance, organic carbon in biological waste treatment and resource recovery) or a growth-limiting micronutrient (such as nitrogen or phosphorus in an estuary or lake). As illustrated in Figure 5.9a, the **half-saturation constant** (K_s) is defined as the substrate

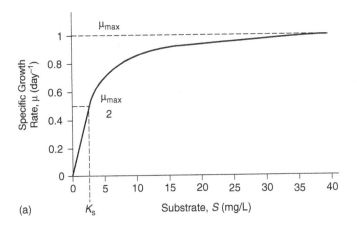

(a)

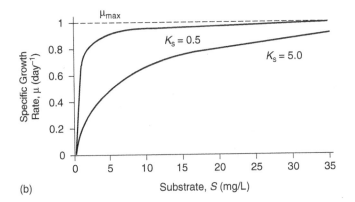

(b)

Figure / 5.9 The Monod Model (a) Basic Monod model, illustrating the relationship between the specific growth rate (μ) and substrate (S) concentration. At high substrate concentrations ($[S] \gg K_s$), μ approaches its maximum value μ_{max}, and the growth is essentially independent of substrate concentration (that is, zero-order kinetics). At low substrate concentrations ($[S] \ll K_s$), μ is directly proportional to substrate concentration (that is, first-order kinetics). (b) Application of the Monod model illustrating the effect of variation in K_s. Organisms with a low K_s approach their maximum specific growth rate at lower substrate concentrations and thus may have a competitive advantage. Note this figure is drawn for the situation where $\mu_{max} = 1$ /day.

(From Mihelcic (1999). Reprinted with permission of John Wiley & Sons, Inc.).

concentration (S) at which the growth rate is one-half of its maximum value, that is, $\mu = \mu_{max}/2$.

The magnitude of K_s reflects the ability of an organism to consume renewable resources (substrate) at different substrate levels. Organisms with a low K_s approach the maximum specific growth rate (μ_{max}) at comparatively low substrate concentrations, while those with high K_s values require higher levels of substrate to achieve the same level of growth. Figure 5.9b illustrates how variability in the half-saturation constant affects growth rate. The physiological basis for this phenomenon lies in the role of enzymes in catalyzing biochemical reactions; low half-saturation constants reflect a strong affinity of the enzyme for substrate.

The Monod model (Equation 5.11) can be substituted into Equation 5.6 (the exponential model) to yield

$$\frac{dX}{dt} = \left(\mu_{max} \frac{S}{K_s + S} - k_d \right) X$$

(5.12)

example/5.3 Resource-Limited Growth

Figure 5.10 shows population density as a function of time for two species of *Paramecium* (a protozoan) grown separately and in mixed batch culture. Grown separately, both species do well, acquiring substrate and achieving high biomass densities. In mixed culture, however, one species dominates, eliminating the other species. Organisms with a small K_s have a competitive advantage because they can reach a high growth rate at lower substrate levels. This can be demonstrated by inspection of the Monod model.

A basic concept of ecology, the **principle of competitive exclusion**, states that two organisms cannot coexist if they depend on the same growth-limiting resource. How then do these two species of *Paramecium* manage to coexist in the natural world? Why isn't the poor competitor extinct?

solution

The answer lies in another ecological principle, *niche separation*. The term *niche* refers to the unique functional role or "place" of an organism in the ecosystem. Organisms that are poorly competitive from a purely kinetic perspective (for example, K_s) can survive by exploiting a time or place where competition can be avoided.

Figure / 5.10 **Two Species of Paramecium Grown Separately and in Mixed Culture** In separate culture, both species do well, acquiring substrate and achieving high biomass densities. In mixed culture, however, one species dominates and eliminates the other species.

(Based on Ricklefs (1983); figure from Mihelcic (1999). Reprinted with permission of John Wiley & Sons, Inc.).

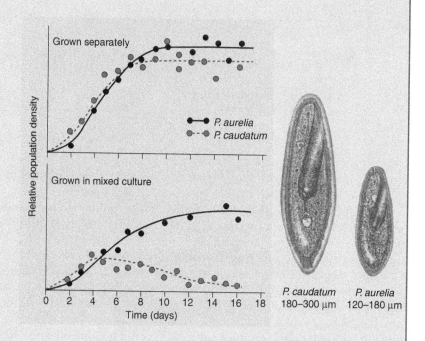

YIELD COEFFICIENT: RELATING GROWTH AND SUBSTRATE UTILIZATION   While attention has been largely devoted here to tracking biomass, substrate fate may be of more interest in many engineering applications. To model substrate concentrations, or to relate substrate consumption to organism growth, we apply the **yield coefficient (*Y*)**, defined as the quantity of organisms produced per unit substrate consumed:

$$Y = \frac{\Delta X}{\Delta S} \qquad (5.13)$$

Y has units of biomass produced per mass of substrate consumed. A yield coefficient value of $Y = 0.2$ indicates that 20 mg of biomass are produced for every 100 mg of substrate consumed. Note that Y for organic carbon is always less than 1, because organisms are not 100 percent efficient in converting substrate to biomass and because some energy must be expended for cell maintenance.

The yield coefficient is also commonly applied for a batch reactor to relate the rate of substrate utilization (dS/dt) to the rate of organism growth (dX/dt):

$$\frac{dS}{dt} = -\frac{1}{Y}\left(\frac{dX}{dt}\right) \qquad (5.14)$$

Substitute the Monod growth limitation model (Equation 5.12) for dX/dt in Equation 5.14:

$$\frac{dS}{dt} = -\frac{1}{Y}\left(\mu_{max}\frac{S}{K_s + S}\right)X \qquad (5.15)$$

We ignore the decay coefficient (k_d) in this expression because substrate utilization is only tied to how fast the organisms grow and not how they die off. This expression is used in a variety of engineering applications, for example, to develop the mass balances on organism growth and substrate utilization that support the design and operation of a wastewater treatment and resource recovery facility.

example/5.4   Yield Coefficient

The organic matter present in domestic wastewater is removed at a rate of 25 mg BOD_5/L-h in a batch aerated biological reactor. BOD (biochemical oxygen demand), defined in Section 5.4, refers to the amount of oxygen consumed in oxidizing a given amount of organic matter, here a representation by effect of substrate concentration.

Use the yield coefficient to compute the mass of microorganisms (measured as VSS) produced daily due to the consumption of organic matter by microorganisms in the aeration basin. Assume that the biological reactor has a volume of 1.5×10^6 L and the yield coefficient Y equals 0.6 mg VSS/mg BOD_5.

solution

The yield coefficient Y relates the rate of substrate (in this case, organic matter) disappearance to the rate of cell growth. This relationship (Equation 5.14) is written in a batch reactor as follows:

$$\frac{dS}{dt} = -\frac{1}{Y}\frac{dX}{dt}$$

Therefore,

$$Y\frac{dS}{dt} = -\frac{dX}{dt}$$

Substitute the given values for Y and the rate of substrate depletion:

$$\frac{0.6 \text{ mg VSS}}{\text{mg BOD}_5} \times \frac{25 \text{ mg BOD}_5}{\text{L-h}} = \frac{15 \text{ mg VSS}}{\text{L-h}}$$

Next, convert this value to a mass per day basic:

$$\frac{15 \text{ mg VSS}}{\text{L-h}} \times 1.5 \times 10^6 \text{ L} \times \frac{24 \text{ h}}{\text{day}} = \frac{5.4 \times 10^8 \text{ mg VSS}}{\text{day}}$$
$$= \frac{540 \text{ kg VSS}}{\text{day}}$$

Note that a lot of biological solids are produced at a wastewater treatment and resource recovery plant each day. This explains why engineers spend so much time designing and operating facilities to handle and dispose of the residual biosolids (sludge) generated at a wastewater treatment and resource recovery plant.

BIOKINETIC COEFFICIENTS   The terms μ_{max}, K_s, Y, and k_d are commonly referred to as biokinetic coefficients because they provide information about the manner in which substrate and biomass change over time (kinetically). Values for these coefficients may be derived from thermodynamic calculations or through field observation and laboratory experimentation; literature compilations of coefficients derived in this fashion are available. Table 5.3 provides some representative values for the biokinetic coefficients as applied in municipal wastewater treatment and resource recovery.

Table / 5.3

Typical Values for Selected Biokinetic Coefficients for the Activated-Sludge Wastewater Treatment Process

Coefficient	Range of Values	Typical Value
μ_{max}	$0.1-0.5 \text{ h}^{-1}$	0.12 h^{-1}
K_s	$25-100 \text{ mg BOD}_5/\text{L}$	$60 \text{ mg BOD}_5/\text{L}$
Y	$0.4-0.8 \text{ VSS/mg BOD}_5$	0.6 VSS/mg BOD_5
k_d	$0.0020-0.0030 \text{ h}^{-1}$	0.0025 h^{-1}

SOURCE: Tchobanoglous et al., 2003.

BATCH GROWTH: PUTTING IT ALL TOGETHER Respiration and the growth-mediating mechanisms introduced earlier may be integrated into a single expression describing population growth in batch culture:

$$\frac{dX}{dt} = \left(\mu_{max}\frac{S}{K_s + S} - k_d\right)\left(1 - \frac{X}{K}\right)X \qquad (5.16)$$

Then we can relate substrate utilization to Equation 5.16 through the yield coefficient:

$$\frac{dS}{dt} = -\frac{1}{Y}\left(\mu_{max}\frac{S}{K_s + S}\right)\left(1 - \frac{X}{K}\right)X \qquad (5.17)$$

Again, we ignore the decay coefficient (k_d) in this expression because substrate utilization is only tied to how fast the organisms grow and not how they die off. Although of considerable importance in natural systems, the carrying-capacity term is not typically included in bio-kinetic models for municipal wastewater engineering, because these systems are designed to operate below their maximum sustainable biomass.

Figure 5.11 illustrates substrate utilization and the attendant phases of population growth in batch culture (no inflow or outflow), according to Equations 5.16 and 5.17. For simplicity, it is assumed that no substrate recycle occurs. Three phases of growth are described in Table 5.4: the exponential or log growth phase, the **stationary phase**, and the death phase. Certain simplifying assumptions regarding growth conditions during the exponential and death phases permit the calculation of substrate and biomass changes at those times.

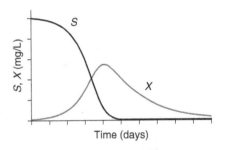

Figure / 5.11 Population Growth in Batch Culture This graph illustrates an exponential-growth phase where initially substrate (S) is abundant and biomass (X) is low. This is followed by a stationary phase where substrate levels support a growth rate equal to the respiration rate and then a death phase where substrate is exhausted and the population is in decline due to unsupported respiratory demand.

Table / 5.4

Phases of Population Growth in a Batch Reactor

Phase of Growth	Description	dX/dt
Exponential or log growth phase	Substrate uptake and growth are rapid. Growth pattern is well approximated by the exponential model (Equation 5.4).	$dX/dt > 0$
Stationary phase	Growth slows due to substrate depletion (or crowding). For a brief period, gains through growth are exactly balanced by losses to respiration and death.	$dX/dt = 0$
Death phase	Substrate is no longer available to support growth and losses to respiration and death. Approximated as an exponential decay (Equation 5.5).	$dX/dt < 0$

example/5.5 Simplified Calculations of Substrate and Biomass

The differential equations that describe biomass and substrate dynamics (Equations 5.16 and 5.17) contain nonlinear terms that require specialized numerical methods for their solution. By applying certain simplifying assumptions, however, we can learn quite a bit about the dynamics of microbial populations and communities. Consider a population of microorganisms with the following characteristics growing in a batch culture: initial biomass $X_0 = 10$ mg DW/L; maximum specific growth rate $\mu_{max} = 0.3$/day; half-saturation constant $K_s = 1$ mg /L; carrying capacity $K = 100,000$ mg DW/L; respiration rate coefficient $k_d = 0.05$/day; initial substrate concentration $S_0 = 2,000$ mg S/L; and yield coefficient $Y = 0.1$ mg DW/mg S.

1. Determine whether this population will ever approach its carrying capacity.

2. Calculate the population biomass after the first 3 days of growth.

3. Calculate the substrate concentration after the first 3 days of growth.

4. If the population peaks at 100 mg DW/L when the substrate runs out, calculate the biomass 10 days after the peak.

solution

1. Changes in substrate and biomass concentrations over time are related by the yield coefficient as given by Equation 5.14. The maximum attainable biomass of this population, based on substrate availability, is given as the product of the maximum potential change in substrate concentration and the yield coefficient:

$$dX = dS \times Y = \frac{2,000 \text{ mg } S}{L} \times \frac{0.1 \text{ mg DW}}{\text{mg } S} = 200 \text{ mg DW}$$

This is well below the carrying capacity of 100,000 mg DW/L; therefore, the population will not run out of substrate and never approach the carrying capacity.

2. Early in the growth phase when substrate concentrations are high (Monod term, $S/(K_s + S)$, approaches 1) and biomass concentrations are low (carrying-capacity term, $1 - X/K$, approaches 1), Equation 5.16 reduces to

$$\frac{dX}{dt} = (\mu_{max} - k_d)X$$

Integrate:

$$X_t = X_0 e^{(\mu_{max} - k_d)t}$$

$$X_3 = \frac{10 \text{ mg DW}}{L} \times e^{(0.3/\text{day} - 0.05/\text{day})3 \text{ days}} = \frac{21 \text{ mg DW}}{L}$$

3. The change in substrate concentration over the 3-day period is given by Equation 5.14:

$$\frac{dS}{dt} = -\frac{1}{Y}\left(\frac{dX}{dt}\right)$$

From the previous calculation, dX/dt over 3 days is $X_3 - X_0 = 21 - 10 = 11$ mg DW/L, and

$$\frac{dS}{dt} = -\left(\frac{0.1 \text{ mg DW}}{\text{mg S}}\right)^{-1} \times \frac{11 \text{ mg DW}}{L} = \frac{-110 \text{ mg S}}{L}$$

The substrate concentration after 3 days of growth is given by

$$S_3 = S_0 - \frac{dS}{dt} = 2{,}000 - 110 = \frac{1{,}890 \text{ mg S}}{L}$$

4. When substrate is exhausted, the Monod term equals 0, and Equation 5.16 reduces to Equation 5.5 and its analytical solution, $X_t = X_0 e^{-k_d t}$. In this case, the peak population decays according to first-order kinetics, so that 10 days after the peak,

$$X_t = X_0 e^{-k_d t} = 100 \times e^{(-0.05/\text{day} \times 10 \text{ days})} = \frac{61 \text{ mg DW}}{L}$$

GROWTH MODELS AND HUMAN POPULATION   Despite the complexity of their reproduction, populations of humans can be simulated using the types of models described in this chapter. Human populations remained relatively unchanged for thousands of years, increasing much more rapidly in modern times (see Figure 5.12). In 2007, the average growth rate coefficient for the world's population was about 0.012 year^{-1} (1.2 percent per year, or 12 births per 1,000 people per year).

Class Discussion
Having recently completed a study of population growth, we might ask ourselves which model best fits the data presented in Figure 5.12 and what that model suggests is likely to happen next.

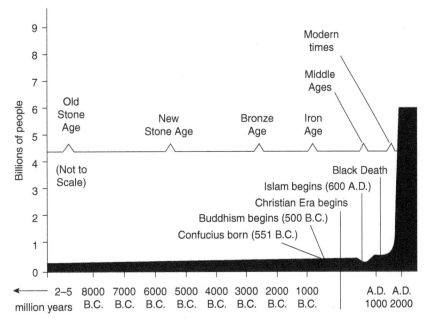

Figure / 5.12   Human Population Growth over Time.

(Adapted from World Population Growth through History with permission from Population Reference Bureau, Washington, D.C.).

Class Discussion

Engineers have a role to play in mitigating current threats and eliminating future insults with designs that improve the quality of life for Earth's population without the associated historical adverse human health and environmental impacts. How do you see your role as a global citizen and as an engineering professional?

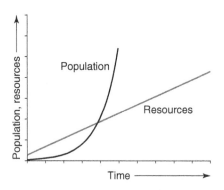

Figure / 5.13 **Malthus's Predictions of Population and Resources Trends** At some point, according to Malthus, demand will outstrip supply.

At this rate, Earth's population will double every 60 years, with most of this population growth occurring in urban areas.

The 18th-century British economist Thomas Malthus considered the questions of what model describes population data and what it predicts. Recognizing that population increased exponentially, Malthus concluded that such growth was checked only by "misery or vice," meaning war, pestilence, and famine. He asserted that while populations increased exponentially, the "means of subsistence" (food) increased in a linear fashion. Therefore, it would be only a matter of time until demand outstripped supply, as shown in Figure 5.13, an event with catastrophic implications.

Two centuries following Malthus's predictions, we have not expended food resources, and the demand and supply curves shown in Figure 5.13 have not yet intersected. This is primarily from the anthropogenic production of nitrogen (N) and phosphorus (P) that have provided numerous social and economic benefits, primarily through meeting demands for food production as population and affluence have increased. Unfortunately, the disturbance of the Earth's equilibrium from excessive inputs of N and P to the environment has resulted in numerous human and ecosystem health concerns. These concerns include freshwater N and P imbalances, greenhouse gas emissions (that is, nitrous oxide), and acidification and eutrophication of terrestrial and aquatic ecosystems (EPA's Office of Water (EPA, 2007b)). In addition, for the two major ingredients in fertilizer, while we require energy to obtain N from the air, we must mine P. And while there is abundant N in the Earth's atmosphere, the amount of readily available P that is mined is expected to run out in this century (Mihelcic et al., 2011).

The Limits to Growth was discussed in Chapter 1. Remember that it warned of the finite limitations of the world's resources. In ***The Limits to Growth***, the Club of Rome warned:

> If "present growth trends in world population, industrialization, pollution, food production, and resource depletion continue unchanged, the limits to growth will be reached sometime within the next one hundred years." They further predicted there would be an adverse impact on industrial capacity.

This somewhat broader perspective on the threats that prescribe limits to growth is more consistent with our observations of the impacts of nutrient (i.e., nitrogen and phosphorus) runoff, acid rain, release of heavy metals and toxic organic chemicals, depletion of the ozone layer, and carbon emissions that lead to climate change. Malthus might have been surprised to see that we would "soil our nest" well before famine, the "last and most dreadful check to population," spread across our planet.

Another way to describe this phenomenon is through a relationship developed in the 1970s and known as the **IPAT equation**:

$$I = P \times A \times T$$

(5.18)

In Equation 5.18, I is the environmental impact, P is population, A is affluence, and T is technology. Environmental impact (I) may be expressed in terms of resource depletion, waste accumulation, or global warming potential. Population (P) refers to the size of the human population, affluence (A) to the level of consumption by that population, and technology (T) to the processes used to obtain resources and transform them into useful goods and wastes.

In addition to highlighting the contribution of population to environmental problems, the IPAT equation makes it clear that environmental problems involve more than pollution and are driven by multiple factors acting together to yield a compounding effect. The product of the affluence (A) and technology (T) terms in Equation 5.18 can be visualized as the per capita demand on ecosystem resources. This demand is sometimes quantified as an ecological footprint (see Application 5.2).

Class Discussion

While ecological footprint does provide general information about the impacts of consumption, what are some of the limitations to this approach? For example, is land area a good surrogate for all environmental impact? Is land contaminated during a process or at end of life considered differently than land used for growing organic crops? What challenges does this pose in using ecological footprint as an indicator of environmental sustainability?

Calculate Your Personal Footprint
www.myfootprint.org/en/

Application / 5.2   The Ecological Footprint

An **ecological footprint** is a determination of the biologically productive land area required to provide an individual's (or country's or city's) resource supplies and absorb the wastes their activities produce.

Another way to think of ecological footprint is as the ecological impact corresponding to the amount of nature an individual (or country or city) needs to occupy to keep intact his or her daily lifestyle (Wackernagel et al., 1997). It is assumed that, if the world's resources were allocated equally among the world's population, there would be 1.8 ha of productive land allocated per person. This value is important because it serves as a benchmark for comparing the ecological footprint of the world's population. This benchmark number is derived from the fact that there is an amount of determined arable land, pasture, forest, ocean, and built-up environment.

Figure 5.14 shows that since 1970, the demand by humans on the natural world has exceeded what the Earth can replenish. This excess demand for Earth's biological capacity is called an **ecological overshoot**, which has grown steadily over the past 40 years. The overshoot reached a 50 percent deficit in 2008, meaning it now takes the Earth 1.5 years to regenerate the renewable resources that humans use and adsorb the waste (for example, carbon dioxide) they produce (WWF, 2012). This is similar to a person withdrawing funds from a bank account faster than the rate at which interest returns funds into the account.

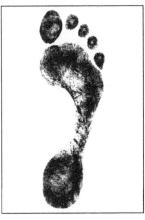
© Juri Samsonov/iStockphoto.

Table 5.5 gives the ecological footprint and available ecological capacity (both as hectares per capita) of the world and selected nations. The table also provides the difference, or *ecological deficit*, which is determined by subtracting the footprint from the available ecological capacity of an individual country. Negative numbers indicate a deficit (that is, ecological overshoot), and positive numbers indicate that some ecological capacity remains within the country's borders. The table shows that many of the world's nations are not currently sustainable if the rest of the world is to share their same level of current consumption of natural resources. This analysis assumes that the impact of an individual is felt within the physical land and ecosystems occupied by that person's home country. Of course, national borders do not bind many environmental emissions, so the impacts will most likely be felt beyond the home country.

On a global basis, the average footprint of the current world is 2.7 ha per person, resulting in an ecological deficit of 0.9 ha per person. It is clear that, to be ecologically sustainable, the world must either decrease its population or decrease the burden each person places on the environment (through more equitable sharing of the world's resources or wider use of green policies and technology), especially in developed countries that consume an unfair share of the world's resources.

Table / 5.5

Ecological Footprints Around the World

	Ecological Footprint (ha per capita)	Biological Capacity (ha per capita)	Difference (ha per capita)
World	2.7	1.8	−0.9
Bangladesh	0.7	0.4	−0.3
Brazil	2.9	9.6	+6.7
Canada	6.4	14.9	+8.5
China	2.1	0.9	−1.2
Germany	4.6	1.9	−2.7
India	0.9	0.5	−0.4
Japan	4.2	0.6	−3.6
Jordan	2.1	0.2	−1.9
Mexico	3.3	1.4	−1.9
New Zealand	4.3	10.2	+5.9
Nigeria	1.4	1.1	−0.3
Russian Federation	4.4	6.6	+2.2
South Africa	2.6	1.2	−1.4
United Kingdom	4.7	1.3	−3.4
United States	7.2	3.9	−3.3

SOURCE: Data from Living Planet Report, WWF, 2012.

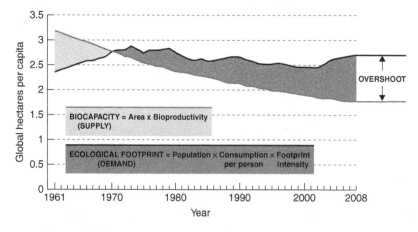

Figure / 5.14   Trends in Ecological Capacity and Biocapacity Showing the Increase in Ecological Overshoot over the Last 40 Years.

In fact, as the global population approaches 10 billion people in the upcoming decades, the footprint allocation of 2.1 ha of productive land per person will approach 1 ha, placing even more attention on issues of fairness and green policies and technologies.

In the United States, while the available ecological capacity is relatively large (3.9 ha per person), there is an ecological footprint of 7.2 ha, resulting in an overall deficit of 3.3 ha per person. China's available ecological capacity is much lower than that of the United States (0.9 ha per person), and its current footprint is much lower as well (2.1 ha per person).

Note also that China has no additional "footprint" of its own to utilize. Imagine what China's ecological overshoot will be if its population of more than 1 billion attempts to emulate the current resource consumption patterns of the United States (remember the IPAT equation). In addition, you might ask yourself whether it is ethically responsible for citizens of the United States and other developed countries to consume the world's resources at current nonsustainable rates with the result that less-developed countries will lack resources to support future development.

From Equation 5.18, it is apparent that only two options exist for reducing environmental impact:

1. Reduce population numbers (P) or
2. Reduce the magnitude of the per capita demand ($A \times T$).

The issues of population size (P), perhaps of paramount importance, and of affluence (A) are policy-oriented issues many engineers work on. Other efforts of the engineer in reducing environmental impact and advancing sustainability focus on (T), the design of greener less resource intensive technologies and use of a systems approach and life cycle thinking.

5.3   Energy Flow in Ecosystems

The character of Earth's many and varied ecosystems is determined to a large extent by their physical setting. Consider the changes in flora (plants) and fauna (animals) observed over the course of a long car trip, especially if traveling north–south or through dramatic changes in elevation. The physical setting includes climatic factors such as temperature (extreme values and duration of seasons), sunlight (day length and annual variation), precipitation (extremes and annual distribution), and wind. Other significant features of the physical setting include soil physics (particle size) and chemistry (pH, organic content, nutrients).

Given an appropriate physical setting, organisms require only two things from the environment: (1) energy to provide power and (2) chemicals to provide substance. Chemical elements are cycled within an ecosystem that could be regional or global, so continued function does not require that they be imported. Energy flows through and propels ecosystems; that is, it does not cycle but rather is converted to heat and lost for useful purposes forever.

5.3.1   ENERGY CAPTURE AND USE: PHOTOSYNTHESIS AND RESPIRATION

The sun is responsible, directly or indirectly, for virtually all of Earth's energy. Sunlight incident on an aquatic or terrestrial ecosystem is

Class Discussion

The IPAT equation also addresses the issue of how equity and sharing are critical factors in sustainable development. For example, as the 2 billion people currently living in the world who live in poverty (less than $2 per day) increase their affluence through increased personal income, other wealthy parts of the world will either have to share some of their affluence or provide access to green technology to those parts of the world to lessen the overall environmental impact of increases in population and affluence. What do you see as the global answer to this complex problem?

Class Discussion

Is an ecological footprint the same as a carbon footprint? (The answer is no). Carbon footprints are part of the ecological footprint. However, an ecological footprint accounts for the demand that carbon emissions have on biological capacity in terms of how many hectares are required to sequester the carbon emissions. One question to consider. Will a shift from consumption of fossil fuels to biomass fuels decrease or increase the overall demand on the world's biological capacity?

© Michal Krakowiak/iStockphoto.

trapped by plant pigments, primarily chlorophyll, and that light energy is converted to chemical energy through a process termed **photosynthesis**.

Artificial photosynthesis is the term often used to describe engineered solar or photovoltaic cell systems designed to capture light energy and convert it into electrical energy. The use of solar energy to power engineered systems rather than the burning of fossil fuels can address many of the current environmental challenges including air pollution, climate change, and depletion of finite resources.

Chemical energy stored through photosynthesis is subsequently made available for use by organisms through respiration. Figure 5.15a provides a simplified representation of photosynthesis, represented as follows:

$$CO_2 + H_2O + \Delta \rightarrow C(H_2O) + O_2 \tag{5.19}$$

where Δ is the sun's energy and $C(H_2O)$ is a general representation of organic carbon (for example, glucose, which is $C_6H_{12}O_6$ or $6\,C(H_2O)$). The free-energy change (ΔG) for photosynthesis (Equation 5.19) is positive, so the reaction could not proceed without the input of energy from the sun.

Chlorophyll acts as an antenna, absorbing the light energy, which is then stored in the chemical bonds of the carbohydrates produced by this reaction. Oxygen is an important by-product of the process. In relation to wastewater treatment, this **photosynthetic source** is considered to be a natural method for providing oxygen to wastewater, as happens in the aeration of lagoon-based treatment systems.

Respiration is the process by which the chemical energy stored through photosynthesis is ultimately released to do work in plants and other organisms (from bacteria to plants and animals):

$$C(H_2O) + O_2 \rightarrow CO_2 + H_2O + \Delta \tag{5.20}$$

Figure / 5.15 **Photosynthesis and Respiration** (a) Simplified version of photosynthesis, the process in which the sun's energy is captured by pigments such as chlorophyll and converted to chemical energy stored in the bonds of simple carbohydrates, for example, $C(H_2O)$. More complex molecules (for example, sugars, starches, cellulose) are then formed from simple carbohydrates. (b) Simplified version of respiration, the reverse of the photosynthetic process. The energy stored in chemical bonds (for example, carbohydrates) is released to support metabolic needs.

(Reprinted from Mihelcic (1999); with permission of John Wiley & Sons, Inc.).

$$C_2O \quad + \quad H_2O \xrightarrow[\text{Chlorophyll}]{\text{Light energy } (\Delta)} C(H_2O) \quad + \quad O_2$$

(a) Photosynthesis

$$O_2 \quad + \quad C(H_2O) \longrightarrow CO_2 + H_2O \quad + \text{ energy } (\Delta)$$

(b) Respiration

Figure 5.15b provides a simplified representation of respiration. The reverse of photosynthesis, this reaction releases stored energy, making it available for cell maintenance, reproduction, and growth. The energy, denoted Δ in Equation 5.20, is equal to the free energy of reaction. Organisms are able to capture and utilize only a fraction (5–50 percent) of the total free energy of this reaction. Thus, all forms of life are by nature rather inefficient.

Respiration is what may be described chemically as an oxidation–reduction or **redox reaction**, which can be written in terms of the following two half-reactions. First, the oxidation of the organic carbon:

$$C(H_2O) + H_2O \rightarrow CO_2 + 4H^+ + 4e^- \tag{5.21}$$

where the valence state of carbon goes from (0) in $C(H_2O)$ to (4+) in CO_2, yielding four electrons. And second, the reduction of oxygen:

$$O_2 + 4e^- + 4H^+ \rightarrow 2H_2O \tag{5.22}$$

where the valence state of oxygen goes from (0) in O_2 to (2–) in H_2O, gaining four electrons. The two half-reactions can be added to yield the overall reaction presented in Equation 5.20. Note that there is no net change in electrons; they simply are redistributed.

Microbial ecologists refer to the respiration described in Equation 5.20 as **aerobic respiration**, because oxygen is utilized as the electron acceptor. Some bacteria, termed obligate or *strict aerobes* (or simply aerobes), rely exclusively on oxygen as an electron acceptor and cannot grow in its absence. At the opposite extreme are microbes that cannot tolerate oxygen, termed obligate or *strict anaerobes*. Facultative microbes can switch their metabolism between aerobic and anaerobic pathways, depending on the presence or absence of oxygen.

When oxygen is absent, **anaerobic respiration** takes place, utilizing a variety of other compounds as electron acceptors. Many bacteria can utilize oxygen as an electron acceptor but in its absence may utilize either nitrate or sulfate. Such bacteria are *facultative aerobes* and have a distinct ecological advantage over strict or obligate anaerobes or aerobes in environments that may be periodically devoid of oxygen. The terms *anaerobic* and *anoxic* are often used synonymously. In wastewater treatment and some natural-systems applications, **anoxic** refers to the case where oxygen is absent and respiration proceeds with nitrate as the electron acceptor.

Table 5.6 presents redox reactions for oxidation of organic matter using a variety of alternate electron acceptors, such as nitrate, manganese, and ferric iron. In the environment, these reactions are thought to take place in the sequence listed—the order of their favorability from a thermodynamic perspective. Thus, reduction of oxygen proceeds first, followed by nitrate, manganese, ferric iron, and sulfate, and finally fermentation occurs. This order is termed the **ecological redox sequence**, with each process carried out by different types of bacteria (for example, nitrate reducers and sulfate reducers). Note that in all of these redox reactions, CO_2 is produced from the degradation of organic manner. These *biogenic* CO_2 emissions are distinguished as a different greenhouse gas emission (but are still important) from the CO_2 emitted from burning of fossil fuels.

Managing Ecosystems to Fight Poverty
http://pdf.wri.org/wrro5_full_hires.pdf

Table / 5.6

Redox Reactions for Oxidation of Organic Matter Using Various Alternate Electron Acceptors

Electron Acceptor	Redox Reaction*	Equation No.
Nitrate	$C(H_2O) + NO_3^- \rightarrow N_2 + CO_2 + HCO_3^- + H_2O$	(5.23)
Manganese	$C(H_2O) + Mn^{4+} \rightarrow Mn^{2+} + CO_2 + H_2O$	(5.24)
Ferric iron	$C(H_2O) + Fe^{3+} \rightarrow Fe^{2+} + CO_2 + H_2O$	(5.25)
Sulfate	$C(H_2O) + SO_4^{2-} \rightarrow H_2S + CO_2 + H_2O$	(5.26)
Organic compounds	$C(H_2O) \rightarrow CH_4 + CO_2$	(5.27)

*Equations are not stoichiometrically balanced, so that participating species in the reactions may be more clearly emphasized.

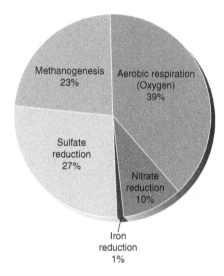

Figure / 5.16 **Contribution of Various Terminal Electron Acceptors to the Oxidation of Organic Matter in the Bottom Waters of Onondaga Lake, New York.**

(Data from Effler (1996); figure from Mihelcic (1999). Reprinted with permission of John Wiley & Sons, Inc.).

Fermentation (Equation 5.27) is an anaerobic process mediated by yeasts and certain bacteria and differs from the other reactions in that organic matter is oxidized without an external electron acceptor. Here, organic compounds serve as both the electron donor and the electron acceptor, resulting in two end products, one of which is oxidized with respect to the substrate and the other of which is reduced. In the production of alcohol, for example, glucose ($C_6H_{12}O_6$, with C in the (0) valence state) is fermented to ethanol (CH_2CH_3OH, with C reduced to the (2−) valence state) and carbon dioxide (CO_2, with C oxidized to the (4+) valence state). **Methanogenesis** is a type of fermentation in which methane (CH_4) is an end product. Recovery of methane from domestic wastewater is part of a waste to energy strategy implemented at domestic wastewater treatment and resource recovery facilities.

Organic matter produced in lakes and wetlands is broken down to carbon dioxide and stable, peat-like end products through aerobic and anaerobic respiration. Figure 5.16 illustrates the relative contribution of oxygen and various alternate electron acceptors to the oxidation of organic matter in the bottom waters and sediments of Onondaga Lake, located in New York. Approximately one-third of the organic matter decomposition during the summer period was accomplished aerobically, that is, having oxygen as the terminal electron acceptor, with the balance utilizing the alternate electron acceptors identified in Table 5.6.

5.3.2 TROPHIC STRUCTURE IN ECOSYSTEMS

In addition to energy, organisms require a source of carbon. Organisms that obtain their carbon from inorganic compounds (for example, CO_2 in Equation 5.19) are called **autotrophs**, loosely translated as self-feeders. This category includes photosynthetic organisms (green plants, including algae, and some bacteria) that use light as their energy source and nitrifying bacteria that use ammonia (NH_3) as their energy source. The simple carbohydrates produced through photosynthesis $C(H_2O)$ in Equation 5.19 and the more complex organic chemicals synthesized later (for instance, starch, cellulose, fats, and protein) are collectively termed **organic matter**.

Organisms that depend on organic matter produced by others to obtain their carbon are termed **heterotrophs**, loosely translated as other feeders. This carbon source could be a simple molecule such as methane (CH_4) or a more complex chemical such as those listed previously. Animals and most bacteria derive both their carbon and energy from organic matter and thus are categorized as heterotrophs.

The amount of organic matter present at any point in time is the system's **biomass** (g C/L or DW/L, g C/m² or DW/m²), and the rate of production of biomass is the system's **productivity** (g C/L-day or DW/L-day, g C/m²-day or DW/m²-day). *Primary production* refers to the photosynthetic generation of organic matter by plants and certain bacteria—for example, algae in lakes and field crops on land. *Secondary production* refers to the generation of organic matter by nonphotosynthetic organisms—that is, those that consume the organic matter originating from primary producers to gain energy and materials and in turn generate more biomass through growth. Secondary producers include zooplankton in aquatic systems and cattle on land.

The trophic, or feeding structure, in ecosystems is composed of the abiotic environment and three biotic components: producers, consumers, and decomposers. Producers, most often plants, assimilate simple chemicals and utilize the sun's energy to produce and store complex, energy-rich compounds that provide an organism with substance and stored energy. Organisms that eat plants, extracting energy and chemical building blocks to make more complex substances, are primary consumers or **herbivores**. Those that consume herbivores are called secondary consumers or **carnivores**. Additional carnivorous trophic levels are possible (tertiary and quaternary consumers). Consumers that eat both plant and animal material are termed **omnivores**.

Figure 5.17 illustrates the various nutritional or trophic levels in a simple aquatic **food chain**. This is a linear subset of the more complex relationships and interactions that make up **food webs** such as the one shown in Figure 5.18. Likewise, the simple terrestrial food chain illustrated in Figure 5.19 is a linear subset of the corresponding food web in Figure 5.20.

5.3.3 THERMODYNAMICS AND ENERGY TRANSFER

The first law of thermodynamics states that energy cannot be created or destroyed, but it can be converted from one form to another. Applied to an ecosystem, this law suggests that no organism can create its own energy supply. For example, plants rely on the sun for energy, and grazing animals rely on plants (and thus indirectly on the sun). Organisms use the food energy they produce or assimilate to meet metabolic requirements for the performance of work, including cell maintenance, growth, and reproduction. Thus, ecosystems must import energy, and the needs of individual organisms must be met by transformations of that energy.

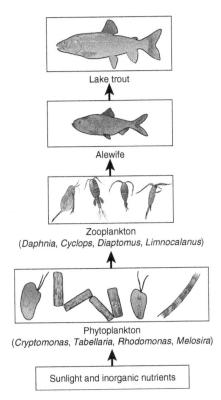

Lake trout

Alewife

Zooplankton
(*Daphnia, Cyclops, Diaptomus, Limnocalanus*)

Phytoplankton
(*Cryptomonas, Tabellaria, Rhodomonas, Melosira*)

Sunlight and inorganic nutrients

Figure / 5.17 **Aquatic Food Chain for Lake Superior** Phytoplankton are primary producers, zooplankton are primary consumers or herbivores, and the alewife and lake trout are secondary and tertiary consumers, respectively, both carnivores.

(From Mihelcic (1999). Reprinted with permission of John Wiley & Sons, Inc.).

Discussion

Switchgrass is being considered as a substrate to produce cellulosic ethanol biofuels. Large areas of states such as Kansas, Nebraska, South Dakota, and North Dakota are prime locations that can support switchgrass production. Based on a careful review of Figures 5.19 and Figure 5.20, what might be some environmental concerns you would consider if assessing the conversion of native prairie or farmed land to support production of switchgrass-derived biofuels?

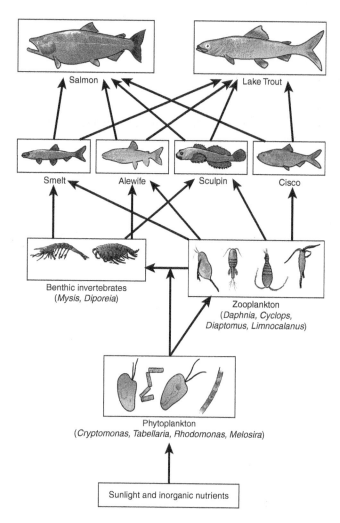

Figure / 5.18 **Food Web for Lake Superior**
This illustration shows the more complex interrelationships commonly found in an ecosystem.

(From Mihelcic (1999). Reprinted with permission of John Wiley & Sons, Inc.).

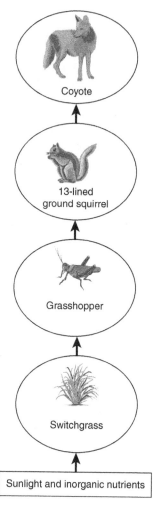

Figure / 5.19 **Simple Food Chain for a Prairie Ecosystem** This representation includes a primary producer (switchgrass) and primary (grasshopper), secondary (13-lined ground squirrel), and tertiary (coyote) consumers.

The second law of thermodynamics states that in every energy transformation, some energy is lost to heat and becomes unavailable to do work. In the food web, the inefficiency of energy transfer is reflected in losses (Figure 5.21) (potentially recycled through the microbial loop) and respiration (heat). Because of this inefficiency, less energy is available at the higher levels of the energy pyramid (Figure 5.22). This explains why it takes a large amount of primary producers to support a single organism at the top of the food chain. (For example, a top predator may require a very large range or territory to support its energy needs.) The inefficiencies of energy transfer also have great bearing on our ability to feed an increasing global population that is also consuming more calories and meat because of increases in economic wealth.

Food and Climate
http://www.fao.org/climatechange/en/

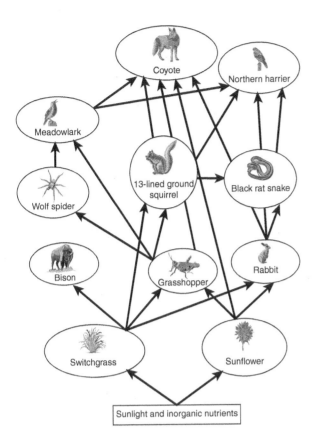

Figure / 5.20 **Simplified Food Web for a Prairie Ecosystem** The coyote, identified as a tertiary consumer and carnivore in the simple food chain (Figure 5.19), is seen within the context of the entire food web to be an omnivore. While the food web may seem complex, it is actually a simplification of the true ecosystem. That food web includes over 1,000 species of plants and animals, underscoring the true complexity of an ecosystem that seems, at first glance, quite simple.

(Based on an above-ground food web developed for the Konza Prairie in Kansas by Dr. Anthony Joern of Kansas State University).

Class Discussion

Based on the information provided in Application 5.3, what type of diet would minimize the environmental impact associated with food consumption? What type of diet results in greater use of water and fossil fuel–based energy? Does feeding cattle via natural grasslands or corn consume less water or result in less soil erosion? Are these types of foods also healthier? Why or why not? Using a life cycle approach, what are the sustainability benefits of purchasing food from local sources?

Application / 5.3   The Food We Eat: Transfer of Energy up the Human Food Chain

Each person in the United States annually consumes an average of 190 lb of meat. More than 60 million people worldwide could be fed on the grain saved if Americans reduced their meat intake by just 10 percent. Where we live on the food chain has other impacts as well. Consider the following (Goodland, 1997):

- Seven pounds of cattle feed is required to produce a pound of beef, compared with 2 lb of fish feed for some aquaculture species. Cattle are also one of the biggest producers of methane, a potent greenhouse gas.

- In the United States, 104 million cattle are the country's largest user of grain.

- Growing an acre of corn to feed cattle takes 535,000 gallons of water. Currently in the United States, 70–80 percent of the corn and soybeans are grown to produce meat.

- Agriculture consumes more freshwater than any other human activity (excluding electricity production). Worldwide, about 70 percent of freshwater is consumed (not recoverable) by agriculture. In the western United States, the figure is about 85 percent.

- Worldwide, food crops are grown on 11 percent of Earth's total fertile land area.

- Another 24 percent of the land is used as pasture to graze livestock for meat and milk products. Marginal land for pastures makes possible the production of meat and milk products on land unsuitable for food crops.

- Most cropland is threatened by at least one type of degradation (including erosion, salinization, and waterlogging of irrigated soils), and 10 million hectares of productive land are severely degraded and abandoned each year. Replacing agricultural land accounts for 60 percent of deforestation now occurring worldwide.

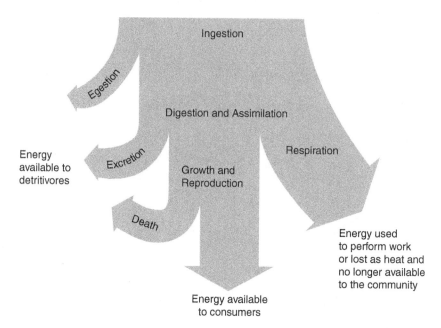

Figure / 5.21 **Energy Losses in a Food Web** A substantial part of the energy ingested by organisms is lost to egestion, excretion, death, and respiration. This inefficiency of energy transfer has a bearing on issues ranging from wastewater treatment microbiology to world population growth.

(From Mihelcic (1999). Reprinted with permission of John Wiley & Sons, Inc.).

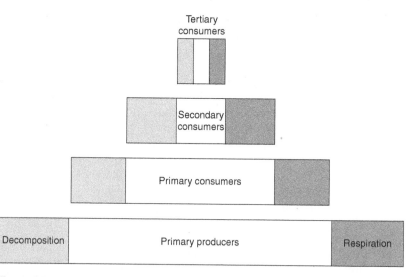

Figure / 5.22 **Energy Pyramid Showing Loss of Energy to Detritivory and Respiration with Movement up the Food Chain** A surprisingly small fraction of the energy originally fixed remains available to transfer to higher trophic levels.

(From Mihelcic (1999). Reprinted with permission of John Wiley & Sons, Inc.).

5.4 Oxygen Demand: Biochemical, Chemical, and Theoretical

The topic of **oxygen demand** involves a significant amount of new terminology and notation, as summarized in Table 5.7.

5.4.1 DEFINITION OF BOD, CBOD, AND NBOD

Organisms derive the energy required for maintenance of metabolic function, growth, and reproduction through the processes of fermentation and respiration. Both organic and inorganic matter may serve as sources of that energy. *Chemoheterotrophs* are organisms that utilize organic matter—$C(H_2O)$—as a carbon and energy source and, under aerobic conditions, consume oxygen in obtaining that energy:

$$C(H_2O) + O_2 \rightarrow CO_2 + H_2O + \Delta \qquad (5.28)$$

Chemoautotrophs are organisms that utilize CO_2 as a carbon source and inorganic matter as an energy source, and usually consume oxygen in obtaining that energy. An example of chemoautotrophy is nitrification, the microbial conversion of ammonia to nitrate (with bicarbonate ion contributing CO_2):

$$NH_4^+ + 2HCO_3^- + 2O_2 \rightarrow NO_3^- + 2CO_2 + 3H_2O + \Delta \qquad (5.29)$$

Table / 5.7

Oxygen Demand: Definition and Notation All terms have units of mg O_2/L.

BOD	*Biochemical oxygen demand*—the amount of oxygen utilized by microorganisms in oxidizing carbonaceous and nitrogenous organic matter.
CBOD	*Carbonaceous biochemical oxygen demand*—BOD where the electron donor is carbonaceous organic matter.
NBOD	*Nitrogenous biochemical oxygen demand*—BOD where the electron donor is nitrogenous organic matter.
ThOD	*Theoretical oxygen demand*—the amount of oxygen utilized by microorganisms in oxidizing carbonaceous and/or nitrogenous organic matter, assuming that all of the organic matter is subject to microbial breakdown, that is, it is biodegradable.
BOD$_5$	*5-day biochemical oxygen demand*—the amount of oxygen consumed (BOD exerted) over an incubation period of 5 days; the standard laboratory estimate of BOD. The BOD$_5$ utilizes the notation y_5, referring to the BOD exerted (y) over 5 days of incubation.
BOD$_U$	*Ultimate biochemical oxygen demand*—the amount of oxygen consumed (BOD exerted) when all of the biodegradable organic matter has been oxidized. The BOD$_U$ utilizes the notation L_o, referring to its potential for oxygen consumption when proceeding to complete oxidation.
COD	*Chemical oxygen demand*—the amount of chemical oxidant, expressed in oxygen equivalents, required to completely oxidize a source of organic matter; COD and ThOD should be near equal.

In these microbially mediated redox reactions, the electron donors are $C(H_2O)$ and NH_4^+, and the electron acceptor is O_2.

Oxygen is consumed in both reactions. **Biochemical oxygen demand (BOD)** can thus be defined as the amount of oxygen utilized by microorganisms in performing the oxidation. BOD is a measure of the "strength" of a water or wastewater: the greater the concentration of ammonia–nitrogen or degradable organic carbon, the higher the BOD.

The reactions described by Equations 5.28 and 5.29 are differentiated based on the source compound for the electron donor: *carbonaceous* and *nitrogenous*. Chemical strength (mg $C(H_2O)$/L or mg NH_3–N/L) is expressed here in terms of its impact on the environment (oxygen consumed, in mg BOD/L). This is representation by effect, as discussed in Chapter 2.

Dissolved oxygen is a critical requirement of the organism assemblage associated with a diverse and balanced aquatic ecosystem. Domestic and industrial wastes often contain high levels of BOD, which, if discharged untreated, would seriously deplete oxygen reserves and reduce the diversity of aquatic life. To prevent degradation of receiving waters, systems are constructed where the supply of BOD and oxygen, the availability of the microbial populations that mediate the process, and the rate at which the oxidations themselves (Equations 5.28 and 5.29) proceed may be carefully controlled. The efficiency of BOD removal is a common performance characteristic of wastewater treatment plants, and BOD is a major feature of treatment plant discharge permits.

5.4.2 SOURCES OF BOD

The simple carbohydrates produced through photosynthesis are used by plants and animals to synthesize more complex carbon-based chemicals such as sugars and fats. These compounds are utilized by organisms as an energy source, exerting a **carbonaceous biochemical oxygen demand (CBOD)**, Equation 5.28). In addition, plants utilize ammonia to produce proteins, that is, complex, carbon-based chemicals with amino groups ($-NH_2$) as part of their structure. Proteins are ultimately broken down (proteolysis) to peptides and then amino acids. The process of deamination then further breaks down the amino acids, yielding a carbon skeleton (CBOD) and an amino group. Conversion of the amino group to ammonia (ammonification) completes the degradation process. The ammonia is then available to exert a **nitrogenous biochemical oxygen demand (NBOD)**, Equation 5.29, when utilized by microorganisms. Figure 5.23 illustrates the chemical structure of some representative carbonaceous and nitrogenous compounds.

Domestic wastewater and many industrial wastes are highly enriched in organic matter compared with natural waters. Proteins and carbohydrates constitute 90 percent of the organic matter in domestic wastewater. Sources include feces and urine from humans; food waste from sinks; soil and dirt from bathing, washing, and laundering; plus various soaps, detergents, and other cleaning products. Wastes from

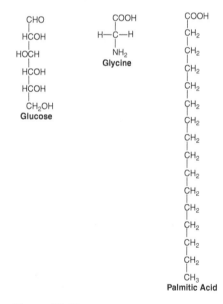

Figure / 5.23 **Chemical Structure of Representative Carbonaceous and Nitrogenous Compounds** Municipal wastewater contains a vast number of different organic chemicals, including sugars (20–25%), amino acids (40–60%), and fatty acids (10%).

(Based on Metcalf and Eddy, Inc. (1989); illustration from Mihelcic (1999). Reprinted with permission of John Wiley & Sons, Inc.).

certain industries, such as breweries, canneries, and pulp and paper producers, also have elevated levels of organic matter.

Table 5.8 presents BOD values for some representative wastes. Even unpolluted natural waters contain some BOD, associated with the carbonaceous and nitrogenous organic matter derived from the watershed and from the waters themselves (for example, from decaying algae and macrophytes, leaf litter, and fecal matter from aquatic organisms). Dissolved-oxygen levels in surface waters—excluding those with excessive algal photosynthesis and attendant O_2 production—often are below the saturation level due to this "natural" BOD.

5.4.3 THEORETICAL OXYGEN DEMAND

Theoretical oxygen demand (ThOD), given in mg O_2/L, is calculated from the stoichiometry of the oxidation reactions involved. A general approach for calculation of *carbonaceous ThOD* (Equation 5.28) is offered by a three-step process provided in Table 5.9.

The calculation is similar for the oxidation of ammonia (NH_4^+ in Equation 5.29) to nitrate: 2 moles (or 64 g) of oxygen are consumed for every mole (or 14 g) of ammonia–nitrogen oxidized. Note that ammonia is reported as mg N/L at 14 g/mole, not as ammonia (17 g/mole) or ammonium (18 g/mole). The stoichiometric coefficient for oxidation of nitrogenous wastes is thus 64/14 or 4.57. A waste containing 50 mg/L of NH_3–N would have a *nitrogenous ThOD* of 229 mg/L.

Table / 5.8

BOD of Selected Waste Streams

Origin	BOD$_5$ (mg O$_2$/L)
River	2
Domestic wastewater	200
Pulp and paper mill	400
Commercial laundry	2,000
Sugar beet factory	10,000
Tannery	15,000
Brewery	25,000
Cherry-canning factory	55,000

SOURCE: Nemerow, 1971.

Table / 5.9

Steps to Calculate the Carbonaceous ThOD

Step	Description of Step	Example
Step 1	Write the equation describing the reaction for oxidation of the carbon-based chemical of interest to carbon dioxide and water (for example, for benzene, C_6H_6).	$C_6H_6 + O_2 \rightarrow CO_2 + H_2O$
Step 2	Balance the equation in the following sequence: (a) balance the number of carbon atoms; (b) balance the number of hydrogen atoms; and (c) balance the number of oxygen atoms.	For benzene, (a) place a 6 in front of CO_2 to balance the carbon; (b) place a 3 in front of H_2O to balance the hydrogen; and (c) place a 7.5 in front of the oxygen to balance the oxygen: $$C_6H_6 + 7.5O_2 \rightarrow 6CO_2 + 3H_2O$$
Step 3	Use the stoichiometry of the balanced chemical reaction, applying unit conversions, to determine the carbonaceous ThOD.	Assume the initial concentration of benzene = 156 mg/L: $$\frac{156 \text{ mg benzene}}{L} \times \frac{1 \text{ mole benzene}}{78 \text{ g benzene}} \times \frac{7.5 \text{ mole } O_2}{\text{mole benzene}} \times \frac{32 \text{ g } O_2}{\text{mole } O_2} = \frac{480 \text{ mg } O_2}{L}$$

example/5.6 Determination of Carbonaceous, Nitrogenous, and Total ThOD

A waste contains 300 mg/L of $C(H_2O)$ and 50 mg/L of NH_3–N. Calculate the carbonaceous ThOD, the nitrogenous ThOD, and the total ThOD of the waste.

solution

Refer to Table 5.9 if you need a review of the process to write the balanced equation describing the oxidation of $C(H_2O)$ to CO_2 and water.

$$C(H_2O) + O_2 \rightarrow CO_2 + H_2O$$

The reaction shows that 1 mole of oxygen is required to oxidize each mole of $C(H_2O)$. The carbonaceous ThOD is determined from the stoichiometry:

$$300 \text{ mg C}(H_2O) \times \frac{g}{1,000 \text{ mg}} \times \frac{1 \text{ mole C}(H_2O)}{30 \text{ g C}(H_2O)} \times \frac{1 \text{ mole O}_2}{\text{mole C}(H_2O)} \times \frac{32 \text{ g O}_2}{\text{mole O}_2} \times \frac{1,000 \text{ mg}}{g} = 320 \text{ mg/L}$$

Next, write the balanced equation describing oxidation of ammonia–nitrogen to nitrate:

$$NH_3 + 2O_2 \rightarrow NO_3^- + H^+ + H_2O$$

This reaction shows that 2 moles of oxygen are required to oxidize each mole of NH_3. Be aware that the ammonia concentration is reported as mg N/L, not mg NH_3/L. The nitrogenous ThOD is determined from the stoichiometry:

$$50 \text{ mg NH}_3 - N/L \times \frac{g}{1,000 \text{ mg}} \times \frac{1 \text{ mole NH}_3 - N}{14 \text{ g NH}_3 - N} \times \frac{2 \text{ mole O}_2}{\text{mole NH}_3 - N} \times \frac{32 \text{ g O}_2}{\text{mole O}_2} \times \frac{1,000 \text{ mg}}{g} = 229 \text{ mg/L}$$

The total ThOD of the waste equals $320 + 229 = 549$ mg/L.

To determine the *total ThOD* of a waste stream containing multiple chemicals (for example, ammonia and organic matter), add the contributions of the component compounds (see Example 5.6).

The stoichiometry for oxidation of ammonia does not vary (that is, Equation 5.29 holds in all cases) because ammonia–nitrogen occurs in only one form with nitrogen in a single valence state. This is not the case for carbon-based compounds (that is, Equation 5.28 does not hold in all cases), as organic carbon exists in a wide variety of chemical species with carbon in several valence states. For this reason, the reaction stoichiometry for each carbon-based compound must be inspected and the oxidation equations balanced individually.

5.4.4 BOD KINETICS

The ThOD calculation defines the oxygen requirement for complete oxidation of ammonia–nitrogen to nitrate–nitrogen or a carbon-based compound to carbon dioxide and water. ThOD does not, however, offer any information regarding the likelihood that the reaction will proceed to completion. For NBOD, this is not an issue, because ammonia–nitrogen is readily oxidized and the ThOD and actual NBOD are identical.

Carbonaceous compounds, in contrast, are not all easily or completely oxidized by microorganisms (**biodegradation**), and the rate of that oxidation may vary widely among different sources of organic matter. For example, the carbon-containing compounds present in a Styrofoam cup are not as biodegradable as those found in tree leaves, and both are less biodegradable than the carbon compound that constitutes sugar. Thus, while a mass of carbon present as sugar or tree leaves or Styrofoam may have the same ThOD, their actual oxygen demand may be substantially different. Further, most wastes are a complex mixture of chemicals (say, Styrofoam + tree leaves + sugar), present in varying amounts, each with a different level of biodegradability. This characteristic of oxygen-demanding wastes is addressed through BOD kinetics.

Consider the oxidation of organic matter as a function of time, as shown in Figure 5.24. In Figure 5.24a y_t is the CBOD exerted (oxygen consumed, in mg O_2/L), and L_t is the CBOD remaining (potential to consume oxygen, in mg O_2/L) at any time, t. At $t = 0$, no CBOD has been exerted ($y_{t=0} = 0$), and all of the potential for oxygen consumption remains ($L_{t=0} = L_0$, the ultimate CBOD). As the oxidation process begins, oxygen is consumed (CBOD is exerted and y_t increases), and the potential to consume oxygen is reduced (CBOD remaining, L_t, decreases). The rate at which CBOD is exerted is rapid at first, but later

© Edfuentesg/iStockphoto.

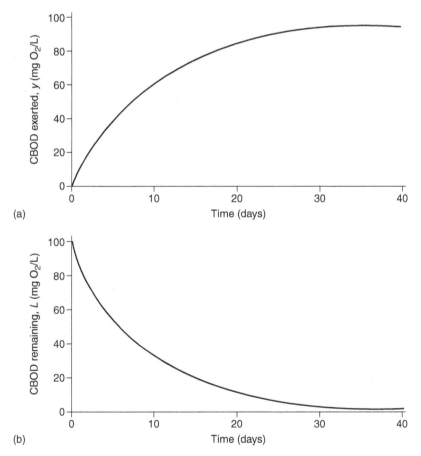

(a)

(b)

Figure / 5.24 Biochemical Oxidation of Organic Matter as a Function of Time (a) CBOD exerted, y, and (b) CBOD remaining, L, as a function of time.

(Mihelcic (1999). Reprinted with permission of John Wiley & Sons, Inc.).

it slows and eventually approaches zero as all of the biodegradable organic matter has been oxidized. The total amount of oxygen consumed in oxidizing the waste is the **ultimate CBOD** (L_0).

The exponential decline in CBOD remaining (L, the potential to consume oxygen), illustrated in Figure 5.24b, can be modeled as a first-order decay:

$$\frac{dL}{dt} = -k_L L \tag{5.30}$$

where k_L is the CBOD reaction rate coefficient (day^{-1}). Equation 5.30 can be integrated to yield the analytical expression depicted in Figure 5.24b:

$$L_t = L_0 e^{(-k_L t)} \tag{5.31}$$

Note that y_t, the CBOD exerted at any time t (see Figure 5.24a), is given by the difference between the ultimate CBOD and the CBOD remaining:

$$y_t = L_0 - L_t \tag{5.32}$$

Substituting Equation 5.31 into Equation 5.32 yields:

$$y_t = L_0(1 - e^{-k_L t}) \tag{5.33}$$

Rearrange to yield an expression for the ultimate CBOD:

$$L_0 = \frac{y_t}{(1 - e^{-k_L t})} \tag{5.34}$$

Equation 5.31 is applied in predicting the change in CBOD over time in natural and engineered systems. Equation 5.34 can be used to convert laboratory measurements of CBOD ($CBOD_5$, discussed subsequently) to ultimate CBOD.

NBOD behaves in an almost identical fashion. The exertion of NBOD follows first-order kinetics, and Equations 5.30–5.34 apply, substituting n, N, and k_N for y, L, and k_L, respectively. The ultimate NBOD (N_0) can be calculated from the ammonia–nitrogen content of the sample, based on the stoichiometry of Equation 5.28 (4.57 mg O_2 consumed per mg NH_3–N oxidized). As shown in Figure 5.25, NBOD exertion begins well after that of CBOD due to differences in the growth rates of the mediating organisms.

Figure / 5.25 **First-Stage (CBOD) and Second-Stage (NBOD) Biochemical Oxygen Demand** Exertion of nitrogenous demand lags that of carbonaceous demand because the nitrifying organisms grow more slowly than microorganisms that derive their energy from organic carbon.

(Mihelcic (1999). Reprinted with permission of John Wiley & Sons, Inc.).

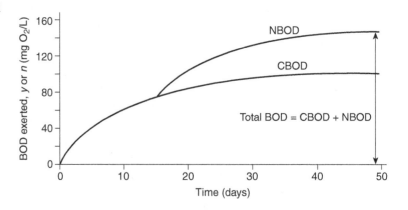

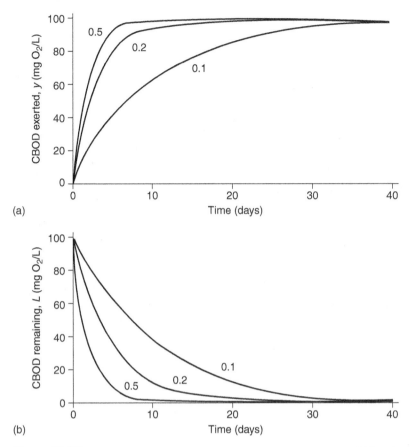

(a)

(b)

Figure / 5.26 **Variations in Rate at Which Organic Matter Is Stabilized, Reflected in the Reaction Rate Constant, k_L** (a) CBOD exerted, y (mg/L), showing effect of variation in k_L. (b) CBOD remaining, L (mg/L), showing effect of variation in k_L.

(Mihelcic (1999). Reprinted with permission of John Wiley & Sons, Inc.).

5.4.5 CBOD RATE COEFFICIENT

The reaction rate coefficient, k_L, utilized in CBOD calculations, is a measure of the biodegradability of a waste. The relationship between k_L and the rates of CBOD exertion (dy/dt) and consumption (dL/dt) is illustrated in Figure 5.26a and 5.26b, respectively. Typical ranges for this coefficient are presented in Table 5.10. Reductions in the magnitude of k_L when moving from untreated domestic wastewater to treated domestic wastewater to unpolluted river water reflect progressive reductions in labile (biodegradable) organic carbon. As with other microbially mediated processes, values for the CBOD and NBOD reaction rate coefficients vary with temperature.

Values for the CBOD reaction rate coefficient can be determined experimentally in the laboratory using the *Thomas slope method*. Measurements of CBOD exerted (y_t) are made daily for 7–10 days. Values for the parameter $(t/y)^{1/3}$ are then calculated, and k_L is determined from the slope and intercept of a plot of $(t/y)^{1/3}$ versus t, according to Equation 5.35:

$$k_L = 6.01 \frac{\text{slope}}{\text{intercept}} \qquad (5.35)$$

Table / 5.10

Ranges of Values for the CBOD Rate Constant

Type of Sample	k_L (day^{-1})
Untreated municipal wastewater	0.35–0.70
Treated municipal wastewater	0.10–0.35
Unpolluted river water	<0.05

SOURCE: Values from Davis and Cornwell, 1991; Chapra, 1997.

5.4.6 BOD: MEASUREMENT, APPLICATION, AND LIMITATIONS

No chemical method of analysis is available to measure BOD directly. It is necessary to call upon the same organisms that exert oxygen demand in nature to perform a bioassay. In the bioassay, water samples are incubated in a controlled environment for several days while oxygen consumption is monitored. As described in Table 5.11, conditions for the test are established so the microorganisms experience an environment favorable for the oxidation of ammonia and organic carbon (exertion of BOD).

The cumulative exertion of BOD over the course of the assay is well described by the curve presented in Figure 5.25. The rate of BOD exertion gradually slows over time (as the organic matter is used up), and cumulative consumption asymptotically approaches a maximum value (the ultimate BOD or BOD_U, in mg O_2/L) after several days. As shown in Figure 5.25, it takes about 30 days to reach 95 percent of the BOD_U (with $k_L = 0.1$/day). Such a wait is impractical where results are required for wastewater plant operation, so a shorter 5-day assay has become standard. Here, only about 60 percent of the BOD_U is exerted, but an accurate estimate of the BOD_U can be obtained by applying Equation 5.33 and a value for k_L to the 5-day measurement.

Table / 5.11

Favorable Conditions for the BOD Test

Condition	Description
Presence of appropriate microorganisms	Typically abundant in untreated and treated domestic wastewater and most natural waters, microbial populations are absent or present in low numbers in many industrial or disinfected wastes and in some natural waters. In these cases, microbes may be purchased or obtained from biological treatment plants and added to the sample as "seed."
Favorable and consistent incubation conditions	Samples are incubated at 20°C to encourage microbial activity (respiration is temperature-dependent) and to facilitate comparison of results among sampling locations and the laboratories performing the assays. To ensure that oxygen is not completely depleted before the end of the incubation period, aerated dilution water is added. The dilution water may also contain inorganic nutrients (for example, Fe, N, and P) required by the microbes. Some wastes may be toxic to microorganisms due to extreme pH or the presence of chemicals such as heavy metals. Sample pH can be adjusted and dilution water added to reduce or eliminate toxicity. The BOD bioassay is conducted in the dark to inhibit oxygen production through photosynthesis should algae be present, as is the case for many samples from natural waters.
Separating CBOD and NBOD	The standard BOD test measures both CBOD and NBOD. On occasion, it is necessary to separate the two processes to support plant design or operation. A chemical may be added to the water sample to inhibit nitrification, yielding CBOD as the sole result of the assay. NBOD may then be determined by difference from an analysis in which no inhibitor was added. In an inhibited assay, results are reported as CBOD; without inhibitor addition, results are reported as BOD.

example/5.7 Laboratory Determination of CBOD

A 15 mL wastewater sample is placed in a standard 300 mL BOD bottle, and the bottle is filled with dilution water. The bottle had an initial dissolved-oxygen concentration of 8 mg/L and a final dissolved-oxygen concentration of 2 mg/L. A blank (a BOD bottle filled with dilution water) run in parallel showed no change in dissolved oxygen over the 5-day incubation period. The BOD reaction rate coefficient for the waste is 0.4/day. Calculate the 5-day (y_5) and ultimate (L_0) BOD of the wastewater.

solution

The BOD_5 (y_5) is the amount of oxygen consumed over the 5-day period corrected for the dilution of the original sample. This can be written as follows:

$$y_5 = \frac{[DO_{initial} - DO_{final}]}{\left[\dfrac{mL\ sample}{total\ test\ volume}\right]} = \frac{\left[\dfrac{8\ mg\ O_2}{L} - \dfrac{2\ mg\ O_2}{L}\right]}{\left[\dfrac{15\ mL}{300\ mL}\right]} = \frac{120\ mg\ O_2}{L}$$

Equation 5.33, with $t = 5$ and $k_L = 0.4$/day, is then applied to determine the ultimate BOD:

$$L_0 = \frac{y_5}{(1 - e^{(-k_L t)})} = \frac{120}{(1 - e^{(-0.4/day \times 5\ days)})} = \frac{138\ mg\ O_2}{L}$$

example/5.8 Determination of Sample Size for the BOD Test

Remember from Chapter 3 that the amount of oxygen that can be transferred to water from the atmosphere is not very high (that is, maximum solubility of 8–12 mg O_2/L for water equilibrated with the atmosphere). Most wastewater samples thus have more BOD to exert than the amount of oxygen available in the BOD sample bottle. Because the BOD test depends on the laboratory observing a measurable decrease in dissolved oxygen over time, the water sample is "diluted" with BOD dilution water. This is so the microorganisms in the BOD sample bottle do not use up all the dissolved oxygen in the system.

The criteria for the test states there should be ≥ 2 mg/L of dissolved oxygen removed over the 5-day incubation period and ≥ 1 mg/L of dissolved oxygen should remain in the sample at the end of the incubation. Thus, a laboratory needs to make an estimate of how many mL of sample should be added to the BOD bottle so both of these criteria are met. Too little sample could result in negligible depletion of oxygen, while too much sample could result in over-depletion of oxygen to below the 1 mg/L minimum final reading.

In some circumstances, you will know from experience what your expected range of BOD_5 is. For example, if you have experience with the waste stream entering the local water reclamation plant and you know it does not change that much over the day, or if you have experience with a particular river that again has a BOD that remains relatively constant over time. In other cases, you may have to make up several BOD tests with different dilutions to ensure that at least one of the dilutions meet the criteria for a valid BOD test.

Suppose the estimated BOD of an influent sample is 400 mg/L and assume the dissolved oxygen of oxygen-saturated dilution water is 8.0 mg/L. You are using a 300 mL BOD bottle. Remember that the criteria for most BOD tests requires that the depletion of oxygen at the end of the 5-day incubation period should be at least 2.0 mg/L and the residual dissolved-oxygen remaining be at least 1.0 mg/L.

example/5.8 (continued)

solution

The minimum and maximum estimated dilution can be determined as follows:

minimum dilution(ensures at least 2 mg/L of oxygen is depleted) is determined as follows.

mL sample added to the BOD bottle = minimum allowable oxygen depletion(in mg/L)
$$\times \text{ volume of BOD bottle(in mL)/estimated BOD(in mg/L)}$$

$$\text{minimum sample (in mL)} = [(8\,\text{mg O}_2/\text{L} - 6\,\text{mg O}_2/\text{L}) \times 300\,\text{mL}]/400\,\text{mg O}_2/\text{L}$$

$$= (2\,\text{mg O}_2/\text{L} \times 300\,\text{mL})/400\,\text{mg O}_2/\text{L} = 600/400$$

$$= 1.5\,\text{mL sample added to } 300\,\text{mL bottle}$$

Maximum dilution (ensures that 1 mg/L of oxygen remains in the bottle at the end of the test) is determined as follows.

mL sample added to the BOD bottle = maximum allowable oxygen depletion(in mg/L)
$$\times \text{ volume of BOD bottle(in mL)/estimated BOD(in mg/L)}$$

$$\text{maximum sample(in mL)} = [(8\,\text{mg O}_2/L - 1\,\text{mg O}_2/L) \times 300\,\text{mL}]/400\,\text{mg O}_2/\text{L}$$

$$= (7\,\text{mg O}_2/\text{L} \times 300\,\text{mL})/400\,\text{mg O}_2/\text{L} = 2,100/400$$

$$= 5.25\,\text{mL added to the } 300\,\text{mL bottle}$$

Because the BOD value used to estimate the sample volume is only an estimate, and BOD bottles do not always have a volume of exactly 300 mL, several bottles with different volumes of sample are typically set up to ensure that test requirements are met. For example, in this example, four bottles might be set up that use sample size of 1, 3, 4, and 6 mL. The results would then be averaged to determine the final BOD$_5$. Those sample dilutions that deplete less than 2 mg/L over the 5-day period, or have a final dissolved-oxygen reading of less than 1 mg/L after the 5-day incubation period, would not be used.

example/5.9 CSI for Environmental Engineers: Wastewater Forensics

A "midnight dumper" discharged a tank truck full of industrial waste in a gravel pit. The truck was spotted there 3 days ago, and a pool of pure waste remains. A laboratory technician determined that the waste has a BOD$_5$ of 80 mg/L with a rate constant of 0.1/day. Three factories in the vicinity generate organic wastes: a winery (ultimate BOD$_U$ = 275 mg/L), a vinegar manufacturer (ultimate BOD$_U$ = 80 mg/L), and a pharmaceutical company (BOD$_U$ = 200 mg/L). Determine the source of the waste.

solution

The BOD$_U$ of the waste may be calculated as follows:

$$L_0 = \frac{y_5}{(1 - e^{(-K_L t)})} = \frac{80\,\text{mg/L}}{(1 - e^{(-0.1/\text{day} \times 5\,\text{days})})} = 203\,\frac{\text{mg O}_2}{\text{L}}$$

This value most closely matches the pharmaceutical manufacturer, but it fails to take into account the fact that the waste has been away from its source (and decaying) for 3 days. The original ultimate BOD_U may be calculated by

$$L_t = L_0 \times e^{-k_L t}$$

Rearrange and solve:

$$L_0 = \frac{L_t}{e^{-k_L t}} = \frac{203 \text{ mg/L}}{e^{(-0.1/\text{day} \times 3 \text{ days})}} = 274 \frac{\text{mg O}_2}{\text{L}}$$

The "midnight dumper" culprit appears to be the winery.

5.4.7 BOD TEST: LIMITATIONS AND ALTERNATIVES

While the BOD_5 test remains a fundamental tool in waste treatment and water quality assessment, concerns regarding its logistics and accuracy have led to proposals for replacement by other measures. While relatively simple to perform, the test has three major shortcomings: (1) the time required to obtain results (5 days is almost unthinkable in today's world of real-time data acquisition); (2) the fact that it may not accurately assay waste streams that degrade over a time period longer than 5 days; and (3) the inherent inaccuracy of the procedure, largely due to variability in seed (bacteria). Table 5.12 provides some troubleshooting advice when conducting the BOD test.

Table / 5.12

Troubleshooting Guide for the BOD$_5$ Test

Symptom	Possible Cause and Corrective Action
DO readings drift downward	Weak batteries for stirring unit result in inadequate flow across membrane—replace batteries
BOD$_5$ demand in dilution water is greater than the acceptable 0.2 mg/L	Deionized water contains ammonia or volatile organic compounds—increase purity of dilution water or obtain from another source. Age water for 5–10 days before use. Deionized water contains semivolatile organic compounds leached from the resin bed—increase purity of dilution water or obtain from another source. Age water for 5–10 days before use. Bacterial growth in reagents and poorly cleaned glassware—more vigorous cleaning of glassware, including washing followed by a 5 to 10 percent HCl rinse followed by 3–5 rinses with deionized water. Discard reagents properly.
Sample BOD values are unusually low in the diluted sample (BOD$_5$ dilution water is within the acceptable range)	Dilution water contains interferences inhibiting the biochemical oxidation process—increase purity of dilution water or obtain from another source. Use deionized water that has been passed through mixed-bed resin columns. Never use copper-lined stills. Distilled water may be contaminated by using copper-lined stills or copper fittings—obtain from another source.

SOURCE: Provided by the U.S. Geological Survey, from Delzer and McKenzie (2007).

Other analytes, such as total organic carbon (TOC), provide greater accuracy and precision but fail to readily distinguish biodegradable and nonbiodegradable organic carbon—precisely the objective of the BOD test in the first place. Further, the BOD test cannot be used to evaluate treatment efficiency for wastes that are poorly biodegradable or toxic. Here a different test is applied, one for **chemical oxygen demand (COD)**.

In this test, a sample containing an unknown amount of organic matter is added to a 250 mL flask. Also added to the flask are $AgSO_4$ (a catalyst to ensure complete oxidation of the organic matter); a strong acid (H_2SO_4), dichromate ($Cr_2O_7^{2-}$, a strong oxidizing agent); and $HgCl_2$ (to provide a Hg^{2+} ion that complexes the chloride ion, Cl^-). The chloride ion interferes with the test and is present in high amounts in many wastewater samples. This is because it can be oxidized to Cl^0 by dichromate as well as by organic matter. However, the complexed form of Cl^- is not oxidized. Thus, if uncomplexed Cl^- is allowed to be oxidized to Cl^0, it can result in a false-positive COD value.

The sample and all the reagents are combined, and the sample is refluxed for 3 h. The organic matter is oxidized (donates electrons), and the chromium is reduced (accepts electrons) from the hexavalent form (Cr^{6+}) to the trivalent form (Cr^{3+}). Thus, the COD test determines how much of the hexavalent chromium is reduced during the COD test. After the sample is cooled to room temperature, the dichromate that remains in the system is determined by titration with ferrous ammonium sulfate. The amount of hexavalent chromium is then related to the amount of organic matter that was oxidized.

Results are expressed in oxygen equivalents (mg O_2/L), that is, the amount of oxygen required to completely oxidize the waste. The test is relatively quick (3 h), and correlation to BOD_5 is easy to establish on a particular waste stream. For example, municipal wastewater has a ratio of 0.4–0.6.

Comparison of BOD and COD results can help identify the occurrence of toxic conditions in a waste stream or point to the presence of biologically resistant (refractory) wastes. For example, a BOD_5/COD ratio approaching 1 may indicate a highly biodegradable waste, while a ratio approaching 0 suggests a poorly biodegradable material.

5.5   Material Flow in Ecosystems

The natural passage of chemicals, as mediated by organisms, occurs within **biogeochemical cycles**. Five chemicals are of particular importance in environmental engineering: C, O, N, P, and S. In addition, the **hydrologic cycle** (see Chapter 8) is of interest because it plays an important role in moving chemical elements through the ecosphere. This section considers each of these key cycles.

Note that humans use and cycle a much greater number of chemical elements in industrial applications than are found in living organisms. This can affect the environment through the mining of exotic elements, concentrating these once-dispersed chemicals, and exposing humans and natural systems to elevated concentrations, often damaging healthy function.

5.5.1 OXYGEN AND CARBON CYCLES

The **oxygen** cycle and **carbon cycle** are closely linked through the processes of photosynthesis (Equation 5.19) and respiration (Equation 5.20). Photosynthesis is the primary source term in the oxygen cycle and the origin of the organic carbon converted to **carbon dioxide** in the carbon cycle. Respiration is the major sink term in the oxygen cycle and is responsible for the conversion of organic carbon to carbon dioxide in the carbon cycle. Photosynthesis is carried out by plants and some bacteria, and respiration is carried out by all organisms, including those that photosynthesize. The interplay of photosynthesis and respiration plays a key role in regulating ecosystem energy balances and in maintaining the oxygen levels required by life in aquatic environments.

Figure 5.27 depicts how the natural carbon cycle has been altered by humans. Note in the top figure how the natural carbon cycle has maintained a constant reservoir of carbon in the air with balanced transfer between the air, oceans, and land. However, anthropogenic activities such as burning of fossil fuels are displacing carbon that was once stored in the land to the atmosphere. The natural carbon cycle has not been able to assimilate this released carbon into existing ocean and land reservoirs. This has resulted in large increases in

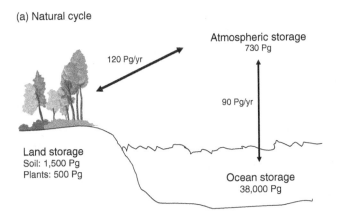

(a) Natural cycle

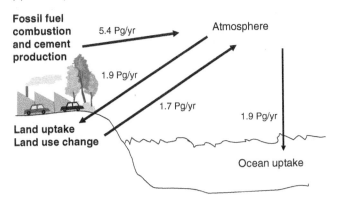

(b) Human perturbations

Figure / 5.27   Carbon Cycle under Natural Conditions and as Modified by Human Perturbation   (a) The natural carbon cycle maintains relatively constant reservoirs of carbon in the air with balanced transfers among compartments. (b) Anthropogenic carbon emissions add to the atmospheric reservoir at rates that are not balanced through uptake by the land and oceans. The result is an increase in atmospheric CO_2 concentrations.

Table / 5.13

Greenhouse Gas Emissions Associated with Particular Economic Activities

Economic Activity and Its Contribution to Global Greenhouse Gas (GHG) Emissions	Explanation
Energy supply accounts for 26% of global GHG emissions	Burning of fossil fuels such as coal, natural gas, and oil to produce electricity and heat.
Industrial activity accounts for 19% of GHG emissions.	Burning of fossil fuels for on-site generation of energy and also emissions from chemical, metallurgical, and mineral production processes.
Land use, land-use change, and forestry accounts for 17% of GHG emissions.	Deforestation results in CO_2 emissions, as does clearing land for agriculture and fire and decay of peat soils.
Agriculture accounts for 13% of GHG emissions.	Agricultural activities that account for most of the emissions are related to the management of soils, livestock, production of rice, and burning of biomass.
Transportation accounts for 13% of GHG emissions.	Transportation currently requires burning oil fossil fuels that are used for road, rail, air, and marine transportation.
Commercial and residential buildings account for 8% of GHG emissions.	On-site energy generation and burning of fossil fuels used for heating and cooking in homes.
Water and wastewater activities account for 3% of GHG emissions.	Methane emissions from landfills account for the majority of these emissions; however, methane, and N_2O emissions produced from treatment of wastewater are also important. Incineration of some solid and industrial wastes also results in emissions.

2004 data from IPCC (2007b).

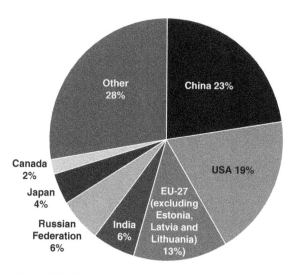

Figure / 5.28 Breakdown of Global CO_2 Emissions from Fossil Fuel Combustion and Some Important Industrial Processes.

(Data from IPCC, 2007b).

global CO_2 concentrations (see Figure 4.14). In fact, Figure 4.14 showed that in the past 50 years, atmospheric global CO_2 concentrations have risen from under 320 ppm to close to 400 ppm.

Table 5.13 describes the particular economic activities that contribute to the majority of global greenhouse emissions. Most global GHG emissions are carbon dioxide from burning of fossil fuels (approximately 57 percent of the total). These emissions are associated with economic activities such as supplying energy, industrial activities associated with chemical, mineral, and metallurgical processes, and transportation. GHG emissions associated with deforestation and decay of biomass are also quite large, accounting for another 17 percent of the total emissions. Methane emissions account for 14 percent of total GHG emissions and nitrogen oxide emissions account for 8 percent of the total are two other major contributors.

Figure 5.28 shows that China and the United States are now the top two producers of CO_2 emissions associated with the consumption of energy. This figure shows other large global emitters of greenhouse gases, including the European Union, India, Russian Federation, Japan, and Canada. The United States' contribution in 2010 for total greenhouse gas emissions was 6,821.8 Tg CO_2e (remember that 1 Tg equals 1 million metric tons) of which 5,706.4 Tg CO_2e were carbon dioxide emissions.

5.5.2 NITROGEN CYCLE

The biogeochemical transformations embodied in the nitrogen cycle (Figure 5.29) are important in both natural and engineered systems. As a result of their association with bacteria and plants, many features of the nitrogen cycle are linked with the oxygen and carbon cycles. Plants take up and utilize nitrogen in the form of ammonia or nitrate, chemicals typically in short supply in agricultural soils, thus leading to requirements for fertilization. Certain bacteria and some plant species (such as legumes and clover) can also utilize atmospheric nitrogen (N_2), converting it to ammonia through a process termed **nitrogen fixation**. Plants incorporate ammonia and nitrate into a variety of organic compounds, such as proteins and nucleic acids, critical to metabolic function. Consumers (both herbivores and carnivores) transfer those nitrogen-rich compounds further up the food chain. The nitrogen species present in organisms are released to the environment through excretion and mortality, whether in nature or in the form of human waste emissions.

The forms of nitrogen depicted in Figure 5.29 are referred as reactive nitrogen. Other reactive nitrogen species include other chemical oxidized inorganic forms such as peroxyactyl nitrate (PAN) and organic compounds such as urea, amines, amino acids, and proteins. **Reactive nitrogen** consists of all the biologically active, chemically reactive, and radiatively active nitrogen compounds that are found in the Earth's atmosphere and biosphere. Figure 5.30 shows the

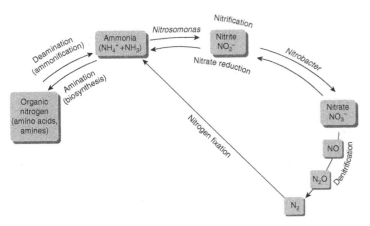

Figure / 5.29 The Nitrogen Cycle Snoeyink and Jenkins, Water Chemistry, 1980, reprinted with permission of John Wiley & Sons, Inc.

(From Mihelcic 1999, reprinted with permission of John Wiley & Sons, Inc.).

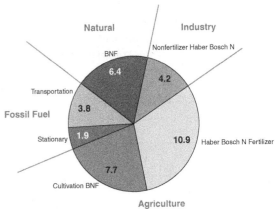

Figure / 5.30 Sources of Reactive Nitrogen Introduced into the United States in 2002 (Tg N/year).

(From EPA, 2011). BNF refers to biological nitrogen fixation.

breakdown of the sources of reactive nitrogen in the United States. Note the dominant roles played by anthropological activities of agriculture, industry, and combustion of fossil fuels. The production of anthropogenic reactive nitrogen through fertilizer production (referred to as Haber Bosch N fertilizer in Figure 5.30) and planting agricultural fields with nitrogen fixing legumes (referred to as cultivation BNF in Figure 5.30) have brought many economic and social benefits to humans, especially through increases in food production that sustain billions of people worldwide. Unfortunately, almost all reactive nitrogen that has been created by humans is ultimately released to the environment.

This major disturbance in the natural nitrogen cycle has brought many adverse public health and environmental effects. Table 5.14 summarizes many of these impacts, including reactive nitrogen's role in production of photochemical smog, decreased atmospheric visibility, acidification of terrestrial and aquatic ecosystems, eutrophication of inland and coastal waters, negative impacts on our drinking water, and contributing to greenhouse gas emissions (EPA, 2011). As we will learn in later chapters, environmental engineers are involved in activities to mitigate and control release of reactive nitrogen into the environment.

Figure 5.31 describes the nitrogen cycle that is commonly observed in a municipal wastewater treatment plant. In the first step, **ammonification**, organic N (for example, protein) is converted to ammonia (NH_4^+/NH_3). The next step involves the transformation of ammonia to nitrite (NO_2^-) (by bacteria of the genus *Nitrosomonas*) and then to nitrate (NO_3^-) (by bacteria of the genus *Nitrobacter*), a process termed **nitrification**. This step requires the presence of oxygen. Nitrate may be transformed to nitrogen gas through the process of **denitrification**, where it is converted to nitrogen gas with subsequent release to the atmosphere. Some of these emissions remain as the greenhouse gas, N_2O (refer back to Figure 5.29). This microbially mediated transformation proceeds under anoxic conditions.

Figure / 5.31 Configuration of a Conventional Municipal Wastewater Treatment Plant for Removal of Nitrogen.

(Mihelcic (1999). Reprinted with permission of John Wiley & Sons, Inc.).

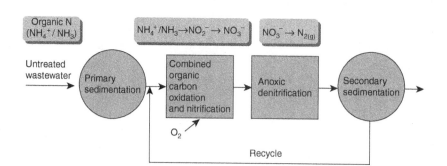

Table / 5.14

Examples of Impacts of Excess Reactive Nitrogen on Human Health and Environment

Impact	Cause(s)	Example Location(s)	Source(s)
Decrease in visibility	Fine particulate matter	National parks and wilderness areas	NO_y and NH_x from burning fossil fuels and agriculture activities
Loss of biodiversity	Nitrogen deposition	Grasslands and forests in the United States receiving N deposition in excess of critical loads	Electric utilities, traffic, and animal focused agriculture
Decline in forests	Ozone and acid deposition	Eastern and western United States	Electric utilities, traffic, and animal-focused agriculture
Reduction of crop yields	Ozone	Eastern and western United States	Electric utilities and traffic
Acidification of surface waters, loss of biodiversity	Acidification of soils, streams, and lakes is caused by atmospheric deposition of sulfur, HNO_3, NH_3, and ammonium compounds	Primarily mountainous regions of the United States	Fossil fuel combustion and agriculture activities
Hypoxia of coastal and inland waters	Excessive nutrient loading, eutrophication, variable freshwater runoff	Gulf of Mexico, Chesapeake Bay, other estuarine and coastal waters	Nitrogen and phosphorus from energy and food production
Harmful algal blooms	Excessive nutrient loading, climate variability	Inland and coastal waters	Excessive nutrient loading of nitrogen and phosphorus
Human mortality	$PM_{2.5}$, O_3, and related toxins	U.S. urban and nearby areas	NO_y and NH_x from fossil fuels and agriculture activities
Damage to public health and the environment	NO_x into air	Chesapeake Bay Watershed	Mobile sources
Damage to public health and the environment	NH_x and nitrate released into air and water	Chesapeake Bay Watershed	Agriculture activities

Adapted from EPA (2011).

5.5.3 PHOSPHORUS CYCLE

Phosphorus-bearing minerals are poorly soluble, so most surface waters naturally contain very little of this important plant nutrient. In addition, readily available sources of this mined material are expected to run out during this century. When phosphorus is mined and incorporated into cleaning agents and fertilizers, biogeochemical cycling (the routing of the element through the environment) is vastly accelerated. Subsequent discharges to lakes, estuaries, and rivers, where phosphorus is the limiting nutrient, can stimulate nuisance algal growth and **eutrophication** (discussed further in Chapter 7), making lakes unpleasant and unavailable for a variety of uses.

Approximately half of the phosphorus excreted by humans is found in urine. The remainder is found in the feces. In contrast, approximately 75 percent of N excreted by humans is in urine. The global demand of P demand is approximately 14 million metric tons. However, currently only 1.5 million metric tons are recovered annually from human waste (through water reuse, biosolids). Thus, integrating the **phosphorus cycle** with wastewater treatment and resource recovery process streams is very important.

5.5.4 SULFUR CYCLE

Like the oxygen and nitrogen cycles, the **sulfur cycle** (Figure 5.32) is to a large extent microbially mediated and thus linked to the carbon cycle. Sulfur reaches lakes and rivers as organic S, incorporated into materials such as proteins, and as inorganic S, primarily in the form of sulfate (SO_4^{2-}).

Hydrogen sulfide (H_2S) is malodorous and toxic to aquatic life at very low concentrations. Pyrite (FeS_2) is often found in and around geologic formations that are mined commercially, as in the case of coal or metals such as silver and zinc. Exposure of pyrite to the atmosphere initiates a three-step oxidation process catalyzed by bacteria including *Thiobacillus thiooxidans*, *Thiobacillus ferrooxidans*, and *Ferrobacillus ferrooxidans*:

$$4FeS_2 + 14O_2 + 4H_2O \rightarrow 4Fe^{2+} + 8SO_4^{2-} + 8H^+ \qquad \textbf{(5.36)}$$

$$4Fe^{2+} + 8H^+ + O_2 \rightarrow 4Fe^{3+} + 2H_2O \qquad \textbf{(5.37)}$$

$$4Fe^{3+} + 12H_2O \rightarrow 4Fe(OH)_{3(s)} + 12H^+ \qquad \textbf{(5.38)}$$

This process yields acid mine drainage rich in sulfate, acidity, and ferric hydroxides (a yellowish-orange precipitate or floc termed "yellow boy").

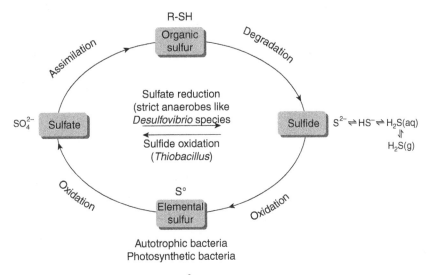

R-SH

Organic sulfur

Assimilation

Degradation

SO_4^{2-} Sulfate

Sulfate reduction
(strict anaerobes like
Desulfovibrio species)

Sulfide oxidation
(*Thiobacillus*)

Sulfide $S^{2-} \rightleftharpoons HS^- \rightleftharpoons H_2S(aq)$

$H_2S(g)$

Oxidation

$S°$

Elemental sulfur

Oxidation

Autotrophic bacteria
Photosynthetic bacteria

Figure / 5.32 **Sulfur Cycle** Reduction of sulfate SO_4^{2-} to hydrogen sulfide (H_2S) can lead to odor problems for wastewater collection systems and treatment plants. Oxidation of reduced sulfur can lead to acidification and discoloration of surface waters.

(From Mihelcic (1999). Reprinted with permission of John Wiley & Sons, Inc.).

While the sulfate is rather innocuous, acidity lowers the pH of the surface waters (often to levels severely impairing water quality), and the floc covers stream beds, eliminating macroinvertebrate habitat. In addition, the low pH of the water dissolves rocks and minerals, releasing hardness and total dissolved solids.

Acid Mine Drainage
http://water.epa.gov/polwaste/nps/acid_mine.cfm

5.6   Ecosystem Health and the Public Welfare

All engineered projects should be designed, constructed, and operated in an environmentally benign manner that will ultimately serve society equitably and protect human and ecosystem health for future generations. This section introduces two topics that are important from this perspective: toxic substances and biodiversity.

5.6.1   TOXIC SUBSTANCES AND ECOSYSTEM AND HUMAN HEALTH

Toxic substances may influence ecosystem health directly through effects manifested at the population or community level and indirectly by initiating an imbalance in ecosystem function (reducing or eliminating the role of a species or group of species). In addition, human health may be harmed through consumption of fish and wildlife contaminated with toxic substances.

Bioconcentration is the direct absorption of a chemical into an individual organism. Examples include mercury moving from water into phytoplankton across the cell wall or into fish via the gills.

Bioaccumulation (also termed biomagnification) refers to the accumulation of chemicals both by exposure to contaminated water (bioconcentration) and ingestion of contaminated food. For example, bioconcentration of a contaminant by plankton results in bioaccumulation in the next trophic level, fish.

Although there are significant losses in transferring energy and biomass up the food chain, as in the case of oxidation and excretion (refer back to Figures 5.21 and Figure 5.22), some chemicals (for example, mercury, PCBs, DDT, and some flame retardants) are retained by organisms. This retention, coupled with the loss of biomass, produces a concentrating effect in each successive level up the food chain.

Figure 5.33 depicts this concentrating effect for mercury in an aquatic system. Note that the seemingly small water concentration of 0.01 μg/L (ppb$_m$) increases by five orders of magnitude to 2,270 ppb$_m$ at the top predator level through bioconcentration and bioaccumulation. Is there a higher trophic level than fish? Yes, potentially humans, other mammals (for example, bears, seals, beluga whales) and birds such as gulls and eagles. Elevated levels of bioaccumulative

Ecosystem Services

http://www.epa.gov/research/ ecoscience/eco-services.htm

| **Application / 5.4** | **Ecosystem Services and Ecosystem Health** |

Society derives many essential goods from natural ecosystems, including seafood, game animals, fodder, fuelwood, timber, and pharmaceutical products. These goods represent important and familiar parts of the economy. What has been less appreciated until recently is that natural ecosystems also perform fundamental life support services, including regulation of climate, water storage, flood control, buffering against extreme weather events, purification of air and water, regeneration of soil fertility, detoxification and decomposition of wastes, and production and maintenance of biodiversity.

Ecosystem services can be subdivided into five categories: (1) *provision*, such as the production of food and water; (2) *regulation*, such as the control of climate and disease; (3) *support*, including nutrient cycles and crop pollination; (4) *culture*, such as spiritual and recreational amenities; and (5) *preservation*, which includes the maintenance of diversity (Daily, 2000; Millennium Ecosystem Assessment, 2005). Such processes are worth many trillions of dollars annually. Yet because most of these benefits are not traded in economic markets, they carry no price tags that could alert society to changes in ecosystem health influencing the supply of those benefits or deterioration of the systems that generate them.

To understand the magnitude of the economic implications of services provided by natural ecosystems, consider the following example. In New York City, where the quality of drinking water had fallen below standards required by the U.S. Environmental Protection Agency (EPA), authorities opted to restore the polluted Catskill Watershed, which had previously provided the city with the ecosystem service of water purification. Once the input of domestic wastewater and pesticides to the watershed area was reduced, natural abiotic processes such as soil adsorption and filtration of chemicals, together with biotic recycling via root systems and soil microorganisms, improved water quality to levels that met government standards. The cost of this investment in natural capital was estimated to be between $1 billion and $1.5 billion, contrasting dramatically with the estimated $6 billion to $8 billion cost of constructing a new water filtration plant with annual operating costs of $300 million (Chichilnisky and Heal, 1998).

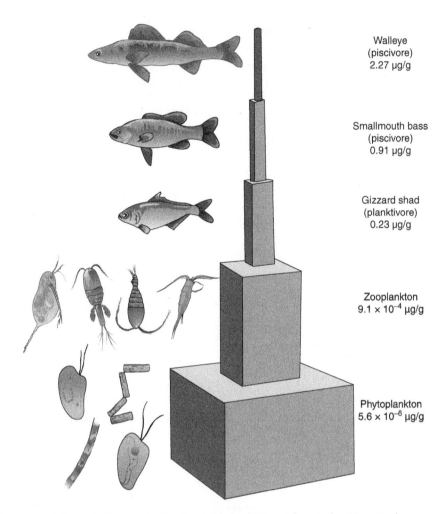

Figure / 5.33 **Bioaccumulation of Mercury in the Food Chain of Onondaga Lake, New York**
Application size represents biomass (decreasing up the food chain through inefficiency of energy
transfer), and shading represents the mercury concentration of the biomass (increasing up the
food chain because it is retained as biomass is reduced). The concentration of mercury in the
water column is ≈ 0.01 μg/L.

(Data from Becker and Bigham (1995); figure from Mihelcic (1999). Reprinted with permission of John Wiley & Sons, Inc.).

substances are routinely observed in wildlife relying on fish as a
significant portion of their diet.

Bioaccumulation of toxic substances has led to severe impacts on
many species of wildlife. Table 5.15 illustrates the impact of chemicals
such as DDT and lead bioaccumulating in bald eagles. In addition,
bioaccumulation can contribute significantly to total human exposure,
and thus environmental risk, for a particular chemical, as discussed in
Chapter 6. Because of these effects and the threat posed to human
populations, there is a pressing need for better understanding of the
dynamics of bioaccumulation and its potential impact on humans and
the environment.

The **bioconcentration factor (BCF)** is the ratio of the concentration of
a chemical in an organism to that in the surrounding environmental

Table / 5.15

Historical Numbers of Nesting Pairs of Bald Eagles in the Contiguous ("Lower 48") States The ban on lead shot in hunting intended to reduce toxicity to waterfowl that often digest the pellets additionally reduced toxicity to eagles, which feed on dead or crippled waterfowl. Land use also has had a strong impact on successful nesting of eagles.

Period	Number of Nesting Pairs	Explanation
1700s	50,000	Presettlement estimated population.
1940	Number unknown; "threatened with extinction"	Bald Eagle Protection Act prohibited killing, selling, or possessing the species.
1960s	400	Habitat loss, lead poisoning, and reproductive failure due to bioaccumulation of the pesticide DDT; Rachel Carson's *Silent Spring* is published.
1972	<800	DDT use banned in the United States.
1991	3,000	Lead shot banned for waterfowl hunting.
2007	10,000	Removal from Endangered Species List by U.S. Fish and Wildlife Service.

SOURCE: U.S. Fish and Wildlife Service.

medium (generally air or water) when uptake directly from that medium is the only mechanism considered.

The **bioaccumulation factor (BAF)** is the ratio of the concentration of a chemical in an organism to that in the surrounding medium when all potential uptake mechanisms (such as through food and water) are included.

Because the BCF addresses only passive uptake (absorption independent of organism-specific feeding patterns), it provides a means of comparing the potential risk of chemicals to organism and ecosystem health. Organisms with high lipid content tend to exhibit greater BCFs; for example, PCB concentrations are typically higher in fatty fish like trout and salmon than in largemouth and smallmouth bass, which are leaner. The phenomenon of chemical bioaccumulation has led to programs of fish consumption advisories in many states, from Florida to northern Minnesota. Specific recommendations are made for removal of fatty tissue when cleaning and preparing fish to minimize human consumption (and bioaccumulation) of contaminants. However, exposure to some contaminants, including mercury, is not reduced by selective trimming of fatty tissue, because this contaminant is uniformly distributed throughout the fish. In this case, the only way consumers can limit exposure is to control the amount of fish eaten per week.

Class Discussion

What fish advisories exist in your state or a neighboring state? What are the chemicals of concern? What is the origin of these chemicals (point discharge, nonpoint discharge, air)? What populations are most at risk? What sustainable solutions (policy and technological) can you implement to solve the problem?

5.6.2 BIODIVERSITY AND ECOSYSTEM HEALTH

The term **biodiversity**, a contraction of *biological diversity*, refers to the great variety present in all forms of life. While the concept of biodiversity originally focused on individual species, many scientists now consider *genetic diversity* (that within a species) and *ecological diversity* of great importance. It is known that significant variation can exist in the genetic makeup of populations of the same species that are separated in time (for instance, by season) or in space. This variation may lead to differences in the response of populations to environmental stress or even the function of the species within an ecosystem. For this reason, fisheries biologists seeking to reestablish a species in a formerly degraded habitat seek young fish from local or regional populations to maintain the characteristics of the original gene pool. At the other end of this spectrum, perhaps it is wise to view ecosystems as representing a diversity of equal or greater value than that of the individual species, thus meriting the attention of those seeking to maintain biodiversity.

THREATS TO BIODIVERSITY Proceeding at a background rate of one species per million species per year, extinction has been outpaced by evolution, and Earth's biodiversity has experienced a steady increase in biodiversity over the past 600 million years. This increase has been punctuated by five episodes of mass extinction thought to be related to meteorite impacts, volcanic eruption, and climate change. But today, humankind has initiated its own meteorite impact, its own volcanic eruption:

> Because of the magnitude and speed with which the human species is altering the physical, chemical and biological world, biodiversity is being destroyed at a rate unprecedented in recent geologic time.
>
> —Thorne-Miller, 1999

While it is the species that becomes extinct, it is important to consider threats to biodiversity within the context of ecosystem structure and function. The Intergovernmental Panel on Climate Change (IPCC) reports:

> The resilience of many ecosystems is likely to be exceeded this century by an unprecedented combination of climate change, associated disturbances (e.g., flooding, drought, wildfire, insects, ocean acidification), and other global change drivers (e.g., land use change, pollution, over-exploitation of resources).
>
> —IPCC, 2007a

Certainly it is the species that is often targeted (for example, by hunting or poaching), but human stress may be manifested as well through negative effects on the ecosystem.

Threats to biodiversity can be organized into five categories: (1) overharvest, (2) habitat destruction, (3) species introductions, (4) chemical pollution, and (5) global atmospheric change. Of these, overharvest has perhaps the longest history. The hunting of waterfowl, harvest of wading birds for their feathers, and the relentless pursuit and killing

Urban Biodiversity
http://www.unep.org/
urban_environment/issues/
biodiversity.asp

Biodiversity in Africa
http://www.eoearth.org/article/
Biodiversity_in_Africa

Class Discussion
We now live in one of the greatest mass extinctions in Earth's history. Discuss the role of engineers in this. Are they or aren't they partially responsible? How can they be involved in solutions to this problem?

of bison are well known from 19th-century America. More recently, overharvest from nonsustainable overfishing is threatening the future of marine fisheries.

Grassland and forest ecosystems that once covered thousands of square miles are now absent or so altered in composition as to be unrecognizable. Together with urban development, these land use changes have led to habitat fragmentation, with the formerly contiguous habitats required by some species now reduced to small patches. The process of species introduction is a natural feature of ecosystem evolution as conditions change and ranges expand. However, human acceleration (accidental or intentional) of this process can be critically damaging to native species and their ecosystem. (Consider, for example, the introduction of an assortment of mammals into New Zealand.)

The vulnerability of an ecosystem to introduction of new species increases through damage to its health (for instance, physical or chemical effects) or the loss of key biotic components (species). Alien **invasive species** are those that become established in a natural ecosystem and threaten native biological diversity. Invasive species have been introduced to aquatic environments intentionally (certain species of trout in Western rivers) and accidentally (sea lamprey through canal construction, zebra mussels in ballast water, and Asian carp through aquaculture escape).

In terrestrial ecosystems, land disturbance during construction of the built environment and the entry of motor vehicles into previously roadless areas have been documented to spread invasive species. The rapid spread and aggressive nature of invasive species is due, in part, to the absence of competition and predation pressure common to their home ecosystem. Invasive species contribute significantly to species extinction and loss of biodiversity, a problem that is increasing due to globalization.

In terms of chemical pollution, oil spills, nutrients input from wastewater treatment plants, agricultural and urban runoff, and heavy metals such as mercury and pesticide residues have had documented negative effects on ecosystems. Global climate change is also predicted to have a dramatic effect on ecosystem health. The IPCC reports, "Approximately 20–30% of plant and animal species assessed so far are likely to be at increased risk of extinction if increases in global average temperature exceed 1.5–2.5°C" (IPCC, 2007a).

SEEING VALUE IN BIODIVERSITY   Since passage of the Endangered Species Act in 1973, most Americans have become familiar with the concept and intrinsic value of biodiversity. However, the application of a philosophy supporting biodiversity may become complex in some management and regulatory arenas. In the case of the northern spotted owl in the old-growth forests of the Pacific Northwest, the timber industry took a strong stand against listing the species, arguing that ensuing joblessness and related socioeconomic impacts far outweighed the value of protecting the animal and biodiversity.

Three reasons are often supplied for encouraging programs supporting biodiversity. The first evolves from the concept that plants and animals may play a critical role in the development of more-productive food supplies and medicines that improve human health and prolong life.

Invasive Species
http://water.epa.gov/type/oceb/
habitat/invasive_species_index.cfm

Aldo Leopold (1887–1948) was a U.S. ecologist, forester, and environmentalist. He was influential in the development of modern environmental ethics and in the movement for wilderness preservation. Aldo Leopold is considered to be a pioneer of wildlife management in the United States and was a lifelong fisher and hunter. In his *Sand County Almanac*, published shortly after his death, Leopold put forth the "Land Ethic," counseling, "A thing is right when it tends to preserve the integrity, stability, and beauty of the biotic community. It is wrong when it tends otherwise."

Leopold went on to state

Photo courtesy of the Aldo Leopold Foundation Archives.

*All ethics so far evolved rest upon a single premise: that the individual is a member of a community of interdependent parts. His instincts prompt him to compete for his place in that community, but his ethics prompt him also to co-operate (perhaps in order that there may be a place to compete for). The **land ethic** simply enlarges the boundaries of the community to include soils, waters, plants, and animals, or collectively: the land. In short, a land ethic changes the role of Homo sapiens from conqueror of the land-community to plain member and citizen of it. It implies respect for his fellow-members, and also respect for the community as such.*[*]

Leopold calls upon us to recognize our role as members of a diverse but fragile environmental community and to use our engineering skills and training in such a way as to protect the land and find value in biodiversity.

[*] Quoted from *A Sand County Almanac: With Essays on Conservation* by Aldo Leopold (2001) Oxford University Press, Inc.

For example, a species of maize discovered in Mexico several decades ago possesses disease resistance and growth patterns that may revolutionize the corn industry. In the field of medicine, examples include aspirin, a major component of which is derived from willow bark or the herb meadowsweet, and anticancer agents produced by the rosy periwinkle, a flowering shrub found only in Madagascar.

The second reason often given in support of biodiversity is that ecosystem structure is determined by the interactions of its components, so the loss of a single component (such as a species) could permanently and fatally disrupt ecosystem function. For example, invasive species can decimate native plants and animals, completely altering the face of an ecosystem. The United States is now experiencing such ecosystem disruption; for example, proliferation of the nonnative spotted knapweed is crowding out native species and reducing livestock forage. Purple loosestrife is becoming dominant in many wetlands, eliminating native plants that are more nutritive to wildlife.

The final reason is an ethical one related to the role of humans as stewards of vast and complicated ecosystems. Our society's environmental ethic has undergone an evolution of thought, moving from the apparently limitless resources of colonial America, through periods of expansion and industrialization to times when we find ourselves bumping up against the limits of the ecosphere and suffering the deterioration and loss of social, economic, and environmental amenities.

Key Terms

- abiotic
- absorption
- aerobic respiration
- ammonification
- anaerobic respiration
- anoxic
- artificial photosynthesis
- autotrophs
- bioaccumulation
- bioaccumulation factor (BAF)
- biochemical oxygen demand (BOD)
- bioconcentration
- bioconcentration factor (BCF)
- biodegradation
- biodiversity
- biogeochemical cycles
- biomass
- biosphere
- biota
- biotic
- carbon cycle
- carbon dioxide
- carbonaceous oxygen demand (CBOD)
- carnivores
- carrying capacity
- chemical oxygen demand (COD)
- climate
- community
- death phase
- denitrification
- ecological footprint
- ecological overshoot
- ecological redox sequence
- ecology
- ecosystem
- engineered system
- eukaryotes
- eutrophication
- exponential-growth model
- exponential or log growth phase
- food chains
- food webs
- half-saturation constant
- herbivores
- heterotrophs
- hydrologic cycle
- International Panel on Climate Change (IPCC)
- invasive species
- IPAT equation
- land ethic
- Leopold, Aldo
- *The Limits to Growth*
- logistic-growth model
- methanogenesis
- Monod model
- natural systems
- nitrification
- nitrogen fixation
- nitrogenous biochemical oxygen demand (NBOD)
- omnivores
- organic matter
- oxygen demand
- phosphorus cycle
- photosynthesis
- photosynthetic source (of oxygen)
- population
- population dynamics
- principle of competitive exclusion
- productivity
- prokaryotes
- reactive nitrogen
- redox reaction
- renewable resources
- respiration
- rotifers
- species
- stationary phase
- substrate
- sulfur cycle
- theoretical oxygen demand (ThOD)
- ultimate CBOD
- yield coefficient (Y)

5.1 The World Health Organization (WHO) reports that diarrhea causes 4 percent of all deaths worldwide, killing 2.2 million people every year, mostly children in developing countries. Diarrhea is a symptom of infection caused by members of which four organism groups listed in Figure 5.4 that can be spread by contaminated water?

5.2 Figure 5.4 described the major organism groups important to environmental engineering, many of which cause water-related diseases. Water-related diseases include those diseases derived from exposure to microorganisms or chemicals that are found in water humans drink. Other water-related diseases include those that have part of their life cycle in water (for example, schistosomiasis), diseases like malaria that have water-related vectors due sometimes to improper management of stormwater and solid waste, and other diseases such as legionellosis, which is carried by aerosols that contain disease causing organisms. Use Figure 5.4 and research you perform on the World Health Organization's web site on "water-related diseases" to answer the following questions: (a) *Giardia intestinalis* and *Cryptosporidium* are members of what group, (b) *Legionella* is a member of what group, (c) Hepatitis A and E are caused by members of what group, (d) typhoid fever is caused by *Salmonella typhi* and *Salmonella paratyphi*, which are part of what group, (e) cholera is caused by *Vibrio cholera*, which is a member of what group, (f) Ascariasis is an infection of the small intestine caused by the large roundworm, *Ascaris lumbricoides*, which is a member of what group?

5.3 Figure 5.4 described the major organism groups important to environmental engineering, many of which are used in treatment of domestic, agricultural, and industrial wastes. Identify the major organism groups listed in this figure (for example, viruses, bacteria, algae, protozoa, rotifers) that: (a) use solar energy to transfer oxygen into wastewater stabilization ponds (that is, lagoons), (b) are key organisms in the removal of the organic matter that makes up BOD in wastewater, (c) are single-cell organisms found in biological wastewater treatment and resource recovery systems that feed on bacteria and algae, (d) are multicellular organisms found in biological wastewater systems, (e) include the gram-positive organism, *Nocardia*, which is normally found in wastewater treatment plants, but if it experiences excessive growth, can result in foaming and poor settling of solids in the secondary settling reactor (that is, clarifier).

5.4 The WHO reports that malaria is the most important parasitic infectious disease. The WHO estimates there are 300–500 million cases of malaria globally with the primary burden assumed by those living in sub-Saharan Africa. Humans contract malaria after being bitten by a malaria-infected mosquito. These mosquitoes breed in fresh and sometimes brackish water, which may be because of improper management of irrigation water, stormwater, and solid waste. Malaria is known to be caused by four species of Plasmodium parasites (*P. falciparum*, *P. vivax*, *P. ovale*, *P. malariae*). Are these parasites prokaryotes or eukaryotes?

5.5 Mathematical models are used to predict the growth of a population, that is, population size at some future date. The simplest model is that for exponential growth. The calculation requires knowledge of the organism's maximum specific growth rate. A value for this coefficient can be obtained from field observations of population size or from laboratory experiments where population size is monitored as a function of time when growing at high substrate concentrations ($S \gg K_s$) (Table 5.16):

Table / 5.16

Field Observations of Population Size Over Time

Time (days)	Biomass (mg/L)
0	50
1	136
2	369
3	1,004
4	2,730
5	7,421

Calculate μ_{max} for this population, assuming exponential growth; include appropriate units.

5.6 Once a value for μ_{max} has been obtained, the model may be used to project population size at a future time. Assuming that exponential growth is sustained, what will the population size in Problem 5.5 be after 10 days?

5.7 Consider a population with an initial biomass (X_0) of 5 mg VSS/L, a maximum specific growth rate (μ_{max}) of 0.9/day, and a respiration rate coefficient of 0.15/day. Determine the biomass concentration (mg VSS/L) at the end of (a) 5 days and (b) 20 days.

5.8 Consider a population with the following characteristics; initial biomass (X_0) of 200 mg TSS/L, maximum specific growth rate (μ_{max}) of 1/day, and a respiration rate coefficient of 0.05/day. (a) Assume logistic growth. Determine the population's biomass (in units of TSS/L) after 2, 10, 100, and 10,000 days if the carrying capacity (K) is reported to be 5,000 mg TSS/L. (b) Assume exponential growth. Determine the population's biomass (in units of TSS/L) after 2, 10, 100, and 10,000 days.

5.9 Table 5.17 provides the U.S. Census Bureau estimate of the population of the city of Boston from 1680 to 2010.

Enter this data into a spreadsheet and make a graph of population over time. (a) Does the population growth follow exponential or logistic growth? (b) How many years did it take for Boston to double its population from 12,000 to 24,000 and 2 million to 4 million? How many years did it take to double the population to 8.8 million? (c) Using your knowledge of half-life from Chapter 3, determine the rate constant for this population growth. Is the rate of growth changing over time? If so, by what percent?

Table / 5.17

Population of Boston from 1680 to 2010

Year	Population	Year	Population
1680	4,500	1920	1,366,000
1720	12,000	1940	1,746,000
1775	16,000	1950	2,301,000
1800	24,900	1980	3,064,000
1830	85,600	1990	3,355,000
1860	374,000	2000	4,032,000
1900	1,009,000	2010	4,407,000

5.10 This chapter described that in 2007 the average rate growth of the global population was 1.2 percent per year. Determine the doubling time expected for the global population using this information.

5.11 The rate of increase for the human population in Mexico was 1.5 percent. (a) How long would you expect their population to double from its current level of 116 million people to 232 million in 2058? (b) What might prevent the population from reaching this level?

5.12 China and the United States were the top two producers of CO_2 emissions associated with energy consumption in 2009. Table 5.18 provides data of CO_2

Table / 5.18

Carbon Dioxide Emissions in China and the United States (in million metric tons)

Year	China	United States	Year	China	United States	Year	China	United States
1980	1,448.46	4,776.57	1990	2,269.71	5,041.00	2000	2,849.75	5,861.82
1981	1,439.86	4,646.85	1991	2,369.25	4,997.69	2001	2,969.58	5,753.70
1982	1,506.94	4,410.83	1992	2,449.16	5,093.53	2002	3,464.84	5,801.17
1983	1,593.39	4,388.02	1993	2,626.64	5,188.87	2003	4,069.24	5,850.63
1984	1,724.49	4,618.83	1994	2,831.55	5,261.43	2004	5,089.78	5,968.49
1985	1,857.81	4,604.84	1995	2,861.68	5,319.89	2005	5,512.70	5,991.47
1986	1,970.82	4,612.97	1996	2,893.38	5,506.37	2006	5,817.14	5,913.68
1987	2,102.78	4,769.96	1997	3,081.74	5,578.43	2007	6,260.03	6,018.13
1988	2,240.37	4,989.55	1998	2,967.26	5,617.03	2008	6,803.92	5,833.13
1989	2,275.34	5,069.96	1999	2,885.72	5,677.10	2009	7,710.50	5,424.53

emissions from energy consumption (in millions of metric tons) by the United States and China from 1980 to 2009 (data from the Energy Information Administration). (a) Enter this data into a spreadsheet and make a graph of CO_2 emissions from 1980 to 2009 for these two countries. (b) What is the percent increase in emissions for each country since 2000? (c) In a half-page, discuss how technology efficiency, price of fossil fuels, and growth of the economy would have impacted U.S. CO_2 emissions in 2008 and 2009? (d) If the population of China was approximately 1.28 billion and the population of the United States was approximately 301 million people, what were their emissions of CO_2 on metric tons per person basis? Compare your results to the 2009 global value of 4.49 metric tons of CO_2 per person.

5.13 Exponential growth cannot be sustained forever because of constraints placed on the organism by its environment, that is, the system's carrying capacity. This phenomenon is described using the logistic-growth model. (a) Calculate the size of the population in Problem 5.5 after 10 days, assuming that logistic growth is followed and that the carrying capacity is 100,000 mg/L. (b) What percentage of the exponentially growing population size would this be?

5.14 As reported by Mihelcic et al. (2009) "water demand is a function of the design population, minimum personal water requirements, and factors such as seasonal activities and infrastructural demands (e.g., from schools, churches, and clinics). Design population (P_N) is the projected population in the last year of the design life." It can be calculated as follows:

$$\text{for populations} < 2{,}000 \quad P_N = P_O\left(1 + \frac{r \times N}{100}\right)$$

or

$$\text{for populations} > 2{,}000 \quad P_N = P_O\left(1 + \frac{r}{100}\right)^N$$

Assume two rural communities in Honduras have initial populations of 1,500 and 2,200, respectively. If you are designing a water system with an expected life of 15 years (N), and the percentage rate of growth is expected to be 3 percent, what is the community size you would design for in 15 years for each community?

5.15 Using information provided in Problem 5.14, (a) determine the expected population for the community of 1,500 people after 15 years for estimated population rate of growths of 1, 2.5, and 5 percent. (b) Assume this rural community is losing population at a rate of 1 percent per year because of the global phenomenon of rural migration to urban areas. What is the expected population of the community in 15 years?

5.16 Food limitation of population growth is described using the Monod model. Population growth is characterized by the maximum specific growth rate (μ_{max}) and the half-saturation constant for growth (K_s). (a) Calculate the specific growth rate (μ) of the population in Problem 5.5 growing at a substrate concentration of 25 mg/L according to Monod kinetics if it has a K_s of 50 mg/L. (b) What percentage of the maximum growth rate for the exponentially growing population size would this be?

5.17 Laboratory studies have shown that microorganisms produce 10 mg/L of biomass in reducing the concentration of a pollutant by 50 mg/L. Calculate the yield coefficient, specifying the units of expression.

5.18 A pilot scale facility maintained under aerobic conditions has monitored the rate of removal of pollutant as 10 mg/L-h. What is the rate of growth of the microorganisms oxidizing the pollutant (mg cells/L-h) if their yield coefficient is reported to equal 0.40 lb cells/lb substrate?

5.19 When food supplies have been exhausted, populations die away. This exponential decay is described by a simple modification of the exponential-growth model. Engineers use this model to calculate the length of time that a swimming beach must remain closed following pollution with fecal material. For a population of bacteria with an initial biomass of 100 mg/L and a $k_d = 0.4$/day, calculate the time necessary to reduce the population size to 10 mg/L.

5.20 A population having a biomass of 2 mg/L at $t = 0$ days reaches a biomass of 139 at $t = 10$ days. Assuming exponential growth, calculate the value of the specific growth coefficient.

5.21 Fecal bacteria occupy the guts of warm-blooded animals and do not grow in the natural environment. Their population dynamics in lakes and rivers—that is, following a discharge of untreated domestic wastewater—can be described as one of exponential decay or death. How many days would it take for a bacteria concentration of 10^6 cell/mL to be reduced to the public health standard of 10^2 cell/mL if the decay coefficient is 2/day?

5.22 The 2012 Living Planet Report from the World Wildlife Fund (WWF, 2012) reported that in the year 2008, the Earth's total biocapacity was 12.0 billion global hectares, which equates to 1.8 global hectares per person. In contract, humanity's ecological footprint was reported to be 18.2 billion global hectares, which equates to 2.7 global hectares per person. (a) Using these values, how many years would it take the Earth to fully regenerate the renewable resources that humanity consumed in 1 year? (b) "Ecological Overshoot" is a term that describes when the global ecological footprint is larger than the Earth's biocapacity. What was the "ecological overshoot" in 2008 reported in global hectares per person? (c) Review Chapter 1 of the Living Planet Report, by what percent did the global living planet index decline between 1970 and 2008? What percent did the freshwater index decline over the same time period?

5.23 According to the web site maintained by Redefining Progress its latest footprint analysis indicates that humans are exceeding their ecological limits by 39 percent. Go to the following web site and determine your own ecological footprint. Record your value and compare it with those for your country and the world. Identify some changes you can make in your current lifestyle and then rerun the footprint calculator to reflect those changes. Summarize the changes you make and how they affect your ecological footprint. The web site is www.rprogress.org/ecological_footprint/about_ecological_footprint.htm.

5.24 Remediation of toluene in a contaminated groundwater aquifer has been found to have the following biokinetic coefficients for microbial growth. $\mu_{max} = 1.2$/day and $K_s = 0.31$ mg/L. What is the growth rate of the microorganism (day^{-1}) removing the toluene if the concentration of the pollutant is 1 ppb and 1 ppm?

5.25 A biological treatment process used to treat wastewater was found to have the following biokinetic coefficients: yield coefficient = 0.52 mg VSS/mg COD, half-saturation constant = 60 mg COD/L, and maximum specific growth rate = 0.96/day. What is the growth rate of the organisms (units of day^{-1}) if the organic matter in the reactor is: (a) a low-strength wastewater with 125 mg COD/L, (b) a high-strength wastewater with COD = 325 mg/L? (c) If the concentration of microorganisms in the biological reactor is 1,000 mg VSS/L, what is the rate of COD utilization?

5.26 What is the ThOD of the following chemicals? Show the balanced stoichiometric equation with your work: (a) 5 mg/L C_7H_3; (b) 0.5 mg/L C_6Cl_5OH; (c) $C_{12}H_{10}$.

5.27 A waste contains 100 mg/L ethylene glycol ($C_2H_6O_2$) and 50 mg/L NH_3–N. Determine the theoretical carbonaceous and the theoretical nitrogenous oxygen demand of the waste.

5.28 Calculate the NBOD and ThOD of a waste containing 100 mg/L isopropanol (C_3H_7OH) and 100 mg/L NH_3–N.

5.29 A waste contains 100 mg/L acetic acid (CH_3COOH) and 50 mg/L NH_3–N. Determine the theoretical carbonaceous oxygen demand, the theoretical nitrogenous oxygen demand, and the total ThOD of the waste.

5.30 A waste has an ultimate CBOD of 1,000 mg/L and a k_L of 0.1 /day. What is its 5-day CBOD?

5.31 A new manufacturing facility is being located in your town. It plans to produce 2,000 m^3/day of a wastewater that consists primarily of water and the chemical phenol dissolved in it at a concentration of 5 mg/L. Phenol has a chemical formula of C_6H_5OH. The company has asked the municipal wastewater treatment to consider treating this industrial waste. Your plant currently treats 30,000 m^3/day with an average influent of 350 mg COD/L. (a) Estimate the increase in COD loading (kg COD/day) if you accept the industrial waste discharge? (b) Estimate the additional amount of oxygen (in kg O_2/day) needed to oxidize the phenol at the treatment plant.

5.32 Untreated municipal wastewater in Europe may average 600 mg/L for the carbonaceous BOD$_5$, while in the United States this average value can be as low as 200 mg/L. One reason for this is because the United States has a greater water use per capita in the home than in Europe and also has problems associated with the infiltration/inflow of water into their wastewater collection system. (a) If the BOD rate constant for untreated wastewater is 0.35/day, calculate the BOD$_U$ of the untreated European and U.S. wastewaters. (b) Assume the dissolved-oxygen concentration of oxygen saturated dilution water used in the BOD test is 8 mg and you are using a 300 mL BOD bottle. Estimate the volume of sample you would add to the BOD bottle to ensure satisfactory test results for the European and U.S. samples (mL).

5.33 (a) Calculate the BOD$_U$ of a waste that has a measured BOD$_5$ of 20 mg/L, assuming a BOD rate coefficient of 0.15/day measured at 20°C. (b) Estimate the rate coefficient and resulting BOD ultimate if the temperature of the waste is raised to 30°C.

5.34 A 5 mL wastewater sample is placed in a standard 300 mL BOD bottle, and the bottle is filled with dilution water. The bottle had an initial dissolved-oxygen concentration of 9 mg/L and a final dissolved-oxygen concentration of 3.5 mg/L. A blank (a BOD bottle filled with dilution water) run in parallel showed no change in dissolved oxygen over the 5-day incubation period. The BOD reaction rate coefficient for the waste is 0.3/day. Calculate the 5-day (y_5) and ultimate (L_0) BOD of the wastewater.

5.35 A city has a population of 50,000 people, an average household generated wastewater flow of 430 L/day-person, and the average BOD_5 of the untreated wastewater in population equivalents is 0.1 kg BOD_5/day-person. If the BOD reaction rate coefficient for the waste stream is 0.4/day, determine the ultimate (L_0) BOD of the wastewater.

5.36 A 10 mL sample is added to a 300 mL BOD bottle. Dilution water is added to the sample bottle and the initial dissolved-oxygen concentration is measured as 8.5 mg/L. After the sample is sealed, the laboratory incorrectly takes a measurement of dissolved-oxygen reading on day 6 of 3 mg/L. If the BOD reaction rate coefficient for the sample is 0.30/day, (a) estimate what the BOD_5 should have been. (b) Estimate what the dissolved-oxygen reading should have been on day 5. (c) Determine the BOD_U of this sample.

5.37 You are provided the following BOD data collected over a 10-day period. Day 1: BOD = 28 mg/L, day 2: BOD = 45 mg/L; day 5: BOD = 89 mg/L; day 6: BOD = 100 mg/L; day 9: BOD = 120 mg/L. Calculate the BOD rate constant and the BOD_U of the sample.

5.38 Suppose the estimated BOD of an influent sample is expected to be 150 mg/L and the dissolved oxygen of oxygen saturated dilution water used in the BOD test is 8.5 mg/L. If you are using a 300 mL BOD bottle, estimate the maximum and minimum amount of sample you would add to the BOD bottle to ensure satisfactory test results.

5.39 If the BOD rate constant at 20°C is 0.12/day, what is the BOD rate constant at 10°C? What fraction of BOD_U would remain in a sample that has been incubated for 3 days (a) at 20°C and (b) at 10°C? (c) Solve for the fraction of BOD_U remaining at 20°C and 10°C, but after 6 days of incubation.

5.40 Suppose the estimated BOD of an influent sample is expected to be 150 mg/L and the dissolved oxygen of oxygen saturated dilution water used in the BOD test is 8.5 mg/L. If you are using a 300 mL BOD bottle, estimate the maximum and minimum amount of sample you would add to the BOD bottle to ensure satisfactory test results.

5.41 Excess nitrogen inputs to estuaries have been scientifically linked to poor water quality and degradation of ecosystem habitat. The nitrogen loading to Narragansett Bay was estimated to be 8,444,631 kg N/year and to Chesapeake Bay is 147,839,494 kg N/year. The watershed area for Narragansett Bay is 310,464 ha and for Chesapeake Bay is 10,951,074 ha. The nitrogen loading rates are estimated for Galveston Bay to be 16.5 kg N per ha per year, 26.9 kg N per ha per year for Tampa Bay, 49.0 kg N per ha per year for Massachusetts Bay, and 20.2 kg N per ha per year for Delaware Bay. (a) Rank the loading rates from lowest to highest for these six estuaries.

5.42 Excess nitrogen inputs to estuaries have been scientifically linked to poor water quality and degradation of ecosystem habitat. Perform a library search for the paper title "Nitrogen inputs to seventy-four southern New England estuaries: Application of a watershed nitrogen model (J. S. Latimer, and M. A. Charpentier, 2010. *Estuarine, Coastal and Shelf Science*, 89: 125–136). Based on this article, what is the percent contribution of the following four sources of nitrogen to the watershed of the New England estuaries? (a) direct atmospheric deposition to the estuaries, (b) wastewater, (c) indirect atmospheric deposition to the watershed of the estuary, (d) fertilizer runoff from lawns, golf courses, and agriculture.

5.43 Humans produce 0.8–1.6 L of urine per day. The annual mass of phosphorus in this urine on a per capita basis ranges from 0.2 to 0.4 kg P. (a) what is the maximum concentration of phosphorus in human urine in mg P/L? (b) what is the concentration in moles P/L? (c) most of this phosphorus is present as HPO_4^{2-}. What is the concentration of phosphorus in mg HPO_4^{2-}/L?

5.44 Assume 50 percent of phosphorus in human excrement in found in urine (the remaining 50 percent is found in feces). Assume humans produce 1 L of urine per day and the annual mass of phosphorus in this urine is 0.3 kg P. If indoor water usage is 80 gallons per day in a single individual apartment, what is the low and high range of phosphorus concentration (in mg P/L) in the wastewater that is discharged from the apartment unit? Make sure you account for phosphorus found in urine and feces.

5.45 Fish that reside in the Potomac River Estuary have BAFs reported as 26,200,000 (units of L/kg) for largemouth bass and 10,500,000 for white perch. If the concentration of PCB116 dissolved in the water column is 0.064 ng/L, what is the estimated concentration of PCB116 in the fish (ng/kg).

5.46 Concentrations of one particular PCB dissolved in the waters of Lake Washington near Seattle were reported to be 42 pg/L. Estimate the fish-specific BAF (in units of L/kg) for (a) cutthroat with a measured fish concentration of 375 ppb, (b) yellow perch with a measured fish concentration of 191 ppb, and (c) pikeminnows which have a measured fish concentration of 1,000 ppb.

5.47 Go to the IPCC web site (www.ipcc.ch). Choose a specific ecosystem to study, and use information from the web site to research the impact of climate change on that ecosystem. Write a one-page essay (properly referenced) summarizing your findings.

5.48 Biofuels are being suggested as a method to close the carbon loop. Do a library and Internet search on a particular type of biofuel. Write a one-page essay that addresses the link between biofuels and the global carbon cycle. Also address the impact that your particular biofuel may have on water quality, food supply, biodiversity, and air quality.

5.49 According to the U.S. Environmental Protection Agency, mountaintop coal mining consists of removal of mountaintops to expose coal seams and subsequent disposal of the associated mining overburden in adjacent valleys (see http://www.epa.gov/region3/mtntop/). In your own words, discuss the environmental impacts associated with mountain top mining. How does this relate to EPA's Healthy Waters Priority Program, and does this method of providing energy fit into a sustainable future that considers social, environmental, and economic balance?

5.50 The U.S. Environmental Protection Agency defines a TMDL as the calculation of the maximum amount of a pollutant that a waterbody can receive and still meet water quality standards. The TMDL approach is a way to apply carrying capacity to a particular waterbody. Under section 303(d) of the Clean Water Act, states, territories, and authorized tribes are required to develop lists of impaired waters. Go the following web site (http://water.epa.gov/lawsregs/lawsguidance/cwa/tmdl/index.cfm) that lists the states which EPA is under court order or agreed in a consent decree to establish TMDLs for. Produce a clear and easy-to-read map of all 50 states that shows the locations of these states.

References

Becker, D. S., and G. N. Bigham, 1995. Distribution of mercury in the aquatic food web of Onondaga Lake, NY. *Water, Air and Soil Pollution, 80*: 563–571.

Chapra, S. C., 1997. *Surface Water-Quality Modeling*. New York: McGraw-Hill.

Chichilnisky, G., and G. Heal, 1998. Economic returns from the biosphere. *Nature, 391*: 629–630.

Daily, G. C., 2000. Management objectives for the protection of ecosystem services. *Environmental Science & Policy, 3*: 333–339.

Davis, M. L., and D. A. Cornwell, 1991. *Introduction to Environmental Engineering*. New York: McGraw-Hill.

Delzer, G. C., S. W. McKenzie, 2007. Five-Day Biochemical Oxygen Demand, U.S. Geological Survey Techniques of Water-Resources Investigations, book 9, chap. A7, sec. 7.0, November 2007, accessed September 24, 2012, from http://pubs.water.usgs.gov/twri9A7/.

Effler, S. W., 1996. *Limnological and Engineering Analysis of a Polluted Urban Lake: Prelude to Environmental Management of Onondaga Lake, New York*. New York: Springer.

Enger, E. D., and B. F. Smith, 2010. *Environmental Science: A Study of Interrelationships*, 12th ed. Boston: McGraw-Hill.

EPA, 2007b. Memorandum: Nutrient Pollution and Numeric Water Quality Standards, May 25, 2008, www.epa.gov/waterscience/criteria/nutrient/files/policy20070525.pdf, Washington, D.C.

Environmental Protection Agency (EPA), 2011. Reactive Nitrogen in the United States: An Analysis of Inputs, Flows, Consequences, and Management Options. A Report of the Science Advisory Board, August 2011, EPA-SAB-11-103.

Goodland, R., 1997. Environmental sustainability in agriculture: Diet matters. *Ecological Economics, 23*(3): 189–200.

Intergovernmental Panel on Climate Change (IPCC), 2007a. Summary for policy makers. In *Climate Change 2007: Impacts, Adaptation and Vulnerability*. Contribution of Working Group II to the Fourth Assessment Report of the Intergovernmental Panel on Climate Change, M. L. Parry, O. F. Canziani, J. P. Palutikof, P. J. van der Linden, and, C. E. Hanson, Eds. Cambridge: Cambridge University Press, 7–22.

Intergovernmental Panel on Climate Change (IPCC), 2007b. Contribution of Working Group I to the Fourth Assessment Report of the Intergovernmental Panel on Climate Change, 2007.

Kupchella, C., and M. Hyland, 1986. *Environmental Science. Living within the System of Nature*, Needham Heights: Allyn and Bacon.

Metcalf and Eddy, Inc., 1989. *Wastewater Engineering: Treatment, Disposal, Reuse*, 2nd ed. New York: McGraw-Hill.

Mihelcic, J. R., 1999. *Fundamentals of Environmental Engineering*. New York: John Wiley & Sons.

Mihelcic, J. R., E. A. Myre, L. M. Fry, L. D. Phillips, and B. D. Barkdoll. *Field Guide in Environmental Engineering for Development Workers: Water, Sanitation, Indoor Air*. American Society of Civil Engineers (ASCE) Press, Reston, VA, 2009.

Mihelcic, J. R., L. M. Fry, and R. Shaw, 2011. Global potential of phosphorus recovery from human urine and feces. *Chemosphere, 84* (6): 832–839.

Millennium Ecosystem Assessment (MEA), 2005. *Ecosystems and Human Well-Being: Synthesis*. Washington, D.C.: Island Press.

Nemerow, N. L., 1971. *Liquid Waste of Industry: Theories, Practices, and Treatment*. Reading: Addison-Wesley.

Ricklefs, R. E., 1983. *The Economy of Nature*, 2nd ed. New York: Chirun Press.

Snoeyink, V. L., and D. Jenkins, 1980. *Water Chemistry*. New York: John Wiley & Sons.

Solomon, S., D. Qin, M. Manning, Z. Chen, M. Marquis, K.B. Averyt, M. Tignor, and H.L. Miller (eds.), *Contribution of Working Group I to the Fourth Assessment Report of the Intergovernmental Panel on Climate Change*. Cambridge University Press, Cambridge, United Kingdom and New York, NY, USA. http://www.ipcc.ch/publications_and_data/ar4/wg1/en/contents.html

Tchobanoglous, G., F. L. Burton, and H. D. Stensel, 2003. *Wastewater Engineering*, 4th ed. Boston: Metcalf & Eddy; New York: McGraw-Hill.

Thorne-Miller, B., 1999. *The Living Ocean: Understanding and Protecting Marine Biodiversity*. Washington, D.C.: Island Press.

Wackernagel, M., L. Onisto, A. Callejas Linares, I. S. López Falfán, J. Méndez García, A.I. Suárez Guerrero, and M. Guadalupe Suárez Guerrero, 1997. *Ecological Footprints of Nations: How Much Nature Do They Use? How Much Nature Do They Have?* Mexico: Centre de Estudios para la Sustentabilidad, Universidad Anáhuac de Xalapa. Commissioned by the Earth Council for the Rio+5 Forum.

World Wildlife Foundation (WWF), 2012. *Living Planet Report*. Gland: WWF International.

chapter/Six Environmental Risk

James R. Mihelcic and
Julie Beth Zimmerman

In this chapter, readers will learn about the distinction between a hazardous and a toxic chemical/material, the meaning of environmental risk, and methods for assessing environmental risk and incorporating it into engineering practice. Toxicity is a complex topic with many contributing factors. Thus, this chapter emphasizes that one of the best strategies for mitigating risk is to reduce or eliminate the use or generation of hazardous chemicals or materials through design. Readers will be introduced to methods to reduce hazard such as green chemistry, the toxic release inventory, and the pollution prevention hierarchy. Readers will also learn about the four components of a complete environmental risk assessment (hazard assessment, dose–response assessment, exposure assessment, and risk characterization), the impact of site-specific conditions on exposure to chemicals and ways that land use ultimately affects environmental risk. Finally, the chapter demonstrates the differences in developing a risk characterization for carcinogenic and noncarcinogenic compounds. The emphasis is on risk determination due to exposure to contaminated chemicals found in water, air, and food. The chapter also demonstrates the method used to determine allowable chemical concentrations in groundwater and contaminated soil.

Learning Objectives

1. Describe how to minimize or eliminate risk by designing for reduced hazard and/or reduced exposure.

2. Summarize the different types of hazard and their potential adverse impacts on human health and the environment.

3. Articulate the meaning of green chemistry, the toxic release inventory, and the pollution prevention hierarchy, and how these three items can be used in engineering practice to reduce environmental risk.

4. Describe the pollution prevention hierarchy, apply it to engineering practice, and describe its relationship to sustainability.

5. Define the terms *environmental justice* and *susceptible populations* in relation to risk assessment, and explain the roles engineers can play in addressing these topics.

6. Articulate the limitations of the risk assessment paradigm for the protection of human health and the environment and the factors that affect the toxicity of a given chemical, including uncertainty associated with data collection and interpretation.

7. Define the four components of a risk assessment and discuss the difference between risk assessment and risk perception.

8. Distinguish between chemical concentration, exposure, and dose.

9. Calculate the acceptable concentration and acceptable risk associated with exposure to a carcinogenic and noncarcinogenic chemical for various and multiple exposure pathways, including chemical partitioning between soil, air, and water phases.

10. Understand the relationship of bioaccumulation/bioconcentration, food web cycles, and toxicity.

11. Perform a basic risk assessment for carcinogens and noncarcinogens, given appropriate data including interpretation of a dose–response curve.

FISH CONTAMINATED DO NOT EAT

6.1 Risk and the Engineer

For the last 60 years of the 20th century, global chemical production increased several hundred times, so that by the end of the first decade of the 21st century, approximately 90,000 chemicals are commonly sold in commerce. Additionally, hundreds of new chemicals are introduced into the market each year. Individuals may be voluntarily or involuntarily exposed to these chemicals at home, at school, in the workplace, while traveling, or simply while jogging in a large urban area. The indoor environment is also becoming an important place for exposure to chemicals, because Americans now spend 85 percent of their time indoors. Therefore, indoor environments, particularly those that are poorly ventilated and have chemical-emitting carpeting, coatings, and adhesives, can have a large impact on human health. Many companies are making efforts to evaluate the potential risks of the chemicals and materials in their products and to use tools such as green chemistry to reduce the inherent risk throughout supply chain.

Risk is the likelihood of injury, disease, or death. In general terms,

$$\text{Risk} = f(\text{hazard}, \text{exposure}) \qquad (6.1)$$

Environmental risk is the risk resulting from exposure to a potential environmental hazard. Environmental hazards can be specific chemicals or chemical mixtures such as secondhand smoke and automobile exhaust. They can also be other hazards such as biological pathogens, stratospheric ozone depletion, climate change, and water scarcity. In this chapter, we will focus on environmental risk to humans derived from exposure to chemicals or materials. However, the concept of environmental risk can be applied to the health of plants, animals, and entire ecosystems—known as **ecotoxicity**—which support human livelihood and enhance our quality of life.

Table 6.1 summarizes many types of hazard, including physical hazards, toxicological hazards, and global hazards. It is important to note that the adjective *hazardous* does not only imply cancer-causing, but also includes any adverse impact to humans or to the environment as a result of exposure to a chemical or material. Further, there are many hazards beyond toxicological ones.

The risk of a chemical may involve its toxic effects or the hazard it presents to workers or to a community—for example, by causing an explosion. Risk has been historically managed by addressing the issue of **exposure**. For example, exposure may be limited by requiring that workers wear protective clothing or by developing warning signs for trucks that transport hazardous chemicals.

Because risk is the product of a function of hazard and of exposure, two implications become clear. As hazard approaches infinity (i.e., maximum toxicity), risk can only be reduced to near zero by reducing exposure to near zero. Conversely, as hazard approaches zero (i.e., inherently benign), exposure can approach infinity without significantly affecting risk. Green chemistry and engineering are methods aimed at minimizing hazard toward zero.

Table / 6.1

Hazard Categories and Examples of Potential Hazard Manifestations

Human Toxicity Hazards		Environmental Toxicity Hazards	Physical Hazards	Global Hazards
Carcinogenicity	Immunotoxicity	Aquatic toxicity	Explosivity	Acid rain
Neurotoxicity	Reproductive toxicity	Avian toxicity	Corrosivity	Global warming
Hepatoxicity	Teratogenicity	Amphibian toxicity	Oxidizers	Ozone depletion
Nephrotoxicity	Mutagenicity (DNA toxicity)	Phytotoxicity	Reducers	Security threat
Cardiotoxicity	Dermal toxicity	Mammalian toxicity (nonhuman)	pH (acidic or basic)	Water scarcity/flooding
Hematological toxicity	Ocular toxicity		Violent reaction with water	Persistence/bioaccumulation
Endocrine toxicity	Enzyme interactions			Loss of biodiversity

In green chemistry, risk is minimized by reducing or eliminating the hazard. As the intrinsic hazard is decreased, there is less reliance on exposure controls and therefore less likelihood for failure. The ultimate goal would be to use completely benign materials or chemicals such that there is no need to control exposure. That is, the chemicals and materials would not cause harm if they are released to the environment or humans are exposed to them.

These relationships among risk, hazard, and exposure are extremely important because current methods to protect human health and the environment are closely tied to the risk paradigm and depend almost exclusively on controlling exposure. There are efforts under way to develop a complimentary paradigm based on a foundation of sustainability to provide not only protection of human health of the environment but also consideration of social and economic benefits.

Green Chemistry
http://www.epa.gov/greenchemistry

Application / 6.1   Green Chemistry

Green chemistry, which emerged as a cohesive area of research in 1991, is defined as the design of products and processes that reduce or eliminate the use and generation of hazardous substances. The green chemistry approach was outlined in the framework of the *12 Principles of Green Chemistry* (Anastas and Warner, 1998), which has served as the guiding document for the field. Green chemistry is one of the most fundamental fields related to science and technology for sustainability in that it focuses at the molecular level to design chemicals and materials to be inherently nonhazardous.

The fundamental research of green chemistry has been brought to bear on a diverse set of challenges, including energy, agriculture, pharmaceuticals and health care, biotechnology, nanotechnology, consumer products, and materials. In each case, green chemistry has been successfully demonstrated to reduce intrinsic hazard, to improve material and energy efficiency, and to ingrain a life cycle perspective. Table 6.2 provides examples of green chemistry that illustrate its breadth of applicability.

Examples of Green Chemistry   Green chemistry reduces toxicity, minimizes waste, saves energy, and cuts down on the depletion of natural resources. It allows for advances in chemistry to occur in a much more environmentally benign way. In the future, when green chemistry is practiced by all chemists and all chemical-related companies, the term "green chemistry" will ideally disappear as all chemistry becomes green.

Polymers. Synthetic polymers or plastics are everywhere. They are used in cars, computers, planes, houses, eyeglasses, paints, bags, appliances, medical devices, carpets, tools, clothing, boats, batteries, and pipes. More than 60 million pounds of polymers are produced in the United States each year. The feedstocks that are used to produce these polymers are virtually all made from petroleum, a nonrenewable resource. Approximately 2.7 percent of all crude oil is used to generate chemical feedstocks.

In order to decrease human consumption of petroleum, chemists have investigated methods for producing polymers from renewable resources such as biomass. NatureWorks polylactic acid (PLA) is a polymer of naturally occurring lactic acid (LA), and LA can be produced from the fermentation of corn. The goal is to eventually manufacture this polymer from waste biomass. Another advantage of PLA is that, unlike most synthetic polymers that litter the landscape and pack landfills, it is biodegradable. PLA can also be easily recycled by conversion back into LA. It can replace many petroleum-based polymers in products such as carpets, bags, cups, and textile fibers.

Computer chips. The manufacture of computer chips requires excessive amounts of chemicals, water, and energy. Estimates indicate that the weight of chemicals and fossil fuels required to make a computer chip is 630 times the weight of the chip, as compared to the 2:1 ratio for the manufacture of an automobile. Scientists at the Los Alamos National Laboratory have developed a process that uses supercritical carbon dioxide in one of the steps in chip preparation, and it significantly reduces the quantities of chemicals, energy, and water needed to produce chips.

Dry cleaning. Condensed phase carbon dioxide is also used as a solvent for the dry cleaning of clothes. Although carbon dioxide alone is not a good solvent for oils, waxes, and greases, the use of carbon dioxide in combination with a surfactant allows for the replacement of perchloroethylene (which is the solvent used most often for dry cleaning clothes, although it poses hazards to the environment and is a suspected human carcinogen).

Other examples. Some other examples of green chemistry include the following:

- taking chromium and arsenic, which are toxic, out of pressure-treated wood
- using new less toxic chemicals for bleaching paper
- substituting yttrium for lead in auto paint
- using enzymes instead of a strong base for the treatment of cotton fibers

Read more: http://www.chemistryexplained.com/Ge-Hy/Green-Chemistry.html

Within the risk paradigm, the resulting engineering efforts to lower the probability of exposure to a wide range of hazards including toxicants, reactive substances, flammables, and explosives have been significant. However, this strategy is tremendously expensive. Worst of all, such an approach can and, as a probability function, will eventually fail. When exposure controls fail, risk is then equal to a function of hazard (see Equation 6.1). This relationship argues for the sustainability paradigm. That is, there is a need to design molecules, products, processes, and systems, and integrate understanding of societal behavior and economic goals into technical solutions, so health and safety do not depend on controls or systems that can fail or be sabotaged (either intentionally or accidentally) but instead rely on the use of benign (minimally hazardous) chemicals and materials.

During the early stages of design of a manufacturing process, there is a great amount of flexibility in developing solutions that prevents or minimizes risk by making decisions that eliminate the use and production of hazardous chemicals. An engineer would not intentionally design a manufacturing process if a direct result were that the process would cause plant workers and community members to contract cancer or that fish located in a local stream would die after exposure to wastewater discharge. Sadly, though, these adverse human and ecosystem impacts are the unintended consequences of many of our current practices of engineering design. However, engineers, with new sustainability awareness, are now embracing "green chemistry" and "green engineering" as a means to develop chemicals, materials, processes, and services that reduce or eliminate the use and generation of hazardous substances, leading to reduced risk to human health and the environment by reducing hazard.

Take buildings, for example. Reflect for a few minutes on the large variety of materials used in building construction, the large list of materials and coatings used to decorate, furnish, and insulate a building, and the large number of chemicals used during operation and maintenance of the building. How many of these structural

example/6.1   Limitations of Controlling Exposure

Describe a past or current event where the failure of an engineered control system allowed for the exposure of humans and the environment to a hazardous release.

solution

Many answers are possible. One example occurred in the early-morning hours of December 3, 1984, when a holding tank containing 43 tons of stored methyl isocyanate (MIC) from a Union Carbide factory in Bhopal, India, overheated and released a toxic, heavier-than-air MIC gas mixture. MIC is an extremely reactive chemical and is used in production of the insecticide carbaryl.

The post-accident analysis of the process showed the accident started when a tank containing MIC leaked. It is presumed that the scientific reason for the accident was that water entered the tank where about 40 m^3 of MIC was stored. When water and MIC mixed, an exothermic chemical reaction started, producing a lot of heat. As a result, the safety valve of the tank burst because of the increase in pressure. This reaction was so violent that the coating of concrete around the tank also broke. It is presumed that between 20 and 30 tons of MIC were released during the hour that the leak took place.

The gas leaked from a 30 m high chimney, and this height was not enough to reduce the effects of the discharge. The reason was that the high moisture content (aerosol) in the discharge evaporated and gave rise to a heavy gas that rapidly sank to the ground, where people had their residences. According to the Bhopal Medical Appeal, around 500,000 people were exposed. Approximately 20,000 are believed to have died as a result; on average, roughly one person dies every day from the effects. Over 120,000 continue to suffer from effects including breathing difficulties, cancer, serious birth defects, blindness, and other problems.

The U.S. Green Building Council has developed a system to certify professionals and the design, construction, and operation of green buildings. This rating system is referred to as **Leadership in Energy and Environmental Design (LEED)**. Building owners, professionals, and operators see advantages to having their building LEED certified. As an engineer (even as a student), you can take an exam to become a *LEED Accredited Professional*. It does not require any experience but does require that you study for and pass a written examination.

LEED scoring exists for new construction, existing buildings, and commercial interiors. Currently under development are methods for core and shell, homes, and neighborhood development. Table 6.3 provides the scoring for **LEED certification** related to new commercial or major renovation projects. Note that the categories involve many things that engineers deal with. These include issues such as site management, storm water management and water/wastewater use, specification of building materials, indoor air quality, and energy conservation.

materials, adhesives, sealants, floor and wall coverings, furniture components, and cleaning agents are selected based on the criteria to maximize the health and productivity of the building's inhabitants by minimizing potential adverse impact (the risk) to humans or the environment? Unfortunately the answer to this question is "very few." Green building design takes into consideration the health of building occupants along with the impact on the environment associated with material choices.

Poorly designed and managed indoor environments have a large adverse economic impact on society that is associated with increased health costs and lower worker productivity. As given in Table 6.4, huge savings could result from reducing that impact in the United States. In high-mortality developing countries, indoor air pollution is now responsible for up to 3.7 percent of the burden of disease. This is because a significant percentage of the environmental risk that leads to loss of disability-free days in the world is due to indoor air pollution from burning solid fuels. Engineers are beginning to address this in the design of innovative cookstoves that can significantly reduce indoor air emissions while considering cultural acceptance.

6.2 Risk Perception

Risk perception examines the judgments people make when they are asked to characterize and evaluate hazardous activities and technologies. People make qualitative or quantitative judgments about the current and desired riskiness of many different hazards through everyday choices and behaviors. These decisions are based on the perceived likelihood (probability) of an injury by a specific hazard and the severity of consequences associated with that injury.

Our judgments about risks are based on several considerations. One important factor is how familiar we are with the hazard. If we believe we know a lot about a hazard because we are often exposed to it, we often underestimate the degree of risk. Another factor is whether or not we are voluntarily being exposed to the hazard. When a person

© Skip O'Donnell/iStockphoto.

LEED Credits Associated with New Commercial Construction and Major Renovation   A project can obtain a maximum of 69 points. There are seven prerequisites (referred to as *Prereq.* in table) that all buildings must meet. No points are assigned for meeting prerequisites. Points (referred to as *credits*) are obtained in five categories, which are not equally balanced: (1) sustainable sites; (2) water efficiency; (3) energy and atmosphere; (4) materials and resources; (5) indoor air quality; and (6) innovation and design process. Certification is granted at four levels, based on the number of points awarded: Certified, 26–32 points; Silver, 33–38 points; Gold, 39–51 points; and Platinum, 52–69 points.

Sustainable Sites (14 Possible Points)		
Prereq. 1	Construction Activity Pollution Prevention	(required)
Credit 1	Site Selection	(1 point)
Credit 2	Development Density & Community Connectivity	(1 point)
Credit 3	Brownfield Redevelopment	(1 point)
Credit 4.1	Alternative Transportation, Public Transportation Access	(1 point)
Credit 4.2	Alternative Transportation, Bicycle Storage & Changing Rooms	(1 point)
Credit 4.3	Alternative Transportation, Low Emitting & Fuel Efficient Vehicles	(1 point)
Credit 4.4	Alternative Transportation, Parking Capacity	(1 point)
Credit 5.1	Site Development, Protect or Restore Habitat	(1 point)
Credit 5.2	Site Development, Maximize Open Space	(1 point)
Credit 6.1	Stormwater Design, Quantity Control	(1 point)
Credit 6.2	Stormwater Design, Quality Control	(1 point)
Credit 7.1	Heat Island Effect, Non-Roof	(1 point)
Credit 7.2	Heat Island Effect, Roof	(1 point)
Credit 8	Light Pollution Reduction	(1 point)
Water Efficiency (5 Possible Points)		
Credit 1.1	Water Efficient Landscaping, Reduce by 50%	(1 point)
Credit 1.2	Water Efficient Landscaping, No Potable Use or No Irrigation	(1 point)
Credit 2	Innovative Wastewater Technologies	(1 point)
Credit 3.1	Water Use Reduction, 20% Reduction	(1 point)
Credit 3.2	Water Use Reduction, 30% Reduction	(1 point)
Energy and Atmosphere (17 Possible Points)		
Prereq. 1	Fundamental Commissioning of the Building Energy Systems	(required)
Prereq. 2	Minimum Energy Performance	(required)
Prereq. 3	Fundamental Refrigerant Management	(required)
Credit 1	Optimize Energy Performance	(1–10 points)
Credit 2	On-Site Renewable Energy	(1–3 points)
Credit 3	Enhanced Commissioning	(1 point)
Credit 4	Enhanced Refrigerant Management	(1 point)
Credit 5	Measurement & Verification	(1 point)
Credit 6	Green Power	(1 point)
Materials and Resources (13 Possible Points)		
Prereq. 1	Storage & Collection of Recyclables	(required)
Credit 1.1	Building Reuse, Maintain 75% of Existing Walls, Floors & Roof	(1 point)

(continued)

Credit 1.2	Building Reuse, Maintain 95% of Existing Walls, Floors & Roof	(1 point)
Credit 1.3	Building Reuse, Maintain 50% of Interior Non-Structural Elements	(1 point)
Credit 2.1	Construction Waste Management, Divert 50% from Disposal	(1 point)
Credit 2.2	Construction Waste Management, Divert 75% from Disposal	(1 point)
Credit 3.1	Materials Reuse, 5%	(1 point)
Credit 3.2	Materials Reuse, 10%	(1 point)
Credit 4.1	Recycled Content, 10% (post-consumer +1/2 pre-consumer)	(1 point)
Credit 4.2	Recycled Content, 20% (post-consumer +1/2 pre-consumer)	(1 point)
Credit 5.1	Regional Materials, 10% Extracted, Processed & Manufactured Regionally	(1 point)
Credit 5.2	Regional Materials, 20% Extracted, Processed & Manufactured Regionally	(1 point)
Credit 6	Rapidly Renewable Materials	(1 point)
Credit 7	Certified Wood	(1 point)
Indoor Environmental Quality (15 Possible Points)		
Prereq. 1	Minimum IAQ Performance	(required)
Prereq. 2	Environmental Tobacco Smoke (ETS) Control	(required)
Credit 1	Outdoor Air Delivery Monitoring	(1 point)
Credit 2	Increased Ventilation	(1 point)
Credit 3.1	Construction IAQ Management Plan, During Construction	(1 point)
Credit 3.2	Construction IAQ Management Plan, Before Occupancy	(1 point)
Credit 4.1	Low-Emitting Materials, Adhesives & Sealants	(1 point)
Credit 4.2	Low-Emitting Materials, Paints & Coatings	(1 point)
Credit 4.3	Low-Emitting Materials, Carpet Systems	(1 point)
Credit 4.4	Low-Emitting Materials, Composite Wood & Agrifiber Products	(1 point)
Credit 5	Indoor Chemical & Pollutant Source Control	(1 point)
Credit 6.1	Controllability of Systems, Lighting	(1 point)
Credit 6.2	Controllability of Systems, Thermal Comfort	(1 point)
Credit 7.1	Thermal Comfort, Design	(1 point)
Credit 7.2	Thermal Comfort, Verification	(1 point)
Credit 8.1	Daylight & Views, Daylight 75% of Spaces	(1 point)
Credit 8.2	Daylight & Views, Views for 90% of Spaces	(1 point)
Innovation and Design Process (5 Possible Points)		
Credit 1.1	Innovation in Design	(1 point)
Credit 1.2	Innovation in Design	(1 point)
Credit 1.3	Innovation in Design	(1 point)
Credit 1.4	Innovation in Design	(1 point)
Credit 2	LEED Accredited Professional	(1 point)

SOURCE: Version 2.2, from U.S. Green Building Council.

Table / 6.4

Estimated Economic Benefits to U.S. Society if Architects and Engineers Design and Operate Buildings with Consideration of Health

$6 billion–$14 billion from reduced respiratory disease
$1 billion–$4 billion from reduced allergies and asthma
$10 billion–$30 billion from reduced sick-building syndrome
$20 billion–$160 billion from increased worker productivity unrelated to health

SOURCE: Fisk (2000).

© Fotobacca /iStockphoto.

voluntarily takes on a risk, he or she also usually underestimates the chances of a resulting injury. This may have to do with how much control individuals feel they have over the situation. Examples of voluntary risk include smoking, driving a car faster than the speed limit, and participating in activities such as mountaineering or skydiving. Also, individuals often feel it is more acceptable to choose a risk than to be put at risk by government or industry.

This attitude toward involuntary versus voluntary risk explains why there is usually a public outcry when a factory contaminates local drinking water or air quality is found to be unsafe. In these cases, the added risk from exposure to contaminated water and air is not voluntary. Individuals feel as if they are being subjected to hazards beyond their control and without their knowledge. Examples of involuntary risk are inhaling secondhand smoke, having a highway or high-voltage power lines placed in your community, and having pesticide residue on the outside of your produce.

6.3   Hazardous Waste and Toxic Chemicals

Exposure to a toxic or hazardous chemical can result in death, disease, or some other adverse impact such as a birth defect, infertility, stunted growth, or a neurological disorder. For humans, this contact with a chemical is typically through ingestion, inhalation, or skin contact. Exposure to the chemical can be associated with drinking water, eating food, ingesting soil and dust, inhaling airborne contaminants that could be in a vapor or particulate form, and contacting chemicals that are transported through the skin. Exposure can also be acute or chronic.

The **Toxics Release Inventory (TRI)** provides information to the public about hazardous waste and toxic chemicals. This inventory was established under the **Emergency Planning and Community Right-to-Know Act (EPCRA)** of 1986 and expanded by the Pollution Prevention Act of 1990. The TRI is a publicly available database, published by the Environmental Protection Agency (EPA) that contains information on releases of nearly 650 chemicals and chemical categories, submitted by

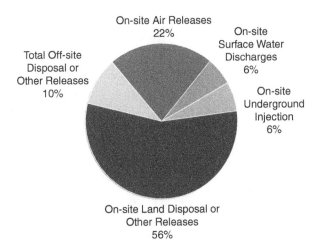

Figure / 6.1 TRI Disposal and Other Releases Reported to the EPA in 2010 Total was 3.93 billion pounds (data from EPA, 2011a).

over 23,000 industrial and federal facilities. The TRI tracks disposal or other releases both on-site and off-site, including wastes directly to air, land, surface water, and groundwater. It also provides information on other waste management strategies, such as recycling, energy recovery, treatment, and discharge to wastewater treatment plants. EPA released the 2010 TRI National Analysis in January 2012. Figure 6.1 shows the breakdown of the 3.93 billion pounds of toxic chemicals disposed of, or released, in 2010. Note that the largest amount of toxic chemicals is disposed/released to land, followed by air, and surface water and underground injection.

The TRI database can be searched by year, geographical location (ZIP code), chemical released, or industry type. Citizens and emergency-response personnel can look up emissions of toxic chemicals in their community. The TRI provides the public with unprecedented access to information about toxic chemical releases and other waste management activities on a local, state, regional, and national level under "the right to know" paradigm. One goal of the TRI is to empower citizens, through information, to hold companies and local governments accountable for how toxic chemicals are managed in their community.

Figure 6.2 shows the total mass of TRI emissions since 2001 (data is available back to 1988), along with the number of facilities reporting releases. The TRI data help the public, government officials, and industry meet three objectives: (1) identify potential concerns and gain a better understanding of potential risks; (2) identify priorities and opportunities to work with industry and government to reduce toxic chemical disposal or other releases and potential risks associated with them; and (3) establish reduction targets and measure progress toward reduction goals.

Figure 6.3 shows the location of facilities that must report the release or disposal of toxic chemicals in the Denver, St. Louis, and Philadelphia metropolitan areas. Note in this figure that in Denver the facilities are concentrated closer to the urban center, while in St. Louis, and especially in Philadelphia, the facilities that emit toxic chemicals are more widely distributed. As seen in Table 6.5, each urban area has over one

Identify TRI Facilities by ZIP Code
http://www.scorecard.org/

Toxics Release Inventory
http://www.epa.gov/tri

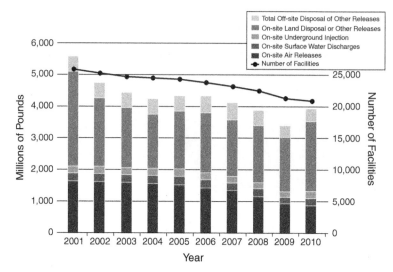

Figure 6.2 TRI Emissions (in million pounds) since 2001 and Number of Facilities Reporting Emissions.

(Adapted from www.epa.gov/tri.).

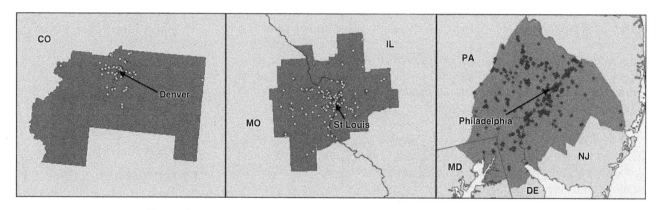

Figure 6.3 TRI Facilities Located in Urban Areas (left to right): Denver, Greater St. Louis, and Philadelphia Metropolitan Areas (2010 TRI data from EPA, 2011a).

hundred facilities that must report their chemical releases under the TRI. Also, each of the urban areas reports well over 1 lb of reportable toxic chemicals emitted per person. Note also how only two of the locations (Greater St. Louis and Philadelphia) reported decreases in release over the past 10 years. You can see from Table 6.5 that the ranking of industrial sectors that release toxic chemicals are different in each area, but generally consist of economic activities that produce chemicals, petroleum, electricity, metals, and food. In addition, observe how the majority of releases may be to air, water, or land depending on the geographical location (EPA, 2011a).

6.3.1 HAZARDOUS WASTE

In the United States, a **hazardous waste** is a regulatory subset of a solid waste. Solid wastes are defined under the **Resource Conservation and Recovery Act (RCRA)**. This regulatory definition says nothing about

Class Discussion

Why do you think that TRI emissions have declined since 1988? What other factors could be impacting these discharges besides community pressure as a result of the information being publicly available?

Table / 6.5

TRI Releases (left to right): Denver, Greater St. Louis, and Philadelphia Metropolitan Areas (2010 TRI data from EPA, 2011a).

	Denver	Greater St. Louis	Philadelphia
Approximate population	2.5 million	2.8 million	6 million
Number of TRI facilities	100	191	314
Total on-site and off-site disposal and Releases	4.4 and 1.2 million pounds	27.2 and 3.7 million pounds	12.7 and 5.8 million pounds
Total on-site releases to air	0.9 million pounds	5.3 million pounds	4.8 million pounds
Total on-site releases to water	0.2 million pounds	1.2 million pounds	7.2 million pounds
Total on-site releases to land	3.4 million pounds	20.7 million pounds	0.7 million pounds
Change in on-site and other releases (2001–2010)	+85%	−41%	−17%
Top five industrial sectors contributing to TRI	(1) Hazardous waste management, (2) metal mining, (3) electric utilities, (4) fabricated metals, (5) food/beverages/tobacco	(1) Metals, (2) electric utilities, (3) petroleum, (4) chemicals, (5) fabricated metals	(1) Chemicals, (2) petroleum, (3) primary metals, (4) food/beverages/tobacco, (5) electric utilities

the waste's physical state, so some "solid" wastes are in liquid form. In the United States, **solid wastes** are legally defined as any discarded material not excluded by 40 C.F.R. 261.4(a). *Excluded wastes* include items such as domestic sewage, household hazardous waste, fly ash and bottom ash from coal combustion, and manure returned to soil. C.F.R. is the abbreviation for Code of Federal Regulations, the document in which federal regulations are published. The number 40 indicates the section of the C.F.R. related to the environment. The C.F.R. can be accessed via the Internet.

Thus, a hazardous waste denotes a regulated waste. Only certain waste streams are designated as hazardous under federal regulations. Wastes are classified as hazardous based on: (1) physical characteristics such as reactivity, corrosivity, and ignitability; (2) toxicity; (3) the quantity generated; and (4) the history of the chemical in terms of environmental damage it caused and the likely environmental fate. Hazardous wastes thus may or may not exhibit toxicity.

6.3.2 TOXICITY

Environmental toxicology, also known as environmental health sciences, is an interdisciplinary field dealing with the effects of chemicals on living organisms. Because energy and material are distributed and cycled through food webs, it is likely that an impact on one level will be reflected in other levels as well. For example, there is evidence that elevated polychlorinated biphenyl (PCB) levels in fish results in

Access the Code of Federal Regulations
http://www.gpoaccess.gov/cfr/

Construction and Demolition Debris
http://www.epa.gov/wastes/conserve/imr/cdm/index.htm

adverse health effects to children born from mothers who included contaminated fish in their diet. While **bioaccumulation** (concentration of a chemical builds up in an organism over time) of PCBs may have had no direct adverse effect on adult fish, there was an impact on some fish offspring and the next trophic level (humans).

Toxic effects can be divided into two types: **carcinogenic** and **noncarcinogenic**. A carcinogen promotes or induces tumors (cancer), that is, the uncontrolled or abnormal growth and division of cells. Carcinogens act by attacking or altering the structure and function of DNA within a cell. Many carcinogens seem to be site-specific; that is, a particular chemical tends to affect a specific organ. In addition, carcinogens may be categorized based on whether they cause direct or indirect effects: *primary carcinogens* directly initiate cancer; *pro-carcinogens* are not carcinogens but are metabolized to form carcinogens and thus indirectly initiate cancer; *co-carcinogens* are not carcinogens but enhance the carcinogenicity of other chemicals; and *promoters* enhance the growth of cancer cells.

Classification of a chemical as being carcinogenic to humans requires sufficient evidence that human exposure leads to a significantly higher incidence of cancer. Such evidence is often collected from workers in job environments where there is prolonged contact with a chemical. (This is called epidemiological data.) While there are few *known* human carcinogens (for example, benzene, vinyl chloride, arsenic, and hexavalent chromium), many chemicals are *probable* human carcinogens (for example, benzo(*a*)pyrene, carbon tetrachloride, cadmium, and PCBs), and hundreds of chemicals have *suggestive evidence* that they are carcinogens. As we will discuss later, chemicals are listed as **suspected carcinogens** when experimental evidence indicates increased cancer risk in test animals and insufficient information is available to show a direct cause–effect relationship for humans.

Noncarcinogenic effects include all toxicological responses other than carcinogenic, of which there are countless examples: organ damage (including kidney and liver), neurological damage, suppressed immunity, and birth and developmental (harming an organism's reproductive ability or intelligence) effects. For example, elevated

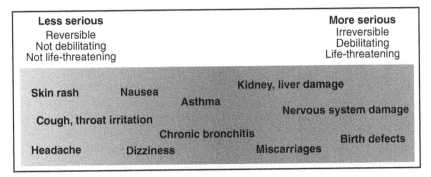

Figure / 6.4 Continuum of Health Risks Due to Exposure to Noncarcinogens Ranging from Less Serious to More Serious.

lead levels in children have been shown to cause learning disorders and lower IQs. The toxic effects manifested following exposure to a chemical often result from interference with enzyme (catalyst) systems that mediate the biochemical reactions critical for organ function. Figure 6.4 depicts the continuum of risks due to exposure to noncarcinogens as ranging from less serious to more serious. Risks that are reversible, not debilitating, and/or not life-threatening are considered to be of less concern than those that are irreversible, debilitating, and/or life threatening.

Chemicals collectively known as **endocrine disruptors** exert their effects by mimicking or interfering with the actions of hormones, biochemical compounds that control basic physiological processes such as growth, metabolism, and reproduction. Endocrine disruptors may exert noncarcinogenic or carcinogenic effects. They are believed to contribute to breast cancer in women and prostate cancer in men. Chemicals identified as endocrine disruptors include pesticides (such as DDT and its metabolites), industrial chemicals (such as some surfactants and PCBs), some prescription drugs, and other contaminants, such as dioxins (National Science and Technology Council, 1996).

The likelihood of a toxicological response is determined by the exposure to a chemical (one factor in Equation 6.1): a product of the chemical dose and the duration over which that dose is experienced. In humans, there are three major **exposure pathways**: ingestion (eating and/or drinking), inhalation (breathing), and dermal (skin) contact. Table 6.6 lists important factors that affect the toxicity of a chemical or material.

Some chemicals (for example, dioxin) can be lethal to test animals in very small doses, whereas others create problems only at much higher levels. Table 6.7 lists chemical compounds with widely varying toxicities. Here, **toxicity** is defined as causing death, an experimental end point that (for test animals) is more readily determined than, for example, lung cancer.

A common method of expressing toxicity is in terms of the **median lethal dose (LD$_{50}$)**, which is the dose that results in the death of 50 percent of a test organism population. The LD$_{50}$ is

Endocrine Disruptors

http://ww2.setac.org/node/100
http://www.who.int/ipcs/assessment/en/

Table / 6.6

Factors That Affect Toxicity of a Chemical or Material

Form and innate chemical activity

Dosage, especially dose–time relationship

Exposure route and timing

Duration of exposure

Species

Ability to be absorbed

Metabolism

Distribution within the body

Excretion

Presence of other chemicals

Susceptibility of receiving organism

Endocrine-disrupting chemicals are chemicals that, when absorbed into the body, either mimic or block hormones and disrupt the body's normal functions. This disruption can happen through altering normal hormone levels, halting or stimulating the production of hormones, or changing the way hormones travel through the body, thus affecting the functions of these hormones control.

These chemicals and substances are accumulating in fish and wildlife, and the number of warnings about eating of fish and wildlife due to endocrine disruptors is increasing and has reached over 30 percent of U.S. lakes and 15 percent of U.S. river miles. Studies document that these chemicals are accumulating in fish and wildlife to levels that are causing serious hormonal and reproductive effects in fish and wildlife at the top of the food chain, including wading birds, alligators, Florida panthers, minks, polar bears, seals, and beluga and orca whales. Many subpopulations with significant exposure are experiencing major reproductive effects, resulting in infertility and reproductive failures.

In humans, several health problems possibly linked to endocrine-disrupting chemicals have been recorded: (1) declines in sperm count in many countries; (2) a 55 percent increase in incidence of testicular cancer from 1979 to 1991 in England and Wales; (3) increases in prostate cancer; and (4) an increase in breast cancer in women, including an annual increase of 1 percent in the United States since the 1940s (Friends of the Earth, 2009).

In wildlife, the following are examples of effects that have been linked to endocrine-disrupting chemicals: (1) masculinization of female dog whelks (a type of shellfish); (2) eggs found in testes of roach fish in many rivers in the United Kingdom; (3) low egg viability, enlarged ovaries, and reduced penis size in Florida alligators; and (4) eggshell thinning and female–female pairing in birds (Friends of the Earth, 2009).

The risks associated with endocrine-disrupting chemicals are only just beginning to be discovered and quantified, because the doses that cause the effects are much lower than those traditionally tested in toxicity studies.

typically presented as the mass of contaminant dosed per mass (body weight) of the test organism, using units of mg/kg. Thus, a rodenticide with an LD_{50} of 100 mg/kg would result in the death of 50 percent of a population of rats, each weighing 0.1 kg, if

Table / 6.7

Oral Median Lethal Dose for Various Organisms and Chemicals

Chemical	Organism	LD_{50} (mg chemical/kg body weight)
Methyl ethyl ketone	Rat	5,500
Fluoranthene	Rat	2,000
Pyrene	Rat	800
Pentachlorophenol	Mouse	117
Lindane	Mouse	86
Dieldrin	Mouse	38
Sarin (nerve gas)	Rat	0.5

SOURCE: Values from Patnaik, 1992.

example/6.2 Chromium Toxicity

Which form of chromium, Cr(III) or Cr(VI), is toxic?

solution

The toxicity of chromium varies greatly depending on which oxidative state it is in. Cr(III), or Cr^{+3}, is relatively nontoxic, whereas Cr(VI), or Cr^{+6}, causes skin or nasal damage and lung cancer. Of course, chemicals can undergo oxidation and reduction reactions in environmental conditions, so release of the lower-toxicity form does not mean that the chromium will pose no risk to human health or the environment.

applied at a dose of 10 mg per rat. A dose of 20 mg per 0.1 kg rat should result in the death of more than 50 percent of the population, and a dose of 5 mg per 0.1 kg rat would result in the death of less than 50 percent.

A similar term, the **median lethal concentration (LC_{50})**, is typically used in studies of aquatic organisms and represents the ambient aqueous contaminant concentration (as opposed to injected or ingested dose) at which 50 percent of the test organisms die.

To identify LD_{50} or LC_{50}, a series of experiments at various concentrations yields a **dose–response curve** as depicted in Figure 6.5. More subtle (behavioral or developmental) changes can also reflect a toxic response but are difficult to assess. These nonlethal end points are measured as an effective concentration that affects 50 percent of the population (EC_{50}).

Recall that what determines toxicity is not only the dose, but also duration of exposure to a chemical or substance. **Acute toxicity** refers to death (or some other adverse response) resulting from short-term (hours to days) exposure to a chemical. **Chronic toxicity** refers to a response resulting from long-term (weeks to years) exposure to a chemical.

Acute effects are typically experienced at higher contaminant concentrations than are chronic effects. For example, the EPA has established acute (1.7 µg/L) and chronic (0.91 µg/L) water-quality criteria for mercury (II) to protect aquatic life in the Great Lakes from toxic effects. Here, the acute criterion is higher than the chronic value. As duration increases, the concentrations that can be tolerated without adverse effect are lower. Acute copper toxicity for rainbow trout has been shown to decrease from an LC_{50} of 0.39 mg/L at a 12 h duration to 0.13 mg/L at 24 h to 0.08 mg/L at 96 h. The toxicity of a specific chemical may also vary among species. Table 6.8 demonstrates this effect, comparing 48 h LC_{50} values for 2,4-dichlorophenoxyacetic acid (2,4-D), a common herbicide used on farms and household lawns, for various aquatic organisms.

While the concentrations of 2,4-D listed in Table 6.8 are not likely to be encountered in surface waters (although levels of agricultural

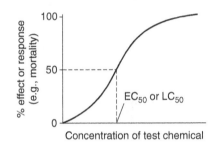

Figure / 6.5 Typical Form of Dose–Response Curve Used in Identifying EC_{50} and LC_{50} for Chemicals and Test Organisms.

Table / 6.8

48-Hour LC_{50} Values for 2,4-D for Selected Organisms

Species	LC_{50} (mg/L)
Daphnia magna (zooplankton)	25
Fathead minnow	325
Rainbow trout	358

SOURCE: Patnaik, 1992.

chemicals in runoff do increase following spring rains and snowmelt), the observed variation in LC_{50} values suggests a scenario in which sediment-living microcrustacean populations would be affected while fish populations would not. Such a scenario could potentially alter and disrupt the food web, with ecosystem-wide impacts. An understanding of food web function and the bioaccumulation and toxicity of contaminants (at each trophic level) is necessary to adequately assess the risk posed by the myriad chemical contaminants introduced into our environment (see Chapter 5).

The species-specific nature of toxicity presents a fundamental shortcoming in procedures commonly applied to estimate effects on humans based on experiments with test animals. Individual humans may be substantially more or less susceptible to the toxic effects of a specific compound at a given dose than are laboratory surrogate organisms. When animal studies are used to determine standards for human exposure, the uncertainties involved in utilizing the results are accounted for through use of conservative assumptions and application of safety factors that may result in an estimate that is conservative by several orders of magnitude—an approach based on a "better safe than sorry" philosophy. In addition, the fact that some wildlife may be more sensitive to toxic chemicals than humans has led to the promulgation of water-quality criteria in which the more stringent of wildlife- or human-health-based standards govern discharge limits. For example, the maximum contaminant level (MCL) for allowable chromium in drinking water is 0.1 mg/L, while the acute criterion for freshwater aquatic life is 21 μg/L. In this case, the wildlife standard is approximately one-fifth of the human-health-based value.

Sensitive segments of a population, known as **susceptible populations**, must receive distinct consideration in determining toxic effects of chemicals or substances. The embryonic, juvenile, elderly, and/or ill segments of any population (human or environmental) are likely to be more susceptible to adverse effects from chemical exposure than are healthy young adults. In some cases, the sex of an individual may also influence its susceptibility.

Synergistic toxicity, resulting from the exposure to multiple chemicals, is a phenomenon that is receiving increased attention. For example, consider two compounds with LC_{50} values of 5 and 20 mg/L, respectively. When present together, their individual LC_{50} values might drop to 3 and 10 mg/L, levels that are lower than the individual LC_{50} values. In some cases, chemicals may have the opposite (antagonistic) effect, resulting in a combination that is less toxic than when present separately.

Table 6.9 provides an example of the possible effects chemical mixtures could have on combined toxicity. Note that the combined effect of the two chemicals (A and B) can be greater or less. This is an example of the difficulty in assessing the risks of chemical mixtures. Unfortunately, scientific studies of chronic synergistic effects are lacking, largely because countless numbers of chemicals and combinations of chemicals exist and because experiments requiring exposure to chemicals of potential concern involve inherent difficulties.

Mercury Report to Congress
http://www.epa.gov/mercury/
reportover.htm

Table / 6.9

Potential Combined Toxicity Resulting from Exposure to a Mixture of Chemicals A and B

Type of Interaction	Toxic Effect, Chemical A	Toxic Effect, Chemical B	Combined Effect, Chemicals A + B
Additivity	20%	30%	50%
Antagonism	20%	30%	5%
Potentiation	0%	20%	50%
Synergism	5%	10%	100%

6.3.3 POLLUTION PREVENTION

Pollution prevention is focused on increasing the efficiency of a process to reduce the amount of pollution generated. This is the idea of incrementalism or **eco-efficiency**, where the current system is tweaked to be better than before. This does not take into account that the current design may not be the best or most appropriate for the current application. That is, the current product, process, or system was not designed with the intent of reducing waste and/or environmental impact. Instead, it is being improved within its current constraints, taking these considerations into account after the fact, after the design has been completed and often has been implemented.

The Pollution Prevention Act of 1990 (see Application 6.5) was passed to *encourage* (not regulate) pollution prevention in the United States. It establishes a **pollution prevention hierarchy** (Figure 6.6) as follows:

- **Source reduction**—Waste (hazardous substance, pollutant, or contaminants) should be prevented at the source (prior to recycling, treatment, or disposal).

- **Recycling**—Waste generated should be reused either in the process that created it or in another process.

- **Treatment**—Waste that cannot be recycled should be treated to reduce its hazard.

- **Disposal**—Waste that is not treated should be disposed of in an environmentally safe manner.

In the case of wastewater treatment, the pollution prevention hierarchy suggests that we may want to focus efforts on identifying ways to eliminate waste materials from being sewered and transported to a treatment plant, rather than devoting all of our efforts to improving the design of treatment facilities. In terms of solid-waste management, it is

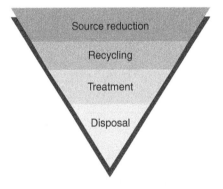

Figure / 6.6 Pollution Prevention Hierarchy.

Class Discussion

The pollution prevention hierarchy clearly shows the source reduction is favored over the other three aspects of pollution prevention. Disposal is the least preferred alternative. How does the pollution prevention hierarchy relate to industrial wastewater treatment and solid-waste management?

The Pollution Prevention Act focused industry, government, and public attention on reducing the amount of pollution through cost-effective changes in production, operation, and raw-materials use. Opportunities for source reduction are often not realized because of existing regulations, and the industrial resources required for compliance focus on treatment and disposal. Source reduction is fundamentally different from and more desirable than waste management or pollution control (2 U.S.C. 13,101 and 13,102, s/s et seq., 1990).

The Congress hereby declares it to be the national policy of the United States that pollution should be prevented or reduced at the source whenever feasible; pollution that cannot be prevented should be recycled in an environmentally safe manner, whenever feasible; pollution that cannot be prevented or recycled should be treated in an environmentally safe manner whenever feasible; and disposal or other release into the environment should be employed only as a last resort and should be conducted in an environmentally safe manner.

(2 U.S.C. 13,101b)

Where Can I Recycle My Stuff?
http://earth911.com

clear that landfill disposal is not the recommended alternative for managing most components of a waste stream. In this case, an engineer would think beyond design of a landfill and focus on broader initiatives to reduce the amount of waste that is generated and discarded.

Allowing more degrees of design freedom and moving upstream for opportunities to redesign the product, process, or system offer greater opportunity for **waste minimization** or even waste elimination. While there may be current barriers, including scientific, technical, or economic, to zero-waste design, it is important to note that the concept of waste is human. In other words, there is nothing inherent about materials, energy, space, or time that makes it waste. It is waste only because no one has yet imagined or implemented a defined use for it.

If the creation of waste cannot be avoided under given conditions or circumstances, designers and engineers can consider alternative mechanisms to effectively exploit these resources for value-added purposes. For example, the waste could be beneficially used as a feedstock by capturing it and recycling/reusing it within the process, the organization, or beyond. This turns a cost and liability into a savings and benefit. Or perhaps construction waste could be captured on site, rather than discarding it to a landfill, so that it can be repurposed for other building applications.

It is important to consider that materials and energy that were utilized and are now "waste" have embedded entropy and complexity, representing an investment in cost and resources. This indicates that the recovery of waste as a feedstock represents both potential environmental and economic benefits.

Zero Waste
http://www.sierraclub.org/
committees/zerowaste/

6.4   Engineering Ethics and Risk

Engineers must understand environmental risk in order to protect all segments of society and all inhabitants of ecosystems. These individuals include the residents of communities in which the engineer resides, aquatic life residing in a river downstream of a construction site or treatment plant, and the global community of more than 7 billion people.

All too often, engineers work to minimize or eliminate the risk of an *average* member of society or an ecosystem inhabitant that is valued for recreational sport or commercial profit. This average societal member is someone that becomes part of a numerical equation that determines risk, and if this risk to be managed through a regulation, there is also an economic value associated with promulgating the regulation (and therefore avoiding the harm). Given the limitations of risk assessment and the multitude of uncertainties, engineers need to carefully consider all segments of society as well as ecosystem health (for example, biodiversity and endangered species). For these reasons, it is always desirable to minimize or eliminate the use and generation of hazardous chemicals and materials wherever possible.

It is important to recognize the susceptible segments of any population that may be significantly more sensitive to environmental exposure to a chemical or substance. For example, the impact of a chemical will vary with a person's age, gender, health status, occupation, and lifestyle. In the phenomenon of **environmental justice**, certain segments of society that are socially economically disadvantaged may be burdened with a greater amount of environmental risk. An environmental justice issue is apparent in Santa Clara County, California, where facilities that are required to list their toxic emissions in the EPA's TRI are located in communities with lower median incomes (Figure 6.7). Economically

Global Burden of Disease

http://www.who.int/
quantifying_ehimpacts/en/

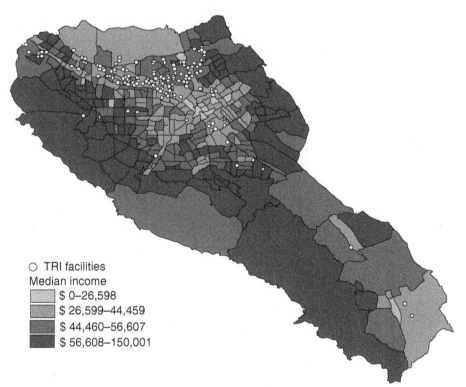

○ TRI facilities
Median income
$ 0–26,598
$ 26,599–44,459
$ 44,460–56,607
$ 56,608–150,001

Figure / 6.7 **Location of TRI Facilities in Santa Cruz (California) and Median Household Income** Note that the facilities that emit toxic chemicals are located primarily in areas with lower median incomes.

(SAGE, Szasz, A., Meuser, M., and A. Szaz, 2000. "Unintended, Inexorable: The Production of Environmental Inequalities in Santa Clara County, California." *American Behavioral Scientist* 43(4) 602–632.

Environmental Justice

http://www.epa.gov/compliance/
environmentaljustice/

Class Discussion

Economically disadvantaged individuals could also be living in a location in the developing world where they are exposed to disease-causing pathogens in unsafe drinking water and bear the added burdens of large-scale impact of HIV/AIDs and chronic effects of malaria. Climate change melting arctic ice is disturbing the subsistence lifestyles of the Inuit people, who live in arctic regions, as well as polar bears that hunt in these areas. Is it fair that a greater risk is assumed by these segments of the global community?

disadvantaged people tend to inhabit places that expose them to a greater number or higher concentrations of toxic chemicals (for example, next to highways that contribute air pollutants, next to industry that emits chemicals of concern, downwind from incinerators); they live in buildings that have hazardous materials associated with older construction or are served by aging infrastructure; or they have employment that results in increased exposure to hazardous materials.

These higher-risk individuals may live in urban or rural areas, typically have little political clout, and often are members of economically disadvantaged minority groups. They could be segments of a population that have elevated exposure levels because of their hunting or fishing habits due to a more subsistence lifestyle (as in the case of Native Americans who eat more fish or parts of fish that contain greater concentrations of toxics than in the average American's diet). They include the African American communities located near the intensive number of oil- and chemical-processing facilities located along the lower Mississippi River.

Engineers have a responsibility to consider these at-risk individuals and the communities they inhabit and to minimize, or eliminate, the likelihood that they bear a greater proportion of environmental risk than wealthier, better educated, or more politically powerful segments of society. With our new and increasing knowledge of sustainable design and green engineering, we have the ability to meet these challenges while continuing to improve quality of life for *all* segments of society in both the developed and developing world by employing more benign chemicals and materials, reducing energy and material consumption, and taking a systems perspective.

Figure 6.8 presents the rates of several waterborne diseases reported in the United States within Hispanic and non-Hispanic populations. While the data does not show a significant difference for diseases related to *Escherichia coli* and *cryptosporidiosis*, there is a large difference in rates of reported infections in the Hispanic population for diseases such as hepatitis A and salmonellosis, and shigellosis are higher for Hispanic than for other parts of the population. All these diseases are indicators of poor water quality, sanitation, and hygiene and in some cases are transmitted by food.

Application / 6.6   Environmental Justice and Human Rights

The EPA defines environmental justice as the "fair treatment and meaningful involvement of all people regardless of race, color, national origin, or income with respect to the development, implementation, and enforcement of environmental laws, regulations, and policies."

The World Health Organization's constitution (first written in 1946) states, "The enjoyment of the highest attainable standard of health is one of the fundamental rights of every human being." In 2002, water was recognized as a basic right when the United Nations Committee on Economic, Social, and Cultural Rights agreed, "The right to water clearly falls within the category of guarantees essential for securing an adequate standard of living, particularly since it is one of the most fundamental conditions for survival."

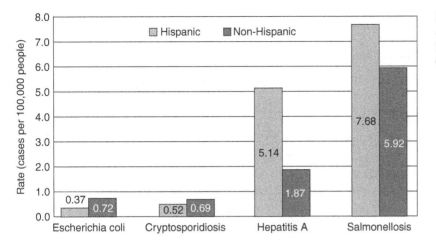

Figure / 6.8 Incidence of water borne diseases in Hispanic and non-Hispanic Populations (figure redrawn from CDC, 2001).

All a frequent consequence of living in conditions with poor access to water, sanitation, and hygiene (Quintero-Somaini and Quirindongo, 2004).

6.5 Risk Assessment

Risk assessments address questions such as these: What health problems are caused by chemicals and substances released into the home, workplace, and environment? What is the probability that humans will experience an adverse health effect when exposed to a specific concentration of chemical? How severe will the adverse response be? The remainder of this chapter focuses primarily on how to quantify the risks associated with exposure to chemicals and other environmental agents and the subsequent impacts on human health.

The four components of a complete **risk assessment** are: (1) hazard assessment; (2) dose–response assessment; (3) exposure assessment; and (4) risk characterization. Figure 6.9 depicts how these four items are integrated. A risk assessment organizes and analyzes a large set of information embedded in the four components to determine whether some environmental hazard will result in an adverse impact on humans or the environment. The environmental hazard could be exposure to a specific chemical or a broader issue such as climate change.

6.5.1 HAZARD ASSESSMENT

A hazard assessment is not a risk assessment. A **hazard assessment** consists of a review and analysis of toxicity data, weighing evidence that a substance causes various toxic effects, and evaluating whether toxic effects in one setting will occur in other settings. The hazard assessment determines whether a chemical or substance is, or is not, linked to a particular health concern, whereas a risk assessment will take into account the hazard assessment as well as the exposure assessment.

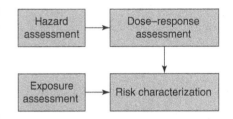

Figure / 6.9 Components of a Complete Risk Assessment The dose–response and exposure assessments are combined to yield a risk characterization.

Risk Assessment
http://www.epa.gov/risk/

Class Discussion

The statements provided in Application 6.6 suggest that the economically disadvantaged should not be burdened with a disproportionate percentage of environmental risk and that access to a basic level of freshwater and sanitation, a healthy workplace, and healthy environment are legal entitlements, instead of commodities or services that should be marginalized or privatized. Do you agree with these statements? Should all people in the world be guaranteed some basic access to water that would guarantee some specified level of health? How about ecosystems?

Integrated Risk Information System (IRIS)

The IRIS is an electronic database that contains human health effect information for hundreds of chemicals (see www.epa.gov/iris/). IRIS provides information for two components of a risk assessment: the hazard assessment and the dose–response assessment. IRIS was developed by the EPA and is written for professionals involved in risk assessments, decision making, and regulatory activities. Thus, it is intended for use by individuals who do not have training in toxicology.

IRIS contains descriptive and quantitative information on hazard identification, oral slope factors, oral and inhalation unit risks (IURS) for carcinogenic effects, as well as oral reference doses (RfDs) and inhalation reference concentrations (RfCs) for chronic noncarcinogenic health effects. These topics are discussed later in this chapter.

Ecotoxicology

The EPA maintains the ECOTOX database as a source for locating single-chemical toxicity data for aquatic life and terrestrial plants and wildlife (see www.epa.gov/ecotox/). This database can be used to assist ecological-hazard assessments and evaluate the potential hazard associated with wastewater effluent and/or leachate.

Sources of toxicity data include test tube studies, animal studies, and human studies and increasingly computational data. Test tube studies are fast and relatively easy, so they are commonly used to screen chemicals. Animal studies may measure acute or chronic effects. They could investigate a general end point (for instance, death) or a more specialized end point (say, a birth defect). Controlled laboratory studies are commonly employed to determine the toxicity of specific chemicals to aquatic life. Human studies typically consist of case studies that alert society to a problem and more extensive controlled epidemiologic studies.

The best study to determine the impact on humans is the epidemiology study. **Epidemiology** is the study of diseases in populations of humans or other animals, specifically how, when, and where they occur. Epidemiologists attempt to determine what factors are associated with diseases (risk factors), and what factors may protect people or animals against disease (protective factors). In many cases, the epidemiological data only arises after there is sufficient evidence in the general population to suggest that there may be a chemical on the market that is cause for concern.

Epidemiological studies can be divided into two basic types depending on whether the events have already happened (retrospective) or whether the events may happen in the future (prospective). The most common studies are retrospective studies that are also called case-control studies. A case-control study may begin when an outbreak of disease is noted and the causes of the disease are not known, or when the disease is unusual within the population studied.

These types of studies have difficulties, however, as summarized in Table 6.10. With epidemiological studies, it is extremely difficult to prove causation, meaning proof that a specific risk factor actually caused the disease being studied. Epidemiological evidence can, however, readily show that this risk factor is associated (correlated) with a higher incidence of disease in the population exposed to that risk factor. The higher the correlation is, the more certain the association.

Table / 6.10

Difficulties of Epidemiology Studies

Matching control groups is difficult, because factors that lead to exposure to a chemical may be associated with other factors that affect health.

Society has become more mobile, so individuals may no longer live in the same community all of their life.

Death certificates typically measure only the cause of death, so they miss health conditions that individuals had over the course of their life.

Other toxicity end points besides death (e.g., miscarriages, infertility, learning disorders) might not be measured with use of death certificates.

Accurate exposure data can be difficult to obtain for a large group of individuals.

Large populations are required for these studies so that rigorous statistical analysis can be applied to the data.

Many diseases can take years to develop.

Weight of evidence is a brief narrative that suggests the potential for whether a chemical or substance can act as a carcinogen to humans. Currently, weight of evidence is categorized by one of five descriptors listed in the left column of Table 6.11. Scientists who analyze the available data obtained from animal or human studies develop these descriptors for carcinogens. The right column of Table 6.11 summarizes how these descriptors are related to the quality and quantity of

Table / 6.11

Explanation of Weight of Evidence Descriptors

Weight of Evidence Descriptor	Relationship of Descriptor to Scientific Evidence
Carcinogenic to humans	Convincing epidemiologic evidence demonstrates causality between human exposure and cancer, or evidence demonstrates exceptionally when there is strong epidemiological evidence, extensive animal evidence, knowledge of the mode of action, and information that the mode of action is anticipated to occur in humans and progress to tumors.
Likely to be carcinogenic to humans	Tumor effects and other key data are adequate to demonstrate carcinogenic potential to humans but do not reach the weight of evidence for being carcinogenic to humans.
Suggestive evidence of carcinogenic potential	Human or animal data are suggestive of carcinogenicity, which raises a concern for carcinogenic effects but is judged not sufficient for a stronger conclusion.
Inadequate information to assess carcinogenic potential	Data are judged inadequate to perform an assessment.
Not likely to be carcinogenic to humans	Available data are considered robust for deciding that there is no basis for carcinogenic human hazard concern.

SOURCE: EPA, Guidelines for Carcinogenic Risk Assessment, March 29, 2005.

available data. IRIS (described in Application 6.7) provides information on the descriptor associated with particular chemicals.

6.5.2 DOSE–RESPONSE ASSESSMENT

DOSE A **dose** is the amount of a chemical received by a subject that can interact with the subject's metabolic process or other biological receptors after it crosses an outer boundary. Depending on the context, the dose may be: (1) the amount of the chemical that is administered to the subject; (2) the amount administered to the subject that reaches a specific location in the organism (for example, liver); or (3) the amount available for interaction within the test organism after the chemical crosses a barrier such as a stomach wall or skin.

To calculate the dose associated with a chemical, determine the mass of the chemical administered per unit time and divide that by the weight of the individual. In the case of an adult or child who is drinking water that contains the chemical of concern, the dose can be determined as shown in Example 6.3.

$example/6.3$ **Determining Dose**

Assume that the chemical of concern has a concentration of 10 mg/L in drinking water, and that adults drink 2 L of water per day, and children drink 1 L of water per day. Assume also that an adult male weighs 70 kg, a female weighs 50 kg, and a child weighs 10 kg. What is the dose for each of these three members of society?

solution

To find the dose associated with a chemical, determine the mass of the chemical taken in per unit time and divide this by the weight of the individual. In this situation, the only route of exposure is from drinking contaminated water. The dose for the three segments of society can be determined as follows:

$$\text{adult female dose: } \frac{10\frac{mg}{L} \times 2\frac{L}{day}}{50\ kg} = 0.40\frac{mg}{kg\text{-}day}$$

$$\text{adult male dose: } \frac{10\frac{mg}{L} \times 2\frac{L}{day}}{70\ kg} = 0.29\frac{mg}{kg\text{-}day}$$

$$\text{child dose: } \frac{10\frac{mg}{L} \times 1\frac{L}{day}}{10\ kg} = \frac{mg}{kg\text{-}day}$$

Note that, in this example, the dose received by the child and adult female is greater than that received by the adult male. This is one reason certain segments of society may be at greater risk when exposed to a specific chemical. Another reason is that, in most situations, children, older adults, and the sick are harmed to a greater extent by exposure to toxic chemicals and pathogens than are young, healthy adults. The same would hold true for plants and animals that inhabit ecosystems.

When determining the dose, scientists can account for absorption of the chemical. For example, the stomach wall may act as a barrier to the absorption of some chemicals that are ingested, while the skin may act as a barrier to chemicals that contact the hands. To account for the fact that 100 percent uptake (or absorption) of some chemicals does not occur, multiply the dose by the percent absorbed (termed f in Example 6.4). The value of f is typically 0 to 0.1 (1 to 10 percent) for metals and 0 to 1.0 (0 to 100 percent) for many organics.

In many states, the absorption efficiency applicable to dermal contact is considered to be 10 percent ($f = 0.10$) for contact with volatile or semivolatile organic chemicals and 1 percent ($f = 0.01$) for inorganic chemicals. For ingestion of contaminated soil and dust, it is assumed to be 100 percent ($f = 1.0$) for volatile organic chemicals and 100 percent for chemicals that sorb more strongly to soil (for example, PCBs and pesticides). However, the medium in which the chemical is present (water versus lipid, air versus water) may determine the extent of absorption. Also, dose is no longer calculated for inhalation risk assessments because inhalation toxicities are now derived as reference concentrations (RfCs) and inhalation unit risks (IURs) (note that risk from inhalation is not covered in this text).

DOSE RESPONSE   In animal laboratory studies, health damage is typically measured over a range of doses (minimum of three). Because sample populations are kept low during these studies to

---

example/6.4   Accounting for Absorption Efficiency When Determining Dose

Assume scientists know that only 10 percent of the chemical discussed in Example 6.3 is absorbed through the stomach wall. In this case, the exposure to the chemical was only through drinking contaminated water. What is the dose for the three target populations?

solution

Since 10 percent of the chemical is transported through the stomach wall, $f = 0.10$. The doses that account for incomplete transport of the chemical of concern for our three segments of society are as follows:

$$\text{adult female dose: } \frac{10\frac{\text{mg}}{\text{L}} \times 0.10 \times 2\frac{\text{L}}{\text{day}}}{50 \text{ kg}} = 0.04\frac{\text{mg}}{\text{kg-day}}$$

$$\text{adult male dose: } \frac{10\frac{\text{mg}}{\text{L}} \times 0.10 \times 2\frac{\text{L}}{\text{day}}}{70 \text{ kg}} = 0.029\frac{\text{mg}}{\text{kg-day}}$$

$$\text{child dose: } \frac{10\frac{\text{mg}}{\text{L}} \times 0.10 \times 1\frac{\text{L}}{\text{day}}}{10 \text{ kg}} = 0.10\frac{\text{mg}}{\text{kg-day}}$$

Here the dose is much lower than when the effect of absorption was neglected; however, children and adult females still receive a greater dose than adult males.

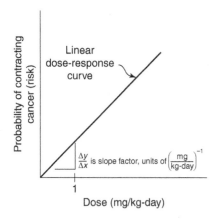

Figure 6.10 **Linear Dose–Response Relationship for a Carcinogenic Chemical or Substance** The zero intercept indicates that, according to this model, there is no threshold effect, so the probability of contracting cancer is zero only if the exposure to a carcinogen is zero. The y-axis can be thought of as the probability of contracting cancer at a given dose. The x-axis is the dose (mg chemical per kg body weight per day). The slope of the dose–response curve near the intercept for a dose of 1 mg/kg-day is referred to as the potency factor or slope factor, with units of inverse mg/kg-day.

save time and money, applied doses must be at relatively high concentrations, that is, at concentrations higher than what are typically observed in the environment (for example, in the workplace and home). Accordingly, a **dose–response assessment** is performed to allow extrapolation of data obtained from laboratory studies performed at higher doses to lower doses that are more representative of everyday life. Because of this extrapolation process, the assessment may overlook hazards such as endocrine-disrupting effects, which can occur at extremely low doses. However, as described in this section, dose–response assessments are performed differently for carcinogens and noncarcinogens.

Carcinogens Scientists have knowledge of the carcinogenic effect of chemicals primarily through laboratory test animal studies. Laboratory studies are conducted at higher doses so scientists can observe statistical changes in response with dose. In this case, the adverse response is formation of a tumor or some other sign of cancer. Carcinogens are treated as having **no threshold effect**, that is, under the assumption that any exposure to a cancer-causing substance will, with some degree of uncertainty, result in the initiation of cancer.

Figure 6.10 shows an example of a dose–response assessment for a carcinogenic chemical. Because a conservative scientific approach is used and, as stated previously, evidence suggests there is no threshold effect, the intersection of the dose–response curve at low doses is through the zero intercept. Application of such a dose–response model implies that the probability of contracting cancer is zero only if the exposure to the carcinogen is zero. The slope of the dose–response curve at very low doses is called the **potency factor** or **slope factor**. The slope factor is used in risk assessments for carcinogens, as we will show later in this chapter. The slope factor is an upper-bound estimate of risk per increment of dose that can be used to estimate carcinogenic risk probabilities for different exposure levels.

As shown in Figure 6.10, the slope factor has units of inverse mg/kg-day, or $(mg/kg\text{-}day)^{-1}$. It equals the unit risk for a chronic daily intake of 1 mg/kg-day. Values of the slope factor for many carcinogenic chemicals are available in the IRIS database (see Application 6.7). As we will see later in several examples, to obtain the overall risk, we multiply the slope factor by the calculated dose.

For most risk assessments, we determine the average daily dose by assuming that an individual is exposed to the maximum concentration of the carcinogen over his or her lifetime. In this case, the individual adult is assumed to live 70 year and weigh 70 kg. This 70 year averaging time (AT) may be different than the actual time of exposure, as we will discuss later when we put everything together to conduct a risk assessment to determine a risk-based cleanup level of contaminants found in drinking water and soil.

Noncarcinogens Noncarcinogenic chemicals do not induce tumors. In this case, the adverse end point would be a health impact such as liver disease, learning disorder, weight loss, or infertility. Obviously many end points do not result in cancer or death. An important point to understand is that, compared with carcinogens, noncarcinogens are assumed to have a **threshold effect**. That is, there is a **dose limit** below which it is believed there is no adverse impact. Figure 6.11 shows the dose–response assessment for a noncarcinogen.

Several new terms are defined in Figure 6.11. First, a **no observable adverse effect level (NOAEL)** is present. The NOAEL is the dose (units of mg/kg-day) at which no adverse health effect is observed. A dose less than or equal to this level is considered safe. However, because there is uncertainty in this safe dose of a noncarcinogen, scientists apply safety factors to the NOAEL to determine the **reference dose (RfD)**.

RfD is defined by EPA as an estimate, with uncertainty spanning perhaps an order of magnitude, of a daily oral exposure to the human population (including sensitive subgroups) that is likely to be without an appreciable risk of deleterious effects during a lifetime.

The RfD can be expressed mathematically:

$$RfD = \frac{NOAEL}{UF} \qquad (6.2)$$

Note that Equation 6.2 and Figure 6.11 both show that the RfD is lower than the NOAEL. The **uncertainty factor (UF)** typically ranges from 10 to 1,000. Application of the UF accounts for numerous uncertainties in applying NOAEL values to estimate RfD values. (Later, Application 6.9 discusses these uncertainties and how the UF values account for these uncertainties in risk characterization.) Inclusion of the UF in determination of a *safe* dose shows that the RfD (units of mg/kg-day) was developed to account for the uncertainty associated with conducting dose–response studies on small homogenous test animal populations for application to humans. The RfD should also account for societal groups (such as children) who may be more sensitive to a chemical. As we will see later, RfDs are used in risk assessments for oral intake of noncarcinogens through drinking water or eating food.

The **reference concentration (RfC)** was developed as an estimate of an inhalation exposure (from breathing) for a given duration that is likely to be without an appreciable risk of adverse health effects over a lifetime. The RfC can be thought of as an estimate (with uncertainty of one order of magnitude or greater) of a continuous inhalation exposure to a noncarcinogen that is likely to be without significant risk to human populations. The IRIS database provides separate values for RfD and RfC. Chapter 8 (Application 8.3) provides an example of how EPA uses the RfD to obtain a drinking water standard for the chemical perchlorate.

6.5.3   EXPOSURE ASSESSMENT

The purpose of the **exposure assessment** is to determine the extent and frequency of human exposure to target chemicals. Some of the questions that are answered during the exposure assessment are listed in Table 6.12.

The exposure assessment can also determine the number of people exposed and the degree of absorption by various routes of exposure. Remember that the exposure assessment study should also determine the exposure of average individuals in society and high-risk groups (for example, workers, children, women, economically disadvantaged

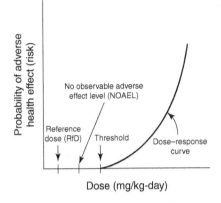

Figure / 6.11   **Dose–Response Relationship for a Noncarcinogenic Chemical or Substance**   Note the presence of a threshold dose below which no adverse response is observed for the response being assessed. The y-axis can be thought of as the probability of contracting an adverse effect at a given dose. The x-axis is the dose (mg chemical per kg body weight per day). The NOAEL is the dose at which no adverse health effect is observed. Doses less than or equal to this level can be considered safe. The RfD is an estimate of a lifetime dose that is likely to be without significant risk. RfDs are used for oral ingestion through exposure routes such as drinking water or eating food.

Table / 6.12

Some Questions Answered during the Exposure Assessment

What are the important sources of chemicals (e.g., pesticide application)?

What are the pathways (e.g., water, air, food) and routes of exposure (e.g., ingestion, inhalation, dermal contact)?

What amount of the chemical are people exposed to?

How often are people exposed?

How much uncertainty is associated with the estimates?

What segments of society (or ecosystem) are more at risk?

Table / 6.13

Some Key Barriers to Brownfield Redevelopment

Issues of liability

Differences in cleanup standards (can vary between state and federal governments)

Cost uncertainty associated with assessing contamination and cleanup

Obtaining financing, because lenders may want the government to waive liability associated with these sites

Community concerns

groups, older adults, area residents). Children typically have a more limited diet that may lead to relatively high but intermittent exposures. They also engage in behaviors such as crawling and mouthing (placing hands and objects in their mouth), which result in an increased exposure of chemicals via oral ingestion. The elderly and disabled may have sedentary lifestyles, which change their exposure. Pregnant and lactating women typically consume more water, which may lead to a different exposure assessment. Lastly, the many physiological differences between men and women, such as body weight and inhalation rates, could lead to important differences in exposures. The EPA has a handbook that provides guidance on what specific values to use in the exposure assessment (EPA, 2011b).

Exposure assessment can also be applied to a specific location. As mentioned earlier in this chapter, additional exposure could be associated with living next to a highway, incinerator, landfill, or factory. It could also be associated with living or working in a particular type of building, drinking a particular water supply, or eating a particular type and amount of food. Many details are considered, and a scientific study goes along with each of these scenarios.

Due to space limitations, we will focus our discussion on how exposure assessment is related to usage of land for residential, commercial, and industrial purposes. Much of this activity is associated with decision making related to engineering abandoned or idle properties (termed **brownfields**, discussed in Application 6.6) into something that is beneficial to society and the environment. Many times, brownfields are contaminated from past activities at the site. Brownfield redevelopment requires that an engineer work with a diverse group of stakeholders—

Application / 6.8   Brownfields

According to the EPA, brownfields are "abandoned, idled, or under-used industrial and commercial sites where expansion or redevelopment is complicated by real or perceived environmental contamination." In the United States, there are an estimated half million brownfields, primarily located in urban areas. Concerns of environmental and economic justice are associated with these sites, because many are located in poorer communities.

Unfortunately, brownfields typically lie idle because purchasers, lenders, and developers stay away for liability reasons and seek out greenfield sites, that is, open spaces typically located on the edge of towns and cities. However, development of greenfield space is not desirable to society and the environment because of issues such as loss of farmland and its associated way of life, loss of open space and wildlife habitat, and problems of flooding associated with stormwater management and paving, which causes increased runoff.

Another set of undesirable impacts associated with developing green spaces is that no infrastructure is present (unlike in an urban area), so it must be built and paid for (which consumes energy and raw materials). In addition, the location of employment away from the urban core can isolate employers from workers who cannot afford ownership of a vehicle or might have to take several bus transfers to reach a job site. Also, development of green space usually results in future problems of sprawl and congestion.

In contrast, brownfields are typically located near existing built environment infrastructure, mass transportation, and labor. Thus, there are clear economic, social, and environmental benefits of redeveloping a brownfield instead of developing a greenfield site.

Table 6.13 lists some key barriers to brownfield redevelopment. Listening to the concerns of the community, engaging stakeholders, and working with local units of government in the planning process are critical components of the engineer's job in successful brownfield redevelopment.

You can learn more about brownfields by visiting the EPA's web site, www.epa.gov/brownfields.

community members, nongovernmental organizations, government officials, financial lenders, real estate agents, and developers—in order to achieve a value-added use for the site.

Table 6.14 gives three types of land usage and associated parameters that might be used in an exposure assessment. The three types

Green Buildings on Brownfields
http://www.epa.gov/brownfields/sustain.htm

New Jersey Brownfields
http://www.nj.gov/dep/srp/brownfields

Table / 6.14

Land Uses and Examples of Exposure Assessment Associated with Each Use   The EPA publishes an *Exposure Factors Handbook* (EPA, 2011b) that provides more detail on specific values used in exposure assessment (EPA/600/R-09/052F, 2011).

Land Use	Examples of This Land Use	Example IR for Drinking Water; Air Inhalation, and Soil Ingestion	Example Exposure Frequency (EF) (days per year) and Exposure Duration (ED) (years)
Residential (primary activity is residential)	Single-family dwellings, condominiums, apartment buildings	**Children drink** 1 L/day **Adults drink** 2 L/day **Adults inhale** 20 m³/day **Children age 1–6 consume** 200 mg soil/day **Adults consume** 100 mg soil/day	**For drinking water** EF: 350 days/year ED: 30 years **For air inhalation** EF: 350 days/year ED: 30 years **For soil ingestion** ED: 6 years for children 1–6 ED: 24 years for adults EF: 350 days for children and adults
Industrial (primary activity is industrial, or zoning is industrial)	Manufacturing, utilities, industrial research, and development, petroleum bulk storage	**Adults drink** 1 L/day **Adults inhale** 10 m³/day	**For drinking water** EF: 245 days/year ED: 21 year **For air inhalation** EF: 245 days/year ED: 21 years **For soil ingestion** ED: 21 years for adults EF: 245 days for children and adults
Commercial (use is a business or is intended to house, educate, or provide care for children, the elderly, the infirm, or other sensitive subpopulations)	Day-care centers, educational facilities, hospitals, elder-care facilities and nursing homes, retail stores, professional offices, warehouses, gas stations, auto services, financial institutions, government buildings	**Adults drink** 1 L/day **Adults inhale** 10 m³/day	**For drinking water** EF: 245 days/year ED: 21 years **For air inhalation** EF: 245 days/years ED: 21 years **For soil ingestion** ED: 21 years for adults EF: 245 days for children and adults

*Recall that the average weight for a male, female, and child are 70 kg, 50 kg, and 10 kg, respectively.

of land usage considered in Table 6.14 are residential, industrial, and commercial. Examples of specific activities that constitute each usage are also provided in the table. The commercial land use category is extremely varied, so it can be split into several subtypes. For example, commercial usage may encompass day-care centers, schools, gas stations, lumberyards, government buildings, professional offices, and commercial businesses that serve food. In all of these cases, there are different levels of restrictions on public access and different exposure levels to workers and customers.

Table 6.14 also estimates human intake of chemicals through mechanisms such as drinking water, breathing air, and ingesting soil (dust). There may also be dermal contact by direct contact with chemicals or contaminated soils. The estimates of these intake rates (IR) are based on scientific studies, the type of individual, and the activity that takes place at the site.

As you can imagine, site specificity is related to exposure assessment. For example, Table 6.15 gives the amount of soil that adheres to the body and is ingested daily for specific human populations based on land use and associated employment. To determine soil adherence, scientists need to know how much skin is exposed for potential dermal (skin) contact with chemicals and contaminated soil. As an example, scientists assume that an adult worker wears a short-sleeved shirt, long pants, and shoes. The amount of skin surface area exposed to dust and dirt for these assumptions of clothing is also a function of the worker's body weight. The total dermal area available for contact is thus assumed to be 3,300 cm^2. This assumes exposed skin consists of the head (1,200 cm^2), hands (900 cm^2), and forearms (1,200 cm^2).

Table / 6.15

Amount of Soil Assumed to Adhere to Surface of Skin and Taken in Daily for Specific Populations Based on Land Use and Associated Employment

Target Population	Soil Adherence (mg soil/cm^2 skin)	Mass of Soil Taken in Daily (mg/day)
Adult living in residential area	0.07	50
Child living in residential area	0.2	200 for ages 1–6; 100 for all others
Adult worker at commercial III	0.01	50
Adult worker at commercial IV	0.1	50
Industrial worker	0.2	50

Commercial III refers to gas stations, auto dealerships, retail warehouses. The worker population is engaged in activities at the property that are of a low soil-intensive nature.

Commercial IV refers to hotels, professional offices, banks. A groundskeeper worker population has been identified as an appropriate receptor population. They engage in activities at the property that are of a high soil-intensive nature.

Climatic conditions such as snow cover and freezing conditions are not assumed to affect the amount of soil ingested by humans, because studies suggest that up to 80 percent of indoor air dust is from outdoor soils. It is believed that outdoor soil is transported inside buildings by air deposition, heating, ventilation and air-conditioning systems, and foot traffic. The indoor air environment clearly affects health, especially because, as stated earlier, Americans now spend 85 percent of their day inside some type of building. However, assessment of dermal exposure to contaminated soil would also consider the climatic conditions in northern areas (snow cover and frozen soil for a particular period of the year) that limits direct contact between soil and skin.

For engineers, knowledge of risk and exposure assessment provides information for determining whether a contaminated site needs to be remediated, and if so, to what level. Contaminated soil and groundwater at brownfield sites can be remediated using engineering technology. Alternatively, technological and institutional barriers can be used to minimize or prevent exposure. For example, paving a parking lot may prevent direct dermal contact with contaminated soils that lie underneath. Another example, in this case to prevent exposure to contaminated groundwater, would be for the property deed to place a restriction on placement of wells if a property is served by a municipal water supply.

6.5.4 RISK CHARACTERIZATION

As shown in Figure 6.9, the **risk characterization** takes into account the first three steps in risk assessment (hazard assessment, dose–response assessment, and exposure assessment). The risk characterization is specifically determined by integrating information from the dose–response and exposure assessments. The process is performed differently for carcinogens and noncarcinogens.

An important question is, What is an acceptable level of risk? Policymakers and scientists have determined that an acceptable environmental risk is a lifetime risk of 1 chance in a million (10^{-6}) of an adverse effect, and an unacceptable risk is 1 chance in 1,000 (10^{-3}) of an adverse effect.

A 10^{-6} risk means that if 1 million individuals were exposed to a toxic chemical at the same level and exposure, then 1 individual would have an adverse effect from this exposure. A 10^{-3} risk means that 1 individual would suffer an adverse effect if 1,000 people were exposed under the same conditions. Typically, state and federal governments have set the acceptable risk between 10^{-4} and 10^{-6}, with 10^{-5} and 10^{-6} being the most commonly used values in policies set by states and the federal government. These values represent the increased risk due to exposure to the hazard over the background risk.

In the examples that follow in the next two subsections, a risk characterization can be used to determine an allowable concentration of a chemical in air, water, or soil for an acceptable risk. It can also be used to determine the resulting environmental risk for a particular chemical at a given concentration and the exposure scenario for that chemical in a particular environmental medium.

In the first scenario, policymakers would fix the acceptable risk at a predetermined level (say, 10^{-4} to 10^{-6}), and the allowable concentration of the chemical in a particular medium that would result in that risk would be estimated. In the second situation, the concentration of the chemical in a particular medium is known, and the risk is determined.

CARCINOGENS   When developing a risk characterization for carcinogens, an important point is that the dose is assumed to be an average daily dose received by a subject over a lifetime of exposure. For carcinogens, this lifetime of exposure is assumed to be 70 years. Later in this section, we will describe how this lifetime exposure is accounted for.

In simple terms, the risk associated with carcinogenic chemicals is equal to the dose (or intake) multiplied by the unit risk associated with a dose of 1 mg/kg-day:

$$\boxed{\text{risk} = \text{dose} \times \text{risk per unit dose}} \tag{6.3}$$

Remember that for carcinogens, the unit risk associated with a dose of 1 mg/kg-day is called the slope factor.

Example 6.5 assumed that individuals were exposed to the chemical carcinogen for their 70-year lifetime. What happens in the case in which exposure is actually less than an individual's entire lifetime? For example, assume that exposure occurred over a 30-year period of employment when a worker was exposed to the chemical only at work. In this case,

$$\boxed{\text{risk} = \text{dose} \times \frac{\text{risk}}{\text{unit dose}} \times \frac{\text{time of exposure}}{\text{lifetime length}}} \tag{6.4}$$

example/6.5   Determining Risk

In Example 6.3, we determined that the dose for an adult male exposed to a chemical found in drinking water at 10 mg/L was 0.29 mg/kg-day. What is the risk associated with this exposure? Is this risk within acceptable guidelines?

Assume this dose is applied over a 70-year lifetime and the chemical found in the water is benzene, a known carcinogen. The IRIS database provides an oral slope factor of 0.055 $(\text{mg/kg-day})^{-1}$ for oral ingestion of benzene.

solution

Remember that previously we learned that the slope factor equals the unit risk for a chronic daily intake of 1 mg/kg-day. To determine the risk characterization, multiply the dose by the slope factor:

$$\text{risk} = 0.29 \frac{\text{mg}}{\text{kg-day}} \times 0.055 \frac{\text{kg-day}}{\text{mg}} = 1.59 \times 10^{-2}$$

This solution means that if 100 individuals were exposed to benzene at a concentration of 10 mg/L over their lifetime, 1.59 individuals would develop cancer. Extrapolated to a population of 10,000, this means that if all of them had similar exposure to benzene as this adult, 159 individuals would develop cancer. Extrapolated to a population of 1 million, we would expect 15,950 individuals to develop cancer.

This is well above acceptable risks of 1 in 10,000 (10^{-4}) and 1 in 1 million (10^{-6}). This is one reason the maximum contaminant level (MCL) for benzene in drinking water is 0.005 mg/L (or 5 μg/L), much lower than the 10 mg/L value used in this example.

Remember that for carcinogens, we assume the exposure takes place over a lifetime (70 years); thus, the lifetime length is set at 70 years. The *lifetime length* term in Equation 6.4 is referred to as the averaging time (AT) and typically has units of days. The AT for carcinogens is assumed to be 25,550 days (70 years × 365 days/year).

The *time of exposure* in Equation 6.4 is the exposure frequency (EF) multiplied by the exposure duration (ED). EF is the number of days an individual is exposed to the chemical per year. ED is the number of years an individual is exposed to the chemical.

Table 6.14 provided examples of EDs and EFs for different land use situations. For example, in the case of residential use and an exposure route of drinking water, the EF is assumed to be 350 days/year (50 weeks), and the ED is assumed to be 30 years. Application of this EF value assumes that an individual spends 2 weeks away from his or her house every year for vacation or other professional or family activities. Application of this ED value assumes that an individual resides in a house for only 30 years of his or her life. Note the differences for other land uses. For example, Table 6.14 gives that, in an industrial setting, an average worker is on the job site only 245 days/year (so EF = 245 days/year) and has an average employment history of 21 years (ED = 21 years).

The assumed values of EF and ED can be used to develop an expression to determine the acceptable concentration of a chemical in drinking water for a stated acceptable risk:

$$\text{acceptable concentration} = \frac{\text{acceptable risk} \times BW \times AT}{SF \times IR \times EF \times ED} \qquad (6.5)$$

In Equation 6.5, if the concentration is known (that is, change term on the left side to "measured concentration"), you can determine the risk associated with that concentration by substituting for the term acceptable risk with the "risk associated with the measured concentration." In Equation 6.5, BW is the average body weight of the target population and IR is the ingestion rate, in this case, 2 L of water per day (2 L/day).

Careful examination of Equation 6.5 shows that it is similar to the simpler-looking Equation 6.4. Some parameters were added to define the terms in Equation 6.4, and the equation was rearranged to set up the problem to compute the acceptable drinking-water concentration rather than the risk. The dose is also hidden in Equation 6.5. Here dose equals the IR multiplied by the acceptable concentration divided by the BW.

How would the acceptable risk in Example 6.6 change if the exposure assessment also showed that the target population consumed 30 g fish per day? The answer is simple. There would be no change in the acceptable risk unless for some reason you had knowledge that toxaphene was found in the fish. In this case, the toxaphene is found in groundwater below this residential neighborhood. We have no information to suggest that the fish these individuals consume came into contact with the contaminated groundwater. If the fish did contain the chemical, the ingestion of toxaphene in the fish would be added to the calculation. That is, the exposure would be from drinking 2 L water per day and eating 30 g fish per day. Section 6.6 will provide an example problem that includes exposure from both water and fish.

Class Discussion

When determining an acceptable level of risk, keep in mind that many individuals who are associated with an individual whose health is harmed also are indirectly harmed. The death or illness of an individual takes an emotional and financial toll on the individual's family members, friends, and coworkers. Also, broader economic and societal costs are associated with death and illness of an individual. Unfortunately, a typical risk characterization does not capture these broader societal and economic impacts. What are your personal and professional feelings about this issue? Are they the same, or do they differ?

example/6.6 Determining an Allowable Concentration of a Carcinogenic Chemical in Drinking Water

Calculate an acceptable groundwater concentration for the chemical toxaphene if a residential development is placed above a groundwater aquifer contaminated with toxaphene. Assume you determine the risk for an adult who weighs 70 kg and consumes 2 L water per day from the contaminated aquifer. The state you work in has determined that an acceptable risk is 1 cancer occurrence per 10^5 people. Use values from Table 6.14 for exposure frequency (EF) and exposure duration (ED) provided for residential land use.

solution

The IRIS database provides an oral slope factor for toxaphene of 1.1 per mg/kg-day. Remember that for carcinogens, AT is assumed to be 70 years. Using Equation 6.5 and exposure data from Table 6.14, solve for the acceptable concentration of toxaphene in the groundwater (assuming the only route of exposure is from drinking contaminated water):

$$\text{Concentration} = \frac{70 \text{ kg} \times 10^{-5} \times 70 \text{ years} \times \dfrac{365 \text{ days}}{\text{year}} \times \dfrac{1{,}000 \text{ } \mu g}{1 \text{ mg}}}{\dfrac{1.1 \text{ kg-day}}{\text{mg}} \times \dfrac{350 \text{ days}}{\text{year}} \times 30 \text{ year} \times \dfrac{2 \text{ L}}{\text{day}}}$$

$$= 0.77 \mu g/L \text{ or } 0.77 \text{ ppb}_m$$

Note that if the acceptable risk were 1 in 1 million (10^{-6}), the allowable toxaphene concentration would decrease to 0.077 µg/L (or 0.077 ppb$_m$).

Another engineering solution may be to investigate whether there is a municipal water supply close to this community that could serve as their source of drinking water. In this case, a deed restriction would be placed on the property so individual property owners could not install a drinking-water well. Furthermore, a hydrogeological study may have to be conducted to assess whether the contaminated groundwater recharges into a stream or river, where the chemical could exert toxicity to aquatic life or perhaps contaminate a downstream drinking-water intake. This option is likely to be costly and will clean up only the contamination currently on site. It will not prevent new quantities of this or other toxic chemicals from being introduced to the site.

NONCARCINOGENS   As stated previously, risk characterizations performed for noncarcinogens are handled differently than for carcinogens. Recall that noncarcinogens have a threshold dose, below which no adverse effect is estimated to occur. A safe dose referred to as the RfD estimates (with an uncertainty of one order of magnitude or greater) a lifetime dose of a noncarcinogen that is likely to be without significant risk to human populations. The IRIS database, discussed previously, provides values for RfDs.

The acceptable risk from exposure to a noncarcinogenic chemical is determined by calculating a **hazard quotient (HQ)**. For the exposure to a carcinogen, the dose is assumed to apply over a 70-year lifetime.

In the dose–response assessment process, numerous sources of uncertainty cause uncertainty in estimates of the risk per unit dose used in Equation 6.3 for carcinogens (the slope factor) and noncarcinogens (the hazard quotient). However, public policy has been developed that accounts for these uncertainties and leads to conservative estimates of the values used in risk characterization.

In the case of carcinogens, the linear dose–response model that is commonly applied (Figure 6.10) is conservative because this model leads to a higher response (risk) estimate at low doses than other models, such as the S-shaped multi-hit model and the U-shaped dose–response curve. It is usually the case that the dose–response data generated using test animal studies must be extrapolated to significantly lower doses in human risk assessments. Furthermore, lack of a threshold level in the dose–response relationship for carcinogens provides a conservative estimate of risk. Even if the dose were one *molecule* per kilogram per day, some risk of getting cancer is estimated when no threshold level is assumed.

For noncarcinogens, application of UFs in determining RfD and RfC (and thus HQ) values provides conservative estimates of HQ values. These UFs account for variation in susceptibility among the members of the human population (inter-individual or intraspecies variability), uncertainty in extrapolating animal data to humans (interspecies uncertainty), uncertainty in extrapolating from data obtained in a study with less-than-lifetime exposure (extrapolating from subchronic to chronic exposure), uncertainty in extrapolating from a lowest observative adverse effect level (LOAEL) rather than from a NOAEL value, and uncertainty associated with extrapolation when the database is incomplete.

The EPA has begun to recommend use of a *benchmark dose* concept to improve the quality of the RfD and RfC values and to reduce the number of uncertainty factors used. This approach uses all available data to estimate the NOAEL, rather than relying on a single point. This development provides an example of means by which public policy is continuously being improved in estimation of risk per unit dose values.

According to EPA, noncarcinogenic effects are evaluated by comparing the estimated daily intake of a chemical over a specific time period with the RfD for the same chemical that was derived over a similar period of exposure. Thus, the HQ is the average daily dose of a chemical received by a subject divided by the RfD:

$$HQ = \frac{\text{average daily dose}}{\text{RfD}} \qquad (6.6)$$

HQs that are less than or equal to 1 mean there is no appreciable or adverse risk; HQs greater than 1 mean there is a possibility that some noncancer effects may occur. This should make sense from careful study of Equation 6.6. If the HQ equals 1, the average daily dose to which an individual or community is exposed equals the RfD (the safe dose).

Equation 6.6 and the information discussed previously can be used to develop an expression to determine the acceptable concentration of a noncarcinogenic chemical in drinking water:

$$\text{acceptable concentration} = \frac{HQ \times RfD \times BW}{IR} \qquad (6.7)$$

All the terms in Equation 6.7 have been defined previously: HQ is the hazard quotient, RfD is the reference dose, BW is the body weight of the target population, and IR is the intake rate (in this case, the water ingestion rate).

example/6.7 Determining the Risk of Noncarcinogenic Chemicals

Though banned in the European Union, atrazine is a widely used as a herbicide in the United States, especially in large-scale production of corn. It is a health concern because of possible adverse impacts on the cardiovascular systems and reproduction. The U.S. Geological Survey has reported atrazine concentrations in the Arkansas River as high as 14 ppb. If the reference dose for atrazine is reported in IRIS as 0.035 mg/kg-day, would a 50 kg female be at risk if she drank 2 L of untreated Arkansas River water per day?

solution

Determine the hazard quotient using Equation 6.6. Remember from Chapter 2 that 1 ppb equals 1 µg/L.

$$HQ = \frac{dose}{RfD} = \frac{14\,\mu g/L \times \dfrac{mg}{1{,}000\,\mu g} \times \dfrac{2\,L}{day} \times \dfrac{1}{50\,kg}}{0.035\,mg/kg\text{-}day} = 0.016$$

Because the HQ is less than 1, we can assume there is no appreciable risk from the atrazine for the individual drinking the river water. Note that the maximum contaminant level goal (MCLG) for atrazine is 3 ppb which a community drinking water treatment plant cannot exceed. In addition, an ecological risk assessment could be performed for the aquatic life living in the river.

example/6.8 Determining Acceptable Concentrations of a Noncarcinogenic Chemical in Drinking Water

Calculate the acceptable groundwater protection standard for an unnamed chemical, classified as a noncarcinogen. An industrial development is placed above a groundwater aquifer contaminated with the chemical, and commercial establishments will use the groundwater as a source of drinking water for their workers. Assume you determine risk for an average adult male who weighs 70 kg and consumes 2 L water per day.

solution

Assume the IRIS database states the chemical's RfD for oral ingestion is 0.01 mg/kg-day. The acceptable concentration of the chemical in the groundwater (assuming the only route of exposure is from drinking contaminated water) is found from Equation 6.7 (and when the HQ is less than or equal to 1):

$$acceptable\ concentration = \frac{HQ \times RfD \times BW}{IR}$$

$$= \frac{1 \times 0.01\,\dfrac{mg}{kg\text{-}day} \times 70\,kg}{2\,\dfrac{L}{day}}$$

$$= 0.35\,mg/L\,(or\ 350\,ppb\ or\ 0.35\,ppm)$$

6.6 More Complicated Problems with at Least Two Exposure Routes

Exposure assessments can become more complicated than shown in the examples presented previously. Up to this point, the dose determination assumed only one route of exposure. This final section investigates how environmental risk can consider several routes of exposure at one time and how some of the environmental partitioning processes studied in Chapters 3 and 5 are incorporated into more complex, multimedia problems in which a chemical partitions between the water, air, and/or solid phases.

The first example will use a risk characterization to determine a surface water-quality standard for which the exposure assessment identified ingestion from eating contaminated fish that live in those waters in addition to drinking the contaminated water. The second example will use a risk characterization to determine acceptable cleanup standards for contaminated soil that can potentially leach vertically into the subsurface and contaminate underlying groundwater. In this situation, the chemical is not only dissolved in the pore water of the soil, but also may be sorbed to organic coatings on the soil particles and/or partitioned into air space voids found in the soil structure. Both examples will require us to use our knowledge of how a chemical partitions in the environment.

6.6.1 SETTING WATER-QUALITY STANDARDS BASED ON EXPOSURE FROM DRINKING WATER AND EATING FISH

Table 6.16 provides information from three different hypothetical states we could independently use to set a surface water-quality standard for a hypothetical chemical. Careful examination of this table shows that the three states would obtain different water-quality standards for the same chemical. How can that be? A further look at the information in Table 6.16 indicates that the states use different assumptions in their exposure assessment.

Each hypothetical state assumes the same weight of an adult (70 kg), the same ingestion rate of water (2 L/day), and the same cancer slope

Class Discussion

What are some socio or economic segments of society that are exposed to higher risk as a result of eating contaminated fish?

Table / 6.16

Exposure Assessment Assumptions Used by Three Hypothetical States in Setting a Water-Quality Standard for a Chemical That Bioaccumulates in Fish

	State 1	State 2	State 3
Ingestion rate of water (L/day)	2	2	2
Body weight of adult (kg)	70	70	70
Ingestion rate of fish (g/day)	6.5	30	15
Bioaccumulation factors, that is, measured concentration in fish divided by measured concentration in water (L water/kg fish)	51,500	336,000 (cold water) 84,086 (warm water)	7,310

factor. However, each state assumes different rates of fish consumption by the adult population and a different magnitude of partitioning behavior of the chemical from water into the fish (the bioaccumulation factor). Note that state 2 even assumes that the type of fish found in cold water has a greater fat (lipid) content, so the chemical of concern (which is very hydrophobic) partitions to a greater extent into cold-water fish than warm-water fish.

The total dose in this case comes from drinking water and eating fish, and in this case, most of the dose results from eating contaminated fish. The high bioaccumulation factors in Table 6.16 indicate that the concentration of the chemical in fish is much greater than in the surface water to which the fish is exposed. When multiplied by the daily ingestion rate for fish, quite a large value results. This is because, in this case, the chemical is very hydrophobic, so it does not tend to dissolve to a great extent in water but instead partitions to a great extent into the fat (lipid) of the cold-water fish.

Related to the discussion topic, the segments of society exposed to higher risk from fish consumption are probably children (higher dose given low body weight), pregnant women (perhaps the chemical is even suspected to impair fetal development), or recreational or subsistence anglers who consume more fish or fish parts that contain higher lipid content than the general population (increased exposure due to increased consumption). This last group could be vacationers who are consuming a large amount of fish for a short duration during a fishing trip. More likely it is individuals who depend on these fish for subsistence. For example, in many parts of the United States, Native American and immigrant populations are known to consume more fish or to eat fish parts that contain higher contaminant concentrations than others consume.

6.6.2 HOW TO DETERMINE ALLOWABLE SOIL CLEANUP STANDARDS THAT PROTECT GROUNDWATER

Now let us turn our attention to determining an appropriate cleanup level to which an engineer should remediate at a site containing contaminated soil. For this particular problem, assume an exposure assessment indicates that the contaminated soil does not pose a direct threat to adults or children who ingest the contaminated soil or breathe vapors that may emit from the contaminated soil. This could be the case for leaking underground storage tanks where the contamination is below the soil surface and the resulting soil vapors do not reach the soil surface. In this case, the problem is that the contaminated soil acts as a source of pollution, and the chemical of concern may leach from the soil and contaminate underlying groundwater. The groundwater may serve as a source of drinking water for a home or municipality, or perhaps the groundwater recharges a stream, where the chemical can then exert toxicity to aquatic life. Figure 6.12 shows the complexity of this problem.

Solving this type of problem requires several steps, detailed in Table 6.17. We will focus our efforts on only one of several hundred chemicals found in petroleum products, benzene. We will also assume that the groundwater does not recharge a stream, so there is no concern about risk to an aquatic ecosystem.

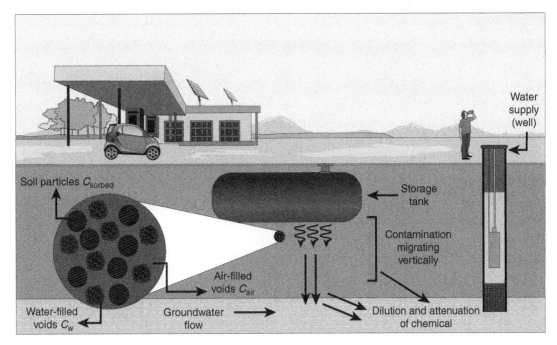

Figure / 6.12 **Complexity of a Situation in Which a Leaking Underground Storage Tank Has Discharged a Chemical That Is Contaminating Subsurface Groundwater.** The chemical may partition between air voids in the soil, water-filled voids in the soil, and organic coatings on soil. Henry's constant is used to relate the gaseous and aqueous concentrations of the chemical at equilibrium. A soil–water partition coefficient is used to relate the concentrations of the aqueous and sorbed phases for sorptive equilibrium. The chemical dissolved in water can leach vertically to the groundwater. In the process, it may chemically or biologically attenuate or be diluted by clean, upgradient groundwater.

We assumed residential use of the groundwater, and Example 6.6 provided us information on the exposure frequency, exposure duration, and averaging time for this particular type of land use. The allowable benzene concentration is determined as follows (using the appropriate slope factor of 0.055 $(mg/kg\text{-}day)^{-1}$ from the IRIS database).

$$\text{concentration} = \frac{70 \text{ kg} \times 10^{-5} \times 70 \text{ years} \times \dfrac{365 \text{ days}}{\text{year}} \times \dfrac{1{,}000 \text{ μg}}{\text{mg}}}{0.055 \dfrac{\text{kg-day}}{\text{mg}} \times \dfrac{350 \text{ days}}{\text{year}} \times 30 \text{ years} \times \dfrac{2 \text{ L}}{\text{day}}} \quad (6.8)$$

The allowable concentration is determined to be 15 μg/L (15 ppb_m).

This value is a factor of three higher than the maximum contaminant level (MCL) for benzene, which is 5 μg/L. The MCL is the enforceable standard set by the EPA, so this default value will be used to compute the soil cleanup standard in subsequent steps. MCLs are based upon treatment technologies, affordability, and other feasibility factors, such as availability of analytical methods, treatment technology, and costs for achieving various levels of removal. The EPA guidance for establishing an MCL states that MCLs are enforceable standards and are to be set as close to the maximum contaminant level goals (MCLGs) (health goals) as is feasible.

Steps to Solve the More Complex Environmental Risk Problem That Occurs in Soil and Groundwater Media

Step	Procedure
Step 1: Determine land use and routes of exposure.	Assume there is underlying groundwater that is currently used (or might be used in the future) as a source of drinking water for a home or community. It was assumed that the groundwater did not recharge a stream or river and that drinking contaminated groundwater is the only way people may be exposed to the contamination.
Step 2: What is the acceptable risk?	Assume that a regulatory body has set the acceptable risk at 10^{-5}.
Step 3: What is acceptable level of benzene in the groundwater?	Perform a risk characterization to determine the concentration of benzene in the contaminated groundwater that will not exceed the acceptable risk stated in step 2. The acceptable concentration of benzene in the groundwater is found from Equation 6.5 (repeated here): $$\text{acceptable concentration} = \frac{\text{acceptable risk} \times \text{BW} \times \text{AT}}{\text{SF} \times \text{IR} \times \text{EF} \times \text{ED}}$$
Step 4: Determine how the concentration of the chemical is changed as it moves vertically through the unsaturated zone to the groundwater.	This is determined either by a detailed hydrogeological study or by making an assumption that provides insight into the dilution and attenuation of the chemical.
Step 4a: Determine the allowable concentration in the soil pore water that surrounds the contamination site from the acceptable concentration in the groundwater.	Use results from the hydrogeological study or a dilution attenuation factor (DAF).
Step 4b: Estimate an allowable soil concentration (total benzene per total mass of wet soil) from the allowable pore water concentration.	Apply knowledge of mass balance and chemical partitioning between air, soil, and aqueous phases.

Step 4 from Table 6.17 is to determine how the concentration of the chemical is changed as it moves vertically through the unsaturated zone to the groundwater. An engineer could perform sophisticated hydrological modeling to determine the vertical movement of the benzene from the contaminated soil found in the unsaturated zone to the underlying saturated zone. For simplicity, let us assume that the migration of a contaminant from soil to underlying groundwater has two stages. First, the chemical must partition (from sorbed or gaseous phases) into the pore water that surrounds the contaminated area. Then the resulting leachate must be transported vertically to the underlying groundwater. As the dissolved chemical is transported downward with the infiltrating water, it may be transformed to a lower concentration by naturally occurring chemical or biological processes. It may also be diluted with the uncontaminated water it meets that is found upgradient in the groundwater.

For simplicity, assume that a value of 16 accounts for all dilution and attenuation processes. That is, the chemical concentration in the pore water that surrounds the contamination divided by 16 will equal the concentration in the groundwater that someone might eventually be exposed to.

Step 4a is to determine the allowable concentration in the soil pore water that surrounds the contamination site from the acceptable concentration in the groundwater. Moving from the allowable concentration in the groundwater to the pore water, the allowable benzene concentration in the pore water within the contaminated soil zone that will leach vertically to not adversely affect the groundwater is

$$0.005\,\text{mg/L} \times 16 = 0.080\,\text{mg/L}$$

Remember that this value of 0.080 mg/L is not the concentration that someone would be exposed to if he or she drank the contaminated water. The value of 0.080 mg/L is the allowable concentration in the pore water of the contaminated soil. It will be decreased by dilution and attenuation so that the concentration that is eventually found in the groundwater is 0.005 mg/L.

Step 4b is to estimate an allowable soil concentration (total benzene per total mass of wet soil) from the allowable pore water concentration. The final step is to take the allowable aqueous phase concentration in the pore water (mg benzene per L water) and convert it to a soil concentration (mg benzene per kg wet soil). To make the problem easier, assume the benzene–air–water–soil system is at equilibrium. The amount of benzene that is partitioned at equilibrium among the pore water, air voids, and the sorbed soil phase is determined.

This total mass of the benzene (from the aqueous pore water, air, and sorbed phases) that is contained in a unit mass of soil is what is measured if you collect a subsurface soil sample and request a laboratory analysis of the soil. This value can also be thought of as the allowable concentration of benzene that can left in the soil (units of mg benzene per kg soil) that, if it leached vertically, would not result in a subsequent groundwater concentration that would harm human health at a stated acceptable risk.

To carry out this computation, we need subsurface property information. Assume the subsurface system is homogenous, the soil porosity is 0.3 percent, soil organic carbon content is 1 percent, and soil bulk density is 2.1 gm/cm³. Assume also that one-third of the void spaces of the soil are filled with air and the remaining two-thirds of the void spaces are filled with water.

The total mass of chemical is equal to the mass of chemical sorbed to the soil plus the mass of chemical in the water-filled voids plus the mass of chemical in the air-filled voids. Mathematically, this statement can be written as follows, using information about equilibrium partitioning (Chapters 3 and 5):

allowable soil concentration

$$= (C_{\text{water}} \times K) + \left(\frac{C_{\text{water}} \times \theta_{\text{water}}}{\rho_b}\right) + \left(\frac{C_{\text{water}} \times \theta_{\text{air}} \times K_H}{\rho_b}\right) \quad \textbf{(6.9)}$$

Here, C_{water} is the pore water concentration of the chemical determined in step 4a (0.08 mg/L), K is the soil–water partition coefficient for our chemical of concern, K_H is the dimensionless Henry's law constant for the chemical of concern, θ_{water} is the soil porosity filled with water, θ_{air} is soil porosity filled with air, and ρ_b is the soil bulk density (assumed to be 2.1 g/cm³).

In this case, we have the following values for these variables:

- C_{water} was determined in step 4a to be 0.080 mg/L.

- θ_{water}, the soil porosity filled with water, is equal to the porosity (0.3) multiplied by the fraction of voids that are filled with water (2/3 in this problem). In this case, θ_{water} is equal to 0.2 (expressed in units of L_{water}/L_{total}).

- θ_{air}, the soil porosity filled with air voids, is equal to the porosity (0.3) multiplied by the fraction of voids that are filled with air (1/3 in this problem). In this case, θ_{air} equals 0.1 (expressed in units of L_{air}/L_{total}).

- K, the soil–water partition coefficient, can be estimated from the octanol–water partition coefficient, as discussed in Chapter 3. Assume benzene has a log K_{ow} of 2.13. For benzene, the correlation from Figure 3.11 can be used to estimate the soil–water partition coefficient normalized to organic carbon (log K_{oc}) as 2.02 (K_{oc} equals $10^{2.02}$ L/kg organic carbon). Because the organic carbon content of the soil was stated to be 1 percent, Equation 3.32 can be used to determine that K is 1.05 L/kg soil.

- The K_H for benzene is 0.18 at 20°C. This value is for reaction of a chemical in the air phase going to the water phase. The units of K_H in this situation are thus L_{water}/L_{air}. To simplify the problem, we will assume that the cooler temperature of the subsurface does not greatly affect the air–water partitioning behavior of benzene.

Plugging in these values for the unknowns in Equation 6.9 results in allowable soil concentration

$$= \left(0.08\,\frac{mg}{L} \times 1.05\,\frac{L}{kg\ soil}\right) + \left(\frac{0.08\,\dfrac{mg}{L} \times 0.2\,\dfrac{L_{water}}{L_{total}}}{2.1\,\dfrac{g\ soil}{cm^3} \times 1{,}000\,\dfrac{cm^3}{L} \times \dfrac{kg}{1{,}000\ g\ soil}}\right)$$

$$+ \left(\frac{0.08\,\dfrac{mg}{L} \times 0.18\,\dfrac{L_{water}}{L_{air}} \times 0.1\,\dfrac{L_{air}}{L_{total}}}{2.1\,\dfrac{g\ soil}{cm^3} \times 1{,}000\,\dfrac{cm^3}{L} \times \dfrac{kg}{1{,}000\ g\ soil}}\right) \qquad \textbf{(6.10)}$$

Solving for the concentrations of benzene in the sorbed phase, pore water phase, and air voids, respectively, results in

allowable soil concentration

$$= 0.084\ mg/kg + 0.0076\ mg/kg + 0.00069\ mg/kg$$

$$\quad (sorbed) \qquad\qquad (water) \qquad\qquad (air)$$

$$= 0.092\ mg\ benzene/kg\ soil$$

(6.11)

This allowable soil concentration of 0.092 mg/kg (0.092 ppm$_m$) is the concentration of benzene that can remain in the soil and still protect the groundwater resource to drinking-water standards. Any areas of the contaminated soil greater than this value would have to be removed or remediated to this lower cleanup level.

Key Terms

- acute toxicity
- bioaccumulation
- brownfield
- carcinogenic
- chronic toxicity
- disposal
- dose
- dose limit
- dose–response assessment
- dose–response curve
- eco-efficiency
- ecotoxicity
- Emergency Planning and Community Right-to-Know Act (EPCRA)
- endocrine disruptors
- environmental justice
- environmental risk
- epidemiology
- exposure
- exposure assessment

- exposure pathways
- green chemistry
- hazard assessment
- hazard quotient (HQ)
- hazardous waste
- Leadership in Energy and Environmental Design (LEED).
- LEED certification
- median lethal concentration LC^{50}
- median lethal dose (LD$_{50}$)
- no observable adverse effect level (NOAEL)
- no threshold effect
- noncarcinogenic
- pollution prevention
- pollution prevention hierarchy
- potency factor
- recycling
- reference concentration (RfC)
- reference dose (RfD)

- Resource Conservation and Recovery Act (RCRA)
- risk
- risk assessment
- risk characterization
- risk perception
- slope factor
- solid waste
- source reduction
- susceptible populations
- suspected carcinogens
- threshold effect
- toxicity
- Toxics Release Inventory (TRI)
- treatment
- uncertainty factor (UF)
- waste minimization
- weight of evidence

6.1 What is the regulatory difference in RCRA between a hazardous and toxic substance?

6.2 Identify several types of risk during your commute to school. Which of these risks would classify as environmental risk?

6.3 Rank these scenarios in order of their environmental risk (low to highest): (a) A factory worker has been provided no protective breathing equipment, and the chemical being emitted has been judged to have zero hazard. (b) A factory worker has been provided protective breathing equipment that removes 99 percent of a hazardous chemical. (c) A factory worker has been provided protective breathing equipment that removes 100 percent of a hazardous chemical. (d) A factory worker has been provided protective breathing equipment that removes 100 percent of a chemical that has been judged to have zero hazard. (e) A toxic chemical was identified in a factory's drinking water supply. The worker you are evaluating has a desk job and is not exposed to any of a toxic chemical emitted in the air of the factory manufacturing area. This worker also brings all her water and other beverages from home in reusable containers.

6.4 Rank these three scenarios in order of their environmental risk (low to highest): (a) Customers visit a bar 6 h per week in a location where the state has passed regulations that prevent customers from smoking inside restaurants and bars. (b) Wait staff are exposed to secondhand tobacco smoke 8 h per day during work. (c) Customers are exposed to secondhand tobacco smoke 2 h per week while dining at the same restaurant as the wait staff in part (b). (d) Wait staff work 8 h per day in an establishment located in a state that has passed regulations that prevent customers from smoking in restaurants and bars.

6.5 What are three considerations besides toxicity that contribute to a chemical being labeled "hazardous"?

6.6 Identify the top three chemical releases in your hometown or university community using the EPA's Toxic Release Inventory database. What information can you find about the toxicity of these chemicals? Is it easy or difficult to find this information? Is the information consistent, or is it conflicting? Does it vary with source (government versus industry)?

6.7 The EPA released the 2011 Toxic Release Inventory National Analysis in January of 2013. Locate this information and fill in Table 6.18 for toxic chemical disposal and releases (in the year 2011) for the following three large aquatic ecosystems: Long Island Sound, the Gulf of Mexico, and the San Francisco Bay Delta Estuary. All these water bodies of water are recognized as being important for ecological, economic, and social reasons.

Table / 6.18

Toxic Release Inventory Data Associated with Three Large Aquatic Ecosystems

	San Francisco Bay Delta Estuary	Gulf of Mexico	Long Island Sound
Number of TRI Facilities			
Total on-site and off-site disposal or other releases			
Total on-site releases to air			
Total on-site releases to water			
Total on-site releases to land			
Underground injection			
Top five industrial sectors contributing to the TRI			

6.8 Define pollution prevention and describe why it is the preferred approach to addressing the challenge of waste.

6.9 What is the difference between pollution prevention and sustainability?

6.10 Use the pollution prevention hierarchy to rank the following scenarios from **least to most preferred**. In addition, label each scenario as an example of source reduction, recycling, treatment, or disposal: (a) Ammonia nitrogen is transformed to less toxic nitrate nitrogen at the wastewater treatment plant and then discharged to a receiving water body, (b) Urine (which contains 75 percent of the nitrogen excreted by the human body) is collected in the household and applied to a backyard garden as a fertilizer, (c) A homeowner decides to no long place food scraps in a garbage disposal connected to the sink and instead sets up a backyard composting machine, (d) Nitrogen in the wastewater is precipitated out and recovered for fertilizer at a centralized treatment plant, as struvite (magnesium ammonium phosphate, $NH_4MgPO_4 \cdot 6H_2O$).

6.11 Label each scenario as an example of source reduction, recycling, treatment, or disposal: (a) The community collects household solid waste and disposes all the waste in a sanitary landfill. (b) The community implements a yard waste collection program to address this component of the wastestream (assume it makes up 14 percent of the total wastestream). The yard waste is then composted and reused in the community. (c) The community changes its billing plan from one flat rate charged per household to a new plan that charges households for each bag (or trash can) of solid waste placed on the curb for pickup. Their idea is this will cause homeowners to reduce the amount of waste they produce and discard. (d) A national policy is instituted to reduce the amount of packaging associated with consumer products. (e) Households begin to purchase locally grown food from local outlets so the packaging associated with food distribution is reduced. (f) The community solid-waste authority requires separation of glass, paper/cardboard, and metal by households in a new curbside recycling program. (g) The community burns solid waste at a high temperature to recover energy, releasing some toxic chemicals into the air and producing an ash product, but reducing the volume of waste that needs to be disposed of in a landfill.

6.12 (a) List a different environmental risk associated with an indoor environment in the developed world and developing world. (b) What particular building occupants are at the greatest risk for the items you have identified?

6.13 Go to the web site of the World Health Organization (www.who.org). Based on the information there, write a referenced two-page essay on global environmental risk. How much environmental risk is from factors such as unsafe water and sanitation, indoor air, urban air, and climate?

6.14 Recalling that the EPA defines environmental justice as the "fair treatment and meaningful involvement of all people regardless of race, color, national origin, or income with respect to the development, implementation, and enforcement of environmental laws, regulations, and policies," research an issue of environmental justice in your hometown or state. What is the environmental issue? What groups of society are being harmed by the environmental issue? What injustice is taking place?

6.15 Online at www.scorecard.org, you can search for the location and number of hazardous-waste sites by location. Use it to search for hazardous-waste sites in your hometown or a city close to your university. Comment on whether the number and location of hazardous-waste sites pose any environmental injustice to residents of the community you are investigating.

6.16 List the four components of a complete risk assessment.

6.17 The Integrated Risk Information System (IRIS), an electronic database that identifies human health effects related to exposure to hundreds of chemicals, is available at www.epa.gov/iris. Go to IRIS and determine (a) the weight of evidence descriptor; (b) the reference dose (RfD); and (c) the slope factor, if available, for the following six chemicals/substances: arsenic, methylmercury, ethylbenzene, methyl ethyl ketone, naphthalene, and diesel engine exhaust.

6.18 Why must children especially be protected from environmental contaminants? Use the term lethal dose (LD) in your answer.

6.19 A study of the potential of acrylonitrile to produce brain tumors in rats was conducted by administering the carcinogen in drinking water for 24 months. The results of the study for female rats are tabulated below.

Dose (mg/kg-day)	Brain tumor incidence
0	1/179
0.12	1/90
0.36	2/91
1.25	4/85
3.65	6/90
10.89	23/88

(a) Determine the slope factor (SF) of the dose–response relationship (assume it is linear). (Don't forget to account for the one rat that had a brain tumor even though he wasn't exposed to the chemical). (b) How accurate is the linear model of the data?

6.20 The EPA maintains a comprehensive site of chemical risk information, called the Integrated Risk Information System (IRIS: http://www.epa.gov/iris/)

Visit the page describing the assessment of the Reference Dose (RfD) for chronic oral exposure: (http://www.epa.gov/ncea/iris/subst/0209.htm#umforal) for the chemical atrazine. Atrazine is a popular herbicide: tens of millions of pounds of atrazine are applied to vegetation in the United States annually, and it is a widespread drinking water contaminant.

After reading through the brief provided by IRIS, please answer the following: (a) What tests did the EPA/Ciba-Geigy perform in order to assess the toxicity of atrazine? Summarize in tabular format the test subjects, time period, and overall key toxic responses to be observed. (b) What uncertainty factor is used, and how was that derived (what are its components)? (c) What are the NOAEL and RfD for atrazine, and to which response do these doses refer?

6.21 Visit the page describing the assessment of the Reference Dose (RfD) for chronic oral exposure: (http://www.epa.gov/ncea/iris/subst/0209.htm#umforal) for the chemical atrazine. Atrazine is a popular herbicide: tens of millions of pounds of atrazine are applied to vegetation in the United States annually, and, is a widespread drinking water contaminant. (a) What are the NOAEL and RfD for atrazine, and to which response do these doses refer? (b) Assume the following additional toxicity data points: LOAEL—5 mg/kg/day (affects 5% of population); LD50—15 mg/kg/day; other data points—22 mg/kg/day (affects 75% of population), 30 mg/kg/day (affects 95% of population). Draw a dose–response curve for these data. (c) What is the approximate slope of the curve? (d) Now consider applying atrazine to your lawn and assume that there are babies in the household that like to eat grass. What is the maximum amount of grass that a baby can eat safely in a day? Assume the following conditions: baby weight ~10 kg, residual atrazine concentration on grass from one application is ~0.01% (where 1% = 10,000 ppm_m). (e) Given these results from part (d), does spraying your lawn with atrazine present any serious risk in this case? Use the "hazard quotient" to make this determination.

6.22 The sigmoid curve used in a dose–response analysis (and many other engineering applications) has the form

$$\text{response}(x) = \frac{1}{1 + e^{LD50-x}}$$

where x is the dose in mg/kg/day.

(a) Show mathematically that the inflection point of the dose–response curve occurs when $x = LD50$. What does this mean in practical terms? (b) Using the information provided in part (a), provide the dose–response equation for atrazine. What is the dose necessary to produce a lethal response in 90% of the test population?

6.23 Consider a pharmaceutical product with two dose–response curves, one that shows the effective dose (ED) and a second that shows the lethal dose (LD). Assume that LD50 = 2 × ED50 and is equal to 28.27 mg/kg/day. This pharmaceutical company makes money from selling each course of this drug (D dollars per *effective* dose) but must pay damages for each death that it causes (1,000D dollars per *lethal* dose). (a) Putting aside questions of ethics and market acceptability, what should the company set as its recommended dose in order to maximize profits? Show all your work. Hint: You will need to utilize the equation in problem 6.22.

6.24 Visit the following EPA Office of Pesticides webpage that provides information on the state of the U.S. federal scientific debate on atrazine: http://www.epa.gov/oppsrrd1/reregistration/atrazine/atrazine_update.htm

Write a one-page memo from the perspective of the strategy team at Syngenta, one of the main global manufacturers of atrazine, giving a summary of the EPA activities, findings, and rulings. From a strategy perspective, what overall recommendations would you give, knowing what you do about how the EPA is considering atrazine?

6.25 Assume an adult female who weighs 50 kg drinks 2 L of water every day and the absorption factor for the chemical of concern is 75 percent (so 25 percent of the chemical is secreted). The concentration of the chemical in the drinking water is 55 ppb. Determine the dose in mg/kg-day.

6.26 (a) Determine the dose (in mg/kg-day) for a bioaccumulative chemical with BCF = 10^3 that is found in water at a concentration of 0.1 mg/L. Calculate your dose for a 50 kg adult female who drinks 2 L lake water per day and consumes 30 g of fish per day that is caught from the lake. (b) What percent of the total dose is from exposure to the water, and what percent is from exposure to the fish?

6.27 Calculate a risk-based groundwater protection standard (in ppb) for the chemical 1,2-dichloroethane for a residential homeowner where the person's well used for drinking water is contaminated with 1,2-dichloroethane. Assume you are determining risk for an average adult who weighs 70 kg. The state where you work has determined that an acceptable risk is 1 cancer occurrence per 10^6 people. Use the values for route of intake, exposure frequency, exposure duration, and averaging time provided for residential use in Table 6.14. Assume an oral slope factor for 1,2-dichloroethane of 9.2×10^{-2} per (mg/kg)/day.

6.28 Determine if exposure by oral ingestion to the chemicals xylene, toluene, arsenic, and hexavalent chromium poses a noncarcinogenic health risk. The chemical-specific reference doses (mg/kg-day) obtained from IRIS are xylene (0.2), toluene (0.8), arsenic (0.0003), and hexavalent chromium (0.003). Assume a 70 kg individual consumes 2 L of water per day with these chemicals dissolved at a concentration of 1 mg/L.

6.29 A commercial area had its own separate groundwater supply system that provided drinking water. Unfortunately the groundwater was contaminated with arsenic at a concentration of 10 ppb. The property owner placed a deed restriction on accessing the groundwater and also contacted the city to complete a hookup with the city water supply. Determine if exposure by oral ingestion to the chemical arsenic poses a noncarcinogenic health risk for individuals consuming drinking water after the property owner' actions have taken place. Assume a 70 kg individual consumes 2 L of water per day and the RfD for arsenic is 0.0003 mg/kg-day.

6.30 (a) Calculate a risk-based groundwater protection standard for the chemical benzo(*a*)pyrene. Assume you are determining risk for an average adult female who weighs 50 kg and consumes 2 L water and eats 30 g of fish per day. The state has determined that an acceptable risk is 1 cancer occurrence per 10^5 people. Use the values for exposure frequency, exposure duration, and averaging time provided for residential land use. (b) According to EPA's Integrated Risk Information System (IRIS), what type of carcinogen is benzo(*a*)pyrene, using the weight of evidence of human and animal studies? (c) Assuming the chemical is leaching from some contaminated soil, estimate the allowable concentration of benzo(*a*)pyrene in the pore water of contaminated soil.

6.31 Is there an unsafe risk associated with a 70 kg adult eating 15 g of fish every day that contains 1 mg/kg of methylmercury? Methylmercury has been shown to cause developmental neuropsychological impairment in human beings. The RfD for methylmercury is 1×10^{-4} mg/kg-day.

6.32 Is there an unsafe risk associated with a 70 kg adult eating 15 g of fish every day that contains 9.8 µg/kg of Arochlor 1254? Arochlor 1254 can exhibit noncarcinogenic effects in humans. Use the IRIS database to find any other information required to solve this problem.

6.33 Concentrations of toxaphene in fish may impair human health and fish-eating birds (such as bald eagles) that feed on the fish. (a) If the log of the octanol–water partition coefficient (log K_{ow}) for toxaphene is assumed to be equal to 4.21, what is the expected concentration of toxaphene in fish? (Assume that the equilibrium aqueous phase toxaphene concentration is 100 ng/L.) (b) If it is assumed that an average person drinks 2 L untreated water daily and consumes 30 g of contaminated fish, what route of exposure (drinking water or eating fish) results in the greatest risk from toxaphene in 1 year? (c) What route of exposure is greatest for a higher-risk group that is assumed to consume 100 g fish per day? Support all of your answers with calculations. Assume the following correlation applies to toxaphene and our problem's specific fish:

$$\log BCF = 0.85 \log K_{ow} - 0.07$$

6.34 Identify a brownfield in your local community, hometown, or a nearby city. What was specifically done at the site? What are several social, economic, and environmental issues associated with restoring the brownfield site?

6.35 The Code of Federal Regulations (CFR) is the codification of the general and permanent rules published in the Federal Register by the executive departments and agencies of the Federal Government. It can be accessed at http://www.gpoaccess.gov/cfr/. What CFR number is associated with the following sections? (for example, 50 CFR for Wildlife and Fisheries). (a) Protection of Environment; (b) Transportation, (c) Conservation of Power and Water Resources, (d) Public Health, and (e) Highways.

6.36 Research the safety of your personal care and household cleaning products using a web site such as http://lesstoxicguide.ca/index.asp?fetch=personal. Develop a table that lists seven current personal care or household cleaning products used in your apartment, home, or dormitory. Add a second column that lists a less hazardous alternative for each of the seven products.

6.37 Do some background research. Some good places to look for this and related information include EXTOXNET, the National Toxicology Program (NTP), the Agency for Toxic Substances and Disease Registry (ATSDR), and the National Library of Medicine (choose one). (a) Is atrazine bioaccumulative and/or persistent in the environment? Explain your answer. (b) Now consider applying atrazine to your lawn and assume that there are children in the household and they like to eat grass, dirt, and worms while playing on grass. Assume the atrazine is both bioaccumulative and persistent. How does this new information about the partitioning and persistence behavior of atrazine affect your consideration of the potential toxicity of atrazine to humans?

References

Anastas, P. T., and J. C. Warner, 1998. *Green Chemistry: Theory and Practice*. Oxford: Oxford University Press.

Centers for Disease Control and Prevention (CDC), 2001 Summary of Notifiable Diseases—United States, 2001, http://www.cdc.gov/mmwr/PDF/wk/mm5053.pdf, accessed October 26, 2012.

Environmental Protection Agency (EPA), 2005. "Fact Sheet: EPA's Guidelines for Carcinogen Risk Assessment" March 29, 2005. http://www.epa.gov/cancerguidelines/cancer-guidelines-factsheet.htm, accessed June 18, 2013.

Environmental Protection Agency (EPA), 2011a. 2010 Toxics Release Inventory National Analysis Overview, http://www.epa.gov/tri/tridata/tri10/nationalanalysis/index.htm, accessed October 22, 2012.

Environmental Protection Agency (EPA), 2011b. Exposure Factors Handbook (EFH): 2011 Edition. U.S. Environmental Protection Agency, Washington, D.C., EPA/600/R-09/052F, 2011.

Fisk, W. J., 2000. Health and productivity gains from better indoor environments and their relationship with building energy efficiency. *Annual Review of Energy and the Environment*, 25: 537–566.

Friends of the Earth, 2009. Endocrine Disrupting Pesticides, http://www.foe.co.uk/index.html, accessed February 21, 2009.

National Science and Technology Council, 1996. The Health and Ecological Effects of Endocrine Disrupting Chemicals: A Framework for Planning. Committee on Environment and Natural Resources, http://www.epa.gov/edrlupvx/Pubs/framewrk.pdf accessed June 18, 2013.

Patnaik, P. 1992. *A Comprehensive Guide to the Hazardous Properties of Chemical Substances*. New York: Van Nostrand Reinhold.

Quintero-Somaini, A., and M. Quirindongo, 2004. *Environmental Health Threats in the Latino Community*. New York: National Resources Defense Council.

Szasz, A., and Meuser, M., 2000. Unintended, inexorable: the production of environmental inequalities in Santa Clara County, California. *American Behavioral Scientist*, 43(4): 602–632.

chapter/Seven Water: Quantity and Quality

James R. Mihelcic, Brian E. Whitman, Martin T. Auer, and Michael R. Penn

Two closely allied engineering disciplines are associated with water: water resources and water quality. Water resources engineering deals with water quantity (for example, its storage and transport), and water quality engineering is concerned with the biological, chemical, and physical nature of water. This chapter provides key concepts, principles, and calculations that support a more sustainable approach to water management. Four hydrologic systems are considered: rivers, lakes/reservoirs, wetlands, and groundwater. Methods to estimate river water quality downstream of pollutant inputs are covered along with management strategies to restore lakes polluted by human activities. Also covered are ways to estimate the flow of water and transport of chemicals in groundwater. Stormwater is considered in a following chapter. Readers will learn how to delineate a watershed and how land use, geographical location, social–economic demographics, and other human activities impact hydrologic cycles and the availability, sources, use and reuse, and quality of water. Concepts for estimating water demand and wastewater generation and sizing water distribution systems and wastewater collection systems are also introduced.

© Charles Taylor/iStockphoto

Learning Objectives

1. Describe the components of the hydrologic cycle and then, specifically, the major components of groundwater systems.
2. Delineate a watershed and estimate runoff within a watershed.
3. Estimate how changes in land use and protection of wetlands and green space impact the hydrologic cycle, runoff, and mass loadings of pollutants (including nutrients) to a watershed.
4. Identify the quantities, sources, and geographic distribution of freshwater on a global basis.
5. Identify the major users of water and the percent of water use associated with types of users.
6. Empathize with the global population that lives in areas not equally served by global water and sanitation and understand challenges experienced by people living in these areas.
7. Associate specific sources of water with water quality and usage.
8. Identify a local or regional water reuse project and describe its environmental and social benefits.
9. Articulate how water demand and energy use are integrated.
10. Estimate water and wastewater flow rates for residents and communities.

11. Distinguish daily demand cycles for industry, residential, and commercial uses.

12. Estimate demand factors and household water usage rates from historical records.

13. Determine water demand associated with fire protection and losses due to items such as leakage and unmetered use.

14. Calculate wet-weather flows based on inflow and infiltration.

15. Project future water demand using extrapolation methods and the type of customer.

16. Lay out a water distribution system.

17. Size a wastewater collection pipe based on the full-pipe design flow velocity and design carrying capacity.

18. Size a wet well based on pump characteristics.

19. Apply mass balance concepts and knowledge of plug flow reactors to investigate issues related to surface water quality.

20. Determine the oxygen deficit in a river.

21. Describe features of the dissolved oxygen (DO) sag curve and determine the location of the critical point and the oxygen concentration at the critical point for a given flow and discharge scenario.

22. Describe the process of lake and reservoir stratification and relate it to issues of water quality such as excess nutrient addition and oxygen depletion.

23. Explain nutrient enrichment of freshwater and coastal waters and the adverse human activities that lead to eutrophication.

24. Develop a systems approach to reducing nutrient loads to a watershed that considers point and nonpoint sources of pollution.

25. Describe eight methods of engineered lake management that controls nutrient inputs.

26. Define a wetland, their importance in buffering nutrient, and other pollutant loadings, and describe the most common contributors to wetland loss and methods to restore them.

27. Describe major point and nonpoint sources of groundwater pollution that are derived from human activities and natural processes.

28. Apply understanding of Darcy's law and the retardation factor to estimate the velocity of groundwater and groundwater pollutants.

29. Describe several methods to remediate soil and groundwater contamination.

30. Describe a regional, national, and global water quality challenge, and present a solution that moves society toward sustainable management of the Earth's water resources.

7.1 Introduction to Water Resources and Water Quality

The fundamental science that deals with the occurrence, movement, and distribution of water on the planet is **hydrology**. The **hydrologic cycle** (Figure 7.1) is defined as the pathways for how water moves and is distributed above, on, and below the surface of the Earth. The degree to which precipitation provides freshwater to the surface and evaporation and evapotranspiration return water to the atmosphere change with geographical location, time of the year, and year. Changes in land use and climate also influence the hydrologic cycle. Important to our discussion in this chapter, the quantity and quality of water also vary as it moves through the hydrologic cycle. Water resources engineering includes the management of hydrological cycle to transport water for supply and wastewater for collection, prevent flooding, and provide water transportation networks.

Pollution may be defined as the introduction of a substance to the environment at levels leading to a loss of a beneficial use of a water, air, or land resource or degradation of the health of humans, wildlife, or ecosystems. Pollutants are discharged to aquatic systems from **point sources** (stationary locations such as an effluent pipe) and from **nonpoint sources** (also called *diffuse*) such as land runoff and the atmosphere. The degree of pollution in water is typically described by units of concentration or the mass flux (or loading) of a pollutant discharged to a water body (units of mass per unit time).

Engineered approaches to pollution management vary with the type of material in question. Pollutants such as nitrogen, phosphorus, organic matter, and suspended solids (SS) are discharged to rivers and estuaries of the world by the tens of millions of tons per year. Figure 7.2a shows the biochemical oxygen demand (BOD) loading (in megatons per year) to global waterways for agricultural, domestic, and industrial sectors in 1995, as well as discharges expected in 2010 and 2020 for countries that are members and nonmembers of the Organization for Economic Cooperation and Development (OECD). Figure 7.2b shows the agricultural nitrogen loading for OECD and non-OECD countries for the same time periods. Note the large contribution of the agricultural sector to global

Figure / 7.1 **The Hydrologic Cycle**
Units of water transfer are 10^{12} m^3/year.

(Data from Budyko (1974); from Mihelcic (1999).
Reprinted with permission of John Wiley & Sons, Inc.).

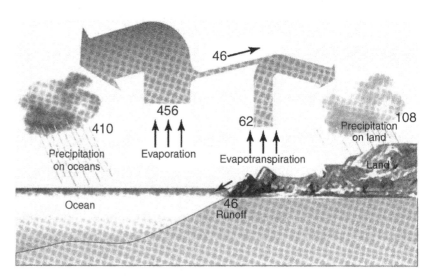

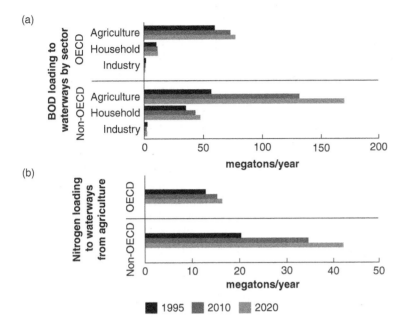

Figure / 7.2 **Pollutants in Global Waterways** (a) Annual BOD loading (megatons) into global waterways (in OECD and non-OECD countries) for agricultural, domestic, and industrial sectors for 1995 and estimated for 2010 and 2020. (b) Nitrogen loading (megatons) for OECD and non-OECD countries for agricultural sectors into global waterways for 1995 and estimated for 2010 and 2020.

(Data from UNESCO, 2003).

BOD and nitrogen loadings. This is one reason the National Academy of Engineering has designated managing the nitrogen cycle as one this century's grand challenges.

The **Clean Water Act** calls for the maintenance of fishable–swimmable conditions in U.S. waters. The Environmental Protection Agency (EPA) has set standards to achieve this goal, retain beneficial uses, and protect human and ecosystem health. Some standards are technology based, requiring a particular level of treatment regardless of the condition of the receiving water. Other standards are water quality based, calling for additional treatment where conditions remain degraded following implementation of standard technologies. Under the National Pollutant Discharge Elimination System (NPDES), permits are required for all those seeking to discharge effluents to surface water or groundwater. Standards may then be met by regulating the conditions of the permit, that is, the load that may be discharged. In cases where controls are not stringent enough to maintain the desired water quality, an analysis is conducted to establish the *total maximum daily load (TMDL)* that may be discharged to a water body, and permits are set accordingly.

Engineering's Grand Challenges
www.engineeringchallenges.org

Visualize Florida Water Issues
www.wateratlas.org

Class Discussion
What methods will protect water ecosystems for future generations as population and urbanization increase, changes occur in land use, and population and increased affluence drive increases in demand for food and biofuels.

7.2 Surface Water, Groundwater, Watersheds

7.2.1 SURFACE WATER AND GROUNDWATER

Surface water occurs as freshwater and seawater in streams, rivers, reservoirs, wetlands, bays, estuaries, and oceans. It also appears in solid form as snow or ice. When precipitation falls to the ground surface, some will run off to surface waters and some will infiltrate into the ground's surface. The water that infiltrates the ground's surface is referred to as **groundwater.** It exists below the land surface and consists of water and air that fills pores and factures that exist underground. The solid materials that providing this pore structure consist of sand, clay, and rock formations.

Some groundwater remains close to the surface and quickly reappears above the surface, providing an important flow to surface water by recharging streams and rivers in what is called **baseflow**. Other groundwater travels vertically through the subsurface due to gravity. Eventually, it meets the water table, where its direction changes to a more horizontal movement. The term **aquifer** describes this underground soil or rock through which groundwater travels. The **saturated zone** is used to describe the aquifer if is saturated with water. This is the zone from which groundwater is extracted to the surface. The **unsaturated zone** is the zone where pores are filled with air and water. In this zone, fluctuations in the moisture content of the pores can occur daily and seasonally through intermittent or longer-term precipitation and other climate events.

Figure 7.3 shows the difference between a **confined aquifer** and **unconfined aquifer**. As seen in the figure, the water in the confined aquifer is separated from atmospheric pressure by an impermeable material. A confined aquifer is also under pressure that is above atmospheric. If penetrated with a well, the water will rise in the well. In

Confined/Unconfined Aquifer

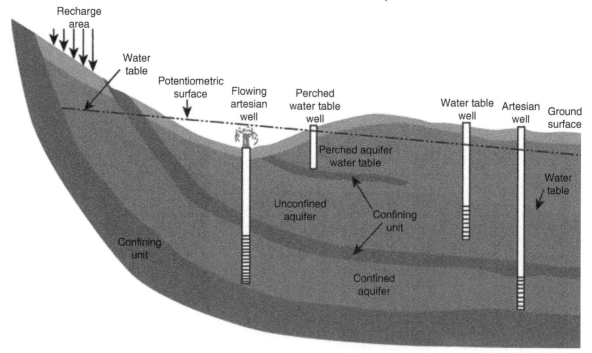

Figure / 7.3 Confined and Unconfined Aquifers.

Table / 7.1

Confined and Unconfined Aquifers (adapted from U.S. Geological Survey Water Science School) (http://ga.water.usgs.gov/edu/watercyclegwdischarge.html)

Unconfined aquifers	In an unconfined aquifer, water has infiltrated from the surface and saturated the subsurface material. If a well is placed in an unconfined aquifer, a pump will be required to lift the water to the surface.
Confined aquifers	A confined aquifer has a layer of rock or a confining layer of clay above and below it that are not very permeable to water. Natural pressure in the confined layer can thus exist, and this pressure may be enough to push water in a well to the surface (noted as an "artesian well' in Figure 7.3). However, not all confined aquifers produce this effect, so pumping may still be required to lift the water to the surface.

contrast, the unconfined aquifer's upper boundary is not a confining layer, but is the top of the water table. Table 7.1 provides additional explanations of these two types of aquifers.

7.2.2 WATERSHEDS

A **watershed** is defined as the land area that drains to a point of concern. Accordingly, lakes and rivers have watersheds. Drainage in the watershed is due to gravity; thus watershed boundaries are defined (or delineated) by topographical ridges as shown in Figure 7.4. Precipitation that falls within a watershed must thus drain somewhere.

Groundwater Assessment in Africa

http://www.unep.org/
groundwaterproject/

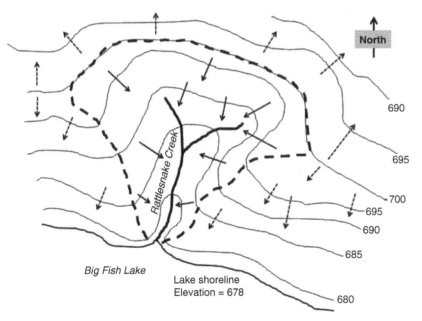

Figure / 7.4   A Small Watershed Defined for a Small Creek and the Lower Boundary of Big Fish Lake (shoreline elevation, 678 ft above sea level)   The dashed line delineates the watershed boundary of Rattlesnake Creek. The solid lines are topographic lines of equal elevation (in this case, feet above sea level). Solid arrows depict rainfall runoff within the watershed of Rattlesnake Creek, and dashed arrows show runoff outside of the Rattlesnake Creek watershed (either to the north to another watershed or southward (east and west of the boundary) into Big Fish Lake.

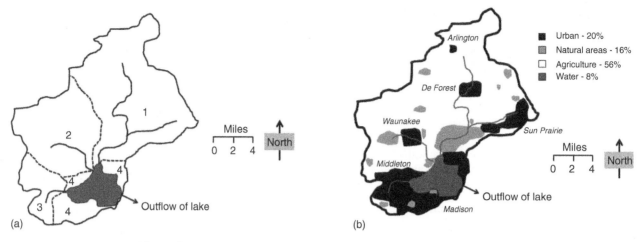

Figure / 7.5 **The Lake Mendota (WI) Watershed, Which Includes Portions of the City of Madison**   (a) The four sub-watersheds of the Lake Mendota watershed are delineated by dashed lines: (1) Yahara River, (2) Six-Mile Creek, (3) Pheasant Branch, and (4) direct runoff into Lake Mendota. The sub-watersheds numbered 1 through 3 drain to streams that flow into the lake. (b) Land use within the Lake Mendota watershed. Natural areas include woodlands, parks, grasslands, and wetlands. Water includes the surface area of the lake and the streams feeding into it.

Larger watersheds may also be divided into sub-watersheds as depicted in Figure 7.5.

Rainfall (or snowmelt) within a watershed has the potential to drain to the point of concern if it does not either infiltrate into the subsurface and become groundwater or enter the atmosphere as evapotranspiration. Rainfall or snowmelt that occurs outside of a given watershed is accounted for in another adjacent watershed.

Understanding watersheds is important for issues of water quantity and quality. For example, the amount of stormwater runoff within a watershed determines the potential for flooding (management of stormwater is discussed in Chapter 9). The topography (slope) and soil (potential for infiltration into the ground versus running off along the surface) are both key factors influencing the amount of runoff or infiltration in a watershed. Other important factors that influence the hydrology of a watershed and movement of pollutants within a watershed is the extent that human activities such as agriculture and urbanization influence the production of pollutants and the presence (or absence) of wetlands and impermeable surfaces. The extent to which urbanization and the construction of impermeable surfaces influence the degree of surface runoff and groundwater infiltration is depicted in Figure 7.6. These same factors, and others, are also important factors influencing the amount of pollution in the runoff.

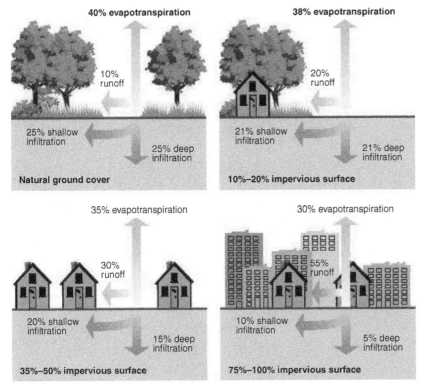

40% evapotranspiration

10% runoff

25% shallow infiltration

25% deep infiltration

Natural ground cover

38% evapotranspiration

20% runoff

21% shallow infiltration

21% deep infiltration

10%–20% impervious surface

35% evapotranspiration

30% runoff

20% shallow infiltration

15% deep infiltration

35%–50% impervious surface

30% evapotranspiration

55% runoff

10% shallow infiltration

5% deep infiltration

75%–100% impervious surface

Figure / 7.6 Nonpermeable Surfaces Change Natural Hydrological Cycles   As natural ground cover is removed and replaced with nonpermeable surfaces such as buildings, roads, and parking lots, there is increased runoff and significantly less recharge of groundwater. Also, evapotranspiration is reduced in an area with large amounts of nonpermeable coverings. The process of evapotranspiration results in a cooling process (much as your skin cools when you perspire), which can negate the impact of urban heat island. (EPA (2000)).

7.2.3   ESTIMATING SURFACE RUNOFF FROM LAND USE

A common approach to estimate runoff is the **Rational method**. The peak runoff flow rate can be estimated as a function of precipitation intensity, land use, and watershed area:

$$Q = \sum C_j i A_j \qquad (7.1)$$

where Q is the peak runoff flow rate (ft^3/s), C_i is a runoff coefficient for a particular land use of type j (dimensionless), i is rainfall intensity (in/h), and A_i is the area within the watershed for a particular land use type j (acres).

Note that in Equation 7.1, the product of C, i, and A using the units provided results in a flow rate with units of acre in/h. By coincidence, 1 acre in/h is approximately equal to 1 ft^3/s. The error of this unit conversion approximation is considered minimal (less than 1 percent) for most estimations. Also, as shown in Equation 7.1, a higher value for a runoff coefficient (C_i) will result in more runoff for a given storm intensity (that is, i, the precipitation) and watershed area, A.

Runoff coefficients are tabulated in many sources. Typical values are provided in Table 7.2. Runoff coefficients can theoretically range from 0 to 1 and are influenced by the particular manner of land use, the type of soil or engineered cover, and the slope of the land. A runoff coefficient closer to zero implies that more rainfall is infiltrating into the subsurface, while runoff coefficients closer to one imply that the

Table / 7.2

Typical Runoff Coefficients and Percent Impervious Area Values for Various Land Uses Runoff coefficients are given for two land slope categories for three types of soils.

Land Use	Percent Nonpermeable Area	Runoff Coefficient for Sandy Soils		Runoff Coefficient for Sandy/Silty Soils		Runoff Coefficient for Clayey Soils	
		Land Slope (%)		Land Slope (%)		Land Slope (%)	
		0–2	2–6	0–2	2–6	0–2	2–6
Industrial	90	0.67	0.68	0.68	0.68	0.69	0.69
Commercial	95	0.71	0.71	0.71	0.72	0.72	0.72
High-density residential (15 homes/acre)	60	0.47	0.49	0.48	0.50	0.51	0.53
Medium-density residential (5 homes/acre)	30	0.25	0.28	0.27	0.30	0.33	0.36
Low-density residential (2 homes/acre)	15	0.14	0.19	0.17	0.21	0.24	0.28
Agricultural	5	0.08	0.13	0.11	0.15	0.18	0.23
Open space (parks, grass, pasture)	2	0.05	0.10	0.08	0.13	0.16	0.21

SOURCE: Wisconsin Department of Transportation (2012).

majority of rainfall is running off to a local water source. As expected, sandy soils have the lowest runoff coefficients and clayey soils have the highest runoff coefficients. Also, as land slope increases, the runoff coefficient increases.

As was depicted in Figure 7.6, land use has a large impact on runoff. For example, a high percentage of roofs and impermeable pavements will prevent or limit infiltration of water to the subsurface. This occurs in industrial, commercial, and high-density residential uses of land. These uses of land accordingly have higher runoff coefficients than land uses with less impervious surfaces such as low-density residential, agricultural, and open green space.

Agricultural practices on the land also influence runoff. Poor agricultural practices such as overgrazing of land result in less vegetative land cover and compact soils. This limits infiltration of precipitation into the ground, which can lead to excessive runoff. Interestingly, with proper management of stormwater using technologies and strategies that mimic natural conditions that enhance infiltration (referred to as low-impact development, which is discussed in Chapter 9), it is possible to design a new development where the amount of post-development stormwater runoff from a residential subdivision may actually be less than the pre-development runoff from poorly managed agricultural land.

example/ 7.1   Use of Rational Method to Determine Runoff from Changes in Land Use

An agricultural watershed that is managed to minimize runoff consists of 100 acres that has a gentle land slope (1–2%) and silty/sandy soils. The land is planned to be developed into a residential subdivision (60% as a low-density residential area and 40% as a medium-density residential area). Estimate the pre- and post-development peak runoff flow rate for a storm with rainfall intensity of 0.5 in/h. Also determine the percent change in runoff between the two land use scenarios.

solution

This problem requires the use of the Rational method to determine to flow rate during the peak storm intensity. Equation 7.1 is written as

$$Q = C \times i \times A$$

Using information provided in this problem and values of the runoff coefficient (C) provided in Table 7.2, we can determine the flow rate (Q) for the pre- and post-development land use scenarios.

Pre-development scenario: $Q = 0.11 \times 0.5 \text{ in./h} \times 100 \text{ acres} = 5.5 \text{ ft}^3/\text{s (cfs)}$

Post-development scenario: $Q = (0.17 \times 0.5 \text{ in./h} \times 60 \text{ acres}) + (0.27 \times 0.5 \text{ in./h} \times 40 \text{ acres}) = 10.5 \text{ ft}^3/\text{s (cfs)}$

Percent change: [(Post-development runoff − pre-development runoff)/ pre-development runoff] × 100%

(10.5 cfs − 5.5 sfs)/5.5 cfs × 100% = 91% increase in runoff

Note the large increase in runoff due to the change in land use from open agricultural to provision of residential housing. This results in less recharge of any precipitation to groundwater where it could be stored for future use. The resulting runoff must not only be managed to prevent flooding of local and downstream properties but also may contain pollutants. This analysis also suggests that the planner and designer could employ protection of wetlands and green space or low impact development strategies (for example, permeable pavements, green roofs, and bioretention cells) that mimic nature's hydrologic processes, thus minimizing runoff and runoff pollution.

7.2.4 ESTIMATING POLLUTANT LOADINGS IN RUNOFF FROM LAND USE

While we will cover water quality in depth later in this chapter, it is worth mentioning at this point that land use impacts both water quantity and quality. High amounts of runoff result in higher flow rates and can cause erosion of soil. Likewise, greater rates of runoff result in higher flow rates in streams and rivers that can cause erosion of streambanks and resuspension of bottom sediment, which can increase loadings of suspended solids (and pollutants attached to these particles) to downstream water bodies. This can have large adverse impact on social, economic, and environmental systems that depend on water quality.

In many cases, the majority of pollutants that contribute to surface water quality problems originate within the watershed. Notable exceptions are mercury that originates from combustion of fossil fuels like coal and could originate locally or hundreds or thousands of miles away.

Table / 7.3

Typical Values for Pollutant Export Coefficients from Runoff (pounds/acre/year) Actual values vary greatly, often by an order of magnitude, depending on hydrology, land slope, and other factors. Agricultural values are especially variable due to different crops, cultivation, and fertilizer application practices.

Land Use	Suspended Solids	Chloride	Phosphorus	Nitrogen
Commercial	1,000	420	1.5	9.8
Industrial	500	25	1.3	4.7
Parking lot	400	300	0.7	8.0
Freeways	880	470	0.9	12.1
High-density residential	420	54	1.0	6.2
Medium-density residential	250	30	0.3	3.9
Low-density residential	10	9	0.04	0.4
Parks	3	–	0.03	–
Agriculture –Cultivated land –Pasture	 2,000–20,000 200–2,000		 0.06–3 0.05–0.6	 2–80 3–14

Data sources include Burton and Pitt (2001); Loehr et al. (1989); and USDA (2009).

Watershed delineation and land use determination can also provide estimates of pollutant loading to receiving waters. Unit area pollutant loadings, also referred to as export coefficients or yield coefficients, are reported in units of mass of pollutant in runoff from a unit area of land surface per time. These units are typically reported as pounds/acre/year (or kg/hectare/year) and vary with land use as given in Table 7.3.

The annual mass loading of a pollutant, L, into a surface water body can be estimated as

$$L = \sum A_i C_{e,i} \tag{7.2}$$

where L is the annual loading of the pollutant (mass/year), A_i is the surface area within the watershed of a particular land use type i, and $C_{e,i}$ is the export coefficient for the pollutant for land use type i. In Equation 7.2, the sum of all A_i values must equal the watershed area.

---

example/ 7.2   Estimating Pollutant Loading to a Watershed from Changes in Land Use

An agricultural watershed that is managed to minimize runoff consists of 100 acres of cultivated land that has a gentle land slope (1–2 percent) and silty/sandy soils. The land is planned to be developed into a residential subdivision (60 percent as a low-density residential area and 40 percent as a medium-density residential area). Estimate the pre- and post-development annual loading of suspended solids (SS) and phosphorus (P) that are in runoff from the land.

example/ 7.2 (continued)

solution

In the absence of specific measurements of flow and concentration at a particular location that would allow us to determine pollutant loadings from field measurements, we can estimate the loading using information provided in Table 7.3 along with Equation 7.2. When there is a range, we will use the lower values in the table for this example.

The loadings for SS and P prior to development can be estimated as

$$L = \sum A_i C_{e,i}$$

$$L_{SS} = 100 \text{ acres} \times 2{,}000 \text{ pounds SS/acre/year} = 200{,}000 \text{ pounds SS/year}$$

$$L_P = 100 \text{ acres} \times 0.06 \text{ pounds P/acre/year} = 6 \text{ pounds of P/year}$$

Using similar methods, we can estimate the loadings for SS and P after development takes place as

$$L_{SS} = (60 \text{ acres} \times 10 \text{ pounds SS/acre/year}) + (40 \text{ acres} \times 250 \text{ pounds/acre/year})$$
$$= 1{,}600 \text{ pounds SS/year}$$

$$L_P = (60 \text{ acres} \times 0.04 \text{ pounds P/acre/year}) + (40 \text{ acres} \times 0.3 \text{ pounds/acre/year})$$
$$= 14 \text{ pounds P/year}$$

The percent change in pollutant loadings from the change in land use can also be determined.

$$[(\text{Post-development load} - \text{pre-development load})/\text{pre-development load}] \times 100\%$$

For SS, the percent change is

$$[(16{,}000 - 200{,}000)/200{,}000] \times 100\% = -92\% \text{ (a 92\% reduction)}$$

For P, the percent change is

$$[(14 - 6)/6] \times 100\% = +133\% \text{ (a 133\% increase)}$$

Note how changes in land use that impact pollutant discharges to surrounding surface water are impacted not only by the particular land use, but also by the particular pollutant.

7.3 Water Availability

The total volume of the world's water is estimated to be $1.386 \times 10^9 \text{ km}^3$. Oceans hold 96.5 percent of this total volume, and the atmosphere contains only $1.29 \times 10^4 \text{ km}^3$ of water (which is only 0.001 percent of the total hydrosphere). Table 7.4 gives the world's freshwater reserves.

The total amount of **freshwater** on our planet is approximately $3.5 \times 10^7 \text{ km}^3$. In terms of freshwater availability, only 2.5 percent of the world's total water budget is estimated to be freshwater, and of this, almost 70 percent is currently present as glaciers and ice sheets. As given in Table 7.4, a large percentage of the world's freshwater is available as a groundwater resource, much of which has a renewal period of over 1,000 years. All of this information shows that very little of the total freshwater budget is available as surface water (lakes, rivers) or as groundwater that is recharged over a short duration.

Figure 7.7 shows the relationship of global water availability to population. The Americas are relatively rich in available water

Table / 7.4

Percent of World's Total Freshwater in Different Locations   The total amount of freshwater on Earth is approximately 3.5×10^7 km^3

Location	Percent of World's Freshwater
Glaciers and permanent snow cover	68.7
Groundwater	30.1
Lakes	0.26
Soil moisture	0.05
Atmosphere	0.04
Marshes and swamps	0.03
Biological water	0.003
Rivers	0.006

SOURCE: Data from UNESCO–WWAP, 2003.

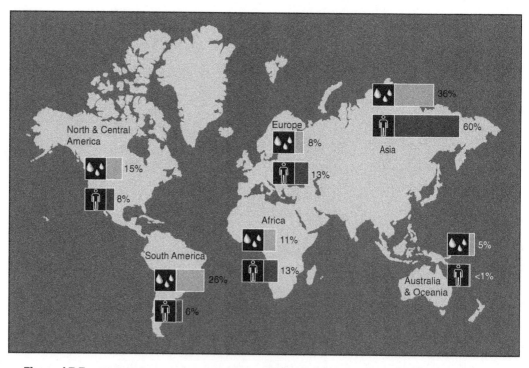

Figure / 7.7   **Global Overview of Water Availability versus Population**   Continental disparities exist, particularly on the Asian continent. About 60 percent of the world's population resides in Asia, yet only 36 percent of the world's water resources are located there.

(Figure redrawn with permission from "The United Nations World Water Development Report: Water for People, Water for Life", Figure 4.2 "Water availability versus population", p. 69, copyright UNESCO-WWAP 2003).

resources relative to their population. North and Central America combined have 8 percent of the world's population and 15 percent of the world available water resources. In contrast, Asia contains 36 percent of the world's available water resources but houses 60 percent of the global population.

7.4 Water Usage

Water is required by a wide variety of human users, including residential homes, commercial entities, industry, agriculture, and importantly, ecosystems. The concept of the ecological footprint (discussed in Chapter 5) assumes that some of the world's ecological capacity should be preserved for biodiversity protection. This ecological capacity requires water. In addition, while everyone in the world depends on ecosystems for their social and economic well-being, much of the world's poor (those living on less than $1 or $2 per day) depend even more on ecosystems for their economic livelihood. The water requirements of ecosystems must therefore be accounted for when managing how water is distributed among various users.

Globally, 3,800 km³ of water are withdrawn every year. Of this, 2,100 km³ are consumed. **Consumed water** is evapotranspirated or incorporated into products or organisms (UN-Habitat, 2003). The difference of 1,700 km³ is returned to local water bodies, usually as wastewater that comes primarily from domestic and industrial users. This large volume returned to the local water system may not be available for easy reuse, though, depending on its next use and, importantly, whether it has been contaminated and/or treated prior to discharge.

Similar to the **ecological footprint** that calculated the land area required to support human activities, a **water footprint** determines the water required to support human activities. Eight countries (in order of consumption) are responsible for half of the world's water footprint: India, China, the United States, Russia, Indonesia, Nigeria, Brazil, and Pakistan. On a per capita basis, for 1997–2001, the United States had the highest footprint: 2,483 m³/capita-year. In comparison, for the same period, the global water footprint was 1,243 m³/capita-year. Footprints of other countries (m³/capita-year) are Australia (1,393), Brazil (1,381), China (702), Germany (1,545), India (980), and South Africa (931) (Hoekstra and Chapagain, 2007).

As stated previously, much of the domestic water in urban areas is discharged back into the environment. Think about the multiple defacto reuses of water along a long river such as the Colorado, Ohio, or Mississippi as drinking water is obtained from a source with multiple upstream wastewater discharges. Historically urban areas met their water needs from local surface waters. Cities thus have an important role to play in ensuring that water they replace back to the environment does not harm ecological systems or downstream users. They are also becoming increasingly dependent on the *interbasin transfer* of water, which requires tremendous amounts of infrastructure investment and associated energy for collection, storage, and transfer.

Class Discussion
Discuss some regional and global challenges you expect over the next century related to the distribution of population and water. How do demographics (for example, income, level of education, age, and gender) relate to your discussion?

World Water Assessment Programme
http://www.unesco.org/water/wwap

Class Discussion
Prior to class, visit www.waterfootprint.org/and calculate your personal water footprint and the footprint of several countries including the United States. Discuss how changes in technology, policy, and human behavior can reduce the water footprint at the household and country level. Which changes are most effective at the household and national level? Which are most equitable?

Calculate Your Water Footprint
http://www.waterfootprint.org

Table / 7.5

Percent of Annual Water Withdrawals Associated with Agricultural, Industrial, and Domestic Sectors

	Agriculture (%)	Industry (%)	Domestic (%)
World	70	20	10
North America	39	47	13
Latin American and Caribbean	73	9	18
Europe	36	49	15
Africa	85	6	9
West Asia and Asia and Pacific	86–90	4–8	6

SOURCE: UN–Habitat, 2003.

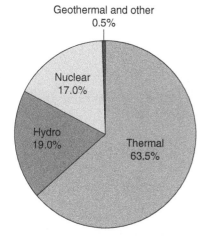

Figure / 7.8 **Breakdown of Global Electricity Production** Thermal electricity generation accounts for two-thirds of electricity production worldwide. Hydropower is the most widely used renewable source of electricity.

World Commission on Dams

http://www.internationalrivers.org/campaigns/the-world-commission-on-dams

Class Discussion

What are some of the social and environmental impacts of large dams? The World Commission on Dams has further information at its web site, www.dams.org.

7.4.1 PRIMARY USE OF WATER IN THE WORLD

Table 7.5 gives the primary uses of water throughout the world (excluding the use of water in production of electricity). Most of the usage listed in Table 7.5 involves **agricultural use** (globally at 70 percent). There are regional differences, however, particularly in the **industrial use** of water in higher-income areas such as Europe and North America. The percent of annual water withdrawals for **domestic use** ranges from 6 to 18 percent of the total withdrawals. Note how the level of development for a particular region of the world affects the distribution of water use.

Table 7.5 excludes the use of water for the energy sector. When the water demands of electricity generation are considered, over half of the water use is for power generation. Figure 7.8 shows that thermal electricity production accounts for two-thirds of global electricity production. Hydropower provides 19 percent of the total electricity generation and nuclear 17 percent. Other sources such as geothermal, tidal, wave, solar, and wind energy (which are not associated with large water usage) account for less than 0.5 percent of the world's electricity production.

A major benefit of **hydropower** is that each additional terawatt of hydropower produced per hour that displaces coal-generated electricity annually offsets 1 million tons of CO_2 equivalents. Hydropower has other benefits, such as low operation and maintenance costs, few atmospheric emissions, and no production of hazardous solid wastes. However, large-scale hydropower has problems, including large investment costs, issues related to fish entrainment and restriction of passage, loss and modification of fish habitat, and displacement of human and wildlife populations.

Many of the people displaced by large-scale hydropower systems are poor, less educated, and indigenous. Remember our discussion on **environmental justice** in Chapter 6? As just one example, the 18.2-gigawatt (GW) Three Gorges dam project in China is estimated to have already displaced more than 1 million people that reside in over 1,200 villages and many cities. The dam has submerged 632 km², which includes burial sites, historic sites of cultural significance, and environmental treasures. Three hundred species of fish live in the Yangtze River, and many have been separated from their spawning grounds. Completed in 2012, the dam is estimated to provide one-ninth of China's electricity needs.

In contrast, **micro-hydropower** systems (generating less than 100 kW) and *mini-hydropower* systems (100 kW to 1 MW) have a much lower negative impact on the environment and society than large hydropower systems. They are usually decentralized and not connected to the electric grid. In terms of environmental benefits, compared with an equivalent coal plant, a 1 MW mini-hydropower system that produces 6,000 MWh every year would supply the electricity needs of 1,500 families and avoid emissions of 4,000 tons of carbon dioxide and 275 tons of sulfur dioxide (UNESCO-WWAP, 2003).

7.4.2   U.S. WATER USAGE

Total water withdrawals in the United States exceed 400,000 million gallons per day (gpd). Table 7.6 gives the breakdown (by use) of fresh and saline water withdrawals in the United States. The largest use of water is for the production of electricity.

Figure 7.9 shows the volume of *freshwater* withdrawals in the United States since 1950, broken down by withdrawals from groundwater and surface water. California and Texas withdraw the most surface water. California and Florida withdraw the most groundwater. The total volume of withdrawals has remained relatively constant since the mid-1980s, varying by less than 3 percent, even though the population has increased over this time period. Surface water has made up 80 percent of the total, and groundwater has made up 20 percent of the

Public Water Information
http://water.usgs.gov/

Table / 7.6

Total Fresh and Saline Water Withdrawals in the United States   These withdrawals total 408,000 million gpd. Freshwater accounts for 85% of this total, and surface water accounts for 79% of the total.

User	Percent of Total Fresh and Saline Water Withdrawals
Thermoelectric	48
Irrigation	34
Public supply	11
Industrial	5
Domestic	<1
Livestock	<1
Aquaculture	<1
Mining	<1

SOURCE: Data from Hutson et al., 2004.

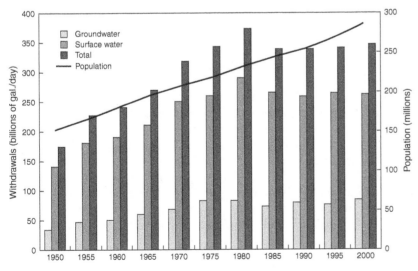

Figure / 7.9   U.S. Freshwater Withdrawals by Source and Population, 1950–2000.

(Courtesy of U.S. Geological Survey; Hutson et al. (2004)).

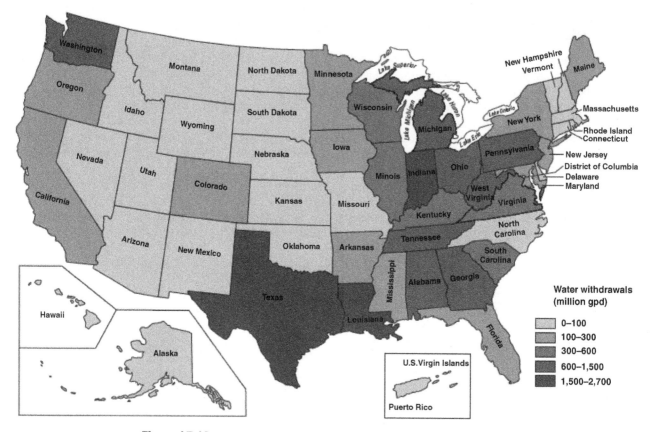

Figure / 7.10 Geographic Distribution of Industrial Water Withdrawals in the United States and Its Territories.

(Courtesy of U.S. Geological Survey; Hutson et al. (2004)).

total over the past 50 years. In contrast, the percent of withdrawals associated with public water supplies has tripled since 1950.

In terms of industrial withdrawals, Figure 7.10 shows the geographic distribution of industrial water withdrawals in the United States. Louisiana, Texas, and Illinois account for 38 percent of total industrial water withdrawals. Over 80 percent of this is from surface waters. The states of Georgia, Louisiana, and Texas account for 23 percent of groundwater withdrawals associated with industrial use.

Table 7.7 reviews various sources of water. Most domestic and industrial users obtain their water from **surface waters** (streams, rivers, lakes, reservoirs) and **groundwater**. However, desalination plants allow **seawater** to be used. Providing safe **reused water** (reclaimed water) is technically feasible. Reclamation is becoming an increasingly important source of water and **nutrients** and is now employed by a wide range of users—for domestic use, agriculture, landscaping, and recharging groundwater.

7.4.3 PUBLIC WATER SUPPLIES

In the United States, **public water supplies** are those that serve at least 25 people and have a minimum of 15 connections. They can be owned by the public or a private organization. This water can serve domestic, commercial, industrial, and even thermoelectric users. Figure 7.11 shows the total water withdrawals associated with public water supplies for every state and several territories of the United States. Public water usage is strongly dependent on population. For example, large states that account for 38 percent of the U.S. population (for example, California, Texas, New York, Florida, and Illinois) account for 40 percent of the total withdrawals from public water supplies.

Water Sources

The largest desalination plant in the United States provides 10 percent of Tampa's (FL) water needs. Capacity is up to 25 mgd.

Table / 7.7

Sources of Water and Issues Associated with the Source

Source of Water	Issues
Surface water	High flows, easy to contaminate, relatively high suspended solids (total suspended solids, TSS), turbidity, and pathogens. In some parts of the world, rivers and streams dry up during the dry season.
Groundwater	Lower flows but natural filtering capacity that removes suspended solids (TSS), turbidity, and pathogens. May be high in dissolved solids (total dissolved solids, TDS), including Fe, Mn, Ca, and Mg (hardness). Difficult to clean up after contaminated. Renewal times can be very long.
Seawater	Energy-intensive to desalinate, so costly compared with other sources, and disposal of resulting brine must be considered. Desalination can occur by distillation, reverse osmosis, electrodialysis, and ion exchange. Of these, multistage distillation and reverse osmosis are the two technologies most commonly used (they account for approximately 87 percent of worldwide desalination capacity). There are more reverse osmosis plants in the world; however, they are typically smaller in capacity than distillation plants.
Reclaimed and reused	Technically feasible. Currently used for irrigating agricultural crops, residential and commercial landscaping, groundwater recharge, and potable water through direct and indirect use. Includes decentralized use of gray water (wastewater produced from baths and showers, clothes and dishwashers, lavatory sinks, and drinking-water fountains). When used for irrigation, nutrients present in reclaimed water can reduce fertilizer usage.

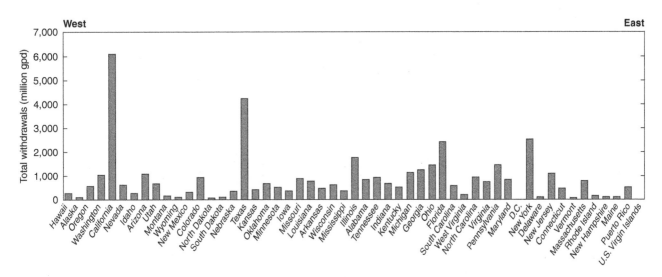

Figure / 7.11 **Public Supply Water Withdrawals in the United States, 2000** The states and territories in the figure are arranged from west to east.

(Courtesy of U.S. Geological Survey; Hutson et al. (2004)).

Learn More About Desalination
http://ga.water.usgs.gov/edu/
drinkseawater.html

Class Discussion

What type of water-efficient features are present in or missing from your house, your apartment or dormitory, and your university campus? Does use of this technology require any behavioral changes by users or maintenance staff? Does it require consideration of gender or cultural differences between users?

Single-family U.S. households average 101 gallons of water use per day per capita (gpdc), indoor and outdoor. In multifamily dwellings such as apartments, water use can be as low as 45 to 70 gpdc, because these households use less water, have fewer fixtures and appliances, and use little or no water outdoors (Vickers, 2001). Outdoor use may exceed natural rainfall in some locations. It ranges from 10 to 75 percent of total residential demand, depending on location. Table 7.8 gives a breakdown of this **household use** by different activities.

The table also gives what the same breakdown would be if households installed more efficient water fixtures and performed regular leak detection. If every U.S. household installed water-efficient features, water use would decrease by 30 percent. This would not only save money but would also eliminate the demand to identify and secure new water sources, saving energy and materials associated with collecting, storing, transporting, and treating water.

7.4.4 WATER RECLAMATION AND REUSE

Because of items such as increasing population, demographic shifts in where people live and industry locates, and current and future issues of climate change and water scarcity, **water reclamation** and **water reuse** have become commonplace in many states as a means to expand the water supply portfolio and provide an additional drought-resistant supply. Reclaimed water also takes advantage of beneficial nutrients found in reclaimed water. In fact Florida, which reuses more reclaimed water than any other state, initially launched its water reuse program to address nutrient pollution concerns in its streams, lakes, and estuaries (NAS, 2012).

States such as California, Arizona, Georgia, and Florida already incorporate water reclamation and reuse into the water management

Table / 7.8

Water Usage in U.S. Households: Typical and Efficient Alternatives Typical water usage in a U.S. household is far greater than when water-efficient fixtures are installed and households pay attention to leak detection. Percentages are based on total use.

Activity	Typical Water Usage, gpdc (% total use)	Water Usage with Water-Efficient Fixtures and Leak Detection, gpdc (% total use)
Showers	11.6 (16.8%)	8.8 (19.5%)
Clothes washing	15.0 (21.7%)	10.0 (22.1%)
Dishwashing	1.0 (1.4%)	0.7 (1.5%)
Toilets	18.5 (26.7%)	8.2 (18.0%)
Baths	1.2 (1.7%)	1.2 (2.7%)
Leaks	9.5 (13.7%)	4.0 (8.8%)
Faucets	10.9 (15.7%)	10.8 (23.9%)
Other domestic uses	1.6 (2.2%)	1.6 (3.4%)
Total	69.3 gpdc	45.3 gpdc

SOURCE: Data from Vickers, 2001.

decisions. This is due to their climate, expanding populations, and extensive water-dependent agricultural production. However, with water scarcity now becoming a common occurrence, areas that were once thought to be water rich *should also consider incorporating* water reuse and reclamation into water management decisions.

Wastewater treatment plants have typically been sited to take advantage of gravity to transport collected wastewater and to make the disposal of treated effluent easy. These are two reasons why many treatment plants are located near surface water bodies. Treatment plants are also situated closer to urban areas, where the vast majority of wastewater is generated. One key to successful water reclamation and reuse is to pair the quality of wastewater effluent with the water quality requirements of new users.

Most reclaimed water today is used in industrial or agricultural settings that may be located far away from a large wastewater treatment plant. Thus, any reclaimed water needs to be transported over large distances before it can be reused. To counter this problem, small decentralized **satellite reclamation plants** are being designed and constructed. These plants combine primary, secondary, and/or tertiary treatment processes to treat a portion of a wastewater stream close to where it can be used, thus eliminating the need to transport reclaimed water over large distances.

Table 7.9 provides several examples of successful water reclamation and reuse. In each case, the source of the reclaimed water is treated

Water Reclamation and Reuse
http://www.epa.gov/region09/water/recycling/

Table / 7.9

Examples of Water Reclamation and Reuse All sources of reclaimed water are treated as domestic wastewater. On a global scale, water reuse capacity is expected to increase from 19.4 to 54.5 million m^3/day by 2015.

Location	Use of Reclaimed Water	Issues Solved by Engineers through Technical, Policy, and Outreach Solutions
Hampton Roads Sanitation District, Virginia	Service water and boiler feed water at oil refinery	Needed to treat ammonia during cold weather and produce water with more consistent turbidity levels.
Irvine Ranch Water District, California	Landscape irrigation for public and businesses; dual-plumbed office buildings use water for toilet and urinal flushing; commercial office cooling towers; agricultural irrigation	TDS builds up in recycled water. Seasonal demands for landscape irrigation need to be balanced with storage limitations of urban environment.
San Antonio Water System, Texas	Power plant cooling water; industrial cooling; river maintenance; landscape irrigation	Water quality deterioration in distribution system can occur from higher solids content of reclaimed water. Concerns of cross connection with potable water and impact of higher TDS levels on vegetation.
South Regional Water Reclamation Facility, Florida	Agricultural and landscape irrigation; freeze protection of citrus crops; groundwater recharge	Possible impacts on irrigating with reclaimed water.
Orange County Water District, California	Groundwater recharge that subsequently supplements potable water supply; landscape irrigation	Emerging contaminants such as low-molecular-weight organics, pharmaceuticals, and endocrine-disrupting chemicals in reclaimed water.

SOURCE: Crook, 2004; GWI, 2005.

domestic wastewater. The reclaimed water has a variety of uses, including domestic, industrial, and agricultural use, as well as ground-water recharge and landscape irrigation. You may want to look up a recent report by the National Research Council (NRC) titled *Water Reuse: Potential for Expanding the Nation's Water Supply through Reuse of Municipal Wastewater Committee on the Assessment of Water Reuse as an Approach to Meeting Future Water Supply Needs* (National Academy of Sciences, Washington, D.C., 2012).

Class Discussion
Many technical issues associated with water reclamation and reuse have been successfully addressed. Are energy and material inputs high or low for this technology? What are some social challenges to reusing water for domestic use? What challenges might you encounter as an engineer working with a community on a water reuse project? How would you overcome these challenges in a fair and equitable manner?

7.4.5   WATER SCARCITY AND WATER CONFLICT

One of the most pressing global security problems in the future is likely to be **water scarcity**, a situation where there is insufficient water to satisfy normal human requirements. A country is defined as experiencing **water stress** when annual water supplies drop below $1{,}700\,\text{m}^3$ per person. When annual water supplies drop below $1{,}000\,\text{m}^3$ per person, the country is defined as **water scarce**. By one measure, nearly 2 billion people currently suffer from severe water scarcity. This number is expected to increase substantially as population increases and as standards of living (and therefore consumption) rise around the world.

Climate change is expected to have an impact on precipitation (see Table 7.10). Some areas may benefit from 10 to 40 percent increases in rainfall, but others are likely to suffer from 10 to 30 percent decreases

Table / 7.10

Example of Possible Impacts of Climate Change on Water Resources Projected for the Mid to Late Century

Phenomenon and Direction of Trend	Likelihood of Future Trends Based on Projections for 21st Century	Major Impact(s)
Over most land areas, warmer and fewer cold days and nights, warmer and more frequent hot days and nights	Virtually certain	Effects on water resources relying on snowmelt; effects on some water supplies
Warm spells/heat waves; frequency increases over most land areas	Very likely	Increased water demand; water quality problems, for example, algal blooms
Heavy precipitation events; frequency increases over most areas	Very likely	Adverse effects on quality of surface water and groundwater; contamination of water supply; water scarcity may be relieved
Increase in area affected by drought	Likely	More widespread water stress
Increase in intense tropical cyclone activity	Likely	Power outages, causing disruption of public water supply
Increased incidence of extreme high sea level (excludes tsunamis)	Likely	Decreased freshwater availability due to saltwater intrusion

SOURCE: Used with permission of the Intergovernmental Panel on Climate Change, *Climate Change 2007: Impacts, Adaptation and Vulnerability*, Summary for Policymakers, from Table SPM.1.

in rainfall. Some regions that will see an increase in rainfall will also become vulnerable to extreme rain events associated with flooding and erosion. The population that will be most vulnerable to climate change is poor and depends on rain for agricultural water and on local water resources for health and economic livelihood. These people also tend to live in areas that are prone to water-associated disasters of drought and flooding.

Water is expected to be a source of both tension and cooperation in the future. This is because more than 215 major rivers and 300 ground-water aquifers are shared by two or more countries. The Organisation for Economic Co-operation and Development (OECD) consists of 30 member countries. The OECD Development Assistant Committee writes that "Water-related tensions can emerge on various geographical scales. The international community can help address factors that determine whether these tensions will lead to violent conflict. Water can also be the focus of measures to improve trust and cooperation."

The following web site chronicles water conflict going back to 3,000 B.C. (http://www.worldwater.org/conflictchronology.pdf). History shows that most water conflict is resolved peacefully. In fact, there have been 507 recorded water conflicts and 1,228 recorded water cooperative events. However, there have been less than 40 recorded reports of violence over water. This shows that water conflict is perhaps not as sensational as popularized in movies like *Chinatown* and books like *Cadillac Desert*. Figure 7.12 shows the specific events that historic water conflict or cooperation was related to. As seen in this figure, most of the documented events associated with water conflict and cooperation are related to changes in the quantity of water flow and design and construction of infrastructure like dams and canals.

7.5 Municipal Water Demand

The amount of water used (or needed) is critical in the planning and design of a municipal water system. The estimated water usage rate is commonly called the **municipal water demand**. In general, the source(s), water facility location and size, and the piping to connect these facilities to the customers all depend upon demand. Although estimating water demand is critical to the planning of a system, there is no single method to measure or estimate it.

The amount of municipal water use is based on land use and the type and number of customers in the system. The design and sizing of a water (or wastewater) treatment plant is based on an estimate of the current and potential future water usage by the customers served by the system. Other factors, such as additional water for fire protection, increase the actual volume of water to be treated.

The design and sizing of the piping network to deliver water or collect wastewater is based not only on the estimated water usage but also on the location of the specific customers relative to the treatment facilities. For example, the location of a large industrial user may not greatly affect the total water to be processed by the treatment plant, but the sizing of the piping network to connect the user to the treatment plant will be greatly affected. An industrial user very near the treatment plant would need a much shorter length of large-diameter pipe than if the user were far from the treatment plant.

Class Discussion
How do your individual and professional actions related to energy use affect water supply and use in ways that ultimately affect future generations of the world's poor and native ecosystems? What decisions made by engineers have broader impacts beyond the local area where they are implemented?

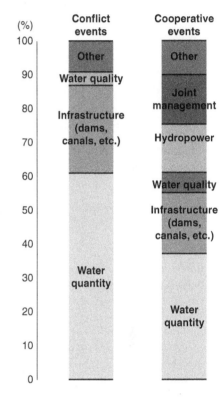

Figure / 7.12 **Water Cooperation or Water Conflict?** Shown are the percent of events that have either caused water conflicts or lead to water cooperation.

(Redrawn with permission from UNEP/GRID-Arendal, 2009. Water—cooperation or conflict? *Vital Water Graphics* 2, http://www.grida.no/graphicslib/detail/water-cooperation-or-conflict_16f8).

Class Discussion
Research and discuss a specific water conflict in your region or globally.

"Dam Hetch Hetchy! As well dam for water-tanks the people's cathedrals and churches, for no holier temple has ever been consecrated by the heart of man."
—John Muir

John Muir at the Merced River with Royal Arches and Washington Column in background, Yosemite National Park, California.

(Reproduced with permission of John Muir Papers, Holt-Atherton Special Collections, University of the Pacific Library. Copyright 1984 Muir-Hanna Trust).

Figure / 7.13   Hetch Hetchy Valley as It Looked before It Was Dammed.

(SOURCE: Sierra Club Bulletin, Vol. VI. No. 4, January, 1908, p. 211.)

Hetch Hetchy Valley is located in the less traveled northwest corner of Yosemite National Park. Yosemite National Park is the United States' second national park, created in 1890. It attracts 3.5 million visitors every year. The Tuolumne River runs through the valley floor and the granite outcroppings that enclose the valley are similar in breadth and beauty to those in the heavily traveled Yosemite Valley.

John Muir (author, preservationist, and founder of the Sierra Club) described Hetch Hetchy as "one of nature's rarest and most precious mountain temples." The U.S. National Park Service writes that

"as early as 1882, Hetch Hetchy Valley had been considered a potential site for a new reservoir. Preservationists, led by John Muir, wanted the valley to remain untouched. They maintained that a dam could be secured outside our wild mountain parks.

Muir and his followers launched a campaign to praise the virtues of Hetch Hetchy. For the first time in the American experience, a national audience considered the competing claims of wilderness versus development. Until the early 1900s, Americans viewed wilderness as something to conquer and natural resources as infinite.

Dam supporters were convinced that a reservoir could offer tremendous social and economic benefits. The fastest-growing city in the West, San Francisco was facing a chronic water and power shortage. In 1906, an earthquake and fire devastated San Francisco, adding urgency and public sympathy to the search for an adequate water supply. Congress passed the Raker Act in 1913, authorizing the construction of a dam in Hetch Hetchy Valley as well as another dam at Lake Eleanor.

The first phase of construction on the O'Shaughnessy Dam (named for the chief engineer) was completed in 1923 and the final phase, raising the height of the dam, was completed in 1938. Today the 117-billion-gallon reservoir supplies water to 2.4 million Bay Area residents and industrial users. It also supplies hydroelectric power generated by two plants downstream. The reservoir is eight miles long and the largest single body of water in Yosemite."

Extracted from "Yosemite," US National Park Service, US Department of Interior http://www.nps.gov/yose/planyourvisit/upload/hetchhetchy-sitebull.pdf

7.5.1 CREATING MODELS TO ESTIMATE DEMAND

Estimating the water demand generally involves creating a model of the system that mimics the real system. Decisions made to create an accurate model depend upon the intended purpose of the model and the information required from it. Therefore, the purposes for the model are defined at the outset, so that the appropriate type of model is selected. In general, **modeling objectives for water demand** are twofold:

1. *Existing systems*: Develop a model to accurately simulate the operation of the existing system.

2. *Proposed systems*: Develop a model that will become a planning tool that will guide the design of a future system.

The type of model and the associated detail are defined based on the specific model objectives. The detail of a model can be classified as either a *macroscale model* or *microscale model*. Macroscale models are used to estimate the overall water demand, sizing of treatment facilities, and system storage required to account for daily water usage cycles. The detail that a specific-size pipe is connected to five commercial customers residing in a Leadership in Energy and Environmental Design (LEED)–certified green building on Main Street would not be required in this case. However, a microscale model of the specific pipe diameter and nearby surrounding system might be used to size a pump in a pumping station used for fire protection. For this case, the water demand would be the water needed to put out the fire.

Table 7.11 summarizes the variety of data used in **estimating water demand**. Depending on the model's objectives and details, the data

Class Discussion

What social, environmental, and economic benefits does Hetch Hetchy Valley provide in its current engineered state? What social, environmental, and economic benefits would be gained if Hetch Hetchy Valley was restored back to its natural state?

Restore Hetch Hetchy
http://www.hetchhetchy.org/

© Eric Delmar/iStockphoto.

Table / 7.11

Types of Data That May Be Needed to Create a Water Demand Model

Type of Data	Description of Data
System data	This is the physical layout of the system. Examples of data include process drawings for a treatment plant, piping network for water distribution or sewer collection, or the layout of a new development. This includes the system dimensions (length, width, height) and elevations.
Operational data	This is the information about the system when it is in operation. Examples of data include water levels in the treatment plant tanks or storage tanks, pumping rates for pumps, or wet-well water levels. Much of this information is known by the system operators.
Consumption data	This is estimated water use by the customers. Data includes the per capita daily water demand, water demand value for specific customers such as a large building and fire protection demand. Also there are typically estimates of changing water use patterns, such as water conservation strategies.
Climate data	This consists of the seasonal temperature and rainfall data. Temperature and rainfall can have a great influence on the estimated water demand. Other climate data could be used if needed, and climate forecasting also may be used.
Demographic and land use data	This is data about the customers and how the customers use their property. This includes population numbers and expected future growth (or decline), the types of customers (residential, commercial, industrial, etc.), and the location of these customers. Transportation planning has large influence on this.

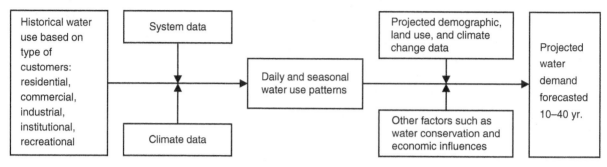

Figure / 7.14 General Process Used to Create a Model to Estimate Future Water Demand.

actually required may be just one or two types, or additional data may needed beyond what is listed in Table 7.11. The availability and accuracy of the data will vary greatly. Maps and drawings of the existing system, historical climate data, and operational data are usually readily available and generally accurate. Data required for future planning, such as demographic changes, future land use, and projected climate, can have significant uncertainty. Once any data are located, they must be critically reviewed for suitability to meet the model's needs. Any historical documents must be evaluated and verified to assure the accuracy of the data.

The general **modeling process to estimate water demand** is provided in Figure 7.14. The process begins with collecting and evaluating historical information about the type of customers served. From this, the daily and seasonal cycles can be determined. Including expected future demographic and land use data (and even climate change forecasting) makes it possible to estimate the future water demand.

Because of data availability and uncertainty, it is common to use different methods to estimate future demand. For example, when estimating the water demand for a residential area to size a treatment plant, a simple analysis of the current water use per household multiplied by the projected number of households might be as good an estimate as a complicated, detailed model of the entire system.

7.5.2   ESTIMATING WATER (AND WASTEWATER) FLOWS

Though we cover wastewater treatment and resource recovery in a later chapter, we will cover estimation of water and wastewater flow together, because they are closely linked. Obtaining water and wastewater data is a fundamental step in designing a water distribution system or sewer collection system, or in sizing a treatment plant. Flow rates and patterns vary greatly from system to system and are highly dependent on the type and number of customers served, climate, and local economics. Figure 7.15 shows the daily water flow cycles depending on the type of customer. Water use (and wastewater production) is time dependent as households and other users of the system incorporate water into their daily lifestyle. Demographics specific to a region will change the shape of this figure. For example, in the United States, the morning peak is usually higher than the afternoon peak. And in bedroom communities, there is usually a very early morning peak as people wake up early for long commutes into urban work areas.

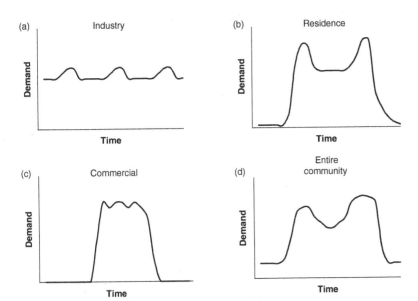

Figure / 7.15 **Daily Demand Cycles**
Cycles differ, depending on the type of customers: (a) industry, (b) residential, (c) commercial, and (d) the entire community.

The best source of information for estimating demand is usually recorded flow data. For existing systems, historical records of water usage are generally available. Generally, water supply and treatment facilities have information on water level changes in reservoirs and the rate of water pumped and treated (so water balances can be used to estimate hourly flows). For customers, there are typically billing records, which may have metered flows associated with them, but these will provide only averages over 1 or 2 months.

Typically, more detailed information is available about **water usage rates** than **wastewater generation rates**. One approach to estimating wastewater generation is to estimate the water usage rate and then assume that between 50 and 90 percent of the water becomes wastewater. However, this range will change greatly with climate, season, land use restrictions, and type of customer.

Indoor water use is generally equivalent to wastewater generation, because water used outdoors generally does not enter the wastewater collection system. A simple model to estimate indoor residential water use is expressed as follows (Mayer et al., 1999):

$$Y = 37.2X + 69.2 \qquad (7.3)$$

where Y is the indoor water use per household (gpd) and X is the number of people per household.

For new customers, historical water use or wastewater generation rates can be used to estimate the new water demand. The historical records are used to estimate values for per unit water use or wastewater generation. Table 7.12 provides typical values and expected ranges. An estimate of the number of new units added to the system is determined based on the information for the projected type of new customers. Then, to determine the additional water demand in gallons per day, the typical flow values are multiplied by the number of additional units.

Table / 7.12

Typical Water Use and Wastewater Generation Values Values are shown for different customers and can be used for estimating future scenarios.

Source	Water Usage Flow (gallons/unit day)			Wastewater Generation Flow (gallons/unit day)	
	Unit	Range	Typical	Range	Typical
Apartment	Person	100–200	100	35–80	55
	Bedroom			100–150	120
Department store	Restroom	400–600	550	350–600	400
Hotel	Guest	40–60	50	65–75	70
Individual residence:					
Typical home	Person	40–130	95	45–90	70
Luxury home	Person			75–150	95
Summer cottage	Person			25–50	40
Office building	Employee	8–20	15	7–16	13
Restaurant	Customer	8–10	9	7–10	8
School:					
With cafeteria, gym, and showers	Student	15–30	25	15–30	25
With cafeteria only	Student	10–20	15	10–20	15

SOURCE: Data obtained from Tchobanoglous and Burton, 1991; Tchobanoglous et al., 2003.

Water usage rates for industrial sources are highly site-specific and should be based on historical records or the design flow rates for new industrial customers. For example, a Ford Taurus (including tires) requires over 147,000 L of water for production and delivery to the marketplace, a pair of blue jeans requires over 6,800 L, and a Sunday newspaper requires 568 L. The wastewater generated from industrial sources also greatly varies depending upon how much water is used and what the water is used for. For many industries, water is used in processes where much of the water can be lost as evaporation. When possible, industries should be metered to determine the actual flows.

Water usage and wastewater generation rates can also be estimated using **land use** data. Although this method is primarily used when designing water distribution or wastewater collection systems for future development (Walski et al., 2003; 2004; AWWA, 2007), it can also be used to estimate the expected flows for sizing water and wastewater treatment facilities. The land use is classified based on customer type (residential, commercial, industrial) and customer density (households per area, light commercial, dense commercial, and so on). From analyzing historical records or using assumed values, a water usage rate or wastewater generation rate per land area can be determined. Then the local zoning regulations for the proposed development provide information for determining a water usage rate and wastewater generation rate.

7.5.3 TIME-VARYING FLOWS AND SEASONAL CYCLES

The methods described in the previous section generally provide average flow rates for water use and wastewater generation. The average flow rate provides an idea of the amount of water that needs to be treated or transported in the pipe network, but the actual design needs to be able to handle the expected daily and seasonal variations in water flow. A properly designed treatment plant must be able to handle a range of expected flows. Storage facilities such as storage tanks for water distribution or wet wells for wastewater collection can be used to minimize the daily flow fluctuations into a treatment plant, but seasonal variations can have a great impact on the treatment facility and how it is operated. Also, the piping network should be sized to handle the expected maximum flow rate but also work effectively for very small flow rates.

The **variation in flows** for municipal systems typically follows a 24 h cycle. However, this cycle can gradually change throughout the week (weekday flow versus weekend flow) and seasonally. The water use in the afternoon of a hot summer weekend day can be enormous due to outside water use such as watering lawns and filling pools. At the same time, however, wastewater generation may follow a typical day pattern because the inside water use would be typical. Alternatively, on a cool, rainy day, water use would be typical, but wastewater flows could be high because of stormwater entering the collection system. For most days in communities, there is very little flow during the night, an increase in flow during the morning hours, and close to average flow during the day, followed by a second increase in flow during the evening hours.

The daily cycle for individual users can also be determined. Figure 7.15 shows examples of the daily water usage rate for different types of customers. Note how the individual demand pattern can be very different from the demand pattern for the entire community. Most of the time, an individual demand pattern is insignificant and has little effect on the entire community's pattern. However, a large water user (such as a large industry) may affect the demand pattern in the local water distribution or sewer collection system, especially in a small community with a large single user.

For municipal systems, a **demand factor (DF)** is determined from historical records to estimate the typical maximum and minimum daily flow rates. Determining demand factors for entire communities is relatively easy, because flow rate records exist at the treatment facilities. The demand factor for different conditions is determined from the average flow rate and extreme-condition flow rate:

$$DF = \frac{Q_{\text{event}}}{Q_{\text{average}}} \qquad (7.4)$$

where Q_{event} is the event flow rate (volume/time), Q_{average} is the average flow rate (volume/time), and DF is the demand factor (unitless). Table 7.13 provides how demand factors are associated with particular events. Historical records can be used to determine the annual average, maximum, and minimum recorded water usage rates or wastewater generation values. Since these values are system-specific, the actual demand factors should be determined for each evaluated system.

Table / 7.13

Commonly Determined Demand Factor Events for Communities

Event	Description	Demand Factor Range
Maximum-day demand	The average rate of all recorded annual maximum-day demand	1.2–3.0
Minimum-day demand	The average rate of all recorded annual minimum-day demand	0.3–0.7
Peak-hour demand	The average rate of all recorded annual maximum-hour demand	3.0–6.0
Maximum day of record	The highest recorded maximum-day demand	<6.0

SOURCE: Adapted from WEF, 1998; Walski et al., 2003.

To determine the **maximum and minimum design flow rates** for a treatment facility or piping network, a peaking factor (similar to the demand factor) is applied to the average daily flow rate. The **peaking factor (PF)** is a multiplier that is used to adjust the average flow rate to design or size components in a water or wastewater treatment plant, or components of a water distribution or wastewater collection system (pipes, pumps, storage tanks, and so forth). Equation 7.5 can be used to determine these design flow rates:

$$Q_{design} = Q_{average} \times PF \tag{7.5}$$

where Q_{design} is the design flow rate (volume/time), $Q_{average}$ is the average flow rate (volume/time), and PF is the peaking factor for design (unitless). Table 7.14 provides peaking factors or design flows used for water and wastewater treatment facility processes.

Details on using peaking factors for the design of treatment facilities and piping networks are discussed elsewhere for water treatment facilities (Crittenden et al., 2005), wastewater treatment facilities (Chen, 1995; Tchobanoglous et al., 2003), water distribution systems (Walski et al., 2003), and wastewater collection systems (Walski et al., 2004).

Table / 7.14

Design Flows and Peaking Factors Used for Sizing Drinking-Water and Wastewater Treatment Plants

Treatment Process of Facility Operation	Water Treatment Plant	Wastewater Treatment Plant
Plant hydraulic capacity	$Q_{max\ day} \times (1.25\ to\ 1.50)$	$Q_{maximum\ instantaneous}$
Treatment processes	$Q_{max\ day}$	$Q_{average} \times (1.4\ to\ 3.0)$
Sludge pumping	$Q_{max\ day}$	$Q_{average} \times (1.4\ to\ 2.0)$

SOURCE: Adapted from Crittenden et al., 2005; Chen, 1995.

example/ 7.3 Using Historical Records to Estimate Demand Factors and Household Water Usage Rate

Estimate the maximum- and minimum-day demand factors, using data gathered from the annual water reports for a small water treatment plant. Then estimate the average household usage rate, using the data for all years. The metered flow values for each year are summarized in Table 7.15.

Table / 7.15

Metered Flow Values for Example 7.3

Year	Average (gpd)	Maximum Day (gpd)	Minimum Day (gpd)	Households Served
2001	834,514	1,325,486	324,851	5,567
2002	843,842	1,354,826	314,584	5,603
2003	854,247	1,334,287	300,145	5,671
2004	837,055	1,341,024	365,454	5,789
2005	828,103	1,362,487	298,764	5,894
2006	858,076	1,356,214	325,141	5,969
2007	861,003	1,384,982	336,954	6,002
2008	868,150	1,368,920	310,247	6,048

solution: Demand Factors

Determine the demand factor every year for the extreme events. For year 2001:

$$DF_{\text{max day}} = \frac{Q_{\text{max day}}}{Q_{\text{average}}} = \frac{1,325,486 \text{ gpd}}{834,514 \text{ gpd}} = 1.59$$

$$DF_{\text{min day}} = \frac{Q_{\text{min day}}}{Q_{\text{average}}} = \frac{324,851 \text{ gpd}}{834,514 \text{ gpd}} = 0.39$$

In the same way, the average demand factor can be determined for the other years, using all the data. From the annual averages, overall average can be determined as follows:

Year	$DF_{\text{max day}}$	$DF_{\text{min day}}$
2001	1.59	0.39
2002	1.61	0.37
2003	1.56	0.35
2004	1.60	0.44
2005	1.65	0.36
2006	1.58	0.38
2007	1.61	0.39
2008	1.58	0.36
Average	1.60	0.38

The maximum-day demand factor is 1.60 and the minimum-day demand factor is 0.38.

Our results compare very well with the established DF values provided in Table 7.13, where the maximum-day demand factor ranged from 1.2 to 3.0 and the minimum-day demand factor ranged from 0.3 to 0.7.

solution: Usage Rates

Next, we estimate the average household usage rate, using the data for all years. The household water usage rate for each year can be determined for each year of data. For year 2001:

$$\text{usage rate} = \frac{Q_{\text{average}}}{\text{metered households}} = \frac{834{,}514 \text{ gpd}}{5{,}567 \text{ households}} = 150 \text{ gpd/household}$$

Using the same formula, the average household user rate can be determined for the remaining years, as given in the following table:

Year	Household Usage Rate (gpd/household)
2001	150
2002	151
2003	151
2004	145
2005	140
2006	144
2007	143
2008	144
Average	**146**

The average household water usage rate is 146 gpd/household.

Remember that average residential water usage is approximately 101 gpdc. So it appears this community is averaging approximately 1.5 individuals per household. Leak detection, incorporation of water-saving technologies, public reminders to conserve water, and promotion of the use of native vegetation that requires little water are some methods that can reduce water usage and eliminate the need to develop additional sources of water that are expensive and ecologically or socially destructive.

7.5.4   FIRE FLOW DEMAND AND UNACCOUNTED-FOR WATER

A water system must be able to supply water quickly for societal needs to ensure adequate protection due to fire emergencies. Also, a portion of the supplied water will be lost due to system leakage, unmetered use (fire protection and maintenance), theft, or other causes.

During a fire emergency, the **fire protection water demand** can have a large effect on supply and distribution. In a community, water used for fire protection is generally pulled from nearby hydrants, which can greatly lower the available water pressure to local customers. For large industries, water is sometimes stored on site for fire protection. Generally, the amount of water required for fire protection depends on the size of the burning structure, the way the structure was constructed, the amount of combustible material in the structure, and the proximity of other buildings.

In the United States, community fire protection is rated and evaluated by the Insurance Services Office (ISO) using the Fire Protection Rating System (ISO, 1998; summarized in AWWA, 1998). For a municipal system, the ISO will evaluate the water supply source, treatment plant and pumping capacities, water distribution piping network, and placement and spacing of hydrants. The water system should have the available storage, pumping capacity, and piping to deliver the maximum daily demand plus the fire flow demand at any time during the day. In many situations, the **fire flow demand** is equal to the ISO's determined **needed fire flow** (NFF) for residential, commercial, and industrial properties. Table 7.16 provides the NFF for small family residences. The NFF is determined based on the spacing of the residential dwellings.

For commercial or industrial structures, the NFF is based on the size of the building, construction class (for example, wood frame), type of occupancy (for example, department store), exposure to adjacent buildings, and what is known as a communication factor (location and types of fire protection doors). The general ISO equation for a minimum NFF being 500 gallons per minute (gpm) is

$$NFF = 18 \times F \times A^{0.5} \times O \times (1 + \sum(X + P)) \qquad \textbf{(7.6)}$$

where NFF is the needed fire flow (gpm), F is the construction factor (0.6–1.5), A is the building effective area (ft^2), O is the occupancy factor (0.75–1.25), X is the exposure factor (0–0.25), and P is the communication factor (0–0.25). The full procedure to determine the NFF for a structure can be found in ISO's *Fire Suppression Rating Schedule* (ISO, 1998) and the AWWA's Manual M-31 (AWWA, 1998). In addition to the NFF, Table 7.17 provides values for the recommended storage for fire protection along with the duration that water should be supplied.

Water that is produced by a treatment facility is delivered to a user in a water distribution system. However, a portion of that water does not make it to the customers or is used as unmetered flow. This **unmetered flow** includes: (1) what is lost in the system due to leaks and breaks in pipes and joints; (2) unmetered uses such as fire protection and maintenance; (3) water theft; and (4) a variety of other minor water losses. Generally, in all cases, more water will be produced and enter the distribution system than is delivered to users. To determine this unaccounted-for water, subtract the sum of the metered water for each user from the metered water leaving the water treatment facilities:

$$\text{unaccounted-for water}(\%) = \frac{\text{water produced} - \text{metered use}}{\text{water produced}} \times 100$$

(7.7)

Many times, the unaccounted-for water is used to gauge the performance of a water distribution system. On an annual basis, it is expected that less than 10 percent of the water produced will be lost as unaccounted-for water. However, for older systems, the unaccounted-for water can be much higher due to the aging pipe network, which can have a significant amount of leakage. Closely monitoring the unaccounted-for water can also be used as an indicator of when something is wrong in the water distribution system. An increase in unaccounted-for water indicates a leak in the piping network, which should be repaired, or significant loss due to another problem, such as water theft. With enough internal metering (meters for water mains, pumping station records), it is possible to determine the general location of needed repairs or problems.

Table / 7.16

NFF for Small Family Residences

Distance Between Buildings (ft)	NFF (gpm)
<11	1,500
11–30	1,000
31–100	750
>100	500

SOURCE: Values from ISO, 1998.

Table / 7.17

Recommended Needed Fire Flow (NFF) Duration for Fire Protection

NFF (gpm)	Duration (h)	Storage (gallons)
<2,500	2	~300,000
3,000–3,500	3	540,000–630,000
>3,500	4	>840,000

SOURCE: Values from Walski et al., 2003.

7.5.5 DEMAND FORECASTING

Estimating future scenarios is a major part of designing a treatment facility, water distribution system, or sewer collection system. In almost all situations, there will be a level of uncertainty regarding how much water is required and wastewater is generated. The long-term planning of a community usually includes the estimation of the **future water demand** for 5, 10, 20, or more years. It is common to estimate the future demand for several different scenarios before deciding on the actual values for design. Also, the comparison of alternative future projections provides a way to understand what effects the input data or assumptions can have on the future water demand. This can be used as a sensitivity analysis to guide community leaders in the decision-making process for how to plan future developments, influence water use practices, or understand the impact of large water users on the system.

Determining the specifics of future scenarios requires a consensus that incorporates environmental, economic, and social needs for current and future generations and is developed by engineers, planners, utilities, and community stakeholders. Population and demographic trends and the location of any new users will greatly influence future water demand. Other important issues could include impact of climate on water availability, climate change, and issues of land use within the watershed. Often more data are collected than are actually used in the analysis.

Figure 7.16 shows some of the potential scenarios that can be evaluated to estimate the future water demand. This approach analyzes and extrapolates historical data to future scenarios. Caution should be used when simply extrapolating forward (linear growth), because past influences may not hold in the future.

Rather than basing projections using an extrapolation method, the forecaster can develop a more in-depth analysis of the potential causes for projected changes in water demand. This analysis is based on estimating the future population, demographics, water use, land use, number of large users (for example, industry), future climate scenarios, and technological and societal issues that affect water conservation. In this case, the water demand is separated into **disaggregated segments** to determine

Class Discussion

Develop future scenario (s) specific to your location that can be added to Figure 7.16 for sustainable water use. In your scenario(s), incorporate water reuse as a source of your regions' water portfolio and account for your location's specific population growth; changes in demographics such as age, education, and wealth; changes in climate; estimation of future precipitation; and expected changes in use (shift from agriculture to domestic uses for example).

Figure / 7.16 **Extrapolated Growth Scenarios of Water Use Based on Historical Demand.**

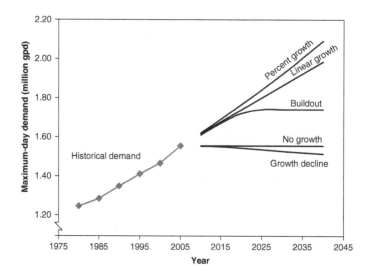

Figure / 7.17 Daily Activity of Collecting Water That Is Seen in Much of the World.

(Photo courtesy of James R. Mihelcic).

the volume of water used (or wastewater generated) per unit. Then the projected change for each segment is predicted, and the resulting new water demand is determined (AWWA, 2007; Walski et al., 2003).

These disaggregated segments are typically based on population estimates or equivalents, or on land use designations. A **population equivalent** is a method of converting the water use (or wastewater generation) of commercial or industrial users into the equivalent amount of water used by a population number. For example, an industrial unit may use the water equivalent of 150 people in a residential area. The projected equivalent population (real population plus population equivalents) is estimated to determine the future water demand.

---

example / 7.4 Projecting Future Water Demand, Using Extrapolation Methods

Project the future average water demand in years 8, 14, and 24, using the following historical records. Use an extrapolation method that includes **linear growth** and **buildout**, shown in Figure 7.16.

Year	Average Metered Water Demand (gpd)
2003	1,797,895
2004	1,843,661
2005	1,907,000
2006	1,813,000
2007	1,890,000
2008	1,901,145
2009	1,891,860
2010	2,012,201
2011	2,058,492
2012	2,051,339

solution

First, graph the recorded values to visualize the historical trend (see Figure 7.18a). Based on these observations, we can make assumptions of expected growth (or decline).

Figure / 7.18   Historical and Projected Water Demand
(a) Recorded values of water demand graphed to visualize the historical trend of data used in Example 7.4. (b) Projected future trend of water demand using the extrapolation method and data for Example 7.4.

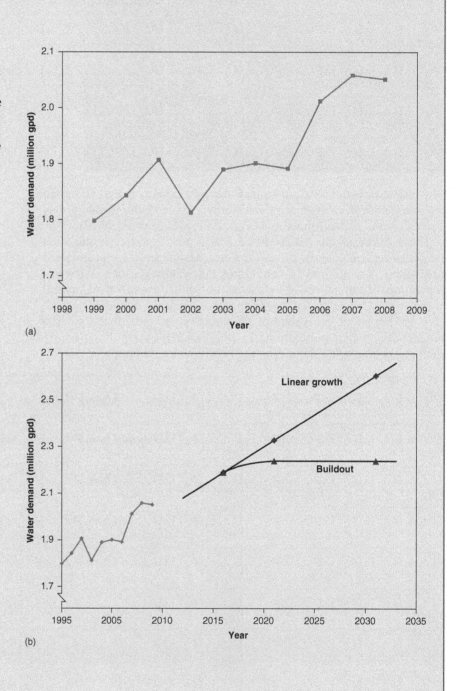

In this case, we will project the future trend using the extrapolation method (Figure 7.16). Both future scenarios are shown in Figure 7.18b. The following table contains the details of the projected water usage. For linear growth, linear regression analysis was used to project the future water demand. The buildout extrapolation method requires additional assumptions. In Figure 7.18b, we assume 8 years of linear growth followed by buildout starting in year 9. Note that many other assumptions can be made to extrapolate the future water demand.

Projected Average Water Demand		
Year	Linear Growth Assumption	Buildout Assumption
2020	2,190,000 gpd	2,190,000 gpd
2026	2,328,000 gpd	2,239,000 gpd
2036	2,604,000 gpd	2,239,000 gpd

Water use (or wastewater generation) can also be categorized based on land use such as light residential, dense residential, heavy commercial, and industrial. After determining the water use for each category and knowing any proposed new development, the forecaster determines a projected water demand. Another important issue is that users do not all require the same quality of water in relation to the source and level of treatment.

7.6 Water Distribution (and Wastewater Collection) Systems

Again, because water production and wastewater generation are closely tied and yet different, we will discuss the collection of resource-laden wastewater in this section along with the piped provision of water. Before the design of a **water distribution** or **wastewater collection** system, a comprehensive investigation of the proposed service area takes place. Part of this investigation includes forecasting water demand throughout the service area, as previously discussed. Other factors such as pipe material, location of appurtenances (manholes, junctions, inlets, and other structures), hydrant placement, type and location of valves, design and locations of pumping stations, and water storage locations all need to be determined and integrated to design a working efficient system.

This process typically uses computer software. (See Walski et al., 2003, 2004 for a full description of such modeling software.) The remainder of this section will discuss system layout, design flow velocities for estimating pipe size, and system needs related to pumping and storage.

7.6.1 SYSTEM LAYOUT

A water distribution system and a wastewater collection system both carry water. However, they do so very differently. Generally, the piping

Figure / 7.19 Examples of a Water Distribution and Wastewater Collection System Layout for a Residential Area (a) This "looped" system is typical for a water distribution system. (b) This "branched" system is typical for a wastewater collection system.

(a) (b)

is placed under or along roadways dedicated for public use or through lands where the utility has a right-of-way easement. It is common practice for the wastewater piping be at least 10 ft from, and 18 in. below, the water service lines, to minimize the possible contamination of potable water (Hammer and Hammer Jr., 1996). Large piping systems are connected to each customer by small laterals or supply pipes. The piping capacity should be designed to meet needs of the customer without excessive costs. Customers should therefore not notice or have to worry about a well-designed and well-maintained system.

A water distribution system has a layout that contains many loops where pressurized flow occurs throughout the system (see Figure 7.19). For large systems, many **looped subsystems** are connected to large pipe force mains or transmission lines. These loops allow water to be delivered to the customers by many different routes. When demand at a specific location is high, as during a fire emergency, water needs to be delivered to that location as efficiently as possible. Hydrants are placed along roads at intersections and spaced so fire personnel can pull water from multiple hydrants if necessary. Also, many shutoff valves are placed throughout the system. Thus, if a break occurs or maintenance is scheduled, isolating the problem area by using shutoff valves will minimize the number of customers that have to go without water. If designed correctly, a water distribution system should provide the needed water for a variety of demand scenarios with adequate pressure and good water quality throughout the whole system.

A wastewater collection system has a layout that is **branched** (or dendritic), with very few or no loops (see Figure 7.19). Most of the time, wastewater flows by gravity in one direction. The smallest pipes are located at the "end" of the branches and progressively get larger as the wastewater flows toward the treatment plant. Since the wastewater flows by gravity, an adequate pipe slope must be maintained throughout the system. There can be sections where wastewater is pumped in a force main to lift the wastewater to a higher elevation to again flow by gravity, or to force wastewater uphill due to the topography. Generally manholes are placed at every pipe diameter change, slope change, junction (two or more pipes joining), and upstream pipe ends. Manholes are spaced no more than 300–500 ft. (90–150 m) between each other to provide enough entry points for maintenance (ASCE, 2007).

© Lacy Rane/iStockphoto.

7.6.2 DESIGN FLOW VELOCITIES AND PIPE SIZING

The analysis of water flow in pipes is based on conservation of mass, conservation of energy, and conservation of momentum. Fluid flow in pipes can be classified as either *full pipe* (pressure flow) or *open channel* (gravity flow). In both cases, modeling software is used to analyze the flow in pipes for water distribution and wastewater collection systems.

Pressure flow conditions are found in water distribution systems where the delivered water must have both the capacity and the pressure to suit the needs of the customer. The **design water velocity in a pipe** is typically between 2.0 ft./(0.6 m/s) to 10 ft./s (3.1 m/s) for peak flow conditions. Smaller design velocities tend to size pipes to be bigger than what is really economically needed. Higher velocities tend to cause excessive head loss throughout the system and to increase the potential for water transients (for example, water hammer).

Typical service pressure requirements are provided in Table 7.18. In residential communities, it is necessary to maintain a minimum pressure for each customer that allows for "normal" water use when more than one water-using device is in service on the second story. Also, the pressure at the residence should not exceed 80 psi; otherwise, the potential for in-house leaks increases, excessive flows in showers and faucets could occur, and the hot-water pressure relief valve could discharge.

Open-channel conditions are commonly found in wastewater collection systems where the water flows by gravity. Since the piping is sized for infrequent high-flow conditions, the pipes are only partially filled most of the time. During low-flow conditions, this causes the wastewater velocity to be too small to move solids, which then can be deposited in the sewer. It is usual practice to design sewer slopes so that

Table / 7.18	
Typical Required Service Pressures in Residential Communities	
Condition	Service Pressure (psi)
Maximum pressure	65–75
Minimum pressure during maximum-day demand	30–40
Minimum pressure during peak-hour demand	25–35
Minimum pressure during fires	20

SOURCE: Values from Mayer et al., 1999.

Table / 7.19

Minimum and Maximum Slope for Gravity Flow in Circular Concrete Sewer Pipe for a Manning's Roughness Value Equal to 0.013

Pipe Diameter		Minimum Slope (2 ft/s, 0.6 m/s)	Maximum Slope (10 ft/s, 3.1 m/s)
in.	mm		
4	100	0.00841	0.21030
6	150	0.00490	0.12247
8	200	0.00334	0.08345
10	250	0.00248	0.06197
12	300	0.00194	0.04860
15	375	0.00144	0.03609
18	450	0.00113	0.02830
21	525	0.00092	0.02304
24	600	0.0008*	0.01929
27	675	0.0008*	0.01648
30	750	0.0008*	0.01432
36	900	0.0008*	0.01123
42	1,050	0.0008*	0.00655

*Minimum practical slope for construction.

the minimum full-pipe velocity is 2.0 ft./s (0.6 m/s) or greater to ensure that solids are flushed out during peak flow hours of the day. Also, it is common practice for the maximum velocity to be about 10 ft./s (3.0 m/s) to avoid damaging the sewer (ASCE, 2007) and causing excessive "sewer spray" in manholes. However, for simple long-pipe sections with straight-through manholes, higher velocities (15–20 ft./s) can be tolerated (Walski et al., 2004). Table 7.19 provides the minimum slope for circular concrete sewer pipe based on the minimum and maximum full-pipe velocity for a Manning's roughness value equal to 0.013.

Equation 7.8 can be used to make an initial estimate of a pipe size based on the full-pipe design flow velocity and design carrying capacity (design peak flow):

$$D = k\sqrt{\frac{Q}{v}} \qquad (7.8)$$

where D is the pipe diameter (in., mm), Q is the design flow rate (gpm, L/s), v is the design velocity (ft./s, m/s), and k represents a constant that is 0.64 for U.S. customary units or 35.7 for SI units.

For example, if the design flow rate is 6,000 gpm (380 L/s) and the design velocity is 5 ft./s (1.5 m/s), then the estimated pipe diameter would be 22.2 in. (568 mm). The next largest nominal pipe size would be selected as a starting point—in this case, a 24 in. (600 mm) pipe. Then a model would be used to evaluate whether this pipe is the best choice for a variety of different scenarios.

7.6.3 PUMPING STATIONS AND STORAGE

A **pumping station** is located where wastewater needs to be lifted or increased pressure is required. Pumps are characterized by their capacity, pumping head, efficiency, and power requirements. The design of a pumping station requires that each pump's performance curve (capacity versus head curve) match the system head curve. The system curve is developed by adding the static head to the head loss (friction plus minor losses) of the system for varied flow rates. The pump characteristic curves should be available from the utility for existing pumps or can be obtained from the manufacturer of the pump. An old pump may not perform as it did when it was new. In fact, older pumps should be checked to verify that the real pump performance matches what is shown on the pumping head performance curve.

Like most mechanical machines, a pump operates best at its **best efficiency point (BEP)**. As a pump operates further from its BEP, the life of the pump or impeller decreases considerably, and the energy costs (and resulting CO_2 emissions) to run the pump greatly increase. The typical operating range of a centrifugal pump is the flow rate between 60 and 120 percent of the flow rate at the BEP. When designing and sizing a pumping station, it is common to have multiple pumps working together to maximize the pumping efficiency and minimize the pumping costs while enabling the facility to deliver the required water for a wide range of flow conditions.

In the example shown in Figure 7.20, the actual pump efficiency is near the BEP, which is at the peak of the pump efficiency curve. Therefore, this

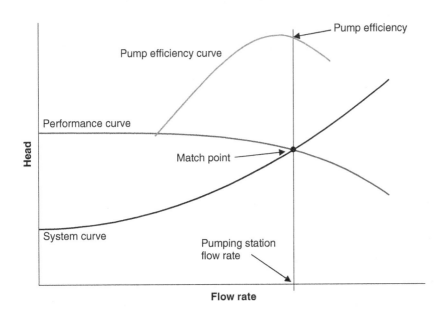

Figure / 7.20 **Pump Characteristic Curves** The match point between a pumping station's performance curve and system curve indicates the flow rate and pumping head leaving the pumping station.

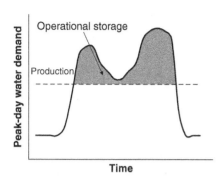

Operational storage

Production

Peak-day water demand

Time

Figure / 7.21 Operational Storage for a Storage Tank in a Water Distribution System.

© Steve Shepard/iStockphoto.

pump configuration is working near its maximum efficiency. As the system curve shifts because of changes in the system (water elevation changes or required pumping rate changes), the match point would move right or left along the performance curve. If it moves too far to the right or left, the pump efficiency can decrease to the point where this configuration may no longer be very efficient or may be outside the working range of the pumps. In this case, additional pumps could be turned on or off to better match the needed pumping rate while maximizing efficiency, or different pumps should be installed that have efficiency curves where the match point is closer to the pump's BEP. These changes will lessen the amount of associated energy consumption.

Often a pumping station is near and controlled by a storage tank or wet well. The water levels in the tank or wet-well control when pumps are on or off and how many pumps are needed. In the case of a variable-speed pump, the water level controls the speed of the pump motor.

A **storage tank** is commonly used to set an elevated hydraulic grade line across a water distribution system. For flat topography, storage tanks are easily seen as elevated water towers. For areas with hilly terrain, many times the storage tanks are placed at ground level in the hills above the community. The water level in a storage tank should be sized so the tank will routinely fill and drain during the daily demand cycle. This is because stagnant water in a tank becomes "stale" as it ages.

The volume of a storage tank is determined as the sum of the operational (equalization) storage, fire flow storage, and other emergency storage if needed. As shown in Figure 7.21, the *operational storage* is determined by estimating the additional volume of water needed for the maximum-day demand above the volume of water produced at the tank location. The *estimated fire flow storage* was provided in Table 7.17. Again, modeling software is used to complete the actual design of storage tanks and the pumping stations that feed them.

A **wet well** provides storage so the actual pumping rate does not need to match the inflow rate into the pumping station. For constant-speed pumps, the wet-well volume is determined based on the **pump cycle time**, which is defined as the time between successive pump starts. It is generally recommended that pumps be started six or fewer times per hour; the pump manufacturer should be consulted. The minimum wet-well volume can then be estimated as

$$V_{min} = \frac{Q_{design} \times t_{min}}{4} \tag{7.9}$$

where V_{min} is the active wet-well volume, Q_{design} is the pump's design flow rate, and t_{min} is the minimum pump cycle time. Derivation of Equation 7.9 is provided by Jones and Sanks (2008).

For example, if the pump's design flow rate is 500 gpm and the pump cycle is six times per hour (so that each pump's cycle time would be 10 min), the estimated minimum wet-well volume would be 1,250 gallons:

$$V_{min} = \frac{500 \text{ gpm} \times 10 \text{ min}}{4} = 1,250 \text{ gallons} \tag{7.10}$$

This value is known as the active volume in the wet well or the amount of water between the "pump on" and "pump off" water levels in the wet well. An additional amount of volume is needed in the wet well to maintain a suction head for the pumps, and more volume is added as a margin of safety.

7.7 River Water Quality

The treatment of river water quality in this section focuses on the management of **dissolved oxygen (DO)** in relation to the discharge of oxygen-demanding wastes. This is a classic issue in surface water quality that remains of interest today with respect to the issuance of discharge permits and the setting of TMDLs for receiving waters.

7.7.1 DISSOLVED OXYGEN AND BOD

Dissolved oxygen is required to maintain a balanced community of organisms in lakes, rivers, and the ocean. When an oxygen-demanding waste (measured as BOD) is added to water, the rate at which oxygen is consumed in oxidizing that waste (**deoxygenation**) may exceed the rate at which oxygen is resupplied from the atmosphere (**reaeration**). This can lead to depletion of oxygen resources, with concentrations falling far below saturation levels (Figure 7.22). When oxygen levels drop below 4–5 mg O_2/L, reproduction by fish and macroinvertebrates is impaired. Oxygen depletion is often severe enough that anaerobic conditions develop, with an attendant loss of biodiversity and poor aesthetics (turbidity and odor problems). Figure 7.22 also illustrates the response of stream biota to BOD discharges.

Consideration of the fate of BOD following discharge to a river is a useful starting point for examining the impact of oxygen-demanding wastes on oxygen resources. Example 7.5 applies the concepts of the mixing basin (Chapter 4) and BOD kinetics (Chapter 5) to examining the oxidation of an organic waste following discharge to and mixing with a river.

In Example 7.5, more than 23 mg O_2/L of ultimate carbonaceous biochemical oxygen demand (CBOD) is exerted over the 50 km stretch downstream of the discharge. To appreciate the impact of this demand on a river's oxygen resources, it is necessary to understand the capacity of water to hold oxygen (saturation) and the rate at which oxygen can be resupplied from the atmosphere (reaeration).

Clean Water Act
http://www.epa.gov/regulations/laws/cwa.html

Surf Your Watershed
http://cfpub.epa.gov/surf/locate/index.cfm

© Galyna Andrushko/iStockphoto.

7.7.2 OXYGEN SATURATION

The amount of oxygen that can be dissolved in water at a given temperature (its equilibrium or **saturation concentration**) may be determined through the Henry's law constant, K_H:

$$DO_{sat} = K_H \times P_{O_2} \qquad (7.11)$$

where DO_{sat} is the saturation DO concentration (in moles O_2/L), K_H is the Henry's law constant (1.36×10^{-3} moles/L-atm at 20°C), and P_{O_2} is the partial pressure of oxygen in the atmosphere (~21 percent or 0.21 atm).

Figure / 7.22 DO Sag Curve (a) and Associated Water Quality Zones ((b)–(d)) Reflecting Impacts on Physical Conditions and the Diversity and Abundance of Organisms.

(From Mihelcic (1999). Reprinted with permission of John Wiley & Sons, Inc.).

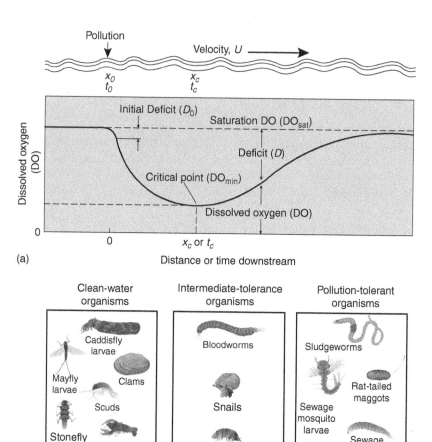

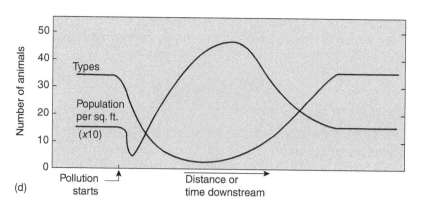

	Stream Zones				
	Clean water	Degradation	Damage	Recovery	Clean Water
Physical condition	Clear water; no bottom sludge	Floating solids; bottom sludge	Turbid water; malodorous gases; bottom sludge	Turbid water; bottom sludge	Clear water; no bottom sludge
Fish species	Cold or warm water game and forage fish; trout, bass,	Pollution-tolerant fish; carp, gar, buffalo	None	Pollution-tolerant fish; carp, gar, buffalo	Cold or warm water game and forage fish; trout, bass,
Benthic invertebrate	Clean water	Intermediate tolerance	Pollution-tolerant	Intermediate tolerance	Clean water

example/7.5 Mixing Basin Calculation for CBOD

A waste with a 5-day CBOD (y_5) of 200 mg O_2/L and a k_L of 0.1/day is discharged to a river at a rate of 1 m^3/s. Calculate the ultimate CBOD (L_0) of the waste before discharge to the river. Assuming instantaneous mixing after discharge, calculate the ultimate CBOD of the river water after it has received the waste. The river has a flow rate (Q) equal to 9 m^3/s and a background ultimate CBOD of 2 mg O_2/L upstream of the waste discharge. Also calculate the ultimate CBOD (L_0) and CBOD$_5$ (y_5) in the river 50 km downstream of the point of discharge. The river has a width (W) of 20 m and a depth (H) of 5 m.

solution

This problem has several steps. First, calculate the ultimate CBOD of the waste before discharge:

$$L_{0,\text{waste}} = \frac{y_5}{(1 - e^{-k_L \times 5 \text{ days}})} = \frac{200 \text{ mg } O_2/\text{L}}{(1 - e^{-0.1/\text{day} \times 5 \text{ days}})} = 508 \text{ mg } O_2/\text{L}$$

Next, perform a mass balance mixing basin calculation to determine the ultimate CBOD after the waste has been discharged and mixed with the river. The general relationship for calculation of the concentration of any chemical in a mixing basin (C_{mb}) is

$$C_{\text{mb}} = \frac{C_{\text{up}} \times Q_{\text{up}} + C_{\text{in}} \times Q_{\text{in}}}{Q_{\text{mb}}}$$

Here the total flow, Q_{mb}, equals $Q_{\text{up}} + Q_{\text{in}}$, and the ultimate CBOD equals

$$L_{0,\text{mb}} = \frac{2 \text{ mg } O_2/\text{L} \times 9 m^3/s + 508 \text{ mg } O_2/\text{L} \times 1 m^3/s}{10 \, m^3/s}$$

$$= 52.6 \text{ mg } O_2/\text{L}$$

This is the value of the ultimate CBOD of the river water after it has received the waste.

To answer the last two questions concerning the ultimate CBOD and 5-day CBOD 50 km downstream of discharge, first calculate the 5-day CBOD of the river water after it has received the waste:

$$y_t = L_0 \times (1 - e^{-k_L \times t})$$
$$y_{5,\text{mb}} = 52.6 \text{ mg } O_2/\text{L} \times (1 - e^{-0.1/\text{day} \times 5 \text{ days}})$$
$$y_{5,\text{mb}} = 20.7 \text{ mg } O_2/\text{L}$$

Next, calculate the ultimate CBOD 50 km downstream of the point of discharge. As the waste travels downstream, it will decay and deplete oxygen according to first-order kinetics. The river downstream of the mixing zone can be modeled as a plug flow reactor (PFR). Therefore,

$$L_t = L_0 \times e^{-k_L \times t}$$

However, the time of travel needs to be calculated. The river velocity (U) is given by

$$U = \frac{Q}{A} = \frac{Q}{W \times H} = \frac{10 \, m^3/s}{20 \text{ m} \times 5 \text{ m}} = 0.1 \text{ m/s} \times \frac{86,400 \text{ s}}{\text{day}} \times \frac{\text{km}}{1,000 \text{ m}}$$

$$= 8.64 \text{ km/day}$$

Then, to determine the time of travel, divide the distance by the river velocity:

$$t = \frac{x}{U} = \frac{50 \text{ km}}{8.64 \text{ km/day}} = 5.78 \text{ days}$$

This value can then be used to determine the ultimate CBOD 5.78 days downriver:

$$L_{0,50km} = L_{0,mb} \times e^{-k_L \times t} = 52.6 \times e^{-0.1/\text{day} \times 5.78 \text{ days}} = 29.5 \text{ mg O}_2/\text{L}$$

and a 5-day CBOD of

$$y_t = L_0 \times \left(1 - e^{-k_L \times t}\right)$$

$$y_{5,50 \text{ km}} = 29.5 \times \left(1 - e^{-0.1/\text{day} \times 5 \text{ days}}\right) = 11.6 \text{ mg O}_2/\text{L}$$

The Henry's law constant varies with temperature (see Chapter 3), so the saturation concentration of DO varies as well. Example 7.6 illustrates the calculation of the saturation DO concentration.

The value for DO_{sat} ranges from approximately 14.6 mg O_2/L at 0°C to 7.6 mg O_2/L at 30°C. These are typical temperature extremes for natural and engineered systems. This shows why fish with high oxygen requirements are associated with colder waters and why the impacts of oxygen-demanding wastes on water quality may be greatest in the summer. In the warmer months of summer, stream flow is typically lower as well, offering less dilution of the waste. The concentration of oxygen in water also decreases as the salinity increases, which becomes important in estuarine and ocean conditions.

example/ 7.6   Determination of Saturation DO Concentration

Determine the saturation DO concentration, DO_{sat}, at 20°C.

solution

Determine DO_{sat} from the appropriate temperature-dependent Henry's law constant and the oxygen partial pressure:

$$DO_{sat} = \frac{1.36 \times 10^{-3} \text{ mole}}{\text{L-atm}} \times 0.21 \text{ atm} = \frac{2.85 \times 10^{-4} \text{ mole O}_2}{\text{L}}$$

Convert to mg O_2/L:

$$DO_{sat} = \frac{2.85 \times 10^{-4} \text{ mole O}_2}{\text{L}} \times \frac{32 \text{ gO}_2}{\text{mole O}_2} \times \frac{1,000 \text{ mg O}_2}{\text{gO}_2}$$

$$= \frac{9.1 \text{ mg O}_2}{\text{L}}$$

Note that the phrases *dissolved-oxygen saturation concentration* and the *solubility of oxygen* are used interchangeably.

Determine the dissolved-oxygen deficit, D, at 20°C for a river with an ambient dissolved-oxygen concentration of 5 mg O_2/L.

solution

From Example 7.6, the DO_{sat} at 20°C was determined to be 9.1 mg O_2/L. Applying Equation 7.12 yields the deficit:

$$D = 9.1 - 5 = 4.1 \text{ mg } O_2/L$$

The actual DO in this case is 5 mg O_2/L, which is below the saturation level. Microbial oxidation of organic matter or ammonia–nitrogen in this river may be leading to oxygen depletion.

7.7.3   THE OXYGEN DEFICIT

The **oxygen deficit** (D, expressed in mg O_2/L) is defined as the departure of the ambient DO concentration from saturation.

$$D = DO_{sat} - DO_{act} \qquad (7.12)$$

DO_{act} is the ambient or measured dissolved-oxygen concentration (mg O_2/L).

Note that negative deficits may occur when ambient oxygen concentrations exceed the saturation value. This happens in lakes and rivers under quiescent, nonturbulent conditions when algae and macrophytes are actively photosynthesizing and producing dissolved oxygen. This **oversaturation** is eliminated when sufficient turbulence is available—for example, due to rapids, waves, and waterfalls.

7.7.4   OXYGEN MASS BALANCE

Examples 7.5–7.7 demonstrated that BOD exertion (e.g., 29.5 mg O_2/L over a 50 km stretch) may exceed a river's oxygen resources, even at saturation. The shortfall (oxygen present minus oxygen required) must be made up through atmospheric exchange, that is, reaeration. Where the demand by deoxygenation exceeds the supply from reaeration, oxygen levels fall, and anaerobic conditions may develop. The dynamic interplay between the oxygen source (reaeration) and sink (deoxygenation) terms can be examined through a mass balance on oxygen in the river. Deoxygenation occurs as BOD is exerted and is described by Equation 7.13. The rate of reaeration is proportional to the deficit and is described using first-order kinetics:

$$\frac{dO_2}{dt} = k_2 \times D - k_1 \times L \qquad (7.13)$$

Here the *in-stream* deoxygenation rate coefficient (k_1, day^{-1}) is comparable to (and, for the purposes of this chapter, the same as) the *laboratory* or *bottle*

CBOD reaction rate coefficient (k_L) discussed in Chapter 5 but also includes in-stream phenomena, such as sorption and turbulence and roughness effects. The reaeration rate coefficient (k_2, day^{-1}) varies with temperature and turbulence (river velocity and depth) and ranges from ~0.1 to 1.2/day.

In practice, the mass balance is written in terms of deficit:

$$\frac{dD}{dt} = k_1 \times L - k_2 \times D \tag{7.14}$$

Note how Equation 7.14 is a simple reversal of the order of the source–sink terms presented in Equation 7.13. Equation 7.14 can be integrated, yielding an expression that describes the oxygen deficit at any location downstream of an arbitrarily established starting point, such as the point where a waste is discharged to a river:

$$D_t = \frac{k_1 \times L_0}{(k_2 - k_1)} \times \left(e^{-k_1 \times t} - e^{-k_2 \times t}\right) + D_0 \times e^{-k_2 \times t} \tag{7.15}$$

where L_0 is the ultimate CBOD and D_0 is the oxygen deficit at the starting point ($x = 0$, $t = 0$), and D_t is the oxygen deficit at some downstream location ($x = x$, $t = t$). The notation t refers to time of travel, defined here as the time required for a parcel of water to travel a distance x downstream. Therefore, $t = x/U$, where x is distance downstream and U is the river velocity.

The time–distance relationship permits expression of the analytical solution for the oxygen deficit in terms of x, the distance downstream of the starting point:

$$D_x = \frac{k_1 \times L_0}{(k_2 - k_1)} \times \left(e^{-k_1 \times x/U} - e^{-k_2 \times x/U}\right) + D_0 \times e^{-k_2 \times x/U} \tag{7.16}$$

Equation 7.16 is called the **Streeter–Phelps model** and was developed in the 1920s for studies of pollution in the Ohio River.

7.7.5 DISSOLVED-OXYGEN SAG CURVE AND CRITICAL DISTANCE

The discharge of oxygen-demanding wastes to a river yields a characteristic response in oxygen levels that is termed the DO sag curve (Figure 7.22). Figure 7.22a demonstrates that a typical DO sag curve has three phases of response:

1. An interval where DO levels fall because the rate of deoxygenation is greater than the rate of reaeration ($k_1 \times L > k_2 \times D$).

2. A minimum (termed the **critical point**) where the rates of deoxygenation and reaeration are equal ($k_1 \times L = k_2 \times D$).

3. An interval where DO levels increase (eventually reaching saturation) because BOD levels are being reduced and the rate of deoxygenation is less than the rate of reaeration ($k_1 \times L < k_2 \times D$).

The location of the critical point and the oxygen concentration at that location are of principal interest, because this is where water quality conditions are at their worst. Design calculations are based on this location because if standards are met at the critical point, they will be met elsewhere. To determine the location of the critical point, first use Equation 7.17 to determine the *critical time* and then multiply the critical time by the river velocity to determine the *critical distance*:

$$t_{crit} = \frac{1}{k_2 - k_1} \times \ln\left(\frac{k_2}{k_1} \times \left(1 - \frac{D_0 \times (k_2 - k_1)}{k_1 \times L_0}\right)\right) \tag{7.17}$$

To find the oxygen deficit at the critical distance, substitute the critical time into Equation 7.15. Knowledge of DO_{sat} then provides the actual DO concentration at the critical distance. Example 7.8 illustrates this approach and suggests opportunities for its application in river management.

---

example/ 7.8 Determining Features of the DO Sag Curve

After receiving the discharge from a wastewater treatment plant, a river has a dissolved-oxygen concentration of 8 mg O_2/L and an ultimate CBOD of 20 mg O_2/L. The saturation dissolved-oxygen concentration is 10 mg O_2/L, the deoxygenation rate coefficient k_1 is 0.2/day, and the reaeration rate coefficient k_2 is 0.6/day. The river travels at a velocity of 10 km/day. Calculate the location of the critical point (time and distance) and the oxygen deficit and concentration at the critical point.

solution

First determine the initial DO deficit at the point of discharge, using Equation 7.12:

$$D_0 = DO_{sat} - DO_{act}$$
$$= 10 - 8 = 2 \text{ mg } O_2/L$$

Then use Equation 7.17 to determine the critical time and knowledge of the river's velocity to determine the critical distance:

$$t_{crit} = \frac{1}{k_2 - k_1} \times \ln\left(\frac{k_2}{k_1} \times \left(1 - \frac{D_0 \times (k_2 - k_1)}{k_1 \times L_0}\right)\right)$$

$$t_{crit} = \frac{1}{0.6/\text{day} - 0.2/\text{day}} \times \ln\left(\frac{0.6/\text{day}}{0.2/\text{day}} \times \left(1 - \frac{2 \text{ mg } O_2/L \times (0.6/\text{day} - 0.2/\text{day})}{0.2/\text{day} \times 20 \text{ mg } O_2/L}\right)\right)$$

$$t_{crit} = 2.2 \text{ days}$$

$$x_{crit} = 2.2 \text{ days} \times 10 \text{ km/day} = 22 \text{ km}$$

Finally, use Equation 7.15 to determine the oxygen deficit and Equation 7.12 to determine the actual dissolved-oxygen concentration for the critical time as just calculated:

$$D_t = \frac{k_1 \times L_0}{(k_2 - k_1)} \times \left(e^{-k_1 \times t} - e^{-k_2 \times t}\right) + D_0 \times e^{-k_2 \times t}$$

$$D_t = \frac{0.2/\text{day} \times 20 \text{ mg } O_2/L}{(0.6/\text{day} - 0.2/\text{day})} \times \left(e^{-0.2/\text{day} \times 2.2 \text{ days}} - e^{-0.6/\text{day} \times 2.2 \text{ days}}\right) + 2 \text{ mg } O_2/L \times e^{-0.6/\text{day} \times 2.2 \text{ days}}$$

$$D_t = 4.3 \text{ mg } O_2/L$$

$$DO = 10 - 4.3 = 5.7 \text{ mg } O_2/L$$

In this example, the deficit occurs 22 km downstream from the point of initial discharge.

7.8 Lake and Reservoir Water Quality

Water quality conditions in lakes and reservoirs are influenced by the magnitude and routing of the chemical and energy fluxes passing through biogeochemical cycles. The cultural perturbations of two of those cycles, phosphorus and nitrogen, result in a water quality issue of widespread interest: *eutrophication*.

7.8.1 THERMAL STRATIFICATION OF LAKES AND RESERVOIRS

A major difference between lakes and rivers lies in the means of mass transport. Rivers are completely mixed, while in temperate latitudes, lakes undergo *thermal stratification*, dividing the system into layers and restricting mass transport. Periods of stratification alternate with periods of complete mixing with mass transport at a maximum. The restriction of mass transport during stratification influences the cycling of many chemical species (such as iron, oxygen, and phosphorus) and can have profound effects on water quality.

The process of thermal stratification is driven by the relationship between water temperature and density. Figure 7.23 shows the maximum density of water occurs at 3.94°C. Ice thus floats and lakes freeze from the top down, instead of from the bottom up, as they would if the maximum density were at 0°C. (Consider the implications of the opposite situation.) During summer stratification, an upper layer of warm, less dense water floats on a lower layer of cold, denser water.

The layers are assigned three names as shown in Figure 7.24: (1) the **epilimnion**, a warm, well-mixed surface layer; (2) the **metalimnion**, a region of transition where temperature changes at least 1°C with every meter of depth; and (3) the **hypolimnion**, a cold, well-mixed bottom layer. The plane in the metalimnion where the temperature–depth gradient is steepest is termed the **thermocline**.

The stratification and destratification processes (mixing) follow a predictable seasonal pattern, as shown in Figure 7.25. In winter, the lake is thermally stratified with cold (~0°C) water near the surface and warmer (2–4°C), denser waters near the bottom. As the surface waters warm toward 4°C in the spring, they become denser and sink, bringing colder waters to the surface to warm.

Lake Tahoe Planning Agency
http://www.trpa.org

Figure / 7.23 **Maximum Density of Water** The maximum density occurs at 3.94°C. Thus, water at approximately 4°C will be found below colder waters (ice at 0°C) in winter and warmer waters (20°C) in the summer.

(From Mihelcic (1999). Reprinted with permission of John Wiley & Sons, Inc.).

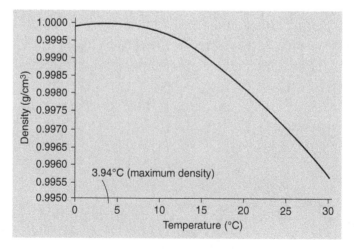

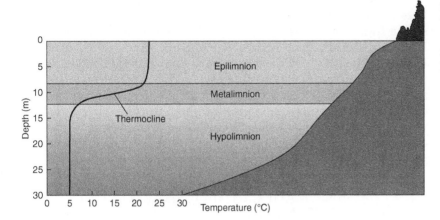

Figure / 7.24 Midsummer Temperature Profile for a Thermally Stratified Lake   Note the epilimnion, metalimnion (with a thermocline), and hypolimnium.

(Adapted from Mihelcic (1999). Reprinted with permission of John Wiley & Sons, Inc.).

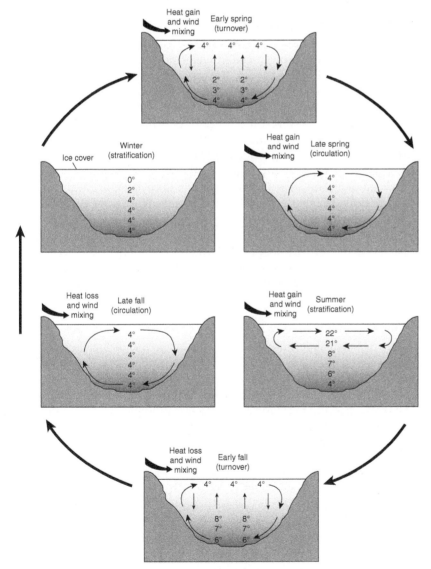

Figure / 7.25 Annual Cycle of Stratification, Overturn, and Circulation in Temperate Lakes and Reservoirs   Variation in meteorological conditions (temperature, wind speed) may cause significant variation to the timing and extent of these events.

(From Mihelcic (1999). Reprinted with permission of John Wiley & Sons, Inc.).

The process of mixing by convection, aided by wind energy, circulates the water column, leading to an isothermal condition termed **spring turnover**. As the lake waters continue to warm above 4°C, the lake thermally stratifies. Surface waters are significantly warmer and less dense than the lower waters during **summer stratification**. In the fall, the input of solar energy decreases, and heat is lost from the lake more rapidly than it is gained. As the surface waters cool, they become denser, sink, and promote circulation through convection, aided by wind. This phenomenon, called **fall turnover**, again leads to isothermal conditions. Finally, as the lake cools further, cold, low-density waters gather at the surface, and the lake reenters **winter stratification**.

7.8.2 ORGANIC MATTER, THERMAL STRATIFICATION, AND OXYGEN DEPLETION

The internal production of organic matter in lakes, resulting from algal and macrophyte growth and stimulated by discharges of growth-limiting nutrients (phosphorus and nitrogen), can dwarf that supplied externally, for example, from wastewater treatment plants and surface runoff. Organic matter produced in well-lit upper waters settles to the bottom, where it decomposes, consuming oxygen. There is little resupply of oxygen under stratified conditions, and if algal and/or macrophyte growth yields a great deal of organic matter, hypolimnetic oxygen depletion may result. Oxygen concentrations in the bottom waters of productive lakes are lower than those in surface waters, and the opposite is true in unproductive waters, where cold bottom waters have a higher oxygen saturation than the warmer surface waters.

Oxygen depletion leads to acceleration in the cycling of chemicals that reside in lake sediments (especially iron and phosphorus), the generation of several undesired and potentially hazardous chemical species (NH_3, H_2S, CH_4), and the extirpation of fish and macroinvertebrates. Oxygen depletion is one of the most important and commonly observed water quality problems in lakes, bays, and estuaries. It is also important in drinking-water reservoirs, where intakes may encounter nuisance algal growth near the top and accumulations of noxious chemicals near the bottom.

7.8.3 NUTRIENT LIMITATION AND TROPHIC STATE

Trophy is defined as the rate at which organic matter is supplied to lakes, both from the watershed and through internal production. The growth of algae and macrophytes in lakes is influenced by conditions of light and temperature and by the supply of growth-limiting nutrients. Because levels of light and temperature are more or less constant regionally, trophy is determined primarily by the availability of growth-limiting nutrients. As mentioned previously, phosphorus is generally found to be the nutrient limiting plant growth in freshwater environments. In the many bays and estuaries such as Chesapeake Bay and Tampa Bay, nitrogen is the limiting nutrient. Because naturally occurring phosphorus minerals are sparingly soluble, anthropogenic inputs can dramatically affect the rate of algal and macrophyte growth and attendant production of organic matter. Table 7.20 shows how

Water Implications of Biofuels
http://dels.nas.edu/dels/rpt_briefs/biofuels_brief_final.pdf

© Jean Schweitzer/iStockphoto.

lakes can be classified into three groups according to their trophic state: **oligotrophic**, **mesotrophic**, and **eutrophic**.

The process of nutrient enrichment of a water body, with attendant increases in organic matter, is termed **eutrophication**. This is considered to be a natural aging process in lakes. Figure 7.26 shows the succession of newly formed water bodies to dry land. Addition of phosphorus through human activities and the resultant aging of the lake are termed **cultural eutrophication**. Variation in land use and population density can lead to a range of trophic states within a given region, for example, from oligotrophic Lake Superior to eutrophic Lake Erie.

Table / 7.20

Classification of Water Bodies Based on Their Trophic Status

Oligotrophic	Nutrient poor; low levels of algae, macrophytes, and organic matter; good transparency; abundant oxygen
Eutrophic	Nutrient rich; high levels of algae, macrophytes, and organic matter; poor transparency; often oxygen-depleted in the hypolimnion
Mesotrophic	Intermediate zone; often with abundant fish life because of elevated levels of organic-matter production and adequate supplies of oxygen

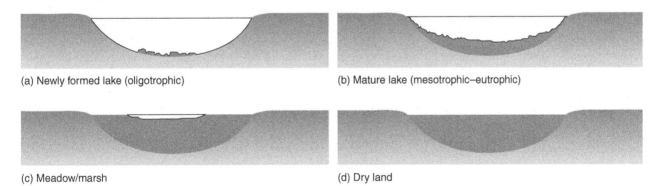

(a) Newly formed lake (oligotrophic)

(b) Mature lake (mesotrophic–eutrophic)

(c) Meadow/marsh

(d) Dry land

Figure / 7.26 **Natural Succession in Lakes** One concept of natural succession in lakes suggests that these systems pass through a series of stages as they become enriched with nutrients and organic matter, eventually being transformed to dry land. The rate of lake aging is importantly influenced by local meteorological conditions, the depth of the lake, and the size and fertility of the drainage basin.

(From Mihelcic (1999). Reprinted with permission of John Wiley & Sons, Inc.).

Over 400 coastal areas in the world are reported to experience some form of eutrophication (Figure 7.27). Of these, 169 are reported to experience hypoxia. These so-called dead zones experience very low oxygen levels (less than 2 mg/L) that can be seasonal or continual.

A small dead zone may occupy 1 km² and occur in a bay or estuary. The large dead zone located in the Gulf of Mexico has been measured at over 22,000 km² (the size of Massachusetts). This location, off the Louisiana shore, contains the most important commercial fishery in the lower 48 states and is fed by runoff from the Mississippi River.

The Mississippi drains 41 percent of the landmass of the lower 48 states and includes Corn Belt states such as Ohio, Indiana, Illinois, and Iowa. Besides using excessive amounts of fertilizers, these states have drained up to 80 percent of their wetlands, which serve as nutrient buffers. In fact, 65 percent of the nutrient input to the Gulf Coast dead zone originates in the Corn Belt, an area that provides food for a growing population and is now viewed by some as a source of energy independence through biofuels. Other nutrient inputs (that contain nitrogen and/or phosphorus) are associated with municipal wastewater and industrial discharges, urban runoff, and atmospheric deposition associated with fossil fuel combustion.

Atmospheric inputs can be large contributors. For example, atmospheric nitrogen inputs that originate with fossil fuel combustion contribute 25 percent of the nitrogen input to the Chesapeake Bay dead zone.

Left untouched, dead zones can cause the collapse of ecosystems and the economic and social systems that depend on them. Fortunately, dead zones can be reversed. For example, the Black Sea once had a dead zone that occupied 20,000 km². After the 1980s collapse of many centralized economies in countries located in the watersheds that drain to the Black Sea, nitrogen inputs fell by 60 percent. This eventually resulted in the dead zone shrinking and in fact disappearing in 1996 (Larson, 2004; Selman et al., 2008).

Figure / 7.27   Coastal Eutrophic and Hypoxic Zones.

(Identified by Selman et al. (2008)).

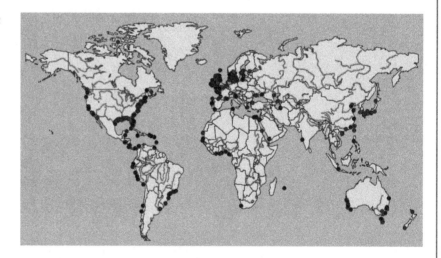

7.8.4 ENGINEERED LAKE MANAGEMENT

The preferred option in quality management of surface water is always to prevent or eliminate discharges through management strategies that would emphasize source reduction, recycling, and reuse of water and nutrients. Figure 7.28 summarizes eight methods of lake management. The focus of lake management is typically on controlling the nutrient phosphorus; however, most of the solutions presented here apply to other pollutants. (Remember that in Application 7.3, management of dead zones would also consider minimizing nitrogen inputs.) In the case of phosphorus, great strides in treatment technologies have reduced phosphorus concentrations in municipal wastewater effluent more than two orders of magnitude from the influent.

Currently methods are being developed to separate and reuse phosphorus that can be implemented at the building or treatment plant level. For example, urine can be collected at the house or building level while struvite (a phosphorus containing precipitate) can be recovered at the treatment plant. A variety of land management practices can also reduce phosphorus and nitrogen loads from watersheds. Fertilizer sales bans and application blackout periods have been used as a policy tool in some locations to minimize nutrient loadings from residential land use while attempts are being made to better manage nutrients related to agricultural practices. Finally, stormwater detention basins, artificial wetlands, and low-impact development may be employed to trap phosphorus (and nitrogen and other materials, such as sediment and trace metals) washed from the land and paved surfaces.

7.9 Wetlands

Wetlands are transitional lands between true aquatic environments and dry terrestrial land (referred to as uplands). They occur where water saturation controls: (1) how soil develops and (2) the types of animal and plant communities that live within or on the surface of the soil. A wetland is defined under the **Clean Water Act** as "those areas that are inundated or saturated by surface or groundwater at a frequency and duration sufficient to support, and that under normal circumstances do support, a prevalence of vegetation typically adapted for life in saturated soil conditions. Wetlands generally include swamps, marshes, bogs and similar areas." Note in all cases a wetland does not have to be saturated the whole year. In fact, it is hydrology, and not necessarily the presence of plants, that controls how a wetland is defined.

Wetlands exist in a variety of areas as discussed in Table 7.21. They provide many economic, social, and environmental benefits. For example, wetlands provide recreational benefits related to fishing, canoeing, and birdwatching. They also reduce property damage and loss of life, as they are critical for controlling flooding. Wetlands also lead to improved water quality because of their assimilative capacity that allows them to reduce loadings of nutrients, sediments, and other pollutants. Many engineering infrastructure projects (especially those related to transportation and site development) have filled in wetlands or adversely altered their hydrology and biogeochemistry.

Class Discussion

How would you address nutrient management using a systems approach? The average population density in coastal areas is twice the global average, while the biodiversity of coastal and freshwater aquatic ecosystems continues to decline. Increases in population and urbanization concentrate nutrients (N and P) in urban areas, where discharges from wastewater treatment plants and urban runoff can create havoc on freshwater and coastal ecosystems. Yet these nutrients are required in rural areas, where crop production is the greatest.

Track Coastal Areas Impacted by Eutrophication and/or Hypoxia
http://www.wri.org/project/eutrophication/map

Wetland Facts
http://www.epa.gov/owow/wetlands

(a) Point source control

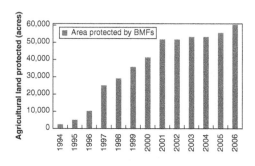

The discharge of phosphorus from a municipal wastewater treatment plant resulted in a hypereutrophic state, manifested in nuisance algal blooms, poor water clarity, and severe hypolimnetic oxygen depletion. Implementation of advanced waste treatment, in multiple stages, has reduced effluent P concentrations and led to a decline in the rate at which oxygen is consumed in the bottom waters (AHOD, areal hypolimnetic oxygen depletion).

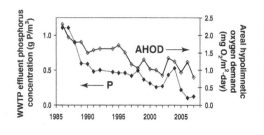

Onondaga Lake, NY

(b) Nonpoint source control

This reservoir, which provides 70% of the water supply for the city of Wichita, is polluted by phosphorus and sediment originating on cropland and rangeland in its watershed. Increases in the land area protected by conservation practices in the watershed are being implemented to reduce phosphorus and sediment loading.

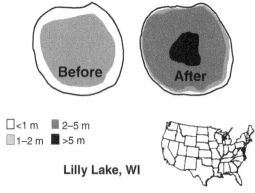

Cheney Lake, KS

(c) Diversion

Algal blooms with attendant elimination of sensitive benthic macroinvertebrates was caused by phosphorus discharges from four municipal wastewater treatment plants. Point source loads were diverted to land application and other uses, including irrigation of citrus groves and groundwater recharge. In-lake phosphorus and chlorophyll levels fell by 50 and 30%, respectively, and Secchi disk transparency increased by 50%.

Lake Tohopekaliga, FL

(d) Dredging

Prolific macrophyte growth and attendant deposition resulted in a water column of 1.4 m overlying 10 m of decomposing plant material. Dredging removed 665,000 m^3 of sediment from this 37 ha lake, increasing the water volume by 128% and the maximum depth to 6.6 m. Severe oxygen depletion and fish kills in winter were eliminated. Macrophytes no longer grew to nuisance levels because increased depth reduced the amount of well-lit plant habitat available. Dredging is expensive and carries with it potential side effects, largely related to sediment resuspension.

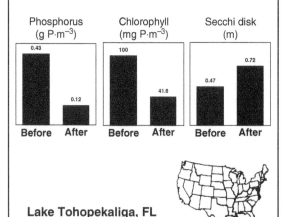

□ <1 m	■ 2–5 m
▨ 1–2 m	■ >5 m

Lilly Lake, WI

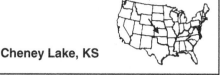

Figure / 7.28 Examples of Engineered Lake Management.

(e) Chemical inactivation

The addition of alum (aluminum sulfate) reduced sediment P release by a factor of 10. Bottom water P concentrations dropped dramatically, and surface water P levels decreased 0.013 to 0.005 g P/m³, establishing conditions of oligotrophy. The treatment persisted for over a decade.

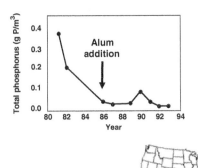

Lake Morey, VT

(f) Hypolimnetic aeration

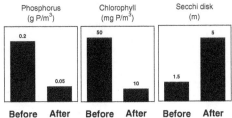

Oxygen was bubbled into the bottom waters of this eutrophic reservoir, maintaining dissolved-oxygen levels above 5 mg/L. Reductions in phosphorus, ammonia-nitrogen, and chlorophyll were achieved, and Secchi disk transparency was increased. Aeration eliminated sulfide from the lake, protecting the water supply for a downstream Chinook salmon and steelhead trout hatchery.

Camanche Reservoir, CA

(g) Herbicides/harvesting

Treatment with the herbicide fluridone eliminated the exotic invasive species Eurasian watermilfoil for four consecutive summers. Some native species decreased in abundance following treatment, while others increased. Secchi disk transparency was reduced following treatment because of reduced competition for phosphorus between algae and macrophytes.

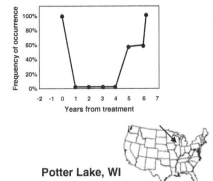

Potter Lake, WI

(h) Biomanipulation

For many years, this urban lake was characterized by high transparency, low algal biomass and a fish population dominated by piscivores (fish-eating species, i.e., largemouth bass). A change to planktivores (plankton-eating species, i.e., bluegill and crappie) and benthivores (bottom-feeding species, i.e., bullheads) increased predation pressure on zooplankton, resulting in increased algal growth and reduced water clarity. The lake was treated with rotenone to remove planktivores and benthivores and restocked with piscivores (largemouth bass and walleye). Reintroduction of piscivores allowed zooplankton populations to rebound, reducing algal biomass and improving water clarity.

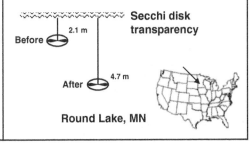

Round Lake, MN

Figure / 7.28 (Continued)

Table / 7.21

Wetlands Include the Following Areas That Fell into One of Five Categories   (from Dahl, 2011).

Areas with hydrophytes and hydric soils, such as those commonly known as marshes, swamps, and bogs

Areas without hydrophytes but with hydric soils—for example, flats where drastic fluctuation in water level, wave action, turbidity, or high concentration of salts may prevent the growth of hydrophytes

Areas with hydrophytes but nonhydric soils, such as margins of impoundments or excavations where hydrophytes have become established but hydric soils have not yet developed

Areas without soils but with hydrophytes, such as the seaweed-covered portions of rocky shores

Areas without soil and without hydrophytes, such as gravel beaches or rocky shores without vegetation

The EPA and the U.S. Army Corps of Engineers have established standards for reviewing permits for discharges that may affect a wetland. These discharges could be associated with construction of residential development, roads, and levees. Section 404 of the Clean Water Act allows the Army Corps to issue permits. Figure 7.29 shows that just by preserving wetlands in an urban environment, one can reduce nitrogen loadings to the urban watershed.

The U.S. Fish and Wildlife Service reports on the current amount of wetlands in the continuous United States (Dahl, 2011). There are currently an estimated 110.1 million acres of wetlands, of which 95 percent are freshwater and the remaining 5 percent are marine or estuarine (saltwater) systems. Overall wetland establishment is increasing through restoration efforts. Figure 7.30 provides an example of such a restoration activity. Note that in this figure, the restoration effort consists of preserving some existing wetlands while reestablishing historical wetlands that had been previously

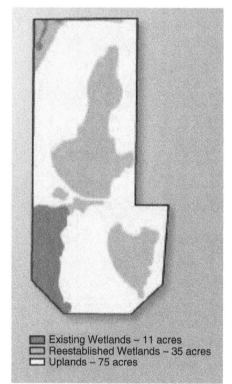

■ Existing Wetlands – 11 acres
□ Reestablished Wetlands – 35 acres
□ Uplands – 75 acres

Figure / 7.30   Wetland Restoration   A 121-acre wetland reestablishment project in southern Wisconsin consists of preserving 11 acres of an existing wetland and reestablishing 35 acres of wetland. The remaining 75 acres is upland forest that is not considered a wetland (image redrawn from Dahl, 2011).

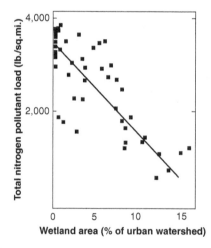

Figure / 7.29   Impact of Amount of Urban Wetland Area on Nitrogen Loading to Urban Watershed   (From EPA (2001)).

drained. Unfortunately wetland loss rates are still increasing with an overall consequence that national losses are outdistancing gains. Loss of wetlands results in habitat loss and fragmentation, limited opportunities for later reestablishment of wetlands, and reductions in future generations ability to protect and restore the beneficial uses of a watershed. The most common causes of wetland loss are listed in Table 7.22.

Table / 7.22

Most Common Contributors to Wetland Loss

Agricultural activities

Residential and commercial development (urbanization)

Construction of roads and highways

National Wetlands Inventory

http://www.fws.gov/wetlands

Louisiana Coastal Land Loss

http://www.nwrc.usgs.gov/special/landloss.htm

Class Discussion

How would you manage the Everglades while balancing social, economic, justice, and environment issues that will not lead to loss of opportunity for current and future generations? What are some key issues associated with water storage and quality, biodiversity, and the economy in this area? How do engineers participate in developing solutions to problems related to these issues? (For further information, see www.evergladesplan.org.)

Application / 7.4   The Florida Everglades

The Everglades in the State of Florida is one of the world's truly unique ecosystems. It includes wetlands and a river once estimated to be 50 miles wide. More than 50 percent of the original wetlands have been lost to conversion to agriculture and urbanization. Efforts to protect this resource have strengthened but often conflict with increasing development demands for rapidly growing resident and tourist populations. Since 1930, the population of southern Florida has increased 25-fold, from 200,000 to more than 5 million, a growth rate approximately 10 times faster than that of the United States. In 1947, Everglades National Park was formed. It now covers 1.4 million acres (nearly 5 percent of the land area of Florida) and has tripled in size since it was established.

In 2000, the U.S. Congress authorized the Comprehensive Everglades Restoration Plan, the largest environmental restoration project in history, with a 30-year time frame and a $10.5 billion budget. As part of the restoration plan, 400 km of canals and levees (installed in the 1940s to control and divert water) will be removed. Measures will be taken to control invasive and exotic species, and 6.4 billion liters a day of runoff will be treated to remove nutrients and other contaminants. Runoff will be stored and redirected to more closely resemble pre-settlement (natural) flow patterns to Florida Bay, as shown in Figure 7.31. More than 200,000 acres of land

have been purchased (50 percent of the project goal) to control land use.

At a web site devoted to reporting on this plan (www.evergladesplan.org), the Army Corps of Engineers spells out its promise and importance:

Implementation of the restoration plan will result in the recovery of healthy, sustainable ecosystems in south Florida. . . . The plan will redirect how water is stored in south Florida so that excess water is not lost to the ocean, and instead can be used to support the ecosystem as well as urban and agricultural needs. . . . The ability to sustain the region's natural resources, economy, and quality of life depends, to a great extent, on the success of the efforts to enhance, protect and better manage the region's water resources (USACE, 2008a).

One key element of the management plan is the Florida panther, a subspecies of mountain lion. The panther once ranged in eight states over the entire southeastern United States, but now it has a range reduced to approximately 10 counties in southern Florida (less than 5 percent of its native range). The current population is estimated at approximately 90 animals. The population decreased dramatically over the past century due to hunting and habitat loss. Habitat loss is primarily from urban sprawl and the conversion of woodlands to agriculture. While loss of habitat is significant, habitat degradation and fragmentation also pose major threats.

The preferred prey for the panther is white-tailed deer, the population of which has also declined. Secondary food sources are feral hogs released for hunting, armadillos, and raccoons. In areas with low deer populations, raccoons are an increasingly important part of the panther diet. Because raccoons feed on fish and crayfish, they have increased mercury content from bioaccumulation. Increased mercury levels in panthers may be linked to decreased health and reproductive success, but conclusive scientific studies have neither proven nor disproven this hypothesis. Vehicle collisions also have been a significant source of panther mortality.

Figure 7.32 shows how integrating wildlife corridors into transportation systems not only protects wildlife but also links ecological areas and wildlife habitat.

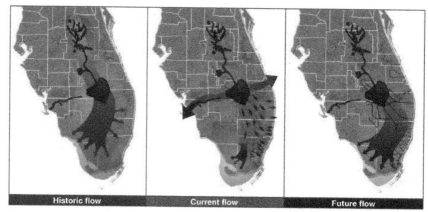

Figure / 7.31   The Florida Everglades, Showing Flow of Water from Historic, Current, and Future Scenarios.

(From www.evergladesplan.org, an effort of the U.S. Army Corps of Engineers in partnership with the South Florida Water Management District and many other federal, state, local and tribal partners). USACE (2008b)).

Figure / 7.32   Use of Wildlife Corridors to Link Ecological Areas and Wildlife Habitat   Compare (a) the existing crossing alternative and (b) proposed crossing alternative for Interstate 90 Snoqualmie Pass East Project, which links Puget Sound to eastern Washington. This area has been recognized as a significant north–south wildlife corridor for animals in the Cascades. Note how the proposed crossing provides a much wider stretch of land for terrestrial and aquatic wildlife crossing. Here wildlife habitats are connected in the proposed structure, and both large and small animals are allowed to cross under the road safely.

(Adapted from U.S. Department of Transportation Federal Highway Administration, Exemplary Ecosystem Initiatives, http://www.fhwa.dot.gov/environment/ecosystems/wa05.htm).

7.10   Groundwater Quality and Flow

Groundwater quality is an important topic because of the large number of individuals and business that use groundwater as a water source and the number of ecosystems that depend on groundwater baseflow. **Groundwater pollution** is the degradation of existing groundwater quality. It can occur from anthropogenic activities and also through natural processes where pollutants leach into the groundwater from surrounding subsurface minerals. These human activities and natural processes can result in pollutant levels that pose a potential risk to human health. Sometimes these natural leaching processes are accelerated through human activities that change the redox chemistry of the subsurface or of metals that are moved to the surface during mining. Changes in redox conditions can cause metals to be converted from an oxidized state to a reduced state where they are then more mobile. Elevated metal concentrations that occur from natural process occur in many parts of the United States and world and can result in unsafe levels of metals such as arsenic, fluoride, and uranium in groundwater.

Global Arsenic Crisis

http://www.who.int/topics/arsenic/en/

7.10.1   SOURCES OF GROUNDWATER POLLUTION

Figure 7.33 shows examples of several human activities and how they can impact groundwater quality. These activities include intentional and unintentional inputs directly to land or the subsurface. Also note in this figure how the pollutants can originate from point sources and nonpoint sources. The types of pollutants that degrade groundwater quality are limitless because of all the different activities that take place on the surface and can introduce pathogens, inorganic chemicals

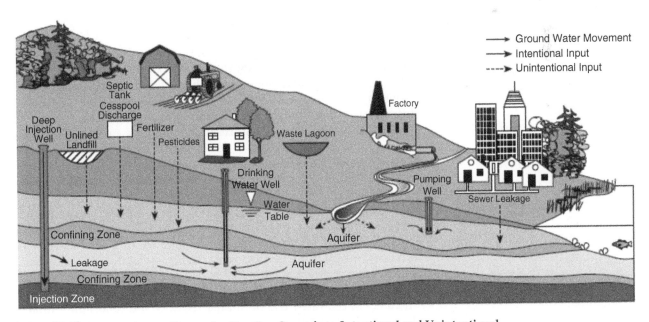

Figure / 7.33   **Groundwater Contamination Can Occur from Intentional and Unintentional Inputs of Point and Nonpoint Sources of Pollution**   (Adapted from EPA, 2000).

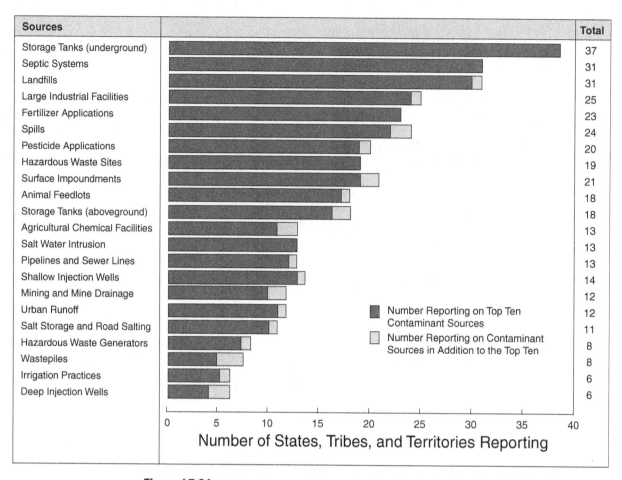

Sources		Total
Storage Tanks (underground)		37
Septic Systems		31
Landfills		31
Large Industrial Facilities		25
Fertilizer Applications		23
Spills		24
Pesticide Applications		20
Hazardous Waste Sites		19
Surface Impoundments		21
Animal Feedlots		18
Storage Tanks (aboveground)		18
Agricultural Chemical Facilities		13
Salt Water Intrusion		13
Pipelines and Sewer Lines		13
Shallow Injection Wells		14
Mining and Mine Drainage		12
Urban Runoff		12
Salt Storage and Road Salting		11
Hazardous Waste Generators		8
Wastepiles		8
Irrigation Practices		6
Deep Injection Wells		6

■ Number Reporting on Top Ten Contaminant Sources

□ Number Reporting on Contaminant Sources in Addition to the Top Ten

Number of States, Tribes, and Territories Reporting

Figure / 7.34 Major Sources of Groundwater Contamination Identified by States, Tribes, and Territories and Reported to EPA (From EPA, 2000).

(e.g., metals, arsenic, fluoride, nitrate), radionuclides (e.g., uranium), and organic chemicals (for example, fuel products, solvents, pesticides, and herbicides), and emerging chemicals like pharmaceuticals. Another important groundwater contaminant is salinity. **Saltwater intrusion** occurs from the overpumping of freshwater from wells located in coastal areas. This process of pumping more water from a coastal aquifer at a rate that is faster than it is recharged by rainfall draws seawater into the aquifer. This turns the water saline, making it not acceptable for crop irrigation, and requiring energy and materials to turn it back to freshwater for human consumption.

Figure 7.34 provides a ranking of potential activities that can have an adverse impact on groundwater quality. As shown in this figure the greatest threats identified are: (1) storage of fuel in underground storage tanks, (2) waste disposal activities related to on-site use of septic tanks to treat wastewater and landfills where municipal and hazardous wastes are disposed of, (3) agricultural activities that include concentrated use of animal feedlots and land application of fertilizers and pesticides, and (4) industrial practices that release metals and organic chemicals.

Underground Storage Tanks in Your Area

http://www.epa.gov/OUST/wheruliv.htm

State-Specific Septic Tank Information

http://www.nesc.wvu.edu/septic_idb/idb.cfm

Approximately one-third of the wastewater in the United States is treated using on-site systems. These systems are designed to reduce the risks of exposure to pathogens and other environmental pollutants, but are not optimized for nutrient removal. Because of evidence that shows that **on-site systems** can contaminate groundwater and springs and other surface waters (Figure 7.35), nitrogen loadings that originate from on-site wastewater systems are receiving increased attention.

A septic tank functions under oxygen-limited conditions. Pollutants in a septic tank may be removed by gravitational settling or through anaerobic digestion of organic carbon. The pretreated wastewater then flows by gravity to a drainfield. Here a series of perforated pipes distribute the wastewater to the subsurface soil. The water percolates downward toward the groundwater table and contaminants can be retrained in the soil matrix or treated by biological, chemical, and physical removal processes in a more aerobic environment.

On-site septic tanks and drainfields are not appropriate in all locations. They are also not designed to specifically remove nitrogen and may not be appropriate in areas that lack appropriate soil infiltration and attenuation properties for the drainfield or where the drainfield discharge is located to close to the groundwater table. The tanks also require regular inspection and cleaning that maintains the tank volume and design hydraulic residence time. Some on-site systems employ mechanical systems that are similar to the activated sludge process described in Chapter 9. They can theoretically obtain nitrogen removal through nitrification and denitrification processes. However, even mechanical systems show only at best 60 percent reduction in nitrogen. This is because of poor operation and maintenance by homeowners and a lack of carbon to support denitrification.

The number of on-site treatment systems continues to grow as urban areas expand beyond the reach of centralized services and as remote and coastal properties are further developed for human habitation. Unfortunately lakes, rivers, groundwater, and other water bodies are susceptible to discharges of nutrients from on-site systems. Research is therefore going on to develop on-site treatment systems that will achieve significant nutrient removal. In addition, some are also thinking how to reclaim water (and the nutrients dissolved in the treated wastewater) that originates from on-site treatment systems, at the residential or community level.

Figure / 7.35   Nitrate Concentrations Detected in Shallow Groundwater under Four Minnesota Residential Communities Served by On-site Septic Systems   These concentrations are higher than in shallow groundwater under residential areas served by sewers (between 1.5 and 2.0 mg NO_3^-/L) or undeveloped (less than 1.0 mg NO_3^-/L) areas (MPCA, 1997).

7.10.2 GROUNDWATER FLOW AND POLLUTANT TRANSPORT

Groundwater pollutants are transported by the processes of advection and dispersion. In Chapter 4 we defined **advection** as transport with the mean fluid flow. This causes pollutants to move vertically during the infiltration process and horizontally with the direction of groundwater flow. Chapter 4 also defined **dispersion** as the transport of chemicals through the action of random motions. Go back to that chapter and review Figure 4.21, which depicted the process of mechanical dispersion that occurs during groundwater flow.

The flow of a groundwater is a similar process as flow through a sand filter or sand column and is thus based on **Darcy's law**. Henry Darcy was a French hydraulic engineer who reported results from sand filter experiments that were conducted in a column. Darcy found that the flow rate (Q) through a porous media (like a sand filter or groundwater) was proportional the head loss (h_L) and inversely proportional to the length of the flow path ($1/L$).

If a proportionality constant is introduced, in this case, K or the aquifer's **hydraulic conductivity** (units of m/day), the flow rate (Q) through a cylinder with cross-sectional area, A, can be written as

$$Q = -KA\frac{h_L}{L} \qquad (7.18)$$

The negative sign in Equation 7.18 accounts for the fact that the head is decreasing in the direction that the water flows.

Hydraulic conductivity is a function of the aquifer material, and Table 7.23 provides typical values of K for a wide range of geological materials. It can range from 150 to 270 m/day for medium gravels, 2.5

Table / 7.23

Representative Values for Hydraulic Conductivity and Porosity for Soils and Other Geologic Materials   (data from Todd, 1980).

Material	Particle Size (mm)	Hydraulic Conductivity (m/day)	Porosity
Coarse gravel	16–32	150	0.28
Medium gravel	8–16	270	0.32
Fine gravel	4–8	450	0.34
Coarse sand	0.5–1	45	0.39
Medium sand	0.25–0.5	12	0.39
Find sand	0.125–0.25	2.5	0.43
Silt	0.004–0.062	0.08	0.46
Clay	<0.004	0.0002	0.42
Fine-grained sandstone	Not applicable	0.2	0.33
Medium-grained sandstone	Not applicable	3.1	0.37
Limestone	Not applicable	0.94	0.30

to 12 m/day for fine to medium sand, 0.08 m/day for silt, and 0.0002 m/day for clay. Equation 7.18 can be expressed in more general terms for the head loss, written as head loss per unit length (dh/dl, dimensionless). This results in the following equation:

$$Q = -KA\frac{dh}{dl} \qquad (7.19)$$

Equation 7.19 can be rearranged to solve for the **Darcy velocity**, v,

$$v = \frac{Q}{A} = -K\frac{dh}{dl} \qquad (7.20)$$

The Darcy velocity is only valid for Reynold's numbers < 1 and assumes the groundwater flow is taking place across the entire cross section (A) of the aquifer. In the subsurface though, there are solids and pore spaces present. Therefore, the actual flow of water is limited to the pore space and the average velocity (v_a) through these pores can be determined by dividing the Darcy velocity by the porosity of the aquifer material:

$$v_a = \frac{v}{\eta} \qquad (7.21)$$

The **porosity** of the aquifer material is defined as

$$\eta = \frac{\text{volume of the void spaces}}{\text{bulk volume}} \qquad (7.22)$$

Table 7.23 provides representative values for porosity. The porosity is a function of the aquifer material; for example, it can equal 0.32 for medium gravels, 0.43–0.39 for fine to medium sand, 0.42 for clay.

Assuming that the chemical–solid interactions are at equilibrium, the effects of sorption on the velocity of a chemical pollutant can be quantified through the **retardation coefficient** (R_f):

$$R_f = 1 + \frac{\rho_b}{\eta}K_p \qquad (7.23)$$

In Equation 7.23, K_p is the soil–water partition coefficient (L/kg or cm^3/mg) that was discussed in Chapter 3, η is the porosity (unitless), and ρ_b is bulk density of the soil or aquifer material (cm^3/g). The retardation coefficient implies that a chemical that is undergoing sorption with the surrounding soil or aquifer material will travel at a slower rate than the average velocity of the groundwater. Thus, a chemical with a retardation coefficient equal to 5 will travel at a velocity that is five times slower than the velocity of the groundwater flow determined in Equation 7.21.

7.10.3 SUBSURFACE REMEDIATION

Remediation of subsurface soils and groundwater takes place after it is determined that there is an unacceptable risk to human health or the environment. In many cases, groundwater recharges a spring, stream, or other surface water body, where contaminants can then exert toxic effects to aquatic organisms. In other case, humans depend on

example/ 7.9   Determining Travel Times of Groundwater and Subsurface Pollutants

A leaking underground storage tank has discharged the solvent, trichloroethene (TCE) into groundwater. A water well used for drinking is located 120 m downgradient from the leaking tank. To ensure the safety of the water supply, the tank is removed to stop the source and a monitoring well is drilled halfway between the water well and the solvent discharge. The difference in hydraulic head between the source and the monitoring well is 35 cm (with the head in the monitoring well higher). A site investigation shows the subsurface aquifer material consists primarily of medium sand. (a) How long does it take groundwater underneath the leaking tank to reach the monitoring well? Assume the TCE moves at the same velocity as the groundwater. (b) If the aquifer material has an organic carbon content of 1 percent and you now account for sorption between the TCE and aquifer material, how long will it take the TCE to reach the monitoring well? Assume the bulk density of the aquifer material is 2.1 gm/cm^3.

solution

For part (a), the travel time of the groundwater and chemical (assuming its movement is not retarded by sorption to aquifer materials) is determined from the point of discharge to the point where the monitoring well is located. It is found by first calculating the Darcy velocity (Equation 7.20) and then dividing this value by the soil porosity (Equation 7.21). Table 7.23 provides information on the hydraulic conductivity (K) and porosity (η) for medium sand ($K = 12$ m/day and $\eta = 0.39$).

$$v_a = \frac{v}{\eta} = -\frac{K}{\eta}\frac{dh}{dl} = -\left(\frac{12\,\text{m/day}}{0.39}\right) \times \left(-\frac{0.35\,\text{m}}{60\,\text{m}}\right) = 0.18\,\text{m/day}$$

The travel time of the groundwater from the pollutant source to the monitoring well is

$$t = \frac{L}{v_a} = \frac{60\,\text{m}}{0.18\,\text{m/day}} = 330\,\text{days}$$

For part (b) we need to determine how sorption impacts the velocity of the TCE. In Example 3.13, we determined that the soil–water partition coefficient for TCE in a soil matrix with an organic carbon content of 1 percent was 1.9 cm^3/gm. Equation 7.23 allows us to estimate the effect that sorption has on the velocity of a chemical pollutant relative to the movement of the groundwater. In this case, the retardation coefficient is determined as

$$R_f = 1 + \frac{\rho_b}{\eta} K_p$$

$$R_f = 1 + \frac{2.1\,\text{gm/cm}^3}{0.39} 1.9\,\text{cm}^3/\text{gm} = 10.2$$

Remember that R_f is unitless. The value of 10.2 implies that the movement of TCE is slower relative to the movement of the water. Thus, we can expect that the TCE would arrive at the monitoring well in (330 days × 10.2 = 3,370 days. This may seem like a long time, but in many cases spills occurred years and decades before they were detected or better managed and regulated. Also observe in this example how the movement of groundwater and the chemical are strongly influenced by subsurface properties (hydraulic gradient, hydraulic conductivity, porosity, organic carbon content) and chemical properties (hydrophobicity of the chemical). All these parameters can result in groundwater and chemicals that move at a rate in the subsurface similar to or slower than the rate at which the groundwater does.

The principles and equations that govern fluid flow through porous media can be applied to a wide variety of challenges related to sustainable development. One example is the design and assessment of a popular "point of use" water treatment technology for the developing world—the ceramic or clay filter (see Figure 7.36), which is now used in over 20 countries.

One advantage of these filters is that they can be produced using locally available materials (e.g., clay, sawdust, water). This mixture is formed and fired in a kiln. Firing in the kiln forms the ceramic materials and combusts the organic material, making the filter porous and permeable to water. The pores are sized to remove larger waterborne particles and associated microorganisms to improve water quality and subsequently reduce health risks. A typical filter is shaped like a bowl or a pot that can be nested within a storage receptacle. The filter pots are suspended in a larger container (for example, a bucket) so that when water is poured into the filter, it flows by gravity through the filter and into the lower container, where treated water can be accessed.

The acceptability of this technology depends not only on the quality of the filtered water but also on the quantity of water produced by the filter (desired flow rates range from 1 to 2 L/h). More information about modeling flow through a porous ceramic filter can be obtained from Schweitzer et al. (2013). For more information on this and design and application of other appropriate technology methods to manage water, wastewater, and indoor pollution, see Mihelcic et al. (2009).

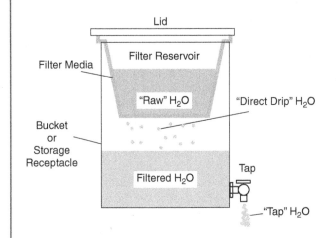

Figure / 7.36   Clay Filter (Porous Medium) Used to Remove Turbidity and Associated Microorganisms from Water   (Reproduced with permission of Ryan W. Schweitzer)

groundwater as a source of drinking water or for supporting agricultural activities.

Remediation or control of subsurface pollutants can prove to be difficult and costly because of the difficulty in not only reaching the subsurface but also visualizing it. Subsurface contaminants can be in diluted forms that are more costly to treat or sorbed to soil particles, which slows their ability to be pumped to the surface for treatment. In addition, site investigations have to make determinations of how much data is required to determine the type and extent of contamination, the movement of chemicals, and the risk they pose, all with the certainty that the subsurface is not homogenous.

Strategies that minimize risk include reducing the hazard (through removing the source of contamination or remediating the chemical concentrations to lower acceptable levels) or reducing or eliminating exposure. Strategies that eliminate or reduce exposure can include requiring placement of barriers between contaminated soil and the public (e.g., layers of pavement or soil). Other strategies to reduce or eliminate risk include placing deed restrictions on usage of contaminated groundwater or provision of a different piped water supply. In Chapter 6, we also provided information on methods to determine what would be the allowable concentration of a particular contaminant that can occur in soil or groundwater that would not result in excessive risk to the public or the environment if they were exposed to the contaminant.

Saltwater intrusion can be reversed or prevented by decreasing the withdrawal rate of groundwater to allow for natural recharge rates to restore groundwater. In other cases, reclaimed wastewater can be injected into the subsurface to replenish an aquifer while also providing a cost-effective method to store water for future use. In many other cases, remediation is applied to contaminated soil or groundwater. Figure 7.37 provides a description of several remediation technologies. Note that they employ biological, physical, and chemical treatment processes.

As shown in this figure, in pump and treat strategies, groundwater is pumped to the surface where it is then treated above ground using established methods used for drinking water or wastewater treatment that include use of air stripping and granular activated carbon. Contaminated soil can be excavated and disposed of in a landfill or treated above ground using chemical, biological, or thermal process. In some cases, the contaminants are treated in place or in situ. The word *in situ* implies the contaminants are treated in place versus excavating them or pumping them to the surface.

In many cases combination of technologies and strategies are employed. For example, at the Baird and McGuire Superfund site (Massachusetts), soil and groundwater were contaminated by chemicals stored in tanks that leaked. The chemicals migrated downward from the source and reached an underlying aquifer, which flowed off-site and contaminated a municipal water supply. To remediate the site, engineers installed a pump and treat system that consisted of eight wells that pumped a total of 127 gpm. This was to contain the contaminated groundwater plume from spreading and also transport contaminated groundwater above ground for treatment. A small treatment plant was designed and constructed to provide treatment of the groundwater for heavy metals. A set of air strippers were also included to remove volatile organic chemicals and activated carbon was used to remove any remaining contaminants. An extensive groundwater monitoring program was also integrated with the remediation plan to demonstrate the remediation strategies were working and a reduction in the risk posed to human health and the environment was also occurring.

EPA provides many online resources to learn about existing and new remediation technologies. This includes access to information about where new technologies were demonstrated, how they performed, and regulatory guidance in using a particular technology (e.g., http://www.epa.gov/superfund/remedytech/remed.htm).

Innovative Remediation Technologies
http://www.epa.gov/tio/

Technology Description	What the Technology Looks Like
Pump and treat consists of installing one or more wells in the subsurface. These extraction wells pump contaminated groundwater to an aboveground treatment system that removes the contaminants. Treatment aboveground can consist of air stripping, activated carbon, or biological treatment. Pump and treat can also be used to contain the contaminated plume from spreading in the groundwater.	
In situ chemical oxidation means that chemical oxidants are injected into the subsurface. There they can oxidize organic chemicals to less hazardous end products. This technology is typically used to used to treat chemicals found in the source area where the initial contamination occurred.	
Natural attenuation relies on several natural processes to decrease (i.e., attenuate) the concentrations of pollutants found in the subsurface. The site is monitored to determine whether contaminants are being diluted, retained by sorption, or degrading under natural conditions. The microorganisms that mediate the biochemical reactions are those that naturally exist in the subsurface.	
Phytoremediation consists of planting specific plants over contamination that occurs in layers of soil close to the surface or shallow groundwater. The plant roots can either: (1) store the contaminants in their roots, stems, or leaves; (2) convert them to less hazardous chemicals in the root zone; (3) transform pollutants to the vapor form which releases them to the air, or (4) sorb contaminants to the roots where microorganisms are present that mediate biodegradation.	

Figure / 7.37 Example of Groundwater Remediation Technologies.

(Adapted from EPA Series: A Citizen's Guide to Cleanup Technologies (EPA, 2012)).

Key Terms

- advection
- agricultural use
- aquifer
- baseflow
- best efficiency point (BEP)
- branched
- Clean Water Act
- confined aquifer
- consumed water
- critical point
- cultural eutrophication
- Darcy velocity
- Darcy's law
- demand factor (DF)
- deoxygenation
- design water velocity in a pipe
- disaggregated segments
- dispersion
- dissolved oxygen (DO)
- domestic use
- ecological footprint
- environmental justice
- epilimnion
- estimating water demand
- eutrophic
- eutrophication
- fall turnover
- fire flow demand
- fire protection water demand
- freshwater
- future water demand
- groundwater

- groundwater pollution
- household use
- hydraulic conductivity (K)
- hydrology
- hydrologic cycle
- hydropower
- hypolimnion
- industrial use
- in situ chemical oxidation
- John Muir
- land use
- looped subsystems
- maximum and minimum design flow rates
- mesotrophic
- metalimnion
- micro-hydropower
- modeling objectives for water demand
- modeling process to estimate water demand
- municipal water demand
- natural attenuation
- needed fire flow (NFF)
- nonpoint source
- nutrients
- oligotrophic
- on-site (wastewater treatment) systems
- open-channel conditions
- oversaturation
- oxygen deficit
- peaking factor (PF)
- permeable (or porous) pavement

- phytoremediation
- point source
- pollution
- population equivalent
- porosity
- public water supplies
- pump and treat
- pump characteristic curves
- pump cycle time
- pumping station
- Rational method
- reaeration
- Retardation coefficient (R_f)
- reused water
- runoff coefficients
- saltwater intrusion
- satellite reclamation plants
- saturated zone
- saturation (oxygen) concentration
- seawater
- septic tank
- spring turnover
- storage tank
- Streeter–Phelps model
- subsurface remediation
- summer stratification
- surface water
- thermocline
- trophy
- unconfined aquifer
- unmetered flow
- unsaturated zone

- variation in flows
- wastewater collection
- wastewater generation rates
- water distribution
- water footprint
- water reclamation
- water reuse
- water scarce
- water scarcity
- water stress
- water usage rates
- watershed
- wet well
- wetlands
- winter stratification

7.1 The average annual rainfall in an area is 60 cm. The average annual evapotranspiration is 35 cm. Thirty percent of the rainfall infiltrates and percolates into the underlying aquifer; the remainder is runoff that moves along or near the ground surface. The underlying aquifer is connected to a stream. Assuming there are no other inputs or outputs of water to the underlying aquifer and the aquifer is at steady state (neither gains nor loses water), what is the amount of baseflow contributed to the stream from the groundwater?

7.2 An agricultural watershed that is managed to minimize runoff consists of 150 acres that has a land slope of 1–2 percent and sandy soils. The land is planned to be developed into a residential sub-division (30 percent as a low-density residential area, 30 percent as a medium-density residential area, and the remaining land preserved as green space. (a) Calculate the peak runoff flow rate (ft^3/min) before and after the development takes place for a storm with rainfall intensity of 0.35 in/h. (b) Determine the percent change in peak runoff between the two land use scenarios. (c) If the green-space was equally developed between low- and medium-density development, how does your answer to part (b) change? (d) How do your answers for the peak runoff flow rate in part (a) change if 100 percent of the land is developed in a high-density residential?

7.3 A rural watershed that is managed to minimize runoff consists of 120 acres of cultivated land that has a gentle land slope (1–2 percent) and silty/sandy soils. The land is planned to be developed into a residential housing with plans for 25 percent low-density residential area, 25 percent for medium-density residential area, and 50 percent set aside as green space. (a) Estimate the annual mass loading due to runoff for suspended solids (SS), phosphorus (P), and nitrogen (N) (lb/year). (b) How does your answer to part (a) change if the slope of the land is 3 percent? (c) How does your answer to part (a) change if the greenspace is developed into commercial development to serve social and economic needs of the community, without leaving any natural areas as open green space?

7.4 Peak rainfall intensities vary by geographical region. Assume the peak rainfall (inches of rain) reported in different states over a 30 minute period are Florida—2.8; Iowa—1.8; Arizona—1.6; and Montana—0.8. (a) What is the peak runoff flow rate (ft^3/min) for an undisturbed open area of 10 acres with a slope of 1.75% that consists of sandy soil? Perform the same calculations for clayey soils. (b) What is the peak runoff flow rate for a similar topography and soil type, but if the 10 acres of open area are commercially developed?

7.5 Western Australia is expected to see a 10–20 percent increase in peak rainfall intensity from the effects of climate change by the year 2030. If the current peak rainfall intensity is 200 mm of rainfall falling over an 18 h period, what is the peak runoff flow rate you expect for a densely populated urban area of 20 acres on sandy soils with a slope of 1.75 percent, that consists of one-half high-density residential development and one-half commercial development? Report your results in cm^3/min and ft^3/min.

7.6 A small public well is used to supply drinking water to a small residential community located in a 26 km^2 watershed. For the month of June, the measured rainfall was 12 cm, the estimated evapotranspiration was 7.5 cm and the surface water runoff entered a small stream with an average flow of 0.32 m^3/s that leaves the watershed. Estimate the average flow (m^3/day) from the public well without depleting the underlying aquifer (neither gains nor loses water). Assume that all water that infiltrates will percolate to the aquifer.

7.7 Go to the United Nations Environment Programme's Global Environment Outlook web page, http://geodata.grid.unep.ch/. Look up two countries located in different hemispheres. (a) What are their current amounts of water withdrawals and freshwater withdrawals? (b) Are these countries currently experiencing water scarcity or expected to experience water scarcity?

7.8 Go to the web site of the U.S. Geological Survey (USGS), www.usgs.gov, and navigate to "Water Use in the United States." Look up the total water withdrawals associated with the following uses in your state: thermoelectric, irrigation, public supply, industrial, domestic, livestock, aquaculture, and mining. Place the eight uses in a table in order of greatest to least water withdrawals. Determine the percent of the total water withdrawals associated with each of these uses. Compare these percentages to the national percentages listed in Table 7.6. Discuss how your state compares with the national average.

7.9 Go to the web site of the U.S. Geological Survey (USGS), www.usgs.gov, and navigate to "Water Use in the United States." Look up the total surface water and groundwater withdrawals associated with your state. Determine the percent of surface water and groundwater withdrawals relative to total withdrawals in your state. Compare these percentages with the national distribution of surface water and groundwater use. Discuss how your state compares with the national average.

7.10 Contact your local water or wastewater authority. Ask for the annual water usage rates (average, maximum-day, minimum-day, and so on). For a water authority, ask how much is unaccounted-for water; for a wastewater authority, ask for how much is wet-weather flows. Use these numbers to estimate a demand factor and per capita (or metered connections) water usage rates. Discuss how your local values compare with the expected range of values described in this chapter.

7.11 **(a)** Estimate your own actual water use during a typical day. List your water use activities and estimate the volume of water used for each activity. **(b)** Compare your water usage rate with that of an average per person rate such as 101 gpdc. **(c)** Explain why your rate may be more or less than the average rate. **(d)** How much of your water use do you think was discharged as wastewater? **(e)** Did you do any water use activity that did not create any collected wastewater?

7.12 Estimate the daily water demand and wastewater generation for a department store that has six floors. On each floor are two sets of men's and women's lavatories. The men's have two toilets, two urinals, and three sinks; the women's have four toilets and three sinks. Assume that each lavatory will be used by 35 people per day.

7.13 Estimate the daily water demand and wastewater generation for a small commercial area that has the following buildings. Clearly indicate all assumptions and the estimated water usage for each building: (a) a 200-room hotel with 35 employees and one kitchen; (b) three restaurants, one being an organic restaurant with regionally produced foods, another an all-you-can-eat buffet (dinner only), and the third a vegan deli open from 5:00 A.M. to 3:00 P.M.; (c) a newsstand that sells magazines, refreshments, and snacks with one lavatory used only by the employees; and (d) a three-story office building with basement employing 140 people and with two sets of men's and women's lavatories per floor.

7.14 Estimate the maximum-day demand plus fire flow for a residential area. The residential area is 400 acres divided into 25-acre lots with yards that are 75 ft wide. Assume the average population density is 2.8 people per home and the maximum-day demand factor is 2.1.

7.15 You are working on a project to put in a new set of townhouse apartments. Estimate the daily and yearly water demand given 30 apartment units with an average of three people living in each unit.

7.16 A 2.5 MGD wastewater treatment plant is currently running at 80 percent capacity during the annual maximum day, servicing a city of 38,500 people with 26.7 mi. sewers. During the next 10 years, it is expected that new residential developments for 15,000 people will be built, along with an additional 6.5 mi. sewers. The sewer is assumed to leak 8,500 gpd/mi. (a) Project the maximum daily demand for the wastewater treatment plant after the new development is built. (b) Should the wastewater treatment plant's capacity be increased?

7.17 You are hired to upgrade the existing water treatment plant for Nittany Lion City. Using the historical records provided in Table 7.24, forecast the water demand through 2024. The population is expected to increase about 1.8 percent per year. (a) Create a graph that has the historical average, minimum-day, and maximum-day water demand in gpd for each year. Extrapolate trend lines for the projected future water demand through 2015. (b) Use your graph to predict the average, minimum-day, and maximum-day water demand for years 2014, 2019, and 2024. (c) Estimate the 2009, 2014, 2019, and 2024 per capita

Table / 7.24

Historical Records Used to Solve Problem 7.17

Year	Water Demand (gpd)			Population Served
	Average	Minimum	Maximum	
2003	1,707,190	1,018,655	2,624,414	14,251
2004	1,713,230	1,086,201	2,817,674	14,352
2005	1,820,602	1,094,415	3,003,411	14,354
2006	1,901,145	1,248,011	2,945,221	14,598
2007	1,891,860	1,068,574	3,038,157	14,587
2008	1,948,648	1,124,125	3,076,542	14,684
2009	1,923,458	1,184,214	3,067,821	14,857

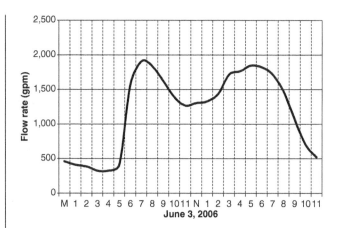

Figure / 7.38   Wilkes City Wastewater Metered Flow Rate (Used in Problem 7.19).

water use by calculating the average water use divided by the population served. (d) Determine a demand factor for the minimum-day and maximum-day demand, using the historical records.

7.18   You are hired to upgrade the existing water treatment plant for USF City. Using the historical records in Table 7.25, forecast the water demand through 2024. A new industry is expected to require 65,000 gpd starting in 2016. (a) Create a graph that has the historical domestic, commercial, and industrial demand for each year. Extrapolate trend lines for the projected future water demand through 2024 for each category. Take into account the additional industry water demand in 2016. (b) Estimate the percent of produced water that is unaccounted-for water based on the historical records. (c) Use the graph and estimated percent unaccounted-for water to predict the total water demand for years 2014, 2019, and 2024.

7.19   The recorded metered flow for June 3 at the Wilkes City wastewater treatment plant is shown in Figure 7.38. (a) Estimate the average flow rate for June 3.

7.20   A storage tank is designed to supply water for fire protection at a small industry. The NFF for this industry is 3,400 gpm. (a) Estimate the volume of water that would be needed for fire protection. (b) Estimate the nominal pipe size of a single pipe supplying the fire protection water from the tank if the design velocity for the pipe is 9.5 ft./s.

7.21   Estimate the size of a storage tank to supply water for fire protection to a 65,000 ft^2 department store ($O = 1.0$). The building is constructed from fire-resistant materials ($F = 0.8$), with a total exposure and communication factors equal to 0.45.

7.22   A pumping station with wet well is to be sized in a wastewater collection system for a design pumping rate of 1,200 gpm. (a) Estimate the minimum active wet-well volume with a pump cycle of four times per hour. (b) Size the force main (pumping station discharge pipe) with a design velocity of 7.5 ft./s.

Table / 7.25

Historical Records Used to Solve Problem 7.18

Year	Metered Flow from Treatment Plant (gpd)	Metered Flow Based on Billing Records (gpd)		
		Domestic	Commercial	Industrial
2003	1,687,517	824,247	423,229	92,676
2004	1,789,453	837,055	465,232	102,707
2005	1,745,658	828,103	476,429	76,916
2006	1,728,750	858,076	454,928	79,029
2007	1,779,854	861,003	461,669	87,422
2008	1,826,650	875,548	475,254	91,214
2009	1,872,456	899,545	479,451	90,248

7.23 Identify one regional and one global water scarcity issue. Develop a long-term sustainable solution that protects future generations of humans and the environment.

7.24 Go to the U.S. Green Building Council web site (http://www.usgbc.org) and research the LEED credits associated with new commercial construction and major renovation (Version 2.2, from U.S. Green Building Council). A project can obtain a maximum of 69 points. (a) How many possible points directly relate to the category water efficiency? (b) What are the specific credits provided for the category water efficiency?

7.25 Go the following web site (http://www.unesco.org/new/en/natural-sciences/environment/water/wwap/) and look up the report, *The 1st UN World Water Development Report: Water for People, Water for Life.* (a) Of the 11 challenge areas, list the ones related to "life and well being" and the ones related to "management." (b) Access the link on "facts and figures on securing the food supply." Develop a table with columns of product, unit equivalent, and water in m^3 per unit for the following products: cattle, sheep and goats, fresh beef, fresh lamb, fresh poultry, cereals, citrus fruits, palm oil, and roots and tubers. Use this table to answer the question whether, on a per kg basis, providing meat or grains/fruits uses more water.

7.26 Go to the following web site to learn how you can save water at home (http://www.epa.gov/WaterSense/pubs/simple_steps.html). For the following three areas (in the bathroom, in kitchen/laundry, outdoors) list a minimum of three items you can do at home to conserve water.

7.27 If the Henry's constant (K_H) for dissolved oxygen is 0.00136 moles/L-atm at 20°C and the concentration of carbon dioxide in the atmosphere is 390 ppm$_v$, what is the concentration of oxygen dissolved in water equilibrated with the atmosphere in: (a) moles/L, (b) mg/L, (c) μg/L, and (d) ppm$_m$? (e) How does your answer to part (b) change if the Henry's constant is reported in different units (K_H equal to 735.3 L-atm/moles).

7.28 A stream at 25°C has a dissolved-oxygen concentration of 4 mg/L. What is the dissolved-oxygen deficit in: (a) mg/L, (b) ppm, (c) ppb, and (d) moles/L?

7.29 The oxygen concentration of a stream is 4 mg/L, and DO saturation is 10 mg/L. What is the oxygen deficit?

7.30 The measured dissolved oxygen concentration using a DO meter of a river is 6 mg/L. The oxygen deficit is 2 mg/L at the same location. What is the saturation concentration of dissolved oxygen in: (a) mg/L, (b) ppm, (c) ppb, and (d) moles/L?

7.31 Dr. Mihelcic is canoeing on the Hillsborough River in Florida, just upstream of Trout Creek Wilderness Area. He is collecting dissolved oxygen readings with his DO meter. He spots several roseate spoonbills, woodstorks, and green herons feeding near the water's edge, gets excited, and drops his oxygen meter into the water. Assuming he is afraid to gather the meter from the bottom of the river because of the presence of alligators, what is the river's dissolved oxygen at this point if he learns later that the water temperature is 20°C and the oxygen deficit at this point is 3 mg/L at the time he was paddling.

7.32 Calculate the dissolved-oxygen deficit for a river at 30°C and a measured dissolved-oxygen concentration of 3 mg/L. The Henry's law constant at that temperature is 1.125×10^{-3} mole/L-atm and the partial pressure of oxygen is 0.21 atm.

7.33 A wastewater treatment plant discharges an effluent containing 2 mg/L of dissolved oxygen to a river that has a dissolved-oxygen concentration of 8 mg/L upstream of the discharge. Calculate the dissolved-oxygen deficit at the mixing basin if the saturation dissolved oxygen for the river is 9 mg/L. Assume that the river and plant discharge have the same flow rate.

7.34 A combined sewer overflow (CSO) discharges an effluent containing 0 mg/L of dissolved oxygen to a stream that has a dissolved-oxygen concentration of 7 mg/L upstream of the discharge. Calculate the dissolved-oxygen deficit at the mixing basin if the saturation dissolved oxygen for the river is 9 mg/L. Assume that the CSO flow rate is one half of the stream flow rate.

7.35 A river traveling at a velocity of 10 km/day has a dissolved-oxygen content of 5 mg/L and an ultimate CBOD of 25 mg/L at distance $x = 0$ km, that is, immediately downstream of a waste discharge. The waste has a CBOD decay coefficient k_1 of 0.2/day. The stream has a reaeration rate coefficient k_2 of 0.4/day and a saturation dissolved-oxygen concentration of 9 mg/L. (a) What is the initial dissolved-oxygen deficit? (b) What is the location of the critical point, in time and distance? (c) What is the dissolved-oxygen deficit at the critical point? (d) What is the dissolved-oxygen concentration at the critical point?

7.36 The wastewater treatment plant for Pine City discharges 1×10^5 m^3/day of treated waste to the Pine River. Immediately upstream of the treatment plant, the Pine River has an ultimate CBOD of 2 mg/L and a flow of 9×10^5 m^3/day. At a distance of 20 km downstream of the treatment plant, the Pine River has an ultimate CBOD of 10 mg/L. The state's Department of Environmental Quality (DEQ) has set an ultimate CBOD discharge limit for the treatment plant of 2,000 kg/day. The river has a velocity of 20 km/day. The CBOD decay coefficient is 0.1/day. Is the plant in violation of the DEQ discharge limit?

7.37 An industry discharges 0.5 m^3/s of a waste with a 5-day CBOD of 500 mg/L to a river with a flow of 2 m^3/s and a 5-day CBOD of 2 mg/L. Calculate the 5-day CBOD of the river after mixing with the waste.

7.38 A high strength waste having an ultimate CBOD of 1,000 mg/L is discharged to a river at a rate of 2 m^3/s. The river has an ultimate CBOD of 10 mg/L and is flowing at a rate of 8 m^3/s. Assuming a reaction rate coefficient of 0.1/day, calculate the ultimate and 5-day CBOD of the waste at the point of discharge (0 km) and 20 km downstream. The river is flowing at a velocity of 10 km/day.

7.39 A new wastewater treatment plant proposes a discharge of 5 m^3/s of treated waste to a river. State regulations prohibit discharges that would raise the ultimate CBOD of the river above 10 mg/L. The river has a flow of 5 m^3/s and an ultimate CBOD of 2 mg/L. Calculate the maximum 5-day CBOD that can be discharged without violating state regulations. Assume a CBOD decay coefficient of 0.1/day for both the river and the proposed treatment plant.

7.40 A river flowing with a velocity of 20 km/day has an ultimate CBOD of 20 mg/L. If the organic matter has a decay coefficient of 0.2/day, what is the ultimate CBOD 40 km downstream?

7.41 A river traveling at a velocity of 10 km/day has an initial oxygen deficit of 4 mg/L and an ultimate CBOD of 10 mg/L. The CBOD has a decay coefficient of 0.2/day, and the stream's reaeration coefficient is 0.4/day. What is the location of the critical point: (a) in time; (b) in distance?

7.42 A paper mill discharges its waste ($k_L = 0.05/$day) to a river flowing with a velocity of 20 km/day. After mixing with the waste, the river has an ultimate carbonaceous BOD of 50 mg/L. Calculate the 5-day carbonaceous BOD at that location and the ultimate carbonaceous BOD remaining 10 km downstream.

7.43 For each of the following cases, assuming all other things unchanged, describe the effect of the following parameter variations on the magnitude of the maximum oxygen deficit in a river. Use the following symbols to indicate your answers: increase (+), decrease (−), or remain the same (=).

Parameter	Magnitude of the Deficit
Increased initial deficit	_____
Increased ultimate CBOD @ $x = 0$	_____
Increased deoxygenation rate	_____
Increased reaeration rate	_____
Increased ThOD @ $x = 0$	_____

7.44 Humans produce 0.8–1.6 L of urine per day. The annual mass of phosphorus in this urine on a per capita basis ranges from 0.2 to 0.4 kg P. (a) What is the maximum concentration of phosphorus in human urine in mg P/L? (b) What is the concentration in moles P/L? (c) Most of this phosphorus is present as HPO_4^{2-}. What is the concentration of phosphorus in mg HPO_4^{2-}/L?

7.45 Assume 66% of phosphorus in human excrement in found in urine (the remaining 34% is found in feces). Assume humans produce 1 L of urine per day and the annual mass of phosphorus in this urine is 0.3 kg P. If indoor water usage is 80 gallons per capita per day in a single individual apartment, what is concentration (in mg P/L) in the wastewater that is discharged from the apartment unit? Account for phosphorus in urine and feces.

7.46 Contact your local wastewater treatment plant to find out the average daily treated flow rate and the average concentration of phosphorus in the untreated influent and treated effluent. Use census data to determine the current population of your area and assuming a growth rate of 3%, the population in 2025 and 2050. (a) If nothing is done in how the plant treats phosphorus and how each human discharges phosphorus, what is the current and future P loading (kg P/day) to the local surface water that takes the plant effluent? (b) Identify one technical and two nontechnical solutions to reduce future phosphorus loading to the wastewater treatment plant. (c) If 50 percent of the treated wastewater is reclaimed and applied to land for residential and agricultural purposes, how would the current loading of phosphorus to local water change? (Assume all the reclaimed water infiltrates to groundwater.)

7.47 Use the library or Internet to research one dead zone located in the United States and one located overseas, such as the Baltic Sea, northern Adriatic Sea, Yellow Sea, or Gulf of Thailand. Write a two-page report discussing environmental, social, and economic issues associated with the dead zones. What management solutions would you propose to reverse the dead zones?

7.48 Go to the EPA's Region 9 web page devoted to Lake Tahoe (http://www.epa.gov/region9/water/watershed/tahoe/). Lake Tahoe (California and Nevada) can be differentiated into deep water and near shore zones. The EPA reports that "in addition to being a scenic and ecological treasure, the Lake Tahoe Basin is one of the outstanding recreational resources of the United States. The communities and the economy in the Lake Tahoe Basin depend on the protection and restoration of its stunning natural beauty and diverse recreational opportunities in the region." (a) List the states which EPA's Region 9 serves. (b) Lake Tahoe is listed under Clean Water Act Section 303(d) as impaired by input of what three pollutants? (c) What one pollutant has the greatest impact on the lake's deep water quality as measured by water transparency? (d) What two additional pollutants play an important role regarding water quality of the near shore zone?

7.49 The EPA reports that "in addition to being a scenic and ecological treasure, the Lake Tahoe Basin is one of the outstanding recreational resources of the United States. Go to the following web site to obtain Secchi depth data at Lake Tahoe (http://terc.ucdavis.edu/research/SecchiData.pdf). (a) Describe how a Secchi depth measurement is performed. (b) Produce one properly labeled and captioned graph that provides the summer average, winter average, and annual average Secchi depth of Lake Tahoe from 1968 to 2011. Use this figure to answer the question whether water quality in the lake has improved since 1968.

7.50 Go to the web site of the Mississippi River/Gulf of Mexico Watershed Nutrient (Hypoxia) Task Force. (http://water.epa.gov/type/watersheds/named/msbasin/index.cfm). On this web site, the EPA reports that "hypoxia can be caused by a variety of factors, including excess nutrients, primarily nitrogen and phosphorus, and waterbody stratification due to saline or temperature gradients." Under the Hypoxia 101 web site, hypoxic waters have DO concentrations of less than 2–3 mg/L. (http://water.epa.gov/type/watersheds/named/msbasin/hypoxia101.cfm). (a) List the four sources of nutrients that impact this water body.

7.51 (a) What was the size (in square miles) of the 2012 Gulf of Mexico hypoxic zone reported by NOAA scientists on July 27, 2012? (b) Was this larger or smaller than in 2011? (c) What caused the change in the dead zone from 2011 to 2012?

7.52 List the three most common contributors to wetland loss in the United States.

7.53 Using Figure 7.29, estimate the nitrogen loading (lb N/square mile) to a urban watershed if the watershed protects (a) 5 percent of its wetlands and (b) 15 percent of its wetlands.

7.54 Two groundwater wells are located 100 m apart in permeable sand and gravel. The water level in well 1 is 50 m below the surface, and in well 2 the water level is 75 m below the surface. The hydraulic conductivity is 1 m/day, and porosity is 0.60. What is: (a) the Darcy velocity; (b) the true velocity of the groundwater flowing between wells; and (c) the period it takes water to travel between the two wells, in days?

7.55 The hydraulic gradient of groundwater in a certain location is 2 m/100 m. Here, groundwater flows through sand, with a hydraulic conductivity equal to 40 m/day and a porosity of 0.5. An oil spill has caused the pollution of the groundwater in a small region beneath an industrial site. How long would it take the polluted water from that location to reach a drinking-water well located 100 m downgradient? Assume no retardation of the pollutant's movement.

7.56 An underground storage tank has discharged diesel fuel into groundwater. A drinking-water well is located 200 m downgradient from the fuel spill. To ensure the safety of the drinking-water supply, a monitoring well is drilled halfway between the drinking-water well and the fuel spill. The difference in hydraulic head between the drinking-water well and the monitoring well is 40 cm (with the head in the monitoring well higher). If the porosity is 39 percent and hydraulic conductivity is 45 m/day, how long after the contaminated water reaches the monitoring well would it reach the drinking-water well? Assume the pollutants move at the same speed as the groundwater.

7.57 Spills of organic chemicals that contact the ground sometimes reach the groundwater table, where they are then carried downgradient with the groundwater flow. The rate at which they are transported with the groundwater is decreased by sorption to the solids in the groundwater aquifer. The contaminated groundwater can reach wells, which is dangerous if the water is used for drinking. (a) For a soil having ρ_b of 2.3 g/cm^3 and a porosity of 0.3, and for which the percent organic carbon equals 2 percent, determine the retardation factors of trichloroethylene $(\log K_{ow} = 2.42)$, hexachlorobenzene $(\log K_{ow} = 5.80)$, and dichloromethane $(\log K_{ow} = 1.31)$. (b) Which compound would be transported farthest, second farthest, and least far with the groundwater if these chemicals entered the same aquifer?

References

American Society of Civil Engineers (ASCE), 2007. *Gravity Sanitary Sewer Design and Construction*, ASCE MOP #60 and WEF MOP #FD-5. Reston: American Society of Civil Engineers; Alexandria: Water Environment Federation.

American Water Works Association (AWWA), 1998. *Distribution System Requirements for Fire Protection*. AWWA Manual M-31. Denver: AWWA.

American Water Works Association (AWWA), 2007. *Water Resources Planning*, 2nd ed. AWWA Manual M-50. Denver: AWWA.

Budyko, M. I., 1974. *Climate and Life*. New York: Academic Press.

Chen W. F., 1995. *The Civil Engineering Handbook*. Boca Raton: CRC Press, Inc.

Burton and Pitt, 2001. *Stormwater Effects Handbook: A Toolbox for Watershed Managers, Scientists, and Engineers*, CRC Press, Boca Raton.

Crittenden, J. C., R. R. Trussell, D. W. Hand, K. J. Howe, and G. Tchobanoglous, 2005. *Water Treatment: Principles and Design*, 2nd ed. Hoboken: John Wiley & Sons, Inc.

Crook, J., 2004. *Innovative Applications in Water Reuse: Ten Case Studies*. Alexandria: WaterReuse Foundation.

Dahl, T. E., 2011. *Status and Trends of Wetlands in the Conterminous United States 2004 to 2009*. Washington, D.C.: U.S. Department of the Interior; Fish and Wildlife Service, 108 pp.

EPA, 2000. National Water Quality Inventory, 1998 Report to Congress, Groundwater and Drinking Water Chapters, EPA 816-R-00-013.

EPA, 2001. Out Built and Natural Environments: A Technical Review of the Interactions Between Land Use, Transportation, and Environmental Quality, EPA 231-R-002.

EPA, 2012. Groundwater Remediation Technologies EPA Series: A Citizen's Guide to Cleanup Technologies.

Global Water Intelligence (GWI), 2005. Water Reuse Markets 2005–2015: A Global Assessment & Forecast. http://www .globalwaterintel.com, accessed November 18, 2008.

Hammer, M. J., and M. J. Hammer Jr., 1996., *Water and Wastewater Technology*, 3rd ed. Englewood Cliffs: Prentice Hall.

Hoekstra, A. Y., and A. K. Chapagain, 2007. Water footprints of nations: Water use by people as a function of their consumption pattern. *Water Resource Management*, 21: 35–48, DOI 10.1007/s11269-006-9039-x.

Hutson, S. S., N. L. Barber, J. F. Kenny, K. S. Linsey, D. S. Lumia, and M. A. Maupin. 2004. *Estimated Use of Water in the United States in 2000*. U.S. Geological Survey Circular 1268, U.S. Geological Survey, Denver.

Insurance Services Office (ISO), 1998. *Fire Suppression Rating Schedule*. New York: ISO. http://www.iso.com.

Intergovernmental Panel on Climate Change (IPCC), 2007. Summary for policymakers. In: *Climate Change 2007: Impacts, Adaptation and Vulnerability. Contribution of Working Group II to the Fourth Assessment Report of the Intergovernmental Panel on Climate Change*, M. L. Parry, O. F. Canziani, J. P. Palutikof, P. J.van der

Linden, and C. E. Hanson, Eds. Cambridge, UK: Cambridge University Press, 7–22.

Jones, G. M., and R. L. Sanks, 2008. *Pumping Station Design*, 3rd ed. Woburn: Butterworth Heinemann.

Larson, J. 2004. Dead Zones Increasing in the World's Coast Waters. Earth Policy Institute (June 16). http://www.earth-policy. org/plan_b_updates/2004/update41 accessed June 27, 2013.

Loehr, R. C., S. O. Ryding, and W. C. Sonzogni, 1989. Estimating the nutrient load to a waterbody. In: *The Control of Eutrophication of Lakes and Reservoirs*, Volume I, *Man and the Biosphere Series*, S. O. Ryding and W. Rast Eds.Parthenon Publishing Group, Nashville, 115–146.

Mayer, P. W., W. B. DeOreo, E. M. Opitz, J. C. Kiefer, W. Y. Davis, B. Mays, and L. W., 1999. *Water Distribution Systems Handbook*. New York: McGraw-Hill, Inc.

Mihelcic, J. R., 1999. *Fundamentals of Environmental Engineering*. New York: John Wiley & Sons, Inc.

Mihelcic, J. R., E. A. Myre, L. M. Fry, L. D. Phillips, and B. D. Barkdoll, 2009. *Field Guide in Environmental Engineering for Development Workers: Water, Sanitation, Indoor Air*. Reston: American Society of Civil Engineers (ASCE) Press.

Minnesota Pollution Control Agency (MPCA), 1997. Septic Systems and Ground Water Quality. http://www.pca.state.mn. us/water/groundwater/gwmap/septic.pdf, accessed January 15, 2013.

National Research Council (NRC), 2012., *Water Reuse: Potential for Expanding the Nation's Water Supply through Reuse of Municipal Wastewater Committee on the Assessment of Water Reuse as an Approach to Meeting Future Water Supply Needs*. Washington, D.C.: National Academy of Sciences.

Schweitzer, R. W., J. C. Cunningham, and J. R. Mihelcic, 2013. Hydraulic modeling of clay ceramic water filters for point-of-use water treatment, *Environmental Science & Technology*, 47(1): 429–435.

Selman, M., S. Greenhalgh, R. Diaz, and Z. Sugg, 2008. *Eutrophication and hypoxia in coastal areas: A global assessment of the state of knowledge*. WRI Policy Note. Washington, D.C.: World Resources Institute.

Tchobanoglous, G., and F. L.Burton, 1991. *Wastewater Engineering: Treatment, Disposal, and Reuse*, 3rd ed. New York: McGraw-Hill, Inc.

Tchobanoglous, G., F. L. Burton, and H. D. Stensel, 2003. *Wastewater Engineering, Treatment and Reuse*, 4th ed. New York: Wiley Interscience.

Todd, D. K., 1980. *Groundwater Hydrology*, 2nd ed. New York: John Wiley & Sons, Inc.

United Nations Educational, Scientific, and Cultural Organization (UNESCO) and World Water Assessment Programme (WWAP), 2003. *Water for People, Water for Life: The United Nations World Water Development Report*. New York: UNESCO/Berghahn Books.

United Nations Human Settlements Programme (UN-Habitat), 2003. *Water and Sanitation in the World's Cities: Local Action for Global Goals*. London: Earthscan.

U.S. Army Corps of Engineers (USACE), 2008a. FAQs: What You Should Know About the Comprehensive Everglades Restoration Plan (CERP). Comprehensive Everglades Restoration Plan Web site, www.evergladesplan.org, accessed November 18, 2008.

U.S. Army Corps of Engineers (USACE), 2008b. Water Flow Maps of the Everglades: Past, Present & Future. Comprehensive Everglades Restoration Plan Web site, www.evergladesplan.org, accessed September 12, 2013.

U.S. Department of Agriculture (USDA). 2009 Summary Report: 2007 National Resources Inventory, Natural Resources Conservation Service, Washington, DC, and Center for Survey Statistics and Methodology, Iowa State University, Ames, Iowa. 123 pages. http://www.nrcs.usda.gov/technical/NRI/2007/2007_NRI_Summary.pdf, accessed June 27, 2013.

Vickers, A., 2001. *Handbook of Water Use and Conservation*. Denver: American Water Works Association.

Walski, T. M., D. V. Chase, D. A. Savic, W. Grayman, S. Beckith, and E. Koelle. 2003., *Advanced Water Distribution System Modeling*. Waterbury: Haestad Press.

Walski, T. M., T. E. Barnard, E. Harold, L. B. Merritt, N. Walker, and B. E. Whitman, 2004. *Wastewater Collection System Modeling and Design*. Waterbury: Haestad Press.

Water Environment Federation (WEF), 1998. *Design of Municipal Wastewater Treatment Plants*. MOP no. 8, vol. 1, 4th ed. Alexandria: Water Environment Federation.

Wisconsin Department of Transportation, Facilities Development, Manual. http://roadwaystandards.dot.wi.gov/standards/fdm/13-10-005att.pdf#fd13-10a5.2, accessed December 15, 2012.

chapter/Eight Water Treatment

David W. Hand, Qiong Zhang, and James R. Mihelcic

In this chapter, readers will learn about the constituents in untreated water, their concentration, and water quality standards associated with them. Mass balance concepts, stoichiometry, and kinetics are employed to develop expressions describing physical–chemical treatment unit processes used to remove constituents. These unit processes include coagulation and flocculation, sedimentation, granular filtration, disinfection, hardness removal by lime–soda softening, and removal of other dissolved organic and inorganic chemicals by activated carbon and membrane techniques. Readers will also learn about the energy requirements for treating water and other considerations in determining energy requirements over the life cycle of provision of water.

Chapter Contents

Learning Objectives

1. Identify physical, chemical, and biological constituents that exist in untreated water and typical concentration ranges for the major constituents.
2. Match major raw-water constituents with the unit process(es) that remove a significant amount of each constituent.
3. Identify the difference between maximum contaminant level goals (MCLGs) and maximum contaminant levels (MCLs) and relate these values to toxicity data and treatment objectives at a water treatment plant.
4. Develop a drinking-water standard from toxicity data.
5. Relate specific biological pathogens to specific human health impact.
6. Develop viable solutions that include water reuse to address the magnitude of global water provision related to improving human health.
7. Use a systems approach to discuss the relationship between watershed management and land use issues associated with water quality, and the design and performance of water treatment plants.
8. Apply Stokes' law and the concept of overflow rate for design of a sedimentation basin.

9. Size and understand the operation of water treatment unit processes used for coagulation and flocculation, sedimentation, granular filtration, disinfection, hardness removal by lime–soda softening, and removal of dissolved organic and inorganic chemicals by activated carbon and membrane techniques.

10. Write the chemical equation(s) to describe the application of different disinfectants and differentiate between residual chlorine and the three chemical species that make up combined chlorine.

11. Demonstrate an in-depth understanding of disinfection processes for different disinfectants used in a developed and developing world setting.

12. Differentiate between the different types of membrane filters and the size and type of constituents they are designed to remove.

13. Identify the magnitude of energy use during plant operation for different unit processes while also being aware of energy considerations associated with the life cycle of particular unit processes.

8.1 Introduction

Freshwater is a finite resource, and readily accessible supplies are becoming less abundant. With water scarcity a reality in many parts of the world, increases in population and income along with the impacts of climate change are expected to further exacerbate this issue. Achieving sustainable solutions is compounded by the energy demands of obtaining, storing, and producing a safe water supply. Approximately 1.4 kWh of energy are needed to collect and treat 1,000 gallons of surface water, and 1.8 kWh are needed to collect and treat a similar volume of groundwater (Burton, 1996; Elliot et al., 2003). For water treatment, most of this requirement is from pumping either raw or treated water or some concentrated waste stream. There is, of course, energy associated with manufacturing and delivering materials and chemicals used during pumping and treatment. Therefore, the distances to a water source and its quality could have a large energy implication from a life cycle perspective (Mo et al., 2011).

And as society develops less desirable sources of water to meet increasing demand, the amount of embodied energy in our water supply is expected to increase. Consequently, there is a need to develop an integrated systems approach to water management strategies (for example, sustainable watershed management, water conservation, and water reuse practices) to meet the global demand for safe drinking water. In terms of reuse, approximately 12 billion gallons of municipal effluent is discharged daily to an ocean or estuary (out of 32 billion gallons discharged every day in the United States). Much of this discharge still contains nutrients that harm coastal ecosystems. If society would only reuse these coastal discharges, we could augment 6 percent of the estimated total water use in the United States, which is equivalent to 27 percent of public water supply (NRC, 2012).

The purpose of **water treatment** is to provide potable water that is palatable. **Potable water** refers to water that is healthy for human consumption and free of harmful microorganisms and organic and inorganic compounds that either cause adverse physiological effects or do not taste good. **Palatable** describes water that is *aesthetically* acceptable to drink or free from turbidity, color, odor, and objectionable taste. Water that is palatable may not be safe.

In developed countries, water is treated to be both potable and palatable. However, some people do not like the palatability of municipal waters, and this has given rise to the increased use of household point-of-use treatment systems and bottled water. Bottled water has added an additional layer of embodied energy to water, because petroleum is used to produce the water container, and there are recycling or disposal costs for the bottles during their end-of-life life stage.

Table / 8.1

Global Definitions of Improved and Unimproved Water Supplies

Improved Water Supplies	Unimproved Water Supplies
Household connections	Unprotected well
Public standpipes	Unprotected spring
Boreholes	Vendor-provided water
Protected dug wells	Bottled water
Protected springs Rainwater collection	Tanker truck–provided water

For those working on engineering projects in the developing world, a water supply can be improved with many types of projects and appropriate technology, from protecting a water source to building a distribution system. Table 8.1 describes how the World Health Organization (WHO) defines *unimproved* and *improved water supplies*. Bottled water is considered unimproved because of possible problems of sufficient quantity, not quality).

8.2   Characteristics of Untreated Water

Most consumers expect drinking water to be clear, colorless, odorless, and free of harmful chemicals and pathogenic microorganisms. Natural waters usually contain some degree of dissolved, particulate, and microbiological constituents, which are obtained from the surrounding environment. Table 8.2 summarizes many of the important chemical and biological constituents found in water. Chapter 7 discussed how precipitation and land use impact the quantity and quality of runoff that enters a watershed. Figure 8.1 shows specifically how several

Table / 8.2

Concentration of Major Constituents Found in Water

General Classification	Specific Constituents	Typical Concentration Range
Major inorganic constituents	Calcium (Ca^{2+}), chloride (Cl^-), fluoride (F^-), iron (Fe^{2+}), manganese (Mn^{2+}), nitrate (NO_3^-), sodium (Na^+), sulfur (SO_4^{2-}, HS^-)	1–1,000 mg/L
Minor inorganic constituents	Cadmium, chromium, copper, lead, mercury, nickel, zinc, arsenic	0.1–10 µg/L
Naturally occurring organic compounds	Naturally occurring organic matter (NOM) that is measured as total organic carbon (TOC)	0.1–20 mg/L
Anthropogenic organic constituents	Synthetic organic chemicals (SOCs) and emerging chemicals of concern used in industry, households, and agriculture (e.g., benzene, methyl tert-butyl ether, tetrachloroethylene, trichloroethylene, vinyl chloride, alachlor)	Below 1 µg/L and up to the tens of mg/L
Living organisms	Bacteria, algae, viruses	Millions

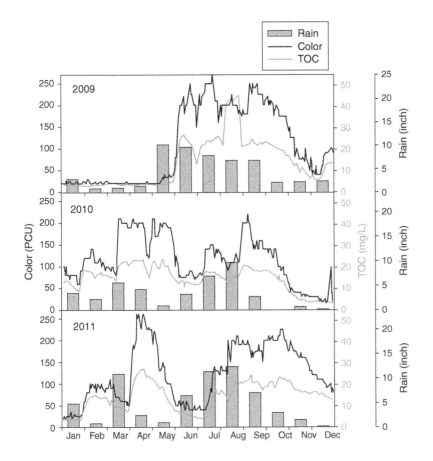

Figure / 8.1 Seasonal Differences in TOC and Color of Hillsborough River Raw Water That Serves the City of Tampa (FL) Water Treatment Plant (2009–2011) The water quality fluctuates largely over the course of the year, impacting the treatment process in many ways. The river's source water is primarily the Green Swamp located in Central Florida. During the wet season (June–September), TOC and color concentrations spike from the large amount of organic matter flushed out of the swamp and river tributaries by heavy rains.

(Courtesy of Dustin Bales, 2012; with permission).

important water quality constituents (specifically TOC and color) change over the season because of seasonal changes in precipitation.

Natural geologic weathering processes can impart dissolved inorganic ions into water, and these can cause problems related to color, hardness, taste, odor, and health. Dissolved organic matter in water, which is derived from decaying vegetation, can impart a yellowish or brownish color to the water. The surrounding terrestrial environment can cause small or colloidal clay particles to be suspended in the water, which impart a turbid or cloudy appearance of the water. Naturally occurring microorganisms such as bacteria, viruses, and protozoa can make their way into natural waters and cause health issues. Synthetic organic chemicals (SOCs) can be released into the environment and cause chronic or acute health problems to humans and aquatic life. Consequently, a water's physical, chemical, and microbiological characteristics need to be considered in the design and operation of a supply and treatment system.

8.2.1 PHYSICAL CHARACTERISTICS

Several aggregate physical characteristics of natural water (also referred to as *raw* or *untreated water*) are used to quantify the appearance or aesthetics of the water. These parameters, described in Table 8.3, are turbidity, number and type of particles, color, taste and odor, and temperature.

Test Your Knowledge of Water-Using Behaviors and Water-Saving Opportunities
http://epa.gov/watersense/
test_your_watersense.html

Drinking Water at the World Health Organization
http://www.who.int/topics/
drinking_water/en/

Table / 8.3

Physical Characteristics of Natural Water

Turbidity	Turbidity measures the optical clarity of water. It is caused by the scattering and absorbance of light by suspended particles in the water. A turbidimeter is used to measure the interference of the light passage through the water. Turbidity is reported in terms of **nephelometric turbidity units (NTU)**. The World Health Organization reports that a turbidity of <5 NTU is usually acceptable but may vary depending upon the availability and resources for treatment. In the United States, many water utilities aim to treat the water to <0.1 NTU.
Particles	Particles in natural waters are solids larger than molecules but generally not distinguishable by the unaided eye. They may adsorb toxic metals or synthetic organic chemicals. Water treatment considers particles in the size range 0.001–100 μm. Particles larger than 1 μm are called **suspended solids**, while particles between about 0.001 and 1 μm can be considered **colloidal particles** (though some researchers go as low as 0.0001 μm). Constituents smaller than 0.001 μm are called **dissolved particles**. **Natural organic matter (NOM)** comprises colloidal particles and **dissolved organic carbon (DOC)**. The DOC is the portion of NOM that can be filtered through a 0.45 μm filter. It is not classified in terms of size.
Color	Color is imparted to water by dissolved organic matter, natural metallic ions such as iron and manganese, and turbidity. Most people can detect color at more than 15 true color units for water in a glass.
Taste and odor	Taste and odor can originate from dissolved natural organic or inorganic constituents and biological sources present in raw waters. They can also be an outcome of the water treatment process.
Temperature	Surface water temperatures may vary from 0.5°C to 3°C in the winter and 23°C to 27°C in the summer. Groundwater can vary from 2.0°C to 25°C depending upon location and well depth.

Turbidity measurements of natural waters vary depending upon the water source. Low turbidity measurements (less than 1 NTU) are typical for most groundwater sources, while surface water turbidity varies depending upon the source. In lakes and reservoirs, turbidity is usually stable and ranges from 1 to 20 NTU, but some waters can vary seasonally due to turnover, storms, and algal activity. Turbidity in rivers is highly dependent on precipitation events and can range from less than 10 NTU to more than 4,000 NTU. Because climate change is expected to change weather events in some parts of the world (including the United States), the resulting runoff and erosion may result in decreased or increased seasonal turbidity of some raw-water supplies. We also discussed in the previous chapter how land use impacts the quality of surface runoff.

Turbidity measurements are primarily used for process control, regulatory compliance, and comparison of different water sources. They are also used as an indicator of increased concentrations of microbial water constituents, such as bacteria, *Cryptosporidium* oocysts, and *Giardia* cysts.

Particles found in natural waters can be measured in terms of their numbers and size. Particle counters can measure the number of suspended particles in size ranges generally from 1.0 to 60 μm. Particle removal is important because it has been suggested as an indicator of removal of *Giardia* and *Cryptosporidium* cysts from water (LeChevallier and Norton, 1995). Accordingly, many treatment facilities employ online particle counters to evaluate process performance and to aid in process-control decisions.

Suspended particles such as algae, organic debris, protozoa cysts, and silt can be removed by conventional sedimentation and depth filtration methods. Coagulation and flocculation processes can remove colloidal particles. However, many dissolved constituents will remain in solution, such as the lower-molecular-weight natural organic matter (NOM) (for example, humic and fulvic acids) and synthetic organic compounds. Other treatment methods, such as activated carbon adsorption and reverse osmosis, may be used to remove these constituents.

Color is categorized as apparent or true color. *Apparent color* is measured on unfiltered samples, so it includes the color imparted by turbidity. *True color* is measured on a water sample passed through a 40 μm filter, so it is a measure of the color imparted by dissolved constituents. While color is not a regulated health concern, it can be an aesthetic problem for some individuals and communities, and treatment is usually provided.

Taste and odor threshold concentrations have been established as guidelines for determining when constituents can be detected. The most prevalent natural odor-causing compounds in surface waters come from the decay of algae (for example, geosmin and methyl isoborneal, which impart a musty odor at concentrations as low as 0.000005 mg/L). Water treated with excess chlorine will have a chlorine odor, which can be detected as low as 0.010 mg/L. Groundwaters that have a low redox potential may contain a dissolved gas such as hydrogen sulfide, which smells like rotten eggs. Waters containing dissolved inorganic compounds such as iron, manganese, and copper may have a metallic taste. The taste of reduced iron (Fe^{2+}) can be detected at 0.04–0.01 mg/L and the taste of reduced manganese (Mn^{2+}) can be detected at 0.4–30 mg/L. Some natural or synthetic organic compounds will impart an objectionable taste to water. Examples of these include phenol, which can be detected at a concentration of 1 mg/L.

Water **temperature** is very important because it affects many physical and chemical parameters of water, such as density, viscosity, vapor pressure, surface tension, solubility, and reaction rates, which are used in the design and operation of a treatment plant and associated conveyance system.

8.2.2 MAJOR AND MINOR INORGANIC CONSTITUENTS

Table 8.4 summarizes the major dissolved inorganic constituents found in water. Calcium is one of the most abundant cations found in water and is the major constituent of water **hardness** (along with magnesium). Calcium concentrations greater than about 60 mg/L are considered a nuisance by some. Chloride (Cl^-) concentrations in terrestrial waters vary from 1 to 250 mg/L depending on the location, and typical surface water is usually less than 10 mg/L Cl^-. However, waters affected by saltwater intrusion and groundwater that contains trapped brine may have chloride concentrations similar to the ocean. Fluoride exists in natural waters primarily as the anion F^- but can also be

Table / 8.4

Major Dissolved Constituents Found in Water

Constituent	Source	Problem in Water Supply	Range in Natural Waters
Calcium and magnesium	Surface water and groundwater	Above 60 mg/L can be considered nuisance as hardness.	For calcium, less than 1 mg/L to more than 500 mg/L. Surface water concentrations of magnesium are less than 10 mg/L up to 20 mg/L. Groundwater concentrations are less than 30 mg/L up to 40 mg/L.
Chloride	Surface water and groundwater; saltwater intrusion	Above 250 mg/L can impart salty taste. Below 50 mg/L can be corrosive to some metals.	Typical surface water is usually less than 10 mg/L.
Fluoride	Surface water and groundwater. Some water utilities add fluoride in the form of sodium fluoride or hydrofluorosilic acid at doses of about 1.0 mg/L	Toxic to humans at concentrations of 250–450 mg/L; fatal at concentrations above 4.0 g/L.	For surface water with total dissolved solids (TDS) concentrations less than 1,000 mg/L, fluoride is usually less than 1.0 mg/L.
Iron and manganese	Surface water and groundwater	Taste threshold of iron for many consumers is around 0.01 mg/L. Iron can impart a brownish color to laundry and bathroom fixtures. Manganese ion can impart a dark brown color. At concentrations around 0.4 mg/L, manganese can impart an unpleasant taste to water and can stain laundry and fixtures.	In oxygenated surface waters, the concentration of total iron is usually less than 0.5 mg/L. In groundwater that has low bicarbonate and dissolved oxygen, iron concentrations can range from 1.0 to 10.0 mg/L. The concentration of manganese ion in surface water and groundwater may be less than 1.0 mg/L.
Nitrate	Surface water and groundwater can contain high concentrations of nitrate from runoff from fertilizers found in urban and agricultural watersheds.	Very high nitrate concentrations may produce infant methemoglobinemia.	
Sulfur	Surface water and groundwater	Groundwater low in dissolved oxygen can contain reduced sulfur compounds, which impart objectionable odors such as that of rotten eggs. Sulfates are also corrosive in concrete structures and pipes.	Sulfate concentrations in freshwater can approach 10 mg/L.

associated with ferric ion, aluminum, and beryllium. Some water utilities add fluoride in the form of sodium fluoride or hydrofluorosilic acid at concentrations of about 1.0 mg/L.

Iron is abundant in geological formations and is frequently found in water. If not removed, it can impart a brownish color to laundry and bathroom fixtures. At concentrations around 0.2–0.4 mg/L, manganese can impart an unpleasant taste to the water and stain laundry and fixtures.

Surface water can contain high concentrations of nitrate (and other forms of nitrogen) from urban and agricultural runoff. Groundwater can also contain high concentrations of nitrate, especially in agricultural areas, where ammonia fertilizers are biochemically converted to nitrate in the soil or areas impacted by on-site treatment such as faulty septic tanks. Nitrate is regulated because high concentrations may produce infant methemoglobinemia.

Sulfur can occur as sulfates ($CaSO_4$, Na_2SO_4, $MgSO_4$) and reduced sulfides (H_2S, HS^-). Sulfides can be found in water where there is significant organic decomposition that results in anoxic conditions. Groundwater low in dissolved oxygen can contain reduced sulfur that imparts objectionable odors. Sulfates are also corrosive in concrete structures and pipes.

Several minor inorganic constituents are sometimes a significant health concern or diminish water quality. Examples include copper, chromium, nickel, mercury, strontium, and zinc. Some of these constituents are the result of the surrounding natural environment, while others are present due to human activities. For example, some industrial sites that use arsenic as a wood preservative have contaminated water supplies, while naturally occurring arsenic is widespread throughout much of the world. In this latter case, arsenic is found mostly as a solid in the mineral form. However, it can be found dissolved in groundwater in the form of arsenite (H_3AsO_3) and arsenate ($H_2AsO_4^-$, $HAsO_4^{2-}$) species. In contrast, lead contamination is usually associated with human activities that even include leaching from old distribution systems.

Lead in Drinking Water
http://www.epa.gov/safewater/lead

Global Arsenic Crisis
http://www.who.int/topics/arsenic/en/

Application / 8.2 The Global Arsenic Crisis

As Table 8.5 shows, naturally occurring arsenic is widespread in water supplies throughout the world. Unfortunately, when humans are exposed over long periods to low concentrations of arsenic from consuming contaminated water, several forms of cancer can develop. The World Health Organization has

Table / 8.5

Global Locations Where Naturally Occurring Arsenic Has Been Detected in Drinking-Water Supplies

Region	Specific Countries
Asia	Bangladesh, Cambodia, China, India, Iran, Japan, Myanmar, Nepal, Pakistan, Thailand, Vietnam
Americas	Argentina, Chile, Dominica, El Salvador, Honduras, Mexico, Nicaragua, Peru, United States
Europe	Austria, Croatia, Finland, France, Germany, Greece, Hungary, Italy, Romania, Russia, Serbia, United Kingdom
Africa	Ghana, South Africa, Zimbabwe
Pacific	Australia, New Zealand

SOURCE: Petrusevski et al., 2007.

accordingly set a drinking-water guideline for arsenic of 10 μg/L (10 ppb).

The magnitude of the problem is most serious in Bangladesh and West Bengal (India). In the 1970s and 1980s, 4 million hand-pump wells were installed in Bangladesh and India to provide people there with a pathogen-free drinking-water supply. The presence of arsenicosis began to appear in the 1980s, shortly after the well installation program. By the early 1990s, it was determined that the arsenic poisoning was originating from these wells. The arsenic is naturally occurring.

Today it is estimated that, every day in Bangladesh, up to 57 million people are exposed to arsenic concentrations greater than 10 μg/L. In West Bengal, an estimated 6 million people are exposed to arsenic concentrations between 50 and 3,200 μg/L. The magnitude of the problem shows why some have called this the greatest mass poisoning of humans that has ever occurred.

The most commonly used arsenic removal systems in both the developed and developing world are based on coagulation–separation and adsorption processes. Membrane filtration (such as reverse osmosis and nanofiltration) is also effective at removing arsenic from water; however, it is not practical in much of the world because of the high costs involved. Accordingly, appropriate technologies have been developed to treat this water. Figure 8.2 shows one such technology.

This unit is installed directly at the hand-pumped wells that were installed in the 1970s and 1980s. It

requires no electricity or chemical addition. The unit is packed with granular activated alumina, which removes the arsenic from the water. The unit can be regenerated with caustic soda about every 4 months. The community is instructed to dispose of the arsenic-laden sludge in a pit lined with bricks. After 10 years of typical operation, it is estimated that the volume of sludge generated will occupy 56 ft^3.

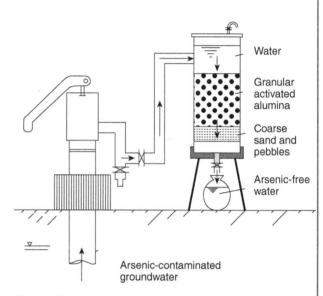

Figure / 8.2   Well Head Arsenic Removal Unit Developed by Dr. Arup Sangupta and Others at Lehigh University.

8.2.3 MAJOR ORGANIC CONSTITUENTS

Organic constituents found in water can either be naturally occurring or associated with human activities. Natural organic matter (NOM) in water is the result of the complexation of soluble organic material derived from biochemical degradation of vegetation in the surrounding environment. NOM occurs in all waters and is measured as total organic carbon (TOC). Typical TOC concentrations in natural waters range from less than 0.1 to 2.0 mg/L in groundwater, 1.0–20 mg/L in surface waters, and 0.5–5.0 mg/L in seawater. Table 8.6 summarizes the impact NOM can have on drinking-water treatment processes.

Table / 8.6

Effect of Natural Organic Matter (NOM) on Water Treatment Processes

Water Treatment Process	Effect
Disinfection	NOM reacts with, and consumes, disinfectants, which increase required dose to achieve effective disinfection.
Coagulation	NOM reacts with, and consumes, coagulants, which increase required dose to achieve effective turbidity removal.
Adsorption	NOM adsorbs to activated carbon, which depletes adsorption capacity of the carbon.
Membranes	NOM adsorbs to membranes, clogging membrane pores, and fouling surfaces. This leads to decline in water passed through the membrane.
Distribution system	NOM may lead to corrosion and slime growth in distribution systems (especially when oxidants are used during treatment).

SOURCE: Adapted from Crittenden et al., 2012.

Anthropogenic organic constituents found in water are associated with industrial activity, land use by agriculture, urban runoff, and municipal effluents from wastewater treatment plants. Most of these organic contaminants are classified as **synthetic organic chemicals (SOCs)**. Representative SOCs are found in fuels, cleaning solvents, chemical feedstocks, and herbicides and pesticides. **Emerging chemicals of concern** are now found in water and wastewater from the use of personal-care and pharmaceutical products.

8.2.4 MICROBIAL CONSTITUENTS

Potable water must be free from pathogenic microorganisms. As just one example of the global magnitude of the problem, the WHO reports that diarrhea contributed 4.7 percent of the global burden of disease in 2011. Of that 4.7 percent, approximately 88 percent was caused by unsafe water, sanitation, and hygiene.

Pathogens are microorganisms that cause sickness and disease. Pathogens include many classes of microorganisms, among them viruses, bacteria, protozoa, and helminths.

Details about some representative pathogenic organisms found in untreated water and associated health effects are provided in Table 8.7. Note the small size of pathogens that can prevent their removal via conventional gravity sedimentation processes. Because there are many different water-based pathogens, monitoring and detecting all of them would require a prohibitive amount of resources. Consequently, **indicator organisms** (such as **coliforms**) have been identified and are used to monitor the microbial water quality.

At present, the EPA requires water utilities to monitor their water distribution system monthly for **total coliforms**. The total coliform rule maximum contaminant level is based on frequency of detection (no

Pharmaceuticals and Personal-Care Products
http://epa.gov/ppcp/faq.html

How's Your Local Drinking Water?
http://water.epa.gov/drink/local/index.cfm

Table / 8.7

Representative Pathogenic Organisms in Raw-Water Supplies

Pathogen(s)	Type	Health Effects in Healthy Persons	Normal Habitat
Vibrio cholera Shape: wormlike Size: 0.5 by 1–2 μm	Bacteria	Classic cholerae—explosive diarrhea and vomiting without fever followed by dehydration; abnormally low blood pressure and temperature; muscle cramps; shock; coma followed by death	Human stomach and intestines
Salmonella (several species) Shape: rod Size: 0.6 μm	Bacteria	*S. typhi* species causes enteric fever, headaches, malaise, and abdominal pain	Intestines of warm-blooded animals
Shigella dysenteriae Shape: round Size: 0.4 μm	Bacteria	Bacillary dysentery: abdominal pain, cramps, diarrhea, fever, vomiting, blood, and mucus in stools	Human stomach and intestines
Escherichia coli Shape: rod Size: 0.3–0.5 by 1–2 μm	Bacteria	Diarrhea	Intestines of warm-blooded animals
Poliovirus types 1, 2, 3 Shape: round Size: 28–30 nm	Virus	Fever, severe headache, stiff neck and back, deep muscle pain, and skin sensitivity	Human intestinal tract
Human adenovirus type 2 Shape: 12 vertices Size: 70–90 nm	Virus	Severe infections in lungs, eyes, urinary tract, genitals; some strains affect intestines	Human intestinal tract
Rotavirus A Shape: round Size: 80 nm	Virus	Severe diarrhea and dehydration	Human intestinal tract
Cryptosporidium parvum Type 1 oocyst Shape: ellipsodial Size: 3–5 μm Sporozoite and merozoites Shape: wormlike Size: 10 by 1.5 μm	Protozoa	Severe diarrhea, abdominal pain, nausea or vomiting, and fever	Human intestinal tract
Giardia lamblia Shape: single-celled flagellated protozoa Size: 9–15 μm long, 5–15 μm wide, 2–4 μm thick	Protozoa	Sudden diarrhea, abdominal cramps, bloating, cramps, and weight loss	Human intestinal tract
Schistosoma haematobium A wormlike organism	Helminthic	Squamous cell carcinoma of the bladder; urolithiasis; ascending urinary tract infection; urethral and ureteral stricture with subsequent hydronephrosis; renal failure	Blood vessels of the human bladder and in mammals

more than 5 percent for systems collecting at least 40 samples per month) or the combination of a positive *Escherichia coli* sample (or fecal coliforms) with a positive total coliform sample. While the total coliform test can provide a good indication of fecal contamination, it cannot prove that the source water is safe. Other methods must be used to confirm the absence of longer-surviving organisms such as viruses and spores.

8.3 Water Quality Standards

The Safe Drinking Water Act (SDWA) (Public Law 93–523) was passed by the U.S. Congress in 1974 and amended in 1986 and 1996 to protect public health by regulating the public water supply. Under this law, the primary responsibility of setting the water quality regulations was moved from the states to the EPA. To protect public health, EPA established primary drinking-water standards by setting health-based **maximum contaminant level goals (MCLGs)** and **maximum contaminant levels (MCLs)** for a large number of pollutants.

The MCL is the enforceable standard and is based not only on health and risk assessment information, but also on costs and the availability of technology. The MCLGs are based solely on health and risk assessment information.

Table 8.8 provides the MCLGs and MCLs for several important chemicals found in water supplies and their potential health effects. A similar set of MCLs has been established by the World Health Organization (see www.who.org).

Safe Drinking Water Act
http://www.epa.gov/safewater/sdwa/index.html

Interactive Presentation to Understand the Safe Drinking Water Act
http://water.epa.gov/learn/training/dwatraining/training.cfm

Safe Drinking Water Hotline 1-800-426-4791
Information about drinking-water and groundwater programs authorized under the SDWA.

Class Discussion
At present, EPA has developed MCLs for over 90 contaminants, even though there are tens of thousands of commonly used chemicals in commerce. It has thus been a daunting task for government regulators to keep pace with the introduction of new chemicals into commerce and prevent or minimize public exposure to them. How would more widespread implementation of green chemistry and green engineering (discussed in Chapters 3 and 6) that are intended to reduce chemical hazard affect issues of regulation and treatment?

Table / 8.8

Representative Chemicals Found in Water and Their Maximum Contaminant Level Goals (MCLGs) and Maximum Contaminant Levels (MCLs)   MCLGs are based solely on health and risk assessment information. MCLs are based not only on health and risk assessment information, but also on costs and the availability of technology.

Chemical	Maximum Contaminant Level Goal (mg/L)	Maximum Contaminant Level (mg/L)
Synthetic Organic Chemicals (SOCs)		
2,4-D (2,4-dichlorophenoxyacetic acid)	0.07	0.07
Alachor	0	0.002
Atrazine	0.003	0.003
Benzo(*a*)pyrene	0	0.0002
Chlordane	0	0.002
Lindane	0.0002	0.0002
Polychlorinated biphenyl (PCB)	0	0.0005
Volatile Organic Chemicals (VOCs)		
1,1-Dichloroethylene	0.007	0.007
1,1,1-Trichloroethane	0.2	0.2

(continued)

(continued)

1,2-Dichloroethane	0	0.005
Benzene	0	0.005
Carbon tetrachloride	0	0.005
cis-1,2-Dichloroethylene	0.07	0.07
Dichloromethane	0	0.005
Ethyl benzene	0.7	0.7
Toluene	1	1
Tetrachloroethylene	0	0.005
Trichloroethylene	0	0.005
Vinyl chloride	0	0.002
Xylenes (total)	10	10
Inorganics		
Arsenic	0	0.01
Cadmium	0.005	0.005
Chromium (total)	0.1	0.1
Cyanide	0.2	0.2
Fluoride	4	4
Lead (at tap)	0	0.015 (action level)*
Mercury	0.002	0.002
Nitrate (as N)	10	10
Nitrite (as N)	1	1

*The concentration of a contaminant which, if exceeded, triggers treatment or other requirements which a water system must follow.
SOURCE: EPA 2012 Edition of the Drinking Water Standards and Health Advisories.

Application / 8.3   Developing a Drinking-Water Standard from Toxicity Data

In this example we demonstrate one way EPA develops a drinking-water standard from toxicity information. In this case we will use EPA's decision to regulate the noncarcinogen perchlorate (ClO_4^-). EPA reports that perchlorate (ClO_4^-) is both a naturally occurring and manmade chemical. It is used to produce rocket fuel, fireworks, flares, and explosives. It can also be found in bleach and some fertilizers. Monitoring data shows that >4 percent of public water systems have detected perchlorate, and 5–17 million people may be exposed to drinking water containing perchlorate.

Perchlorate is an important issue when protecting human health, because scientific research indicates it can disrupt the thyroid's ability to produce important developmental hormones that are critical for normal brain development and growth of fetuses, infants, and young children. To give you an idea of the effort involved in developing a

drinking-water standard for one chemical, in this process EPA considered input from their Science Advisory Board and almost 39,000 public commenters since 2007.

EPA decided to regulate perchlorate under the SDWA (see http://water.epa.gov/drink/contaminants/unregulated/perchlorate.cfm). In the process of developing a drinking-water standard, EPA first develops a primary drinking-water standard for perchlorate by setting a health-based MCLG. After this is accomplished, they set the enforceable MCL.

In this case, the drinking-water standard is related to toxicity information provided by the **reference dose (RfD)**. In Chapter 6, we defined the RfD as an estimate, with uncertainty spanning perhaps an order of magnitude, of a daily oral exposure to the human population (including sensitive subgroups) that is likely to be without an appreciable risk of deleterious effects during a lifetime. The RfD for perchlorate can be obtained from the data available in the Integrated Risk Information System (IRIS) (http://www.epa.gov/IRIS/). IRIS reports the RfD for perchlorate is 0.7 μg per kg of body weight per day (0.7 μg/kg/day). This RfD is based on the No Observed Effect Level (NOEL) of 7 μg/kg/day and application of an uncertainty factor (UF) of 10 to account for differences in sensitivity between the healthy adults and the most sensitive population, that is, fetuses of pregnant women who might have hypothyroidism or iodide 21 deficient.

So how does EPA go from the RfD to a drinking-water standard? The health-based MCLGs EPA will set for drinking water can be determined as:

$$\text{MCLG } (\mu g/L) = \frac{\text{RfD} \left(\frac{\mu g}{kg - day} \right) \times \text{body weight (kg)}}{\text{drinking water intake} \left(\frac{L}{day} \right)} \times \text{RSC}$$

In this equation, EPA assumes the default values for body weight (70 kg) and drinking-water ingestion rate (2 L/day). The RSC is the relative source contribution and is the percentage of the RfD that remains for drinking water after other sources of perchlorate exposure occurs. Remember that you could intake perchlorate through other routes of exposure besides drinking water. The RSC is based on studies performed by the Food and Drug Administration (FDA) and in this case, EPA is proposing to use an RSC of 62% (so the RSC = 0.62) for a pregnant woman. This means that 62% of a pregnant woman's exposure to perchlorate would be through drinking water.

If you place these values into the above equation, you will obtain an MCLG for perchlorate of 15 μg/L (15 ppb) in drinking water.

Remember from our reading in this chapter that the MCLG is a nonenforceable goal defined under the SDWA as "the level at which no known or anticipated adverse effects on the health of persons occur and which allows an adequate margin of safety." The SDWA specifies that the enforceable MCL be set as close to the MCLG as feasible using the best available technology, treatment techniques, and other means (considering cost). This process was not yet finalized as we were writing this book, but it does show the MCL that EPA will eventually set for perchlorate will be ≥15 μg/L.

Information and text courtesy of U.S. Environmental Protection Agency.

8.4   Overview of Water Treatment Processes

The typical *unit processes* used for the treatment of surface water and brackish waters are shown in Figure 8.3. Table 8.9 summarizes the unit processes associated with significant removal of particular water constituents. Treatment of surface waters (Figure 8.3a) mostly requires the removal of particulate matter and pathogens. Removing particles also

Table / 8.9

Unit Processes That Remove a Significant Amount of Raw-Water Constituents

Constituent	Unit Process(es)
Turbidity and particles	Coagulation/flocculation, sedimentation, granular filtration
Major dissolved inorganics	Softening, aeration, membranes
Minor dissolved inorganics	Membranes
Pathogens	Sedimentation, filtration, disinfection
Major dissolved organics	Membranes, adsorption

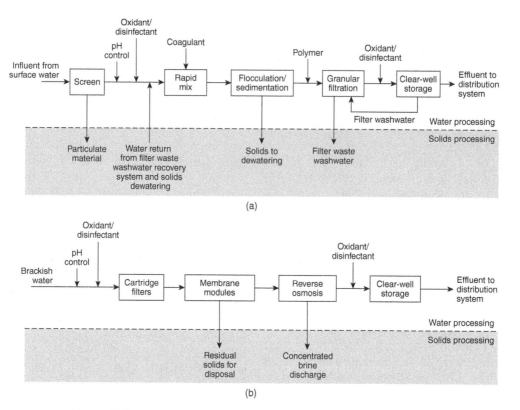

Figure / 8.3 **Typical Water Treatment Unit Processes and Their Arrangement** These processes are typically used for: (a) treatment of surface water and (b) treatment of water with high levels of dissolved constituents.

(From Crittenden et al. (2012). Redrawn with permission of John Wiley & Sons, Inc.).

| **Application / 8.4** | **Visualizing the Water Treatment Process** |

These Internet videos will allow you to tour water treatment plants and better visualize many of the unit processes described throughout this chapter.

Video 1 (5:26): Produced for the Beaufort Jasper Water Authority, this video tours a 39 MGD treatment plant located in South Carolina that extracts and treats water from the Savannah River.
www.youtube.com/watch?v=0bXIqS5NcRY

Video 2 (3:20): Covers the basic processes of treating drinking water.
www.youtube.com/watch?v=9z14el51ISwg&NR=1

Video 3 (5:26): Covers pretreatment and sedimentation (developed to assist water treatment operators obtain certification).
www.youtube.com/watch?v=uqrDdFDUWE8& feature=channel_video_title

Video 4 (4:41): Covers coagulation, flocculation, and sedimentation (developed to assist water treatment operators obtain certification).
www.youtube.com/watch?v=oQOG5ymU4Yg& feature=relmfu

Video 5: Take a virtual tour of a drinking-water treatment plant.
http://water.epa.gov/drink/tour/

assists in pathogen removal, because most pathogens either are particles or are associated with particles. Application 8.5 and Figure 10.4 show the much greater complexity of a specific water treatment plant.

If the water source contains dissolved constituents, then additional unit processes can be added to remove them as well. Membrane processes are now widely used for drinking-water treatment, and a schematic diagram for the treatment of brackish water is shown in Figure 8.3b. This treatment process will become more important in the future as increases in population and demand, along with climate change, force society to search for waters of poorer quality. These sources are sometimes high in TDS and are present as brackish groundwater, seawater, and **reclaimed water**. Reclaimed water is now being considered in many water-scarce areas to provide an alternative source of water. In terms of water treatment, one issue is whether the water should be treated for potable or not potable applications, such as irrigation of residential yards, agriculture, or public green space. These nonpotable uses can take advantage of the valuable nutrients that engineers can leave in reclaimed water.

Application / 8.5   The Real Thing: The Treatment Process at the David L. Tippin Water Treatment Facility

Drinking-water treatment processes depend on a variety of factors, including seasonal factors (remember Figure 8.1) and whether the source is surface water, groundwater, or reclaimed water. Here we discuss a real treatment process for surface water that serves the City of Tampa (FL) so a reader can understand the complexity of water treatment. The David L. Tippin Water Treatment Facility treats surface water from the Hillsborough River Reservoir (Figure 8.1). The local water management district works to protect water quantity and quality through management of the watershed. Some of this includes managing how much groundwater (which provides baseflow) can be withdrawn farther up in the watershed and through managing how land is used (or protected) within the watershed.

The treatment plant is permitted for a maximum flow of 120 MGD and serves approximately 600,000 people. Figure 8.4 shows the treatment process, which consists of coagulation and flocculation, advanced oxidation by ozone, biofiltration, and disinfection by chloramination.

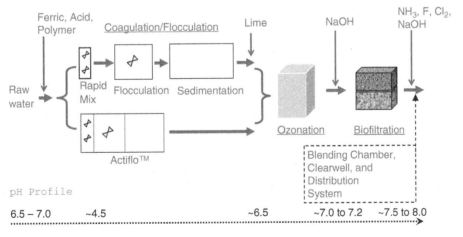

Figure / 8.4   The Current Water Treatment Process at the David L. Tippin Water Treatment Facility (Tampa, FL).

(Courtesy of Dustin Bales, 2012; with permission).

The process begins by pumping surface water into parallel rapid mix and Actiflo™ systems. As the water travels between the water intake and the beginning of the coagulation/flocculation processes, sulfuric acid, polymer, and ferric sulfate are added to the water. Actiflo™ is a high-rate settling and coagulation/flocculation process that uses sand to promote flocculation. The resulting flocs from the two coagulation/flocculation processes settle by gravity in sedimentation basins.

Following this the pH is raised to between 6 and 6.5 by adding lime ($Ca(OH)_2$) prior to ozonation. Ozonation consists of an eight chambered contactor with weirs located between each chamber controlling flow conditions. Between 0.5 and 3 ppm of ozone is diffused into the water in the first two chambers. Any remaining ozone at the end of the chambers is quenched by the addition of hydrogen peroxide (H_2O_2). Caustic (NaOH) is then used to raise the pH to between 7.2 and 8 prior to biofiltration. Biofiltration consists of a 24 in layer of granular activated carbon (GAC) on top of 12 in of sand. Microbes reside in the GAC and increase removal of turbidity and low-molecular-weight organic material that may contribute to formation of biofilms in the water distribution system. Manganese and iron are also removed by the filters.

After filtration, water is directed to a blending chamber. Here chlorine (dosed from chlorine gas dissolved into a sidestream) is added first, followed by anhydrous ammonia (NH_3) after an approximately 15-min. retention time. Fluoride is added at the same spot as ammonia. Chloraminated finished water is then stored in a clearwell until pumped into the water distribution system.

(Text courtesy of Dustin Bales, with permission)

8.5   Coagulation and Flocculation

The most common method used to remove particles and a portion of dissolved organic matter is a combination of coagulation and flocculation followed by sedimentation and/or filtration. **Coagulation** is a charge neutralization step that involves the conditioning of the suspended, colloidal, and dissolved matter by adding coagulants. **Flocculation** involves the aggregation of destabilized particles and formation of larger particles known as floc.

8.5.1   PARTICLE STABILITY AND REMOVAL

Surface charge is the primary contribution to particle stability. Stable particles are likely to remain suspended in solution (and measured as turbidity or TSS). Suspended colloids and fine particles are relatively stable and cannot flocculate and settle in a reasonable period of time. The stability of particles in natural waters primarily depends on a balance of the repulsive and attractive forces between particles. Most particles in natural waters are negatively charged, and a *repulsive electrostatic force* exists between particles of the same charge.

Counteracting these repulsive forces are *attractive forces* between particles, known as *van der Waals forces*. The potential energy from the combined repulsive electrostatic force and attractive van der Waals forces is related to the distance between two particles. Because the net attractive force is very weak at long distances, flocculation usually will not occur. At very short distances, an energy barrier exists, and the kinetic energy arising from Brownian motion of particles is not high enough to overcome the energy barrier. After a coagulant is added, the repulsive forces are reduced, particles will come together, and rapid flocculation can occur. Table 8.10 explains the combined mechanisms of coagulation and flocculation in greater detail.

Table / 8.10

Mechanisms of Coagulation and Flocculation

Compression of the electrical double layer (EDL)	Most particles in water have a net negative surface charge. The EDL consists of a layer of cations bound to the particle surface and a diffuse set of cations and anions that extend out into the solution. When the ionic strength is raised, the EDL shrinks (the repulsive forces are reduced).
Charge neutralization	Since most particles found in natural waters are negatively charged at neutral pHs, they can be *destabilized* by adsorption of positively charged cations or polymers, such as hydrolyzed metal salts and cationic organic polymers. The dose (in mg/L) of such salts or polymers is critical for subsequent flocculation process. With the proper dose, the charge will be neutralized, and particles will come together. However, if the dose is too high, the particles, instead of being neutralized, will attain a positive charge and become stable once again.
Adsorption and interparticle bridging	With the further addition of *nonionic polymers* and long-chain low-surface-charge polymers, a particle can be adsorbed on the chain, and the remainder of the polymer may adsorb on available surface sites of other particulates. This results in the formation of a bridge between particles. Again, an optimum dose (in mg/L) of the nonionic polymer exits. If too much polymer is added, particles will be enmeshed in a polymer matrix and will not flocculate.
Precipitation and enmeshment	Enmeshment (also referred to as *sweep floc*) occurs when a high enough dose of aluminum (and iron salts) is added and they form various hydrous polymers that will precipitate from solution. As the amorphous precipitate forms, particulate matter is trapped within the floc and swept from the water with the settling floc. This mechanism predominates in water treatment applications where aluminum or iron salts are used at high concentrations and pH is maintained near neutral.

8.5.2 CHEMICAL COAGULANTS

A **coagulant** is the chemical that is added to destabilize particles and accomplish coagulation. Table 8.11 provides examples of commonly used coagulants. Selection of the proper coagulant depends upon: (1) the characteristics of the coagulant, (2) concentration and type of particulates, (3) concentration and characteristics of NOM, (4) water temperature, (5) water quality (for example, pH), (6) cost and availability, and (7) dewatering characteristics of the solids that are produced.

Table / 8.11

Types of Coagulants Commonly Used in the Field

Coagulant Type	Examples
Inorganic metallic coagulants	Aluminum sulfate (also referred to as alum, $Al_2(SO_4)_3 \cdot 14H_2O$); sodium aluminate ($Na_2Al_2O_4$); aluminum chloride ($AlCl_3$); ferric sulfate ($Fe_2(SO_4)_3$), and ferric chloride ($FeCl_3$)
Prehydrolyzed metal salts	Made from alum and iron salts and hydroxide under controlled conditions; include polyaluminum chloride (PACl), polyaluminum sulfate (PAS), and polyiron chloride
Organic polymers	Cationic polymers, anionic polymers, and nonionic polymers (for synthetic polymers, molecular weight in the range of 10^4–10^7 g/mole)
Natural plant-based materials	*Opuntia* spp. and *Moringa oleifera* (used in many parts of the world, especially the developing world)

Natural coagulants are being promoted in many parts of the world because they are considered renewable, they can be used as food and fuel, and their production relies on local materials and labor.

Coagulant and flocculant aids are substances that enhance the coagulation and flocculation processes. Coagulant aids are typically insoluble particulate materials, such as clay, diatomite, powdered activated carbon (PAC), or fine sand, that form nucleating sites for the formation of larger flocs. They are used in conjunction with the primary coagulants. Flocculant aids such as anionic and nonionic polymers are used to strengthen flocs. They are added after the addition of coagulants and the destabilization of the particles.

The most commonly used coagulant is aluminum sulphate, commonly referred to as alum (molecular weight of 594 g/mole). Addition of Al^{3+} in the form of alum (or Fe^{3+} in the form of the iron salts such as ferrous sulphate ($Fe_2(SO_4)_3$ or ferric chloride, $FeCl_3$) at concentrations greater than their solubility limits results in the formation of the hydroxide precipitate that is typically used in the sweep floc mode of operation. The overall stoichiometric reaction for addition of alum in the formation of a hydroxide precipitate is as follows:

$$Al_2(SO_4)_3 \cdot 14H_2O + 6(HCO_3^-) \rightarrow 2Al(OH)_{3(s)} + 3SO_4^{2-} + 14H_2O + 6CO_2$$

(8.1)

In Equation 8.1, alkalinity (expressed as HCO_3^-) is consumed with the addition of alum. This is because alum and the other iron salts are weak acids. Based on stoichiometry, 1 mg/L of alum will consume approximately 0.50 mg/L of alkalinity (as $CaCO_3$). If the natural alkalinity of the water is not sufficient, it may be necessary to add lime or soda ash (Na_2CO_3) to react with the alum to maintain the pH in the appropriate range. The pH range for operating region of alum is 5.5–7.7, and for iron salts is 5–8.5.

Jar testing is widely used for screening the type of coagulant and the proper coagulant dosage. A jar test apparatus is shown in Figure 8.5. It consists of six square batch reactors, each equipped with a paddle mixer that can turn at variable speeds. In a jar test, batch additions of various types and different dosages of coagulants are added to the water sample. A rapid-mixing stage is combined with the addition of the coagulant. This stage is followed by a slow-mixing stage to

Figure / 8.5 Jar Test Apparatus Used to Screen Coagulants for Correct Coagulant and Proper Dosage.

(Photo courtesy of David Hand).

enhance floc formation. The samples are then allowed to settle under undisturbed conditions, and the turbidity of the settled supernatant is measured and plotted as a function of coagulant dose in order to determine the proper coagulant dosage.

---

example/8.1   Use of Jar Testing to Determine the Optimal Coagulant Dosage

A jar test was conducted on untreated water with an initial turbidity of 10 NTU and a HCO_3^- concentration of 50 mg/L as $CaCO_3$. Using the following data obtained from a jar test, estimate the optimum alum dosage for turbidity removal and the theoretical amount of alkalinity that will be consumed at the optimal dosage. Alum is added as dry alum (molecular weight of 594 g/mole).

Alum dose, mg/L	5	10	15	20	25	30
Turbidity, NTU	8	6	4.5	3.5	5	7

solution

The data are graphed as shown in Figure 8.6. The graph shows that the turbidity reaches the lowest value when the alum dose is 20 mg/L. This is the optimum alum dosage.

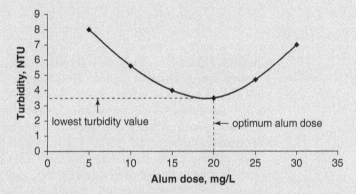

Figure / 8.6   Jar Test Results That Will Aid Identification of Proper Coagulant Dosage.

Next, we must determine the amount of alkalinity consumed and check this value against the naturally occurring alkalinity to determine whether additional alkalinity needs to be added. Use the stoichiometry of Equation 8.1 to determine the alkalinity consumed:

$$
\text{alkalinity consumed} = (20 \text{ mg/L Al}_2(SO_4)_3 \cdot 14H_2O) \times \left(\frac{1g}{1{,}000 \text{ mg}}\right)
$$

$$
\times \left(\frac{1 \text{ mole/L Al}_2(SO_4)_3 \cdot 14H_2O}{594 \text{ g/mole Al}_2(SO_4)_3 \cdot 14H_2O}\right) \times \left(\frac{6 \text{ mole HCO}_3^-}{1 \text{ mole Al}_2(SO_4)_3 \cdot 14H_2O}\right)
$$

$$
\times \left(\frac{1 \text{ eqv alkalinity}}{1 \text{ mole HCO}_3^-}\right) \times \left(\frac{100 \text{ g CaCO}_3}{2 \text{ eqv alkalinity}}\right)
$$

$$
= 0.01 \text{ g/L as CaCO}_3 = 10 \text{ mg/L as CaCO}_3
$$

Thus, 10 mg/L of alkalinity as $CaCO_3$ are consumed, and the water sample had an initial alkalinity of 50 mg/L as $CaCO_3$. Therefore, the alkalinity in the raw water is sufficient to buffer the acidity produced after alum addition.

8.5.3 OTHER CONSIDERATIONS

Coagulants are dispersed into the water stream via **rapid-mixing** systems: (1) pumped mixing (for example, pumped flash mixing, which can be simple and reliable); (2) hydraulic methods (for example, in-line static mixer, which is simple, reliable, and nonmechanical); and (3) mechanical mixing (conventional stirred tanks being the most common). Several of the devices are illustrated in Figure 8.7.

Differential sedimentation and Brownian motion are the two major mechanisms for particle aggregation (Han and Lawler, 1992). Thus, the *gentle mixing* of the water is the key for proper flocculation. To assist particle aggregation, mechanical mixing is typically employed to maintain the particles in suspension. Flocculation systems can be divided into two groups: (1) mechanical flocculators (vertical-shaft turbine, horizontal-shaft paddle) and (2) hydraulic flocculators. Three common types of flocculation systems are illustrated in Figure 8.8.

Rapid-mixing systems and most flocculation units operate under turbulent mixing conditions. Under these conditions, velocity gradients are not well defined, and the **root mean square (RMS) velocity gradient** has been widely adopted for assessing energy input:

$$\overline{G} = \sqrt{\frac{P}{\mu V}} \tag{8.2}$$

where $\overline{G}$ is the global RMS velocity gradient (energy input rate, in s^{-1}; P is the power of mixing input to the vessel (J/s), μ is the dynamic viscosity of water $(N\text{-}s/m^2)$, and V is the volume of the mixing vessel (m^3). The value of $\overline{G}$ assumes that all elemental cubes of liquid in the volume V are sheared at the same rate (on average) and $\overline{G}$ is a time-averaged value.

The mixing performance depends not only on the velocity gradient $\overline{G}$ but also the hydraulic detention time (t) and the product of $\overline{G}$ and t, a measure of degree of mixing. In practice, the values of $\overline{G}$ and $\overline{G}t$ are used as design criteria. Typical values for $\overline{G}$, t, and $\overline{G}t$ in the design of rapid-mix systems and flocculation tanks are provided in Table 8.12. $\overline{G}$ has been found to not be as important as previously thought in the design of flocculation tanks such as horizontal- and vertical-shaft

---

Table / 8.12

Typical Values Used in Design of Rapid-Mixing and Flocculation Systems

System Category	RMS Velocity Gradient, $\overline{G}$ (s^{-1})	Detention Time, t	$\overline{G}t$ Values
Mechanical mixing	600–1,000	10–120 s	5.0×10^4 to 5.0×10^5
In-line mixing	3,000–5,000	1 s	1.0×10^3 to 1.0×10^5
Horizontal-shaft paddle flocculator	20–50	10–30 min	1.0×10^4 to 1.0×10^5
Vertical-shaft turbine flocculator	10–50	10–30 min	1.0×10^4 to 1.0×10^5

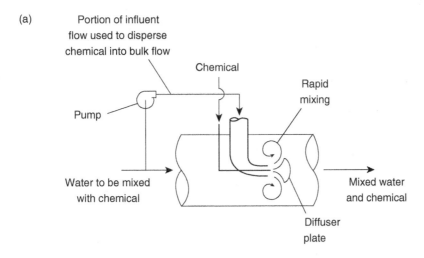

(a)

Portion of influent flow used to disperse chemical into bulk flow

Chemical

Rapid mixing

Pump

Water to be mixed with chemical

Mixed water and chemical

Diffuser plate

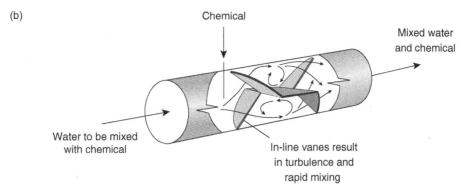

(b)

Chemical

Mixed water and chemical

Water to be mixed with chemical

In-line vanes result in turbulence and rapid mixing

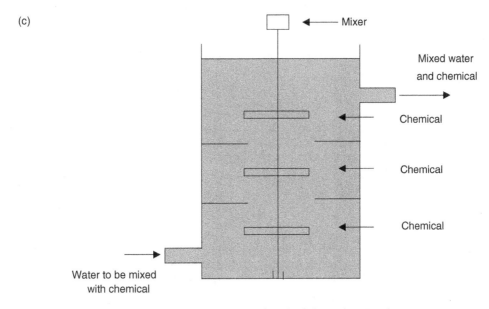

(c)

Mixer

Mixed water and chemical

Chemical

Chemical

Chemical

Water to be mixed with chemical

Figure / 8.7 Rapid-Mixing Approaches Used to Disperse a Chemical Coagulant During Water Treatment   The approaches are: (a) pumped flash mixing, (b) in-line static mixer, and (c) conventional stirred tank.

(From Crittenden et al. (2012). Redrawn with permission of John Wiley & Sons, Inc.).

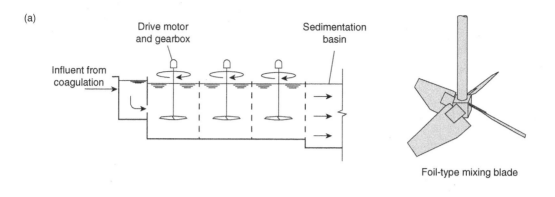

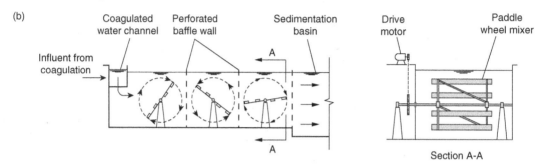

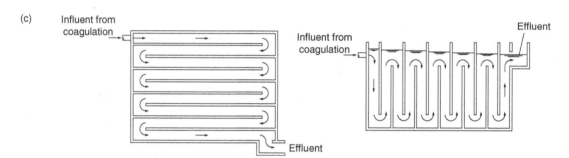

Figure / 8.8 **Common Types of Gentle Mixing Employed in Flocculation Systems** The drawing show (a) a vertical-shaft turbine flocculation system, (b) a horizontal-shaft paddle wheel flocculation system, and (c) a hydraulic flocculation system. Note how the hydraulic flocculation system requires no energy input during the use life stage.

(Crittenden et al. (2012). Redrawn with permission of John Wiley & Sons, Inc.).

flocculators (Han and Lawler, 1992). In these cases, it is important though to design the flocculation tank at the lowest value of $\overline{G}$ that will ensure the particles remain in suspension and under the upper limit to avoid breakup of formed flocs. Using the lowest achievable, $\overline{G}$ will also result in less energy requirements.

Electricity is needed for mixers, chemical pumps, chemical mixers and drainage pumps used in the coagulation and flocculation process. The range of energy use varies widely, though, by treatment plant capacity as given in Table 8.13. With the increase in system capacity, the electricity consumption per m³ water produced decreases. This indicates that the larger centralized treatment system may have lower direct energy requirement for coagulation and flocculation process during the operation phase; however, the energy needed for coagulant manufacturing and delivery is not considered here. For example, 6,290 MJ of energy is consumed to produce 1 metric ton of alum (PRé Consultants, 2004).

Table / 8.13

Electricity Consumption (kWh/m³) for Coagulation and Flocculation for Surface Water Treatment Systems of Different Plant Capacity

Electricity consumption	1 MGD	5 MGD	10 MGD	20 MGD	50 MGD	100 MGD
Rapid Mixing	0.011	0.009	0.008	0.008	0.008	0.008
Flocculation	0.003	0.003	0.002	0.002	0.002	0.002
Alum Feed System	0.0024	0.0005	0.0003	0.0003	0.0002	0.0002
Polymer Feed System	0.0124	0.0025	0.0012	0.0006	0.0002	0.0001

SOURCE: Burton, 1996.

example/8.2   Design of a Mechanical Rapid-Mix Tank

A conventional stirred tank is used for rapid mixing in a water treatment plant with a flow of 100×10^6 L/day. The water temperature is 10°C. Determine the tank volume and power requirement.

solution

The tank volume equals the flow rate (Q) times the hydraulic detention time (θ). Table 8.12 provides appropriate detention times. Conventional stirred tanks are considered a form of mechanical mixing, and we will select a detention time value of 60 s:

$$V = Q \times \theta = \frac{100 \times 10^6 \text{ L}}{\text{day}} \times 60 \text{ s} \times \frac{1 \text{ min}}{60 \text{ s}} \times \frac{1 \text{ day}}{1{,}440 \text{ min}} \times \frac{\text{m}^3}{1{,}000 \text{ L}} = 69 \text{ m}^3$$

To determine the power requirement, use Table 8.12 to select an appropriate RMS velocity gradient. Here we will select a $\overline{G}$ value of 900/s, and the product of $\overline{G}$ and t ($900/\text{s} \times 60 \text{ s} = 5.4 \times 10^4$) is within the range ($5 \times 10^4$ to 5×10^5) provided in Table 8.12. At 10°C, $\mu = 0.001307$ N-s/m². To obtain power consumption, rearrange Equation 8.2 to solve for P:

$$P = \overline{G}^2 \times \mu \times V = (900/\text{s})^2 \times 0.001307 \frac{\text{N-s}}{\text{m}^2} \times 69 \text{ m}^3 \times \frac{1 \text{ kN}}{1{,}000 \text{ N}}$$

$$= 73 \frac{\text{kN} \cdot \text{m}}{\text{s}} = 73 \text{ kW}$$

Northbrook, Illinois, is the first community in Illinois and one of only a few in the United States to offset the energy used to run its water plant. The Northbrook water treatment plant provides 2.1 billion gallons water per year for 34,000 residents. It began purchasing 155 MWh/year in renewable-energy certificates from wind farms in north central Illinois several years ago to offset electricity derived from coal (Figure 8.9).

After seeing the movie *An Inconvenient Truth*, Northbrook's director of public works recommended the village increase its purchase of renewable-energy certificates to 4,500 MWh/year, enough energy to operate the water treatment plant.

Figure / 8.9 **A Wind Farm in North Central Illinois** Wind turbines provide energy that the water treatment plant for Northbrook, Illinois, purchases through renewable-energy certificates. The community purchases enough energy to operate the water treatment plant.

(Photo courtesy of Iberdrola Renewables, LLC).

8.6   Hardness Removal

Water hardness is caused by divalent cations, primarily calcium and magnesium ions (Ca^{2+} and Mg^{2+}). When Ca^{2+} and Mg^{2+} are associated with alkalinity anions (for example, HCO_3^-), the hardness is defined as **carbonate hardness**. The term **noncarbonated hardness** is used if the Ca^{2+} and Mg^{2+} are associated with nonalkalinity anions (for example, SO_4^{2-}). The distribution of hard waters in the United States is shown in Figure 8.10.

Complexation agents can be added to prevent divalent cations from precipitating, or hardness can be removed. A process flow diagram for a two-stage excess **lime–soda ash softening process** is shown in Figure 8.11. Lime is sold commercially in forms of quicklime (90 percent CaO) and hydrated lime (70 percent CaO). Granular quicklime is usually crushed in a slaker and fed to slurry containing about 5 percent calcium hydroxide. Powdered hydrated lime is prepared by fluidizing in a tank containing a turbine mixer. The soda ash (approximately 98 percent Na_2CO_3) is a grayish-white powder that can be added either with lime or following the addition of lime. Carbon dioxide is used for recarbonization to reduce the pH and precipitate excess calcium of the lime-softened water.

When lime slurry ($Ca(OH)_2$) is added to water, it first reacts with free carbon dioxide, because the CO_2 is a stronger acid than HCO_3^-. (Remember that, by definition, HCO_3^- can act as an acid or a base.) The chemical reactions for the removal of carbonate and noncarbonate hardness are provided in Equations 8.3–8.7:

$$CO_2 + Ca(OH)_2 \rightarrow CaCO_{3(s)} + H_2O \qquad (8.3)$$

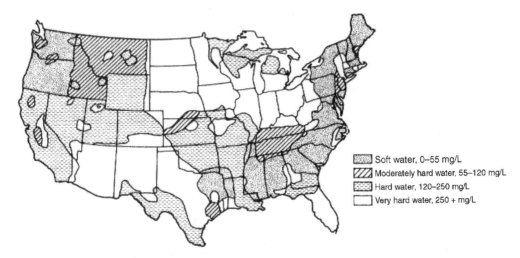

Figure / 8.10 **Distribution of Hard Water in the United States** Units are mg/L as CaCO₃.
The areas shown define approximate hardness values for municipal water supplies.

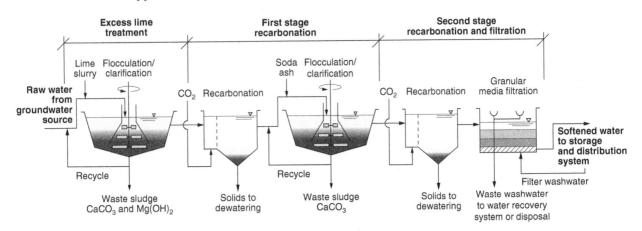

Figure / 8.11 **Process Flow Diagram for a Two-Stage Excess Lime–Soda Treatment Process
Used to Treat Hard Waters** Considerations should be made for reusing all waste streams.

(From Crittenden et al. (2012). Redrawn with permission of John Wiley & Sons, Inc.).

Carbonate hardness removal is given by

$$Ca(HCO_3)_2 + Ca(OH)_2 \rightarrow 2CaCO_{3(s)} + 2H_2O \qquad \textbf{(8.4)}$$

$$Mg(HCO_3)_2 + 2Ca(OH)_2 \rightarrow 2CaCO_{3(s)} + Mg(OH)_{2(s)} + 2H_2O \quad \textbf{(8.5)}$$

Noncarbonate hardness removal is given by

$$MgSO_4 + Ca(OH)_2 \rightarrow Mg(OH)_{2(s)} + CaSO_4 \qquad \textbf{(8.6)}$$

$$CaSO_4 + Na_2CO_3 \rightarrow CaCO_{3(s)} + Na_2SO_4 \qquad \textbf{(8.7)}$$

As shown in Equations 8.4–8.7, lime will remove CO_2 (Equation 8.3) and carbonate hardness (Equations 8.4 and 8.5) and replace magnesium with calcium in solution (Equation 8.6). Equation 8.7 shows that soda ash (Na_2CO_3) is then used to remove calcium noncarbonated hardness, which may be present in the untreated water or may be the result of the precipitation of magnesium noncarbonated hardness (Equation 8.6).

The amount of soda ash required depends on the amount of non-carbonated hardness to be removed.

Complete conversion of bicarbonate to carbonate for calcium precipitation will take place only at a pH above 12. The optimum pH depends on the concentration of both calcium and bicarbonate ions. In practice, the optimum pH for maximum calcium carbonate precipitation may be as low as 9.3, because more carbonate is formed due to the shift in carbonate–bicarbonate equilibrium as calcium carbonate precipitates.

The removal of magnesium as $Mg(OH)_2$ precipitate (Equation 8.6) requires a pH value of at least 10.5. Therefore, extra lime (30–70 mg/L as $CaCO_3$) in excess of the stoichiometric amount is added to raise the pH. Jar testing can be used to determine the amount of excess lime required for a given water source.

Similar to the coagulation and flocculation process, electricity is needed for softening process. The electricity consumption for a lime feed system varies for systems with different capacities. It can range from 0.0024 kWh/m^3 for a 1 MGD treatment system to 0.0002 kWh/m^3 for a 50 MGD treatment system (Burton, 1996). The energy consumption will be much higher if the chemical production and delivery is also included in the system boundary of evaluation which would occur if the system's life cycle was considered.

example/8.3   Lime–Soda Ash Softening

A groundwater contains the following constituents: $H_2CO_3^* = 62$ mg/L, $Ca^{2+} = 80$ mg/L, $Mg^{2+} = 36.6$ mg/L, $Na^+ = 23$ mg/L, alkalinity $(HCO_3^-) = 250$ mg/L as $CaCO_3$, $SO_4^{2-} = 96$ mg/L, and $Cl^- = 35$ mg/L. The facility is to treat 50×10^6 L/day (15 MGD) of water from this source using lime–soda ash to reduce the hardness.

1. Determine the total, carbonate, and noncarbonated hardness present in the raw water.

2. Determine the lime and soda ash dosages for softening (units of kg/day). Assume the lime is 90 percent CaO by weight and the soda ash is pure sodium carbonate.

solution

This problem requires several steps. First construct a table of the chemical constituents and their concentrations in terms of mg/L as $CaCO_3$ (see Table 8.14).

The second step determines the total hardness, carbonate hardness, and noncarbonate hardness. The total hardness is the sum of the calcium and magnesium ions as $CaCO_3$:

$$\text{total hardness} = (200 + 150) = 350 \text{ mg/L as } CaCO_3$$

The carbonate hardness is the sum of the calcium and magnesium ions associated with bicarbonate ions. Because the total hardness (350 mg/L as $CaCO_3$) is greater than bicarbonate alkalinity (250 mg/L $CaCO_3$), all the bicarbonate is associated with calcium (200 mg/L $CaCO_3$) and magnesium (50 mg/L $CaCO_3$). The carbonate hardness is thus equal to bicarbonate alkalinity as $CaCO_3$:

$$\text{carbonate hardness} = 200 + 50 = 250 \text{ mg/L as } CaCO_3$$

The noncarbonate hardness equals the magnesium ions not associated with carbonate hardness ($MgSO_4$):

$$\text{noncarbonate hardness} = (150 - 50) = 100 \text{ mg/L as } CaCO_3$$

example/8.3 (continued)

The second question requests we determine the daily mass of lime and soda ash required for softening. The lime that is required will react with CO_2 ($H_2CO_3^*$), $Ca(HCO_3)_2$, $Mg(HCO_3)_2$, and $MgSO_4$. The stoichiometric amount of $Ca(OH)_2$ can be calculated based on Equations 8.3–8.6.

$$Ca(OH)_2 \text{ required to react with } CO_2 = 100 \text{ mg/L as } CaCO_3$$
$$Ca(OH)_2 \text{ required to react with } Ca(HCO_3)_2 = 200 \text{ mg/L as } CaCO_3$$
$$Ca(OH)_2 \text{ required to react with } Mg(HCO_3)_2 = 2 \times Mg(HCO_3)_2 = 2 \times 50 = 100 \text{ mg/L as } CaCO_3$$

$$Ca(OH)_2 \text{ required to react with } MgSO_4 = 100 \text{ mg/L as } CaCO_3$$

Remember that 30–70 mg/L of extra lime must be added to raise the pH above 10.5 to ensure that magnesium is removed as $Mg(OH)_2$. Assume 30 mg/L (as $CaCO_3$) of additional lime is required in the process.

The total lime requirement is thus determined from the summation of all five individual lime requirements:

$$\text{lime required} = \left(\frac{100 \text{ mg } CaCO_3}{L} + \frac{200 \text{ mg } CaCO_3}{L} + \frac{100 \text{ mg } CaCO_3}{L} + \frac{100 \text{ mg } CaCO_3}{L} + \frac{30 \text{ mg } CaCO_3}{L} \right)$$
$$\times \frac{56 \text{ mg } CaO/\text{mmole}}{100 \text{ mg } CaCO_3/\text{mmole}} \times \frac{kg}{10^6 \text{ mg}} \times \frac{50 \times 10^6 L}{day} \times \frac{kg \text{ bulk lime}}{0.9 \text{ kg } CaO}$$
$$= 16{,}500 \text{ kg/day}$$

The soda ash (Na_2CO_3) required is also determined by reaction stoichiometry:

soda ash = noncarbonate hardness
$$= 100 \text{ mg } CaCO_3/L \times \frac{106 \text{ mg } Na_2CO_3/\text{mmole}}{100 \text{ mg } CaCO_3/\text{mmole}} \times \frac{kg}{10^6 \text{ mg}} \times \frac{50 \times 10^6 \text{ L}}{day} = 5{,}300 \text{ kg/day}$$

Table / 8.14

Table Constructed to Solve Example 8.3

Chemical Constituent	Concentration (mg/L)	Equivalents/ mole	Molecular Weight (g/mole)	Equivalent Weight (g/eqv)	Concentration (meqv/L)	Concentration (mg/L as $CaCO_3$)
$H_2CO_3^*$	62	2	62.0	31.0	2.0	100
Cations						
Ca^{2+}	80	2	40.0	20.0	4.0	200
Mg^{2+}	36.6	2	24.4	12.2	3.0	150
Na^+	23.0	1	23.0	23.0	1.0	50
				Total	9.0	400
Anions						
Alk (HCO_3^-)	250.0	2	100.0	50.0	5.0	250
SO_4^{2-}	96.0	2	96.0	48.0	2.0	100
Cl^-	35	1	35.5	35.5	1.0	50
				Total	9.0	400

Note: Alkalinity is expressed as $CaCO_3$.

8.7 Sedimentation

Sedimentation is the process in which the majority of the particles will settle by gravity within a reasonable time and be removed. Particles with densities greater than 1,000 kg/m^3 will eventually settle, and particles with densities less than 1,000 kg/m^3 will float to the water surface. In water treatment, there are common types of settling: *discrete particle settling* and *flocculant settling*.

8.7.1 DISCRETE PARTICLE SETTLING

Discrete particle settling occurs when particles are discrete and do not interfere with one another as they settle. For this type of settling, the movement of a particle in water is determined by a balance of a downward gravitational force, an upward buoyancy force, and an upward drag force.

The settling velocity of particles in a liquid such as water can be described by either Stokes' law or Newton's law. Table 8.15 describes each of these laws in greater detail. Stokes' law was derived in Chapter 4. It is applicable to spherical particles when the Reynolds number is less than or equal to 1 (laminar flow). Newton's law is used to determine the settling velocity of particles when the Reynolds number is greater than 1 (transition and turbulent flow). The dimensionless **Reynolds number (Re)** is defined as

$$\text{Re} = \frac{\rho d_p v_s}{\mu} = \frac{d_p v_s}{\upsilon} \approx \frac{\text{inertial forces}}{\text{viscous forces}} \tag{8.8}$$

where ρ is the density of the liquid (kg/m^3), d_p is the particle diameter (m), v_s is the settling velocity of the particle at any point in time (m/s), μ is the dynamic viscosity of the liquid (N-s/m^2), and υ is the kinematic viscosity of the liquid (m^2/s).

Table / 8.15

Determination of Settling Velocity of Particles Using Stokes' and Newton's Laws

Applicable Law	Settling Velocity (m/s)	Terms	Drag Coefficient	Applicability
Stokes' law	$v_s = \dfrac{g\left(\rho_p - \rho\right)d_p^2}{18\,\mu}$	g is the acceleration due to gravity (m/s^2); ρ_p is the density of the particle (kg/m^3); ρ is the density of the liquid (kg/m^3); d_p is the particle diameter (m); and μ is the dynamic viscosity of the liquid (N-s/m^2)	For laminar flow: $C_d = \dfrac{24}{\text{Re}}$	Applicable for spherical particles when the Reynolds number ≤ 1 (laminar flow). Has limited application in water treatment because conditions in most treatment facilities are not laminar.
Newton's law	$v_s = \sqrt{\dfrac{4g\left(\rho_p - \rho\right)d_p}{3C_d\rho}}$	g is the acceleration due to gravity (m/s^2); ρ_p is the density of the particle (kg/m^3); ρ is the density of the liquid (kg/m^3); C_d is the drag coefficient.	For the transition regime: $C_d = \dfrac{24}{\text{Re}} + \dfrac{3}{\sqrt{\text{Re}}} + 0.34$ C_d becomes constant in turbulent regime (Re $> 10,000$)	Applicable for particles when the Reynolds number > 1 (transition and turbulent flow).

example/8.4 Application of Stokes' Law

Calculate the terminal settling velocity for a sand particle that has a diameter of 100 μm and a density of 2,650 kg/m³. The water temperature is 10°C.

solution

The terminal settling velocity of the particle can be calculated using Stokes' law (Table 8.15). For water at 10°C, $\rho = 999.7 \, \text{kg/m}^3$, $\mu = 1.307 \times 10^{-3} \, \text{N-s/m}^2$, and $\upsilon = 1.306 \times 10^{-6} \, \text{m}^2/\text{s}$.

$$v_s = \frac{g\left(\rho_p - \rho\right)d_p^2}{18\,\mu}$$

$$= \frac{9.81 \, \text{m/s}^2 \times \left(2,650 - 999.7 \, \text{kg/m}^3\right) \times \left(1.0 \times 10^{-4} \, \text{m}\right)^2}{18 \times 1.307 \times 10^{-3} \, \text{N-s/m}^2} \times \frac{3,600 \, \text{s}}{\text{h}} = 24.8 \frac{\text{m}}{\text{h}}$$

We must verify the flow conditions to ensure Stokes' law is applicable. The Reynolds number is calculated to verify that the particle is settling under laminar conditions:

$$\text{Re} = \frac{d_p v_s}{\upsilon} = \frac{1.0 \times 10^{-4} \, \text{m} \times 24.8 \, \text{m/h} \times (1 \, \text{h}/3,600 \, \text{s})}{1.306 \times 10^{-6} \, \text{m}^2/\text{s}} = 0.53$$

Because Re < 1, laminar flow exists, and Stokes' law is applicable.

example/8.5 Application of Newton's Law

Calculate the terminal settling velocity for a sand particle that has a diameter of 200 μm and a density of 2,650 kg/m³. The water temperature is 15°C.

solution

The terminal velocity of the particle is calculated using Stokes' law (Table 8.15). For water at 15°C, $\rho = 999.1 \, \text{kg/m}^3$, $\mu = 1.139 \times 10^{-3} \text{N-s/m}^2$, and $\upsilon = 1.139 \times 10^{-6} \, \text{m}^2/\text{s}$.

$$v_s = \frac{g\left(\rho_p - \rho\right)d_p^2}{18\,\mu}$$

$$= \frac{9.81 \, \text{m/s}^2 \times \left(2,650 - 999.1 \, \text{kg/m}^3\right) \times \left(2.0 \times 10^{-4} \, \text{m}\right)^2}{18 \times 1.139 \times 10^{-3} \, \text{N-s/m}^2} \times \frac{3,600 \, \text{s}}{\text{h}} = 113.8 \frac{\text{m}}{\text{h}}$$

Check the Reynolds number (Equation 8.8) to verify the particle is settling under laminar conditions:

$$\text{Re} = \frac{d_p v_s}{\upsilon} = \frac{\left(2.0 \times 10^{-4} \, \text{m}\right) \times (113.8 \, \text{m/h}) \times (1 \, \text{h}/3,600 \, \text{s})}{\left(1.139 \times 10^{-6} \, \text{m}^2/\text{s}\right)} = 5.55$$

Because Re > 1, Stokes' law is not valid. The equation for the drag coefficient is provided in Table 8.15, and the settling velocity can be calculated using Newton's law (Table 8.15).

example / 8.5 (continued)

Because v_s cannot be determined explicitly, a trial-and-error solution must be used. Using the value of Re just obtained (that is, 5.55), the drag coefficient can be calculated as

$$C_d = \frac{24}{5.55} + \frac{3}{\sqrt{5.55}} + 0.34 = 5.94$$

The terminal settling velocity also can be recalculated:

$$v_s = \sqrt{\frac{4(9.81 \text{ m/s}^2) \times (2,650 - 999.1 \text{ kg/m}^3) \times 2.0 \times 10^{-4} \text{ m}}{3 \times 5.94 \times 999.1 \text{ kg/m}^3}} \times 3,600 \frac{\text{s}}{\text{h}}$$

$$= 97.1 \frac{\text{m}}{\text{h}}$$

The Reynolds number is calculated again, and then the drag coefficient and terminal settling velocity are recalculated. After several iterations, a convergent answer is obtained, as given in Table 8.16. The settling velocity begins to converge at the sixth or seventh trial and has a value of 86.5 m/h.

Table / 8.16

Iterative Process Used in Example 8.5 to Determine Settling Velocity Using Newton's Law   After several iterations, a convergent answer is obtained, as shown here.

Trial	Re (dimensionless)	C_d (dimensionless)	v_s (m/h)
0	5.55	5.94	97.1
1	4.74	6.78	90.9
2	4.43	7.18	88.3
3	4.31	7.36	87.3
4	4.26	7.43	86.8
5	4.23	7.47	86.6
6	4.23	7.48	86.5
7	4.22	7.49	86.5

8.7.2 PARTICLE REMOVAL DURING SEDIMENTATION

Figure 8.12 shows particle trajectories in a rectangular sedimentation basin. Here it is assumed that particles move horizontally at the same velocity as the water and are removed by gravity once they reach the bottom of the basin. Particle trajectories in the basin depend upon the particle settling velocity (v_s) and the fluid velocity (v_f).

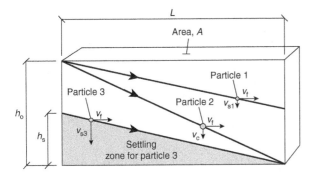

The settling velocity for discrete particles is constant, because particles will not interfere with one another, and the size, shape, and density of particles is assumed to not change as they move through the reactor. A particle (particle 2 in Figure 8.12) that enters at the top of the basin and settles just before it flows out of the basin is called a *critical particle*. Its settling velocity is defined as the **critical particle-settling velocity**, determined as follows:

$$v_c = \frac{h_o}{\theta} \tag{8.9}$$

where v_c is the critical particle-settling velocity (m/h), h_o is the depth of the sedimentation basin (m), and θ is the hydraulic detention time of the sedimentation basin (h).

The critical particle-settling velocity is also called the **overflow rate** (OR) because it is equal to the ratio of process flow rate to surface area:

$$v_c = \frac{h_o}{\theta} = \frac{h_o Q}{V} = \frac{h_o Q}{h_o A} = \frac{Q}{A} = OR \tag{8.10}$$

where A is the surface area of the top of the settling basin (m^2) and Q is the process flow rate (m^3/h). Important to our discussion is the term OR, the overflow rate. Note in Equation 8.10 that the OR is not a function of the tank depth.

The OR (m^3/m^2-h, also written as m/h) is equal to the critical settling velocity, v_c. Any particles with a settling velocity (v_s) greater than or equal to v_c (or the OR) will be removed. Particles with a settling velocity (v_s) less than v_c can also be removed depending on their position at the inlet. For example, assuming particle 3 in Figure 8.12 has a settling velocity, v_{s3}, less than v_c, it can be removed because of its inlet position. Note that in Figure 8.12, particle 1 will not be removed, assuming its settling velocity (v_{s1}) is less than v_c relative to its entry point into the sedimentation basin.

The percentage of particles removed is determined as follows:

$$\text{percent of particles removed} = \frac{v_s}{OR} \times 100 \tag{8.11}$$

example/8.6 Determining Particle Removal

A treatment plant has a horizontal flow sedimentation basin with a depth of 4 m, width of 6 m, length of 36 m, and process flow rate of 450 m³/h. What removal percentage should be expected for particles that have settling velocities of 1.0 and 2.5 m/h? What is the minimum size of particles that would be completely removed? Assume the particle density is 2,650 kg/m³. The water temperature is 10°C.

solution

First, determine the sedimentation basin overflow rate (critical settling velocity), using Equation 8.10:

$$\text{OR} = v_c = \frac{Q}{A} = \frac{450\,\text{m}^3/\text{h}}{36\,\text{m} \times 6\,\text{m}} = 2.1\,\text{m}^3/\text{m}^2\text{-h}$$

The percent removal of particles for each particle size can then be calculated as follows.

For particles with a settling velocity of 1.0 m/h, because v_s equals 1.0 m/h (which is less than the v_c of 2.1 m/h), the percentage of removal is calculated using Equation 8.11:

$$\text{fraction of particles removed} = \frac{v_s}{\text{OR}} = \frac{1.0\,\text{m/h}}{2.1\,\text{m/h}} = 0.48$$

For particles with a settling velocity of 2.5 m/h, because v_s equals 2.5 m/h (which is greater than v_c of 2.1 m/h), all particles with this settling velocity will be removed:

$$\text{fraction of particles removed} = 1.0$$

The final question is to determine the minimum particle size that would result in complete removal. This particle size can be determined using Stokes' law. The settling velocity of the minimum-size particle is equal to the critical settling velocity, $v_s = v_c = 2.1\,\text{m/h}$. Plugging v_s into Stokes' law will provide the correct particle size. For water at 10°C, $\rho = 999.7\,\text{kg/m}^3$, $\mu = 1.307 \times 10^{-3}\,\text{N-s/m}^2$, and $\upsilon = 1.306 \times 10^{-6}\,\text{m}^2/\text{s}$.

$$v_s = \frac{g\left(\rho_p - \rho\right)d_p^2}{18\,\mu} = \frac{9.81\,\text{m/s}^2 \times (2,650 - 999.7\,\text{kg/m}^3) \times (d_p)^2}{18 \times 1.307 \times 10^{-3}\,\text{N-s/m}^2} \times \frac{3,600\,\text{s}}{\text{h}} = 2.1\frac{\text{m}}{\text{h}}$$

$$d_p^2 = 8.48 \times 10^{-10}\,\text{m}^2$$

$$d_p = 2.9 \times 10^{-5}\,\text{m} = 2.9 \times 10^{-3}\,\text{cm}$$

To verify the flow conditions, check the Reynolds number (Equation 8.8) to see if the particle is settling under laminar conditions:

$$\text{Re} = \frac{d_p v_s}{\upsilon} = \frac{8.48 \times 10^{-10}\,\text{m} \times 2.1\,\text{m/h} \times 1\,\text{h}/3,600\,\text{s}}{1.306 \times 10^{-6}\,\text{m}^2/\text{s}}$$

$$= 3.8 \times 10^{-7}$$

The Reynolds number is much less than 1. Therefore, laminar flow exists, and Stokes' law is valid.

Readers should note that horizontal settling tanks can be designed and operated without the use of mechanical rakes and other moving parts that require energy, maintenance, and funds to purchase spare parts. They may be more appropriate technology for many applications throughout the world.

Typical detention times for a rectangular sedimentation basin with horizontal flow are in the range of 1.5–4 h. Poor inlet conditions, density currents, and uneven weirs can decrease the design detention time (Miller and Esler, 2013). Other important design criteria are considered in the design of such basins, including the number of basins (more than 2 so one can be taken offline for maintenance), basin depth (3–5 m), the horizontal mean flow velocity (0.3–1.1 m/min), overflow rate (1.25–2.5 m/h), length-to-depth ratio (15), and length-to-width ratio (4–5) (Crittenden et al., 2012).

Materials can be saved by placing the basins side by side (so they share a common wall) and integrating the flocculation process into the front end of the sedimentation basin. Combining the flocculation and sedimentation basins typically requires a diffuser wall to separate the two processes. The wall has small hole openings of diameter 100–200 mm. The inlet structure to a sedimentation basin is also designed to provide uniform distribution of the water over the entire cross section of the basin and to maintain an appropriate velocity that not only prevents sedimentation at the inlet zone but also limits breaking up the formed flocs. A velocity of 0.15–0.60 m/s will maintain the floc suspension for most drinking-water applications (Crittenden et al., 2012). Circular sedimentation tanks can also be designed and particle size removal evaluated using Equations 8.10 and 8.11.

8.7.3 OTHER TYPES OF SETTLING

Table 8.17 summarizes the various types of settling observed during treatment of water and wastewater. The type of discrete particle settling discussed in Example 8.6 is also referred to as Type I settling. In **Type I settling**, particles settle discretely at a constant settling velocity.

When particles flocculate during settling due to the velocity gradient of fluids or differences in the settling velocities of particles, their size is increasing, and they settle faster as time passes. This type of settling is known as flocculant or **Type II settling**, which is found in coagulation processes and most conventional sedimentation basins. At high particle concentrations (greater than 1,000 mg/L), a blanket of particles is formed, and a clear interface is observed between the blanket and clarified water above it. This type of settling is hindered or **Type III**

Table / 8.17

Types of Particle Settling Encountered During Drinking-Water and Wastewater Treatment

Type of Settling	Description	Where Used in Treatment Process
Type I	Particles settle discretely at a constant settling velocity.	Grit removal
Type II	Particles flocculate during settling due to velocity gradient of fluids or differences in the settling velocities of particles. Their size is increasing, and they settle faster as time passes.	Coagulation processes and most conventional sedimentation basins
Type III	Blanket of particles is formed at high particle concentrations (above 1,000 mg/L), and a clear interface is observed between the blanket and clarified water above it.	Lime-softening sedimentation and in wastewater sedimentation and sludge thickeners

settling, which occurs in lime-softening sedimentation, activated-sludge sedimentation, and sludge thickeners.

Even when particles settle by gravity during the sedimentation process, electricity may still be required needed for agitators, clarifier drives, and sludge pumps. The electricity consumption for sedimentation has been found to range from 0.0037 kWh/m^3 for a 1 MGD treatment system to 0.0023 kWh/m^3 for a 50 MGD treatment system (Burton, 1996). Again, this is only the direct energy used in the process and does not consider the embodied energy, for example, including the additional requirement of handling solid residuals.

8.8   Filtration

Filtration is widely used for removing small flocs or precipitated particles. It may be used as the primary turbidity removal process such as direct filtration of raw water with low turbidity. It is also used for removal of pathogens, such as *Giardia lamblia* and *Cryptosporidium*. Two types of filtration employed in water treatment facilities include **granular media filtration** (discussed next) and membrane filtration (discussed in a later section).

8.8.1   TYPES OF GRANULAR FILTRATION

Granular media filtration can operate at either a high hydraulic loading rate (5–15 m/h) or a low rate (0.05–0.2 m/h). (The units on the hydraulic loading rates are really m^3/m^2-h, which reduces to m/h.) In both processes, the influent water is driven by gravity flow through a bed of granular material, and particles are collected within the bed.

High-rate filtration (also known as **rapid filtration**) is the process used by nearly all U.S. filtration plants. **Slow sand filtration** is an appropriate water treatment technology for rural communities due to its simplicity, land availability, and low energy requirement. It is also commonly employed in community and household point-of-use systems implemented in the developing world. Table 8.18 compares the process characteristics of slow sand filtration and rapid filtration.

Figure 8.13 shows a schematic of a dual-media granular filtration system. Filter media in rapid filtration must be fairly uniform in size to allow the filters to operate at a high loading rate. Coagulation pretreatment is necessary before filtration because particles have to be properly destabilized for effective removal.

Biosand Filter used for household water treatment in many parts of the developing world.

Rapid filtration operates over a cycle consisting of a filtration stage and a backwash stage. During the *filtration stage*, water flows downward through the filter bed, and particles are captured within the bed. During the *backwash stage*, water flows upward and flushes captured particles up and away from the bed. Typically, the filtration step lasts from 1 to 4 days, and the **backwashing** takes 15 to 30 min.

The **head loss** during the filtration and backwashing stages is important for proper design of granular filtration systems. For a given depth of filter media and the effective size of media, the **clean-bed head**

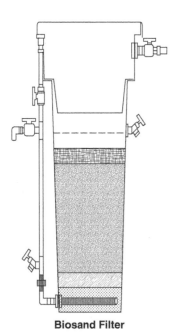

Biosand Filter

Comparison of Typical Ranges for Design and Operating Parameters for Slow Sand Filtration and Rapid Filtration Some filters are designed and operated outside of these ranges.

Process Characteristic	Slow Sand Filtration	Rapid Filtration
Filtration rate	0.08–0.25 m/h	5–15 m/h
	(0.03–0.10 gpm/ft^2)	(2–6 gpm/ft^2)
Media diameter	0.15–0.30 mm	0.5–1.2 mm
Bed depth	0.9–1.5 m (3–5 ft)	0.6–1.8 m (2–6 ft)
Required head	0.9–1.8 m (3–6 ft)	1.8–3.0 m (6–10 ft)
Run length	1–6 months	1–4 days
Pretreatment	None required	Coagulation
Regeneration method	Scraping	Backwashing
Maximum raw-water turbidity	10 NTU	Unlimited with proper pretreatment

SOURCE: Crittenden et al., 2012. Reprinted with permission of John Wiley & Sons, Inc.

loss depends on the bed porosity, filtration rate, and water temperature. The effect of the porosity and filtration rate on head loss is illustrated in Figure 8.14. The clean-bed head loss increases as the porosity decreases or as the filtration rate increases. The figure also shows that the clean-bed head loss is more sensitive to filtration rate at a lower porosity. The clean-bed head loss also increases as temperature decreases, because fluid viscosity increases. For example, the clean-bed head loss at 5°C is 60–70 percent higher than it is at 25°C.

The **backwash rate** must be above the minimum fluidization velocity of the largest media. The largest media is typically taken as the d_{90} diameter. The minimum fluidization velocity can be calculated by inputting the fixed-bed porosity into design equations. Backwash rates range from approximately 20 to 56 m/h, with a typical rate being 45 m/h. The targeted expansion of the bed is approximately 25 percent for anthracite and 37 percent for sand.

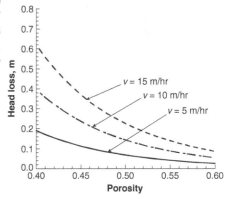

Figure / 8.14 Effect of Porosity and Filtration Rate (v) on Clean-Bed Head Loss through Clean Granular Filter Bed The clean-bed head loss increases as the porosity decreases or as the filtration rate increases.

(Crittenden et al. (2012). Redrawn with permission of John Wiley & Sons, Inc.).

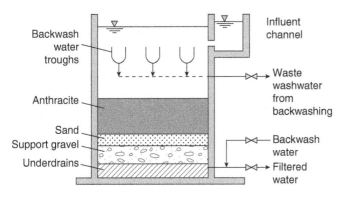

Figure / 8.13 Dual-Media Granular Media Filter.

(Crittenden et al. (2005). Adapted with permission of John Wiley & Sons, Inc.).

8.8.2 MEDIA CHARACTERISTICS

The sand used for slow sand filtration is smaller and less uniform than the media used in rapid filtration. Only washed sand should be used. Beach or riverbed sand cannot be used before processing, because its size and uniformity are usually higher than the criteria for a slow sand filter. The coagulation pretreatment step is not required, because destabilization is not important for slow sand filtration. Community slow sand filters are typically housed in reinforced-concrete structures with graded gravel (0.3–0.6 m) as a support layer and an underdrain system for water collection.

Slow sand filtration operates over a cycle consisting of a filtration stage and a regeneration stage. In the filtration stage, water flows downward by gravity through a submerged sand bed (0.9–1.5 m) at low rate, and head loss builds. When the head loss reaches the available head (typically after weeks or months), the filter is drained, and the top 1–2 cm of sand are scraped off, cleaned, and then stockpiled on-site. The operation and scraping cycle repeats over several years until the sand bed reaches a minimum depth of 0.4–0.5 m. The stockpiled sand is then replaced in the filter to restore the original bed depth.

Filter media and bed characteristics are very important for evaluation of filtration process performance and design of filtration systems. For rapid filtration, sand, anthracite coal, garnet, and ilmenite are commonly used for filter media. Sometimes GAC is used when filtration is combined with adsorption in a single unit process.

Sand is the granular media used in *slow sand filtration*. Important media characteristics include media size (described by the effective size), size distribution (described by a uniformity coefficient), density, shape, and hardness. Table 8.19 compares important filter and bed characteristics for rapid and slow sand filters. Figure 8.15 shows how a sand sample is analyzed by a sieve analysis to determine appropriate size for use in filtration.

In Figure 8.15, a 1,000 g sample of naturally occurring sand was sifted through a stack of sieves, and the weight retained on each sieve was recorded and plotted. Here $d_{10} = 0.43$ mm, and $d_{60} = 1.18$ mm. Therefore, the uniformity coefficient equals 2.7 (see Table 8.19). See how the uniformity coefficient of this naturally occurring sand is much higher than typical values used in rapid filters (1.3–1.7). The higher uniformity coefficient will result in severe media stratification during backwash, causing excessive head loss and reducing overall effectiveness of the filter. Therefore, the sample sifted for this example needs to be processed to a fairly uniform size.

During filtration and the backwash processes, energy is consumed by sludge pumps, process water pumps, and vacuum pumps. The electricity consumption for filtration process does not vary with systems capacity. It is typically 0.0021 kWh/m^3 for filter surface wash pumping and 0.0034 kWh/m^3 for backwash water pumps (Burton, 1996). Energy savings can be realized from proper design (and use) of efficient motors and pumps, adjustable-speed drives, improvements in instrumentation and control, and installation of valves that reduce head loss. For example, though not applicable in all situations, ball and flapper check valves have a much lower head loss than swing check valves. The energy required for filter media production and delivery to the treatment plant along with handling of residuals can also be included in an energy analysis if life cycle thinking is integrated.

Filter Media and Bed Characteristics for Rapid and Slow Sand Filters

	Important Media Characteristics
Rapid filtration Uses granular materials such as sand, anthracite coal, garnet, ilmenite, and granular activated carbon (GAC)	The **effective size (ES or d_{10})** is determined by sieve analysis. It is the media diameter at which 10% of the media by weight is smaller. The effective size is determined by sieve analysis. The typical effective size for rapid filtration media are sand 0.4–0.8 mm, anthracite coal 0.8–2.0 mm, garnet 0.2–0.4 mm, ilmenite 0.2–0.4 mm, GAC 0.8–2.0 mm. The **uniformity coefficient (UC)** is the ratio of the 60th-percentile media diameter (the diameter at which 60% of the media by weight is smaller) to the effective size (d_{10}): $$UC = \frac{d_{60}}{d_{10}}$$ UC is an important parameter in the design of rapid filters because it directly affects the overall effectiveness of the filter bed because of media stratification during backwash. During backwash, coarse grains (more weight) will settle to the bottom of the bed and are difficult to fluidize for effective cleaning. Fine grains (less weight) will collect at the top of the filter, which in turn will cause excessive head loss during the filtration stage. Typical ranges of uniformity coefficients for rapid filtration media are as follows: sand 1.3–1.7, anthracite coal 1.3–1.7, garnet 1.3–1.7, ilmenite 1.3–1.7, GAC 1.3–2.4.
Slow sand filtration Uses sand	Slow sand filter media has a smaller effective size (ES or d_{10}) to achieve a lower filtration rate. It also has a higher uniformity coefficient (UC) because backwash is not involved in slow sand filtration operation; therefore stratification is not a concern. The typical effective size for slow sand filtration is between 0.3 and 0.45 mm, and the uniformity coefficient is less than 2.5. **Filter bed porosity** is like soil porosity. Porosity has a strong influence on the head loss and filtration effectiveness. If the porosity is too small, the head loss will be high, and the filtration rate will decrease rapidly over operation time. If the porosity is too high, the filtration rate will exceed the criterion, and the effluent will not meet the treatment objective. Typical values for porosity are in the range of 40%–60%.

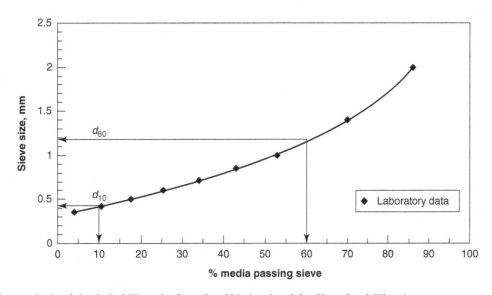

Figure / 8.15   Analysis of the Suitability of a Sample of Native Sand for Slow Sand Filtration.

8.9 Disinfection

Pathogens can either be removed by treatment processes such as granular filtration or inactivated by disinfection agents. The term **disinfection** in drinking-water practice refers to two activities:

1. **Primary disinfection**: the inactivation of microorganisms in the water.

2. **Secondary disinfection**: maintaining a disinfectant residual in the treated-water distribution system (also called residual maintenance).

Disinfection By-products

http://www.epa.gov/safewater/
disinfection/

8.9.1 CURRENT DISINFECTION METHODS

Generally, **disinfectants** can be classified as *oxidizing agents* (for example, chlorine and ozone), *cations of heavy metals* (silver or copper) and *physical agents* (heat or UV radiation). The most commonly used disinfectant is free chlorine. Four other common disinfectants are combined chlorine, ozone, chlorine dioxide and ultraviolet (UV) light. Combined chlorine is often limited to secondary disinfection.

Table 8.20 summarizes the effectiveness, regulatory limits, typical application, and chemical source of the five most common disinfectants. Table 8.21 provides detailed information on the important disinfection chemistry and application considerations for the five common disinfectants (including the importance of pH, as shown in Figure 8.16).

Reactors used for disinfection are usually called *contactors*. Free chlorine, combined chlorine, and chlorine dioxide are most often used in contactors close to ideal plug flow reactors, such as baffled, serpentine contactor chambers. Both types of contactors can be designed so that they are highly efficient, closely approaching ideal plug flow. Ozone is generally introduced in bubble chambers in series. UV light is often applied in proprietary reactors where short-circuiting is a concern because contact times are so short.

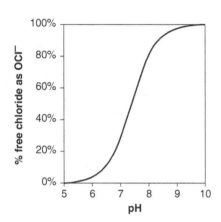

Figure / 8.16 Effect of pH on the Fraction of Free Chlorine Present as Hypochlorite Ion (OCl⁻).

8.9.2 DISINFECTION KINETICS

The mechanisms for pathogen inactivation during disinfection are complex and not well understood. Therefore, kinetic models have been developed that are based on laboratory observations. **Chick's law** (Equation 8.12) is the most straightforward model to describe the disinfection process. It assumes the rate of the disinfection reaction is **pseudo first order** with respect to the concentration of the pathogens being inactivated:

$$\frac{dN}{dt} = -K \times N \qquad (8.12)$$

where dN/dt is the rate of change in the number of organisms with time (organisms/volume/time), N is the concentration of organisms

Effectiveness, Regulatory Limits, Typical Application, and Chemical Source of the Five Most Common Disinfectants

Issue	Disinfectant				
	Free Chlorine	Monochloramine	Chlorine Dioxide	Ozone	UV Light
Effectiveness					
Bacteria	Excellent	Good	Excellent	Excellent	Good
Viruses	Excellent	Fair	Excellent	Excellent	Fair
Protozoa	Fair to poor	Poor	Good	Good	Excellent
Endospores	Poor to good	Poor	Fair	Excellent	Fair
Frequency of use as primary disinfectant	Most common	Common	Occasional	Very common	Increasingly common
Regulatory limit on residuals	4 mg/L	4 mg/L	0.8 mg/L	—	—
Formation of chemical by-products					
Regulated by-products	Forms 5 trihalomethanes (THMs) and 4 haloacetic acids (HAAs)	Traces of THMs and HAAs	Chlorite	Bromate	None
By-products that may be regulated in the future	Several	Cyanogen halides, nitrosodimethylamine (NDMA)	Chlorate	Biodegradable organic carbon	None known
Typical application Dose, mg/L	1–6	2–6	0.2–1.5	1–5	20–100 mJ/cm^2
Chemical source	Delivered as liquid gas in tank cars, 1 ton and 68 kg cylinders as calcium hypochlorite powder for very small applications. On-site generation from salt and water–using electrolysis.	Same sources as for chlorine. Ammonia is delivered as aqua ammonia solution, liquid gas in cylinders, or as solid ammonium sulfate. Chlorine and ammonia are mixed in treatment process.	Same sources as for chlorine. Chlorite as powder or stabilized liquid solution. ClO$_2$ is manufactured with an on-site generator.	Manufactured on-site using a corona discharge in very dry air or pure oxygen. Oxygen is usually delivered as a liquid. Oxygen is also manufactured on-site in some very large plants.	Uses low pressure or low-pressure–high-intensity UV (254 nm) or medium-pressure UV (several wavelengths) lamps in the contactor.
Typical contactor	In the past was added at beginning of plant and residual carried through. Increasingly, individual contactors are used.	In the past was added at beginning of plant and residual carried through. Increasingly, individual contactors are used.	In the past was added at beginning of plant and residual carried through. Increasingly, individual contactors are used.	Has always been added in specially engineered contactors. These contactors are using more compartments.	Lamps are placed in gravity channels or in specially manufactured UV reactors. Because the contact time is so short, reactors must be tested.

SOURCE: Crittenden et al., 2012. Adapted with permission of John Wiley & Sons, Inc.

Table / 8.21

Important Chemical Reactions Associated with Common Disinfectants

Disinfectant	Important Chemical Reactions	Considerations
Free chlorine When gaseous chlorine is added to water, it rapidly reacts with water to form hypochlorous acid (HOCl) and hydrochloric acid (HCl). Hypochlorous acid and hypochlorite ion together are often referred to as *free chlorine* (free chlorine = $HOCl + OCl^-$). Both of these chemical species are active disinfecting agents; however, hypochlorous acid (HOCl) is much more effective than OCl^- for disinfection.	$Cl_{2(g)} + H_2O \rightarrow HOCl + HCl$ **(8.13)** $HOCl \rightarrow H^+ + OCl^-$ $K_a = 10^{-7.5}$ **(8.14)**	As can also be seen from the equilibrium constant for Equation 8.13 and Figure 8.16, HOCl is the predominant form of free chlorine at a pH less than 7. Consequently, a treatment plant operator will attempt to maintain the pH at 7 or slightly lower to increase the disinfection power of the added chlorine. Although chlorine disinfection is very effective and cost-advantageous, use of chlorine has some concerns. One of them is by-product formation; chlorine will react with dissolved organic matter that occurs naturally in waters to form carcinogenic THMs.
Combined chlorine When chlorine and ammonia (NH_3) are both present in water, they react to form three *chloramine compounds* (NH_2Cl, $NHCl_2$, NCl_3) according to the three reactions on the right. These three chloramines together are referred to as **combined chlorine**. (combined chlorine = $NH_2Cl + NHCl_2 + NCl_3$). The **total chlorine residual** is the sum of the combined chlorine and any free chlorine residual.	*Monochloramine formation:* $HOCl + NH_3 \rightarrow NH_2Cl + H_2O$ **(8.15)** *Dichloramine formation:* $HOCl + NH_2Cl \rightarrow NHCl_2 + H_2O$ **(8.16)** *Trichloramine formation:* $NHCl_2 + HOCl \rightarrow NCl_3 + H_2O$ **(8.17)**	The formation of these species depends upon the ratio of Cl_2 to NH_3–N. At a high ratio of Cl_2 to NH_3–N, the oxidation of ammonia to nitrogen gas and nitrate ion occurs. $3HOCl + 2NH_3 \rightarrow N_{2(g)} + 3H_2O + 3HCl$ **(8.18)** $4HOCl + NH_3 \rightarrow HNO_3 + H_2O + 4HCl$ **(8.19)** The reactions are also dependent upon chlorine dosage, temperature, pH, and alkalinity. At low pH values, other reactions become more significant, such as these: $NH_2Cl + H^+ \rightarrow NH_3Cl^+$ **(8.20)** $NH_3Cl^+ + NH_2Cl \rightleftharpoons NHCl_2 + NH_4^+$ **(8.21)**
Chlorine dioxide This chemical does not produce significant amounts of **trihalomethanes (THMs)** as by-products from reactions with organics.	$2NaClO_2 + Cl_{2(g)} \rightarrow 2ClO_{2(g)} + 2NaCl$ **(8.22)** $2NaClO_2 + HOCl \rightarrow 2ClO_{2(g)} + NaCl + NaOH$ **(8.23)** $5NaClO_2 + 4HCl \rightarrow 4ClO_{2(g)} + 5NaCl + 2H_2O$ **(8.24)**	Chlorine dioxide is explosive at elevated temperatures, on exposure to light, or in the presence of organic substances. Chlorine dioxide does not produce THMs; however, it produces inorganic products such as chlorite (ClO_2^-) and chlorate (ClO_3^-), which have health concerns at certain levels of exposure.

(continued)

Disinfectant	Important Chemical Reactions	Considerations
Chlorine dioxide (*continued*) Chlorine dioxide also has a higher oxidizing power than chlorine; however, at neutral pH values typical of most waters, it has only about 70% of the oxidizing capacity of chlorine.		ClO_2 is generated on-site using sodium chlorite with gaseous chlorine (Cl_2), aqueous chlorine ($HOCl$), or acid (usually hydrochloric acid, HCl).
Ozone Ozone is a stronger oxidant than the other three disinfectants.	$3O_2 \rightarrow 2O_3$ **(8.25)** Ozone can decompose to the hydroxyl radical (HO·), which is formed by the reactions with high concentrations of hydroxide ion (OH^-) or natural organic matter (NOM). $3O_3 + OH^- + H^+ \rightarrow 4O_2 + 2H_2O·$ **(8.26)** $O_3 + NOM \rightarrow HO· + \text{by-products}$ **(8.27)**	Ozone is a highly reactive gas and decays rapidly under ambient conditions. Therefore, it has to be generated on-site, commonly by electrical discharges in the presence of O_2. Ozone reacts with microbes by direct oxidation or through the action of hydroxyl radicals generated as in Equations 8.26 and 8.27. Hydroxyl radicals are scavenged by carbonate species (HCO_3^-, CO_3^{2-}) and reduced metal ions (e.g., Fe^{2+}, Mn^{2+}). High pH conditions or high concentrations of organic matter favor hydroxyl radical oxidation reactions. Ozone disinfection is primarily dependent on its direct reactions, and the residual of ozone is important. Low pH, high alkalinity, low concentrations of organic matter, and low temperature will increase the stability of aqueous ozone residuals. Unfortunately, use of ozone will not result in a residual that can continue the disinfection process in the water distribution system. In addition, ozonation of waters containing bromide produces bromate (BrO_3^-), which is believed to be a carcinogen to humans.
UV radiation This is electromagnetic radiation having a wavelength between 100 and 400 nm.	Light in the UV spectrum can be divided into vacuum UV, short-wave UV (UV-C), middle-wave UV (UV-B), and long-wave UV (UV-A). The region of the wavelength of 200–300 nm is a germicidal range where deoxyribonucleic acid (DNA) absorbs UV.	The photons in UV light react directly with the nucleic acids in the form of DNA and damage DNA. This will inhibit further transcription of the genetic code of the cell and prevent its successful reproduction. However, microorganisms may evolve to repair UV-induced damage through photoreactivation and dark repair mechanisms. Therefore, reactivation is an important consideration in UV disinfection. The performance of UV disinfection systems is highly influenced by dissolved substances and particulate matter in waters, which have to be considered in UV reactor design.

(organisms/volume), and K is the Chick's law rate constant (time^{-1}). Integrating Equation 8.12 results in

$$\ln\left(\frac{N}{N_0}\right) = -K \times t \qquad (8.28)$$

where N_0 is the initial concentration of organisms (organisms/volume).

The rate of disinfection can be determined by plotting the log organism concentration ratio (N/N_0) versus time. To better estimate the rate of disinfection, several measurements should be made at each time step. Due to inaccurate measurement, the best fit often may not pass through zero. The disinfection rate from Equation 8.28 is related to disinfectant concentration, and the reaction has a different rate constant for each concentration.

example/8.7 Application of Chick's Law

Given the data in the first three columns of Table 8.22, graph the data for the inactivation of *Poliovirus* type 1, using the disinfectant bromine. Determine the Chick's law rate constant for each of the two disinfectant concentrations.

solution

Calculate the log of the survival value (N/N_0) for each data point. The results are shown in the two right columns of Table 8.22. Then plot $\log(N/N_0)$ as a function of time (t) and fit a linear line through the data. Figure 8.17 shows this graph.

The value of the slope of the line in Figure 8.17 corresponds to the Chick's law rate constant. For a disinfectant concentration of 21.6 mg/L, the rate constant K (base 10) is 1.8333/s; thus, K (base e) is 4.22/s. For the disinfectant concentration of 4.7 mg/L, the rate constant K (base 10) is 0.6667/s; thus, K (base e) is 1.54/s. Remember that, as we discussed previously, this is a pseudo first-order rate constant, so the rate constant differs for different disinfectant concentrations.

Figure / 8.17 Log (N/N_0) as a Function of t for Example 8.7.

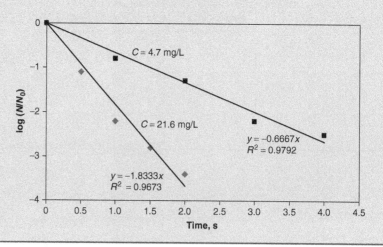

Table / 8.22

Data Used to Solve Problem Applying Chick's Law in Example 8.7

C (mg/L)	t (sec)	N (number of organism/L)	N/N_0	$\log (N/N_0)$
21.6	0.0	500	1	0
21.6	0.5	40	0.080	−1.1
21.6	1.0	3	0.006	−2.2
21.6	1.5	1	0.002	−2.7
21.6	2.0	0.2	0.0004	−3.4
4.7	0.0	500	1	0
4.7	1.0	79	0.158	−0.8
4.7	2.0	25	0.050	−1.3
4.7	3.0	3	0.006	−2.2
4.7	4.0	1.5	0.003	−2.5

The **Ct approach** uses the product $C \times t$, which can be viewed as the dosage of disinfectant. C is the concentration of a chemical disinfectant (mg/L). It is measured after the time segment, t, where t is the time required to achieve a level of inactivation. A similar concept is the product of the UV light intensity (I, mW/cm^2) and the time of exposure, t. $I \times t$ (units of mW/cm$^2 \times$ s or mJ/cm^2) is used during UV disinfection to compute the dose of UV light.

The Ct approach is a useful way to compare the relative effectiveness of different disinfectants and the resistance of different organisms. Figure 8.18 illustrates this by comparing the Ct and It required for a 99 percent inactivation of several microorganisms using the five common disinfectants. Figure 8.18 shows that ozone requires a lower Ct for most microorganisms than the other three chemical oxidants need; therefore, ozone is a stronger disinfectant. Also, the microorganism called *C. parvum* requires the highest Ct among four chemical oxidants, so this microorganism is most resistant to these disinfectants. The It data for UV disinfection are included in Figure 8.18. However, not much can be concluded by comparing the It values with the Ct values for the same organism.

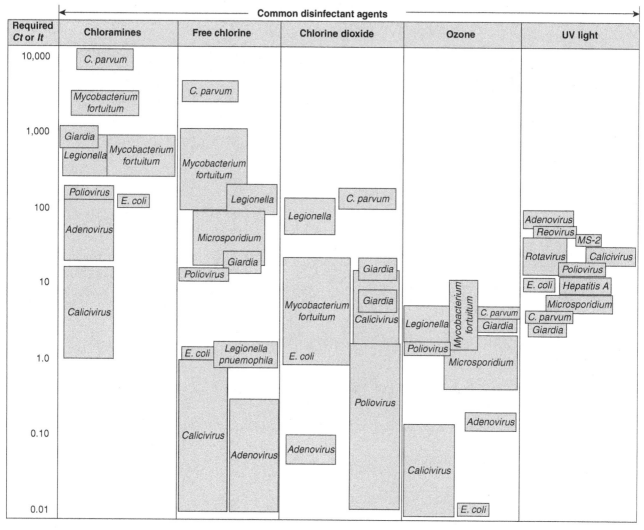

Figure / 8.18 **Overview of Disinfection Requirements for 99% Inactivation** This chart compares the *Ct* or *It* required for a 99% inactivation of several pathogens, using five common disinfectants (listed on the top). Ozone requires a lower *Ct* for most pathogens than the other three chemical oxidants, so ozone is a stronger disinfectant. The microorganism *C. parvum* requires the highest *Ct* among the four chemical oxidants, so this microorganism is most resistant to these disinfectants.

(Crittenden et al. (2012). Redrawn with permission of John Wiley & Sons, Inc.).

During the disinfection process, electricity is used by mixers, process water pumps, and chemical pumps. The production and delivery of disinfecting agents to treatment facilities also requires energy inputs that occur over the life cycle of the material. Electricity consumption has been found to be approximately 2 kWh/m^3 for the operation of a chlorination disinfection process at surface water treatment plants with capacities ranging from 1 to 20 MGD. The electricity requirement for chlorination at groundwater treatment plants, however, was found to be about 9 kWh/m^3 (Burton, 1996). This shows that energy use at a drinking-water treatment plant is influenced by the source of the raw water.

example / 8.8 Application of *Ct* Value

The Surface Water Treatment Rule (SWTR) requires at least 99.99 percent (4-log) removal and/or inactivation of viruses. A water treatment plant provides disinfection using free chlorine with a contact time of 30 min. Determine the concentration of free chlorine required for disinfection at pH of 7 and water temperatures of 5°C and 20°C. The *Ct* values are provided: $Ct = 8$ min-mg/L at pH = 7 and temperature of 5°C; $Ct = 3$ min-mg/L at pH = 7 and temperature of 20°C.

solution

The required residual concentration of free chlorine (mg/L) for the given contact time of 30 min is determined as follows:

$$C_{20} = \frac{3 \text{ min-mg/L}}{30 \text{ min}} = 0.1 \text{ mg/L}$$

$$C_{5} = \frac{8 \text{ min-mg/L}}{30 \text{ min}} = 0.27 \text{ mg/L}$$

Note that at the lower temperature, higher *Ct* values are required. Therefore, a higher concentration of free chlorine may have to be applied during winter months.

Application / 8.7   Pathogen Removal and Disinfection in the Developing World

In the developing world, inactivation of micro-biological contaminants is accomplished through the addition of disinfection agents such as free chlorine, UV light, or heat (Table 8.23). Water storage needs to be carefully considered as well, because water safe at the point of collection is often recontaminated with fecal matter during collection, transport, and use in home. Vessels with narrow mouths and taps reduce the chance of recontamination, and open containers should always be covered (Mihelcic et al., 2009).

Hand-dug wells can be chlorinated by lowering a chlorination pot (Figure 8.19) into the well or directly injecting chlorine into the well daily. Some communities take the intermediate measure of periodically injecting chlorine into a well to reduce contamination. However, disinfection will be short-lived with this method if the pathogens are present in the groundwater source.

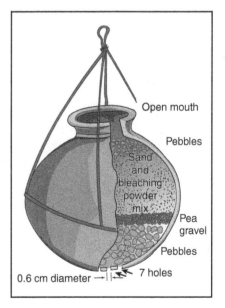

Figure / 8.19 Example of a Chlorination Pot.

(From Mihelcic et al. (2009). Drawing courtesy of Linda D. Phillips).

Disinfection Agents and Processes Used at the Household Level in the Developing World

Disinfection Agent or Process	Examples
Heat	Boiling is still the most common means of treating water in the developing world but requires valuable wood (which can be expensive), produces air pollution, and has safety concerns related to scalding of children.
	Water has a flat, deaerated taste.
	Most experts suggest attaining a "rolling boil" for 1–5 min. The WHO recommends reaching a rolling boil. This gets you in excess of temperatures needed for sterilization.
	Heating to pasteurization temperatures (60°C) for 1–10 min will destroy many waterborne pathogens.
Solar	Solar systems take time and work with limited volumes of water.
	Uses combination of UV and pasteurization processes for disinfection. Heating to pasteurization temperatures (60°C) for 1–10 min will destroy many waterborne pathogens.
	Solar disinfection (SODIS) is a simple treatment method that takes advantage of the bacterial destruction potential of sunlight. Treatment involves placing clear bottles of water to be treated in direct sunlight for a determined amount of time.
	Solar cookers or reflectors (made of cardboard or aluminum foil) can attain pasteurization temperatures of 65°C.
Chlorination	Imparts taste to water. Requires turbidity be below 1–5 NTU. The WHO guideline for chlorine is 5 mg/L, which is above the taste threshold of 0.6–1.0 mg/L.
	Liquid laundry bleach is a readily available source of chlorine in many developing communities. The powder form of chlorine is calcium hypochlorite, which includes chlorinated lime, tropical bleach, bleaching powder, and HTH. Calcium hypochlorite can be found in 30–70% solutions. Liquid bleach is in the form of sodium hypochlorite and contains 1–18% chlorine.
	At the community level, *chlorination pots* are added to hand-dug wells and use bleaching powder or chlorinated lime (Figure 8.19). Pressed calcium hypochlorite tablets can also be added to hand-dug wells.
	Mechanical chlorinators with no moving parts are sometimes constructed on top of a water storage tank and drip a bleach solution into the tank.
	At the household level, a 1% chlorine stock solution can be used. Chlorine is added at 1–5 mg/L to achieve a residual of 0.2–0.5 mg/L after contact time of 30 min. This increases to 1 h contact time for cold waters.
Iodine	Imparts taste to water. The WHO recommends a dose of 2 mg/L with a contact time of 30 min. Iodine is better at penetrating particulate matter than chlorine but is more expensive.
Sedimentation and filtration	The three-pot system uses a series of three pots where water is transferred between pots daily, and gravity settling reduces helminth ova and some protozoans by more than 90%, especially with 1–2 day storage. Virus and bacteria removal are much smaller.
	Filtration can consist of a wide variety of media: granular media, slow sand, fabric, paper, canvas, and ceramics.

SOURCE: Mihelcic et al., 2009.

8.10 Membrane Processes

By 2007, over 20,000 membrane (including reverse osmosis) plants were operating worldwide. This number is expected to grow at a significant pace as population and water consumption increase. **Membrane processes** involve water pumped under pressure, called *feedwater*, into a housing containing a semipermeable membrane, where some of the water filters through the membrane and is called *permeate*.

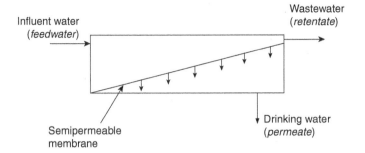

Figure / 8.20   **Diagram of Membrane Separation Process**   The influent is separated into a permeate stream (for drinking water) and a retentate stream, which currently becomes waste. One challenge to the water treatment industry is to find beneficial uses for the retentate.

(Crittenden et al. (2005). Redrawn with permission of John Wiley & Sons, Inc.).

The rest of the water containing the filtered constituents passes by the filter and is called *retentate,* as shown in Figure 8.20.

The pressure required for constant filtration or flux through the membrane is called the *transmembrane* pressure. This pressure provides the driving force for filtration to take place. As the pore size of the membrane decreases, the transmembrane pressure increases. The **flux rate** is the rate at which the permeate flows through the membrane area and is expressed as L/m^2-day (gpm/ft^2). Permeate is usually stabilized if needed, disinfected, and sent to the distribution system.

8.10.1   CLASSIFICATION OF MEMBRANE PROCESSES

Selection of the type of membrane systems depends upon the constituents to be removed. Four types of membrane systems are used in water treatment: **microfiltration (MF), ultrafiltration (UF), nanofiltration (NF),** and **reverse osmosis (RO)**. Figure 8.21 classifies the four membrane processes used in drinking-water treatment. Table 8.24 provides more information about the four types of membrane filters. Microfiltration and ultrafiltration are sometimes classified as membrane filtration because

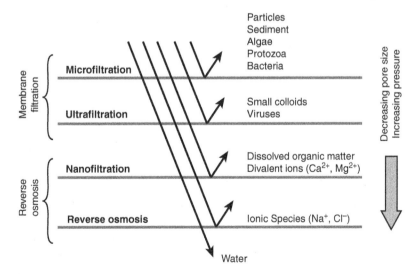

Figure / 8.21   **Classification of Four Pressure-Driven Membrane Processes**   The processes are microfiltration, ultrafiltration, nanofiltration, and reverse osmosis.

(Adapted from Crittenden et al. (2012). Redrawn with permission of John Wiley & Sons, Inc.).

Types of Membrane Filter Systems

Type of Membrane Filter	Considerations
Microfiltration (MF)	Has membrane pore sizes ≈ 0.1 in nominal diameter.
	Removes particles, algae, bacteria, and protozoa that have sizes larger than the nominal diameter.
	Transmembrane pressure operating ranges are 0.2–1.0 bar (2–15 psig).
Ultrafiltration (UF)	Has membrane pore sizes as small as 0.01 μm in nominal diameter.
	Can remove constituents such as small colloids and viruses.
	Operating pressure ranges are 1–5 bar (15–75 psig).
Nanofiltration (NF)	Has membrane pore sizes as small as 0.001 μm in nominal diameter.
	Removes dissolved organic matter and some divalent ions such as calcium and magnesium ions.
	Transmembrane pressure operating ranges are 5.0–6.7 bar (75–100 psig).
Reverse osmosis (RO)	Filters are considered nonporous, and usually only constituents that are the size of water molecules can pass through the filter.
	Transmembrane operating pressure ranges are 13.4–80.4 bar (200–1,200 psig).

they primarily remove particulate constituents. Nanofiltration and reverse osmosis are sometimes classified as reverse osmosis processes because they remove dissolved constituents.

The **energy** use required to pass water through a membrane can be determined as follows:

$$Hp = \frac{Q \times \Delta P}{1,714}$$

(8.29)

where Hp is the power required to pass a million gallons of water per day through the membrane (kWh/MGD), Q is the flow rate (gpm), and ΔP is the required feed pressure (psi).

8.10.2 MEMBRANE MATERIALS

Membranes are manufactured in tubular form, flat sheets or as fine hollow fibers. They are composed of either natural or synthetic materials. Modified natural types consist of cellulose acetate, cellulose diacetate, cellulose triacetate, and a blend of both di- and triacetate materials. Synthetic membrane materials may be composed of polyamide, polysulfone, acrylonitrile, polyethersulfone, Teflon, nylon, and polypropylene polymers.

The membrane thickness can range from 0.1 to 0.3 μm. Some membrane materials, such as cellulose acetate and polyamide, are manufactured as thin film composites (TFC). TFC membranes usually have a very thin active layer that is bonded to a thicker porous support material for strength.

Some membrane materials are sensitive to temperature, pH, and oxidants. Cellulose acetate membranes have an operating temperature range of 15–30°C and are not tolerant above 30°C. They are also sensitive to hydrolysis at high and low pH. The triacetate and di/triacetate blends have improved hydrolytic stability in water with high and low pH. Polyamide, polysulfone, nylon, Teflon, and polypropylene membranes have high physical stability, do not hydrolyze in the pH range of 3–11, and are immune to bacterial degradation. However, some polyamide membranes may be subject to oxidation by disinfectants such as chlorination.

8.10.3 MEMBRANE PROCESS TYPES AND CONFIGURATIONS

The most common types of membrane configurations used in water treatment are *spiral-wound* and *hollow-fiber modules*. A spiral-wound element consists of sheets of the membrane, permeate material, channel spacer, and outer membrane wrap wound around a porous permeate tube. Water enters one end of the element and flows across the wound filter surface area, and the water permeates through the membrane into the collection tube while the concentrate or retentate flows out the other end of the filter element. Spiral-wound elements can range from 100 to 300 mm in diameter and 1 to 6 m in length. Bundles of elements, usually 30–50 in number, are skid mounted (called skids or arrays) and operated in parallel mode.

For RO operation, skid-mounted units or arrays can also be operated in parallel and in series operation. For example, two arrays may operate in parallel (first stage) and the retentate from each array is fed to two other arrays (second stage) to increase water recovery. Figure 8.22 displays an RO plant with four skid-mounted units.

Perhaps the most widely used membrane is the hollow-fiber type. **Hollow fibers** have outside diameters ranging from 0.5 to 2.0 mm wall

Figure / 8.22   Reverse Osmosis Drinking-Water Treatment Plant Showing Four Skid-Mounted Units.

(Photo provided by George Tchobanoglous).

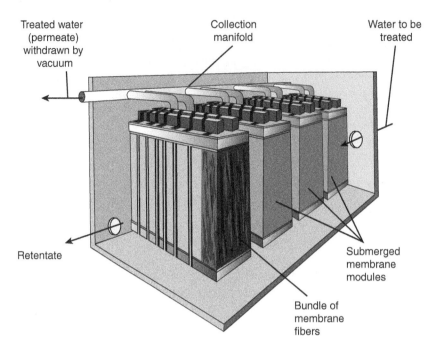

Figure / 8.23 Membranes Submerged in Tanks.

(Crittenden et al. (2012). Redrawn with permission of John Wiley & Sons, Inc.).

thickness ranging from 0.07 to 0.60 mm, and lengths of about 1 m. Typically, 7,000–10,000 fibers are housed in stainless-steel or fiberglass pressurized modules with diameters ranging from 100 to 300 mm and lengths ranging from 1 to 2 m.

In addition to pressure vessels, microfiltration and ultrafiltration membranes can operate submerged in tanks. Figure 8.23 is a schematic of what is called a *submerged-membrane system*. In the submerged mode, the membranes are submerged in a completely mixed tank, where the feed water enters the tank. Permeate is drawn through the membrane by applying a vacuum on the permeate side of the membrane. The retentate is usually drawn off in a semibatch mode or in a continuous mode if the reactor is operating as a completely mixed-flow reactor (CMFR).

8.10.4 MEMBRANE SELECTION AND OPERATION

Membrane operation is highly dependent upon the influent water characteristics and the type of constituents in the water that need to be removed. If fresh surface water requires removal of particulate constituents, then the required process is membrane filtration (microfiltration or ultrafiltration). If the removal of dissolved constituents is required, then the process should be nanofiltration or RO. Figure 8.24 depicts the layouts of microfiltration and RO water treatment plants.

As shown in Figure 8.24a, microfiltration plants typically require a prefiltration step to remove coarse particles with diameters in excess of 200–500 μm. If dissolved constituents (such as manganese, iron, or hardness) or colloidal constituents are present and

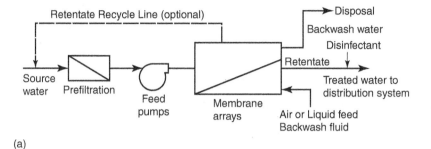

(a)

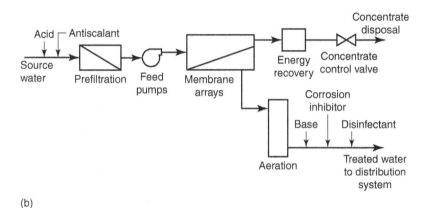

(b)

Figure / 8.24 **Typical Components of a Membrane Drinking-Water Treatment Plant** This schematic compares (a) a microfiltration plant and (b) reverse osmosis plant.

(Crittenden et al. (2012). Redrawn with permission of John Wiley & Sons, Inc.).

interfering with membrane operation, chemical pretreatment steps may be required. Feed pumps are utilized to provide the transmembrane pressure necessary for filtration to occur. Over time, particulate and dissolved constituents build up on the surface of the membrane, causing the permeate flux to decrease or increase the transmembrane pressure.

At some point in the operation, the membranes require backwashing to restore their performance. Backwashing a microfiltration filter is accomplished by reversing fluid flow through the use of air or permeate at a pressure higher than normal operating pressure. Permeate is then disinfected and sent to the distribution system. If there is retentate, it can be returned directly to the feed line or to a mixing basin upstream of the membranes. The backwash water containing cleaning chemicals and solids may be sewered for treatment at the wastewater treatment plant, treated on-site, or permitted for discharge into source water.

For microfiltration membranes, the backwash cycle occurs approximately every 30–90 min, depending upon the influent water quality. After a period of time, chemical cleaning is required to reverse the loss of flux and restore the permeability. The loss of flux is due to fouling of the membrane caused by particulates, dissolved organic matter, and biological fouling from microorganism growth. This is called *reversible fouling*, as the loss of flux can be restored. *Irreversible fouling* is due to membrane compaction during the first year of operation. Compaction occurs when the large voids collapse in the porous membrane support layer because of excessive applied pressure. The applied pressure reduces the size of the support layer voids, causing a reduction in the permeability through the entire membrane cross section.

More on Desalination

http://www.water.ca.gov/desalination/

As shown in Figure 8.24b, reverse osmosis plants require acid and chemical addition to prevent scale formation on the membranes. The scale formation is caused by soluble salts in the feed water concentrating in the membrane to a degree that they exceed their solubility product and begin to precipitate as a solid on the membrane surface. Adding acid will change the solubility of the salts, and the antiscalant will help prevent the formation of the scale or at least slow down the rate of precipitation.

Prefiltration is used to prevent particles from clogging RO membranes. For surface waters, cartridge filters, granular filtration, and/or microfiltration may be necessary to remove particles before RO filtration. Disinfection may also be required to prevent microbial fouling of the RO membranes. Care must be taken to ensure that the RO membranes are not susceptible to oxidation by the disinfectant.

Permeate water quality is usually acidic (pH ≈ 5.0), low in alkalinity, and corrosive, and depending upon the water source, permeate may contain dissolved gas (for example, hydrogen sulfide if the water source is reduced groundwater). Some permeates may require aeration to remove unwanted dissolved gases. The corrosiveness of the water requires adjustments to pH and alkalinity by using a basic solution and adding a corrosion inhibitor (sodium silicate and sodium hexametaphosphate). A disinfectant is usually added to the water before it enters the clear well and distribution system.

Since the retentate stream is under high pressure, a concentrate control valve is used to capture the energy and reduce the pressure of the retentate stream. The concentrate control valve is an energy recovery system and can be used to reduce energy use at the RO plant. The retentate may be further treated, as with the use of evaporation ponds to further concentrate the salt, or may be discharged into the ocean, brackish estuary, or river, or to a municipal sewer. Obviously, the retentate stream is easier to dispose of in coastal areas, where the discharge does not present as many problems, than at inland locations where freshwater is present. However, permitting will be required to ensure protection of coastal ecosystems.

8.10.5 MEMBRANE PERFORMANCE

Table 8.25 summarizes typical operating characteristics and general performance parameters for membrane systems. For microfiltration systems, typical water recovery exceeds 95 percent and suspended-solids removal is around 90–98 percent, depending upon the membrane's nominal pore size. Microbiological removal can be as high as 7-log removal for protozoa such as *Giardia lamblia* cysts and *Cryptosporidium parvum* oocysts. Bacteria removal has been reported from 4- to 8-log removal, and virus removal from 7- to 10-log removal.

For RO systems, water recoveries range from 50 to 90 percent, depending upon the influent water type and the staging. TDS removals can be in the range of 90–99.5 percent removal. Microbiological removal is excellent, with greater than 7-log removal for protozoa and between 4- and 7-log removal for bacteria and virus removal (Asano et al., 2006).

Operational Parameters for Membranes

Parameter	Microfiltration	Ultrafiltration	Nanofiltration	Reverse osmosis
Flux rate, $L/m^2 \cdot d$	400–1,600	400–800	10–35	12–20

Operating pressure, kPa
    Freshwater
    Seawater
	0.0007–0.01	0.007–0.7	350–550
			500–1,000
				1,200–1,800
				5,500–8,500

Energy consumption, kWh/m^3
    Freshwater
    Seawater
	0.4	3.0	0.6–1.2	1.5–2.5
				5–10

SOURCE: Data from Crittenden et al., 2012; and Asano et al., 2006.

RO membranes are also very effective in removing SOCs. Table 8.26 describes several mechanisms that foul the membranes as well as methods to remove or prevent the fouling.

In general, membrane processes have much higher energy consumption requirements compared to other water treatment processes. For example, reverse osmosis used for seawater desalination has reported electricity consumption that ranges from 2.3 to 7 kWh/m^3 water produced (Hancock et al., 2012; Muñoz et al., 2010; Stokes and Horvath, 2009; Lyons et al., 2009; Raluy et al., 2005a, b). The energy consumption for membrane processes varies with the salinity of the raw water and the pore size of membranes. For example, RO used for treatment of brackish groundwater (which has lower salinity compared with seawater) has lower electricity consumption; only 0.55–2 kWh/m^3, which is quite lower than the range reported earlier in this paragraph. In addition, changing the process from RO to NF (which results in a larger pore size) will reduce the electricity consumption by 65% (Vince et al., 2008). Overall energy consumption can be also be reduced through recovering energy contained in the high-pressure rejected stream and transferring it to the low-pressure feed stream (Vince et al., 2008).

Types of Membrane Fouling and Methods to Control   (Crittenden et al., 2012).

Type of Fouling	Description	Controlled by
Fouling by particles	Large particles form a cake on the surface of the membrane.	Backwashing will remove particles that are larger than membrane pore sizes.
Biofouling	Biofilm forms on the membrane surface and may secrete extracellular material which adds to fouling. Attachment is strong so backwashing will not remove.	Chemical disinfectants like chlorine are added to the feed water.
Natural organic matter	Natural organic matter (NOM) adsorbs to the surface of the membrane. Believed to be the higher molecular weight and colloidal fractions of NOM that are of sufficient size to restrict membrane pores.	Careful removal of NOM through coagulation, filtration, or biofiltration.

example/8.9 Calculation of Microfiltration Membrane System

A municipality is upgrading its microfiltration plant by replacing membranes of 0.2 μm pore size with membranes that have 0.1 μm pore size. The plant consists of eight arrays with 90 modules per array, and the plant capacity is 29,214 m^3/day. The modules are 119 mm in inside diameter and 1,194 mm in length, and they have an available filtration surface area of 23.4 m^2. The new hollow fibers have 1.0 mm outside diameter and length of 1,194 mm. Calculate the following:

1. Total surface area available for filtration
2. Membrane flux rate in L/m^2-h and gpm/ft^2.
3. Total number of membrane fibers for the plant and each module

solution

The total surface area available for filtration is determined as follows:

$$\text{total surface area} = 8\,\text{arrays} \times \frac{90\,\text{modules}}{\text{array}} \times \frac{23.4\,m^2}{\text{module}} = 16{,}848\,m^2$$

The membrane flux rate is determined as follows:

$$\text{flux rate} = \frac{\text{total plant flow rate}}{(\text{total surface area available for filtration})}$$

$$= \left(\frac{29{,}214\,m^3/\text{day}}{16{,}848\,m^2}\right) \times \left(\frac{\text{day}}{24\,h}\right) \times \left(\frac{1{,}000\,L}{m^3}\right)$$

$$= 72.25\,L/m^2\text{-h}$$

$$\text{flux rate} = 72.25\,L/m^2\text{-h} \times \frac{\text{gallons}}{3.785\,L} \times \frac{m^2}{(3.28\,\text{ft})^2} \times \frac{h}{60\,\text{min}} = 0.0296\,gpm/ft^2$$

The total number of membrane fibers required for the plant and each module is

$$\text{total number of fibers} = \frac{\text{total plant surface area available for filtration}}{\text{external surface area of a single fiber}}$$

$$= \frac{16{,}848\,m^2}{2\pi r L} = \frac{16{,}848\,m^2}{2\pi \times (0.001\,m) \times (1.194\,m)}$$

$$= 2{,}246{,}000$$

$$\left[\text{number of fibers per module}\right] = \frac{\text{total number of fibers}}{\text{number of modules}}$$

$$= \frac{2{,}246{,}000}{8\,\text{arrays} \times \left(\dfrac{90\,\text{modules}}{\text{array}}\right)} = 3{,}120$$

8.11 Adsorption

8.11.1 TYPES OF ADSORPTION PROCESSES

Adsorption processes are widely used to remove organic and inorganic constituents from water and air. For example, **granular activated carbon (GAC)** and **powdered activated carbon (PAC)** are widely used for removing SOCs and odor compounds from drinking-water supplies. Under the Safe Drinking Water Act, the EPA has designated GAC as the best available technology for removing many organic and inorganic constituents from water supplies (for example, SOCs, arsenic, and radionuclides). GAC is also widely used for treating indoor air (for example, removing formaldehyde, toluene, and radon).

Adsorption is a process whereby molecules are transferred from a fluid stream and concentrated on a solid surface by physical forces. A dissolved molecule in the fluid stream (gas or aqueous) that is attracted to and adsorbs to the solid surface is called *adsorbate*. The solid surface onto which the adsorbate adsorbs is called the *adsorbent*. The physical attraction is controlled by van der Waals–type forces, nonspecific binding mechanisms (attractive forces on a molecular level), or the attractive forces between the adsorbate and the adsorbent surface. SOCs that are amenable to adsorption are those that are hydrophobic (water-fearing). Those that are hydrophilic (water-loving) prefer to stay in water, so their affinity to adsorption is low. Table 8.27 provides examples of SOCs that are favorably and unfavorably adsorbed onto activated carbon. Remember that Chapter 3 discussed the mechanisms of adsorption, including use of the Freundlich isotherm to describe equilibrium between the adsorbed and aqueous phase of water contaminants.

8.11.2 ADSORBENT TYPES

The most widely used adsorbent is activated carbon. Activated carbon comes in both granular (GAC) and powdered (PAC) form. Table 8.28 compares the use of GAC and PAC and a few physical properties of these two activated carbon adsorbents. Another adsorbent is granular ferric hydroxide (GFH), which is used primarily for removing arsenic and selenium. In comparison to GAC and PAC, GFH has a specific surface area of 250–300 m^2/g, with particle sizes ranging from 0.32 to 2.00 mm. These adsorbents are porous and have a large internal surface area for adsorption to occur. The high total specific surface area of these adsorbents provides many sites for adsorption to take place.

Treatment with GAC is carried out in a fixed-bed mode of operation either as gravity feed or under pressure. In some cases, GAC can replace granular media in filter operations. Beds in parallel consist of more than one adsorber where the influent concentration is fed to each bed and the effluent is blended.

PAC is typically added at the raw-water intake before the rapid-mix unit, or at the filter inlet before sand filtration, depending upon the other water treatment processes used. For example, PAC use can interfere with pre-oxidation and/or coagulation processes, resulting in higher PAC dosage requirements. The PAC membrane uses a completely mixed-flow reactor (CMFR) for adequate contact time, followed by PAC/water separation using an ultrafiltration membrane, and the PAC is recycled back to the head

Practical Information on Small Community Water Systems
http://www.nesc.wvu.edu/drinkingwater.cfm

Table / 8.27

Examples of Synthetic Organic Compounds That Are Favorably and Unfavorably Adsorbed onto Activated Carbon from Water

Compound	Favorably Adsorbed	Unfavorably Adsorbed
1-aminobutane		X
Acetone		X
Atrazine	X	
Benzene	X	
Carbon tetrachloride	X	
Chloroform	X	
Ethanol		X
Geosmin	X	
Lindane	X	
Methanol		X
Methyl isoborneal	X	
Tert-butyl alcohol		X
Tetrachloroethylene	X	
Toluene	X	
Trichloroethylene	X	

Table / 8.28

Comparison of Activated Adsorbents Used in Water Treatment Granular ferric hydroxide can be employed to remove arsenic and selenium.

	Primary Use	Total Specific Surface Area (m²/g)	Particle Size (mm)	Design Considerations
Granular activated carbon (GAC)	Removal of organic constituents, inorganic constituents such as mercury, fluoride, perchlorate, and arsenic	500–2,500	0.3–2.4	Carried out in a fixed-bed mode of operation either as gravity feed or under pressure. The three common modes of fixed-bed operation are: (1) single adsorber operation, (2) adsorbers operated in parallel, and (3) adsorbers operated in series.
Powdered activated carbon (PAC)	Primarily for removal of seasonal taste and odor; also used for periodic problems with agricultural pollutants such as pesticides and herbicides (e.g., during spring runoff)	800–2,000	0.044–0.074	Injected at head of the conventional water treatment plant and removed during sedimentation and filtration processes. Key design parameter is required dosage, obtained using jar tests.

of CMFR. The membrane reactor configurations may be used when chronic problems are associated with taste and odor or micropollutants.

Energy consumption for GAC adsorption is similar as in filtration process, primarily from pumping during operation and backwash. The electricity consumption is approximately 0.01 kWh/m^3 water produced (Vincent et al., 2008). From a life cycle perspective, energy consumption for the process should also consider the energy required for adsorbent production and delivery, adsorbent regeneration, and waste handling. The electricity used in the operation can be reduced by improving the energy efficiency of the system as discussed in the filtration section. The embodied energy can be reduced by methods that include using adsorbent regeneration processes that are less energy-intensive.

Application / 8.8   Small-Scale Systems for Water Treatment

By definition, a public drinking-water system regularly supplies drinking water to at least 25 people or 15 service connections. Of the approximately 158,000 public water supply systems inventoried by the Environmental Protection Agency in 2005, almost 95 percent were classified as small (population 501–3,300) or very small (population less than 500) systems. These small systems serve a range of facilities, including small municipalities, mobile-home parks, apartment complexes, schools, and factories, as well as churches, motels, resorts, restaurants, and campgrounds. While the majority of the U.S. population is served by larger water systems, small systems play a vital role in providing safe drinking water.

Figure 8.25 shows the unit processes used to treat water at a resort located in northern Minnesota. Water is pumped from Rainy Lake and flows through several treatment process steps before being sent into the distribution system. Lake water is first chlorinated and then enters a set of pressure sand filters that remove larger particles. A 30 μm bag filter provides additional pretreatment. Water then flows through two ultrafiltration membrane units operating in parallel, providing more than the required 2-log removal of *Cryptosporidium* oocysts. Storage tanks provide chlorine contact time for inactivation of viruses and other pathogens and also provide distribution storage volume. Because of the high organic content in the source water (10–12 mg/L TOC), high levels of disinfection by-products have formed at this point in the process. Water therefore flows through a GAC filter (not shown) to remove the disinfection by-products. Because regulations require that a disinfectant residual be maintained at all points of the distribution system, a flow-paced rechlorination system (not shown) provides the final step of treatment.

Some challenges for this type of small system include the complexity of the treatment coupled with the lack of a full-time water operator, obtaining service in a remote area, and the ever-changing technology market and cost considerations.

Figure / 8.25 **Small-Scale Water Supply System Serving a Seasonal Resort and Restaurant Near Voyageurs National Park (Minnesota)** Average water use for this system is approximately 3,000 gpd. The system ensures that the water consumed by the resort owners and visitors, some as part of the famous Long Island iced teas served at this resort, is safe to drink.

(Photo and information courtesy of Anita Anderson, Minnesota Department of Public Health).

Key Terms

- adsorption
- backwash rate
- backwashing
- carbonate hardness
- charge neutralization
- Chick's law
- clean-bed head loss
- coagulant
- coagulation
- coliform
- colloidal particles
- color
- combined chlorine
- critical particle-settling velocity
- Ct approach
- discrete particle settling
- disinfectants
- disinfection
- dissolved organic carbon (DOC)
- dissolved particles
- effective size (ES or d_{10})
- emerging chemicals of concern
- energy
- enmeshment
- filter bed porosity
- filtration
- flocculation
- flux rate

- granular activated carbon (GAC)
- granular media filtration
- hardness
- head loss
- high-rate filtration
- hollow fibers
- indicator organisms
- interparticle bridging
- jar testing
- lime–soda ash softening process
- maximum contaminant level goal (MCLG)
- maximum contaminant level (MCL)
- membrane processes
- microfiltration (MF)
- nanofiltration (NF)
- natural organic matter (NOM)
- nephelometric turbidity unit (NTU)
- Newton's law
- noncarbonated hardness
- overflow rate
- palatable
- particles
- pathogens
- potable water

- powdered activated carbon (PAC)
- precipitation
- primary disinfection
- pseudo first order
- rapid filtration
- rapid mixing
- reclaimed water
- reference dose (RfD)
- reverse osmosis (RO)
- Reynolds number (Re)
- root mean square (RMS) velocity gradient
- secondary disinfection
- sedimentation
- slow sand filtration
- Stokes' law
- suspended solids
- synthetic organic chemical (SOC)
- taste and odor
- temperature
- total chlorine residual
- total coliforms
- trihalomethanes (THMs)
- Type I, II, and III settling
- turbidity
- ultrafiltration (UF)
- uniformity coefficient (UC)
- water treatment

8.1 The EPA provides reports (sometimes referred to as consumer confidence reports) that explain where your drinking water comes from and whether any contaminants are in the water. Go to this information at the "Local Drinking Water Information" page of EPA's web site, (http://water.epa.gov/drink/local/index.cfm). Look up the utility that serves your university and the largest city near your hometown. (a) What is the source of water? (b) Are there any violations? (c) If so, are they for physical, biological, or chemical constituents?

8.2 Nitrate concentrations exceeding 10 mg NO_3^- as N/L are a concern in drinking water due to the infant disease known as methemoglobinemia. Nitrate concentrations near three rural wells were reported as 5 mg NO_3^-/L, 35 mg NO_3^-/L, and 50 mg NO_3^-/L. Do any of these wells exceed this 10 ppm regulatory standard?

8.3 What are the major differences and similarities between the water quality of a typical surface water and typical groundwater source?

8.4 Jar testing was performed using alum on a raw drinking-water source that contained an initial turbidity of 20 NTU and an alkalinity of 35 mg/L as $CaCO_3$. The optimum coagulant dosage was determined as 18 mg/L with a final turbidity of 0.25 NTU. Determine the quantity of alkalinity consumed as $CaCO_3$.

8.5 Jar tests were performed on untreated river water. An optimum dose of 12.5 mg/L of alum was determined. Determine the amount of natural alkalinity (mg/L as $CaCO_3$) consumed. If 50×10^6 gallons/day of raw water are to be treated, determine the amount of alum required (kg/year).

8.6 A utility is trying to achieve 25 percent removal of TOC and is using jar test to determine the optimal coagulant dose. The following table contains their jar test data (data from EPA 815-R-99-012, 1999). What is the optimal coagulant dose (mg/L)?

Alum Dose (mg/L)	Settled Water TOC (mg/l)	Alum Dose (mg/L)	Settled Water TOC (mg/L)
0	5.45	60	3.60
10	5.50	70	3.24
20	5.50	80	3.00
30	5.00	90	2.78
40	4.78	100	2.53
50	4.52		

8.7 Ferric sulfate is available as a commercial coagulant and is popular at removing turbidity and color. The chemical reaction for its addition to water is

$$Fe_2(SO_4)_3 + 3Ca(HCO_3)_2 \rightarrow 2Fe(OH)_{3(s)} + 3CaSO_4 + 6CO_2$$

Results of a jar test to determine the optimal coagulant dose are provided below. The initial water sample has a pH = 6.5, turbidity of 30 NTU, and alkalinity of 250 mg/L as $CaCO_3$.

Ferric Sulfate dose, mg/L	5	10	15	20	25
Turbidity, NTU		15	5	1	0.9	2

(a) What is the optimal mass of ferric sulfate you would need to purchase every day to treat 1×10^6 gallons/day to a turbidity below 1 NTU (assume 100 percent purity of the coagulant). (b) Do you have to add alkalinity to the system? If so, how much (in concentration units as mg $CaCO_3$/L)?

8.8 A mechanical rapid-mix tank is to be designed to treat 50 m^3/day of water at a temperature of 12°C. Using typical design values in the chapter, determine the (a) tank volume and (b) power requirement.

8.9 An in-line mixer is to be used for rapid mixing. The plant flow is 3,780 m^3/day, the water viscosity is 0.001307 N-s/m^2, and the RMS velocity gradient is 10^4/s. Estimate the daily power requirement for the in-line mixer.

8.10 The city of Melbourne, Florida, has a surface water treatment plant that produces 20 MGD of potable drinking water. The water source has hardness measured as 94 mg/L as $CaCO_3$, and after treatment, the hardness is reduced to 85 mg/L as $CaCO_3$. (a) Is the treated water soft, moderately hard, or hard? (b) Assuming all the hardness is derived from calcium ion, what would the concentration of calcium be in the treated water (mg Ca^{2+}/L). (c) Assuming all the hardness is derived from magnesium ion, what would the concentration of magnesium be in the treated water (mg Mg^{2+}/L).

8.11 A laboratory provides the following analysis obtained from a 50 mL sample of raw water. $[Ca^{2+}] = 60$ mg/L, $[Mg^{2+}] = 10$ mg/L, $[Fe^{2+}] = 5$ mg/L, $[Fe^{3+}] = 10$ mg/L, total solids = 200 mg/L, suspended solids = 160 mg/L, fixed suspended solids = 40 mg/L, and volatile suspended solids $= 120$ mg/L. (a) What is the hardness of this water sample in units of mg/L as $CaCO_3$? (b) What is the concentration of total dissolved solids of this sample?

8.12 A source water mineral analysis shows the following ion concentrations in the water: $Ca^{2+} = 70$ mg/L, $Mg^{2+} = 40$ mg/L, and $HCO_3^- = 250$ mg/L as $CaCO_3$. Determine the water's carbonate hardness, noncarbonated hardness, and total hardness.

8.13 (a) Calculate the lime dosage required for softening by selective calcium removal for the following water analysis. The chemical constituents in the water are $CO_2 = 17.6$ mg/L, $Ca^{2+} = 63$ mg/L, $Mg^{2+} = 15$ mg/L, $Na^+ = 20$ mg/L, Alk $(HCO_3^-) = 189$ mg/L as $CaCO_3$, $SO_4^{2-} = 80$ mg/L, and $Cl^- = 10$ mg/L. What is the finished-water hardness?

8.14 A municipality treats 15×10^6 gallons/day of groundwater containing the following: $CO_2 = 17.6$ mg/L, $Ca^{2+} = 80$ mg/L, $Mg^{2+} = 48.8$ mg/L, $Na^+ = 23$ mg/L, Alk $(HCO_3^-) = 270$ mg/L as $CaCO_3$, $SO_4^2 = 125$ mg/L, and $Cl^- = 35$ mg/L. The water is to be softened by excess lime treatment. Assume that the soda ash is 90 percent sodium carbonate, and the lime is 85 percent weight CaO. Determine the lime and soda ash dosages necessary for precipitation softening (kg/day).

8.15 Water contains 7.0 mg/L of soluble ion (Fe^{2+}) that is to be oxidized by aeration to a concentration of 0.25 mg/L. The pH of the water is 6.0, and the temperature is 12°C. Assume the dissolved oxygen in the water is in equilibrium with the surrounding atmosphere. Laboratory results indicate the pseudo first-order rate constant for oxygenation of Fe^{2+} is 0.175/min. Assuming steady-state operations and a flow rate of 40,000 m^3/day, calculate the minimum detention time and reactor volume necessary for oxidation of Fe^{2+} to Fe^{3+}. Perform the calculations for both a CMFR and a PFR. (You should be able to work this out from information provided in Chapters 3 and 4.)

8.16 Calculate the settling velocity of a particle with 100 μm diameter and a specific gravity of 2.4 in 10°C water.

8.17 Calculate the settling velocity of a particle with 10 μm diameter and a specific gravity of 1.05 in 15°C water.

8.18 A water treatment plant processes 21,000 m^3 of water per day. Assume two types of flocculated particles enter a rectangular sedimentation basin that has dimensions of depth $= 4$ m, width $= 6$ m, and length $= 40$ m. The first type of particle has a settling velocity of 0.5 m/h and the other type has a settling velocity of 1.8 m/h. What percent of particles are removed for each of the two types of particles?

8.19 What percent of particles with diameter of 100 μm and a particle density of 2,650 kg/m^3 are removed in a 1,500 m^2 rectangular sedimentation basin of that contains water at 10°C? Assume the plant flow rate is 1.26×10^6 m^3/day.

8.20 A treatment plant has a horizontal flow sedimentation basin with a depth of 4 m, width of 6 m, and length of 36 m. It has a process flow rate of 400 m^3/h. What is the percent removal for particles entering this basin, assuming they all have a diameter of 0.0029 cm and a particle density of 2,650 kg/m^3, and they are in water at a temperature of 10°C?

8.21 Research the use of a filtration method that provides household (point-of-use) treatment in the developing world. Write a one-page report that is clearly referenced. In your report, describe the technology and address these issues: Is the technology affordable to the local population? Does it use local materials and local labor for its construction? What are the observed health improvements after implementation of the treatment system? What specific training do you believe is required to ensure proper operation of the technology?

8.22 Research the global arsenic problem in the United States and in Bangladesh. In a two-page essay, identify, compare, and contrast the extent of the problem (spatially and in terms of population affected). What are the current methods of treatment employed to remove the arsenic in both countries? What is the current regulatory standard for arsenic set by the EPA and the guideline suggested by the WHO?

8.23 A 1,000 g sample of naturally occurring sand was sifted through a stack of sieves and the weight retained on each sieve is recorded as given in the table below. Determine the effective size and uniformity coefficient for the media.

Sieve Designation	Sieve Opening, mm	Weight of Retained Media, g
10	2.000	140
14	1.400	160
18	1.000	170
20	0.850	100
25	0.710	90
30	0.600	85
35	0.500	80
40	0.425	70
45	0.355	65

8.24 When Cl_2 gas is added to water during the disinfection of drinking water, it hydrolyzes with the water to form HOCl (Equation 8.13). Assume that the disinfection power of the acid HOCl is 88 times greater than the conjugate base, OCl^-. The pK_a for HOCl is 7.5. (a) What percentage of the total disinfection power measured as free chlorine (HOCl + OCl^-) exists in the acid form at pH = 6? (b) At pH = 7?

8.25 Given the following data, graph the data for the *Poliomyelitis* virus, using hypobromite as a disinfectant. Determine the Chick's law rate constant and the time required for 99.99 percent (4-log removals) inactivation of this virus.

Time (s)	N (number of organisms/L)
0.0	1,000
2.0	350
4.0	78
6.0	20
8.0	6
10.0	2
12.0	1

8.26 Graph the following data for the inactivation of a virus using hypochlorous acid (HOCl). Determine the coefficient of specific lethality and the time required to obtain 99.99 percent inactivation using 1.0 mg/L of HOCl.

Time (min)	log (N/N_o)
1.0	−0.08
3.0	−0.64
5.0	−1.05
9.0	−1.87
15.0	−3.23

8.27 (a) Define the meaning of Ct product. (b) In addition to C and t, what factors influence the rate of chemical disinfection? (c) What kind of microorganism is most readily inactivated by free chlorine? (d) What kind is most difficult to inactivate?

8.28 During drinking-water treatment, 17 lb of chlorine are added daily to disinfect 5 million gallons of water. (a) What is the aqueous concentration of chlorine in mg/L? (b) The chlorine demand is the concentration of chlorine used during disinfection. The chlorine residual is the concentration of chlorine that remains after treatment so the water maintains its disinfecting power in the distribution system. If the chlorine residual is 0.20 mg/L, what is the chlorine demand in mg/L?

8.29 Visit the web site for the World Health Organization (www.who.org). Write a referenced essay of up to two pages about a particular pathogen and associated disease that is transmitted through contaminated water. What is the global extent of the public health crisis in terms of spatial and population effects? Does the disease affect wealthy or poor communities? What are the social and engineered barriers that can be used to reduce human exposure to the specific pathogen?

8.30 Research the professional engineering society Water for People that works on water supply in the developing world. Determine how you could contribute to such a professional society as a student and after graduation. Provide specific detail about membership requirements, costs, and how you might be involved.

8.31 (a) Use the Internet to research the number of people in the world who do not have access to an improved water supply. Then look up the Millennium Development Goals (MDGs) at the United Nations web site, www.un.org. MDG 7 states that by 2015 the world will decrease by half the number of people without access to an improved water supply. Select a country in Africa and another country in Asia or Latin America. Compare the progress of these two countries in meeting goal 7 in terms of the number of people still not served by an improved water supply.

8.32 What is the source of drinking water in the town where you currently live (groundwater, surface water, reclaimed water, or a mixture)? Sketch the unit processes used to treat this water in order of occurrence as currently practiced. What water constituent (s) does each unit process remove?

8.33 Identify three significant water users in your town or city that could benefit from the use of reclaimed water. What economic, social, and environmental challenges do you see that might need to be overcome before you can implement your plan for these users to use reclaimed water? What would you do as engineer to overcome these barriers?

8.34 A city is upgrading its water supply capacity to 81,378 m^3/day, using microfiltration. The new plant membrane system will consist of 25 arrays with 90 modules per array. The modules have an inside diameter of 120 mm, a length of 1,200 mm, and an available surface area of 30 m^2. The membranes have an outside diameter of 1.0 mm and a length of 1,200 mm. Determine: (a) the total surface area available for filtration, (b) the membrane flux rate in L/m^2-h, and (c) the total number of membrane fibers required for the plant and each module.

8.35 A municipality uses a microfiltration membrane system to treat 35,000 m^3/day. The membrane system consists of nine arrays with 80 modules per array. The modules are 119 mm in inside diameter and 1,194 mm in length, and have an available filtration surface area of 27 m^2. Determine the (a) total surface area available for filtration, and (b) membrane flux rate in L/m^2-h, and gpm/ft^2.

8.36 Atrazine and trichloroethylene can be removed from water by adsorption to activated carbon. The Freundlich parameters for atrazine are $K = 182$ mg/g $(L/mg)^{1/n}$ and $1/n = 0.18$. The parameters for trichloroethylene are $K = 56$ mg/g $(L/mg)^{1/n}$ and $1/n = 0.48$. What is the adsorbed concentration of both contaminants (units of mg chemical per gram of activated carbon) if you want the aqueous phase concentration at equilibrium to be 10 μg/L? (this problem may require you review information presented in Chapter 3).

8.37 A methyl tert butyl ether (MTBE) adsorption isotherm was performed on an activated carbon at 15°C using 0.250 L amber bottles with an initial MTBE concentration, C_o, of 150 mg/L. The isotherm data for each experimental point is summarized below. Calculate the adsorbed-phase concentration, q_e, for each isotherm point, plot the $\log(q_e)$ versus $\log(C_e)$ and determine the Freundlich isotherm parameters K and $1/n$ (this problem may require you review information presented in Chapter 3).

Mass of GAC, g	MTBE Equilibrium Liquid-Phase Concentration, C_e, mg/L
0.155	79.76
0.339	42.06
0.589	24.78
0.956	12.98
1.71	6.03
2.4	4.64
2.9	3.49
4.2	1.69

8.38 PAC is to be added to a water treatment plant to remove 10 ng/L of methylisoborneol (MIB) which is causing odor problems in the finished water. A standard jar test is performed to evaluate the impact of PAC dosage on the removal of MIB. The results are shown in Figure 8.26. If 60 percent MIB removal is required, determine the dosage of PAC required and quantity of PAC needed for 3 months (90 days) of treatment if the plant flow rate for is 40,000 m^3/day.

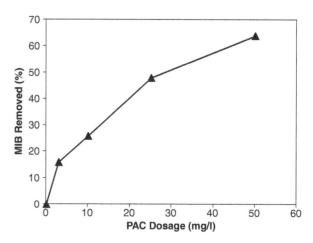

Figure / 8.26 Results from Jar Test Investigating Powdered Activated Carbon Usage at a Water Treatment Plant.

8.39 Reverse osmosis is used to treat brackish groundwater water and requires 1 kWh of energy per 1 m^3 of treated water. In comparison, reverse osmosis of seawater requires 4 kWh of energy per 1 m^3 of treated water (this difference is because of the higher TDS concentration of seawater). According to eGRID, the carbon dioxide equivalent emission rate is 1,324.79 lb CO_2e/MWh in Florida and 727.26 lb CO_2e/MWh in California. Estimate the carbon footprint of using reverse osmosis to desalinate 1 m^3 brackish groundwater and 1 m^3 seawater in Florida and California. Ignore line losses in your estimate (you may have to go back to Chapter 2 to review carbon footprints and eGRID).

8.40 Fill in the rest of this table, providing the electricity requirements to treat 1 MDG of treated water. Fill in the carbon footprint assuming the treatment plant is located in California. According to eGRID, the carbon dioxide equivalent emission rate is 727.26 lb CO_2e/MWh in California (you may have to review Chapter 2 for information on carbon footprints and eGRID).

Unit Process	kWh Required to Treat 1 m^3	kWh Required to Treat 1 MGD	Carbon Footprint to Treat 1 MGD (CO_2e)
Coagulation/ Flocculation			
Sedimentation			
Disinfection with Chlorine			
Adsorption with GAC			

References

Asano, T., F. L. Burton, H. L. Leverenz, R. Tsuchihashi, and G. Tchobanoglous, 2006. *Water Reuse: Issues, Technologies, and Applications*. Wakefield: Metcalf and Eddy.

Bales, D. W., 2012. Optimization of an Advanced Water Treatment Plant: Bromate Control and Biofiltration Improvement, MS Thesis, Civil & Environmental Engineering, University of South Florida, 120 pp.

Burton, F. L., 1996. Water and Wastewater Industries: Characteristics and Energy Management Opportunities. Report prepared for Electric Power Research Institute (EPRI) CR-106941.

Crittenden, J. C., R. R. Trussell, D. W. Hand, G. Tchobanoglous, and K. Howe, 2012. *Water Treatment Principles and Design*, 3rd ed. New York: John Wiley & Sons.

Elliot, T., B. Zeier, I. Xagoraraki, and G. Harrington. 2003. "Energy Use at Wisconsin's Drinking Water Facilities." ECW Report Number 222-1. Wisconsin Focus on Energy & Energy Center of Wisconsin, Madison (July).

Han, M., and D. F. Lawler, 1992. The (relative) insignificance of G in flocculation. *Journal of American Water Works Association*, 84(10): 79–91.

Hancock, N. T., N. D. Black, and T. Y. Cath, 2012. A comparative life cycle assessment of hybrid osmotic dilution desalination and established seawater desalination and wastewater reclamation processes. *Water Research*, 46(4), 1145–1154.

LeChevallier, M. W., and W. D. Norton, 1995. *Giardia and Cryptosporidium* in raw and finished water. *AWWA* 87(9): 54–68.

Lyons, E., P. Zhang, T. Benn, F. Sharif, K. Li, J. Crittenden, M. Costanza, and Y. S. Chen, 2009. Life cycle assessment of three water supply systems: importation, reclamation and desalination. *Water Science & Technology*, 9(4): 439–448.

Mihelcic, J. R., E. A. Myre, L. M. Fry, L. D. Phillips, and B. D. Barkdoll, 2009. *Field Guide in Environmental Engineering for Development Workers: Water, Sanitation, Indoor Air*. Reston: American Society of Civil Engineers Press.

Miller, T., J. Esler. 2013. What every operator should know about secondary clarification. *Water Environment and Technology*, 25(6): 62–64.

Mo, W., Q. Zhang, J.R. Mihelcic, and D. Hokanson. 2011. Embodied energy comparison of surface water and groundwater supply options. *Water Research*, 45(17): 5577–5586.

Muñoz, I.; Mila-i-Canals, L.; Fernández-Alba, and A. R. 2010. Life Cycle Assessment of Water Supply Plans in Mediterranean Spain: The Ebro River Transfer versus the AGUA Programme. *J. Indust. Ecol.* 2010 14(6), 902–918.

National Research Council (NRC), 2012. "Water Reuse: Potential for Expanding the Nation's Water Supply through Reuse of Municipal Wastewater Committee on the Assessment of Water Reuse as an Approach to Meeting Future Water Supply Needs," Washington, D.C.: National Academy of Sciences.

Petrusevski, B., S. Sharma, J. C. Schippers, and K. Shordt, 2007. *Arsenic in Drinking Water*. Thematic Overview Paper 17. Delft, The Netherlands: IRC International Water and Sanitation Centre.

PRé Consultants, 2004, *SimaPro Database Manual: The BUWAL 250 Library*. The Netherlands.

Raluy, R. G., L. Serra, and J. Uche, 2005a. Life cycle assessment of water production technologies part 1: LCA of different commercial technologies (MSF, MED, RO). *International Journal of Life Cycle Assessment*, 10(5): 346–354.

Raluy, R. G., L. Serra, and J. Uche, 2005b. Life cycle assessment of desalination technologies integrated with renewable energies. *Desalination*, 183(1–3): 81–93.

Stokes, J., and A. Horvath, 2009. Energy and air emission effects of water supply. *Environmental Science & Technology*, 43(8): 2680–2687.

Vince, F., E. Aoustin, P. Breant, and F. Marechal, 2008. LCA tool for the environmental evaluation of potable water production. *Desalination*, 220(1–3): 37–56.

chapter/Nine Wastewater and Stormwater: Collection, Treatment, Resource Recovery

James R. Mihelcic, Julie Beth
Zimmerman, David W. Hand,
Brian E. Whitman, and
Martin T. Auer

*In this chapter, readers will learn about
the composition of wastewater and the
various unit processes employed to
protect human health, improve water
quality, and recover resources. Mass
balance and biochemical kinetics are
employed to develop expressions to
size a reactor used to remove biochemical oxygen demand. Less mechanized
and natural treatment processes such
as free surface wetlands, lagoons,
and septic tanks are discussed. The
treatment and recovery of nitrogen
and phosphorus are also discussed,
highlighting conventional removal
approaches as well as strategies to
reclaim these nutrients for value-added
applications, including the processing
of sludge and its role in energy production. Readers will also learn about
energy requirements (and energy sources) in terms of wastewater treatment
technologies and how operation of a
plant influences energy usage. Similar
to nutrient recovery from wastewater
treatment plants, wastewater reuse is
also explored as a means to more
sustainable water resource management. Finally, the relationship of stormwater and wastewater will be explored
to understand how the effective management of stormwater, particularly
through low-impact development,
can mitigate significant wastewater
challenges.*

© Marcus Clackson/iStockphoto

Learning Objectives

1. Identify distinct hydrological, physical, chemical, and
   biological inputs that make up municipal wastewater, list
   typical concentrations of the major constituents, and determine
   recoverable resources associated with particular constituents.

2. Match major wastewater constituents with the unit process(es)
   that remove or recover a significant amount of each constituent.

3. Apply mass balances and other design relationships to design a grit chamber, flow equalization basin, activated-sludge biological treatment system, and waste stabilization pond.

4. Integrate mass balances with biological growth kinetics to develop activated-sludge design equations and relate solids retention time, food-to-microorganism ratio, sludge wasting, and growth kinetics to plant design and operation.

5. Demonstrate in-depth knowledge of differences between aerobic, anoxic, and anaerobic biological processes used to treat wastewater.

6. Empathize with the magnitude of global sanitation coverage and its relationship to human health.

7. List advantages and disadvantages of membrane bioreactors.

8. List the components of wastewater solids management, calculate sludge volume index, and relate this value to the settling characteristics of sludge.

9. Relate the configuration and operation of a wastewater treatment and resource recovery plant to the removal and recovery of nitrogen and phosphorus via biochemical and chemical processes.

10. Describe specific removal processes that occur in different treatment zones in facultative lagoons and free surface wetlands.

11. Apply theoretical oxygen balance and site-specific solar insolation data to design a wastewater lagoon.

12. Identify the magnitude of energy usage during wastewater treatment and the greenhouse gases emitted from different unit processes.

13. Identify sources of nutrients and energy that can be recovered from wastewater treatment and recovery plant as well as opportunities for water reclamation and appropriate reuse.

14. Discuss economic, social, and environmental benefits of water reuse.

15. Demonstrate understanding of how design of unit processes for water reuse must be integrated with intended application(s).

16. Calculate wet-weather flows based on inflow and infiltration.

17. Describe best management practices (BMPs) for controlling urban stormwater that are referred to as low-impact development, such as rain gardens, permeable pavements, green roofs, and bioswales.

18. Differentiate between the components of a green roof and calculate the volume of water a green roof can store.

19. Design a bioretention cell for given precipitation and design scenarios.

9.1 Introduction

Municipal wastewater treatment plants, also referred to as **publicly owned treatment works (POTWs)**, receive inputs from many domestic and industrial sources as well as from combined stormwater sewer systems. These distinct hydrological inputs are depicted in Figure 9.1. There are four components of **domestic wastewater**: (1) wastewater from domestic, commercial, and industrial users; (2) stormwater runoff; (3) infiltration; and (4) inflow. With population increase, climate change, nutrient cycling, and water scarcity becoming common, sustainable treatment of wastewater must address issues beyond treatment performance and cost. Furthermore, it is becoming increasingly clear that more effective management of stormwater is needed, especially given the importance of nutrient management and changes in urbanization and storm intensity and frequency as a result of climate change.

Industrial wastewaters vary in quantity, composition, and strength, depending on the specific industrial source. The Environmental Protection Agency (EPA) has identified 129 *priority pollutants*. Industrial wastes include conventional pollutants found in domestic wastewater but may also contain heavy metals, radioactive materials, and refractory organics. Industries may choose to treat their waste on-site, following specific guidelines, *best available treatment* and *effluent guidelines* for priority pollutants. They may also choose to sewer their waste to a municipal wastewater treatment plant, after first providing pretreatment to protect operation of the municipal treatment plant and prevent discharge of pass-through pollutants.

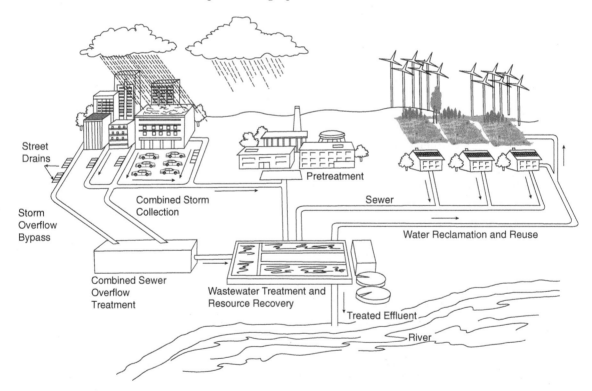

Figure 9.1 **Wastewater Infrastructure Management** This schematic shows the many possible hydrological and chemical and biological pollutant contributors to municipal wastewater.

Table 9.1 provides the global definition for *improved* and *unimproved* sanitation technologies. Currently, 2.5 billion people in the world are without access to improved sanitation technology (including 1 billion who have no facilities at all), and lack of sanitation has a large negative impact on human health and the environment (WHO, 2010).

The perception of sanitation varies significantly from culture to culture. Improvements in health are not the only reasons that communities accept sanitation projects. According to a survey of rural households in the Philippines, the reasons why people were satisfied with their newly built latrines included lack of smell and flies, cleaner surroundings, privacy, less embarrassment when friends visit, and fewer incidences of gastrointestinal disease (Cairncross and Feachem, 1993). For an example of an appropriate latrine design being used in the Pacific island republic of Vanuatu, see Figure 9.2.

Fry et al. (2008) analyzed barriers to global sanitation coverage including inadequate investment, poor or nonexistent policies, governance, too few resources, gender disparities, and water availability. The challenges studied were found to be significant barriers to sanitation coverage, but water availability was not a primary obstacle at a global scale. However, water availability was found to be an important barrier to as many as 46 million people, depending on the sanitation technology selected.

Table / 9.1

Improved and Unimproved Sanitation Technologies: Global Definitions

Improved	Unimproved
Connection to a public sewer	Service or bucket latrines (excreta are removed manually)
Connection to a septic system	Public latrines
Pour-flush latrines	Open latrines
Ventilated improved pit latrines	
Composting latrines Simple pit latrines	

Figure / 9.2   **Double-Vault Compost Latrine Being Constructed in Vanuatu**   These latrines can be designed to separate urine and feces. The urine is either routed to a soak pit or collected in a pot or routed to a garden and used as a fertilizer. One side of the latrine is used for up to 12 months, while the other side is composting. Desiccants such as wood ash and sawdust are added to reduce odors and kill pathogens. The privacy shelter is being constructed of local wood and woven plant material. Compost latrines require no addition of water, unlike other sanitation technologies, and allow the nutrients to be used locally.

(Photo courtesy of Eric Tawney. Built during workshop cosponsored by Dr. Leonie Crennan).

Joint Monitoring Programme for Water Supply and Sanitation
http://www.wssinfo.org/

Two purposes of municipal wastewater treatment are to protect human health and prevent pollution of a receiving surface water or groundwater. Examples of pollutants associated with untreated wastewater include dissolved-oxygen depletion (measured as biochemical oxygen demand (BOD) and chemical oxygen demand (COD)), unsightly and oxygen-depleting solids (total suspended solids, TSS), nutrients that cause eutrophication (N and P), chemicals that exert toxicity (NH_3, metals, organics), emerging chemicals of concern, and pathogens (bacteria and viruses). Aesthetic problems include visual pollution and odor. In terms of pathogens, humans produce on average 10^{11}–10^{13} coliform bacteria per day. While treatment processes are very efficient at removing pathogens and other pollutants, in the near future, treatment plants will need to be concerned with removing other chemicals now found in our wastewater. These include fragrances, surfactants found in soaps and detergents, pharmaceutical chemicals, endocrine-disrupting chemicals, and other **emerging chemicals of concern**.

Through the Federal Water Pollution Control Act of 1972 (commonly known as the **Clean Water Act**), the U.S. Congress established a national strategy to reduce water pollution. The objectives of the Clean Water Act are to restore and maintain the chemical, physical, and biological integrity of the nation's waters by achieving a level of water quality that provides for the protection and propagation of fish, shellfish, and wildlife and for recreation in and on the water, and to eventually eliminate the discharge of pollutants into U.S. waters (zero discharge). This is accomplished through the **National Pollutant Discharge Elimination System (NPDES)**, which issues permits defining the types and amounts of polluting substances that may be discharged. The NPDES permitting system is administered and enforced at the state level. Violation of the technology-based and water-quality-based effluent standards may result in civil penalties (fines) and criminal penalties (imprisonment).

© David Pullicino/iStockphoto.

Class Discussion

How will indoor and outdoor household water conservation influence the characteristics of domestic wastewater and stormwater runoff? Revisit this topic later in the chapter to discuss how water conservation affects collection, treatment, and treatment of wastewater, recovery of resources, and stormwater management.

9.2 Characteristics of Domestic Wastewater

Raw (that is, untreated) wastewater is considered highly polluted, yet the amount of contaminants it contains may seem small. For example, 1 m^3 municipal wastewater weighs approximately 1 million g, yet it may contain only 500 g pollutants. This small fraction of pollution can have serious ecological and health impacts, however, if discharged untreated. It also contains valuable resources that energy and nutrients can be recovered from.

Domestic wastewater appears gray and turbid and has a temperature of 10–20°C. Table 9.2 provides the composition of an **average-strength municipal wastewater** and shows the most common **municipal wastewater constituents**. We will not go into great detail about each constituent at this time because their measurement and importance were described in previous chapters, as noted in Table 9.2. As we discuss the various unit processes for treatment, you may want to refer back to this table, because specific processes remove distinct wastewater constituent(s).

Table / 9.2

Concentration of Major Constituents Found in Average-Strength Wastewater

Constituent	Discussed Previously in	Average Concentration	Comments
Biochemical oxygen demand (BOD)	Chapters 2 and 5	200 mg/L	Oxygen-demanding materials can deplete the oxygen content of receiving waters.
Suspended solids	Chapters 2 and 8	240 mg/L (total solids typically 800 mg/L)	Cause water to be turbid; may contain organic matter and thus contribute to BOD; may contain other pollutants or pathogens. Organic solids can be anaerobically digested to produce energy.
Pathogens	Chapters 5 and 8	3 million coliforms per 100 mL	Disease-causing microorganisms usually associated with fecal matter.
Nutrients such as nitrogen and phosphorus	Chapters 2, 3, and 5	Total nitrogen: 35 mg N/L Inorganic nitrogen: 15 mg N/L Total phosphorus: 10 mg P/L	Can accelerate growth of aquatic plants, contribute to eutrophication; ammonia is toxic to aquatic life, can contribute to NBOD. Value as crop fertilizer during water reuse.
Toxic chemicals	Chapters 3, 5, 6, and 8	Variable	Heavy metals such as mercury, cadmium, and chromium; organic chemicals such as pesticides, solvents, fuel products.
Emerging chemicals of concern	Chapters 6 and 8	Unknown or variable	Pharmaceuticals, caffeine, surfactants, fragrances, perfumes, other endocrine-disrupting chemicals.

9.3 Overview of Treatment Processes

Design and operation of a wastewater and resource recovery treatment facility require an understanding of unit operations that employ fundamental physical, chemical, and biological processes (see Chapters 3–5) to remove and/or recover specific water quality constituents. Assembly of the correct process removal train requires the accomplishment of four tasks: (1) identifying the characteristics of the untreated wastewater, (2) identifying treatment and recovery objectives and assessing community involvement, (3) integrating unit operations into a complete process that recognizes the appropriateness and limits of each unit process and how they complement each other, and (4) integrating concepts of green engineering, life cycle thinking, and sustainability to incorporate issues beyond end-of-pipe treatment standards and capital and operating costs (for example, water reuse and energy consumption).

Figure 9.3 provides an aerial view of a typical municipal wastewater treatment plant. The schematic drawing in Figure 9.4 shows how the different unit processes can be integrated. Different processes occur for the treatment of liquid and solid-waste streams. The steps involved in conventional wastewater treatment are:

Wastewater and Stormwater Fact Sheets
http://water.epa.gov/scitech/wastetech/mtbfact.cfm

Primer for Municipal Wastewater Treatment
http://water.epa.gov/aboutow/owm/upload/2005_08_19_primer.pdf

Figure / 9.3 Overhead View of **Wastewater Treatment Plant** This plant serves approximately 14,000 people.

(Photo courtesy of the Portage Lake Water and Sewage Authority).

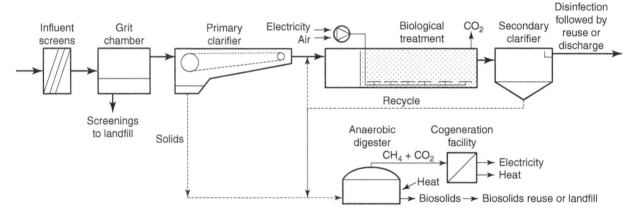

Figure 9.4 **Typical Layout of Conventional Wastewater Treatment** Preliminary treatment with screens and grit removal is followed by a primary clarifier, biological treatment, a secondary clarifier, and anaerobic treatment of the sludge.

(Adapted from figure provided by Dr. Diego Rosso, University of California–Irvine).

(1) pretreatment, (2) primary treatment, (3) secondary treatment, (4) tertiary treatment to remove nutrients (N, P), and (5) disinfection. With wastewater now viewed as a resource by many communities, the flow chart for conventional wastewater treatment is now changing to accommodate issues of energy recovery, nutrient recovery, and water reuse.

Pursuant to Section 304(d) of Public Law 92-500, the EPA published its definition of minimum standards for secondary treatment. Table 9.3 provides an overview of these treatment standards and the specific unit processes that remove significant amounts of specific wastewater constituents.

These Internet videos will allow you to tour wastewater treatment and resource recovery plants and better visualize many of the unit processes described throughout this chapter.

Video 1 (10:29): Produced by the Lake County (Ohio) and shows each unit process of the 38 MGD Gary L. Kron Water Reclamation Facility.
http://www.youtube.com/watch?v=8YMfSR1kD3U

Video 2 (4:10): An animation with text, but no sound. Covers the basic processes of treating wastewater.
http://www.youtube.com/watch?v=Y_1FRWHbz-o

Video 3 (10:01): The Water Environment Federation's narrated video provides a virtual tour that includes treatment, water reuse, management of sludge, and energy generation from anaerobic digestion.
http://news.wef.org/wef-video-teaches-about-operation-of-water-resource-recovery-facility-2/

Video 4 (3:13): Blue Plains Advanced Wastewater Treatment Plant in Washington, D.C., has major efforts to recover energy and nutrients from wastewater.
http://www.werf.org/c/KnowledgeAreas/Resource_Recovery/Latest_News/Video_Wastewater_Plants.aspx

Table / 9.3

Minimum Standards for Treatment and Unit Processes That Remove a Significant Amount of Major Wastewater Constituents

Constituent and EPA Standard for Minimum Treatment	Unit Process(es) That Remove Significant Amount of Constituent
Biochemical oxygen demand (BOD): Allowable 30-day BOD_5 is 30 mg/L, and 7-day BOD_5 is 45 mg/L.	BOD can be in a dissolved or particulate form. Biological reactor; primary and secondary sedimentation.
Suspended solids: Allowable 30-day TSS is 30 mg/L, and 7-day TSS is 45 mg/L.	Primary and secondary sedimentation. Important in production of energy and/or reuse of biosolids.
Pathogens: Depends upon the NPDES permit based on the receiving water (e.g., at the plant depicted in Figure 9.3, fecal coliforms, <200 counts/mL monthly average or <400 counts/mL 7-day average).	Primary and secondary sedimentation; disinfection. Predation also occurs in the biological reactor.
Nutrients such as nitrogen and phosphorus: Depends upon the NPDES discharge permit.	Nutrients can be in a dissolved or particulate form. Sedimentation; biological reactor; addition of chemicals to precipitate and recover phosphorus. Nitrogen recovered as struvite or in biosolids and both nutrients are present in reclaimed water.
Toxic chemicals	Some are removed via sedimentation (if they are sorbed to or complexed by particles), some are biodegradable, and some pass through the treatment plant. Advanced oxidation processes can be used to treat in water reuse scenarios.
pH: Discharge must be in the range of 6.5–10.0.	Not applicable.

9.4 Preliminary Treatment

Preliminary treatment prepares the wastewater for further treatment. It is used to remove oily scum, floating debris, and grit, which may inhibit biological processes and/or damage mechanical equipment. Equalization tanks are employed to balance flows or organic loading. Industrial effluents may additionally require physical–chemical pretreatment for removal of ammonia–nitrogen (air stripping), acids/bases (neutralization), heavy metals (oxidation–reduction, precipitation), or oils (dissolved-air flotation).

9.4.1 SCREENING

Bar racks (parallel bars or rods, 20–150 mm) and **bar screens** (perforated plates or mesh, 10 mm or less) retain the coarse solids (large objects, rags, paper, plastic bottles, etc.) present in wastewater, preventing damage to the piping and mechanical equipment that follows this treatment step (Figure 9.5). They are cleaned by hand in some older, smaller plants, but most are equipped with automatic cleaning rakes. Screenings are typically disposed of by landfilling or incineration.

As an alternative to screens, some plants utilize a **comminutor**, which grinds up (comminutes) coarse solids without removing them from the wastewater flow. This reduction in size makes the solids easier to treat in subsequent operations that employ settling. Comminutors eliminate the need for handling and disposal of coarse solids removed during screening.

9.4.2 GRIT CHAMBERS

Grit consists of particulate materials in wastewater that have specific gravities of approximately 2.65 and a temperature of 15.5°C. Particles with specific gravities between 1.3 and 2.7 have also been removed, based on field data. Grit can consist of inorganic sand or gravel (about 1 mm in diameter), eggshells, bone fragments, fruit and vegetable pieces and seeds, and coffee grounds.

Grit is primarily removed to prevent abrasion of piping and mechanical equipment. During grit removal, some organic materials are removed along with the grit. Sometimes grit-washing equipment is added to remove organic materials and return them to the wastewater.

For horizontal-flow systems, grit is removed through sedimentation by gravity (using Stokes' law or Newton's law). In an aerated **grit chamber**, air is introduced along one side of the tank, which provides a helical flow pattern of the wastewater through the chamber, enabling the grit to settle out while keeping the smaller organic material suspended in the wastewater. The *aerated grit chamber* has the added advantage that it keeps the wastewater fresh by adding oxygen to the wastewater. In a *vortex grit chamber*, the wastewater enters and exits tangentially, creating a vortex flow pattern where the grit settles to the bottom of the tank.

Both aerated and vortex systems are designed based on typical design parameters. Table 9.4 provides design information used to size an aerated grit chamber. These chambers are normally designed to remove particles with diameters of at least 0.21 mm. Detention times range from 2 to 5 min based on peak hourly flow, and air flow rates range from 0.2 to 0.5 m³ of air per minute per length of tank. Example 9.1 illustrates the design of an aerated grit chamber.

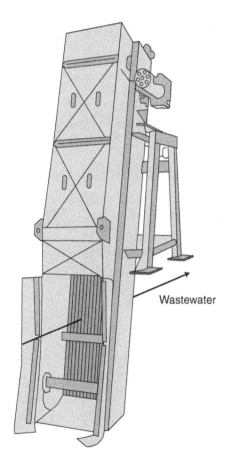

Figure 9.5 **Bar Screens Used to Remove Coarse Solids from Wastewater** If not removed, these solids may damage piping and mechanical equipment later in the treatment process.

Wastewater

Table / 9.4

Design Information Used to Size Aerated Grit Chambers

Parameter	Range
Peak flow detention time (min)	2–5
Depth (m)	2–5
Length (m)	7.5–20
Width (m)	2.5–7
Width-to-depth ratio	1:1 to 5:1
Length-to-width ratio	3:1 to 5:1
Air requirement per length of tank (m^3/m-min)	0.2–0.5
Quantity of grit ($m^3/10^3$ m^3)	0.004–0.20

SOURCE: Tchobanoglous et al. (2003).

example/9.1 Design of an Aerated Grit Chamber

Design an aerated grit chamber system to treat a 1 day sustained peak hourly flow of 1.5 m^3/s with an average flow of 0.6 m^3/s. Determine: (a) the grit chamber volume (assuming two chambers will be used); (b) the dimensions of the two grit chambers; (c) the average hydraulic retention time in each grit chamber; (d) air requirements, assuming 0.35 m^3/m-min of air; and (e) the quantity of grit removed at peak flow, assuming a typical value of 0.015 $m^3/10^3$ m^3 of grit in the untreated wastewater.

solution

Much of this problem can be solved by using the design guidance provided in Table 9.4.

The volume of the grit chambers is determined assuming a detention time of 3 min:

$$\text{total grit chamber volume} = 1.5 \text{ m}^3/\text{s} \times 3 \text{ min} \times 60 \text{ s/min} = 270 \text{ m}^3$$

$$\text{volume of each grit chamber} = \frac{1}{2} \times 270 \text{ m}^3 = 135 \text{ m}^3$$

Assuming a width-to-depth ratio of 1.5:1 and a depth of 3 m, the dimensions of the two grit chambers are

$$\text{grit chamber width} = 1.5 \times 3 \text{ m} = 4.5 \text{ m}$$

$$\text{grit chamber length} = \frac{\text{volume}}{\text{width} \times \text{depth}} = \frac{135 \text{ m}^3}{4.5 \text{ m} \times 3 \text{ m}} = 10 \text{ m}$$

The average hydraulic detention time in each grit chamber is based on the average flow rate:

$$\text{detention time} = \frac{\text{volume}}{\text{flow}} = \frac{135 \text{ m}^3}{(0.6 \text{ m}^3/\text{s})/(2 \text{ tanks})} \times \frac{1 \text{ min}}{60 \text{ s}} = 7.5 \text{ min}$$

The air requirements, assuming 0.35 m^3/m-min of air, are

$$\text{total air requirement} = (2 \text{ tanks}) \times (10 \text{ m length}) \times (0.35 \text{ m}^3 \text{ air/m-min}) = 7.0 \text{ m}^3/\text{min}$$

Finally, solve for the amount of grit to be disposed of, assuming peak flow conditions:

$$\text{grit volume} = (1.5 \text{ m}^3/\text{s}) \times (0.015 \text{ m}^3/10^3 \text{ m}^3) \times (86,400 \text{ s/day}) = 1.94 \text{ m}^3/\text{day}$$

Similar design methods are employed to size horizontal-flow and vortex grit removal devices.

9.4.3 FLOTATION

Flotation is the opposite of sedimentation, utilizing buoyancy to separate solid particles such as fats, oils, and greases, which would not settle by sedimentation. The separation process is accomplished by introducing air at the bottom of a floatation tank. The air bubbles rise to the surface, where they are removed by skimming. A popular variation on this scheme is termed *dissolved-air flotation*. Recycled effluent is retained in a pressure vessel, where it is mixed and saturated with air. The effluent is then mixed with the raw wastewater, and as the pressure returns to atmospheric, dissolved air comes out of solution, carrying floatable solids to the surface, where they may be skimmed and collected.

Today, fats, oil, and grease (termed FOG)—specifically, material generated at local restaurants or campus dining halls—do not have to become part of the wastewater stream. They can be easily converted to biodiesel (which could help fuel your municipality's vehicle fleet) and used to generate energy, when combined with digester gas, or used as a supplementary fuel in solid-waste-to-energy plants.

9.4.4 EQUALIZATION

Flow equalization is implemented to dampen the flow and organic loading rate to a wastewater treatment facility. Remember from Chapter 7 that large variations occur in the flow for many reasons. Implementing flow equalization in some instances can overcome operational problems associated with large flow variations and improve the performance of the downstream unit processes. For example, the biological processes used during wastewater treatment can be more easily controlled with a steady flow rate and near-constant BOD loading. In addition, implementation of flow equalization can reduce the size of the downstream treatment processes and in some cases improve performance at plants that are overloaded.

Figure 9.6 compares the diurnal flow rate and BOD loading variation with an equalized flow and BOD loading pattern. BOD loading is equal to the flow times the concentration of BOD in the wastewater and has units of kg BOD/m^3 of wastewater per day. As Figure 9.6 shows, the dampening of the flow rate and BOD can be considerable.

Flow equalization can be accomplished in two ways: in-line or off-line equalization. *In-line equalization* is the process where all the flow passes through the equalization basin. In contrast, with *off-line equalization*, only a portion of the flow is diverted through the equalization basin. Off-line flow equalization requires that the diverted flow be pumped and mixed with the plant influent when the influent flow rate to the plant is reduced. This typically occurs late at night. In this case, the flow can be equalized, but the change in BOD loading is reduced less than with in-line flow equalization. Therefore, in-line equalization is typically used when stringent dampening of both the flow and organic loading is required.

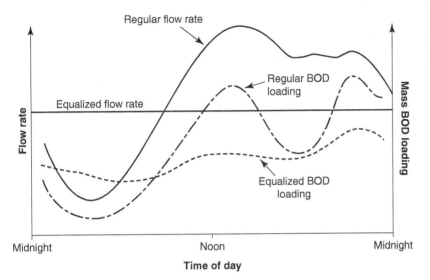

Figure / 9.6 **Changes in Regular Flow Rate and Regular Mass BOD Loading over a Typical Day** The equalized flow rate is shown as a constant, and the equalized BOD loading is dampened, so larger variations over the day are removed.

Mass diagrams like the one shown in Figure 9.7 can be used to determine the required *equalization storage volume*. Figure 9.7 shows the *cumulative influent volume* and the *cumulative average influent volume* as a function of the time of day. To determine the equalization volume required, take the vertical distance between the average volume and the parallel line that is tangent to the cumulative inflow volume curve for Figure 9.7. The tangent point on the cumulative inflow curve is where the equalization tank is empty. As time increases, the slope of the cumulative inflow curve is greater than the average inflow curve. The equalization tank will fill up to about midnight, when the slopes of both curves are about the same. From midnight to a little past noon, the slope of the cumulative inflow curve is less than the average curve, meaning the equalization tank is losing volume (is draining).

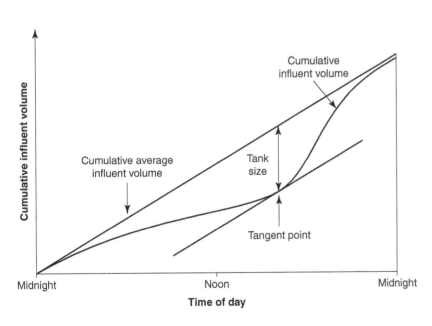

Figure / 9.7 **Cumulative Influent Volume and Cumulative Average Influent Volume as a Function of Time of Day** The cumulative average influent volume and the cumulative influent volume can be plotted to determine the required tank size for equalization storage volume. After the tangent point is determined on the cumulative influent volume curve, a line is drawn parallel to the cumulative average influent volume curve. The distance between the tangent point and the cumulative average influent volume curve is the required storage volume. The curve drawn through the tangent point provides information on when the storage volume is filling and emptying.

For Figure 9.7, the equalization volume would be found at the tangent of both parallel lines on the cumulative influent volume curve. During the time between the two tangent points, approximately 1 P.M. to about midnight, the equalization basin is filling, as the slope of the cumulative influent volume curve is greater than the average influent volume curve. From about midnight to 1 P.M., the slope of the cumulative influent volume curve is less than the average influent volume curve, and the equalization tank is draining.

example/9.2   Sizing a Flow Equalization Tank

Given the data for average hourly flows given in Table 9.5 (in the two left columns), determine the required online flow equalization volume (m^3).

Table / 9.5

Data and Results for Flow Equalization Problem in Example 9.2   The average cumulative influent flow (not shown, units of m^3/h) is determined by dividing the cumulative influent volume by 24 h.

Time Period	Volume of Flow During Period (m^3)	Cumulative Influent Volume (m^3)
Midnight–1 A.M.	1,090	1,090
1–2	987	2,077
2–3	701	2,778
3–4	568	3,346
4–5	487	3,833
5–6	475	4,308
6–7	532	4,840
7–8	838	5,678
8–9	1,375	7,053
9–10	1,565	8,618
10–11	1,630	10,248
11–noon	1,649	11,897
Noon–1	1,640	13,537
1–2	1,545	15,082
2–3	1,495	16,577
3–4	1,490	18,067
4–5	1,270	19,337
5–6	1,270	20,607

(continued)

Table / 9.5

Time Period	Volume of Flow During Period (m³)	Cumulative Influent Volume (m³)
6–7	1,290	21,897
7–8	1,424	23,321
8–9	1,548	24,869
9–10	1,550	26,419
10–11	1,476	27,895
11–midnight	1,342	29,237

solution

This solution requires several steps. First, determine the cumulative hourly flow during the period. The answers are given in the right column of Table 9.5. For one time period, the cumulative hourly flow is

$$\begin{bmatrix} \text{cumulative hourly} \\ \text{flow, } 1-2 \text{ A.M.} \end{bmatrix} = V_{M-1} + V_{1-2} = 1{,}090 + 987 = 2{,}077 \text{ m}^3$$

Then, to determine the cumulative average influent volume (not listed in the table), divide the cumulative flow (listed in the table) by 24 h:

$$\text{average flow} = \frac{\text{cumulative flow}}{24 \text{ h}} = \frac{29{,}237 \text{ m}^3}{24 \text{ h}} = 1{,}218 \text{ m}^3/\text{h}$$

The solution now requires a graph of the cumulative influent volume and cumulative average influent volume. (The figure is not shown, so readers should consult Figure 9.7 and complete this on their own.) From this graph, the required equalized flow volume is found to be approximately 4,100 m³.

9.5 Primary Treatment

The goal of **primary treatment** is to remove solids through quiescent, gravity settling. Typically, domestic wastewater is held for a period of approximately 2 h. **Settling tanks**, also referred to as **sedimentation tanks** or **clarifiers**, can be either rectangular or circular. During sedimentation, solids settle to the bottom of the tank, where they are collected as a liquid–solid sludge. Figure 9.8 shows the inside of a circular clarifier.

Primary treatment removes about 60 percent of the suspended solids (TSS), 30 percent of the BOD, and 20 percent of the phosphorus (P). The BOD and phosphorus removed in this stage are primarily in the particulate phase (that is, part of the TSS). Any dissolved BOD, N, or P will pass through primary treatment and enter secondary treatment. Coagulants can be added to improve the removal of particulate matter. This may reduce the overall energy costs required during second treatment to biologically convert these particles to CO_2, water, and new biomass.

The clarified effluent that exits primary treatment is routed to secondary treatment, and the solids (the sludge) removed during settling are segregated for further treatment and recovery of energy or reuse of biosolids. Primary sludge is malodorous, may contain pathogenic organisms, and has high water content (perhaps less than 1 percent solids). These characteristics make it difficult to dispose of. Secondary clarifiers are designed to remove much smaller particles because, as we will explain in the next section, most of the particulate matter at this point in the treatment plant consists of microorganisms.

This chapter does not go into great detail on the design of settling tanks for wastewater treatment. Chapter 8 provided a discussion of sedimentation theory and design principles, including how to use established **overflow rates** to size settling tanks. Of important note, the actual hydraulic detention time can be much shorter than calculated because of uneven weirs, poor inlet conditions, and density currents, all which can contribute to a reduction in the efficiency of solids removal (Miller and Esler, 2013).

Figure / 9.8 Secondary clarifiers (diameter is 130 ft) with Stamford baffle (clarifier in bottom of photo is empty of water). Clarifiers are located at the Stamford (CT) Water Pollution Control Facility. The plant is designed to treat an average wastewater flow of 24 MGD from the City of Stamford and the Town of Darien. Treated effluent is discharged to the East Branch of Stamford Harbor which is located on Long Island Sound.

(Photo courtesy of Dr. Jeanette Brown, University of Connecticut).

example/9.3 Sizing a Primary Settling Tank

A municipal wastewater treatment plant treats an average flow of 12,000 m^3/day and a peak hourly flow of 30,000 m^3/day. Two circular clarifiers are to be designed, using a depth of 4 m and overflow rate of 40 m^3/m^2-day. Calculate the area, diameter, volume, and detention time required for each clarifier.

solution

To calculate the surface area required for clarification, divide the average flow rate (Q) by the overflow rate (OR):

$$\text{total clarifier area} = \frac{Q}{\text{OR}} = \frac{12,000 \text{ m}^3/\text{day}}{40 \text{ m}^3/\text{m}^2\text{-day}} = 300 \text{ m}^2$$

Since there are two clarifiers, the area for each clarifier would be

$$\text{clarifier area} = \frac{300 \text{ m}^2}{2 \text{ clarifiers}} = 150 \text{ m}^2$$

The tank diameter can be calculated from the area as follows:

$$\text{clarifier diameter} = \sqrt{\frac{\text{clarifier area}}{\frac{\pi}{4}}} = \sqrt{\frac{150 \text{ m}^2}{\frac{\pi}{4}}} = 13.8 \text{ m}$$

The diameter will be rounded up to 14 m for the final design.

The actual area for each clarifier is calculated as follows:

$$\text{clarifier area} = \frac{\pi}{4}(14 \text{ m})^2 = 154 \text{ m}^2$$

The volume of each clarifier is calculated as follows:

$$\text{clarifier volume} = \text{area} \times \text{depth} = \left(\frac{\pi}{4}(14 \text{ m})^2\right) \times (4 \text{ m}) = 616 \text{ m}^3$$

To determine the hydraulic detention time, divide the clarifier volume by the flow rate (Q) to each clarifier:

$$\text{detention time} = \frac{\text{volume}}{Q} = \frac{616 \text{ m}^3 \times 24 \text{ h}/\text{day}}{6,000 \text{ m}^3/\text{day}} = 2.46 \text{ h}$$

The observed overflow rate (OR) is calculated as follows:

$$\text{OR} = \frac{Q}{\text{area}} = \frac{6,000 \text{ m}^3/\text{day}}{154 \text{ m}^2} = 39 \text{ m}^3/\text{m}^2\text{-day}$$

Determine the detention time and overflow rate at peak flow:

$$\text{OR at peak flow} = \frac{(Q \text{ at peak flow})/2}{\text{area}} = \frac{15,000 \text{ m}^3/\text{day}}{154 \text{ m}^2} = 97.4 \text{ m}^3/\text{m}^2\text{-day}$$

$$[\text{detention time at peak flow}] = \frac{\text{clarifier volume}}{Q \text{ at peak flow}} = \frac{616 \text{ m}^3 \times 24 \text{ h}/\text{day}}{30,000 \text{ m}^3/\text{day}/2} = 0.99 \text{ h}$$

At the average flow, the calculated values of detention time and overflow rate are within the ranges we discussed in the previous chapter. At peak flow, the calculated value of detention time is fine, but the overflow rate is slightly lower than desired. The final clarifier design may need to have an increased surface area to provide enough detention time for sufficient solids to settle.

9.6 Secondary Treatment

The wastewater that exits the primary clarifier has lost a significant amount of the particulate matter it contained, but it still has a high demand for oxygen due to an abundance of dissolved organic matter (measured as BOD). **Secondary treatment** (which is a form of biological treatment) utilizes microorganisms to decompose these high-energy molecules.

There are two basic approaches to biological treatment, differing in the manner in which the waste is brought into contact with the microorganisms. In *suspended-growth reactors*, the organisms and wastewater are mixed together, while in *attached-growth reactors*, the organisms are attached to a support structure, and the wastewater is passed over the organisms.

9.6.1 SUSPENDED-GROWTH REACTORS: ACTIVATED SLUDGE

The most common biological treatment system is a **suspended-growth** system called the **activated-sludge** process. Effluent from the primary clarifier is routed to an **aeration tank** (also referred to as an **aeration basin**), usually by gravity, and mixed with a diverse mass of microorganisms comprising bacteria, fungi, rotifers, and protozoa. This mixture of liquid, waste solids, and microorganisms is called the **mixed liquor**. A measurement of TSS obtained from the aeration basin is termed the **mixed liquor suspended solids (MLSS)**, expressed in mg/L. Volatile suspended solids (VSS) can be used as a surrogate to describe the reactor's biomass. This is because most of the solids are microorganisms that have a high carbon content in their cell structure. Typically the volatile fraction of the **mixed liquor volatile suspended solids (MLVSS)**, expressed in mg/L, is 60–80 percent of the MLSS.

The food web of the activated-sludge process is shown in Figure 9.9. The food web is somewhat truncated, both laterally (primary producers are unimportant because the waste provides a source of organic matter) and vertically (higher consumers are absent because the system is engineered to top out at a point where the remaining particulate matter is easily removed by sedimentation).

Different organism groups predominate depending on the degree of stabilization of the waste. At first, amoeboid and zooflagellate protozoans dominate, utilizing the dissolved and particulate organic matter initially present. Next zooflagellate and free-swimming ciliate protozoans increase in numbers, feeding on developing populations of bacteria. Finally, stalked ciliates and rotifers become most abundant, feeding from the surfaces of activated-sludge floc. Plant operating practices—such as the solids retention time (SRT), which will be discussed later—dictate the degree of stabilization and thus the successional position of the microbiology community. Molecular biology techniques are currently being used to further understand the unique microbial ecology of wastewater treatment systems.

The majority of the BOD is degraded in the presence of oxygen, so air is added to the reactor to supply oxygen, which must be transferred to the aqueous phase. This requires energy inputs. In practice,

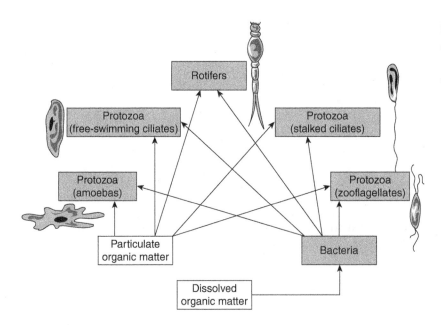

Figure / 9.9 Food Web of the Activated-Sludge Process.

(From Mihelcic (1999). Reprinted with permission of John Wiley & Sons, Inc.).

dissolved-oxygen concentrations in the aeration tank are maintained at 1.5–4.0 mg/L, with 2 mg/L being a common value. Levels greater than 4 mg/L do not significantly improve operation but raise operating costs because of the energy associated with forcing air into the system. Low oxygen levels can lead to *sludge bulking*, an abundance of filamentous organisms with poor settling characteristics.

Bacteria are primarily responsible for assimilating the dissolved organic matter in wastewater, and the rotifers and protozoa are helpful in removing the dispersed bacteria, which otherwise would not settle out. This would cause the plant's effluent to not meet permit requirements for suspended solids. The energy derived from the decomposition process is primarily used for cell maintenance and to produce more microorganisms. Once most of the dissolved organics have been used up, the microorganisms are routed to the secondary (or final) clarifier for separation.

In the secondary clarifier, two streams are produced: (1) a clarified effluent, which is sent to the next stage of treatment (usually disinfection); and (2) a liquid–solid sludge largely comprising microorganisms (but perhaps 2–4 percent solids). Lying at the bottom of the secondary clarifier, without a food source, these organisms become nutrient-starved or *activated*. A portion of the sludge is then pumped to the head of the tank (**return activated sludge**), where the process starts all over again. The remainder of the sludge is removed from the system and is processed for disposal (**waste activated sludge**). As we will see in the next several sections, it is necessary to continuously waste sludge from the systems to balance the gains in biomass that occur through microbial growth.

Key to performance of a secondary settling tank is the quality of the activated sludge produced in the biological reactor. For example, if you have a small or undersized secondary settling tank, a sludge that settles faster and has a lower sludge volume index (SVI) may work best for

The National Small Flows Clearinghouse provides technical assistance to help small communities and homeowners with their wastewater treatment

http://www.nesc.wvu.edu/wastewater. cfm

you. It is also critical that weirs be level to avoid hydraulic short circuiting within the settling tank. Inlets in the settling tank's feedwell can also be adapted by inserting an insert that is designed to dissipate the energy of the incoming flow, thus promoting flocculation and settling of the biological solids, which are small and not very dense (Miller and Esler, 2013).

DESIGN OF THE ACTIVATED-SLUDGE SYSTEM   A set of equations allow sizing of the biological reactor and, importantly, understanding relationships in the activated-sludge system between microorganism concentration, solids removal, and influent organic matter. Figure 9.10 shows a schematic of the activated-sludge process with a control volume added for our mass balance.

In this reactor, the organisms convert dissolved organic matter (measured as CBOD and NBOD) into gaseous CO_2, water, nitrate, and particulate organic matter (more microorganisms). A settling tank (called the *secondary clarifier*) that follows the biological reactor captures the particulate matter (sludge).

The typical life of a microorganism in a wastewater plant is first to feed in the aeration basin for several hours (4–6 h, for example), then flow to the secondary clarifier for several additional hours, where the organism rests while it settles to the bottom of the tank. When the organisms are hungry again, they are recycled back to the biological reactor to seed the biological reactor with a metabolically active (hungry) group of organisms. For a microorganism in the process, this process is repeated several times (feed and rest, feed and rest, feed and rest, and so on).

Because the microorganism population is increasing due to the presence of substrate (CBOD and NBOD) and the plant operator needs to maintain a constant concentration of microorganisms in the aeration basin, some organisms must be removed from the secondary clarifier. These organisms are *wasted* from the process; hence the term **wasting of sludge** is used to describe the removal of solids from the activated-sludge system via the secondary clarifier.

Figure / 9.10   Schematic of the Activated-Sludge Process.

(From Mihelcic (1999). Reprinted with permission of John Wiley & Sons, Inc.).

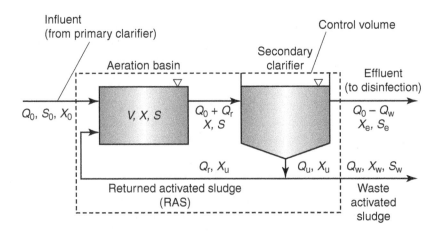

To develop a master design equation, we will first set up and analyze two mass balances, conducted on dissolved organic matter (substrate) and solids (biomass). This analysis, when combined with our understanding of microbial growth, will allow us to determine the volume of the aeration basin. In all these expressions, Q represents flow, expressed in m^3/day; S is substrate concentration (usually measured as mg BOD or COD per L); X is solids (biomass) concentration, measured as mg SS/L or mg VSS/L; and V is the volume of the aeration tank, expressed in m^3. The subscripts refer to influent (o), effluent (e), recycle (r), underflow from clarifiers (u), and wasted solids (w).

Application / 9.3 Should you use time derivatives (e.g., dX/dt and dS/dt) or rates of growth (r_g) and substrate depletion (r_s) in a flow through reactor?

While it was appropriate to use time derivatives in a batch reactor (like we did in Chapters 4 and 5), it is not appropriate to use time derivatives to express rates of biomass growth or substrate depletion in a flow through reactor at steady state. Careful examination of Figure 9.10 shows it is not a batch reactor, but is instead a flow through reactor (that is under steady-state conditions). Thus, in the following analysis where we work to derive a final design equation for the activated-sludge systems depicted in Figure 9.10, we will use the terms rate of growth (r_g) and rate of substrate depletion (r_s) instead of dX/dt and dS/dt, respectively.

We can show how dX/dt and r_g are equal in a batch reactor by setting up a mass balance for biomass.

Accumulation = [flow in] − [flow out]
+ sources/production
− sinks/consumption

Plugging in values for a batch system (there is no flow into or out of the reactor) and assuming there are no sinks/consumption of biomass results in

$$V \, dX/dt = 0 - 0 + (V \times r_g) - 0$$

Dividing both sides of the equation by the volume, V, we can see that in a batch reactor, $dX/dt = r_g$.

You can perform a similar analysis on the substrate depletion.

MASS BALANCE ON SOLIDS (BIOMASS)   A mass balance on microorganisms (solids) within the stated control volume (the dashed line in Figure 9.10) can be expressed as follows:

$$\begin{bmatrix} \text{biomass entering} \\ \text{aeration basin} \end{bmatrix} + \begin{bmatrix} \text{biomass produced} \\ \text{due to growth in} \\ \text{aeration basin} \end{bmatrix} = \begin{bmatrix} \text{biomass leaving} \\ \text{system} \end{bmatrix}$$

(9.1)

Using Equation 9.1, Figure 9.10, and the stated control volume, the mathematical expression that describes the *solids mass balance* is

$$Q_o X_o + V r_g = (Q_o - Q_w) X_e + Q_w X_w \qquad (9.2)$$

Here r_g is the rate of growth of the microorganisms (units for example of mg VSS/L-day). Assuming that **Monod kinetics** describes microbial growth and that first-order decay describes microbial death, the overall rate of biomass growth in the aeration basin (r_g) that results from growth and decay can be written as follows:

$$r_g = \frac{\mu_{max}SX}{(K_s + S)} - k_d X \tag{9.3}$$

This overall growth term (Equation 9.3) can be substituted into the mass balance expression shown in Equation 9.2 for r_g. In addition, we can assume that X_o and X_e are very small in relation to X (that is, $X_o \approx X_e \approx 0$). This is a good assumption because the concentration of biomass in the reactor, X, is maintained at approximately 2,000–4,000 mg TSS/L, while the influent solids into the aeration basin (X_o) might be 100 mg SS/L, and the effluent solids (X_e) less than 25 mg/L. (Note that X is the MLSS or MLVSS, defined previously.)

After performing this substitution and making the assumptions of X_o and X_e being negligible in terms of solids concentration, the resulting expression can be rearranged to yield

$$\boxed{\frac{\mu_{max}S}{K_s + S} = \frac{Q_w X_w}{VX} + k_d} \tag{9.4}$$

Equation 9.4 is important. It will be revisited in the next section, so readers should become familiar with it.

MASS BALANCE ON SUBSTRATE (BOD) Next, a mass balance is performed on the dissolved organic material (the BOD), which is substrate for the organisms. A mass balance on substrate (food) within the stated control volume (the dashed line in Figure 9.10) can be worded as follows:

$$\begin{bmatrix} \text{substrate entering} \\ \text{aeration basin} \end{bmatrix} - \begin{bmatrix} \text{substrate consumed} \\ \text{by microorganisms} \end{bmatrix} = \begin{bmatrix} \text{substrate leaving} \\ \text{system} \end{bmatrix} \tag{9.5}$$

Using Equation 9.5, Figure 9.10, and the stated control volume, we can write the mathematical expression that describes the *substrate mass balance*:

$$Q_o S_o - V r_s = (Q_o - Q_w)S + Q_w S \tag{9.6}$$

In Equation 9.6, r_s is the rate of substrate utilization (units for example of mg BOD_5/L-day). Here the effluent substrate, S_e, and the substrate in the waste sludge, S_w, are assumed to equal S ($S = S_e = S_w$). This should make sense because the secondary clarifier's purpose is to remove solids, not to biologically transform substrate to carbon dioxide and water.

The yield coefficient (Y) relates the change in substrate concentration (r_s) to the change in biomass concentration (r_g). Thus, the change in substrate concentration with time, r_s, can be written as follows:

$$r_s = \left(\frac{1}{Y}\right)\left(\frac{\mu_{max}S}{K_s + S}\right)X \qquad (9.7)$$

This expression for r_s can be substituted into the substrate mass balance (Equation 9.6). Rearranging the overall expression then results in

$$\boxed{\frac{\mu_{max}S}{K_s + S} = \frac{Q_oY}{VX}(S_o - S)} \qquad (9.8)$$

Note that the left side of the two final rearranged expressions obtained from the solids (Equation 9.4) and substrate mass balances (Equation 9.8) are the same. Thus, these two expressions can be set equal to provide a design expression:

$$\frac{Q_wX_w}{VX} = \frac{Q_oY}{VX}(S_o - S) - k_d \qquad (9.9)$$

Equation 9.9 can be used to solve for the volume of the aeration basin (V). This is because all the other terms either can be measured or are fixed. Table 9.6 provides a real-world explanation of the terms in this activated-sludge design expression (Equation 9.9). However, one problem arises with using Equation 9.9 to solve for V; the volume term appears on both sides of the equation. Fortunately, as we will see in the next section, the terms on the left side of Equation 9.9 can be combined into one term, referred to as the solids retention time (SRT).

Table / 9.6

Explanation of Terms Used in Expression for Activated-Sludge Design (Equation 9.9)

Term(s) in Final Design Expression	Explanation
Yield coefficient (Y) and decay coefficient (k_d)	Biokinetic coefficients, which are either measured orestimated (see Chapter 5 for example values)
Substrate concentration in reactor (S)	Plant effluent concentration, which is set by the state through the NPDES permitting process
Initial substrate concentration entering reactor (S_o) and flow entering reactor (Q_o)	Independent parameters that are a function of community demographics such as population, community wealth, water conservation measures, and commercial and industrial activity in a community
Rate of wasting sludge (Q_w) and concentration of solids in the wasted sludge (X_w)	Items a plant operator can control by wasting sludge, especially the rate at which solids are removed (wasted) from the system

SOLIDS RETENTION TIME The term on the left side of Equation 9.9 is an important expression for design and operation of an activated-sludge plant. It is the inverse of a term referred to as the **solids retention time (SRT)**. SRT is sometimes also referred to as **sludge age** and **mean cell retention time (MCRT)**. SRT is defined as

$$SRT = \frac{VX}{Q_w X_w}$$ (9.10)

If you look closely at the units of this expression, you will see that it has units of time (typically days). The sludge age in most treatment plants typically ranges from 2 to 30 days.

The SRT is not the hydraulic retention time (V/Q). Sludge age refers to the average time a microorganism spends in the activated-sludge process before it is expelled, or wasted, from the system.

Remember that the organisms feed in the aeration basin and then rest in the secondary clarifier until they are recycled back into the aeration basin to feed and then are sent back to the secondary clarifier to rest. This process is repeated many times until the organism is finally removed from the process, or wasted. The SRT thus refers to the number of days that an average microorganism undergoes this feed-and-rest cycle.

Equation 9.10 can be substituted into Equation 9.9 to yield our final design equation:

$$\frac{1}{SRT} = \frac{Q_o Y}{VX}(S_o - S) - k_d$$ (9.11)

Equation 9.11 can be used to size the aeration basin (solve for V) for a given (or range of) SRT.

---

example/ 9.4 Design of the Aeration Basin Based on Solids Retention Time

Given the following information, determine the design volume of the aeration basin and the aeration period of the wastewater for an activated-sludge treatment process: population = 150,000; flow rate is 33.75×10^6 L/day (equals 225 L/person-day); and influent BOD_5 concentration is 444 mg/L (note that this is high-strength wastewater). Assume that the regulatory agency enforces an effluent standard of $BOD_5 = 20$ mg/L and a suspended-solids standard of 20 mg/L in the treated wastewater.

A wastewater sample is collected from the biological reactor and is found to contain a suspended-solids concentration of 4,300 mg/L. The concentration of suspended solids in the plant influent is 200 mg/L, and that which leaves the primary clarifier is 100 mg/L. The microorganisms in the activated-sludge process can convert 100 g BOD_5 into 55 g biomass. They have a maximum growth rate of 0.1/day and a first-order death rate constant of 0.05/day, and they reach half of their maximum growth rate when the BOD_5 concentration is 10 mg/L. The design SRT is 4 days, and sludge is processed on the belt filter press every 5 days.

example/9.4 (continued)

solution

For the aeration basin volume. This problem provides a lot of information. To solve for the aeration basin volume (V), you need to know what information is important and what is not required. Look closely at Equation 9.11. Here, S_o equals the substrate (or BOD_5) entering the biological reactor, so assume that some BOD_5 is particulate and is removed in the primary clarifier. Assuming that 30 percent of the plant influent BOD_5 is removed during primary sedimentation, this means that $S_o = 0.70 \times 444\,mg/L = 310\,mg/L$. Accordingly,

$$\frac{1}{SRT} = \left[\frac{Q_o Y}{VX}(S_o - S) \right] - k_d$$

$$\frac{1}{4\,\text{days}} = \left[\frac{\left(33.75 \times 10^6\,\dfrac{L}{\text{day}}\right) \times \left(0.55\,\dfrac{g\,SS}{g\,BOD_5}\right)}{V \times \left(4{,}300\,\dfrac{mg\,SS}{L}\right)} \times \left(310\,\dfrac{mg}{L} - 20\,\dfrac{mg}{L}\right) \right] - \frac{0.05}{\text{day}}$$

Solve for $V = 4.173 \times 10^6\,L$.

For the aeration period. The plant's aeration period is the number of hours that the wastewater is aerated during the activated-sludge process. This equals the hydraulic detention time of the biological reactor:

$$\theta = \frac{V}{Q} = \frac{4.173 \times 10^6\,L}{33.75 \times 10^6\,\dfrac{L}{\text{day}}} = 0.12\,\text{day} = 3\,h$$

example/9.5 Use of Solids Retention Time to Calculate Solids Processing

Using data provided from Example 9.4, how many kg of primary and secondary dry solids need to be processed daily from the treatment plant?

solution

Assume that the amount of solids processed from the primary sedimentation tanks equals the difference in suspended-solids concentrations (influent minus effluent) measured across the sedimentation tanks multiplied by the plant flow rate:

$$33.75 \times 10^6\,\frac{L}{\text{day}} \times \left(200\,\frac{mg\,TSS}{L} - 100\,\frac{mg\,TSS}{L}\right) \times \left(\frac{kg}{10^6\,mg}\right) = 3{,}375\,\text{kg primary solids per day}$$

We are not provided with the concentration difference of suspended solids across the secondary sedimentation tanks, so we cannot determine the amount of secondary solids produced daily in the same manner that we used for primary solids. However, careful examination of the expression of solids retention time (SRT = 4 days) shows the term $Q_w X_w$ equals the answer. Therefore:

$$4\,\text{days} = \frac{VX}{Q_w X_w} = \frac{4.173 \times 10^6\,L \times \left(4{,}300\,\dfrac{mg\,SS}{L}\right)}{Q_w \times X_w}$$

Solve for $Q_w X_w$, which equals 4,486 kg secondary dry solids per day.

RELATING SOLIDS RETENTION TIME TO MICROBIAL GROWTH RATE You may have already noted that the inverse of SRT has the units of day^{-1}. Remembering from Chapter 5 the definition of the *specific growth rate* of microorganisms,

$$r_g = \mu X \tag{9.12}$$

This expression can be arranged to solve for the specific growth rate, μ:

$$\mu = \frac{r_g}{X} \tag{9.13}$$

Equation 9.13 shows that the specific growth rate (units of day^{-1}) equals the mass of biomass produced in the aeration basis (kg MLSS produced per day) divided by the mass of biomass present in the reactor (kg MLSS).

Remember that SRT refers to the average time a microorganism spends in the activated-sludge process before it is expelled, or wasted, from the system. If a treatment plant operator wants to maintain the same concentration of biomass in the biological reactor (that is, X), the operator would have to waste the same volume of solids per day ($Q_w X_w / V$) that are produced by microbial growth (r_g).

Equations 9.13 and 9.10 are thus related. In fact, for a completely mixed activated-sludge process, the SRT (which is controlled by wasting solids) is the inverse of the average of the specific growth rate of the microorganisms:

$$\boxed{\frac{1}{\text{SRT}} = \mu} \tag{9.14}$$

The relationship shown in Equation 9.14 is important for a design engineer and plant operator because it tells us there is a *critical* value of SRT. Below this **critical SRT** value (sometimes referred to as SRT$_{min}$), the microbial cells in the activated-sludge process will be **washed out** or removed from the system faster than they can reproduce. This would not be good, because if specific types of microorganisms wash out of the system, the activated-sludge process will lose its ability to degrade particular pollutants. For example, washing out of nitrifying organisms will result in poor removal of ammonia nitrogen, and washing out of heterotrophic organisms will result in poor removal of BOD.

Fortunately, the SRT$_{min}$ can be approximated as

$$\boxed{\frac{1}{\text{SRT}_{min}} \approx \mu_{max} - k_d} \tag{9.15}$$

μ_{max} and k_d were defined previously in Chapter 5. Never design a biological treatment process where the SRT is equal to the SRT$_{min}$! In fact, many treatment plants are designed for an SRT that is 2–20 times greater than the SRT$_{min}$.

FOOD-TO-MICROORGANISM RATIO The rate of food introduction (BOD loading) is largely fixed by the flow rate (Q_o) and BOD (S_o) of the influent. The size of the microbial population is equal to the product of the MLSS (or MLVSS) concentration in the biological reactor (X) and the reactor volume (V). Earlier we stated that operating experience in waste treatment plants suggests that MLSS concentrations in the reactor should be maintained at levels ranging from 2,000 to 4,000 mg/L. Too low concentrations (less than 1,000 mg/L) may lead to poor settling, and too high concentrations (greater than 4,000 mg/L) may result in solids loss in the secondary clarifier overflow and excessive oxygen requirements.

Another key process design parameter (besides SRT) is referred to as the **food-to-microorganism (F/M) ratio**. It can also be used to estimate the required tank volume. Essentially a feeding rate, the F/M ratio is equivalent to the BOD loading rate divided by the mass of MLSS in the reactor. The **BOD loading rate** (kg BOD/m^3-day) is the mass of food that enters the biological reactor per day divided by the volume of the reactor.

Using the terminology from Figure 9.10

$$F/M = \frac{S_o Q_o}{XV} \tag{9.16}$$

F/M has units of kg BOD/kg MLSS-day. Remembering from Chapter 4 the definition of **hydraulic retention time** ($\theta = V/Q$), the F/M ratio can also be written as

$$F/M = \frac{S_o}{\theta X} \tag{9.17}$$

Referring back to Table 9.6, S_o and Q_o are largely fixed by local demographics such as population, wealth, and the mix of residential and commercial establishments in a community. Note that water conservation measures will not reduce the mass of food entering the system. Water conservation will reduce Q_o, but the resulting S_o will increase proportionally. Remember also that the concentration of microorganisms in the biological reactor (X) is controlled by how much solids the operator wastes. Thus, a particular reactor volume (V) can be selected to achieve the desired F/M ratio.

---

example/ 9.6 Calculating the F/M Ratio

Determine the F/M ratio (in units of lb BOD_5/lb MLSS-day), using data provided from Example 9.4.

solution

Remember that by definition,

$$F/M = \frac{Q \times S_o}{X \times V} = \frac{\left(33.75 \times 10^6 \frac{L}{day}\right) \times \left(310 \frac{mg}{L}\right)}{\left(4,300 \frac{mg\ SS}{L}\right) \times (4.173 \times 10^6\ L)}$$

$$= 0.58 \frac{kg\ BOD_5}{kg\ MLSS\text{-}day} = 0.58 \frac{lb\ BOD_5}{lb\ MLSS\text{-}day}$$

Note that converting units of F/M ratio from metric to English units requires no conversion factor, because the mass unit is in both the numerator and denominator. Also, be careful in your units for F/M, because the denominator can have units of MLSS or MLVSS.

Table / 9.7

Relationship of Solids Retention Time (SRT) and Food-to-Microorganism (F/M) Ratio

SRT (days)	F/M (gm BOD/gm VSS-day)
5–7	0.3–0.5
20–30	0.10–0.05

SOURCE: Values obtained from Tchobanoglous et al. (2003).

Equation 9.16 shows the F/M ratio is really a feeding rate. The lower the F/M ratio, the lower the feeding rate, the hungrier the microorganisms, and the more efficient the removal. Likewise, if the SRT is lowered, the operator will be wasting sludge and decreasing their solids inventory (decreasing X). Examination of Equation 9.16 shows that in this case, with lower SRT and lower X, the F/M ratio will increase.

Our examination of F/M will not go into additional detail. However, readers need to understand that SRT and F/M are related (Table 9.7). This relationship includes the efficiency of the BOD removal and some of microbial parameters that were discussed in Chapter 5 (Y and k_d). Higher SRTs equate to lower F/M, and lower SRTs equate to higher F/M. This should make sense. At a higher SRT, fewer cells are being wasted from the system, so the concentration of microorganisms in the biological reactor (X) will increase. Because the incoming food (S_o times Q_o) is not controlled by the plant operator, examination of Equation 9.16 shows that the F/M ratio would decrease.

At low F/M ratios, the microorganisms are maintained in the **death growth phase** or **endogenous-growth phase**, meaning they are starved and thus very efficient at BOD removal. Because S_o is relatively constant for domestic wastes and because there are limits on the levels of X that a reactor can support, maintenance of a low F/M ratio requires either a very small flow or a very large tank volume. In either case, this leads to a long hydraulic residence (aeration) time.

Operating an activated-sludge plant at low F/M ratios is termed **extended aeration**. The cost of operation and maintenance is high for large tank volumes, so extended aeration is largely limited to systems with small organic loads (for example, at mobile-home parks and recreational facilities).

At high F/M ratios, the microorganisms are maintained in the exponential-growth phase. These organisms are more food saturated, meaning there is an excess of substrate, so BOD removal is less efficient. This approach is termed *high-rate activated sludge*. In this approach, higher MLSS concentrations are employed, so a shorter hydraulic residence time is achieved, and smaller aeration tank volumes are required.

In addition to influencing BOD removal efficiency, the selection of an F/M ratio affects the settleability of the sludge and thus the efficiency of TSS removal. In general, as the F/M ratio decreases, the settleability of the sludge increases. Starving microorganisms flocculate and thus settle well, while those maintained at high F/M ratios form buoyant filamentous growths, which settle poorly, a condition termed *sludge bulking*.

SETTLING CHARACTERISTICS OF ACTIVATED SLUDGE Design of settling clarifiers is similar to that of primary clarifiers, except that the hydraulic detention rates and overflow rates reflect the fact that the particles are smaller than in primary settling. The efficiency of settling in the secondary clarifier is influenced by the degree to which the MLSS flocculates (that is, the ability of the MLSS to stick together to form a larger mass of particles). This **flocculation** capacity is lowest in exponential-growth phase populations (high F/M ratio), increases in declining-growth phase populations (intermediate F/M ratio), and is highest in endogenous-growth phase populations (low F/M ratio).

example/9.7 Relating F/M to Aeration Tank Volume

The suspended-solids concentration is 220 mg/L in the plant influent; 4,000 mg/L in the primary sludge, 15,000 mg/L in the secondary sludge, and 3,000 mg/L exiting the aeration basin. The concentration of total dissolved solids in the plant influent is 300 mg/L, and the concentration of total dissolved solids exiting the aeration basin is 3,300 mg/L. The BOD_5 is 150 mg/L measured after primary treatment and 15 mg/L exiting the plant. Total nitrogen levels in the plant are approximately 30 mg N/L.

If the F/M ratio is 0.33 lb BOD_5/lb MLSS-day, estimate the hydraulic retention time of the aeration basins if the total plant flow is 5 million gallons/day.

solution

Note that this problem statement provides a lot of extra material, so readers must understand the order of various unit processes in a wastewater treatment plant, as well as the definition of F/M ratio. The mass of food that the microorganisms see equals the plant flow rate multiplied by the concentration of BOD_5 exiting the primary sedimentation tank (which thus enters the aeration basin). This value is S_o.

$$F/M = 0.33 \frac{\text{lb BOD}_5}{\text{lb MLSS-day}} = \frac{Q \times S_o}{V \times X} = \frac{Q \times \left(150 \frac{\text{mg}}{\text{L}}\right)}{V \times \left(3,000 \frac{\text{mg MLSS}}{\text{L}}\right)}$$

$Q/V = 6.6$/day, and because the hydraulic detention time equals V/Q, the detention time (θ) equals 0.15 day or 3.6 h. The size of the aeration basins can then be found from knowledge of the design flow (or Q) of the plant: $V = Q \times \theta$.

There are two reasons for this behavior. First, flocculation is aided by the presence of microbial-produced slime (polysaccharide gums), which helps the particles stick together. Slime is produced through the attrition of slime layers on the zoogleal masses. Slime layers are most abundant in populations grown under the endogenous phase and are least abundant in exponential-phase populations. Second, as we described in the previous chapter, flocculation is greatest under conditions where particles can be easily brought together. Inactive, endogenous-phase cells behave as simple colloids and flocculate well. Active, highly motile particles characteristic of exponential-phase populations (high F/M ratio) tend to flocculate poorly.

If you place some activated sludge in a 1,000 mL graduated cylinder and watch the sludge settle over time, you would observe four zones of settling. Near the top of the cylinder, you would observe **discrete settling**, which is typical of low particle concentrations. In this zone, the particles settle alone according to **Stokes' law**. Beneath this, you would observe **flocculent settling**. Here the particles coalesce during sedimentation, and the sample is still relatively dilute. The third zone is called **hindered settling**. Here, as the solids concentrations increase, interparticle forces hinder settling of neighboring particles. You can actually observe the particles settling as a unit, and you observe a solid–liquid interface. The final zone, called **compression settling**, occurs near the bottom of the cylinder and is visible as time passes. Here the concentration of solids is now large, so the downward movement of solids is opposed by the upward movement of water. In this case, no settling can occur until water is compressed from the sludge.

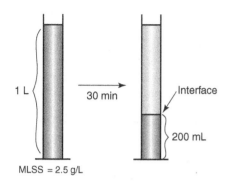

1 L { ... **30 min** → ... **Interface**

200 mL

MLSS = 2.5 g/L

Figure 9.11 Determination of Sludge Volume Index The SVI is measured to determine the settleability of the sludge. The SVI is the volume (in mL) occupied by 1 g MLSS (dry weight) after settling for 30 min in a 1,000 mL graduated cylinder. In the situation depicted here, 2.5 g MLSS occupy 200 mL of volume after 30 min settling, so the SVI is 80.

A test termed the **sludge volume index (SVI)** is performed at the treatment plant to determine the settleability of the sludge. The SVI is the volume (in mL) occupied by 1 g MLSS (dry weight) after settling for 30 min in a 1,000 mL graduated cylinder (Figure 9.11). The units of the SVI are mL/g.

Another way to think of the SVI is that it is the percent volume occupied by sludge in an MLSS sample. It thus tells you how well your sludge is currently settling. Table 9.8 provides a method to interpret the value of SVI in terms of how well sludge will settle.

Good settling characteristics (low SVI) are one indication of a properly operating wastewater treatment and resource recovery plant. A person at such a plant who collects a sample from the biological reactor would first observe a moderate amount of brown **foam** in the sample. After the sample has been allowed to sit for a few minutes, the resulting supernatant should appear clear and have a low BOD. Further examination of the contents of the biological reactor under a microscope would show a well-flocculating sludge with large numbers of free-swimming ciliates and bacteria.

During extended aeration, **endogenous respiration** takes place, because there is not enough food (BOD) to support the population of microorganisms. The microorganisms will thus begin to utilize food they have stored in their cellular structure, and some organisms will begin to die. A dead organism's cells begin to lyse and subsequently become food for other, higher organisms such as stalked ciliates and rotifers. During endogenous respiration, the biological solids appear very dense (so the SVI is very low).

A common occurrence in an activated-sludge plant is the site of a poorly settling brown foam that is high in TSS. The microorganisms found in this foam have hydrophobic cell walls. Most of these organisms belong to a group called **nocardioform** organisms. Because of the hydrophobic cell wall, air bubbles from the aeration process can attach to the microorganism and the associated biological floc. As a result, the biological floc rises to the surface. At the surface, the air bubble will eventually collapse but leave behind the floated solids, which appear as foam.

Fortunately several design and operating techniques can eliminate or reduce the formation of foam. For example, treatment plants need to watch their inputs of toxic chemicals that can change the ecology of the reactor and may favor the presence of these dispersed bacteria. Also, activated-sludge plants can be run at low SRT in an attempt to wash out the nocardioform organisms. Recycling foam that contains these organisms back to the front end of the plant should be discouraged. In addition, it has been found that activated-sludge systems containing subsurface aeration basin draw-off and secondary scum baffles have higher nocardioform levels than systems with aeration basin draw-off and no clarifier surface baffles. Apparently, reactor features that encourage the selection of dispersed nocardioforms over a clumped and settleable biological floc result in greater chance for foaming (Jenkins, 2007).

Table / 9.8

Interpretation of Sludge Volume Index (SVI)

Value of SVI	Settleability of the Sludge
0–100	Good
100–200	Acceptable
>200	Poor

9.7 Modifications to the Activated-Sludge Process

Operation of the activated-sludge process at midrange F/M ratios with microorganisms in the declining-growth phase is termed **conventional activated sludge**. This option offers a balance between removal

Table / 9.9

Examples of Process Modifications Made to the Conventional Activated-Sludge System

Process	Description
Conventional activated sludge	Primary effluent and return activated sludge are introduced at head of aeration basin. The aeration is provided in a nonuniform manner over the length of the tank as more aeration is required at the beginning of the tank, since the organic loading is higher there because the BOD is removed along the length of the aeration basin.
Step feed aeration	Modification where primary clarifier effluent is introduced at several points along the beginning of the aeration basin. The peak oxygen demand is thus more evenly distributed throughout the aeration tank. Aeration is uniform along the length of the aeration basin.
Contact stabilization	The aeration basin is separated into a stabilization zone followed by a small contact zone. Primary clarifier effluent is routed to the contact zone first. Return activated sludge is recycled back into the stabilization zone.
Extended aeration	Similar to conventional activated sludge except primary clarifier is usually eliminated, SRT is very long (20–30 days), and hydraulic detention times are close to 1 day. Used primarily by smaller communities, schools, resorts.
Oxidation ditch	Oval reactor where wastewater moves at relatively high velocities. Return activated sludge is recycled back to beginning of the reactor.
Sequencing batch reactor	Fill and draw reactors where a minimum of two reactors are used. While one reactor is being filled, the other reactor is overseeing the biological reactions, settling of solids, and removal of settled wastewater.

efficiency and cost of operation. Various reactor configurations are available (see Table 9.9), each with its own set of advantages and disadvantages. The two basic types are plug flow and completely mixed flow reactors. Plug flow reactors offer higher treatment efficiency than completely mixed flow reactors but are less able to handle spikes in the BOD load. Other modifications of the process are based on the manner in which waste and oxygen are introduced to the system.

9.7.1 MEMBRANE BIOREACTORS

One of the fastest-growing segments of biological wastewater treatment processes and water reuse is the use of **membrane bioreactors (MBRs)**. They are also appropriate for deployment in sewers where they can be used to extract and reuse water that contains nutrients upstream of a centralized plant, where the water and nutrients are of use. This process is referred to as sewer mining.

MBRs combine the suspended-growth activated-sludge process previously described with the microfiltration membrane process (discussed in Chapter 8). Figure 9.12 displays two types of process configurations: (1) *the submerged-membrane process* and (2) *the external-membrane process*.

For both process flow configurations, the mixed liquor in the aeration basin is filtered through the membrane, separating the biosolids from the effluent water. In the submerged-membrane process, a vacuum of less than 50 kPa is applied to the membrane that

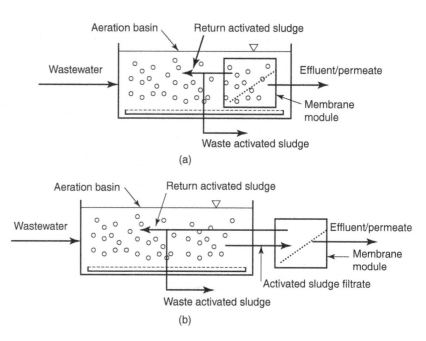

Figure / 9.12 Two Bioreactor Membrane Processes (a) MBR where the membranes are immersed into the aeration basin. (b) MBR where the membranes are external to the aeration basin.

filters the water through the membrane while leaving the biosolids in the aeration basin. For the external-membrane system, a pump is used to pressurize the mixed liquor at less than 100 kPa, and the water is filtered through the membrane while the biosolids are sent back to the aeration basin. In both systems, wasting of solids is done directly from the aeration basin.

MBRs have several advantages over conventional activated-sludge systems (Tchobanoglous et al., 2003). The use of MBRs eliminates the need for secondary clarifiers or filters. The MBRs can operate at much higher MLSS loadings, which decreases the size of the aeration basin; can operate at longer SRTs, leading to less sludge production; and can operate at lower DO concentrations with the potential of nitrification/denitrification at long SRTs. The MBR requires about 40–60 percent less land use footprint than a conventional activated-sludge plant, which is especially important in urban areas, where land is a premium and populations are expanding. Also, the effluent quality is much better in terms of BOD, low turbidity, TSS, and bacteria.

In addition, MBRs do not have the problematic issues such as sludge bulking, filamentous-organism growth, and pinpoint floc, sometimes experienced in many conventional activated-sludge facilities. However, they do contain more dispersed microorganisms, and the resulting biological flocs tend to be smaller than obtained from gravity settling. For an example of a facility that has realized some of these benefits, see Figure 9.13 and Figure 9.14 and Table 9.10.

MBRs do have some disadvantages, including (1) higher capital costs, (2) the potential for short membrane life due to membrane fouling and higher energy costs resulting from module aeration, and (3) the operational issue that membranes need to be cleaned on a cyclic basis. In addition, a more skilled operation staff is needed in case of problems with plant disturbances, which can upset the plant operation very quickly.

Figure / 9.13 Aerial View of the Traverse City, Michigan, Wastewater Treatment Facility, Highlighting Some Major Unit Processes The plant employs membrane bioreactors. The plant consists of eight trains containing 13 cassettes, and each cassette consists of 32 membrane modules. The plant was initially designed to treat 19,000 m³/day (maximum monthly flow) and was upgraded with the bioreactor membrane system to treat 32,000 m³/day (maximum monthly flow) to 68,000 m³/day (peak-day flow) of wastewater. With addition of the MBRs, the physical footprint of the plant was decreased by approximately 40 percent, as the two secondary clarifierswere no longer required.

(Photo courtesy of David W. Hand).

Figure / 9.14 Membrane Cassette Being Lowered into the Aeration Basin at the Traverse City, Michigan, Wastewater Plant.

(Photo courtesy of David W. Hand).

Table / 9.10

Comparison of Effluent Characteristics of Membrane Bioreactor (MBR) and Conventional Wastewater Treatment at the Traverse City (Michigan) Wastewater Treatment Plant

Parameter	Performance of Conventional Activated-Sludge Plant Before Installation of MBR	After Installation of MBR
BOD_5	2–5 mg/L	<2 mg/L
TSS	8–20 mg/L	<1 mg/L
NH_3–N	<0.03–20–mg/L	<0.03 mg/L
PO_4–P	0.6–4.0 mg/L	<0.5 mg/L
Fecal coliforms	50–200 cfu (colony-forming units)/100 mL	<1 cfu/100 mL

9.8 Attached-Growth Reactors

In contrast to suspended-growth treatment systems, discussed in the previous section, microorganisms can also be attached (or fixed) to a surface during biological treatment. Table 9.11 provides an overview of specific types of **attached-growth** processes, identifying some advantages and disadvantages associated with each of them.

In the **trickling filter**, primary effluent is "trickled" over and percolates through a 1–3 m deep tank filled with stone media, slag, or plastic media (termed the filter bed). An active biological growth forms on the solid surfaces (the active biofilm thickness ranges from 0.07 to 4.0 mm) and dissolved organic matter (BOD) diffuses from the water phase into the biofilm as the wastewater trickles down. Design of trickling filters is based on a maximum allowable hydraulic loading (5–10 m^3/m^2-day) and a maximum organic loading (250–500 g BOD m^3/d). The organic loading must be limited so that the substrate uptake capabilities of the system are not saturated (which would result in poor removal efficiency). The hydraulic load must be limited so that the filter is not flooded. In this case, ponding could occur, which would limit oxygen transfer into the system. Trickling filters are a very appropriate technology (along with waste stabilization ponds) in many parts of United States and the world because of their low energy and material requirements and relatively fewer operation and maintenance requirements.

Table / 9.11

Attached-Growth Processes Used to Treat Wastewater   The three processes are configured in several specific types of reactors, which have different advantages and disadvantages.

Attached-Growth Processes	Specific Type	Advantages and Disadvantages
Nonsubmerged attached-growth systems	Trickling filters; biotowers; rotating biological contactors (RBCs)	Less energy required; simpler operation and fewer equipment maintenance needs than suspended-growth systems; better recovery from shock toxic loads than suspended-growth systems; difficult to accomplish biological N and P removal compared with suspended designs.
Suspended-growth processes with fixed-film packing	Submerged RBCs; aeration basins with submerged packing materials	Increased treatment capacity; greater process stability; reduced solids loadings on the secondary clarifier; no increase in operation and maintenance costs.
Submerged attached-growth aerobic processes	Upflow and downflow packed-bed reactors and fluidized-bed reactors that do not use secondary clarification	Small physical footprint with an area requirement one-fifth to one-third of that needed for activated-sludge treatment.

Approximately one-third of the wastewater in the United States is treated using on-site systems. These systems are designed to reduce the risk of exposure to pathogens and other pollutants (for example, TSS and BOD) but are not designed and optimized for nutrient removal.

The management of on-site wastewater treatment systems is undergoing a shift from one based primarily on conventional septic systems to one that accommodates alternative technologies on sites poorly suited for conventional septic systems. Accompanying this shift is a greater emphasis on site-specific designs, for example, for sites with poorly draining soils or near sensitive ecosystems. In addition, on-site treatment systems must be able to accommodate transient loading and long idle times (for example, when homeowners are on vacation or are seasonal residents as is common in some coastal areas). They must also have low complexity because they must be maintained by homeowners who often have little operations and maintenance experience.

A septic tank comprises a sealed tank with an inlet and outlet (Figure 9.15). Wastes flow by gravity into the tank and after several days of hydraulic retention time, partially treated effluent flows from out of the tank, usually to a subsurface leachfield where soil processes of filtration, adsorption, and biodegradation degrade or retain some additional pollution. In the tank, solids settle out by gravity and undergo anaerobic decomposition. This results in the production of water, gas, sludge, and a layer of floating scum. The settled solids gradually build up at the tank bottom and must be periodically removed.

A septic tank can be sized based on the number of bedrooms (not bathrooms) in a home. One way to estimate the size of septic tank necessary for an average household would be to multiply the number of bedrooms by 150 gallons per bedroom per day, and then multiply this number by 2 or 3 to allow for 2 or 3 days retention time in the tank. The use of water-saving devices will improve the system performance because they result in longer residence times for pollutants in the tank and ensure the leachfield is not overloaded.

What size septic tank would you recommend for a three-bedroom house assuming the available tank sizes are 750, 1,000, 1,200, and 1,500 gallons? Assume you wish to have 2 days of residence time for the pollutants in the tank.

solution

3 bedrooms × 150 gallons/bedroom × 2 = 900 gallons

Select the 1,000 gallon tank which is the closest size available to meet our design guidelines.

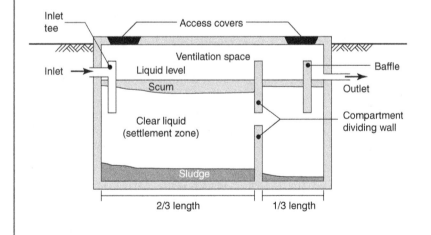

Figure / 9.15   Septic Tank Schematic for On-site Treatment of Wastewater.

Septic Systems
http://water.epa.gov/infrastructure/septic/

9.9 Removal and Recovery of Nutrients: Nitrogen and Phosphorus

Secondary treatment is sometimes inadequate to protect the receiving water. Additional removal or recovery of pollutants, especially nitrogen (N) and phosphorus (P), is accomplished through a variety of physical, chemical, and biological processes collectively termed advanced or **tertiary wastewater treatment**. As we will see, advanced removal of nutrients such as N and P is now being incorporated into the existing biological process. Water reuse takes advantage of the fertilizer potential of nutrients found in wastewater.

Today suspended- or attached-growth biological systems can treat inorganic nitrogen down to 1–1.5 mg/L and phosphorus to as low as 0.1 mg/L (after filtration). Dissolved organic nitrogen concentrations will still remain at the range of 0.5–1.5 mg/L. An excellent review of the history of biological removal of nitrogen and phosphorus is available elsewhere (Barnard, 2006).

A significant fraction of phosphorus was removed in some parts of the world via source reduction by eliminating it from soaps and detergents. Interestingly, up to 90 percent of the total nitrogen and 75 percent of the total phosphorus discharged from a household into a sanitary sewer is now found in urine. Much of this is diluted by the excessive water usage found in North American homes and commercial districts. Some countries are now proposing the development of toilets that separate urine from feces and dual-sewer systems that handle each waste stream.

Composting toilets can also be designed to separate urine from feces. They are used throughout the world and can be constructed from concrete

Class Discussion

Urine contains most of the nitrogen and a good part of the phosphorus in domestic wastewater. Think about the advantages and disadvantages of separating urine from feces at the building and community scale. What environmental, cultural, and economic issues would need to be addressed if the established methods to collect and treat wastewater were changed to one more focused on resource recovery?

Application / 9.5   The Chemistry of Urine

Understanding the chemistry of human urine is central to identifying innovative processes for recovery and reuse of the valuable nutrients found in urine. Human urine contains urea, ammonia, inorganic anions (Cl^-, SO_4^{2-}, PO_4^{3-}), inorganic cations (Na^+, K^+, Ca^{2+}, Mg^{2+}), natural organic metabolites (for example, citrate), and pharmaceuticals.

Reports of urine composition vary widely in the literature. Representative characteristics of fresh urine are pH 6–7, ionic strength (0.1–0.4 M), urea (415 mmol N/L), total NH_3 (18 mmol NH_3–N/L), Cl^- (108 mmol/L), SO_4^{2-} (8 mmol/L), total phosphates (12 mmol P/L), Na^+ (116 mmol/L), K^+ (56 mmol/L), Ca^{2+} (3 mmol/L), Mg^{2+} (3 mmol/L), and natural organic metabolites (<20 mmol/L) (Udert et al., 2003; Saude et al., 2007). Recent research shows that the majority of pharmaceuticals consumed by humans are excreted in urine with concentrations typically <1 mmol/L (Lienert and Larsen, 2010).

After urine leaves the human body and is stored for a period of time, its composition changes due to hydrolysis of urea catalyzed by urease-positive bacteria, which are ubiquitous in collection systems. Hydrolysis converts 1 mole of urea into 2 moles of ammonia and 1 mole of bicarbonate. This increases the pH to 8–9 and total NH_3 concentration to 123 mmole NH_3–N/L. In contrast, total phosphates decrease to 2.5 mmole P/L, Ca^{2+} to 0.7 mmol/L, and Mg^{2+} to 0.04 mmol/L. This is from precipitation of minerals such as struvite and hydroxyapatite (Darn et al., 2006).

$$Mg^{2+} + NH_4^+ + PO_4^{3-}$$
$$+ 6H_2O \rightarrow MgNH_4PO_4 \cdot 6H_2O_{(s)}(\text{struvite})$$

$$10\,Ca(OH)_2 + 6H_3PO_4 \rightarrow Ca_{10}\,(PO_4)_6(OH)_{2(s)}$$
$$+ 18H_2O\ (\text{hydroxyapatite})$$

Thus, urine composition and time-varying changes must be considered when identifying appropriate processes for urine collection and recovery.

Adapted with permission from information provided by Dr. Treavor H. Boyer, University of Florida.

block or modular plastic units that resemble ceramic toilets (Application 9.1). The lead author of this book has a working composting toilet. Both the urine and composted feces can be used to amend agricultural soils.

9.9.1 NITROGEN

North American treatment and resource recovery plants typically receive influent nitrogen in the range of 25–40 mg N/L. This value can reach the low hundreds of mg/L in areas where wastewater is primarily sewage and the sewage is not diluted by excessive water use. At a minimum, treatment plants seek ammonia removal, but it is becoming more common to totally remove nitrogen from a wastewater effluent.

One source reduction strategy is to reduce the amount of nitrogen associated with wastewater influent, removing the burden from the WWTP operations. Separate collection of human urine eliminates nutrients from domestic wastewater and substantially decreases the nutrient load to the WWP. Urine contains most of the nutrients excreted by humans: 85–90% nitrogen, 50–80% phosphorus, and 80–90% potassium (Larsen and Gujer, 1996).

Collecting this highly concentrated waste stream separately enables elimination as well as the reuse of these nutrients (Maurer et al., 2006). This collection can occur at the building or community level, or at a centralized plant. For example, at a centralized treatment and resource recovery plant, struvite ($MgNH_4PO_4 \cdot 6H_2O$) precipitation can efficiently convert two dominate wastewater nutrients into a good slow-release fertilizer (Johnston and Richards, 2003) with the chemical reaction provided in Application 9.5.

Approximately 10 percent of the nonwater portion of a microbial cell is nitrogen. Therefore, the growth of biological solids removes some nitrogen from the dissolved to the particulate phase. However, this removal of nitrogen is nowhere sufficient to protect receiving water bodies from pollution.

Nitrification, the conversion of ammonia (NH_4^+) to nitrite (NO_2^-) and then nitrate (NO_3^-), is accomplished during secondary treatment by specialized genera of lithotrophic bacteria. *Nitrosomonas* (and *Nitrosococcus*) bacteria convert ammonia (NH_4^+) to nitrite (NO_2^-), and *Nitrobacter* (and *Nitrospira*) bacteria convert nitrite to nitrate (NO_3^-). The overall reaction for these two processes can be written as follows:

$$NH_4^+ + 2O_2 \rightarrow NO_3^- + 2H^+ + H_2O \qquad \textbf{(9.18)}$$

If you work out the stoichiometry in Equation 9.18, you will find that 4.57 g of oxygen are required to oxidize every 1 g of nitrogen (as N).

Equation 9.19 includes not only nitrification of ammonia but also the incorporation of dissolved inorganic carbon and some of the ammonia into biomass (written as $C_5H_7NO_2$):

$$NH_4^+ + 1.863O_2 + 0.098CO_2 \rightarrow 0.0196C_5H_7NO_2 + 0.98NO_3^-$$
$$+ 0.0941H_2O + 1.98H^+ \qquad \textbf{(9.19)}$$

In Equation 9.19, only 4.52 g oxygen are required to oxidize every gram of nitrogen (as N). This stoichiometric value is lower than Equation 9.18 because Equation 9.19 accounts for some ammonia being used for synthesis of new cells. Equation 9.19 also shows that lithotrophic microorganisms obtain carbon for their cellular mass not from dissolved organic carbon, but by converting inorganic carbon (dissolved CO_2). Both Equations 9.18 and 9.19 show that some alkalinity is consumed for every mole of ammonia oxidized.

The nitrification reactions proceed slowly, require adequate oxygen (more than 0.5 mg/L) and alkalinity, and are sensitive to temperature, pH (preferring pH near 7), and the presence of toxic chemicals. For complete nitrogen removal, a variety of bacteria, including those of the genus *Pseudomonas*, can convert nitrate to nitrogen gas (N_2). In this reaction, the nitrate serves as the electron acceptor (as oxygen does in carbonaceous oxidation), and the organic material in the wastewater is the electron donor.

Readers should refer back to Chapter 5 and review the nitrogen cycle (Figure 5.29). Observe that in the denitrification step, the end product is N_2 gas. However, we are now finding out that some of the nitrate ends up being released as the intermediate N_2O, which is emitted to the atmosphere and is a **greenhouse gas.** (See how increased populations will produce increased amounts of greenhouse gases because of constituents found in domestic waste streams.) Assuming the biodegradable organic material (measured as CBOD) in wastewater can be written as $C_{10}H_{19}O_3N$, the removal of nitrate to nitrogen gas can be written as follows:

$$C_{10}H_{19}O_3N + 10NO_3^- \rightarrow 5N_{2\ (gas)} + 10CO_2 + 3H_2O + NH_3 + 10OH^-$$

$$(9.20)$$

In Equation 9.20, alkalinity is produced (written as OH^-), and no dissolved oxygen is written in the expression. In fact, the presence of dissolved oxygen will inhibit the nitrate-reducing enzymes required for the denitrification reaction. This occurs at dissolved-oxygen concentrations as low as 0.1 or 0.2 mg/L.

During the design and operation of the biological reactor to denitrify nitrate, the amount of CBOD in the wastewater is a critical design parameter. If organic carbon is limiting in the waste stream, a waste organic stream (for example, dairy waste) or chemicals (for example, methanol) can be added in carefully controlled amounts to support **denitrification**. However, the amount added must be carefully controlled to ensure that no untreated residual BOD remains.

More than nine patented methods are available to configure the biological reactor such that nitrogen can be removed via nitrification and denitrification reactions. Figure 9.16 shows the most commonly used process, termed the **modified Ludzak–Ettinger (MLE) process**. Here an anoxic zone is located at the beginning of the biological reactor. All the processes use a combination of aerobic and anoxic zones where the biological reactor is configured to remove organic carbon (in the aerobic and anoxic zones), convert inorganic ammonia nitrogen (in the aerobic zone), and remove inorganic nitrate nitrogen (in the anoxic zone).

In the anoxic zone, the internal carbon of the wastewater (measured as CBOD) is oxidized to carbon dioxide and new biomass, while the nitrate, serving as the electron acceptor, is reduced to nitrogen gas. The nitrate is produced in the aerobic zone (by nitrification reactions that convert ammonia to nitrate) and is recycled to the anoxic zone of the biological reactor.

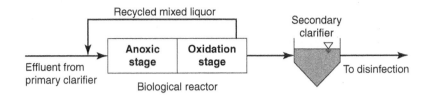

Figure / 9.16 Modified Ludzak–Ettinger (MLE) Process for Configuring a Biological Reactor to Remove Nitrogen   In the second oxygenated compartment (Oxidation), nitrification of ammonia to nitrate occurs and the nitrate contained in the mixed liquor is recycled back to the first anoxic stage (Anoxic) for denitrification.

One key to biological nitrogen removal is to prevent the washout of the slower-growing autotrophic bacteria (for example, *Nitrosomonas*) that convert ammonia nitrogen to nitrate. This requires that the SRT be greater than the inverse of the growth rate of the nitrifying bacteria. The nitrifiers are also more sensitive to inhibition by toxic chemicals that can find their way into a treatment plant.

Compared with traditional wastewater treatment or a plant that treats BOD and ammonia nitrogen, the nitrifying/denitrifying processes have been found to have lower total operating costs if one considers the cost of aeration, sludge disposal costs, and credits for releasing less methane. Table 9.12 summarizes this information in detail.

Sludge disposal costs (Table 9.12) are greater in conventional activated sludge because the SRT is lower. Remember that when the SRT is high, there is greater endogenous respiration, so more of the resulting waste sludge is oxidized during the treatment process, and less sludge is produced. When fewer solids are produced from the biological reactor, there will be less sludge produced that will require sludge digestion (and a corresponding decrease in energy recovery via methane production in the sludge digester).

Table / 9.12

Operational Costs for Activated-Sludge Practices   Assumptions: plant flow of 20,000 m³/day; influent of 350 mg BOD/L; and effluent of 20 mg BOD/L.

	Conventional Activated Sludge	Conventional Activated Sludge with Nitrification	Conventional Activated Sludge with Nitrification and Denitrification
Range of SRT (days)	1.2–8.5	12–21	4.7–22
Sludge disposal cost ($/day)	140	78	69
Oxygen requirement (kg O_2/day)	3,800	5,034	3,469
Aeration cost ($/day)	39	52	36
Methane production credit ($/day)	96	36	32
Total cost (sludge disposal cost + aeration cost − methane production credit) ($/day)	83	94	73

SOURCE: Results from Rosso and Stenstrom (2005).

Class Discussion

Given all the choices in Table 9.12, how would you operate a municipal wastewater treatment and resource recovery plant to balance environmental, social, and economic issues related to protection of human health and water quality as well as resource recovery in a sustainable manner? What technology innovations would be most significant in advancing this goal?

Anaerobic sludge digestion produces **methane**, and if less sludge is produced, less methane also will be produced. This methane can be utilized as an energy source to supply heat or electricity. This is why the conventional process listed in Table 9.12 shows a higher methane production credit than the other processes.

Readers should investigate whether their local wastewater treatment plant recovers methane that is produced to use for heating or electricity generation. With future energy needs and climate change upon us, many treatment and resource recovery plants have implemented the successful recovery and use of generated methane. However, technology to convert methane to electricity still produces CO_2, because it is an end product of methane combustion with oxygen. However, CO_2 has a global warming potential that is 25 times less than CH_4.

In all three processes compared in Table 9.12, oxygen is required to oxidize organic carbon (CBOD) and ammonia nitrogen (NBOD). Table 9.12 demonstrates how conventional activated sludge with nitrification increases the oxygen requirement (and thus the aeration cost) compared with conventional activated sludge. However, with the addition of nitrification and denitrification, the oxygen requirements are lowered, because the nitrate that is produced during the nitrification process can be used as an electron acceptor in the denitrification process (where it displaces oxygen as the electron acceptor).

9.9.2 PHOSPHORUS

Traditionally, chemicals such as alum ($Al_2(SO_4)_3$, ferric sulfate ($Fe_2(SO_4)_3$), and ferric chloride ($FeCl_3$) have been added to remove **phosphorus** by precipitation. All three chemicals precipitate dissolved polyphosphates (as illustrated here for alum):

$$Al^{3+} + PO_4^{3-} \rightarrow AlPO_{4(s)} \qquad \textbf{(9.21)}$$

Chemicals are typically added either during primary or secondary treatment. They thus generate a chemical sludge in addition to the sludge associated with the respective treatment processes. Of course, source reduction activities such as preventing phosphorus use in detergents or recovery of urine can reduce the need for a portion of this treatment requirement. Biological processes can also be used to remove significant amounts of phosphorus and are gaining widespread usage. The phosphorus content of a typical dry cell is approximately 1 percent, so some phosphorus is removed simply by the growth and subsequent wastage of sludge. Typically, some hybrid of chemical addition and enhanced biological uptake is used to reach low P concentration in the effluent and minimize chemical usage.

In the 1970s, it was determined that enhanced biological phosphorus removal was possible by use of organisms called **phosphate-accumulating organisms (PAOs)**. These organisms were found to uptake phosphorus well above the 1 percent typical of most microorganisms.

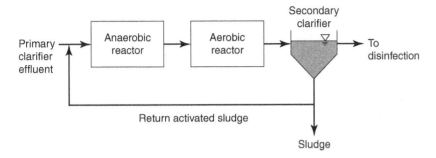

Figure / 9.17 Configuration of a Biological Reactor to Remove Phosphorus In the anaerobic compartment, phosphate is stored internally by phosphate-accumulating microorganisms. Phosphorus is thus removed by converting it from dissolved phosphate to particulate phosphorus stored in biological cells, which are removed in the secondary clarifier.

Fortunately PAOs are present in wastewater treatment systems. They have the ability to take up volatile fatty acids such as acetic acid when oxygen and nitrates are not present. It was found that if PAOs were first exposed to an anaerobic zone (no oxygen present) followed by an aerobic zone, they would take up extra phosphate as they grew on the organic carbon in the aerated zone of the biological reactor (Figure 9.17). This extra phosphate is stored in energy-rich polyphosphate chains, which are subsequently used to take up the volatile fatty acids. This ability to transfer phosphorus in the dissolved aqueous phase to the particular microbial phase does not require chemical addition, and the particulate phosphorus can be removed as sludge and land applied to reuse the nutrient.

Application / 9.6 Phosphorus in Detergents—Application of Policy to Achieve Source Reduction of Nutrient Inputs to the Environment

In the mid 1960s, many of the nation's waterways were rapidly turning green and succumbing to plant and algae growth. A primary reason for this impact on water quality was the high levels of phosphates, one of several major plant nutrients, found in domestic and municipal sewage effluents. Phosphates have traditionally been used as an active agent in laundry and dishwasher detergents as builder and scavenger of cations found in hardwater. They were utilized in concentrations of 30–40 percent of the final product. Because the effluent of the wastewater treatment plants are directly discharged to lakes, ponds, and rivers, the untreated phosphates are available as a nutrient and can promote algae growth, contributing to eutrophication.

Historically, half the phosphorus input to Lakes Erie and Ontario came from municipal and industrial wastewater sources, of which 50–70 percent came from detergents. Over half of the phosphorus input to the Potomac estuary also came from detergents in municipal and industrial effluents (Congressional Report HR 91-1004. April 14, 1970). It was generally agreed that detergents accounted for about 50 percent of the wastewater phosphorus nationwide (Hammond, 1971). There was a growing public consensus that in order to save lakes (like Lake Erie), phosphates must be banned from detergents.

Given the challenges in removing phosphates using secondary wastewater treatment and growing demand for detergent products, it was suggested that the most cost-effective way to manage phosphates in natural systems was to eliminate or minimize them at the source—that is, in the detergents. As such, in 1993, the United States banned the use of phosphates in *laundry detergent*. As can be seen in Figure 9.18, this regulatory driver had a significant impact on the amount of phosphates entering the wastewater plant system.

And in 2010, 16 states (Illinois, Indiana, Maryland, Massachusetts, Michigan, Minnesota, Montana,

New Hampshire, Ohio, Oregon, Pennsylvania, Utah, Vermont, Virginia, Washington, and Wisconsin) banned the sale of *dishwasher detergents* that contain greater than 0.5 percent phosphorus. The European Union is pursuing similar regulatory frameworks to eliminate or reduce the amount of phosphates permitted in laundry and dish detergents. As a result of the ban, familiar brands are offering innovative new dish detergents with few or no phosphates, while traditionally eco-friendly brands such as Seventh Generation and Method have gained in popularity.

Figure / 9.18   phosphorus load for a wastewater treatment facility in Georgia before and after the implementation of a mandatory restriction on the use of phosphate detergents. Note that the total volume of wastewater discharge from the facility did not decline, so the decrease in phosphates can be attributed to the regulations on detergents and not to a decline in wastewater generation.

(Redrawn from U.S. Department of the Interior, U.S. Geological Survey, http://ga.water.usgs.gov/edu/phosphorus.html).

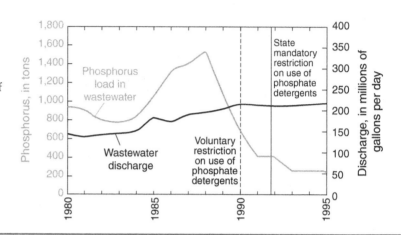

9.10   Disinfection and Aeration

The final step before flow measurement and discharge to the receiving water is **disinfection**. The purpose of disinfection is to ensure removal of pathogenic organisms. This is most commonly accomplished by the addition of liquid sodium hypochlorite, chlorine dioxide, or chlorine gas; on-site hypochlorite generation; ozonation; or exposure to ultraviolet light. Disinfection was covered in Chapter 8, so it is not discussed further in this chapter.

During aeration, oxygen is transferred from a gaseous phase to the liquid phase. Remember in Chapter 3 that we discussed the solubility of oxygen in water. Though oxygen makes up approximately 21 percent of Earth's atmosphere, only 8–11 or so parts per million (mg/L) of oxygen can dissolve into water at equilibrium if air is used as the source.

Table 9.13 compares the three methods commonly used to aerate wastewater: (1) *surface aeration*, (2) *fine-pore diffusion*, and (3) *coarse-bubble diffusion*. Table 9.13 also provides information on the energy usage associated with each of these aeration technologies. Fine-pore diffusers reduce energy costs by 50 percent over coarse-bubble diffusers, but they are more easily fouled by constituents found in wastewater. Figure 9.19 shows examples of fine-pore diffusers placed in an aeration basin.

Disinfecting Combined Sewer Overflows

http://water.epa.gov/scitech/wastetech/upload/2002_06_28_mtb_chlor.pdf

Table / 9.13

Aeration Devices Used during Wastewater Treatment

Aeration Device	Description
Surface aerator	Shear wastewater surface with a mixer or turbine to produce a spray of fine droplets that land on the wastewater surface over a radius of several meters. Can be connected to a solar-powered pump.
Diffusers (fine pore and coarse bubble) Fine bubbles have a diameter less than 5 mm; coarse bubbles have diameters as large as 50 mm.	Nozzles or porous surfaces are placed in the tank bottom, where they release bubbles that travel upward toward the tank's surface. Fine-pore diffusers are used more in the United States and Europe than coarse-bubble diffusers. Fine-pore diffusers reduce energy costs by 50 percent over coarse-bubble diffusers. Fine-pore diffusers can be fouled or experience scale buildup, so they need more cleaning. Cleaning typically consists of either emptying a tank, using a hose to clean the diffusers, or scrubbing with a 10–15 percent HCl solution. Note that emptying a tank works best in the situation where a plant has excess capacity (typically at larger plants). Periodic cleaning will maintain the diffuser's efficiency, which will reduce energy requirements. The transfer efficiency of gas to liquid in the presence of wastewater contaminants (e.g., surfactants, dissolved organic matter) is quantified by the α factor. This α factor is lower for fine-pore diffusers, which suggests that the presence of contaminants inhibits the transfer of oxygen to a greater extent than for coarse bubbles.

SOURCE: Rosso and Stenstrom (2006).

(a)   (b)

Figure / 9.19   **Fine-Pore Diffusers**   (a) Fine-pore diffusers at bottom of rectangular aeration basin. (b) Fine-pore diffusers in operation in same aeration basin.

(Photo courtesy of James R. Mihelcic).

Conservation, improvement in efficiency of use, and **water reclamation and reuse** are all methods that can be used to bridge the gap between demand for and supply of water. Treatment technology is now so advanced that wastewater can be converted to a water source for use in a variety of residential, commercial, industrial, and agricultural applications. Globally, water reuse capacity is increasing rapidly.

One challenge, though, is matching issues of quality, demand, and supply. For example, the extensive water needs of agriculture are typically located far away from treatment plants that collect wastewater, which are located in urban areas. In this case, it makes little sense to use energy to pump treated wastewater back to rural areas. One solution to this problem is the use of small-scale **satellite reclamation processes**. These types of small plants treat and reclaim wastewater where it is needed.

An example of using technology discussed in our drinking-water and wastewater chapters to reclaim wastewater is the Gippsland Water Factory, located in Traralgon (Victoria, Australia). Every day this facility treats 16,000 m^3 of municipal wastewater and 19,000 m^3 of industrial wastewater. Figure 9.20 provides an example of how the Gippsland Water Factory (Australia) has combined treatment processes used to produce the reclaimed water. In this case, the reclaimed water is being used to supplement a current source of groundwater (Daigger et al., 2007).

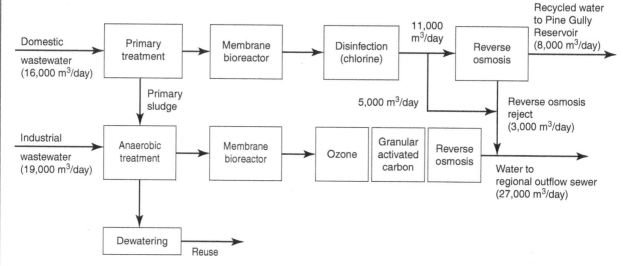

Figure 9.20   **Combination of Drinking-Water and Wastewater Unit Processes Employed to Produce Usable Water from a Domestic and Municipal Wastewater Source**   This example is from the Gippsland Water Factory located in Victoria, Australia.

(Daigger, G.T., A. Hodgkinson, and D. Evans. A Sustainable Near-Potable Quality Water Reclamation Plant for Municipal and Industrial Wastewater. Reprinted with permission from *Proceedings of WEFTEC®.07, the 80th Annual Water Environment Federation Technical Exhibition and Conference*, San Diego, CA, October 13–17, 2007. Copyright © (2007) Water Environment Federation: Alexandria, Virginia).

9.11   End of Life Sludge Management and Energy Recovery

The WateReuse Association
http://www.watereuse.org/

The sludge generated through primary and secondary treatment has three characteristics that make its direct disposal difficult: (1) it is aesthetically unpleasing in terms of odor; (2) it is potentially harmful because of the presence of pathogens; and (3) it contains too much water, which makes it difficult to process and dispose of. The first two problems are often solved by *sludge stabilization* and the third by *dewatering*.

9.11.1 SLUDGE STABILIZATION

The objective of **sludge stabilization** is to reduce problems associated with sludge odor and putrescence and the presence of pathogenic organisms. The first stabilization alternative, **aerobic digestion**, is simply an extension of the activated-sludge process. Waste activated sludge is pumped to dedicated aeration tanks for a much longer time period than with the activated-sludge process. The concentrated solids are allowed to progress well into the endogenous respiration phase, in which food is obtained through the destruction of viable organisms. The result is a net reduction in organic matter.

Another method of sludge treatment is **anaerobic digestion**. It is more commonly employed, because it does not require energy-intensive aeration. It is primarily a three-step biochemical process mediated by specialized groups of microorganisms (Table 9.14). Anaerobic process are now being considered for replacing aerobic processes for several reasons, including the production of methane and its associated energy value as well as the production of fewer biosolids that require extensive management.

Table / 9.14

Three-Step Biochemical Process during Anaerobic Digestion of Wastewater Solids

Step	Description	Example Reactions
Step 1: Hydrolysis	Microorganisms produce extracellular enzymes, which solubilize particulate organics in the presence of water.	Organic waste ($C_6H_{10}O_6$) converted to glucose $$C_6H_{10}O_4 + 2H_2O \rightarrow C_6H_{12}O_6 + 2H_2$$
Step 2: Acidogenesis and acetogenesis	A specialized group of bacteria termed the acid formers use the process of acidogenesis to first convert the soluble organics (things like sugars, amino acids, fatty acids) to volatile fatty acids (e.g., weak organic acids such as propionic acid and acetic acid). Then in a second process termed acetogenesis, hydrogen and carbon dioxide are also formed. The microorganisms that mediate the hydrolysis and fermentation steps are facultative and obligate anaerobic bacteria.	Glucose ($C_6H_{12}O_6$) converted to: propionic acid: $$C_6H_{12}O_6 \rightarrow 2CH_3CH_2COOH + 2CO_2$$ and acetic acid: $$C_6H_{12}O_6 \rightarrow 3CH_3COOH$$ Then compounds such as propionate are converted to hydrogen and acetic acid: $$CH_3CH_2COO^- + 3H_2O \rightarrow CH_3COOH + HCO_3^- + 3H_2$$
Step 3: Methanogenesis	A group of specialized bacteria, the methane formers, convert hydrogen and the organic acids (such as acetic acid here) that the acid formers produced to the end products **methane** and **carbon dioxide**. The methane-forming bacteria are termed strict obligate anaerobes.	$$CO_2 + 4H_2 \rightarrow CH_4 + 2H_2O$$ $$CH_3COOH \rightarrow CH_4 + CO_2$$

A pH near neutral is preferred for anaerobic digestion, and at a pH below 6.8, the methane formers begin to be inhibited. If the digester is not operated properly, the methane-forming bacteria will not be able to use the hydrogen that is produced at a fast enough rate. In this case, the pH of the reactor may drop due to accumulation of volatile fatty acids from the fermentation step. The methane formers may become further inhibited by the low pH; however, the acid formers keep on mediating the second step. This also lowers the pH, which may *sour* the digester and stop the process. Lime is often added to correct this problem.

Anaerobic digesters yield solid residues that can be used to make soil amendments. The gases can be used to produce heat, electricity, or fuel because the resulting gas of an anaerobic digester is comprised of approximately 35 percent CO_2 and 65 percent CH_4. In terms of greenhouse gas emissions, remember that in Chapter 2 we learned that the radiative forcing of gases differs. Thus, 1 ton of methane emissions equates to 25 tons of carbon dioxide emissions in terms of their equivalent global warming potential as greenhouse gases. The methane that results from anaerobic digestion should be viewed as a valuable gas that should not be emitted directly into the atmosphere.

Methane generated at a wastewater treatment and resource recovery plant can be converted to electricity or used for heating. Combusting methane still results in production of a greenhouse gas, according the following equation:

$$CH_4 + 2O_2 \rightarrow CO_2 + 2H_2O \qquad (9.22)$$

Equation 9.22 shows that, on an equivalent basis, 2.75 kg CO_2 are produced for every 1 kg of CH_4 that is combusted. But although some greenhouse gas is produced by combusting methane, there is not only a benefit from converting methane to carbon dioxide, but also some carbon offset associated with generating electricity from the methane.

Application / 9.8   Energy Production from Wastewater and Solid Waste

The potential for electricity production from wastewater treatment is not trivial. The recovered energy can be sold back to the grid as "green" power or can be used to operate pumps and blowers at the WWTP or to maintain optimal digester temperatures, dry the biosolids, and provide space heating for the WWTP. Besides the real possibility to use solar, wind, or microhydro power, wastewater should also be viewed as a source of energy (and water and nutrients). On average, a typical WWTP processes 100 gallons per day of wastewater for every person served. Approximately 1.0 ft^3 of digester gas can be produced by an anaerobic digester per person per day. This volume of gas can provide approximately 2.2 W of power generation. The heating value of the biogas produced by an anaerobic digester is approximately 600 BTU per cubic foot.

As just one example, the Los Angeles Sanitation District treats approximately 520 million gallons wastewater every day and manages the final disposal of half the 40,000 tons/day of nonhazardous solid waste generated in Los Angeles County. From this, the Sanitation District currently obtains 23 MW of electricity from digester gas, 63 MW from landfill gas, and 40 MW from combusting solid waste. In comparison to needs of the Sanitation District, this 126 MW of electricity production dwarfs the 41 MW of electricity required by the district (McDannel and Wheless, 2007). Methane from digester gas is even used in a fuel cell after it has been transformed to hydrogen upstream. Fuel cells generate electricity by using electrochemical reactions between hydrogen and oxygen.

Recall the methane production credits compared in Table 9.12. These are related to the quantity of sludge produced during the different biological configurations of the activated-sludge process used to remove nitrogen.

9.11.2 DIGESTERS

In the past, a two-stage **digester** was employed. In this process, the primary tank was covered, heated to 35°C, and kept well mixed to enhance the reaction rate. The secondary tank had a floating roof cover. The secondary tank would not be mixed or heated and was used for gas storage and for concentrating the solids by settling. Settled solids (termed **digested sludge)** were routed to a dewatering process, and the liquid supernatant was recycled to the beginning of the treatment plant. The reason two-stage digesters have fallen out of favor is the cost associated with building a second tank to be used primarily for storage.

In the single-stage digester (see Figure 9.21a), sludge is pumped to the reactor every 30–120 min to maintain constant conditions within the digester. The digester is typically sized based on a design SRT. In this case, the SRT equals the mass of solids in the reactor divided by the mass of solids removed every day. Typical design SRTs for anaerobic digesters range from 15 to 30 days. Egg-shaped digesters (Figure 9.21b) are now preferred by many practitioners because they require a smaller physical footprint, have lower operation and maintenance costs, provide for better mixing and heating of solids, and do not accumulate scum and grit like the conventional digester shown in Figure 9.21a.

DC Water Blue Plains WWTP Digesters

http://www.dcwater.com/ site_archive/news/documents/ digesteroverview.pdf

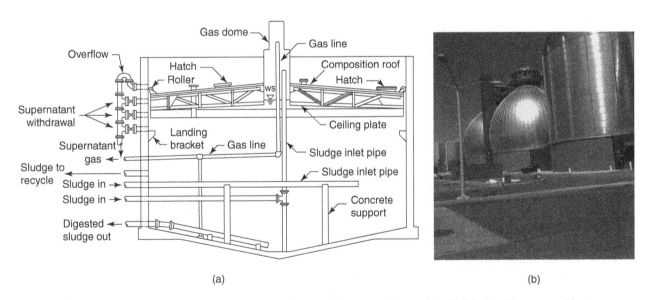

(a)   (b)

Figure / 9.21   (a) Single-Stage Digester Used to Stabilize Primary and Secondary Solids in a Treatment Plant (b) Egg-Shaped Digesters at Newtown Creek (NY), one of New York City's fourteen wastewater plants that treat 1.3 billion gallons of wastewater every day.

(Photo courtesy of Dr. Jeanette Brown, University of Connecticut).

9.11.3 DEWATERING

After stabilization, solids are typically dewatered before disposal. *Dewatering* is generally the final method of volume reduction before ultimate disposal. Sludge pumped from the primary and secondary clarifiers has a solids content of only 0.5 to a few percent. Dewatering can improve this to 15–50 percent.

The simplest and most cost-effective dewatering method if land is available and labor costs are low is to use *drying beds*. The beds consist of tile drains in gravel covered by about 10 in. of sand. Liquid is lost by seepage into the sand and by evaporation. A typical drying time is 3 months. If dewatering by sand beds is considered impractical, mechanical techniques may be employed. One mechanical dewatering method is a belt filter press (Figure 9.22), where the sludge is introduced to a moving belt and squeezed to remove water, producing a sludge cake. A second mechanical method is the centrifuge, a solid bowl where solids are moved to the wall by centrifugal force and scraped out by a screw conveyor. The performance of mechanical dewatering devices may be improved by some type of chemical pretreatment. Here, polymers are added to improve dewatering.

There is great potential to retrofit existing treatment plants for greater efficiency and recovery of nutrients. One such opportunity is to apply dedicated sidestream treatment of the anaerobic sludge digestate. This waste stream is what remains after the anaerobic digestion process and is typically high in N and P and contains a significant fraction of the treatment plant's nutrient loading.

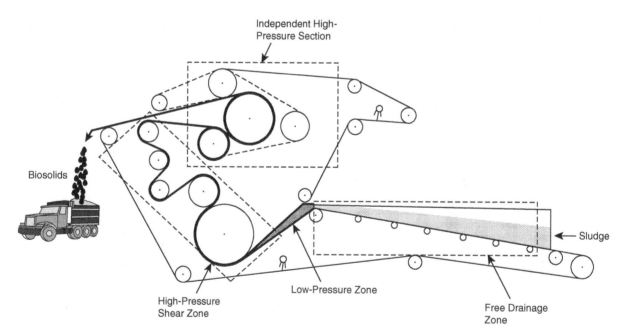

Figure 9.22 **Belt Filter Press** A belt filter applies pressure to sludge, which squeezes out water. The parts include dewatering belts, rollers and bearings, a belt tracking and tensioning system, controls and drives, and a belt-washing system. The sludge is applied to the right and drier biosolids are collected to the left. A polymer is likely applied early on (either at beginning of press or even before that) so that there is enough reaction time to assist with dewatering.

(Redrawn from EPA, 2000a).

Dewatering of anaerobically digested solids also results in the production of "reject" sludge digestate streams that have 15–30 times greater nitrogen concentration than is found in a typical municipal wastewater. (Depending on the means of dewatering the sludge, this stream is also referred to as *filtrate* (for a belt press) or *centrate* (for the centrifuge)). This sludge digestate is a perfect place to focus nutrient recovery efforts, because it can contain 15–40 percent of the total nitrogen load within a wastewater treatment plant.

Sidestream nutrient management from sludge digesters and sludge dewatering unit processes, in particular recovery of the nutrients, may thus be the single most effective strategy to improving nutrient management and helping a wastewater treatment plant meet objectives of resource recovery that are related to sustainable management of nutrients because it: (1) reduces a recurring nutrient burden (25 percent), which means less subsequent requirements for energy, chemicals, and less sludge production, (2) reduces unsteady loading, and (3) generates an economically beneficial product that contains valued nutrients.

9.11.4 DISPOSAL

Dewatered sludge is incinerated, applied to agricultural land, composted, treated with chemicals or heat, provided to the public or a municipality as a soil conditioner, or disposed of in a landfill. Recently, burial in a landfill was not considered a preferred disposal method because of the lack of landfill space. However, burial in a landfill is considered by some as a disposal option in order to sequester carbon. Use of dewatered sludge (also referred to as sludge or biosolids) in agricultural areas can be seen as a way to return organic carbon nutrients back into the environment. The resulting material can then be used to support agricultural production of food or as a soil conditioner that can be used in community gardens. However, there have been health concerns raised about this practice due to the pathogens that may still be present in the biosolids and that may then be dispersed into the environment during land application.

On a global basis, readily available phosphorus (P) that is used to promote agricultural activity on phosphorus-deficient soils is expected to run out in the next 70–100 years. The phosphorus that is present in urine and accumulated in biosolids is one source of P. If collected, the phosphorus available in human urine and feces could account for 22 percent of the total global phosphorus demand (Mihelcic et al., 2011). Of course, the sludge should not be contaminated with industrial and household hazardous wastes. Therefore, an integrated program to use biosolids from a municipal wastewater treatment plant should be coordinated with well-publicized household hazardous-waste collection program and aggressive monitoring and enforcement of industrial pretreatment standards where industries discharge to municipal sewers.

Two concerns about biosolids are runoff and groundwater contamination associated with chemical constituents in the sludge and the presence of pathogens. Table 9.15 gives the survival times of pathogens in soil. The survival times for some pathogens range from days to several years. This is one reason for either treating biosolids before application through composting or minimizing the risk by preventing human exposure by placing the solids in an area with little human contact.

Biosolids

http://www.epa.gov/owm/mtb/biosolids

Table / 9.15

Survival Times in Soil for Commonly Occurring Pathogens Found in Domestic Wastewater

Pathogen	Absolute Maximum	Common Observed Maximum
Bacteria	1 year	2 months
Viruses	6 months	3 months
Protozoa	10 days	2 days
Helminths	7 years	2 years

SOURCE: Data from Kowal (1995).

Sludge that is applied to land is broken down into *Class A* and *Class B biosolids*. Class A solids can be applied in areas open to the public. These biosolids may even be provided (or sold) to the public in a small bag. Therefore, Class A solids must be further treated by heat or chemical treatment to reduce the presence of pathogens to undetectable levels.

Application / 9.9   Dealing with Household Food Wastes

Garbage disposals have become a standard feature in many homes because of consumer perceptions of status and convenience. Garbage disposals result in a larger fraction of food wastes entering the wastewater collection system. This can increase the BOD and TSS of wastewater by 10–20 percent. As explained in this chapter, removal of these pollutants requires additional plant capacity (more or larger reactors) and energy to pump and aerate the wastewater. Also, some of the organic carbon found in the food wastes will be converted to solids, which will require treatment and handling. Another part of the organic carbon will be converted to climate-change-causing greenhouse gases such as CO_2 and CH_4.

Photo courtesy of James R. Mihelcic

Other options to deal with food scraps include depositing them in the solid-waste stream, where they might be landfilled; reusing them as animal feed for local farms; collecting them at the community level for composting, and treating them with backyard composting. Landfills produce the greenhouse gases CH_4 and CO_2; food waste has three times the methane potential of biosolids (376 m^3 gas/ton versus 120 m^3 gas/ton). Composting can occur at the home or be arranged at the community level. But even with composting, biological processes will break down the organic matter and produce CO_2. Some people might have a negative bias against household composting, but backyard composting does not require transportation costs or mechanical energy in the form of mechanically aerating the compost.

Thermal treatment can consist of heat drying or composting. Chemical treatment typically involves a combination of increased pH and temperature.

Class B solids are processed to a point where pathogens may still be present, but land restrictions are in place to limit the public's exposure. The most common example of restricted land use where human exposure is limited would be to land apply Class B solids to an agricultural field. Restrictions are further placed on the timing of when root crops or aboveground crops may be harvested after the final land application of municipal wastewater sludge. There are also restrictions on application of solids to agricultural land to ensure that runoff from the field does not cause problems with water quality in local streams and lakes.

9.12   Natural Treatment Systems

Natural waste treatment systems are discussed in this section, which emphasizes the treatment technologies of lagoons and wetlands. These technologies not only use more natural methods to treat wastewater, but also have lower capital costs because they do not employ aboveground reactors constructed from steel-reinforced concrete, metal, or plastic. They also typically have lower operation costs because they may rely on natural aeration methods (versus mechanical aeration) and may utilize nonoxygenated biological processes. Natural wastewater treatment systems are also employed in decentralized treatment systems. Figure 9.23 shows one such system, the Living Machine®, which can be scaled to homes, dormitories, offices, and schools.

9.12.1   STABILIZATION PONDS

Stabilization ponds are referred to as **lagoons** or **oxidation ponds**. The lagoon is essentially an engineered hole-in-the-ground design to confine wastewater for treatment before discharge to a natural watercourse. Lagoons are typically found in smaller communities and cities.

Figure / 9.23   The Living Machine®
This example of engineered natural treatment processes uses methods that incorporate bacteria, protozoa, plants, and snails to treat wastewater. The EPA reports that the largest of these systems can treat 80,000 gpd. They have produced effluent with BOD_5, TSS, and total nitrogen less than 10 mg/L. Phosphorus removal is reported to be 50 percent with effluents in the range of 5–11 mg/L.

(Adapted from EPA (2001)).

Table / 9.16

Types of Waste Stabilization Ponds and Associated Design Information

Type of Stabilization Pond	Comments	Water Depth (m)	Detention Time (days)
Facultative lagoon	Uses combination of aerobic, anoxic, and anaerobic processes. Not typically mixed or aerated. Does not function well in colder climates.	1.2–2.4	20–180
Aerated lagoon	Typically placed in series, in front of a facultative pond. Aeration consists of either mechanical surface aerators or submerged diffused aeration systems. Requires less area than a facultative lagoon and can operate effectively in the winter.	1.8–6	10–30
Anaerobic pond	Usually used to pretreat high-strength wastewaters. Deep, nonaerated, and no mixing. Performance decreases at temperatures below 15°C.	>8	≤50
Maturation pond	Typically treats effluent from an activated sludge process, trickling filter, or facultative lagoon. Also referred to as a tertiary or polishing pond. Designed to remove pathogens.	<1	10–5

Table 9.16 describes the various types of stabilization ponds. Each type of stabilization pond supports different biological processes. The type of biology is influenced by the depth of the lagoon and whether the lagoon is mixed and aerated.

The capability of lagoon treatment systems to meet water quality guidelines and protect public health and environmental integrity, coupled with their cost-effectiveness and ease of operation and management, make them a highly desirable technology, particularly for smaller communities and developing countries. They are designed in such a way as to remove primary wastewater constituents, including TSS, BOD, nutrients, and pathogens. Major pathogen removal mechanisms in ponds include sedimentation, adsorption to particles, lack of food and nutrients, solar ultraviolet radiation, temperature, pH, predators, straining/filtration in sediments, toxins and antibiotics excreted by some organisms, and natural die-off. In many settings, they can be linked with a water reuse project where the water and the dissolved nutrients are of value to agricultural users.

The main purpose of an **anaerobic lagoon** is to remove BOD and TSS. Anaerobic lagoons are typically the first in a series of several lagoons and are characterized by a lack of oxygen resulting from high organic loading rates. Therefore, BOD removal processes are anaerobic. These lagoons are designed based on the volumetric BOD loading, which is calculated by dividing the product of the influent BOD concentration (S_o in earlier sections of this chapter) and the average flowrate (the numerator in the F/M ratio, Equation 9.16) by the volume of the pond, as shown in Equation 9.23 (Mara, 2004):

$$\text{Volumetric BOD Loading} = \frac{BOD_o \times Q}{V} \qquad \textbf{(9.23)}$$

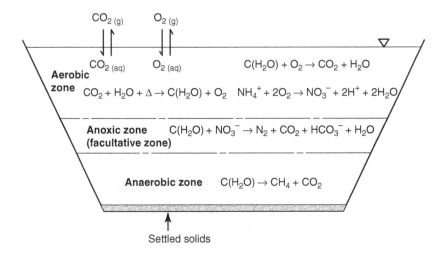

Figure / 9.24 Zones of a Facultative Lagoon.

where BOD_o is the influent BOD (referred to as S_o earlier in this chapter), Q is the average flowrate, and V is the lagoon volume.

At temperatures above 25°C, approximately 70 percent of BOD may be removed in an anaerobic lagoon with a volumetric loading rate of up to 350 g BOD/m³/day. Because the biochemical reactions that occur in the lagoon are temperature-dependent, at temperatures below 10°C, only 40 percent of BOD is removed with volumetric loading rates of up to 100 g BOD/m³/day.

Figure 9.24 shows the various zones found in a **facultative lagoon** and the respective biological processes that occur. An *aerobic zone* is located near the surface. It is aerated due to oxygen transfer from the overlying air to the water and also by algal photosynthesis. The amount of **photosynthetic oxygen** production can be significant and does not require any energy input, except from the sun. In the presence of oxygen, CBOD is converted to CO_2, and NBOD is converted to nitrate (along with the production of biomass solids). An *anaerobic zone* forms on the bottom of a facultative lagoon, where the solids settle. This part of the lagoon supports anaerobic biological fermentation processes discussed in Section 9.11.1 on sludge digestion and converts CBOD into CH_4 and CO_2.

Between these two layers there may be a small *anoxic layer*, also termed the facultative zone. In this zone, denitrification reactions may take place, where nitrate can be reduced to nitrogen gas, oxidizing CBOD in the process. However, several studies have reported that nitrification/denitrification is negligible in stabilization ponds. Therefore, the majority of nitrogen removal is believed to occur via ammonia volatilization to the atmosphere, due to increased pH values when algal activity increases, or from algal and bacteria uptake and incorporation of nitrogen into the sediment.

For a facultative lagoon, the amount of sunlight needed to produce enough oxygen to degrade the CBOD in the wastewater can be determined using the estimated efficiency of algae in the lagoon and monthly solar insolation data (Oakley, 2005). Algae use solar radiation, carbon dioxide, ammonia, and phosphate in the wastewater to produce new algal biomass and oxygen, as shown in Equation 9.24. Lagoons thus can eliminate the need for provision of oxygen to the wastewater by mechanical aeration.

$$106CO_2 + 65H_2O + 16NH_3 + H_3PO_4 \xrightarrow{\text{solar radiation}} C_{106}H_{181}O_{45}N_{16}P + 118O_2$$

(9.24)

In this equation, $C_{106}H_{181}O_{45}N_{16}P$ (MW of 2,428 g/mole) represents algae biomass. Applying mass balance and stoichiometric principles of Equation 9.24 demonstrates that 1.55 kg of oxygen is produced for every kg of algae biomass present in the system. For facultative lagoons located in regions with warm climates, it can generally be assumed that 24,000 kJ of sunlight will produce approximately 1 kg of algae, and the algae will have a conversion efficiency (CE) of the sun's energy of approximately 3.0 percent (so CE = 0.03). Therefore, the *theoretical maximum surface loading rate* (SRL_{max}) (units of kg O_2/ha-day) for providing oxygen to a facultative lagoon can be calculated as

$$SLR_{max} = \frac{I_s \times CE \times \left(1.55 \frac{kg\,O_2}{kg\ \text{algae}}\right)}{24,000 \frac{kJ}{kg\ \text{algae}}}$$

(9.25)

where I_s is the solar insolation (kJ/ha-day) and CE is a conversion efficiency (%) of the sun's energy to algal biomass as described previously. Equation 9.25 is then used to size the surface area requirement of a facultative lagoon along with information on the organic loading, as shown in Example 9.8.

Typical area organic loading values of facultative ponds range from 15 to 80 kg/ha/day (Note how in this loading rate the denominator term has units of area in it, not volume as found in a volumetric loading rate described previously). One possible problem with these ponds is that they cool off quickly and can experience a large reduction in the rate of biological activity during the winter months of northern climates. In addition, algae may accumulate in the pond effluent, which will cause problems with an effluent TSS going well above 20–100 mg/L in a poorly designed lagoon.

---

example/9.8   Sizing a Facultative Lagoon Based on a Theoretical Oxygen Balance

In this problem we will determine the minimum surface area for a facultative lagoon treating municipal wastewater to be sited in Tampa, Florida. We assume the conversion efficiency (CE) of the algae is 3 percent and that 24,000 kJ of sunlight are needed to produce 1 kg of algae biomass. The average daily concentration of CBOD in the influent wastewater is 200 mg/L (0.2 kg/m³), and the average flowrate is 3.5 MGD (13,230 m³/day).

solution

The surface area of the facultative lagoon can be designed using a theoretical oxygen balance, as described above, using site-specific solar insolation data and the estimated efficiency of algae in the lagoon. Monthly solar insolation data that is location specific is available from the National Aeronautics and Space Administration (NASA) web site (http://eosweb.larc.nasa.gov/sse).

The solar insolation at ground level will vary seasonally and is typically the lowest during the winter months. For design purposes, the maximum allowable surface loading rate should be calculated for the month with the lowest solar radiation. For example, in Tampa, Florida (27.95° N, 82.46° W), the lowest average monthly solar insolation incident on a horizontal surface is 3.32 kWh/m^2-day (1.2×10^8 kJ/ha-day) in December.

The maximum surface loading rate of oxygen to a facultative lagoon in Tampa for the month of December is calculated using Equation 9.25 to be 232 kg O$_2$/ha-day, as shown below:

$$\text{SLR}_{max} = \frac{I_s \times CE \times 1.55 \dfrac{\text{kg O}_2}{\text{kg algae}}}{24{,}000 \dfrac{\text{kJ}}{\text{kg algae}}}$$

$$= \frac{1.2 \times 10^8 \dfrac{\text{kJ}}{\text{ha-day}} \times 0.03 \times 1.55 \dfrac{\text{kg O}_2}{\text{kg algae}}}{24{,}000 \dfrac{\text{kJ}}{\text{kg algae}}} = 232 \frac{\text{kg O}_2}{\text{ha-day}}$$

The volumetric organic loading rate of a facultative lagoon is calculated by multiplying the flowrate entering the pond by the concentration of CBOD at the pond influent. The required area is then found by dividing that product by the maximum loading of oxygen to pond (SLR$_{max}$).

Here the volumetric loading rate of CBOD would be equal to 2,646 kg CBOD/day ($Q \times$ CBOD$_o$). When divided by the SRT$_{max}$ for oxygen for a particular location and season, we can estimate that the lagoon should have an area of at least 11.44 ha (that is, 28.3 acres):

$$\text{lagoon area} = \frac{0.2 \dfrac{\text{kgCBOD}}{\text{m}^3} \times 13{,}230 \dfrac{\text{m}^3}{\text{day}}}{232 \dfrac{\text{kg O}_2}{\text{ha-day}}} = 11.44 \text{ ha}$$

Be aware that the shape and hydraulics of a lagoon can affect the lagoon's oxygen balance. For example, a long, narrow facultative lagoon will have higher loading of CBOD at the influent than it will near the effluent.

Figure / 9.25 Facultative Stabilization Ponds in Punata, Bolivia.

(Photo courtesy of Matthew E. Verbyla).

As stated previously, typical organic loading values of facultative ponds in temperate climates range from 15 to 80 kg/ha/day, although in warm tropical climates, surface loading values may range from 100 to 1,000 kg/ha/day. One possible problem with ponds in colder climates is that they cool off quickly and can experience a large reduction in biological activity due to reduced temperature and reduced solar radiation during the winter months. In addition, algae may accumulate in the pond effluent, which will cause problems with an effluent TSS going well above 20–100 mg/L in a poorly designed lagoon.

The primary objective of a **maturation lagoon** is pathogen removal. This is an important step when considering the water reuse potential of the treated wastewater for integration with agricultural uses that place an economic value on the water and nutrients found in lagoon effluent. There are several models that have been proposed to design a maturation lagoon for the removal of bacterial pathogen indicators, all of which are based on pseudo-first-order kinetics. One of the more widely accepted models is a modified version of the Wehner–Wilhelm equation (1956) for dispersed flow for the removal of *Escherichia coli* as is discussed by Mara (2004):

$$C_{\text{eff}} = C_{\text{inf}} \left(\frac{4a}{(1+a)^2} \right) e^{(1-a)/2\delta} \tag{9.26}$$

$$a = \sqrt{1 + 4k_B \theta \delta} \tag{9.27}$$

$$\delta = \left(\frac{L}{W} \right)^{-1} \tag{9.28}$$

Here k_B is a pseudo-first-order rate constant (day^{-1}), θ is the hydraulic retention time (day), L is the length (m), W is the width (m), C_{eff} is the pathogen effluent concentration, and C_{inf} is the pathogen influent concentration.

Values for k_B have generally been reported between 0.7 and 2.6 for a temperature of 20°C. The rate increases at temperatures above 20°C and decreases at lower temperatures. It is important to note that bacterial pathogen indicators, such as *E. coli* or fecal coliform bacteria, do not always successfully predict the removal of other pathogens, such as viruses or protozoan parasites.

9.12.2 WETLANDS

Natural ecosystems such as wetlands are the prototype for raising water quality by using natural energy (sunlight) and ambient temperature, without adding materials or requiring large amounts of human labor. They can also provide the public with open green space. A wastewater treatment system designed with that prototype in mind can be sustainable in terms of energy and material input/output as well as social and environmental benefits. Wastewater treatment technologies that combine the soil–water–air–vegetation environment include constructed wetlands and evapotranspiration beds. Both require pretreatment of the influent solids load with either a septic tank, oxidation pond, or other primary treatment structure for settling solids. The two types of constructed wetlands are the free water surface (FWS) wetland and the subsurface flow (SSF) wetland (Figure 9.26). Only FWS wetlands will be described in detail in this chapter.

Figure / 9.26 Subsurface Flow Wetland (before Planting of Vegetation) That Serves More Than 200 Students at the Pisgah All Age School (Jamaica) SSF wetlands typically use gravel as the aquatic plant rooting media, and the water level is intentionally maintained below the surface of the gravel. Several hydraulic regimes may be employed: horizontal flow, vertical upflow, and vertical downflow. The septic tanks and wetland treat only black water from the toilets and no gray water. The wetland has two parallel plastic-lined rock media beds (separated by the black plastic in center of bed). Each bed has inside dimensions of 19.1 m length by 4.7 m width by 0.5 m depth of media. Media are irregularly shaped washed stones ordered as "1/2 in. river shingle," and the bulk of the stone ranges in size from 1/4 to 1 in. (6–25 mm) diameter. The media porosity was measured to be 37.7 percent without plant roots.

(Photo courtesy of Ed Stewart, Sanitation Districts of Los Angeles County).

FREE WATER SURFACE (FWS) WETLANDS **Free water surface (FWS) wetlands**, also called **surface flow wetlands**, are similar to natural open-water wetlands in appearance and treatment mechanisms (see Figure 9.27). The majority of the wetland surface area has aquatic plants rooted in soil or sand below the water surface. The wastewater travels in deep sheet flow over the soil and through the plant stems (zones 1 and 3). The area (zone 2) with no surface vegetation is exposed to sunlight and open to the air to increase the potential for oxygen transfer from the gaseous to aqueous phase. Zone 2 may also have submerged aquatic plants to enhance the dissolved-oxygen content.

The first vegetated zone (at water depth of approximately 1 ft.) acts as an anaerobic settling chamber so that only a 1 or 2 day hydraulic retention time is necessary to achieve the required reactions. The hydraulic residence time of the open zone (zone 2, at water depth of approximately 3 ft) should be less than the amount of time required for algae to form and will depend on climate and temperature as well as nutrient limitations. In the United States and Canada, this time is typically between 2 and 3 days. The hydraulic residence time for the second vegetated zone (zone 3, at water depth of approximately 1 ft.) is 1 day to achieve denitrification.

Decentralized Wastewater Treatment

http://water.epa.gov/infrastructure/septic/

Figure / 9.27 Zones of a Free Water Surface Wetland.

Table / 9.17

Maximum Wetland Areal Mass Loading Values and Typical Resultant Effluent Concentrations Data were obtained for a variety of applications, from sewage to stormwater, and cover a range of temperate climate locations, from Florida to Canada.

Constituent	Free Water Surface (FWS) Wetland Loading	Subsurface Flow (SSF) Wetland Loading	Effluent Concentration
BOD	60 kg/ha-day	60 kg/ha-day	30 mg/L
TSS	50 kg/ha-day	200 kg/ha-day	30 mg/L
TKN	5 kg/ha-day	Not applicable	10 mg/L

SOURCE: EPA (2000).

There may be multiple vegetated and open zones to achieve the desired treatment objectives. The calculation of head loss along the length of an FWS wetland is usually not necessary, since a typical FWS wetland with a recommended 5:1 to 10:1 aspect ratio (L:W) may have a hydraulic slope gradient of only 1 cm in 100 m (EPA, 2000b). The use of hydraulic residence time and the maximum areal loading guidelines (see Table 9.17) makes sizing for TSS, BOD, and nitrogen an iterative design process. The values provided in Table 9.17 represent maximum monthly mass loading rates that should reliably keep an FWS wetland's effluent below the noted concentration.

Table 9.18 compares FWS wetlands with SSF wetlands. One advantage of FWS wetlands is that they can be designed to provide long-term nitrogen removal because of the aerobic open-water zones that allow biological nitrification. However, this is only the case for wetlands that are shallow or well mixed. In deeper water zones, good top-to-bottom mixing may not occur, so deeper water may not achieve nitrification levels. SSF wetlands will not typically provide long-term nitrogen removal without plant harvesting or some oxygenation of the water by means of cascading or mechanical aeration.

Table / 9.18

Comparison of Free Water Surface and Subsurface Flow Wetlands

Characteristics	Free Water Surface	Subsurface Flow
Wastewater exposure	Open-water aerobic zones enhance biological nitrification and provide wildlife habitat.	Wastewater remains 2–4 in. below the media surface, so there is no surface water to attract aquatic birds, little risk of human exposure, and no mosquito breeding.
Hydraulics	Not likely to overflow from an accumulation of solids at the inlet.	Surface flooding will occur at the inlet if there is excessive accumulation of solids.
Bedding	The sand or sandy loam plant rooting media have a lower materials cost than the gravel media used in SSF wetlands.	The plant rooting rock media should have diameters of 0.25–1.5 in. and be relatively free of fines. This is more expensive than FWS wetland media.
Dimensions	Recommended aspect ratio (L:W) is 5:1 to 10:1. Depth of water may range from a few inches in vegetated zones to 4 ft. in open-water zones.	Recommended aspect ratio is in the range of 1:1 to 0.25:1. Depth of gravel may be 1–2 ft.

9.13 Energy Usage during Wastewater Treatment

Energy use during wastewater treatment is not trivial. In fact, drinking-water and wastewater treatment combined account for 3 percent of electricity use in the United States, and energy costs can account for up to 30 percent of a treatment plant's total operation and maintenance costs. This translates into several percent of the total U.S. greenhouse gas emissions on an annual basis. Mechanical systems have been discussed in much of this chapter and have been preferred in highly populated areas. Land treatment systems utilize soil and plants without significant need for reactors and operational labor, energy, and chemicals. They also allow for keeping nutrients on land versus discharge to surface waters. Lagoon treatment systems are also less mechanized, as was discussed previously. While large treatment plants (in excess of 100 MGD) serve many cities and a large percentage of the U.S. population, most treatment plants in the United States serve small communities. In fact, the EPA reports there are over 16,000 wastewater treatment plants, and over 80 percent of existing plants have a capacity less than 5 MGD.

Operating and maintenance costs associated with wastewater treatment include labor, purchase of chemicals and replacement equipment, and energy to aerate, lift water, and pump solids. Mechanized plants obviously cost more to run than less-mechanized forms of treatment. Table 9.19 gives a breakdown of energy use at a 7.5 MGD treatment plant. As expected, the activated-sludge process accounts for more energy use than gravity settling and sludge dewatering. For smaller plants, the operational life stage of the life cycle has been found to have the highest energy consumption (95 percent), compared with the construction and refurbishment/demolition stages. This is significant, because energy production and its use are associated with many environmental problems, including release of airborne pollutants and global warming.

Figure 9.28 shows an example of energy requirements to treat a million gallons per day (MGD) of wastewater with mechanical, lagoon, and terrestrial treatment systems. Note the much higher energy costs associated with mechanical treatment. This is primarily due to the mechanical aeration of water, which accounts for 45–75 percent of a treatment plant's energy costs. However, mechanical treatment systems are well documented to be very effective at treating wastewater constituents to specified levels, especially given the smaller land area required per unit of

Table / 9.19

Breakdown of Energy Use at a 7.5 MGD Wastewater Treatment Plant

Unit Process/ Activity	Percent of Total Energy Use
Activated sludge	55
Primary clarifier	10
Heating	7
Dewatering solids	7
Pumping raw wastewater	5
Secondary clarifier (returned activated sludge)	4
Other	12

SOURCE: California Energy Commission.

Class Discussion

After visiting your local WWTP, how would you redesign the plant to be more sustainable considering issues of future population growth, cultural preferences of homeowners, carbon neutrality, reuse of nutrients, and minimizing water use? What source reduction would you implement?

Energy Demands on Water Resources: Report to Congress

http://www.sandia.gov/energy-water

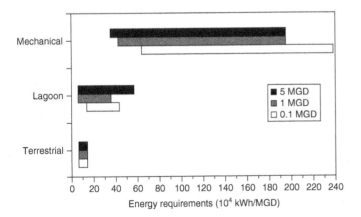

Figure / 9.28 Total Energy Requirements for Various Sizes and Types of Wastewater Treatment Plants Located in Intermountain Areas of the United States Total electricity requirements are measured in kWh/ MGD at flow rates of 0.1, 1, and 5 MGD.

(Reprinted from *Journal of Environmental Management* 88(3), Muga, H. E., and J. R. Mihelcic. Sustainability of wastewater treatment technologies. 437–447. Copyright 2008 with permission of Elsevier).

wastewater treated. Obviously, the future of wastewater treatment needs to look beyond treatment objectives and integrate issues of energy and materials use throughout the process life cycle. For just one small example, pumps selected for wastewater treatment and drinking water are typically purchased based on initial costs, not pumping efficiencies.

9.14   Wastewater Reclamation and Reuse

The use of recycled water offers two highly significant advantages: the reduction of pollutant loading on receiving bodies and the provision of a new source of water. **Wastewater reclamation** is the treatment or processing of wastewater to make it reusable, while **wastewater reuse** is using wastewater in a variety of beneficial ways. The foundation of water reuse requires three fundamentals: (1) providing reliable treatment of wastewater to meet strict water quality requirements for the intended reuse application; (2) protecting public health; and (3) gaining public acceptance (Asano et al., 2007). Whether water reuse is appropriate for a specific community requires a careful review of economic considerations, potential uses for the reclaimed water, and the current regulatory requirements for the level of treatment.

Designing for water reuse requires an understanding of the intended applications, which will govern the degree of wastewater treatment required. The dominant applications for the use of reclaimed water include direct uses such as agricultural irrigation, landscape irrigation, industrial recycling, and reuse and indirect uses such as groundwater recharge. Among them, agricultural and landscape irrigation are widely practiced throughout the world, with well-established health protection guidelines and agronomic practices (Asano and Bahri, 2010).

As water is cycled through water treatment, water use, collection, wastewater treatment, and the quality changes. A conceptual comparison of water quality changes during this cycle is shown in Figure 9.29 and different water quality EPA-recommended guidelines for various water reuse applications can be found in Table 9.20. Note that direct reuse is not included in the table, but that the treatment technologies and standards would be appropriate to drinking water as discussed in Chapter 8. While wastewater reclamation and reuse is a sustainable approach and can be

Figure / 9.29   Water Quality Changes during Municipal Use in a Time Sequence.

(Reprinted from Asano and Bahri, 2010 with permission of Dr. Takashi Asano).

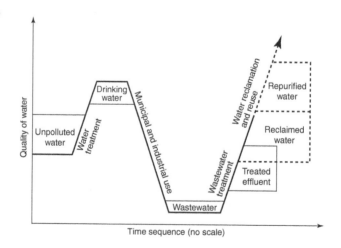

Table / 9.20

Water Treatment, Quality, Monitoring Requirements, and Setback Distances for Various Applications of Reused Water (from EPA, 2012).

Reuse Category and Description	Treatment	Reclaimed Water Quality	Reclaimed Water Monitoring	Setback Distances
Urban Reuse				
Unrestricted The use of reclaimed water in nonpotable applications in municipal settings where public access is not restricted.	• Secondary • Filtration • Disinfection	• pH = 6.0–9.0 • ≤ 10 mg/L BOD • ≤ 2 NTU • No detectable fecal coliform/100 mL • 1 mg/L Cl$_2$ residual (min.)	• pH – weekly • BOD – weekly • Turbidity – continuous • Fecal coliform – daily • Cl$_2$ residual – continuous	• 50 ft. (15 m) to potable water supply wells; increased to 100 ft. (30 m) when located in porous media
Restricted The use of reclaimed water in nonpotable applications in municipal settings where public access is controlled or restricted by physical or institutional barriers, such as fencing, advisory signage, or temporal access restriction	• Secondary • Disinfection	• pH = 6.0–9.0 • ≤ 30 mg/L BOD • ≤ 30 mg/L TSS • ≤ 200 fecal coliform/100 mL • 1 mg/L Cl$_2$ residual (min.)	• pH – weekly • BOD – weekly • TSS – daily • Fecal coliform – daily • Cl$_2$ residual – continuous	• 300 ft. (90 m) to potable water supply wells • 100 ft. (30 m) to areas accessible to the public (if spray irrigation)
Agricultural Reuse for Food Crops				
Food Crops The use of reclaimed water for surface or spray irrigation of food crops, which are intended for human consumption, consumed raw.	• Secondary • Filtration • Disinfection	• pH = 6.0–9.0 • ≤ 10 mg/L BOD • ≤ 2 NTU • No detectable fecal coliform/100 mL • 1 mg/L Cl$_2$ residual (min.)	• pH – weekly • BOD – weekly • Turbidity – continuous • Fecal coliform – daily • Cl$_2$ residual – continuous	• 50 ft. (15 m) to potable water supply wells; increased to 100 ft. (30 m) when located in porous media

2012 EPA Guidelines for Water Reuse

http://nepis.epa.gov/Adobe/PDF/P100FS7K.pdf

cost-effective in the long run, the additional treatment of wastewater beyond secondary treatment for reuse and the installation of reclaimed water distribution systems maybe costly compared to traditional water supply alternatives such as conservation and use of imported water or groundwater. However, the energy savings and other social, economic, and environmental benefits can be much greater.

9.15 Wet-Weather Flow Implications for Wastewater

Given the tremendous energy and resources costs associated with wastewater treatment, there is a significant driver to minimize the water quantity loading and to improve water quality of the influent. This suggests a strong need to more effectively manage stormwater and its interface with the wastewater collection and treatment systems. Runoff water from either rain events or snowmelt can enter the wastewater collection system through manhole covers, designed inlets and catchments, and defects or cracks in the piping network. Domestic wastewater collection systems are designed to carry only domestic sewage (*sanitary system*) or to carry both domestic sewage and runoff (*combined system*). When a sanitary system is used to collect domestic sewage, many times a separate stormwater collection system is designed for runoff events.

Wet-weather flows are defined as the stormwater or snowmelt that directly enters a combined sewer system through designed inlets and catchments or that enters into the sanitary sewer through manholes and pipe defects. The amount of wet-weather flow that enters a sanitary sewer is generally in the same order of magnitude as the domestic (dry-weather) flows. However, the wet-weather flow can be much higher for aged sanitary systems or when there are many "leaky" manhole covers. For combined systems, the wet-weather flow is many times higher than the dry-weather flows and is used to size the pipes in the system. Wet-weather flows are generally subdivided into two categories: **inflow** and **infiltration** (I/I), described in Table 9.21.

When the wet-weather flows are not distinguished from each other, the combination of flows is called inflow/infiltration (I/I). The I/I flow can vary widely, based on the age and condition of the sewer system, climate or season, and groundwater elevation. (If the groundwater table is above the sewer system, there can be infiltration into the sewer.) Recording typical I/I values can be based on the length of sewer, land area drained into the sewer, or the number of manholes.

Table / 9.21

Categories of Wet-Weather Flow

Inflow	Water that enters a collection system through direct connections such as stormwater catchments, roof leaders, sump pumps, yard and foundation drains, manhole covers, and other designed inlets. It is estimated by subtracting the typical dry-weather flow from the total metered flow after a storm event.
Infiltration	Water that enters a collection system through pipe defects, pipe joints, manhole walls, or other nonengineered designed inflows. It is estimated as the metered flow early in the morning when domestic use is relatively small and the sewer system has been drained of domestic sewage. The remaining flow is mostly infiltration.

I/I values for new sanitary sewers are generally between 200 to 500 gpd per inch of pipe diameter and mile of pipe length (Hammer and Hammer Jr., 1996). An aged sewer may have higher values. I/I based on the number of manholes is generally divided into three categories based on the rainfall depth: (1) low, or 3,000 gallons/in.-manhole; (2) medium, or 7,700 gallons/in.-manhole; and (3) high, or 20,000 gallons/in.-manhole (Walski et al., 2004). Common I/I values based on land area range from 20 to 3,000 gallons/acre-day for nonrainy days. During storm events or snowmelt, I/I values could exceed 50,000 gallons/acre-day for older sewer systems with considerable leaks and points for inflow (Tchobanoglous et al., 2003).

The type of sewer system can greatly influence the design I/I value. A sanitary sewer would have a relatively small I/I design flow rate where most is from infiltration. A combined system, which is designed to handle a large amount of runoff, would have a much higher design I/I flow rate. The wastewater collection system and treatment facilities should have the hydraulic capacity to handle the maximum-day domestic wastewater generation plus expected I/I.

Design infiltration values for new sewers vary greatly based on location, type of pipe material, construction practices, and whether the pipe is above or below the groundwater table. Regulatory agencies in most states have a maximum allowance for infiltration. For a sewer system designed to capture stormwater runoff, inflow values are estimated using a runoff model based on a design rain event, drainage area, and land use information. Since this inflow value is based on a design storm event, the potential for hydraulic failure (surcharging in the sewer system or overloading the treatment plant) is high for large-rainfall events.

A **combined sewer** is a type of sewer system that collects sanitary sewage and stormwater runoff in a single pipe system (Figure 9.30). Combined sewers can cause serious water pollution problems due to **combined sewer overflows (CSOs)**, which occur when the capacity of the sewer system is overwhelmed by accommodating a large amount of stormwater (EPA, 2004). CSOs are the direct discharge of untreated wastewater and stormwater to receiving waterbodies. A significant amount of pollutants in a CSO event can be attributed to the stormwater including oil, grease, fecal coliform from pet and wildlife waste, pesticides, as well as pollutants from roadways.

To minimize or eliminate CSOs, which is mandated by the EPA, there are several strategies including sewer separation, CSO storage, expanding capacity, and constructing retention basins. **Municipal storm sewer separation systems (MS4)** involve building a second piping system to manage stormwater eliminating the need for the combined sewer system and the WWTP to accommodate rain water. These projects, while effective at addressing CSOs, are extremely expensive and do not provide any treatment of the pollutants associated with the stormwater that is now directly discharged. **CSO storage facilities** or **retention basins** can also be constructed to hold excess flow that can then be slowly released back into the combined sewer as the treatment and resource recovery plant has capacity after the storm event is over. Retention basins are designed to also provide some level of treatment and disinfection to the stormwater prior to discharge.

example/9.9 Projecting Wastewater Generation Based on Types of Customers

A residential community with a population of 10,000 is planning to expand its wastewater treatment plant and sewer system. In 15 years, the population is estimated to increase to 17,000 residents, and a new 500-person apartment complex is to be constructed. A new industrial park, also planned, will contribute an average flow of 550,000 gpd and maximum-day flow of 750,000 gpd. It is expected that approximately 9 mi new sewer pipe will be needed.

The present average daily flow into the plant is 1.0 MGD with 16.5 mi sewers. The current average inflow and infiltration (I/I) is 2,800 gpd/mi, and maximum-day I/I expected on a rainy day is 53,000 gpd/mi. Residential per capita water use is expected to be 8 percent less in 15 years, due to in-house water-saving strategies. The maximum-day wastewater generation typically occurs on a rainy day. Estimate the future average and maximum-day flow rates.

solution

Based on the information provided, we first need to estimate the various contributions to the overall wastewater generation. We can use information provided in this chapter and from Chapter 7. The current estimated per capita generation of wastewater from different sources is as follows:

Estimate the current average I/I:

$$I/I = 2,800 \text{ gpd/mi} \times 16.5 \text{ mi} = 46,200 \text{ gpd}$$

Determine the current average domestic wastewater generation:

domestic wastewater generation = metered flow $-$ I/I = 1,000,000 gpd $-$ 46,200 gpd = 953,800 gpd

The current per capita wastewater generation can then be calculated as

$$\text{per capita wastewater generation} = \frac{\text{domestic wastewater generation}}{\text{population served}} = \frac{953,800 \text{ gpd}}{10,000 \text{ people}} = 95.4 \text{ gpdc}$$

We can now estimate the future average wastewater flow rate to be 2,140,900 gpd from the following four contributions:

Future average domestic flow (including 8 percent wastewater reduction):

$$95.4 \text{ gpdc} \times 0.92 \times 17,000 \text{ people} = 1,492,000 \text{ gpd}$$

Apartment complex average flow (using values from Table 7.12 in Chapter 7):

$$55 \text{ gpdc} \times 500 \text{ people} = 27,500 \text{ gpd}$$

Industrial park average flow (from problem statement): 550,000 gpd.

Average infiltration/inflow (I/I):

$$(16.5 + 9) \text{ mi} \times 2,800 \text{ gpd/mi} = 71,400 \text{ gpd}$$

In similar fashion, we can also estimate the future maximum-day wastewater flow rate (assuming the dry-day domestic wastewater generation is equal to the wet-day domestic wastewater generation) as 3,621,000 gpd from the following four contributions:

Future domestic maximum flow (including 8 percent wastewater reduction):

$$95.4 \text{ gpdc} \times 0.92 \times 17{,}000 \text{ people} = 1{,}492{,}000 \text{ gpd}$$

Apartment complex maximum flow (using values from Table 7.12):

$$55 \text{ gpdc} \times 500 \text{ people} = 27{,}500 \text{ gpd}$$

Industrial park maximum flow (from problem statement): 750,000 gpd

Maximum infiltration/inflow (I/I):

$$(16.5 + 9) \text{ mi} \times 53{,}000 \text{ gpd/mi} = 1{,}351{,}500 \text{ gpd}$$

Note that the values used in this example are "typical" and not overly aggressive in terms of water conservation. Readers are encouraged to think much bigger in terms of sustainable water management. For example, can the projected 8 percent reduction in wastewater generation be raised much higher to achieve more sustainable use of existing water? What would happen if the apartment complex were designed to be "green" in terms of aggressively reducing water usage, not only through use of water-saving technology and education of the building occupants, but also with innovative use of gray water in cooling and landscaping? How about if the design of the industrial complex selected some businesses that had an integrated reuse of others' generated wastewater? And how will water conservation impact the "strength" of the wastewater?

9.16   Managing Wet-Weather Flows

There are several "gray" strategies as discussed above that have traditionally been utilized to manage stormwater and to limit its undesirable impacts on wastewater treatment and resource recovery. These traditional techniques are intended solely to reduce the peak flow rate of water entering the plant. However, there is a growing awareness that more effective management of stormwater can benefit not only plant operations but provide opportunities for beneficial use of captured stormwater either for on-site applications or for aquifer recharge (which can then be linked with water reuse strategies).

There are further examples where reimagining stormwater management can have broad benefits to sustainability objectives as well. For example, low-impact development mimics the natural hydrology that existed before development occurred. It not only reduces the peak flow rate, but also considers the timing of the discharge off the site as well as retention of rainwater. It also integrates principles of biodiversity, green space, water storage, groundwater recharge, and water quality improvements into the

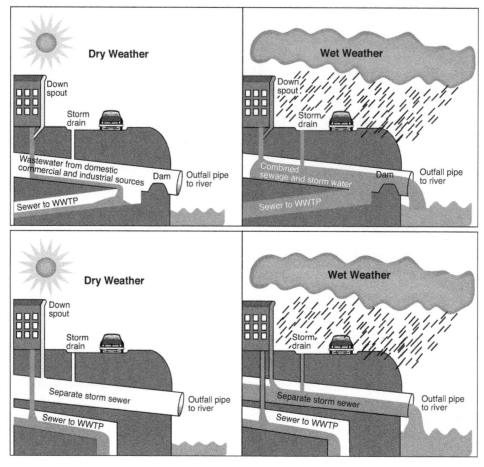

Figure 9.30 **Combined Sewer System** During dry weather (and small storms), all flows are handled by the WWTP. During large storms, the relief structure allows some of the combined stormwater and sewage to be discharged untreated to an adjacent water body. A municipal separate storm sewer system manages runoff through a different pipe, directly discharging rainfall to receiving bodies, eliminating this additional load on the WWTP.

(Redrawn from EPA, 2004).

overall plan. Table 9.22 compares the philosophies of traditional stormwater management with low-impact development.

There are many benefits to pursuing a low-impact development approach. One major problem of the built environment is the impact of *nonpermeable surfaces* on the natural hydrologic cycle and water quality. Early in Chapter 7, Figure 7.6 showed how covering natural surfaces with buildings, roofs, roads, and parking lots decreases the amount of precipitation that infiltrates to groundwater. In natural systems, approximately 50 percent of precipitation recharges to the subsurface, but in a highly urbanized environment, this declines to 15 percent, with only 5 percent reaching deep groundwater.

Because of the water problems associated with nonpermeable surfaces, Phase II of the National Pollutant Discharge Elimination System (NPDES) requires small municipalities with separate storm sewer systems (called MS4s) to address stormwater runoff from their site with the recommended

Comparison of Stormwater Management Philosophies Site plans will typically integrate the use of several best management practices (BMPs) such as rain gardens, permeable pavement (pavers and gravel), green roof, bioswales, and even underground gravel beds for stormwater detention.

Traditional Management	Low-Impact Development
Catchbasins and pipes	Swales
Mowed detention basin	Planted basin as in bioretention cells
Large central detention pond	Small, distributed detention areas
Runoff rate	Runoff rate and volume
Flooding is primary concern	Flooding and water quality are primary concerns
Time of concentration reduced significantly	Time of concentration maintained or extended
Runoff has nutrients, suspended particles, and hazardous materials	Runoff constituents treated by gravity settling, filtration, sorption, and interaction with microorganisms and vegetation

SOURCE: Ward, 2007.

use of structural **best management practices (BMPs)**. Examples of BMPs related to this regulation are grassed swales, permeable pavements, and bioretention cells. Fortunately, BMPs can be easily integrated into a new or existing development, even at the household level.

Beyond water quality issues, the impact of urbanization on runoff also affects a city's ability to store freshwater. For example, in urbanized coastal areas, precipitation that becomes runoff will travel quickly to saline seawater, where it then becomes energy intensive and expensive to treat to drinking or agricultural standards. This problem is significant, because the average population density in coastal areas is twice the global average. Promoting groundwater recharge by preferring permeable surfaces should thus be considered a method to store freshwater for later use by ecosystems and humans. This is especially important because groundwater makes up over 30 percent of the world's freshwater reserves.

9.17 Green Stormwater Management

Stormwater management systems that mimic nature by integrating stormwater into building and site development can reduce the damaging effects of urbanization on rivers and streams. Disconnecting the flow from storm sewers and directing runoff to natural systems like landscaped planters, swales and rain gardens or implementing a green roof reduces and filters stormwater runoff. All of these strategies present benefits to wastewater treatment operation and have the potential to realize multiple sustainability objectives depending on the site, place, and other location-specific considerations. Each of the green stormwater management technologies and strategies discussed present economic costs and benefits depending on capacity, ability to treat contaminants, economics, performance variability with season and age.

9.17.1 GREEN ROOFS

Rooftop runoff can be a substantial contributor to municipal storm sewer systems. To provide an idea of the extent of urban roof areas, the estimated roof area in metropolitan Chicago is 680 km^2, and in cities such as Phoenix, Seattle, and Birmingham, it is estimated that connected residential roofs account for 30–35 percent of annual runoff volume.

Compared with traditional blacktop and metal roofs, **green roofs** provide many private and public benefits, listed in Table 9.23. Some of these benefits include enhanced stormwater management, reduction in building energy costs, increase of green space and habitat, and reduction in urban heat island.

There are three types of green roofs:

1. *Intensive green roofs* have a thicker soil layer (150–400 mm) and weigh more, so they require more structural support. They work well with existing concrete roofs (for example, on parking decks).

2. *Extensive green roofs* have a thinner soil layer (60–200 mm), so they require less structural support.

3. The *semi-intensive green roof* has some components of both the extensive and the intensive roof systems.

Table 9.24 summarizes design and maintenance criteria for each type of green roof.

Native vegetation is always the preferred alternative. Whether or not the plants are native plays an important role in determining the degree of maintenance and irrigation. Extensive green roofs tend to be the more manicured with lots of mulch coverage and sedum vegetation, implying the need to weed regularly throughout the lifetime of the roof. Intensive green roofs tend to incorporate more native vegetation.

Table / 9.23

Private and Public Benefits of Green Roofs

Benefit	Description
Increased roof life	Life expectancy of a "naked" flat roof is 15–25 years because of high surface temperatures and UV radiation degradation. Green roofs increase roof life by moderating these impacts.
Reduced noise levels	Sound is reflected by up to 3 dB, and sound insulation is improved by up to 8 dB.
Thermal insulation	Provides additional insulation, which reduces heating and cooling costs.
Heat shield	Transpiration during growing season results in a cooler building climate.
Use of space	Can be incorporated into personal, commercial, and public space.
Habitat	Provides habitat for plant and animal species.
Stormwater retention	Runoff can be reduced by 50–90%, especially important during peak precipitation events.
Urban heat island	Transpiration results in cooler roof surfaces, which reduces building contribution to heating the local built environment.

SOURCE: Information courtesy of the International Green Roof Association, Berlin.

Table / 9.24

Design and Maintenance Criteria for Green-Roof Types Each has different characteristics in terms of planting depth and structural loading requirements.

	System Buildup Height	Weight	Costs	Possible Use
Extensive green roof	60–200 mm	60–150 kg/m^2 13–30 lb/ft^2	Low	Ecological protection layer
Semi-intensive green roof	120–250 mm	120–200 kg/m^2 25–40 lb/ft^2	Middle	Designed green roofs
Intensive green roof	150–400 mm on underground garages >1,000 mm	180–500 kg/m^2 35–100 lb/ft^2	High	Parklike garden

SOURCE: Information courtesy of the International Green Roof Association, Berlin.

The green roof on Chicago's City Hall is planted with native prairie vegetation. Native vegetation can grow densely (so it requires little to no mulch), and it more closely replicates a native ecosystem. The use of native species in the design of an intensive green roof requires heavy weeding of invasive species during the first 2–3 years but little maintenance after this initial period. In contrast, a manicured sedum-planted extensive green roof will require maintenance over its lifetime. Native vegetation also stands up better to climatic variation than ornamentals and introduced species. This implies a lesser need for irrigation on an intensive green roof planted with native species.

Note that irrigation needs and the relationship to the type of vegetation will depend on the depth of the soil. For example, irrigation requirements for a succulent nonnative plant could be equal to those for native vegetation in the shallow soils of a green roof. This is because the deep-root advantage of some native plants is lessened in the shallow soil layers of green roof.

Figure 9.31 shows the components of a typical green roof. Such a roof consists of a multilayered system that sits above the roofing deck and provides water and root protection (items 3–7 in Figure 9.31). These layers sit under a drainage system (item 9). On top of the drainage layer is some growing material (such as soil), which is planted (items 11 and 12, respectively).

The volume of water that a green roof can store after a rainfall event (V) is determined as

$$V = P \times A \times C \tag{9.29}$$

Chicago Green Roofs
www.chicagogreenroofs.org/

where P is the precipitation (mm), A is the roof area, and C is a measure of the water-holding capacity of the growing medium (ranges from 0 to 1). This volume of stored water can be compared with the volume of stormwater generated by a conventional roof ($P \times A$).

9.17.2 PERMEABLE (OR POROUS) PAVEMENTS

Permeable (or porous) pavement is pavement that allows for the vertical passage of water. This pavement not only reduces runoff by

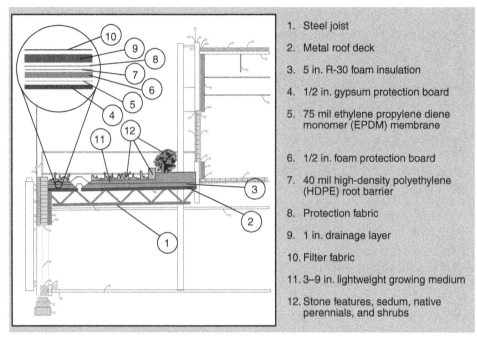

Figure 9.31 **Components of a Green Roof** This particular roof uses a combination of extensive (3–4 in. soil) and intensive (4–9 in. soil) planting areas. It reduced the nonpermeable surface area 3,626 ft², required an engineered support of 62 lb/ft², and cost \$31.80/ft² in 2005.

(Courtesy of Wetlands Studies and Solutions, Inc., Gainesville, Va).

enhancing groundwater recharge but also reduces the urban heat island. Permeable pavements are also believed to be more skid-resistant and quieter than conventional pavements. They also eliminate the need for some drainage systems (as with other BMPs).

Permeable pavement can be constructed from grass, gravel, crushed stone, concrete pavers, concrete, and asphalt (see Table 9.25).

Table / 9.25

Examples of Permeable (Porous) Pavement

Permeable (Porous) Pavement	Examples
Grass	Excellent option for situations where commercial parking is needed in winter months when ground cover is frozen or is only needed several times a year (e.g., outside football stadium or on edge of big box stores)
Gravel or crushed stone	Common on many driveways and roads
Grid	Consists of washed gravel, plastic grid, and filter fabric; has high infiltration rate and removes sediments
Paver stones	High infiltration rates and filter sediments; easy to plow and maintain
Porous concrete or asphalt	For porous concrete, cement, and water with little to no sand or aggregate; high permeability (15–25% voids, flow rates around 480 in water depth per hour); filter sediment; easy to plow and maintain

Porous pavement can be as simple as grassy surfaces and interlocking paving stones that allow vegetation to grow between the block edges. Paver stones can also be placed along trees, because they will not harm root systems like traditional pavement does. It can also refer to a multilayer system that includes a permeable course of paver stones (or gravel) at the surface, a compacted sand subbase immediately below the permeable cover, a fabric filter, and a compacted base on the bottom.

In situations where strength of the paving material is of concern, permeable pavement can be mixed with traditional pavement. In this scenario, parking areas and pedestrian walkways are specified as porous pavement, and traditional paving materials are used in limited-use areas where heavy loads are brought (for example, truck travel to a loading dock). In this case, pavement markings or signage can guide larger delivery trucks so they remain off the porous pavement that surrounds the traditional pavement.

Permeable concrete consists of conventional concrete materials with the coarse aggregate being limited in its range of sizes, and the presence of fine aggregate being minimal or nonexistent (PCA, 2004). *Porous asphalt* is sometimes referred to as open-graded coarse aggregate, bonded together by asphalt cement, with sufficient interconnected voids to make it highly permeable to water (EPA, 1999).

9.17.3   BIORETENTION CELLS

Bioretention cells are shallow depressions in the soil to which stormwater is directed for storage and to maximize infiltration. They are sometimes referred to as *bioinfiltration cells, vegetated biofilters,* and *rain gardens*. They are most often mulched (for aesthetic value and water treatment) and planted with native vegetation that promotes evapotranspiration. The design objective of maximizing infiltration will reduce the volume of water that needs to be stored and/or treated (hence the name bioinfiltration cell).

Figure 9.32 shows the detailed design of a bioretention cell. It includes the use of an underdrain, substantial aggregate backfill, and geotextile lining. Bioretention cells are often incorporated along streets to capture road runoff, which requires the inclusion of a curb cutoff. One novel aspect is that they can be designed to incorporate a wide variety of uses, from high infiltration to pretreatment of urban runoff and removal of nitrogen (see Figure 9.33 for examples). They can also be sized and installed by homeowners (WDNR, 2003).

As the vegetative cover grows, bioretention cells are expected to have increased capacity to accept water as the root network of the plants evolves and increases transpiration. In contrast to septic tank drainage fields, where biomats can develop due to relatively high organic and nutrient loadings, no studies have yet found loss of infiltration performance in bioretention cells. If needed, the soil may be carefully loosened (and aerated) to restore the infiltrative capacity and break up the biomat that may develop on the upper horizontal plane of the cell where water enters.

One common myth is that bioretention cells attract mosquitoes. Mosquitoes require 7–12 days to lay and then hatch their eggs. The

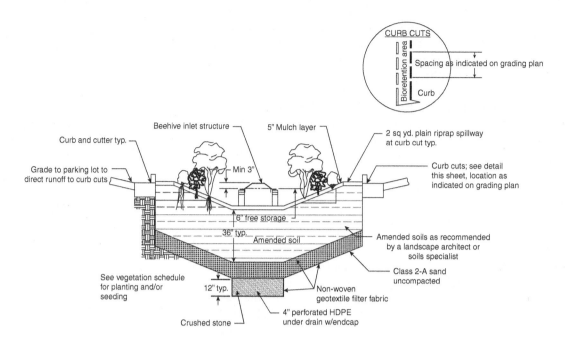

Figure 9.32 **Typical Design of a Commercial Bioretention Cell** Note the location of amended soils, underdrains, outlet structures, vegetation, and inlet protection. These types of bioretention cells typically include the use of an underdrain, substantial aggregate backfill, or geotextile lining.

(Courtesy of Spicer Group, Inc., Saginaw, Mich. Detail developed in 2006).

standing water in a properly designed bioretention cell will be present for only several hours after a rain event. Also, the plants will attract dragonflies, which prey on mosquitoes.

DESIGN AND CONCEPT OF FIRST FLUSH Many states now recommend or require that the first flush of a storm event be captured and treated. Bioretention cells can be sized based on the concept of first flush. **First flush** is defined as the first 0.5–1 in. of runoff that is associated with a rain event and is calculated over the entire nonpermeable area of a site.

In a residential home where a yard slopes toward the bioretention cell, the nonpermeable area (for example, the roof) would be increased by the area of yard that is draining into the bioretention cell. The reason for including the area of the yard that drains to the cell is that there is a misconception that residential yards constitute high-infiltration green space. In fact, the effects of soil compaction during development are substantial, which is why land disturbance should be minimized during any land development. For similar reasons of soil compaction, heavy equipment should never travel across the cell during construction.

Because climates vary among states, some states have guidelines in terms of the release of the first flush. For example, in Michigan, this water volume must be released over a 1- to 2-day period or infiltrated into the ground within 3 days. For regional detention

(a) This bioretention cell facilitates high recharge of groundwater. Here in situ soils are recommended to have infiltration rates of at least 1 in./hr. and a depth of at least 2.5 ft. for adequate filtration.	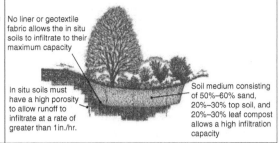
(b) This bioretention cell facilitates high filtration and partial recharge of runoff. The underdrain location ensures a desired rate of drainage. Again, the depth is at least 2.5 ft.	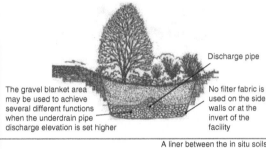
(c) This bioretention cell is designed to handle higher nutrient loadings by facilitating a fluctuating aerobic/anaerobic zone in the layer below the underdrain. The area below the underdrain also provides a storage area and recharge zone.	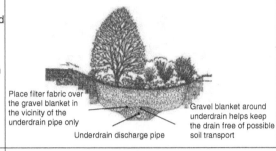
(d) This bioretention cell is designed for pretreatment of highly contaminated water before discharge at an outlet pipe. The liner prevents groundwater contamination.	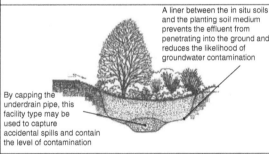

Figure / 9.33 Bioretention Cells Designed for Different Purposes
From top to bottom: (a) infiltration and recharge facility for enhanced infiltration; (b) filtration and partial recharge facility; (c) infiltration, filtration, and recharge facility; and (d) filtration-only bioretention cell.

(Redrawn from *The Bioretention Manual*, developed by Prince George's County Government, Md., 2006).

applications, Michigan suggests treatment of the 90 percent nonexceedance storm (the storm event for which 90 percent of all runoff-producing storms are smaller than or equal to the specified storm) (Ward, 2007).

Bioretention systems are sized based on several different methods, including the Prince George's County manual method, the runoff frequency method, and the rational method. Modeling programs such as EPA SWMM, WIN-TR-55, HEC-HMS, and HydroCad are applicable for modeling stormwater at the small site to regional scale. These tools have limitations in terms of simulating specific hydrologic mechanisms in bioretention at the site scale, but knowledgeable users have applied them to develop conservative designs. Two good examples of widely used models are the Prince George's County (Maryland) BMP module and RECARGA (from the University of Wisconsin–Madison).

Bioretention cells can also be designed based on the first flush and on whether the water is stored below grade or above grade. Sizing the cell based on storage below grade will require an estimate of the cell's porosity. Sizing the cell based on the volume of water that can be stored above grade requires an understanding of how deep a water level the plant species requires to survive for a short period of time, along with accounting for the above-grade volume taken up by the plants.

To estimate the volume of a bioretention cell required to store the first flush below grade, determine the volume of rainwater generated at a site:

$$\text{volume of rainwater to be stored} = \text{first flush} \times \text{nonpermeable area}$$

(9.30)

In Equation 9.30, the first flush is 0.5–1.0 in rainwater generated during a precipitation event, and the nonpermeable area includes the area being drained to the cell. The maximum volume of the bioretention cell below grade can be determined as follows:

$$\text{volume of bioretention cell} = \frac{\text{volume of rainwater to be stored}}{\text{soil porosity}}$$

(9.31)

The porosity is defined as

$$n = \frac{\text{volume of the void space}}{\text{total volume}} = \frac{V_V}{V_t}$$

(9.32)

Equation 9.31 assumes that the soil porosity is saturated with precipitation associated with the first flush. Because soil porosity has units of volume voids divided by total volume, the area of the bioretention cell (A) can be written as

$$A = \frac{\text{volume of bioretention cell}}{\text{depth}}$$

(9.33)

where the below-grade depth is typically 3–8 in.

When designing a bioretention cell to store water above grade, the designer should consider the volume of water generated by nonpermeable surfaces and the depth of water that can submerge a portion of the planted vegetation for a short duration, along with the above-grade volume occupied by the plants. The volume required above grade for the bioretention cell to store the first-flush precipitation event is

$$
\begin{bmatrix} \text{total volume} \\ \text{above grade for the} \\ \text{bioretention cell} \end{bmatrix} = \begin{bmatrix} \text{volume of} \\ \text{rainwater to} \\ \text{be stored} \end{bmatrix} + \begin{bmatrix} \text{volume taken} \\ \text{up by} \\ \text{vegetation} \end{bmatrix}
$$

(9.34)

In Equation 9.34, the volume of rainwater associated with the first flush that will need to be stored is determined from Equation 9.30. The volume above grade taken up by vegetation is

$$
V = \begin{bmatrix} \text{number} \\ \text{of plants} \end{bmatrix} \times \begin{bmatrix} \text{cross-sectional} \\ \text{area of the} \\ \text{plant stem} \end{bmatrix} \times \begin{bmatrix} \text{allowable depth that} \\ \text{a plant can be} \\ \text{submerged for} \\ \text{a short duration} \end{bmatrix}
$$

(9.35)

With the information provided by Equation 9.35, the total volume required above grade for the bioretention cell can be determined from Equation 9.34. This volume can be divided by the vegetative-specific allowable depth that a plant can be submerged for a short duration (variable in Equation 9.35) to determine the required area. This area may be restrained by site constraints. Also, permeable pavements can be used to reduce the volume of the first flush that is generated.

You can easily design and construct a rain garden for your home. A residential bioretention cell—like commercial ones—is usually 4–8 in deep. The area of a residential bioretention cell that treats runoff from a homeowner's roof typically ranges from 100 to 300 ft^2. In all cases, bioretention cells (especially those without an overflow drain) are graded so that when they overflow, water flows away from buildings. Parking lot bioretention systems between rows of parking sometimes have a deeper slope, but the level of such an outlet would still reflect the specific inundation requirements of the plants and/or reflect the desired volume captured (for example, a first-flush design or 90 percent nonexceedance design).

Residential bioretention cells can also be sized based on understanding the specified depth of the cell (based upon vegetation inundation tolerances) along with knowledge of the soil type. In this case, the area of the cell is determined as follows:

$$
\begin{aligned} &[\text{area of residential bioretention cell}] \\ &\quad = \text{nonpermeable area} \times \text{size factor} \end{aligned}
$$

(9.36)

The size factor is related to the porosity of the soil and its ability to infiltrate rainwater. Table 9.26 provides size factors for residential bioretention cells as a function of cell depth and soil type.

Table / 9.26

Size Factors for Sizing Residential Bioretention Cells

Soil Type	Depth of Cell		
	3–5 in.	6–7 in.	8 in.
Sandy soil	0.19	0.15	0.08
Silty soil	0.34	0.25	0.16
Clayey soil	0.43	0.32	0.20

SOURCE: WDNR, 2003.

9.17.4 BIOSWALES AND OTHER LAND USE TECHNIQUES

Bioswales (also referred to as *grass swales* and *infiltration trenches*) are engineered conveyance channels that consist of native vegetation. They are not lined with a material such as concrete. To the untrained eye, a bioswale will appear to be a grassy channel with a ridge of higher soil placed on either side of the channel, perhaps planted with trees or shrubs.

Bioswales are designed with the longest route of conveyance in mind. Thus, as the water flows through the bioswale, it either transpires through plants or infiltrates through the soil. A meandering conveyance channel encourages infiltration. When a meandering flow path is impossible, porous rock check dams may be used at intervals along bioswales for additional filtration and reduction in the rate of runoff. Bioswales are usually installed along highways or between rows of a parking lot. They have sloped walls and are also sloped along the length for conveyance. When they are placed along roads, sometimes curbs are completely removed.

Another way to reduce the impact of nonpermeable coverings is to retain natural vegetation and preserve wetlands. As previously mentioned, grass cover is a permeable pavement, especially in applications where the space is needed for winter months (as in the case of holiday shopping season), when ground is frozen, or is needed in frequently (for example, surrounding sports events and fairs). **Green space** areas can also be preserved to provide seasonal flood storage and many other benefits (Table 9.27). Here the green space may flood during seasonal heavy-rain events or after snowmelt. During these times, the space may also provide wildlife habitat and recreation such as birdwatching. When the space dries out later in summer and fall, it can be used as recreational green space.

Take a virtual tour of a natural drainage system, review the Seattle SEAStreets project

http://www2.cityofseattle.net/util/ tours/seastreet/slide1.htm

Class Discussion

How would your class manage the stormwater generated from the classroom's building roof and an outside paved area? Where does the stormwater currently go? How would you maximize the benefit of low-impact development while accounting for movement of students, faculty, staff, services, water, and biodiversity? How would you design a system that uses the least energy and materials? What economic, social, and environmental benefits would be preserved for future generations?

Table / 9.27

Benefits of Green Space (based on Wright Wendel et al., 2011).

Social Benefits	Environmental Benefits	Economic Benefits
Enhanced recreational opportunities	Improved air quality	Increased property values
Increased levels of physical activity	Water pollutant filtration	Improved ability to attract and retain businesses and residents
Increased sense of security	Increased control of stormwater runoff & flooding	Tourism
Improved mental health	Reduced loads on stormwater systems	Decreased needs for police and prisons
Reduced crime and juvenile delinquency	Groundwater recharge	Reduced pollution prevention measures
	Reduced heat island effect	
	Wildlife habitat	

Key Terms

- activated sludge
- aeration basin
- aeration tank
- aerobic digestion
- anaerobic digestion
- anaerobic lagoon
- attached growth
- average-strength municipal wastewater
- bar rack
- bar screen
- best management practices (BMPs)
- bioretention cells
- bioswales
- BOD loading rate
- carbon dioxide
- clarifiers
- Clean Water Act
- combined sewer
- combined sewer overflow (CSO)
- comminutor
- compression settling
- constructed wetland
- conventional activated sludge
- critical SRT
- CSO storage
- death growth phase
- denitrification
- digested sludge
- digester
- discrete settling
- disinfection
- domestic wastewater
- emerging chemicals of concern
- endogenous-growth phase
- endogenous respiration
- energy
- extended aeration

- facultative lagoon
- first flush
- flocculation
- flocculent settling
- flotation
- flow equalization
- foam
- food-to-microorganism (F/M) ratio
- free water surface (FWS) wetland
- green roof
- green space
- greenhouse gas
- grit
- grit chamber
- hindered settling
- hydraulic retention time
- improved sanitation
- infiltration
- inflow
- lagoon
- maturation lagoon
- mean cell retention time (MCRT)
- membrane bioreactors (MBRs)
- methane
- mixed liquor
- mixed liquor suspended solids (MLSS)
- mixed liquor volatile suspended solids (MLVSS)
- modified Ludzak–Ettinger (MLE) process
- Monod kinetics
- municipal separate storm sewer system (MS4)
- municipal wastewater constituents
- municipal wastewater treatment plants

- National Pollutant Discharge Elimination System (NPDES)
- nitrification
- nocardioform
- overflow rate
- oxidation ponds
- permeable (or porous) pavement
- phosphate-accumulating organisms (PAOs)
- phosphorus
- photosynthetic oxygen
- preliminary treatment
- primary treatment
- publicly owned treatment works (POTWs)
- retention basins
- return activated sludge
- satellite reclamation processes
- secondary clarifier
- secondary treatment
- sedimentation tanks
- settling tanks
- sludge age
- sludge stabilization
- sludge volume index (SVI)
- solids retention time (SRT)
- stabilization ponds
- Stokes' law
- surface flow wetlands
- tertiary wastewater treatment
- trickling filter
- unimproved sanitation
- washed out
- waste activated sludge
- wastewater reclamation
- wastewater reuse
- wasting of sludge
- water reclamation
- water reuse
- wet-weather flows

chapter/Nine Problems

9.1 Research an emerging chemical of concern that might be discharged to a wastewater treatment plant or household septic systems. Examples would include pharmaceuticals, caffeine, surfactants found in detergents, fragrances, and perfumes. Write an essay of up to three pages on the concentration of this chemical found in wastewater influent. Determine whether the chemical you are researching is treated in the plant, passes through untreated, or accumulates on the sludge. Identify any adverse ecosystem or human health impacts that have been found for this chemical.

9.2 Research whether there are state or regional pollution prevention programs to keep mercury out of your local municipal wastewater treatment plant. This mercury might come from your university laboratories or local dental offices and hospitals. What are some of the specifics of these programs? How much mercury has been kept out of the environment since the program's inception?

9.3 A laboratory provides the following solids analysis for a wastewater sample: TS = 225 mg/L, TDS = 40 mg/L, FSS = 30 mg/L. (a) What is the total suspended solids concentration of this sample? (b) Does this sample have appreciable organic matter? Why or why not?

9.4 A 100 mL water sample is collected from the activated-sludge process of municipal wastewater treatment. The sample is placed in a drying dish (weight = 0.5000 g before the sample is added) and then placed in an oven at 104°C until all moisture is evaporated. The weight of the dried dish is recorded as 0.5625 g. A similar 100 mL sample is filtered, and the 100 mL liquid sample that passes through the filter is collected and placed in another drying dish (weight = 0.5000 g before the sample is added). This sample is dried at 104°C, and the dried dish's weight is recorded as 0.5325 g. Determine the concentration (in mg/L) of (a) total solids, (b) total suspended solids, (c) total dissolved solids, and (d) volatile suspended solids. (Assume VSS = 0.7 × TSS.)

9.5 Obtain the World Health Organization (WHO) report on "Urine diversion: Hygienic risks and microbial guidelines for reuse." Review Figure 2 in Chapter 1 of this report (Introduction). (a) How many grams of N, P, and K are excreted every day in a Swedish person's urine?

9.6 Humans produce 0.8–1.6 L of urine per day. The annual mass of phosphorus in this urine on a per capita basis ranges from 0.2 to 0.4 kg P. (a) What is the maximum concentration of phosphorus in human urine in mg P/L? (b) What is the concentration in moles P/L? (c) Most of this phosphorus is present as HPO_4^{2-}. What is the concentration of phosphorus in mg HPO_4^{2-}/L?

9.7 Assume 50 percent of phosphorus in human excrement is found in urine (the remaining 50 percent is found in feces). Assume humans produce 1 L of urine per day and the annual mass of phosphorus in this urine is 0.3 kg P. If indoor water usage is 80 gallons per day in a single individual apartment, what is the low and high range of phosphorus concentration (in mg P/L) in the wastewater that is discharged from the apartment unit? Make sure you account for phosphorus found in urine and feces.

9.8 The following equation shows the stoichiometry for recovery of phosphorus and nitrogen from wastewater through precipitation of struvite.

$$Mg^{2+} + NH_4^+ + PO_4^{3-} + 6H_2O$$
$$\rightarrow MgNH_4PO_4 \cdot 6H_2O_{(s)}$$

If the composition of the wastewater under consideration for struvite recovery is 7 mg P/L, NH_4^+ is 25 mg NH_4^+–N/L, and Mg is 40 mg Mg^{2+}/L. Is there sufficient Mg and NH_4^+ to precipitate all the phosphorus, assuming it all exists as PO_4^{3-}?

9.9 Design an aerated grit chamber system to treat a 1-day sustained peak hourly flow of 1.6 m³/s with an average flow of 0.65 m³/s. Determine: (a) the grit chamber volume (assuming two chambers will be used); (b) the dimensions of the two grit chambers; (c) the average hydraulic retention time in each grit chamber; (d) air requirements, assuming 0.20 m³ of air per m length of tank per minute; and (e) the quantity of grit removed at peak flow, assuming a typical value of 0.20 m³ of grit per one thousand m³ of untreated wastewater.

9.10 A wastewater treatment plant receives a flow of 35,000 m³/day. Calculate the required volume (m³) for a 3 m deep horizontal-flow grit chamber that will remove particles with a specific gravity of more than 1.9 and a size greater than 0.2 mm diameter.

9.11 A wastewater treatment plant will receive a flow of 35,000 m^3/day. Calculate the surface area (m^2), diameter (m), volume (m^3), and hydraulic retention time of a 3 m deep circular, primary clarifier that would remove 50 percent of suspended solids. Assume the surface overflow rate used for the design is 60 m^3/m^2-day.

9.12 Assume a plant flow of 12,000 m^3/day. Determine the actual detention time observed in the field of two circular settling tanks with depth of 3.5 m that were designed to have an overflow rate not to exceed 60 m^3/m^2-day and a detention time of at least 2 h.

9.13 A wastewater treatment plant has a flow of 35,000 m^3/day. Calculate the mass of sludge wasted each day ($Q_w X_w$, expressed in kg/day) for an activated-sludge system operated at an SRT of 5 days. Assume an aeration tank volume of 1,640 m^3 and an MLSS concentration of 2,000 mg/L.

9.14 You are provided with the following information about a municipal wastewater treatment plant. This plant uses the traditional activated-sludge process. Assume the microorganisms are 55 percent efficient at converting food to biomass, the organisms have a first-order death rate constant of 0.05/day, and the microbes reach half of their maximum growth rate when the BOD_5 concentration is 10 mg/L. There are 150,000 people in the community (their wastewater production is 225 L/day-capita, 0.1 kg BOD_5/capita-day). The effluent standard is $BOD_5 = 20$ mg/L and TSS $= 20$ mg/L. Suspended solids were measured as 4,300 mg/L in a wastewater sample obtained from the biological reactor, 15,000 mg/L in the secondary sludge, 200 mg/L in the plant influent, and 100 mg/L in the primary clarifier effluent. SRT is equal to 4 days. (a) What is the design volume of the aeration basin (m^3)? (b) What is the plant's aeration period (days)? (c) How many kg of secondary dry solids need to be processed daily from the treatment plants? (d) If the sludge wastage rate (Q_w) is increased in the plant, will the solids retention time go up, go down, or remain the same? (e) Determine the F/M ratio in units of kg BOD_5/kg MLVSS-day. (f) What is the mean cell residence time?

9.15 Using information provided in Example 9.4, determine the critical SRT value (sometimes referred to as SRT_{min}). This term refers to the SRT where the cells in the activated-sludge process would be washed out or removed from the system faster than they can reproduce.

9.16 If the specific growth rate for an activated-sludge process equals 0.10/day, what is the SRT for this system (units of days). (b) What is the mean cell retention time for the same system (units of day)?

9.17 In the following sentences, circle the correct term in boldface. If the SRT is low (for example, 4 days), which conditions exist? (a) The F/M ratio is **low/high**. (b) The power requirements for aeration will be **less/greater**. (c) The microorganisms will be **starved/saturated** with food. (d) The mean cell retention time is **low/high**. (e) The sludge age is **low/high**. (f) The sludge wastage rate may have been recently **increased/decreased**. (g) The MLSS may have been **increased/decreased**.

9.18 The suspended-solids concentration entering a treatment and resource recovery plant is 200 mg/L in the plant influent; 3,000 mg/L in the primary sludge; 12,500 mg/L in the secondary sludge; and 3,500 mg/L exiting the aeration basin. The concentration of total dissolved solids in the plant influent is 350 mg/L, and the concentration of total dissolved solids exiting the aeration basins is 2,300 mg/L. The BOD_5 is 100 mg/L measured after primary treatment and 3 mg/L exiting the plant. Total nitrogen levels in the plant are approximately 35 mg N/L.

If the F/M ratio is 0.35 g BOD_5/gram MLSS-day, estimate the hydraulic retention time of the aeration basins if the daily plant flow is 15 million liters.

9.19 Determine the sludge volume index (SVI) for a test where 3 g MLSS occupies a 450 mL volume after 30 min settling.

9.20 A 2 g sample of MLSS obtained from an aeration basin is placed in a 1,000 mL graduated cylinder. After 30 min settling, the MLSS occupies 600 mL. Does the following sludge have good, acceptable, or poor settling characteristics?

9.21 Figure 9.16 shows the modified Ludzak–Ettinger (MLE) process, which is used to configure a biological reactor to remove nitrogen. Explain the role of the two compartments in terms of: (a) whether they are oxygenated; (b) whether CBOD is removed in the compartment; (c) whether ammonia is converted in the compartment; (d) whether nitrogen is removed from the aqueous phase in the compartment; and (e) the primary electron donor(s) and electron acceptor(s) in each compartment.

9.22 Investigate the specific mechanisms by which ammonia nitrogen, total nitrogen, and phosphorus are treated or recovered at your local municipal wastewater treatment plant. Are the processes chemical or biochemical (or a combination)? Discuss your answer.

9.23 Investigate the specific mechanisms that your local municipal wastewater treatment plant uses for aeration. Is it surface aeration, fine- or coarse-bubble aeration, or natural aeration (via a facultative lagoon or attached-growth system)?

9.24 A wastewater treatment plant will receive a flow of 35,000 m^3/day (~10 MGD) with a raw wastewater $CBOD_5$ of 250 mg/L. Primary treatment removes ~25% of the BOD. Calculate the volume (m^3) and approximate hydraulic retention time (h) of the aeration basin required to run the plant as a "high rate" facility (F/M = 2 kg BOD/kg MLSS-day). The aeration basin MLSS concentration will be maintained at 2,000 mg MLSS/L.

9.25 Table 9.28 provides suspended solids concentrations in several different wastestreams at a municipal wastewater treatment plant. The BOD_5 is measured in the sewer located just before the treatment plant as 250 mg/L, after primary treatment is 150 mg/L, and after secondary treatment is 15 mg/L. Total nitrogen levels in the plant are approximately 30 mg N/L. (a) If the design hydraulic retention time of each of four aeration basins operated in parallel equals 6 h and the total plant flow is 5 million gallons per day, what is the F/M ratio in units of lb BOD_5/lb MLVSS-day. (b) Suppose the plant engineer wishes to increase the concentration of microorganisms in the biological reactor because she expects the substrate level to increase. What would she command the operator to do to accomplish this goal?

9.26 Determine the minimum surface area for a facultative lagoon treating municipal wastewater to be sited in Tampa, Florida. Assume the conversion efficiency of the algae is 3.5 percent and that 24,000 kJ of sunlight are needed to produce 1 kg of algae biomass. The average daily concentration of CBOD in the wastewater to be treated is 250 mg/L, and the average flowrate is 4 MGD.

9.27 Determine the minimum surface area for a facultative lagoon treating municipal wastewater to be sited in your local community. Assume that the conversion efficiency of the algae is 3% and that 24,000 kJ of sunlight are needed to produce 1 kg of algae biomass. Use the solar radiation characteristics of your local area and obtain an average daily concentration of CBOD in the wastewater and average wastewater flow rate from your instructor that is for your area.

9.28 The community of San Antonio is located in Caranavi Province, Bolivia. According to the 2005 year survey, there are 420 habitants of this community. The population is estimated to increase to 940 by the year 2035. The average peak flow is currently 1.2 L/s and is expected to increase to 2.14 L/s by 2035. The organic load is estimated to be 45 g BOD_5/capita-day. The community is considering a free surface wetland to treat their wastewater. (a) What is the BOD_5 loading generated in the year 2035 (kg/day)? (b) Use the BOD loading to estimate the maximum surface area (ha) required for a free surface wetland that would serve the community in 2035 and remove BOD and TSS to 30 mg/L. (c) Assuming you are now considering sizing a facultative lagoon instead of a free surface wetland, quickly estimate the required surface area (m^2) for a facultative lagoon to handle a peak flow in 2035, assuming a design water depth of 4 m and a hydraulic detention time of 20 days.

9.29 (a) What size septic tank would you recommend for a two-bedroom cottage assuming the available tank sizes are 750, 1,000, 1,200, and 1,500 gallons? Assume you wish to have 3 days of residence time for the pollutants in the tank. (b) How would your problem change for a four-bedroom house with a 2-day residence time?

9.30 Table 9.1 indicated that pit latrines are considered an improved technology for treating wastewater. Determine the depth required for a pit latrine 1 m × 1 m in. area that serves a household of seven people and has a design life of 10 years. Assume the pit is dug above the water table and the

Table / 9.28

Suspended-Solids Concentration for Different Process Streams in Problem 9.25

Process Stream	Suspended-Solids Concentration (mg SS/L)
Plant influent	200
Primary sludge	5,000
Secondary sludge	15,000
Aeration basin effluent	3,000

occupants use bulky or nonbiodegradable materials for anal cleansing (for example, corncobs, stones, newspaper); therefore, the solids accumulation rate is assumed to be 0.09 m^3/person/year. Allow a 0.5 m space between the ground surface and the top of the solids at the end of the design life, which is the point when the pit is filled.

9.31 (a) Estimate the volume of gas production produced from 1 metric ton of food waste and 1 ton of wastewater solids. (b) Based on the results from part (a), assuming the mass of food waste and wastewater solids generated in a community is the same, would you recommend that a municipality develop a program to collect and digest (with energy recovery) food waste or wastewater solids? Explain your answer based on potential energy production but also an implementation standpoint. Assume the methane production potential of wastewater solids is 120 m^3/metric ton, food waste has three times the methane production potential per volume of wastewater solids, and methane makes up 60 percent of the total gas produced from anaerobic digestion.

9.32 (a) If methane has an energy content of 39 MJ/m^3 and digester gas is approximately 60 percent methane, what volume of total digester gas must an anaerobic digester produce annually to provide potable water for a family of six for a year? (b) If the methane is provided from anaerobic digestion of food waste, how many pounds of food waste would a family have to generate per day to provide this energy to heat water (lb/day)? The United Nations states that the minimum potable water requirement to provide drinking water, sanitation, and hygiene is 20 L per person per day. Assume the water has an initial temperature of 25°C and you must raise the temperature to 100°C to produce potable water. The energy required to raise water up 1°C is equal to 4,200 J/L-°C and there is 39 MJ of energy per m^3 of methane. Assume that 1 metric ton of food waste produces 600 m^3 of total gas.

9.33 Assume 1 kg of volatile solids (VS) produces 0.5 m^3 of methane, but only half of the VS added to the digester will be broken down to gaseous compounds. If you want to produce 120 L of methane per day, how many pigs will you need to maintain to contribute waste to the digester? Assume a 60 kg pig produces 5 kg of manure per day with 10 percent being VS.

9.34 Waste must be held in the digester for a period of time for digestion to occur, but the length of time depends on temperature. Using the data in Table 9.29,

Table / 9.29

Data for Problem 9.34

Temperature (°C)	Retention Time (days) Minimum Recommended
10	55
20	20
30	8

calculate the appropriate digester capacity and dimensions (diameter and height) in m for each temperature listed assuming 20 L of input per day. Fix the digester dimensions at 1:5 diameter to height.

9.35 A 2.5 MGD wastewater treatment plant is currently running at 80 percent capacity during the annual maximum day servicing a city of 38,500 people with 26.7 miles of sewers. During the next 10 years, it is expected that new residential developments for 15,000 people along with 6.5 more miles of sewer will be built. The sewer is projected to have an I/I equal to 8,500 gpd/mile. (a) Project the maximum daily demand for the wastewater treatment plant after the new development is built. (b) Should the wastewater treatment plant capacity be increased?

9.36 A residential community with a population of 15,000 is planning to expand its wastewater treatment plant. In 20 years, the population is estimated to increase to 23,000 residents, and 1,000 students per day are expected to commute to proposed junior college from outside the area. A new industry will also move in and contribute an average flow of 350,000 gpd and maximum-day flow of 420,000 gpd. The present average daily flow into the plant is 1.45 million-gpd. The average inflow and infiltration (I/I) is 6 gallons/capita day and maximum day I/I is 42 gallons/capita day (rainy day). Residential per capita water use is expected to be 15 percent less in 20 years due to in-house water-saving strategies. The demand factor for domestic (residential use only) wastewater is determined to be 2.4 for the maximum day. Compute the future average and maximum-day flow rates. Hint: Compute the present per capita flow rates first; [total flow rate − I/I] divided by current population.

9.37 List five advantages to precipitating struvite from the nitrogen and phosphorus found primarily in urine discharged to municipal wastewater.

9.38 For the following influent water quality of a waste-water treatment plant that employs struvite recovery that is 70 percent effective in nutrients, determine which nutrient (N or P) in the influent is limiting for struvite precipitation, and why? The influent contains $[NH_4^+ - N](80.5\,mg\,N/L)$ and $PO_4^- - P](20.7\,mg\,P/L)$.

9.39 Assuming a homeowner installs a 60-gallon rain barrel at their home which has roof with 215 ft^2 of surface area, how much rainfall (in feet) could be stored? Assume that only 90 percent of the rain that falls on the roof enters the rain barrel due to leaky downspouts.

9.40 Assume that the roof area for a residential home is 12 ft × 30 ft. (a) If a green roof is placed on the home, what percentage of a 0.5 in. rain event will be stored on the roof if the growing medium has a water-holding capacity of 0.25? (b) What is the volume of water (in gallons) that is stored during this rain event?

9.41 What area (in ft^2) is needed for two bioreten-tion cells used to collect rainwater coming from a household roof? The roof has dimensions of 30 ft × 40 ft. It drains to two downspouts, each of which will be routed to a bioretention cell. Assume the soil surrounding the home is silty and the cell will be dug to a depth of 6 in.

9.42 A 1-acre paved parking lot measures 50 ft × 20 ft. What volume of bioretention cell is required (in ft^3) that can handle a first flush from the nonpermeable pavement of 0.5 in? Assume the soil porosity is 0.30.

9.43 Select a specific location on your campus that has a building and associated parking lot. Redesign this area, incorporating at least three low-impact development techniques. Besides thinking about management of stormwater, also consider the move-ment of people and vehicles, and the use of native plant species.

9.44 Size a rain garden for your current home, apartment, or dormitory to treat stormwater that originates from the roof.

9.45 The average cost of delivering an acre-foot of treated water in a water scarce region is $5,900 and the cost of delivering an acre-foot of reclaimed water in the same region is $6,400 (due to treatment and transport). Given that the following factors can be credited to reclaimed water, what is the range of cost of reclaimed water as compared to treated water as a percentage?

Increased potable water supply: $300–$1,000/acre-foot

Water supply reliability: $100–$140/acre-foot

Effluent disposal savings: $200–$2,000/acre-foot

Downstream effects: $400–$800/acre-foot

Energy conservation: $0–$240/acre-foot

9.46 Assume the energy requirement to treat wastewater using a mechanical process is 1 million kWh per million gallons of water treated. According to eGRID, the carbon dioxide equivalent emission rate is 1,324.79 lb CO_2e/MWh in Florida and 727.26 lb CO_2e/MWh in California. Estimate the carbon foot-print of treating 50 million gallons of wastewater in Florida and California. Ignore line losses in your estimate (you may have to go back to Chapter 2 to review carbon footprints and eGRID).

References

Asano, T., F. L. Burton, H. L. Leverenz, R. Tsuchihashi, and G. Tchobanoglous, 2007. *Water Reuse: Issues, Technologies, and Applications.* New York: McGraw-Hill.

Asano, T., and A. Bahri, 2010. *Global Challenges to Wastewater Reclamation and Reuse.* On the Water Front, Stockholm International Water Institute, 2010.

Barnard, J. L., 2006. Biological nutrient removal: where we have been, where we are going. *Proceedings of the 79th Annual Water Environment Federation Technical Exhibition and Conference,* Dallas.

Cairncross, S., and R. G. Feachem, 1993. *Environmental Health Engineering in the Tropics: An Introductory Text.* New York: John Wiley.

Congressional Report HR 91-1004. April 14, 1970. Phosphates in Detergents and the Eutrophication of America's Waters. Committee on Government Operations.

Daigger, G. T., A. Hodgkinson, and D. Evans, 2007. A sustainable near-potable quality water reclamation plant for municipal and industrial wastewater. *Proceedings of the 80th Annual Water Environment Federation Technical Exhibition and Conference.* San Diego.

Darn, S.M., R. Sodi, L. R. Ranganath, N. B. Roberts, and J. R. Duffield, 2006. Experimental and computer modeling speciation studies of the effect of pH and phosphate on the precipitation of calcium and magnesium salts in urine. *Clinical Chemistry and Laboratory Medicine,* 44: 185–191.

Environmental Protection Agency (EPA). 1999. *Storm Water Technology Fact Sheet: Porous Pavement.* EPA 832-F-99-023. Washington, D.C.: Environmental Protection Agency, Office of Water.

Environmental Protection Agency (EPA), 2000a. *Biosolids Technology Fact Sheet/Belt Filter Press,* Office of Water, EPA 832-F-00-057.

Environmental Protection Agency (EPA), 2000b. *Manual: Constructed Wetlands Treatment of Municipal Wastewaters.* Cincinnati: Office of Research and Development, EPA 625/R-99/010.

Environmental Protection Agency (EPA), 2001. *The "Living Machine" Wastewater Treatment Technology: An Evaluation of Performance and System Cost,* EPA 832-R-01-004.

Environmental Protection Agency (EPA), 2004. *Report to Congress: Impacts and Control of CSOs and SSOs.* Washington, D.C., EPA-833-R-04-001.

Environmental Protection Agency (EPA), 2012. *Guidelines for Water Reuse,* EPA-600/R-12/618, 643 pp.

Fry, L. M., J. R. Mihelcic, and D. W. Watkins, 2008. Water and non-water-related challenges of achieving global sanitation coverage. *Environmental Science & Technology,* 42(4): 4298–4304.

Hammer, M. J., and M. J. Hammer Jr., 1996. *Water and Wastewater Technology,* 3rd ed. Englewood Cliffs: Prentice Hall.

Hammond, A.L., 1971. Phosphate replacements: problems with the washday miracle. *Science,* 172: 361–363.

Jenkins, D., 2007. From TSS to MBTs and beyond: a personal view of biological wastewater treatment process population dynamics.

Proceedings of the 80th Annual Water Environment Federation Technical Exhibition and Conference, San Diego.

Johnston, A. E., and I. R. Richards, 2003. Effectiveness of the water-insoluble component of triple superphosphate for yield and phosphorus uptake by plants. *Journal of Agricultural Science,* 140: 267–274.

Kowal, N. E., 1985. *Health Effects of Land Application of Municipal Sludge.* Cincinnati: Health Effects Research Laboratory, EPA, EPA/600/1-85/015.

Larsen, T.A., and W. Gujer, 1996. Separate management of anthropogenic nutrient solutions (human urine). *Water Science Technology,* 34(3–4): 87–94.

Lienert, J., T. A. Larsen, 2010. High acceptance of urine source separation in seven European countries: a review. *Environmental Science & Technology,* 44: 556–566.

Mara, D. D., 2004. *Domestic Wastewater Treatment in Developing Countries,* 1st ed. London, UK: Earthscan/James & James.

Maurer, M., W. Pronk, and T. A. Larsen, 2006. Review: treatment processes for source-separated urine. *Water Research,* 40(17): 3151–3166.

McDannel, M., and E. Wheless, 2007. The power of digester gas: a technology review from micro to megawatts. *Proceedings of the 80th Annual Water Environment Federation Technical Exhibition and Conference,* San Diego.

Mihelcic, J. R., 1999. *Fundamentals of Environmental Engineering.* New York: John Wiley & Sons.

Mihelcic, J.R., L. M. Fry, R. Shaw, 2011. Global potential of phosphorus recovery from human urine and feces. *Chemosphere,* 84(6): 832–839.

Miller, T., and J. Esler. 2013. "What every operator should know about secondary clarification." *Water Environment & Technology,* June, 62–64.

Muga, H. E., and J. R. Mihelcic, 2008. Sustainability of wastewater treatment technologies. *Journal of Environmental Management,* 88(3): 437–447.

Oakley, S. M., 2005. *Lagunas de Estabilización en Honduras: Manual de Diseño, Construcción, Operación y Mantenimiento, Monitoreo y Sostenibilidad.* U.S. Agency for International Development—Honduras (USAID—Honduras), Red Regional de Agua y Saneamiento de Centroamérica (RRAS-CA), Fondo Hondureño de Inversión Social (FHIS), Tegucigalpa, Honduras. (*Design and Operations Manual for Wastewater Stabilization Ponds in Honduras,* published in Spanish).

Portland Cement Association (PCA). 2004. "Pervious Concrete Mixtures and Properties." *Concrete Technology Today* 25(3).

Rosso, D., and M. K. Stenstrom, 2005. Comparative economic analysis of the impacts of mean cell retention time and denitrification on aeration systems. *Water Research,* 39: 3773–3780.

Rosso, D., and M. K. Stenstrom, 2006. Surfactant effects on α–factors in aeration systems. *Water Research,* 40: 1397–1404.

Saude, E.J., D. Adamko, B. H. Rowe, T. Marrie, B. D. Sykes, 2007. Variation of metabolites in normal human urine. *Metabolomics*, 3.4: 439–451.

Tchobanoglous, G., F. L. Burton, and H. D. Stensel, 2003. *Wastewater Engineering*. Boston: Metcalf & Eddy/McGraw-Hill.

Udert, K. M., T. A. Larsen, and W. Gujer, 2003. Estimating the precipitation in urine-collecting systems. *Water Research*, 37.11: 2667–2677.

Walski, T. M., T. E. Barnard, E. Harold, L. B. Merritt, N. Walker, and B. E. Whitman, 2004. *Wastewater Collection System Modeling and Design*. Waterbury: Haestad Press.

Ward, A. S. 2007. "A Review of the Practice of Low Impact Development: Bioretention Design, Analysis, and Lifecycle Assessment." (M.S. report, Civil & Environmental Engineering, Michigan Technological University).

Wehner, J. F., and R. H. Wilhelm, 1956. Boundary conditions of flow reactor. *Chemical Engineering Science*, 6(2): 89–93.

Wisconsin Department of Natural Resources (WDNR). 2003. DNR Publication PUB-WT-776. Available from the University of Wisconsin Extension, http://learningstore.uwex.edu/Assets/pdfs/GWQ037.pdf, accessed September 14, 2013.

World Health Organization (WHO) and United Nations Children Fund Joint Monitoring Programme for Water Supply and Sanitation (JMP), 2010. Progress on Drinking Water and Sanitation: Special Focus on Sanitation. UNICEF, New York and WHO, Geneva.

Wright Wendel, H. E., J. A. Downs, and J. R. Mihelcic. 2011. "Assessing equitable access to urban green space: the role of engineered water infrastructure," *Environmental Science & Technology*, 45(16): 728–6734.

chapter/Ten Solid-Waste Management

Mark W. Milke and
James R. Mihelcic

In this chapter, readers will learn about management of municipal solid wastes. The types of solid waste and their quantities, composition, and physical properties are first described. The chapter addresses the storage, collection, transport, treatment, and disposal of solid wastes, including recycling and materials recovery, composting, incineration, and landfilling. Methods are reviewed to estimate the production of greenhouse gas emissions from landfills, including the public benefit the landfill gas can provide. The chapter concludes with an introduction to public consultation, public policy, and cost estimation. Emphasis is placed on the application of basic mass balance concepts to solid-waste problems, with a mixture of quantitative problems and discussion of broader management topics.

© Stephanie DeLay/iStockphoto

Learning Objectives

1. Describe the key components of a solid-waste management system.
2. Integrate principles of the pollution prevention hierarchy into a solid-waste management system.
3. Identify the objectives of solid-waste management.
4. Describe key relevant U.S. legislation related to solid wastes.
5. Distinguish between municipal solid waste and other solid wastes.
6. Calculate dry and wet weight generation rates for specific solid-waste components from available data.
7. Discuss the differences between solid-waste management in developing and developed countries.
8. Explain the issues associated with design and operation of successful solid-waste subsystems (collection, transfer stations, materials recovery facilities, composting facilities, waste-to-energy facilities, and landfills).
9. Solve mixing problems to determine an appropriate C/N ratio for composting.
10. Calculate oxygen requirements for aerobic biological or thermal treatment processes, as well as methane (a greenhouse gas) generation rates from landfills, using stoichiometry and mass composition data.
11. Explain the concerns with landfill gas and leachate and how they are addressed.
12. Discuss the five pathways by which landfills generate greenhouse gases and estimate greenhouse gas emissions from a landfill.
13. Calculate the size of an area landfill based on daily cell construction.

PAPER   PLASTIC   TRASH

14. Empathize with community stakeholders related to siting a landfill or waste-to-energy facility, and carefully decide how to reach consensus.
15. Identify key public consultation methods and their strengths and weaknesses.
16. Identify key public policy options for solid-waste management and summarize their strengths and weaknesses.
17. Estimate costs for different-sized landfills, using economy-of-scale factors.

10.1 Introduction

Solid wastes include paper and plastic generated at home, ash produced by industry, cafeteria food wastes, leaves and cut grass from parks, hospital medical wastes, and demolition debris from a construction site. These materials are considered a **waste** when owners and society believe they no longer have value.

Solid-waste management varies greatly between cultures and countries and has evolved over time. The components of solid-waste management are depicted in Figure 10.1. Solid-waste management requires an understanding of **waste generation**, **storage**, **collection**, **transport**, **processing**, and **disposal**. The end points in Figure 10.1 are recycled materials, compost, and energy recovery; these end points will become more common as society adopts more sustainable waste management practices that consider systems thinking. Referring back to previous chapters, remember that *waste* is a human-derived word; thus, society needs to identify ways to minimize the amount of waste that is generated, transported, processed, and disposed.

Solid wastes differ from liquid or gaseous wastes because they cannot be pumped or flow like fluids. However, solid wastes can be placed into solid forms (including soils) and thus can be contained more easily. These differences have led to different approaches for managing solid wastes than the approaches described in previous chapters for liquid and gaseous waste streams.

Proper management of solid wastes has five main objectives:

1. Follow the **pollution prevention hierarchy**, which prefers source reduction and recycling over treatment and disposal.

2. Protect public health.

3. Protect the environment (including biodiversity) and view the waste material as a resource.

Zero Waste as a Design Principle for the 21st Century
http://www.sierraclub.org/committees/zerowaste/
http://www.zerowaste.co.nz

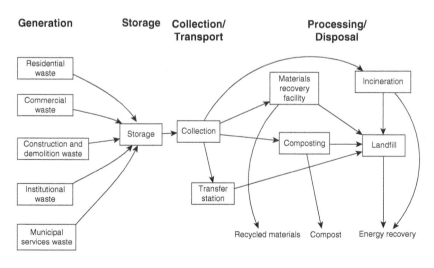

Figure / 10.1 Overview of the Solid-Waste Management System   The system consists of storage, collection and transport, processing, and disposal. Materials recovery, composting, recycling, and energy recovery are important in the processing and disposal stage.

4. Address social concerns (equity, environmental justice, aesthetics, risk, public preferences, recycling, renewable energy).

5. Minimize economic, social, and environmental costs.

It is common practice for individual communities to place varying weights on these objectives. Therefore, many different systems exist to manage solid waste.

Preindustrial communities arranged to have solid waste collected and disposed of in central locations termed *middens*. Some middens can be found today. For example, some coastal areas along Florida's Gulf Coast are home to middens derived from eating shellfish that formed small islands that can now be explored. As population, urbanization, and consumption increased, solid waste became more of a problem. Some wastes were left in streets and alleys, feeding dogs, pigs, and rats. Other wastes were hauled outside of the city and dumped in large mounds. The increasing population of cities and the large death rates experienced in many parts of the world due to plague, cholera, and other infectious diseases led to a demand to clean cities of solid waste. Organized collection, treatment, and disposal of solid waste began in the late 1800s and were closely tied to the objective of improving public health and sanitation. Early approaches to solid-waste management included feeding food wastes to farm animals, burning waste to heat city water, and creating solid-waste **dumps** (often in wetlands) to reclaim land.

During the 1900s, increased industrialization resulted in the production of different wastes—more *hazardous solid wastes*. At the same time, increasing urban populations, along with increasing affluence and consumption, increased the amount of solid waste produced. Nonrecovered solid wastes were placed into engineered facilities, which were called engineered or sanitary **landfills**. The improvement from dumps to engineered sanitary landfills has been a gradual process over the past 100 years that has led to technically advanced and sophisticated systems for protection of human health and the environment.

RCRA

http://www.epa.gov/lawsregs/laws/rcra.html

The major piece of U.S. legislation affecting solid-waste management is the **Resource Conservation and Recovery Act (RCRA)** of 1976. RCRA also mandated tracking and rigorous management of hazardous wastes. It also led to regulations to improve the design of landfills and reduce their risk. The *Pollution Prevention Act* then followed.

In the past 20 years, there has been an increased emphasis on solid-waste management strategies that prefer source **reduction** and **reuse, recycling** and composting, and energy recovery, over treatment and disposal as part of the **pollution prevention** hierarchy discussed in Chapter 6. Numerous local, state, and national regulations have been enacted to develop and promote pollution prevention initiatives. These initiatives have been motivated by a desire to further reduce adverse social and environmental impacts and to conserve natural resources (including water and energy). At the same time, there has been an increased recognition of the importance of wider systems thinking that includes life cycle assessment in evaluating solid-waste management options. Accordingly, recent trends place greater importance on an integrated or systems-based approach to solid-waste management.

10.2 Solid-Waste Characterization

Solid wastes can be characterized by their source, original use (e.g., as glass or plastic), hazard, or underlying physical or chemical composition. Wastes that spread disease are termed **putrescible** wastes. They can spread disease directly (as in the case of dirty diapers) or indirectly by providing a food source for disease vectors such as insects (flies) or animals (rats, dogs, birds).

10.2.1 SOURCES OF SOLID WASTE

The sources of solid waste and typical constituents are identified in Table 10.1. Some solid wastes (e.g., mining wastes and most agricultural and industrial wastes) are managed by the waste generator. Smaller sources are usually managed jointly under one integrated

Table / 10.1

Sources of Solid Waste and Typical Percentage That Makes Up Municipal Solid Waste

Source	Examples	Comments	Typical Percentage of MSW
Residential	Detached homes, apartments	Food wastes, yard/garden wastes, paper, plastic, glass, metal, household hazardous wastes.	30–50%
Commercial	Stores, restaurants, office buildings, motels, auto repair shops, small businesses	Same as above, but more variable from source to source. Small quantities of specific hazardous wastes.	30–50%
Institutional	Schools, hospitals, prisons, military bases, nursing homes	Same as above; variable composition between sources.	2–5%
Construction and demolition	Building construction or demolition sites, road construction sites	Concrete, metal, wood, asphalt, wallboard, and dirt predominate. Some hazardous wastes possible.	5–20%
Municipal services	Cleaning of streets, parks, and beaches; water and wastewater treatment grit and biosolids; leaf collection; disposal of abandoned cars and dead animals	Waste sources vary among municipalities.	1–10%
Industrial	Light and heavy manufacturing, large food-processing plants, power plants, chemical plants	Can produce large quantities of relatively homogeneous wastes. Can include ashes, sands, paper mill sludge, fruit pits, tank sludge.	Not MSW
Agricultural	Cropping farms, dairies, feedlots, orchards	Spoiled food wastes, manures, unused plant matter (e.g., straw), hazardous chemicals.	Not MSW
Mining	Coal mining, uranium mining, metal mining, oil/gas exploration	Can produce vast amounts of solid waste needing specialized management.	Not MSW

SOURCE: Tchobanoglous et al., 1993.

system. The solid wastes jointly managed by a municipality are called **municipal solid waste (MSW)**. The focus of this chapter is on the management of MSW, though many of the principles and processes discussed are also relevant to the management of industrial, agricultural, and mining wastes.

10.2.2   QUANTITIES OF MUNICIPAL SOLID WASTE

Table 10.2 provides the quantities of MSW generated and managed in the United States from 1960 to 2010. Note how generation rates have increased dramatically in just over 50 years. Though recycling rates increased in the past 30 years, over the same period, the rate of waste production steadily increased as well.

The 2010 generation rate of 0.74 Mg per person per year excludes construction and demolition debris and biosolids from wastewater treatment plants (which, as listed in Table 10.1, is often included in the solid waste managed by a municipality). *As a rough rule of thumb, an overall generation rate for MSW in most industrialized countries is now approximately 1 Mg per person per year.* Because residential solid waste makes up 30–50 percent of the MSW, this converts to a *residential waste generation rate of approximately 1 kg per person per day.*

The amount of generated MSW can vary within a year, between urban and rural areas, geographically, with income, and among countries. The fraction managed by recycling, composting, incineration, or landfilling varies even more according to local conditions. It is common to read reports that a particular city or country produces more waste than others. However, these reports need to be read cautiously, because different authorities count different waste streams, and some define *produced* waste as waste that remains after recycling and composting. One key is to understand the differences between *generation rates* and *discard rates* (the difference being the part of the waste stream that is reused or recycled).

Class Discussion

EPA reports the amount of household solid waste in the United States typically increases by 25 percent between Thanksgiving and New Year's Day, from 4 million tons to 5 million tons. What specific technological, policy, and human behavioral changes can a household and community implement to reduce the volume of waste generated during the holiday season?

Table / 10.2

Quantities of Municipal Solid Waste in the United States over Time

	Mg per Person per Year[1]					
	1960	1970	1980	1990	2000	2010
Generation	0.44	0.54	0.61	0.76	0.78	0.74
Recycling	0.03	0.04	0.06	0.11	0.17	0.19
Composting	Negligible	Negligible	Negligible	0.01	0.05	0.06
Incineration	0.00	0.00	0.01	0.11	0.11	0.09
Landfill[2]	0.42	0.50	0.54	0.53	0.45	0.40

[1]These quantities exclude construction and demolition debris and wastewater treatment plant biosolids.
[2]This includes small quantities of waste incinerated without energy recovery and does not include wastes produced during recycling, composting, and incineration (e.g., ashes).
SOURCE: EPA, 2011.

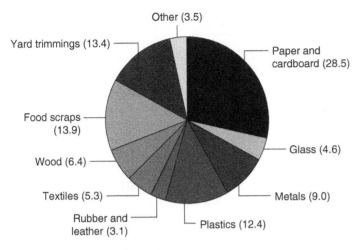

Figure / 10.2 Percentage of Various Materials (on a Mass Basis) That Compose U.S. Municipal Solid Waste, 2010.

(Data from EPA (2011)).

10.2.3 MATERIALS IN MUNICIPAL SOLID WASTE

Figure 10.2 shows the estimated percentage breakdown of the various materials found in MSW at the point of generation in 2010. This information can be used to assess which waste streams can be targeted for composting or materials recovery programs. For example, Figure 10.2 shows that a high percentage of the overall waste is yard waste (13.4 percent) and paper/cardboard (28.5 percent). The types of materials in MSW have also changed over time. Figure 10.3 shows this change over the past 50 years, especially for materials associated with packaging (plastic and paper/cardboard).

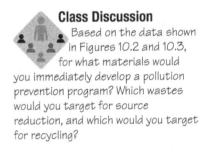

Class Discussion

Based on the data shown in Figures 10.2 and 10.3, for what materials would you immediately develop a pollution prevention program? Which wastes would you target for source reduction, and which would you target for recycling?

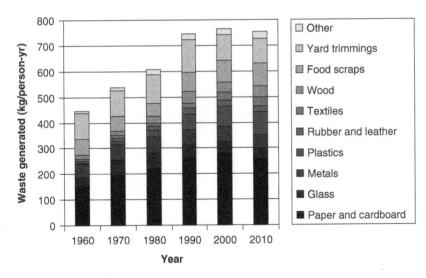

Figure / 10.3 Rate of Generation of Various Materials in U.S. Municipal Solid Waste, 1960–2010.

(Data from EPA, (2011)).

The solid waste generated in developing countries is much less (0.15–0.3 Mg per person per year) than in developed countries (0.7–1.5 Mg per person per year). The composition of the waste also differs throughout the world, as given in Table 10.3. Key differences in waste composition in developing countries include the higher fraction of organic putrescibles and the lower fraction of manufactured products, such as paper, metals, and glass.

Higher-income households tend to generate more inorganic material from packaging waste, whereas lower-income households produce a greater fraction of more organic material through preparing food from base ingredients. However, some high-income households in the developing world may generate the same amount of organic material because they prepare more fresh, unpackaged food. These differences tend to become reduced as countries develop their economies.

Combined with fewer financial resources and skills, the differences in solid-waste production mean that unique solid-waste management practices are required in locations with developing economies.

Table / 10.3

Solid-Waste Composition for Five Cities of the World

Location	Food Waste	Paper	Metals	Glass	Plastic, Rubber, Leather	Textiles	Ceramics, Dust, Ash, Stones	Generation (Mg/person-year)
Bangalore, India	75.2	1.5	0.1	0.2	0.9	3.1	19	0.146
Manila, Philippines	45.5	14.5	4.9	2.7	8.6	1.3	27.5	0.146
Asuncion, Paraguay	60.8	12.2	2.3	4.6	4.4	2.5	13.2	0.168
Mexico City, Mexico	59.8*	11.9	1.1	3.3	3.5	0.4	20	0.248
Bogota, Colombia	55.4*	18.3	1.6	4.6	16	3.8	0.3	0.270

*Includes small amounts of wood, hay, and straw.
SOURCE: Diaz et al., 2003.

Typically, solid-waste **generation rates** (in kg of a specific material generated per day or year) are determined by collecting data for the total waste generated and the percentage of a specific material in the solid waste. It can be misleading to compare percentage composition data between two different months or two different locations, because differences are likely in the total waste generation rate. Instead, one should compare waste generation rates on the basis of kg *per year* for each material of interest. This point is demonstrated in Example 10.1.

10.2.4   COLLECTION OF SOLID-WASTE CHARACTERIZATION DATA

The characterization of MSW is a complex task. Because solid waste varies greatly in composition and quantity within a region and over time, there will always be large uncertainty in estimates of solid-waste composition. The generation of good data assists with making

example/10.1 Calculating Solid-Waste Generation Rates

The solid-waste composition and the quantity of solid waste generated from two cities located in the developing world are as follows:

Component	City 1	City 2
Food waste (%)	47.0	65.5
Paper and cardboard (%)	6.3	6.5
Ash (%)	36.0	10.2
Other (%)	10.7	17.8
Waste generation rate (kg/person-day)	0.38	0.28

Which city generates more ash on a per capita basis? Which city generates more nonash waste on a per capita basis?

solution

To determine which city produces more ash on a per capita basis, multiply the overall waste generation rate by the percentage of the component of interest (in this case ash) in the overall waste stream:

$$\text{City 1:} \quad 0.38 \text{ kg/person-day} \times \frac{36.0}{100} = 0.14 \text{ kg ash waste/person-day}$$

$$\text{City 2:} \quad 0.28 \text{ kg/person-day} \times \frac{10.2}{100} = 0.03 \text{ kg ash waste/person-day}$$

City 1 produces significantly more ash on a per person basis.

To find the nonash waste generation rate, first determine the nonash waste percentage by subtraction and then multiply this value by the total waste generation rate:

$$\text{City 1:} \quad 0.38 \text{ kg/person-day} \times \frac{(100 - 36.0)}{100} = 0.24 \text{ kg nonash waste/person-day}$$

$$\text{City 2:} \quad 0.28 \text{ kg/person-day} \times \frac{(100 - 10.2)}{100} = 0.25 \text{ kg nonash waste/person-day}$$

The nonash waste generation rates are roughly equal.

The greatest difference is that one city generates much more ash than the other. One possible explanation is that it is more common for people in City 1 to burn solid fuels such as wood or coal for cooking and heat. Another possible explanation is that City 2 collects ash waste along with other solid wastes, while City 1 collects only some of the ash waste produced. Other climatic and social–economic reasons are plausible.

In any case, the example shows how waste characterization is an important aspect of a solid-waste management program, because waste generation and composition can differ within a country and around the world.

appropriate management decisions, but reduction of uncertainty in estimated composition can be very expensive.

Three methods are commonly used to characterize solid waste:

1. *Literature review.* This method relies on using past data to characterize the makeup of solid waste. It has some limitations: Definitions used to collect the data can be unclear. Solid-waste generation patterns vary with time and space. Most historical data lack an analysis of the data's uncertainty. And most historical data lack a joint estimate of total waste and percent composition.

2. *Input–output analysis.* This method relies on using data on the consumption of materials to estimate the generation of waste. The EPA data used for Figures 10.2 and 10.3 are examples of this method of analysis. This approach also has weaknesses: It requires clear boundaries so that imports and exports outside the boundary (e.g., outside the United States) can be accounted for. It requires assumptions about storage and consumptive use (e.g., eating) of purchased goods. It requires assumptions about wastes generated without an economic record (e.g., yard wastes).

3. *Sampling surveys.* This method relies on collection of actual data and statistical methods to estimate averages and uncertainty. Many locations require periodic surveys, and a number of methods are available to aid in sampling surveys (ASTM, 1992; New Zealand Ministry for the Environment, 2002). This approach has three weaknesses: Large variability means that a large number of samples are needed, which in turn drives up costs. High variability from one season to another can mean that periodic surveys must be conducted for a year or two before useful data are obtained. Finally, because uncertainty increases as the percent decreases, the method is not effective for relatively uncommon components.

10.2.5 PHYSICAL/CHEMICAL CHARACTERIZATION OF WASTE

The selection of a particular management option for solid wastes depends on specific physical and chemical characteristics of the waste. Data on waste generation rates are provided in terms of mass units (mass/capita-time) and not volume units, because the density of waste can vary greatly among wastes or over time. Estimates of **waste density** are important so that space requirements can be estimated for the waste at various stages (e.g., collection, transport, disposal) within the waste management system.

Table 10.4 provides typical densities for different types of solid waste during different stages of solid-waste management. The large increases in density that come with compaction of waste in a truck or landfill, and from baling recovered materials, are important in evaluating the economics of related solid-waste management options.

Collecting combustible waste for waste-to-energy production in Japan

(Photo courtesy of James R. Mihelcic).

Table / 10.4

Densities of Various Municipal Solid Wastes and Recovered Materials

	Density Range (kg/m³)		Density Range (kg/m³)
Mixed MSW		Plastic containers	32–48
Loose	90–180	Miscellaneous paper	48–64
Loose (developing countries)	250–600	Newspaper	80–110
In compactor truck	300–420	Garden waste	64–80
After dumping from compactor truck	210–240	Rubber	210–260
In landfill (initial)	480–770	Glass bottles	190–300
In landfill (with overburden)	700–1,100	Food waste	350–400
Shredded	120–240	Tin cans	64–80
Baled	480–710	*Recovered materials (densified)*	
Recovered materials (loose)		Baled aluminum cans	190–290
Powdered refuse-derived fuel	420–440	Cubed ferrous cans	1,040–1,500
Densified refuse-derived fuel	480–640	Baled cardboard	350–510
Aluminum scrap	220–260	Baled newspaper	370–530
Ferrous scrap	370–420	Baled high-grade paper	320–460
Cardboard	16–32	Baled PET plastic	210–300
Aluminum cans	32–48	Baled HDPE plastic	270–380

SOURCE: Diaz et al., 2003.

The **moisture content** of solid waste is determined as follows:

$$\text{moisture content} = \frac{\text{mass of moisture}}{\text{total mass of waste}} \quad (10.1)$$

This definition is different from the dry-weight basis, the definition commonly used in geotechnical engineering applications. However, it is the same as the wet weight basis, the definition commonly used in the soil sciences. The dry weight can be found as follows:

$$\text{dry mass} = \text{total mass of waste} \times \frac{100 - \text{moisture content (in \%)}}{100}$$

$$(10.2)$$

Table / 10.5

Common Physical/Chemical Characteristics of Solid-Waste Components

	Moisture (% by wet mass)	Energy Value as Received (MJ/kg)	Energy Value after Drying (MJ/kg)	Carbon (% by dry mass)	Hydrogen (% by dry mass)	Oxygen (% by dry mass)	Nitrogen (% by dry mass)	Sulfur (% by dry mass)	Ash (% by dry mass)
Food wastes	70	4.2	13.9	48	6.4	37.6	2.6	0.4	5
Magazines	4.1	12.2	12.7	32.9	5	38.6	0.1	0.1	23.3
Paper (mixed)	10	15.8	17.6	43.4	5.8	44.3	0.3	0.2	6
Plastics (mixed)	0.2	32.7	33.4	60	7.2	22.8	<0.1	<0.1	10
Textiles	10	18.5	20.5	48	6.4	40	2.2	0.2	3.2
Rubber	1.2	25.3	25.6	69.7	8.7	<0.1	<0.1	1.6	20
Leather	10	17.4	18.7	60	8	11.6	10	0.4	10
Yard wastes	60	6.0	15.1	46	6	38	3.4	0.3	6.3
Wood (mixed)	20	15.4	19.3	49.6	6	42.7	0.2	<0.1	1.5
Glass	2	0.2	0.2	0.5	0.1	0.4	<0.1	<0.1	99
Metals	4	0.6	0.7	4.5	0.6	4.3	<0.1	<0.1	90.6

SOURCE: Some data from Tchobanoglous et al., 1993.

Different amounts of moisture are associated with different solid wastes, and data are typically collected on moist solid waste *as received*. This value is then converted to dry waste mass before further calculations are made. Table 10.5 provides typical values for the moisture content of different solid-waste components, along with information on the energy content and elemental chemical composition.

The design of systems for **energy recovery** requires data on the energy content of the waste. Similarly, evaluation of many solid-waste treatment systems will require information on the elemental composition of wastes. The high-energy components of MSW are plastic and paper. Food waste and yard waste have high moisture content, which limits the energy they release when burned.

The moisture contents provided in Table 10.5 are typical values. These values can vary greatly depending on the specific composition of the waste component or local factors such as the weather. For example, yard waste that is generated during the summer could be predominately grass, and after collection in wet weather, the yard waste could have a moisture content of 80 percent. In late autumn, the yard waste could be predominately leaves, and after collection in dry weather, it could have a moisture content as low as 20 percent. The varying moisture content of a waste material can affect evaluation of the overall composition of a waste stream (as shown in Example 10.2) and estimates of the total waste. It is therefore preferable to work with the dry mass of waste during intermediate calculations.

example/10.2 Adjusting for Varying Moisture Content of Waste

Food wastes can make up a significant fraction of a municipal solid-waste stream. In order to design a food waste collection system, a municipality is interested in determining the amount of food waste generated. The solid waste was analyzed on one particular day when it was found that the total annual waste generation was 700 kg/person-year. The study also showed that the food waste was 20 percent of the total (wet) mass of waste generated (see the following table). Thus, the rate of food waste generation is 140 kg/person-year.

The study did not, however, measure the moisture content of the waste. Assume that the waste percentage data were collected on a dry summer day with a low moisture content. Assuming a typical moisture content, estimate the food waste generation rate and the total waste generation rate.

	Percentage of Total Mass	Low Moisture Content (%)	Typical Moisture Content (%)
Food waste	20	50	70
Paper waste	30	3	10
Yard waste	30	20	60
Other waste	20	2	5

solution

First determine the dry generation rate of the various waste streams on a day when the waste has a low moisture content. This value can be converted to the typical dry generation rate (which is probably more applicable to the whole year). The total generation rate of each component of the solid-waste stream can be calculated and tabulated as given in Table 10.6. To estimate conditions on a day with a typical moisture content, determine the dry generation rate by using an assumed low moisture content (provided in the preceding table). Then add the typical moisture back to the dry generation rate.

Equations 10.1 and 10.2 can be combined to solve for the total mass as a function of dry mass and moisture content:

$$\text{total mass} = \frac{\text{dry mass}}{\dfrac{100 - \text{moisture content (in \%)}}{100}}$$

The results of our analysis are as given in Table 10.6.

The expected typical total generation rate of solid waste with typical moisture added is 1,024.1 kg/person-year, and the wet mass generation rate of food waste is 233.3 kg/person-year. These values are significantly higher than the sampled values (provided in the first column of Table 10.6), because the sampled values were determined on a relatively dry summer day.

Table / 10.6

Results for Example 10.2

	Sampled Total Generation Rate (kg/person-year)	Assumed Moisture Content (%)	Dry Generation Rate (kg/person-year)	Typical Moisture Content (%)	Typical Total Generation Rate (kg/person-year)
Food waste	140	50	70	70	233.3
Paper waste	210	3	203.7	10	226.3
Yard waste	210	20	168	60	420.0
Other waste	140	2	137.2	5	144.4
Total	700		578.9		1,024.1

Table / 10.7

Methods to Classify a Solid Waste as Hazardous

Characteristic of Waste	Question Related to Characteristic
Ignitable	Can the waste create a fire (e.g., waste solvents)?
Corrosive	Is the waste very acidic or basic and so able to corrode storage containers (e.g., battery acids)?
Reactive	Can the waste participate in rapid chemical reactions leading to explosions, toxic fumes, or excessive heat (e.g., lithium that can react with water explosively, explosives, cyanide sludge, strong oxidizing agents)?
Toxic	Can the waste cause internal damage to a person or organism (e.g., poisons causing death or blindness, carcinogens)?
Radioactive	Can the waste release subatomic particles that can cause toxic effects (e.g., some medical and laboratory wastes, wastes associated with nuclear energy production)?
Infectious	Can the waste lead to the transmission of disease (e.g., used syringes, hospital medical wastes)?

SOURCE: Environmental Protection Agency, http://www.epa.gov/osw/hazard/wastetypes/index.htm.

10.2.6 HAZARDOUS-WASTE CHARACTERIZATION

Wastes are considered **hazardous waste** when they pose a direct threat to human health or the environment. Wastes can be classified as hazardous by one of six characteristics: ignitable, corrosive, reactive, toxic, radioactive, or infectious. Table 10.7 provides more detail on each characteristic.

Most countries have laws and regulations related to hazardous wastes and provide a definition that classifies particular wastes as legally hazardous or not. In the United States, the relevant regulations are promulgated under RCRA and focus primarily on large generators of relatively homogeneous wastes. The management of hazardous wastes is a specialized topic that is beyond the scope of this book. We have described in Chapter 6 how green chemistry and green engineering can be deployed to reduce or eliminate the production and use of hazardous chemicals (and their associated risk and wastes).

Small quantities of wastes that exhibit the characteristics of a hazardous waste are often not legally considered a hazardous waste and are managed (and thus disposed of) along with municipal waste. Table 10.8 lists common household hazardous products. The storage and use of these products in the home is a concern (especially because of the poor indoor air environments). Many municipalities provide household hazardous-waste programs that educate consumers on using green

Is It Nonhazardous or Hazardous Waste?

http://www.epa.gov/osw/

Class Discussion

What green alternatives could you use at your house or apartment to eliminate use of household hazardous wastes listed in Table 10.8? How would you effectively communicate this information to the public as an engineer employed by the local municipality?

Table / 10.8

Common Hazardous Products Found in Households

Product	Concern
Household Cleaning Products	
Oven cleaners	Corrosive
Drain cleaners	Corrosive
Pool acids, chlorine	Corrosive
Chlorine bleach	Corrosive
Automotive products	
Motor oil	Ignitable
Antifreeze	Toxic
Car batteries	Corrosive
Transmission and brake fluid	Ignitable
Lawn and garden products	
Herbicides, insecticides	Toxic
Wood preservatives	Toxic
Indoor pesticides	
Flea repellents and shampoos	Toxic
Moth repellents	Toxic
Mouse and rat poisons	Toxic
Home maintenance/hobby supplies	
Oil or enamel-based paints	Flammable
Paint solvents and thinners	Flammable

SOURCE: Adapted from Environmental Protection Agency (http://www.epa.gov/wastes/conserve/materials/hhw.htm) and Tchobanoglous et al., *Integrated Solid Waste Management*, 1993, copyright The McGraw-Hill Companies.

alternatives (see www.care2.com/greenliving/) while also collecting household hazardous waste.

Readers should consider the impact of their purchases on the environment. To varying degrees, household hazardous-waste products will increase environmental impacts in ecosystems and at wastewater treatment plants, materials recovery facilities (MRFs), compost facilities, landfills, and waste-to-energy facilities. In addition, the use of these items in the home or workplace presents an environmental risk to humans who occupy buildings. Specifying and using degradable and nontoxic cleaning products and equipment reduces emissions and operation and maintenance costs, and improves indoor air quality and occupant productivity.

Learn about Pharmaceutical Collection

http://www.epa.gov/wastes/hazard/generation/pharmaceuticals/collection.htm

Pollution Prevention Act

http://www.epa.gov/oppt/p2 home/pubs/laws.htm

Proper solid-waste management is important in all parts of the world.* The components of solid-waste management can look quite different, though. Here we explore solid-waste management in a neighborhood of the city of Sikasso (Mali), which has a population of approximately 150,000. Figure 10.4 provides an overview of the components.

On-Site Storage

Women sweep their small shops and homes every morning and evening and place the waste in a trash can or corner. Residents may purchase a trash can constructed of halved metal drums, which have holes along the walls. Residential solid waste consists mainly of paper, organics (dust, leaves), and some plastic. There is also significant solid waste generated at markets, especially organics and cardboard.

Collection

Typically, solid-waste collection begins with privately owned organizations that collect solid waste.

Transport and Transfer

If your household has an account with the collection organization, a man with a donkey cart will come to your concession/house each day, empty your trash can, and take the waste to a collection area on the periphery of the city. They will not empty your trash without a trash can. Those without trash cans and agreements will ask children to carry trash to the collection areas. In some cases, the mayor or a private collection organization owns a large truck and tractor and can transfer large piles of trash from central areas to a collection area. In Sikasso, there are 10 collection areas on the periphery.

Figure / 10.4 Solid-Waste Management in the City of Sikasso, Mali (Population Approximately 150,000).

(Photo courtesy of Brooke T. Ahrens and Jennifer R. McConville).

Market

Household

Processing

Scavengers, especially children, living near the collection areas may pick through the solid waste before the collection team arrives to take the trash outside the city. Some residents may burn the trash occasionally if the piles become too large.

Disposal

Later on, solid waste is burned or carried from the periphery to rural areas. It may be spread out over fallow fields, like compost. Currently, there are no official landfills for disposal, but plans for construction are pending.

Source Reduction, Recycling, Composting

Much of the refuse is organic. People reuse cans, bags, and other plastics and metals as long as they can. These may be used for containers in the reselling of products (e.g., juice, spices) or recycled into toys and art (e.g., metal milk cans hammered into toy trucks).

Disease Vectors

Rats, mice, maggots, cockroaches, and mosquitoes may be found in collection areas. Those living in proximity to these zones may experience increased health risks. Many people throughout the world still do not understand the connection between these vectors and diarrheal diseases or malaria.

Cultural Considerations

Men and women consider transporting and burning solid waste children's work, unless they have an account with the collection organization.

*This case study courtesy of Brooke T. Ahrens. Photos courtesy of Brooke T. Ahrens and Jennifer R. McConville.

10.3   Components of Solid-Waste Systems

Recall from Figure 10.1 that a solid-waste system comprises waste generation, storage, collection and transport, and processing and disposal. In this section, we consider each responsibility in greater detail, beginning with storage, collection, and transport.

10.3.1   STORAGE, COLLECTION, AND TRANSPORT

Storage, collection, and transport of MSW typically accounts for 40–80 percent of the total cost of solid-waste management. Four questions need to be considered when designing a storage, collection, and transport system:

1. Which wastes should be collected from the generator, and which should the generator transport to a processing facility?

2. To what extent should the generators be asked to separate collected waste into different fractions?

3. Should waste be transported directly to a treatment/disposal facility, or should collection vehicles transfer wastes into a more efficient vehicle first?

4. How will implementation of pollution prevention strategies influence current and future storage, collection, and transport practices.

These questions cannot be considered independently. For example, a community might be considering whether to have house-to-house collection of newspapers versus a system where newspaper is dropped off at recycling centers. If newspaper is collected house to house, an appropriate system will be required for households to collect and deliver newspapers to the curb for collection. An assessment is needed whether a separate vehicle should be used to collect newspapers or whether existing collection for recycled goods or general waste could be used. If a separate vehicle is used for newspaper collection, it might be more efficient for those trucks to drive directly to a dockside storage area for shipping of wastepaper. However, if the same truck is used for collection of both newspaper and general waste, it might be more efficient for the truck to travel to a transfer station, where the newspaper is separated and sent in a special vehicle to the dock, while the general waste is taken to a landfill.

Common options available for the interrelated problems of storage, collection, and transport can be used to devise creative and sustainable solutions. Figure 10.5 provides examples of vehicles and a container used for collection, storage, and transport of specific components of MSW.

For residential MSW, the most common collection method is curbside. Residents are asked to segregate wastes into various types (e.g., recyclables, organic waste, and general waste), using different containers or bags, which are placed at the curb for collection. Collection is performed by one or more trucks. The number of workers per truck varies among communities, with some trucks operated by only one person and residents using large wheeled bins for waste storage. Some communities use collection systems that weigh bins (or charge per bag) and thus bill the generator accordingly. For residential waste collection, the costs increase most rapidly with the number of stops, the number of workers, and the total number of trucks in service. These factors have influenced collection systems, so they now typically use large waste containers for residents, more multiuse trucks, higher density of waste per truck, and fewer workers per truck.

Commercial and institutional waste generators typically use larger storage containers and have a separate system for waste collection. This system uses vehicles specifically designed for collection of large quantities of waste per stop. High-rise commercial or residential buildings are a special case. Many have specialized systems to transport waste to the bottom of the building for storage and compaction prior to collection.

Drop-off stations for recyclables are another valuable part of an MSW collection system. These can be designed for use by people who are walking or riding a bicycle, as well as driving via personal vehicles or using shared public transit. Drop-off systems can also be used for general waste in rural areas without door-to-door collection. Where vehicular access would be inappropriate or difficult (say, in a very old city or a tourist zone), pneumatic systems can be used to convey wastes by vacuum out of the urban core.

Transfer stations are used in larger cities and towns to reduce the costs associated with transport. They have also become common because of the movement toward regional disposal sites (versus the many local disposal sites that existed several decades ago). Collection trucks are specialized vehicles and are most efficiently

Solid-Waste Storage in Fiji

(Photo courtesy of James R. Mihelcic).

(a) This front-end loading vehicle is commonly used for commercial collection.

(Photo courtesy of Heil Environmental).

(b) The side-loading vehicle is commonly used for residential collection.

(Photo courtesy of Heil Environmental).

(c) The rear-loading vehicle is suited for residential collection.

(Photo courtesy of Heil Environmental).

(d) This is a street and footpath sweeper.

(Photo courtesy of Hako GmbH).

(e) This bin offers a means to collect recyclables in a shopping district.

(Photo courtesy of Glasdon Group Limited).

Figure / 10.5 Devices to Aid Solid-Waste Collection, Storage, and Transport.

Figure / 10.6 Recycling Systems: Waste Generators, Materials Recovery Facilities, and Markets for Recovered Materials.

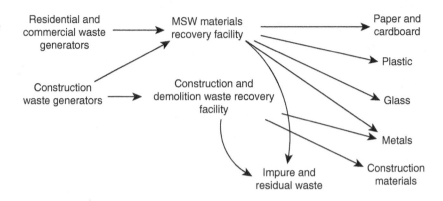

used to collect, rather than transport, waste. While a typical collection truck might hold 4–7 Mg of MSW, a larger truck using more efficient compaction can carry 10–20 Mg of MSW. Stations can also be used to transfer waste to containers that are shipped by rail or sea. With increases in the distance to waste treatment and disposal sites and the amount of waste generated, urban transfer stations generally become more economical.

10.3.2 RECYCLING AND MATERIALS RECOVERY

Recycling requires separation of materials and removal of low-quality wastes. Successful recycling systems use a mixture of separation at the source by the waste generator, by machinery at a central location, and by trained people at a central location. Successful recycling systems require careful consideration of costs involved, and of the markets for recycled goods. Several types of **materials recovery facilities (MRFs)** can be used, with some specializing in processing of separately collected wastes (Figure 10.6).

Expansion in recycling requires the development of new markets. Otherwise, the excess of supply over demand will lead to decreases in the value of the recovered materials, to the point that more resources are used to recover the materials than are saved by the recovery. In some cases, additional recycling can occur by distributing information between waste holders and potential users. Waste exchange systems can be operated to help one small business (or homeowner) solve a waste problem while another finds a valuable input.

Note that in recycling programs, the control of impurities can be critical. For example, a small amount of ceramics in glass can make the glass impractical to recycle. Limiting the number of impurities requires extensive and ongoing communication with waste generators, whether they be children, adults, business owners, or community leaders.

TYPES OF MATERIALS RECOVERED OR RECYCLED Plastics recycling is a challenge partly because the plastics industry has developed and marketed many unique types of plastics that are not necessarily compatible when recycled. To assist with plastics recycling, an international resin code is marked on most plastic consumer products

Table / 10.9

Types of Plastics Found in Commercial Products with Resin Codes Used to Aid in Recovery

Resin Code	Material	Sample Applications
1	Polyethylene terephthalate (PET)	Plastic bottles for soft drinks; food jars
2	High-density polyethylene (HDPE)	Bottles for milk; bags for groceries
3	Polyvinyl chloride	Blister packs; bags for bedding, pipe
4	Low-density polyethylene	Bags for dry cleaning and frozen foods
5	Polypropylene	Containers for takeout meals
6	Polystyrene	Cups and plates; furniture and electronics packaging

(Table 10.9). The most commonly recovered plastics are polyethylene terephthalate (PET) (type 1) and high-density polyethylene (HDPE) (type 2). As society moves away from use of nonrenewable resources such as petroleum plastics, there will greater use of biomaterials for packaging products.

Recovered paper is typically transformed back to new paper products. Wastepaper is of higher value when the paper fibers are longer and there are fewer impurities. Glossy magazines currently have a lower value than office paper because they use minerals that provide the gloss to the paper. Previously recycled paper loses value because the recycling process shortens the fibers.

Because of the high energy requirements to process aluminum ore, aluminum is typically of high value per unit weight of recovered material. Ferrous metals (iron, steel) have been recovered by scrap metal processors for many decades. With a developed market for waste ferrous metal, the recovery of ferrous metals from appliances, vehicles, equipment, cans, and demolition debris is now common.

The system for turning waste glass, called cullet, into new glass is well developed. However, the large cost of transportation to a glass smelter can make it impractical to turn waste glass into new glass. As a result, new markets for this material are under development.

Construction and demolition debris includes metals, wood, stone, and concrete. Some building materials (e.g., tiles and fittings) can be reused, while others are processed for new uses. Broken stone and concrete can become aggregate for new concrete or for other building-fill purposes.

SEPARATION OF MATERIALS A wide variety of mechanical equipment is available for separation of waste materials Magnets can separate ferrous metals, but only after any bags have been opened and the waste has been placed on conveyors. To separate papers and plastics, machinery can exploit the lower density and larger size of these materials. Methods can involve screens, sieve-like inclined shaking tables, bursts of air, and rotating sieves called trommels.

E-Cycle Your Electronic Wastes
http://www.epa.gov/wastes/conserve/materials/ecycling/index.htm

In developing countries, it is common for *scavengers*, the *informal sector*, to participate in solid-waste management activities. This is due primarily to inadequate municipal services, which create a large need for informal waste collection and an opportunity for income among the poor. Medina (2000) writes:

When scavenging is supported—ending exploitation and discrimination—it represents a perfect illustration of sustainable development that can be achieved in the Third World: jobs are created, poverty is reduced, raw material costs for industry are lowered (while improving competitiveness), resources are conserved, pollution is reduced, and the environment is protected.

In some cases, paper and plastic wastes are separated from other low-energy materials and then left in a mixed state to be used as a source of fuel. The paper/plastic mix is termed **refuse-derived fuel (RDF)**, and can be shredded and then compressed to lessen transport costs. Mechanized techniques are also available to separate aluminum from other materials and to distinguish various colors of glass.

In many situations, people are employed to aid in the separation of waste materials. In some cases, people ensure that high-quality recovered goods are produced. In other cases, staff pick specific wastes from a conveyor and place them into separate containers. The health and safety of workers is obviously a primary concern in MRFs.

Design of an MRF is difficult, and creative approaches are valued. Separated materials need to be compressed for transport and safely stored. A typical process flow diagram for an MRF is shown in Figure 10.7. The MRF design needs to manage waste materials that arrive in variable quantities, as well as adapt to markets that vary in the price paid for processed materials.

10.3.3 COMPOSTING

Composting is a microbial process that treats biodegradable wastes. The reactions are similar to those employed in aerobic wastewater treatment (discussed in Chapter 9). Wastes are processed down to a suitable size, water is added, air is allowed to enter to transfer oxygen into the waste pile, and the waste is mixed to ensure even degradation. The microorganisms feed on the organic matter in the waste, producing carbon dioxide and leaving behind a solid (called *compost*) that can be applied to soil. The two most common applications of composting are for: (1) industrial/agricultural wastes, such as wood waste, fish-processing waste, and solids generated at a municipal wastewater treatment plant; and (2) source-separated MSW, such as yard wastes separately collected or a mixture of yard and food wastes separately collected.

Composting has several objectives: (1) to reduce the mass of waste to be managed; (2) to reduce pollution potential; (3) to destroy any pathogens; and (4) to produce a product that can be marketed or used by the local community.

Table 10.10 provides some detail of the two most common types of composting systems in use: windrow and in-vessel. Figure 10.8

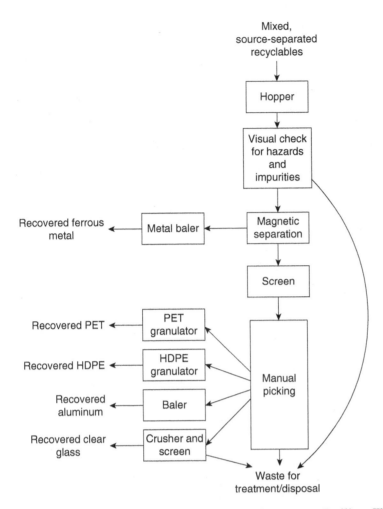

Figure / 10.7 **Process Flow Diagram for a Materials Recovery Facility** This facility is designed to process source-separated PET and HDPE plastics, metal and aluminum cans, and clear glass.

illustrates each of them. A **windrow** is a trapezoidal pile of processed organic matter left in the open air. It must be occasionally turned to ensure that all organic matter spends time inside the pile, where the temperature and moisture content are ideal for decomposition. An **in-vessel** system maintains the processed organic matter in a large container. Although this process is more expensive to construct, it allows for more precise control of the process, more readily manages air emissions, and produces compost faster.

Table / 10.10

Most Common Types of Composting Systems

System Type	Particle Size	Types of Waste	Mixing Frequency	Time to Obtain Compost
Windrows	5–20 mm	Mixed yard	Once per week	2–4 months
In-vessel	5–20 mm	Yard and food	Hourly	1–2 months

(a) Windrow composting of yard waste

(Photo from U.S. Department of Agriculture, Agricultural Research Service Information Staff).

(b) In-vessel composting to treat kitchen and yard waste

(Photo from Magherafelt District Council, Northern Ireland).

Figure 10.8   Two Composting Systems.

Production of compost requires maintenance of temperatures above 40°C for several days or longer. Providing the right conditions for aerobic microbial growth leads to rapid release of energy in the form of heat, and composting systems can commonly reach temperatures of 70°C for short time periods. The high temperature ensures destruction of pathogens and unwanted seeds, and leads to a faster production of the final product. The process is controlled through nutrient (especially nitrogen) content, pH, moisture content, and the air content.

The nutrient content is commonly expressed as a **carbon-to-nitrogen ratio** on a dry-weight basis (the C:N ratio). This ratio should be in the range of 20–40 (20:1 to 40:1) for materials entering a composting process. Table 10.11 lists the nitrogen content and C:N ratio for a variety of materials commonly composted.

When one waste is not compostable on its own, it can be mixed with other materials to ensure the proper nutrient content, pH, moisture content, and air porosity. Example 10.3 demonstrates a typical calculation used to determine the correct composition of the waste added to a composting system.

Composting can be performed by individuals, by individual businesses, or by municipalities with large amounts of organic waste. In all cases, the principles are the same. Backyard composting can reduce the costs (and associated environmental impacts) of collection, transport, and processing and disposal of organic wastes. Many municipalities now provide subsidized or free home composting units (or worm farms) to encourage the practice. To be effective, home composting must have the proper mix of nitrogen- and carbon-rich materials and have adequate airflow and moisture. Poor household composting practices will not reach the temperatures required for effective treatment, can create odor problems, and can attract animals. Municipalities need to find the proper balance of education, subsidies, and enforcement to achieve effective backyard composting systems.

Table / 10.11

Nutrient Content of Various Materials Used in Composting Composting systems perform best when the carbon-to-nitrogen ratio is optimized in the range of 20–40 (C:N of 20:1 to 40:1).

Material	Nitrogen (% dry mass)	C:N Ratio (dry mass basis)
Urine	15–18	0.8
Human feces	5.5–6.5	6–10
Cow manure	1.7–2	18
Poultry manure	5–6.3	15
Horse manure	1.2–2.3	25
Activated sludge	5	6
Nonlegume vegetable wastes	2.5–4.0	11–12
Potato tops	1.5	25
Wheat straw	0.3–0.5	130–150
Oat straw	1.1	48
Grass clippings	2.4–6.0	12–15
Fresh leaves	0.5–1.0	41
Sawdust	0.1	200–500
Food wastes	3.2	16
Mixed paper	0.19	230
Yard wastes	2.0	23

SOURCE: Haug, 1993.

example/ 10.3 Determining Proper Ingredients for Successful Composting

A poultry manure has a moisture content of 70 percent and 6.3 percent N (on a dry mass basis). The manure is to be composted with oat straw with a moisture content of 20 percent. The desired C:N for the mixture is 30. Using Table 10.11 values for the nitrogen composition and C:N ratio for these two materials, determine the kg of oat straw required per kg of manure to attain the desired C:N ratio.

solution

Assume 1 kg of moist poultry manure dry mass. Let X = kg of moist oat straw on a dry mass basis. The mass of carbon and nitrogen obtained from each material in the mixture is

$$\text{Dry mass nitrogen from poultry manure} = 1\,\text{kg} \times (1 - 0.7) \times 0.063 = 0.0189\,\text{kg}$$
$$\text{Dry mass carbon from poultry manure} = 1\,\text{kg} \times (1 - 0.7) \times 0.063 \times 15 = 0.2835\,\text{kg}$$
$$\text{Dry mass nitrogen from oat straw} = X\,\text{kg} \times (1 - 0.2) \times 0.011 \times 0.0088 = X\,\text{kg}$$
$$\text{Dry mass carbon from oat straw} = X\,\text{kg} \times (1 - 0.2) \times 0.011 \times 48 = 0.4224 \times X\,\text{kg}$$

The overall C:N ratio is

$$30 = \frac{\text{(mass carbon from poultry manure + mass carbon from oat straw)}}{\text{(mass nitrogen from poultry manure + mass nitrogen from oat straw)}}$$

$$30 = \frac{(0.2835 + 0.4224 \times X)}{(0.0189 + 0.0088 \times X)}$$

Solving for X, we find $X = 1.8$ kg. Thus, for every 1 kg of poultry manure, 1.8 kg of oat straw must be added to obtain an optimal C:N ratio of 30. The reason for this is that the poultry manure is a better source of nitrogen and the oat straw provides a better source of carbon.

How to Compost
http://howtocompost.org/

It takes 2,000 pounds of MSW to equal the heat energy in 500 pounds of coal.

At the larger scale, composting systems must match the product to suitable markets. Municipalities can return (or sell) compost back to the local residents who generated the waste. Larger markets for compost include municipal parks, golf courses, nurseries, landscapers, landfills (as daily and final cover material), and turf growers. Many users will insist on a lack of impurities in compost, and strict control of the wastes inputted to a composting system will ensure a quality, in-demand product.

10.3.4 WASTE-TO-ENERGY

Waste-to-energy (also called **incineration**) is a combustion process where oxygen is used at high temperatures to liberate the energy in waste. In the United States in 2004, waste-to-energy facilities burned 29 million tons of MSW. In addition, 380 U.S. landfills now recover methane. Waste-to-energy can reduce the amount of waste needing disposal, generate energy for a community, and also reduce MSW transportation costs. It becomes more favorable for wastes that have high-energy content, low-moisture content, and low-ash content. These wastes include paper, plastics, textiles, rubber, leather, and wood (energy values previously listed in Table 10.5).

Table 10.12 describes six incineration systems. The most commonly used method is *mass-burn incineration*. In this process, unsegregated MSW is combusted. Figure 10.9 provides a schematic of a typical mass-burn incinerator. Incineration creates two solid by-products. *Bottom ash* is the unburned fraction of the waste. *Fly ash* is the particulate matter suspended in the combustion air and removed by air pollution control equipment. Both of these ashes have a number of hazardous components and need to be carefully managed; therefore, a mixture of recovery and landfill processes are used for ashes.

Air pollution control is required for incinerators to limit emissions of particulates, volatile metals (such as mercury), nitrous oxides, and incomplete combustion products (such as dioxins). In the late 1980s, incinerators were the major contributor of dioxin emissions in the United States, but since 2000, the contribution of dioxin emissions from incinerators has been less than 6 percent of the total. Informal

Table / 10.12

Common Incineration Systems

Type of Incinerator System	Explanation
Mass-burn	Unsegregated MSW is combusted.
Modular	Small incinerators focus on treatment of specific waste streams (e.g., medical waste).
Refuse-derived fuel (RDF)	Energy-rich waste streams can be separated from other wastes and burned, typically as a substitute for fossil fuels such as coal, in power plants. Wastewater treatment biosolids are one such waste stream.
Co-incineration	Specific postproduction commercial/industrial wastes, such as wood wastes from construction, can be combusted with production wastes, such as paper mill sludge or dried wastewater treatment plant biosolids to produce energy.
Hazardous waste	Hazardous organic wastes (e.g., solvents, pesticides) can be combusted to destroy the wastes, though this requires very close attention to air emissions.
Cement kilns	Cement factories may provide suitable conditions for combustion of many wastes, including waste tires and waste oils, during the production of cement.

backyard burning of solid waste (e.g., burn barrels) is now the major source of U.S. dioxin emissions (EPA, 2006).

Sufficient oxygen must be provided in incineration processes to ensure complete combustion. Carbon, hydrogen, and sulfur in the waste stream are oxidized to CO_2, H_2O, and SO_2 during combustion. Although some oxygen is present in wastes initially, most of this oxygen comes from air. Example 10.4 shows how to estimate the air required in combustion.

Backyard Burning

http://www.epa.gov/wastes/nonhaz/municipal/backyard/

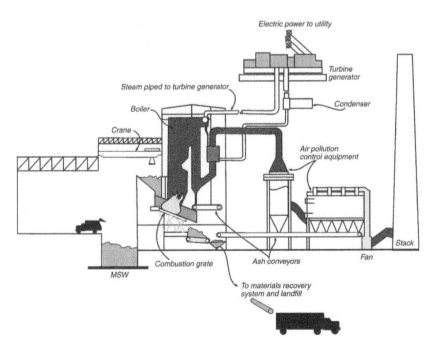

MSW — Combustion grate — Crane — Boiler — Steam piped to turbine generator — Turbine generator — Condenser — Air pollution control equipment — Electric power to utility — Stack — Fan — Ash conveyors — To materials recovery system and landfill

Figure / 10.9 Mass-Burn Incineration System for Treatment of Municipal Solid Waste Programs in reducing and recycling wastes and eliminating household hazardous wastes and heavy metals (e.g., batteries) from a waste stream can be coordinated with incineration systems to minimize potential risk associated with pollutants found in the offgas and ash.

How would you lead a public meeting where you were going to discuss behavioral changes that would decrease use of burn barrels in a suburban or rural community? What stakeholders need to be present at such a meeting? What specific actions might you employ to achieve consensus?

Incineration systems often have high construction and operating costs. However, the costs can be offset by savings in transporting wastes to a disposal site, land requirements for disposal, and recoverable energy. Successful incineration systems are those where the waste is appropriate for incineration and the cost offsets can be found. In addition, incineration systems are complex and require advanced skills in construction and operation. These two factors cause incineration systems to be preferred in more economically advanced regions of the world and where land values, energy costs, and transportation costs associated with other alternatives are high.

example/10.4   Air Requirements for Combustion of Municipal Solid Waste

A refuse-derived fuel (RDF) comprises 60 percent mixed paper, 30 percent mixed plastic, and 10 percent textiles. Assume it is dried before combustion. Determine the volume of air (in L) at 20°C, 1 atmosphere pressure that is required to combust 1 kg of the RDF.

solution

Use values of percent composition of solid waste from Table 10.5 to determine the moles of C, H, S, and O in the waste. (The molecular weights are 12, 1, 32, and 16 g/mole, respectively.) Complete a table for each of the RDF components:

	Dry Mass (g)	Moles C	Moles H	Moles S	Moles O
Paper (mixed)	600	21.7	34.8	0.038	16.6
Plastic (mixed)	300	15.0	21.6	<0.01	4.3
Textiles	100	4.0	6.4	0.006	2.5
Total	1,000	40.7	62.8	0.05	23.4

The moles of O_2 required for combustion are determined for each molecular species. The balanced oxidation reactions are as follows:

$$C + O_2 \rightarrow CO_2$$
$$H + \tfrac{1}{4}O_2 \rightarrow \tfrac{1}{2}H_2O$$
$$S + O_2 \rightarrow SO_2$$

From the balanced reactions, 1, $\tfrac{1}{4}$, and 1 mole of O_2 are required per mole of carbon, hydrogen, and sulfur, respectively. In addition, each mole of oxygen (O) in the waste can offset the requirement for 0.5 mole O_2 gas. The overall moles of O_2 required to incinerate the waste are then

$$\left[\left(40.7 \text{ moles C} \times \frac{1 \text{ mole } O_2}{\text{mole C}} \right) + \left(62.8 \text{ moles H} \times \frac{\tfrac{1}{4} \text{ mole } O_2}{\text{mole H}} \right) + \left(0.05 \text{ mole S} \times \frac{1 \text{ mole } O_2}{\text{mole S}} \right) \right]$$
$$- \left(23.4 \text{ moles O} \times \frac{0.5 \text{ mole } O_2}{\text{mole O}} \right) = 44.8 \text{ moles } O_2$$

This value of O_2 can be converted to liters of oxygen, using the ideal gas law, after which we can determine the volume of air. Rearrange the ideal gas law $(PV = nRT)$ to solve for V. Use a value of $R = 0.082$ L-atm/mole-K, along with the values of 20°C and 1 atmosphere pressure:

$$V = 44.8 \text{ moles } O_2 \times 0.082 \text{ L-atm/mole-K} \times \frac{(273 + 20)K}{1 \text{ atm}} = 1{,}080 \text{ L } O_2/\text{kg}$$

Because air is approximately 20.9 percent O_2, the liters of O_2 can be converted to liters of air:

$$1{,}080 \text{ L } O_2/\text{kg} \times \frac{1 \text{ L air}}{0.209 \text{ L } O_2} = 5{,}200 \text{ L air/kg waste}$$

10.3.5 LANDFILL

Landfills are engineered facilities designed and operated for the long-term containment of solid wastes. Design of the landfill will vary greatly based on the waste and the location of the facility. Based on waste type, the four main types of landfills are listed in Table 10.13, along with the relevant federal regulations under RCRA. In landfills, wastes are placed and compacted into solid forms, then covered to limit exposure to water and air. **Leachate** is water that contacts the wastes and becomes a contaminated wastewater. As biological materials decompose in landfills, oxygen is consumed, and carbon dioxide is produced. Over time, an anaerobic environment evolves that leads to the production of methane gas. Chapter 9 discussed in detail the biochemical processes in which organic wastes are converted to methane.

As shown in Figure 10.10, landfills are technically advanced facilities with sophisticated environmental protection systems. Environmental protection in landfills occurs through a combination of four barriers: (1) appropriate siting; (2) engineered design that is carefully implemented during construction and operation; (3) exclusion of inappropriate wastes; (4) collection and use of landfill gas as a source of energy; and (5) short- and long-term monitoring.

Table / 10.13

Major Types of Landfills Regulatory oversight depends on the type of waste.

Type of Waste	Applicable Regulations in United States
Construction and demolition debris	Regulation of construction and demolition debris landfills is managed at the state level in the United States
Municipal solid waste	40 C.F.R. Part 258
Industrial waste (e.g., for ash and mining waste)	40 C.F.R. Part 257
Hazardous waste	40 C.F.R. Parts 264 and 265

Figure / 10.10 Cross Section of a Modern Landfill, Showing Barriers Incorporated into Engineering Design
The many barriers help protect public health and the environment.

(Redrawn with permission of the National Solid Wastes Management Association (NSWMA), 2003.)

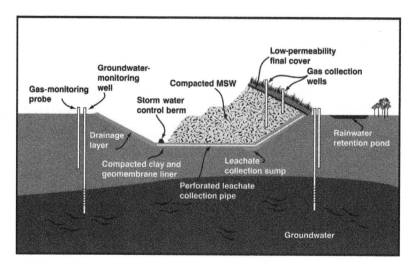

LANDFILL SITING   Landfills need to be located where the risks to the environment and society are low, so that even in the case of poor design, construction, or operation, the resulting risk is minimized. Negating issues associated with social and environmental justice is also critical. Table 10.14 summarizes locations to avoid when siting a landfill and other issues that need to be considered. Landfill siting is a highly contentious social issue, and the engineer's role is to provide input and analysis in an equitable manner.

A common tool for evaluating potential landfill sites is a *geographic information system (GIS)*. GIS is a valuable means of processing large amounts of data and ensuring an evaluation of all options. Figure 10.11

Table / 10.14

Items to Consider When Siting a Landfill

Location-specific items to avoid
  Floodplains
  Active geological faults
  Land prone to slips or erosion
  Wetlands and intertidal zones
  Areas with significant ecosystems and important biodiversity
  Areas of cultural or archaeological significance
  Drinking-water catchments

Other items to consider
  Integrate strategies of pollution prevention such as source reduction and recycling to maximize landfill capacity.
  Minimize costs for transport of wastes by placing the landfill near centers of waste generation.
  Minimize costs required for construction of transportation infrastructure required to access the site.
  Identify sites less prone to extremes of rain or wind.
  Identify sites with soils that can be used during construction.
  Match the potential site to a final use of the landfill that will benefit the local community and use the produced energy.
  Develop equitable solutions to issues of environmental justice and other social objections in the development of a landfill.
  Seek collocation of users that can make beneficial use of waste materials or derived energy.

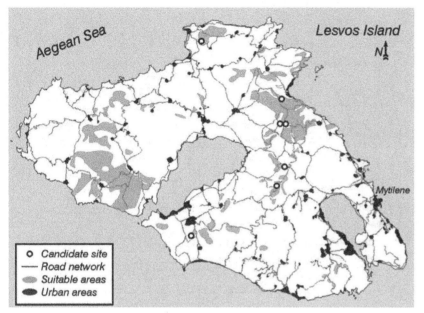

Figure / 10.11 Graphical Information System Evaluation of Potential Landfill Sites Suitable areas are highlighted in the gray shaded areas, which can be compared to existing transportation networks and waste-producing urban areas highlighted in dark shaded areas. The location is Lesvos Island, in the Aegean Sea east of Greece.

(Reproduced from T. D. Kontos et al., *Waste Management and Research*, 21(3): 262, copyright 2003. Reprinted by permission of SAGE Publications, Inc.).

shows the result of a GIS evaluation of potential landfill sites. In this case, the location of existing transportation infrastructure can be integrated with the location of high-volume urban waste generators and other location issues listed in Table 10.14. GIS also allows the engineer to add demographic information that is related to other concerns such as equitable distribution of risk.

LANDFILL DECOMPOSITION Wastes that are disposed into a landfill undergo a series of interrelated chemical and biological reactions. These reactions determine the quantity and composition of the gas and leachate produced by the landfill, and thus determine the management required. In terms of the biological reactions that take place, a landfill is best viewed as a batch biochemical decomposition process.

Figure 10.12 depicts gas and leachate composition over time as the waste of biological origin (e.g., food waste, yard waste, paper) decomposes. In the early stages of decomposition (shown in Figure 10.12), oxygen is consumed, and carbon dioxide and organic acids are produced. Both of these products decrease the pH of the leachate. An increase in the leachate's oxygen demand (measured as COD and BOD) is also observed in the early stages of decomposition as the organics are converted from a particulate to dissolved phase. In later stages, after all the oxygen is consumed and an anaerobic environment is established, microorganisms convert the high-BOD organic acids to methane gas. Here the leachate becomes less concentrated as the dissolved constituents are converted to the gaseous phase and the readily leachable components of the waste material become less prevalent. At later stages of decomposition, leachate is still high in strength and will still require collection and treatment.

The time required to reach a steady state for methane production can vary from 1 or 2 years to as long as 20 years. Methane production may never occur if biological conditions are unfavorable. Reasons for this

Figure / 10.12 Typical Landfill
Decomposition Pathways The upper
figure (a) shows gas composition (and
production) over time. The lower figure
(b) shows the relative leachate
concentration of various constituents. As
the rate of gas production increases, the
leachate becomes less strong in terms of
the concentration of leachate
constituents.

(Based on Farquhar and Rovers (1973)).

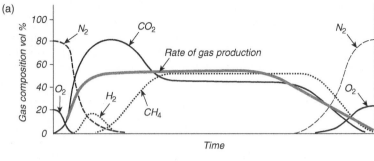

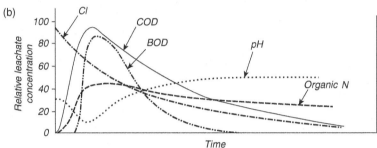

could be the presence of methane-inhibiting chemicals in the waste
stream or infiltration of oxygen from a poorly designed cover. During
steady-state methane production, the gas is roughly 50 percent methane and 50 percent carbon dioxide—but remember that methane has
25 times the global-warming potential of carbon dioxide.

The resulting leachate typically has a higher BOD than domestic
municipal wastewater, and it may be as hazardous as many industrial
wastewaters. The concentrations of constituents in leachate depend
greatly on the type of waste placed in the landfill. The exclusion of
hazardous wastes from landfills improves the likelihood of proper
biological production of methane at the same time that it limits the
hazard associated with landfill leachate. Steady methane production
is important in terms of leachate quality, as methane production
reduces the hazard of leachate because the dissolved organic components of leachate (measured as COD or BOD) are converted to
gaseous methane.

LANDFILL GAS   The production of landfill gas is best viewed as
both a problem and an opportunity. First, consider the reasons why it
is a problem: (1) It can be explosive when mixed with oxygen. (2) It
can be a human health concern for site workers. (3) It can create odors.
(4) It can displace oxygen in soils, which may suffocate nearby plants.
(5) It can emit methane to the atmosphere, which contributes to
greenhouse gas emissions.

The main reason that landfill gas is an opportunity is that it has the
potential to produce an economical and non-fossil-fuel-derived form of
electricity. Table 2.5 showed that in the United States alone, CH_4
emissions from landfills were 107.8 Tg of CO_2 equivalents in 2010.
In Chapter 9, we discussed how the Los Angeles County Sanitation
District currently obtains 63 MW of electricity from landfill gas.

The total amount of **methane** that can be produced will vary depending on the amount of biodegradable material and the suitability of landfill conditions to biological methane production. Typical values are 100 L CH_4 produced per kg MSW landfilled. The maximum amount of methane produced can be estimated from the stoichiometry of the waste decomposition:

$$C_a H_b O_c N_d + \frac{(4a - b - 2c + 3d)}{4} H_2O \rightarrow \frac{(4a + b - 2c - 3d)}{8} CH_4$$
$$+ \frac{(4a - b + 2c + 3d)}{8} CO_2 + dNH_3$$

(10.3)

where a, b, c, and d are stoichiometric coefficients for a specific organic chemical. The actual amount of methane produced may be only 10–50 percent of the maximum amount estimated from Equation 10.3. This is because some organic waste is not degradable under anaerobic conditions and because in some sections of a landfill, effective biodegradation may be hampered by low moisture, presence of toxins, adverse pH, or a lack of nutrients.

Figure 10.12 showed that the rate at which methane is produced (in units of L methane/kg waste-year) is commonly described with a lag phase of zero methane production, followed by exponential decay, as follows (for $t > t_{lag}$):

$$\text{rate of } CH_4 \text{ production} = V_{gas} \times k \times e^{[-k \times (t - t_{lag})]}$$

(10.4)

where V_{gas} is the total volume of gas (L) that can be produced per kg of waste, k is a first-order decay rate (time^{-1}), t is the time measured from the point the waste is disposed, and t_{lag} is the lag time required before the waste begins to produce methane.

To determine the methane produced between time t_1 and t_2, we can integrate this function to provide the following equation (with $t_1, t_2 > t_{lag}$):

$$\text{cumulative } CH_4 = V_{gas} \times \left[1 - e^{[-k(t_2 - t_{lag})]} \right] - V_{gas} \times \left[1 - e^{[-k(t_1 - t_{lag})]} \right]$$

(10.5)

For the situation when $t_1 \ll t_{lag}$, Equation 10.5 reduces to (with $t_2 > t_{lag}$)

$$\text{cumulative } CH_4 = V_{gas} \times [1 - e^{[-k(t_2 - t_{lag})]}]$$ (10.6)

The first-order decay rate is related to the half-life, $t_{1/2}$ (as discussed in Chapter 3):

$$k = \frac{0.693}{t_{1/2}}$$ (10.7)

Half-lives for methane production in a landfill depend on the waste's degradability and can vary from 1 to 35 years (McBean et al., 1995; Pierce et al., 2005). They also vary with the moisture content of the waste. Dry waste will have a longer half-life than moister waste. Accordingly, some landfills recycle leachate into the waste pile or add other sources of water to the waste to increase the rate of decomposition.

example/10.5 Estimating Methane Production

Calculate the volume of methane that can be collected (in m^3/person at 0°C) from landfilling 1 year's waste for each person. Use the landfilling rate for 2010 provided in Table 10.2. Assume the three waste components that produce methane are food wastes (15 percent of total), mixed paper (30 percent of total), and yard wastes (25 percent of total). Assume that 60 percent of the food and paper waste and 40 percent of the yard trimmings decompose.

solution

This problem requires that we determine the volume of methane produced for every 1 kg of landfilled waste and multiply this value by the mass of waste landfilled per person per year.

For each kg of waste, determine the moles of carbon, hydrogen, and oxygen that will degrade to methane, and then use the molar stoichiometry from Equation 10.3 to determine the moles of methane produced. As Table 10.15 shows, the solution uses not only information from Table 10.5, but also values provided in the problem statement and molecular weights, to find the required molar ratios.

Table / 10.15

Results for Example 10.5

	Wet Weight (g)	Moisture Content (%)	Dry Weight (g)	Total Carbon (g)	Total Hydrogen (g)	Total Oxygen (g)	Total Nitrogen (g)
Food wastes	150	70	45	21.6	2.9	16.9	1.2
Paper (mixed)	300	10	270	117.2	15.7	119.6	0.8
Yard wastes	250	60	100	46.0	6.0	38.0	3.4
Other	300	—	—	—	—	—	—
Total	1,000						

	Degraded Carbon (g)	Degraded Hydrogen (g)	Degraded Oxygen (g)	Degraded Nitrogen (g)	Degraded Carbon (moles)	Degraded Hydrogen (moles)	Degraded Oxygen (moles)	Degraded Nitrogen (moles)
Food wastes	13.0	1.7	10.2	0.7	1.08	1.71	0.63	0.05
Paper (mixed)	70.3	9.4	71.8	0.5	5.85	9.32	4.49	0.03
Yard wastes	18.4	2.4	15.2	1.4	1.53	2.38	0.95	0.10
Other	—	—	—	—	—	—	—	—
Total					8.46	13.41	6.07	0.18

From Equation 10.3, the moles of methane produced are determined as $(4a + b - 2c - 3d)/8$. From Table 10.15, $a = 8.46$, $b = 13.41$, $c = 6.07$, and $d = 0.18$. Using this expression, the methane produced per kg of waste is 4.32 moles.

Using the ideal gas law, at 0°C (273 K), there are 22.4 L gas per mole or 0.0224 m^3 of gas per mole of gas. The volume of methane produced per kg waste is then

$$4.32 \text{ moles CH}_4 \times 0.0224 \text{ m}^3/\text{mole} = 0.0967 \text{ m}^3 \text{ per kg}$$

The problem states that 90 percent of the methane was recovered; therefore, $0.087 \, m^3$ methane is produced per kg waste.

From Table 10.2, the U.S. landfilling rate in 2010 was 0.40 Mg (or 400 kg) per person per year. Therefore, the gas production rate is

$$0.0968 \, m^3/kg \times 400 \, kg/person\text{-}year = 38.7 \, m^3 \, methane/person\text{-}year \, at \, 0°C$$

The production of landfill gas leads to pressure in landfills. This results in gas movement. Gas flow in landfill waste (and soil) has many similarities to groundwater flow. To control the movement of gas, gas-impermeable barriers are incorporated into the design, and outside the landfill high-permeability soils are placed, which direct gas flow into trenches. Even in past circumstances when landfill gas wasn't collected for energy, it was common to collect and combust the gas to minimize negative impacts associated with landfill gas.

Full capture of landfill gas requires installing inside the landfill waste, gas wells, and gas-permeable layers, along with pumping and piping systems (refer back to Figure 10.10). New landfills can typically capture more than 90 percent of the methane produced. Landfill gas can then provide heat, steam, or electricity. The most common method of converting gas to electricity is with large, 1 MW modular engines. Because landfill gas is a renewable energy source, many efforts are taking place around the world to expand its use. As one example, the Landfill Methane Outreach Program is a voluntary assistance and partnership program run by the EPA. Readers are encouraged to visit this site or similar ones in their country.

Landfill Methane Outreach Program
http://www.epa.gov/lmop

Application / 10.4   Greenhouse Gas Impacts of Landfills

Landfills provide a good example of the challenges and complexities of assessing the climate-change impacts of an economic or social activity. There are five pathways by which landfills generate greenhouse gases.

1. Methane (CH_4) production leaking to the atmosphere.

2. Storage of biogenic carbon found in waste materials such as paper, food, and yard waste.

3. Direct consumption of fossil fuels over the landfill's life cycle that include the life stages of construction, operation, and end of life (i.e., closure).

4. Indirect greenhouse gas emissions over the landfill's life cycle from the life stages of construction, operation, and end of life (i.e., closure).

5. A reduction in use of fossil fuels for energy because of substitution with landfill methane gas.

Pathways 1, 3, and 4 lead to a worsening of greenhouse gas impacts, while pathways 2 and 5 reduce the overall greenhouse gas impact. The net effect is the sum of the contribution from each of these five pathways. Therefore, landfills can have an overall negative or positive impact on greenhouse gas emissions.

Pathway 1 is a serious concern because each ton of methane released has 25 times the greenhouse gas impact of a ton of carbon dioxide (refer to Table 2.4 and discussion of global warming potential in Chapter 2). New landfills install gas collection systems and these can greatly reduce the methane released to the atmosphere, though some will still escape. In addition, the microorganisms present in the soil directly above a landfill can oxidize a fraction of the methane before it reaches the atmosphere. An estimate of pathway 1 requires an estimate of the gas collection factor, and a soil oxidation factor, neither of which are well known (Levis and Barlaz, 2011).

Pathway 2 applies to food, paper, and yard wastes only. These materials, if not disposed of to a landfill (which is anaerobic), would naturally oxidize to CO_2, in the presence of oxygen. Therefore, each mole of carbon retained (or sequestered) in the landfill reduces the emissions of CO_2. This does not apply to plastics, which are composed of organic carbon, but are considered nonbiogenic carbon because they are manufactured from fossil fuels.

Table 10.15 (in Example 10.5) provided the results of a calculation where we determined the degraded carbon associated with food, paper, and yard wastes. The non-degraded biogenic carbon is the carbon that is relevant to this pathway because it is what is sequestered in the landfill (and therefore does not decompose and produce greenhouse gases). Pathway 2 requires an estimate of the total amount of degradation of these biogenic carbon sources. This varies with waste composition as well as landfill design. The value is currently not well known, and its estimate is closely tied to the estimate of a gas collection factor, which further complicates estimation.

Pathways 3 and 4 have been found to be negligible compared to the other pathways (Camobreco et al., 1999). Pathway 3 is easier to estimate from an examination of the fossil fuel use in vehicles and other direct uses that support the various life stages of disposing of wastes in a landfill. Pathway 4 requires the use of life cycle assessment methods, which introduce assumptions in an estimate of indirect emissions.

Pathway 5 can only be counted when the landfill gas is put to beneficial use. Though it is very common to have a beneficial use of landfill gas associated with new landfills, it is not always the case. The benefit that arises also varies on the form of energy that the landfill gas substitutes for. For example, if landfill methane is burned to produce electricity and that reduces the need for hydroelectric power, little if any greenhouse gas reduction results. However, if the use of the landfill gas reduces the consumption of coal for generation of electricity or heat, then a large benefit is gained.

The benefit thus varies greatly by region, country, and over time (e.g., remember our discussion of eGRID in Chapter 2). However, a general rule of thumb is that 0.8 metric tons of CO_2 equivalents are saved for every MW-h of energy produced from biogenic carbon (USDOE, 2007). This value when combined with a common value for conversion of gas to electricity ($270 m^3/h$ of CH_4 to 1 MW of gas (GMOP, 2012)) results in a savings rate of 0.0030 metric tons CO_2e/m^3 CH_4 gas used (remember that 1 metric ton = 1,000 kg).

example/ 10.6 Determining the Overall Greenhouse Gas Emissions from Landfilling Solid Waste

Use the data and results from Example 10.5 to determine the overall greenhouse gas emissions from landfilling 1 year's worth of waste for one person (in Mg of CO_2e). Assume 80 percent of the landfill gas is collected and combusted for energy and 20 percent of the uncollected methane is oxidized in the soil before it is captured for combustion. We will also assume that 0.003 Mg of CO_2e can be offset per m^3 of CH_4 combusted for energy. In addition, we will neglect the direct and indirect emissions of greenhouse gases from the consumption of fossil fuels during construction, operation, and closure of the landfill.

solution

Our last assumption allows us to ignore the contribution from pathways 3 and 4 that were discussed in Application 10.4. The overall greenhouse gas impact from pathways 1, 2, and 5 will be determined separately and then added up to find the overall greenhouse gas emissions associated with landfilling the person's 1 year of solid waste.

For pathway 1 (the direct methane emission impact), we must first determine the methane produced by the landfilled waste of one person over 1 year. In Example 10.5, this was found to be 38.7 m^3 of CH_4 at 0°C. This value is reduced because we assume that first, 80 percent of the methane produced is collected, and then of the 20 percent that is not collected, 20 percent of it is oxidized by microorganisms in the overlying soil:

$$38.7 \, m^3 \, CH_4 \text{ produced} \times (1 - 0.8)(\text{remains after oxidation})$$
$$= 7.74 \, m^3 \, CH_4 \text{ not collected}(30.96 \, m^3 \text{ collected and burned})$$

$$7.74 \, m^3 \, CH_4 \text{ not collected} \times (1 - 0.2) = 6.19 \, m^3 \, CH_4 \text{ emitted}$$
$$(20 \text{ percent or } 1.55 \, m^3 \, CH_4 \text{ oxidized in the soil})$$

This value is then converted into CO_2 equivalents by applying the ideal gas law and using the molecular weight of methane and a global warming potential of 25 for methane (see Chapter 2).

$$6.19 \, m^3 \, CH_4 \times 1,000 \, L/m^3 \times (1 \text{ mole } CH_4/22.4 \, L \, CH_4) \times (16 \, g \, CH_4/1 \text{ mole } CH_4) \times (1 \, Mg/10^6 \, g)$$
$$\times (25 \, Mg \, CO_2 e/Mg \, CH_4) = 0.11 \, Mg \, CO_2 e$$

For pathway 2, subtract the total degraded biogenic carbon from the total biogenic carbon to determine the residual, nondegraded biogenic carbon that is sequestered in the landfill. The values were generated in Table 10.15 (Example 10.5) for each of the three biogenic sources of carbon (i.e., food, mixed paper, yard wastes).

$$(21.6 \, g \, C - 13.0 \, g \, C) + (117.2 \, g \, C - 70.3 \, g \, C) + (46.0 \, g \, C - 18.4 \, g \, C)$$
$$= 83.1 \, g \text{ sequestered carbon/kg of landfilled waste}$$

This value of sequestered carbon can be converted to $CO_2 e$ by multiplying by the mass of waste that is landfilled per person per year, and then using simple stoichiometry to convert from mass of carbon to mass of CO_2 equivalents. Table 10.2 stated that 0.40 Mg of waste is landfilled per person per year in the United States. This equals 400 kg of waste per person per year.

$$83.1 \, g \, C \text{ sequestered/kg waste landfilled} \times (400 \, kg \text{ waste landfilled/person-year})$$
$$\times (44 \, g \, CO_2 e/12 \, g \, C) \times (1 \, Mg/10^6 \, g) = 0.12 \, Mg \, CO^2 e$$

For pathway 5, we start with the total volume of methane produced, multiply it by the collection efficiency to provide the value for the methane used for energy and then apply a conversion factor to obtain the $CO_2 e$ per volume of methane used for energy.

$$38.7 \, m^3 \, CH_4 \text{ produced} \times (0.8 \, m^3 \, CH_4 \text{ combusted}/1.0 \, m^3 \, CH_4 \text{ produced})$$
$$\times (0.003 \, Mg \, CO_2 e \text{ avoided}/m^3 \, CH_4 \text{ combusted}) = 0.093 \, Mg \, CO_2 e$$

Note how pathway 1 has a negative greenhouse gas impact (it results in production of methane), while pathways 2 and 5 are positive, because they either sequester carbon in the subsurface or utilize the generated methane. Totaling these values provides the net overall effect in a per person per year basis:

$$-0.11 \, Mg \, CO_2 e + 0.12 \, Mg \, CO_2 e + 0.093 \, Mg \, CO_2 e = +0.10 \, Mg \, CO_2 e$$

The result shows that it is possible that modern landfills can help to reduce overall greenhouse gas impacts. Remember, though, that as we discussed previously, this is not the case for all landfills in all settings.

Table / 10.16

Composition of Young and Old Landfill Leachates

Constituent	Units	Young Leachate	Old Leachate
BOD_5	mg/L	10,000	100
COD	mg/L	18,000	300
Organic nitrogen	mg/L as N	200	100
Alkalinity	mg/L as $CaCO_3$	3,000	500
pH	–	6	7
Hardness	mg/L as $CaCO_3$	3,500	300
Chloride	mg/L	500	200

SOURCE: Tchobanoglous et al., 1993.

LANDFILL LEACHATE No matter what controls are implemented to minimize the movement of water into a landfill, some water will enter and produce **leachate**. Control of leachate needs to consider the quantity and quality of the leachate, as well as its potential adverse effects. Leachate concentrations will vary dramatically by location and the life of the landfill (refer back to Figure 10.12). Table 10.16 provides typical concentrations for young and old leachates. Concentrations of leachate constituents are much higher than similar constituents found in untreated municipal wastewater. The information provided in Figure 10.12 and Table 10.16 also shows how the concentration of the constituents decreases as the landfill ages and the readily leachable components are removed.

In addition to pollution prevention strategies, three other strategies are used to control the volume and strength of leachate. A landfill will typically use a combination of these three strategies to limit leachate impacts.

1. *Isolation.* Waste is isolated by limiting the entrance of water and thus the production of leachate. The waste may be bound to some physical or chemical matrix to reduce its leaching potential. This latter option is most suitable for highly hazardous wastes (e.g., waste that is radioactive) but is always part of any overall strategy to limit leachate impact.

2. *Natural attenuation.* The natural physical, chemical, and microbiological properties of soil treat the leachate. The method also relies on the dilution of leachate. This option is very suitable for poorly resourced communities with small quantities of nonhazardous wastes.

3. *Controlled biological degradation.* Landfill conditions are modified to optimize degradation. This option typically involves adding moisture and ensuring suitable mixing, ensuring that toxic chemicals are kept out of the waste, and maintaining proper pH, nutrients, and monitoring. It has been termed a landfill bioreactor strategy

and is most suitable for landfills where advanced technical skills are available and nonhazardous wastes are disposed. This strategy accelerates stabilization of the waste material and so reduces the long-term risk from leachate.

Leachate management requires a series of subsystems: (1) hydraulic barriers to limit the ability of leachate to leave a landfill and for rainfall to enter the landfill; (2) collection systems to convey the leachate from the base of the landfill to an external location; and (3) a leachate treatment system. Figure 10.10 depicts these subsystems.

Hydraulic barriers are constructed from compacted clay, manufactured geomembranes, or geosynthetic clay products. A combination of these can be used to provide a system with multiple benefits. The top barrier is referred to as a **cover** or **final cover**, and the bottom barrier is called a **liner**. Hydraulic barriers can decrease the amount of leachate leaving a site by a factor of 1,000 or more over existing soils. Quality control of the installation of the barriers is critical to ensuring good performance.

Collection systems rely on gravity to convey leachate to a low point, a sump, within the landfill. Highly permeable, rounded gravel and perforated pipes are placed above the liner to ensure rapid movement of leachate and thus reduce the likelihood that pore pressures will increase to the point where leachate can seep out of the landfill or cause problems of geotechnical stability. From a low point, leachate is typically pumped out to a storage location.

Because landfill leachate is similar in many ways to a high-strength industrial wastewater, similar options are considered for treatment. The leachate could be conveyed (by pipe or truck) to a municipal wastewater treatment plant, where it would be carefully metered into the plant flow. Other options are to treat it on-site and then discharge it to land or water, or else to partly treat it on-site before conveyance to a central facility. The choice depends on the nature of the leachate, the environmental effects of discharge of treated leachate at the landfill, and the costs of conveying the leachate to a larger treatment facility.

LANDFILL DESIGN   As shown in Figure 10.13, a landfill is constructed as a series of daily **cells**, where a day's waste is compacted and covered. Waste is delivered at the *working face* and then compacted

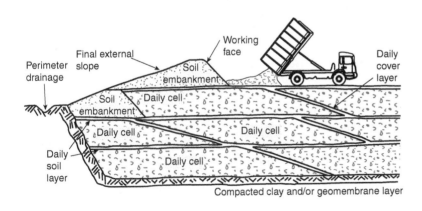

Figure / 10.13   Landfill Construction from Daily Cells and Lifts   A daily cell of solid waste is covered with daily soil; additional cells are added vertically to the landfill, which eventually makes a lift.

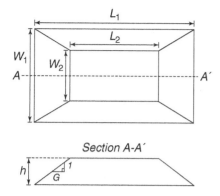

Figure 10.14 **Geometric Design for the Area Landfill** The top figure (a) is a plan view and the bottom figure (b) is a cross section through the upper figure.

against the edge of the landfill or previous daily cell. A horizontal layer of daily cells is called a **lift**. The number of lifts will depend on the topography of the site and the desired working life of the landfill.

The two principal design types are area and valley. An **area landfill** requires a relatively flat area and attempts to fit as much waste into that space as possible (moving vertically upward). This needs to be accomplished without causing geotechnical stability problems or exceeding any height limitation (h in Figure 10.14) set by zoning standards. For a rectangular land area, the relevant solid is a truncated, rectangular-based pyramid as depicted in Figure 10.14.

Using the dimensions provided in Figure 10.14 and letting the slope, G, be the length divided by the height ($0.5 \times [L_1 - L_2]/h$), the volume of this pyramid is given as follows:

$$V = \frac{h}{3} \times \left\{ L_1 \times W_1 + [(L_1 - 2\,Gh)(W_1 - 2\,Gh)] \right.$$
$$\left. + \sqrt{L_1 \times W_1 \times (L_1 - 2\,Gh)(W_1 - 2\,Gh)} \right\}$$

(10.8)

Note that the volume determined in Equation 10.8 is the maximum amount of compacted solid waste that could be contained in the landfill, assuming that no volume is taken up by daily or final cover soil.

The soil located under this type of landfill can be used for covering compacted refuse or for other purposes. It is thus common to excavate soil under an area landfill before construction. Equation 10.8 can also be adapted to estimate the volume of compacted waste placed below grade.

To determine the volume available from a **valley fill landfill** design, compare the topographic contours of a site before filling and estimated contours after filling (Figure 10.15). To maximize the waste placed per unit surface area, the fill needs to increase at maximum slope from the low point of the valley until it reaches its final height. Modern computer-aided design software can quickly assess volumes of valley fills.

Typical densities for compacted solid waste vary from 700 to 1,000 kg/m^3. Previously, Table 10.4 provided how the density of solid waste increases as the waste is collected, transported, and then landfilled. The density is 90–178 kg/m^3 for loose refuse and increases to 475–772 kg/m^3 when first placed in the landfill. This density increases further with pressure from lifts of waste above. The density assumed in design will vary with the composition of the waste (e.g., construction and demolition waste is denser) and the depth of waste (greater depths lead to greater pressure and thus greater density).

In addition to compacted waste, the **landfill volume** will often include substantial amounts of **daily soil cover**. This soil is usually obtained from the landfill site. A typical thickness, T, of daily soil cover is 200 mm. The daily cell is designed based on the daily compacted volume of refuse (V_r), the slope of the cell (G), the daily refuse height (H), the refuse length (L), and the working face width (W) (different than L_1 and W_1 in Equation 10.8). The volume of daily soil cover (V_s) required in an idealized daily cell can be related to the volume of

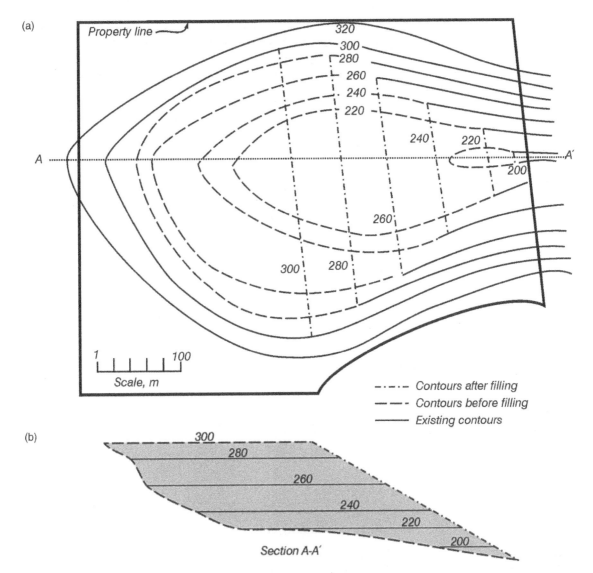

(a)

Property line

320
300
280
260
240
220

240
220
200

A ·········· A'

260

300 280

1 ⊢⊢⊢⊢⊢⊢ 100
Scale, m

–·–·– Contours after filling
––– Contours before filling
——— Existing contours

(b)

300
280
260
240
220
200

Section A-A'

Figure / 10.15 **Topographic Contours Used to Determine Available Volume for Waste Material in a Valley Fill Landfill Design** The top figure (a) shows the topographic contours. The bottom figure (b) shows a section drawn through the upper figure.

(Redrawn with permission of Tchobanoglous et al., *Integrated Solid Waste Management*, 1993, copyright The McGraw-Hill Companies).

compacted refuse (V_r) as follows (Milke, 1997):

$$\frac{V_s}{V_r} = \left[\left(1 + \frac{T}{H}\right) \times \left(1 + \frac{G \times T}{L}\right) \times \left(1 + \frac{G \times T}{W}\right) \right] - 1 \quad \textbf{(10.9)}$$

The height of a daily cell (H) usually depends on site management factors and can vary from 2 to 5 m. G is the grade ratio of length to height and is unitless. The daily volume of compacted refuse, V_r, is given by $H \times L \times W$.

The working face width will depend on the number of vehicles the site can accommodate at any one time under safe conditions.

The working face width can be estimated by dimensional analysis. For example, the working face width can be determined, assuming a daily discard rate of 1,000 Mg/day (equals 1,000 tonnes/day) (and noting the assumptions related to truck size and operation of the landfill incorporated in Equation 10.10) as follows:

$$\frac{1{,}000 \text{ tonnes}}{\text{day}} \times \frac{1 \text{ truck}}{8 \text{ tonnes}} \times \frac{1 \text{ day}}{6 \text{ h operation}} \times \frac{6 \text{ m}}{\text{truck}}$$

$$\times \frac{0.167 \text{ h unloading}}{\text{truck}} = 21 \text{ m} \qquad (10.10)$$

This value of 21 m implies that 3.5 trucks can unload at once, which is not practical. The working face length should be rounded up to the nearest multiple of truck width. In this case, we assumed a 6 m distance is required per truck to ensure safe operation, and the working face should fit four trucks, which would result in $W = 24$ m.

---

example/ 10.7 Sizing a Landfill

A landfill and associated structures are to be built on flat ground. The dimensions for the landfill portion are 1,000 m by 1,000 m. The maximum height allowed (h) is 9 m aboveground (excluding final cover). The landfill will be open 6 days a week and accept 1,000 Mg (equals 1,000 tonnes/day) of waste per day of operation. How many years can the facility operate?

Other assumptions are as follows: thickness of daily soil cover $(T) = 0.2$ m; refuse height in daily cell $(H) = 3$ m; compacted density $= 1{,}000$ kg/m^3; working face width $(W) = 24$ m; and slope $(G) = 3$.

solution

The total landfill volume is obtained from Equation 10.8:

$$V = \frac{9 \text{ m}}{3} \times \Big\{ [1{,}000 \text{ m} \times 1{,}000 \text{ m} + (1{,}000 \text{ m} - 2(3)(9 \text{ m}))(1{,}000 \text{ m} - 2(3)(9 \text{ m}))]$$

$$+ \sqrt{1{,}000 \text{ m} \times 1{,}000 \text{ m} \times 1{,}000 \text{ m} - 2(3)(9 \text{ m})(1{,}000 \text{ m} - 2(3)(9 \text{ m}))} \Big\}$$

$$V = 3 \text{ m} \times \Big\{ \big[10^6 \text{ m}^2 + 894{,}916 \text{ m}^2\big] + 946{,}000 \text{ m}^2 \Big\}$$

$$= 8.5227 \times 10^6 \text{ m}^3$$

The total volume of a daily cell consists of the volume occupied by compacted refuse (V_r) and daily cover soil (V_s). Using this knowledge and Equation 10.9, the volume of the daily cell can be determined as follows:

$$V_{\text{daily cell}} = V_s + V_r$$

Solve Equation 10.9 for V_s and substitute it into the previous equation:

$$V_{\text{daily cell}} = V_r \Big\{ \Big(1 + \frac{T}{H}\Big) \times \Big(1 + \frac{G \times T}{L}\Big) \times \Big(1 + \frac{G \times T}{W}\Big) \Big\}$$

In the above equation, L is determined based on our previous definition of V_r,

$$L = \frac{V_r}{H \times W} = \frac{\frac{1,000\ \text{Mg}}{1,000\ \text{kg/m}^3} \times \frac{1,000\ \text{kg}}{\text{Mg}}}{3\ \text{m} \times 24\ \text{m}} = 13.9\ \text{m}$$

Solving for $V_{\text{daily cell}}$

$$V_{\text{daily cell}} = \frac{1,000\ \text{Mg/day}}{1,000\ \text{kg/m}^3} \times \frac{1,000\ \text{kg}}{\text{Mg}} \times \left\{ \left(1 + \frac{0.2\ \text{m}}{3\ \text{m}} \right) \times \left(1 + \frac{3 \times 0.2\ \text{m}}{13.9\ \text{m}} \right) \times \left(1 + \frac{3 \times 0.2\ \text{m}}{24\ \text{m}} \right) \right\}$$

$$= 1,140\ \text{m}^3/\text{day}$$

Find the volume of all the daily cells per year, assuming 6 days' operation per 7-day week:

$$V_{\text{daily cells}} = 1,140\ \text{m}^3/\text{day} \times \frac{365\ \text{days}}{\text{year}} \times \frac{6\ \text{days}}{7\text{-day week}}$$

$$= 356,700\ \text{m}^3/\text{h}$$

Finally, determine the number of years of capacity in the total volume of the landfill:

$$\text{years} = \frac{8.5227 \times 10^6\ \text{m}^3}{356,700\ \text{m}^3/\text{year}} = 24\ \text{years}$$

Readers may want to resolve this problem assuming that the disposal rate decreases over time, or that the population increases and the waste discard rate remains the same.

LANDFILL MANAGEMENT A landfill requires careful management over its lifetime. The decomposition of waste and the increase in pressure from waste added in higher lifts will cause the waste to settle. This necessitates the repair of roads, gas collection, and water drainage systems. Surface water can be easily contaminated if it contacts the waste; thus, careful separation of rainwater and waste is needed. Landfills require heavy machinery associated with earthworks and waste compaction after the waste is placed on the working face. Roads on a landfill require careful management, not only because of the use of heavy vehicles, but also because of settlement of waste. A detailed and rigorous system for protection of safety and health of workers is a key aspect of good landfill management. Landfills must also be good neighbors to the surrounding community. Therefore, vigilant measures are required to reduce impacts from noise, odor, birds, dust, and litter.

Good management is closely tied to good monitoring. Environmental protection and occupational safety require monitoring of gas, leachate, groundwater, and surface water. In addition, landfill managers will monitor the waste that arrives to ensure that inappropriate materials are not deposited and to provide data on the rate of arrival of waste.

Waste density is also often monitored to allow for planning the landfill's life. Landfills will typically charge for waste on the basis of the mass received. The charge must be high enough to pay for large construction costs and also provide for ongoing monitoring and maintenance after the landfill has completed its useful life.

10.3.6 SOLID-WASTE ENERGY TECHNOLOGIES

Because of an increasing interest in obtaining energy from waste, a number of alternative **energy technologies** are undergoing development. **Anaerobic digestion** is a biological process of converting separated biodegradable solid wastes into methane gas and a solid residue that is suitable for turning into compost. The process works very similarly to those that produce landfill gas. The difference is that in an anaerobic digester (discussed in Chapter 9), the wastes are mixed in a large vessel, and optimal degradation conditions are ensured, giving greater production of energy from the wastes. The technology is currently relatively expensive, and for it to be successful, the wastes must be readily degradable (such as food wastes), and the other alternatives available, such as landfill or incineration, must be expensive.

Gasification is a process, similar to incineration, where less than stoichiometric amounts of oxygen are applied to the waste in the reaction vessel. The high temperatures that result from partial combustion lead to the production of high-energy gases, which in turn can be converted into power, usually in less polluting ways. Pyrolysis is a similar process, where even less oxygen or no oxygen is applied to the reaction, leading to the production of high-energy gases and a solid residue (char) that can be separated to give a high-energy residue for later combustion. Both processes have been most successfully applied to relatively homogeneous wastes, such as tires or plastics (Malkow, 2004).

10.4 Management Concepts

Successful solid-waste management requires a systems approach. Rather than attempt to analyze whether individual system components are better or not, society needs to evaluate, in a holistic, integrated manner, the combination of components that will maximize benefits at a given cost for current and future generations. Figure 10.16 provides

The College and University Recycling Council (CURC) organizes and supports environmental program leaders at institutions of higher education in managing resource, recycling, and waste issues

http://www.nrc-recycle.org/

Figure / 10.16 Example of a Systems Approach to Solid-Waste Management Note the many technologies and waste streams that are integrated in this approach.

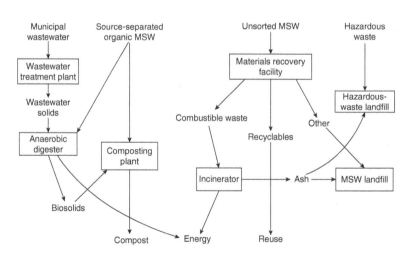

Waste management techniques such as composting, recycling, and source reduction result in the reduction of the quantity of waste bound for disposal options such as landfilling. Three waste reduction strategies were studied in the Parish of St. Ann in the small island nation of Jamaica (Post and Mihelcic, 2010). Factors designated as incentives to waste reduction existed primarily at the household level—specifically, waste segregation, household education, environmental concern, and knowledge—whereas the barriers existed primarily at the national or regional levels, namely government

(Courtesy of James R. Mihelcic).

policies and finances. (More information on the incentives and barriers to recycling programs can be found at Troschinetz and Mihelcic, 2009.)

The greatest potential for initiating waste reduction strategies and diverting waste from a landfill in this setting was thus within the household, specifically by community-based waste reduction initiatives that build upon already existing practices and improve local solid-waste management.

Example of source reduction at the point of collection. Photo taken in Lavena, Island of Taveuni (Fiji), home of the rare orange dove.

---

an example of how municipal solid-waste management can involve a complex mix of technologies appropriate to the types of waste.

Arriving at a system like the one shown in Figure 10.16 requires focus on the overall objectives; creativity in developing new sustainable possibilities, which may include redefining the original problem; and recognition of the impact that decisions on one part of the system can have on the functioning of the overall system. For example, operating a successful incinerator requires a steady supply of high-energy waste, such as paper. If a community chooses to invest in an incinerator, it then becomes difficult to consider other options for managing wastepaper, such as recycling through materials recovery. If a community chooses to separate food wastes and can do so at a low level of contamination, then the possibility of treating the waste along with wastewater-derived solids becomes an option that did not exist before.

Separation of waste types by waste generators, known as **source separation**, is a key part of a good system. If a community is able and willing to separate solid-waste components, the ability to create value from the wastes increases. Evaluation of the economics of source separation should consider the avoided disposal cost of separated wastes.

Class Discussion
What type of meaningful source reduction could be implemented in your own household? How would you implement the plan with all members of your house or apartment?

10.4.1 CONSULTATION

Public concern about solid waste is high, as it is for many other engineering activities. As a result, engineers today must discuss proposed projects and programs with the various **stakeholders** that make up the public—an effort that includes listening to the public's concerns and ideas. A first step in a process of consultation is the identification of stakeholders who have a direct or indirect interest in the project. Stakeholders can be neighbors, the local community, the wider community, news media, elected officials, environmental and social interest groups, and cultural groups.

Table / 10.17

Consultation Methods and Their Potential to Achieve Specific Outcomes   Consultation methods that appear to take more time (e.g., workshops, advisory committees, mediation) typically result in a higher outcome.

Consultation Method	Outcomes							
	Inform Stakeholders	Identify Values	Generate Options	Change Opinions	Resolve Conflict	Change Proposal	Costly Consultation	Lengthy Consultation
Information releases	High	Low	Low	Moderate	Low	Low	Moderate	Moderate
Field trips/site visits	Moderate	Moderate	Moderate	Moderate	Moderate	Moderate	Moderate	Low
Information stands/club visits	Moderate	Moderate	Moderate	Moderate	Low	Low	Low	Moderate
Contact person	Low	High	Low	Low	Low	Low	Low	Moderate
Public meetings	Moderate	High	Moderate	Low	Low	Low	Low	Low
Workshops	Moderate	High	High	Moderate	Moderate	Moderate	Moderate	Moderate
Advisory committees	Low	Moderate	High	High	High	High	High	High
Mediation	Low	Moderate	Moderate	Low	High	High	Moderate	High

Table 10.17 summarizes reasons to consult with stakeholders and identifies methods that have greater potential to match the desired outcome. In general, consultation methods that involve greater cost and longer time periods tend to result in better outcomes.

To be effective, consultation must begin early, be provided with adequate resources, be open and sincere, and involve good listening. In general, giving up a degree of power to stakeholders and allowing modifications to the project will result in the greatest chance of public acceptance. Consultation is therefore a critical part of managing any engineering project.

EPA's WAste Reduction Model (WARM) Tracks Greenhouse Gas Emissions Reductions from Different Waste Management Practices

http://www.epa.gov/climatechange/wycd/waste/calculators/Warm_home.html

10.4.2   POLICY OPTIONS

Table 10.18 provides an overview of the policy options to meet objectives for solid-waste management. Good policy development requires assessments of costs and benefits, a focus on objectives, and a consideration of risks and unintended effects (Australian Productivity Commission, 2006).

10.4.3   COST ESTIMATION

Socially acceptable waste management facilities can be expensive, and equitable compromises between costs and social benefits are needed. An underlying cause of difficulty is the economy of scale of

Table / 10.18

Policy Options for Meeting Objectives for Solid-Waste Management

Policy Option	Description	Example	When to Consider	When to Avoid
Public	Inform the public of preferred behaviors.	Exchanges of commercial waste	Information is in individual's best interest as well as helpful to waste management.	Changing behavior information.
Ecolabeling	Inform the public which consumer goods create fewer waste problems.	Reusable shopping bags; ecolabeled detergents	Consumers lack information on waste impacts of products.	Differences between options are small or difficult to assess.
Waste targets	Government or industry groups set future goals.	50% increase in paper recycling by 2020	Society agrees on direction but lacks a focus.	Target does not consider costs, side effects, or risks.
Government subsidies	Government supports recycling or waste minimization efforts following pollution prevention hierarchy.	Grants to community recycling efforts	Environmental effects of waste are not reflected by costs.	Grants awarded for activities that would happen in any case.
User pays	Producers of waste rather than the government pay the full cost of management.	Weight-based charges for residential solid waste	Cost of charging system is small.	Users avoid charges by illegal practices.
Enforcement	Violators of rules pay a fine.	Tickets for littering	Behavior is clearly negative, and few violators exist.	Many violators and each causes very small impact.
Deposit-refund	Consumers receive refund as incentive for proper management of waste.	Deposit-refund for beverage containers and car batteries	High negative consequences of improper waste management.	Large costs to operate system, and few social and environmental benefits.
Waste taxes	Government imposes taxes on waste.	Landfill tax	Taxes linked to environmental consequences of activities.	Large costs to operate system or undesirable side effects.
Producer responsibility	Producers responsible to take back goods at end-of-use life stage.	Computer return systems	Goods can be reused easily in production of new products.	Large costs in responsibility collection, storage, and transport.
Ban of goods and practices	Government prohibits goods or practices.	Ban of specific pesticides; ban of backyard burning	Goods/practices have high potential for harm, and other policy options are too costly.	Impacts are small or can be managed with other policies.

most waste management facilities, which means that a facility twice as large does not cost twice as much. For example, the costs of a landfill per tonne (Mg) over a year in the European Union (in 2003 euros) had the following form (Tsilemou and Panagiotakopoulos, 2006):

$$\text{total landfill cost} = 5{,}040 \times X^{-0.3} \qquad (10.11)$$

for situations where X was between 60,000 and 1,500,000 Mg/year. From Equation 10.11, the total landfill cost per Mg of discarded solid waste for 60,000 Mg per year can be estimated at 186 euros/Mg. For 10 times the amount of solid waste discarded (600,000 Mg per year), the cost decreases to 93 euros/Mg.

Economy of scale means that larger landfills, incinerators, and compost plants are economically favored. However, they are more likely to face public opposition. Of course, sustainable development has shown that local solutions are often a preferred alternative.

Key Terms

- anaerobic digestion
- area landfill
- carbon-to-nitrogen (C:N) ratio
- cell
- collection
- composting
- construction and demolition debris
- cover
- daily soil cover
- disposal
- dumps
- energy recovery
- energy technology
- final cover
- gasification
- generation rate
- greenhouse gas emissions
- hazardous waste

- incineration
- in-vessel
- landfill
- landfill volume
- leachate
- leachate management
- lift
- liner
- materials recovery facility (MRF)
- methane
- moisture content
- municipal solid waste (MSW)
- pollution prevention
- pollution prevention hierarchy
- processing
- putrescible
- recycling
- reduction

- refuse-derived fuel (RDF)
- Resource Conservation and Recovery Act (RCRA)
- reuse
- solid waste
- solid-waste management
- source separation
- stakeholders
- storage
- transfer stations
- transport
- valley fill landfill
- waste
- waste density
- waste generation
- waste-to-energy
- windrow

10.1 A community with a population of 150,000 has a solid-waste generation rate of 1.5 kg solid waste/day-person. Assume that yard waste makes up 15 percent of the total waste generated (by weight) and yard waste is banned by the state from being disposed of in a landfill; therefore, the community has set up a program to collect and compost yard waste. Assume the density of the loose solid waste is 140 kg/m^3 at the curb is compacted to 340 kg/m^3 in the truck that collects the waste at the home, and is 220 kg/m^3 after the material is removed from the compacter truck at the landfill. (a) Is this generation rate above or below the current value for a U.S. residential community? (b) What is the volume of waste that is discarded every day by the community at the source (m^3)? (c) What is the volume of waste that will removed from the compacter truck at the landfill (m^3)?

10.2 A new solid-waste landfill site is being designed with a projected life of 10 years. The landfill will serve a population of 250,000 that generates 1 kg solid waste/day-person. Assume that yard waste makes up 15 percent of the total waste (by weight), paper makes up 40 percent of the total waste (by weight), and metals make up 10 percent of the waste (by weight). The municipality bans the placement of yard waste in landfill and has a recycling program that collects one-half of all discarded metals. What is the volume of waste that is discarded by the community every day (assume a waste density at the curb of 140 kg/m^3).

10.3 Design and safely perform a waste characterization on the solid waste at your residence and at an office at your university or college. (a) How does your waste characterization compare with the data in Figure 10.2? (b) Which of the following pollution prevention strategies (source reduction, reuse, recycle) would you implement to reduce the discard rate?

10.4 Identify one source of solid waste on your campus that could readily be reduced, one source that could be reused, and one that could be recycled. What social, economic, and environmental benefits would come from implementing a plan to deal with the three items you identified?

10.5 Research the energy and water savings associated with recycling 1,000 kg office paper. Which value do you consider the most reliable of the ones you found? Justify your choice and provide a reference for your preferred source of information.

10.6 Using the values provided in Example 10.2, estimate the low moisture content and typical moisture content for the waste as a whole.

10.7 Waste composition has been measured for two cities. The results are summarized in Table 10.19.

Table / 10.19

Data for Problem 10.7

	City 1	City 2
Wet-weight generation rate (kg/person-day)	2.0	1.8
Wet-weight composition (%)		
Food	15	10
Paper	30	40
Yard	20	15
Other	35	35
Moisture content of fractions (% on wet-weight basis)		
Food	80	50
Paper	10	4
Yard	80	30
Other	5	4

(a) Which city generates more paper on a dry-weight basis? (b) Find the percent moisture (wet-weight basis) for City 1. (c) A nearby disposal site receives all of its MSW from cities 1 and 2. The average moisture content for MSW disposed of at the site is 20 percent. What fraction of the dry-weight refuse comes from City 1?

10.8 What is the dry-weight percent composition for the following combined waste?

Component	% Composition	% Moisture (Wet Weight)
Paper	40	6
Yard/food	30	60
Other	30	3

10.9 The mass composition of dry paper is 43 percent carbon, 6 percent hydrogen, 44 percent oxygen, and 7 percent other. Estimate the liters of air required to burn 1 kg dry paper. Assume carbon dioxide and water are the only products of combustion of carbon, hydrogen, and oxygen. Assume a temperature of 20°C and pressure of 1 atm.

10.10 Estimate the oxygen demand for composting mixed garden waste (units of kg of O_2 required per kg of dry raw waste). Assume 1,000 dry kg mixed garden waste has a composition of 513 g C, 60 g H, 405 g O, and 22 g N. Assume 25 percent of the nitrogen is lost to $NH_{3(g)}$ during composting. The final C:N ratio is 9.43. The final molecular composition is $C_{11}H_{14}O_4N$.

10.11 Waste of the composition shown in the following table is disposed of at a rate of 100,000 Mg/year for 2 years in one section of a landfill. Assume that half of the waste is disposed of at time = 0.5 years, and half at a time = 1.5 years. Assume that gas production follows the first-order relationship used in Equation 10.4, and use the additional information provided in the table. How long until 90 percent of the gas will be produced in this section?

	Initial Mass (Mg)	Half-Life (year)
Slowly biodegrading	10,000	10
Rapidly biodegrading	40,000	3
Nonbiodegrading	50,000	Infinite

10.12 Assume all the waste in one section of a landfill was added at the same time. After 5 years, the gas production rate reached its peak. After 25 years (20 years after the peak), the production rate had decreased to 10 percent of the peak rate. Assume first-order decay in the gas production rate after reaching its peak. Assume no gas is produced prior to the peak of 5 years. (a) What percentage of the total gas production do you predict has occurred after 25 years? (b) How long do you predict until 99 percent of the gas has been produced?

10.13 Equal amounts of two types of waste are disposed into a section of a landfill. They both start producing gas at $t = 0$, and so there is no lag time. Assume first-order decay for gas production. Each type of waste can produce 150 L CH_4/kg of waste. Waste type A produces gas with a half-life of 6 years, and waste type B produces gas with a half-life of

3 years. How long (to the nearest year) until 90 percent of each gas has been produced?

10.14 Determine whether your local (or regional) landfill produces energy from methane gas. If so, what is the mass of solid-waste disposed at the landfill on an annual basis, and what is the amount of CH_4 generated? Relate these numbers to a calculation you can perform with appropriate assumptions.

10.15 (a) Calculate the volume of methane produced (m^3/year) due to landfilling for the years 1970 and 2010. Assuming the landfilled MSW produces gas in a similar fashion between the 2 years. The U.S. Census Bureau reports the U.S. population was 203,392,031 in 1970 and was 308,745,531 in 2010. Use the landfilling and composting rates provided in Table 10.2. Assume the three waste components that produce methane did not change over time and are food wastes (15 percent of total), mixed paper (30 percent of total), and yard wastes (15 percent of total). Assume that 60 percent of the food and paper wastes and 40 percent of the yard trimmings will decompose if placed in a landfill. (b) Determine the energy (in MW) of landfill gas produced in 1970 and 2010. Assume 1 MW of gas is produced for every 270 m^3/h of CH_4 produced at the landfill

10.16 Return to Example 10.6 in this chapter. The overall greenhouse gas effect of a landfill is sensitive to a number of parameters and assumptions. In Example 10.6, an assumption of 80 percent gas recovery is used and leads to an overall greenhouse gas benefit of 0.10 metric tons CO_2e. Leaving all parameters and assumptions in place used to solve Example 10.6, what is the percentage of landfill gas collected that provides an overall benefit of reducing greenhouse gas emissions.

10.17 What percentage reduction in yard waste would be required to reduce the NH_4^+ released in landfill leachate by 1 kg per Mg of MSW? Assume only yard waste contributes to NH_4^+ in leachate. Assume the waste composition provided in Figure 10.2 and Table 10.5. Assume all N in yard waste is eventually released as NH_4^+.

10.18 Daily cells for a landfill are operated so that the following conditions are maintained: thickness of daily cover = 0.2 m; slope (horizontal:vertical) = 3:1; working face for refuse = 30 m; height of refuse = 3 m; and volume of daily refuse = 1,800 m^3/day. The landfill is interested in reducing requirements for daily cover soil over its 20-year life and is considering three options. Which option would be the best? Why?

Option 1	Increase height of refuse to 4 m.
Option 2	Increase daily refuse volume to 2,000 m³/day.
Option 3	Decrease working face to 20 m.

10.19 Estimate the landfill area required, in hectares, given the following specifications: daily cover thickness = 0.2 m; total final cover = 1.0 m (in addition to daily cover); height above ground before biodecay and settlement = 10 m; lift height = 3–5 m; depth below ground level that waste can begin to be placed = 5 m; landfill site area is square; MSW generation rate = 100,000 Mg/year; side slopes at 3 horizontal:1 vertical for daily cells and external slopes; working face width = 8 m; open for disposal 360 days per year; 30-year life; and in-place density of fresh MSW = 700 kg/m³.

10.20 You need to budget for a new transfer station in your district. A similar transfer station cost $1 million, but that station was 50 percent larger than yours. How much money should you budget so that your local government will have enough money to pay for the new transfer station? Assume that the economy-of-scale factor for transfer stations is 0.9.

10.21 Go the World Health Organization web site (www.who.org). Learn about a disease that is transmitted through improper disposal of solid waste. What is the extent of the disease on a global level?

10.22 Identify an engineering professional society that you could join as a student or after graduation that deals with issues of solid-waste management. What are the dues for joining this group? What benefits would you receive as a member while working in practice?

10.23 Determine the number of students currently enrolled full time at your university or college. Then, using the information provided in Figure 10.2 and Tables 10.2 and 10.5, determine the energy content associated with a day's worth of solid waste that would be generated by this population.

10.24 Assume the population of the United States will reach 420,000,000 in 2050. Estimate the annual mass of municipal solid waste that will be generated in the United States in 2050 and the annual amount that will require landfilling (both answers in metric tons). Use information provided in Table 10.2. Justify your assumptions on changes in solid-waste generation per person and landfill disposal per person from now until 2050. HINT: Graph waste generation versus time and also the percent of waste landfilled versus time and observe the trends. Make your own assumptions (e.g., waste generation will be the same in 2050 as 2010; waste generation will decrease back to 1960 levels; and/or the percent of landfill waste will decrease as recycling becomes more mainstream or it will remain the same).

10.25 Nonlegume vegetable wastes have a moisture content of 80 percent and are 4 percent N (on a dry mass basis). The vegetable wastes are to be composted with readily available sawdust. The sawdust has a moisture content of 50 percent and is 0.1 percent N (on a dry mass basis). The desired C:N for the mixture is 20. The C:N ratio for vegetable wastes is 11, and the C:N ratio for sawdust is 500. Determine the kg of sawdust required per kilogram of vegetable waste that results in an initial C:N ratio of 20.

10.26 A mixture of organic materials is to be composted. The mixture begins with 40 percent moisture and 80 percent of the solids are VS. Assume that 50 percent of the VS are lost through composting along with 70 percent of the moisture. What is the moisture content of the final compost?

10.27 EPA provides five methods of composting at this web site: http://www.epa.gov/compost/types.htm. Develop a table that lists the five methods in one column and brief description of the method in a second column.

10.28 Estimate the total landfill costs (in 2003 euros) for a situation where you must landfill (a) 75,000 Mg of solid waste per year and (b) 1,000,000 Mg of solid waste per year.

References

ASTM International, 1992. ASTM Standard D5231-92, *Standard Test Method for Determination of the Composition of Unprocessed Municipal Solid Waste*. West Conshohocken: ASTM International. Available at www.astm.org.

Australian Productivity Commission (APC), 2006. *Waste Management*. Report No. 38. Canberra: APC.

Camobreco, V., R. Ham, M. Barlaz, E. Repa, M. Felker, C. Rousseau, and J. Rathle, 1999. Life-cycle inventory of a modern municipal solid waste landfill. *Waste Management and Research*, 17: 394–408.

Department of Energy (DOE), Form EIA-1605 (2007). Voluntary Reporting of Greenhouse Gases, Appendix F. Electricity Emission Factors.

Diaz, L. F., G. M. Savage, L. L. Eggerth, and C. G. Golueke, 2003. *Solid Waste Management for Economically Developing Countries*, 2nd ed. Concord: Cal Recovery.

Environmental Protection Agency (EPA), 2006. *An Inventory of Sources and Environmental Releases of Dioxin-Like Compounds in the United States for the Years 1987, 1995, and 2000*. Washington, D.C.: EPA, EPA/600/P-03/002f.

Environmental Protection Agency (EPA), 2011. *Municipal Solid Waste Generation, Recycling and Disposal in the United States: Facts and Figures for 2010*. Washington, D.C., EPA530-F-11-005.

Farquhar, G. J., and F. A. Rovers, 1973. Gas production during refuse decomposition. *Water, Air, and Soil Pollution*, 2: 483–495.

GMOP (Global Methane Outreach Program), 2012. *International Best Practice Guide for Landfill Gas Energy Projects* (Chapter 4), USEPA. Available at http://www.globalmethane.org/tools-resources/tools.aspx).

Haug, R. T., 1993. *The Practical Handbook of Compost Engineering*. Boca Raton: Lewis Publishers.

Kontos, T. D., D. R. Komilis, and C. P. Halvadakis, 2003. Siting MSW landfills on Lesvos Island with a GIS-based methodology. *Waste Management and Research*, 21: 262–277.

Levis, J. W., and M. A. Barlaz, 2011. Is biodegradability a desirable attribute for discarded solid waste? Perspectives from a National Landfill Greenhouse Gas Inventory Model. *Environmental Science and Technology*, 45: 5470–5476.

Malkow, T., 2004. Novel and innovative pyrolysis and gasification technologies for energy efficient and environmentally sound MSW disposal. *Waste Management*, 24: 53–79.

McBean, E. A., F. A. Rovers, and G. J. Farquhar, 1995. *Solid Waste Landfill Engineering and Design*. New York: Prentice Hall.

Medina, M., 2000. Scavenger cooperatives in Asia and Latin America. *Resources, Conservation, and Recycling*, 31: 51–69.

Milke, M. W., 1997. Design of landfill daily cells to reduce cover soil use. *Waste Management and Research*, 15: 585–592.

National Solid Wastes Management Association (NSWMA), 2003. *Modern Landfills: A Far Cry from the Past*. Washington, D.C. NSWMA. Available at www.nswma.org.

New Zealand Ministry for the Environment (NZME), 2002. *Solid Waste Analysis Protocol*. March, Ref. ME 430. Wellington: NZME.

Pierce, J., L. LaFountain, and R. Huitric, 2005. *Landfill Gas Generation and Modelling Manual of Practice*. Silver Spring: Solid Waste Association of North America.

Post, J. L., and J. R. Mihelcic, 2010. Waste reduction strategies for improved management of household solid waste in Jamaica. *International Journal of Environment and Waste Management*, 6(1/2): 4–24.

Tchobanoglous, G., H. Theisen, and S. A. Vigil, 1993. *Integrated Solid Waste Management*. New York: McGraw-Hill.

Troschinetz, A. M., and J. R. Mihelcic, 2009. Sustainable recycling of municipal solid waste in developing countries. *Waste Management*, 29(2): 915–923.

Tsilemou, K., and D. Panagiotakopoulos, 2006. Approximate cost functions for solid waste treatment facilities. *Waste Management and Research*, 24: 310–22.

chapter/Eleven Air Quality Engineering

James R. Mihelcic and
Amy L. Stuart

In this chapter, readers will learn about the sources of air pollution, the characteristics of clean and polluted outdoor and indoor air, and the adverse impacts that criteria, hazardous, and odorous air pollutants have on human health, the economy, and the environment. The framework of how air pollutants are regulated is discussed, including recent legal decisions that now allow for regulation of the major greenhouse gas, carbon dioxide. Urban, regional, and global scales of air pollution problems are described and related to issues of transport and impact. The structure of the troposphere and stratosphere is presented and examples of point, area, and mobile sources of air emissions to the troposphere are then described, along with the historical emission trends of criteria and hazardous air pollutants. The control of emissions is described in great detail with emphasis on source reduction and demand management strategies. Four other strategies to control air emissions are discussed (regulatory, market-based, and voluntary approaches, and control technologies) and several control technologies for gases and particles are described with emphasis on applicability and design. The chapter then reviews available methods to measure or estimate air emissions that can include use of mass balances and emissions factors. The chapter concludes with descriptions of how atmospheric and ground-level conditions control the vertical and horizontal movement of air parcels and also provides instruction on how to apply Gaussian dispersion models to estimate downwind concentrations of air pollutants.

© Marcus Lindström/iStockphoto

Learning Objectives

1. Describe how the urban, regional, and global scale of different air pollutant problems influences chemical reaction, transport, and impact.

2. Explain how management of air pollution encompasses not only effect, but, importantly, human demands for goods, services, and activities such as transportation, and apply the IPAT equation to automotive air emissions.

3. Describe important features and air pollutants of the troposphere and stratosphere.

4. Describe the composition of polluted indoor air in a global context and the impact fuel usage has on global indoor air quality while empathizing with at-risk individuals.

5. Identify the six criteria air pollutants and other hazardous and odorous air pollutants found in the ambient and indoor environments, and associate specific health, economic, and environmental impacts with the pollutants.

6. Describe the formation of carbon dioxide during combustion of fossil fuels, the legal justification that allows carbon dioxide to be regulated as an air pollutant, and impact that climate change may have on air quality.

7. Write and explain the chemical reactions that describe the formation of ground-level and stratospheric ozone.

8. Use and apply the air quality index as a measure to understand daily air pollution concentrations.

575

9. Demonstrate empathy for population groups that are unfairly assigned a greater environmental risk due to releases of air pollutants.

10. Differentiate between point, area, and mobile sources of air emissions that can also include fugitive emissions.

11. Explain how engine combustion leads to emissions of nitrogen oxides, carbon monoxide, carbon dioxide, and hydrocarbons.

12. Differentiate between different strategies to control emissions that include source reduction, demand management, regulatory approaches, market-based approaches, voluntary approaches, and control technologies.

13. Innovate different methods of transportation demand management to reduce vehicle congestion and air emissions.

14. Design and/or apply the following air pollutant control technologies for specific air pollutants: adsorption, baghouse, biofilter, cyclone, electrostatic precipitator, thermal oxidizer, ventilation hood, and venturi scrubber.

15. Integrate mass balances with chemical kinetics to size a thermal oxidizer.

16. Estimate air emissions using methods such as monitoring, modeling, mass balance approaches, and emission factors.

17. Explain how meteorological and ground-level features influence atmospheric stability and the vertical and horizontal movement of air parcels.

18. Apply Gaussian dispersion models to estimate downwind concentrations of air pollutants.

11.1   Introduction

In Chapter 2, we presented a table that showed the composition of the atmosphere (Table 2.3). Though almost every element listed in the periodic table is found in the atmosphere, if you go back and review Table 2.3, you will notice the atmosphere is composed mostly of nitrogen (78.1 percent), oxygen (20.9 percent), and argon (0.93 percent). For comparison, carbon dioxide now makes up over 0.039 percent of the atmosphere. However, anthropogenic activities cause the release of many different compounds into the air at concentrations that are high enough to cause adverse impacts on human health, crops and other vegetation, building materials, climate, and even habitants of aquatic ecosystems. The movement of air also does not respect the geopolitical boundaries of a county, state, or nation. Therefore, transboundary air pollution problems are common and the solutions can be quite complex.

The **Tragedy of the Commons** (discussed in Chapter 1) also applies directly to air pollution. Remember that the Tragedy of the Commons describes the relationship where individuals or organizations consume a shared resource (like air) and then return their wastes back into the shared resource. In this way, we learned that the individual or organization receives the benefit of the shared resource but distributes the cost to others who also use that resource. The social, economic, and environmental costs associated with air pollution are large and could result in increased health costs, lessening of life span, degradation of historic monuments and buildings, and crop damage, loss of ecosystem productivity, and consequences associated with change in climate.

Figure 11.1 shows the number of people living in U.S. counties with air quality concentrations that exceed the national-based standards set

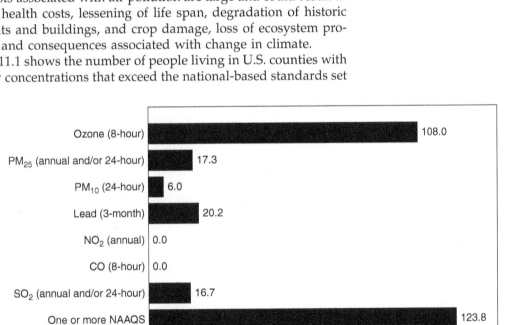

Figure / 11.1   Number of People (in millions) Living in U.S. Counties with Air Quality Concentrations Greater Than the Level of the Primary (Health-Based) National Ambient Air Quality Standards (NAAQS) in 2010   The time period listed refers to the averaging time basis for the standard.

(Figure redrawn from EPA, 2012a).

to protect human health. Note the number of people exposed to high levels depends on a particular air pollutant, but ranges from tens to hundreds of millions (the population of the United States is slightly over 300 million). Historically air pollution has been associated with human communities and industrialized activities. We now understand that the consequences of air pollution are found in every location of the Earth. Air pollution is found in cities from the combustion of air and fossil fuels in a coal-fired power plant and vehicles' engines, in rural areas from the particulates generated from abrasive agricultural activities, and in the indoor environment from evaporation of chemicals from building materials or surrounding soil (including naturally occurring radon). Emissions from combustion of fossil fuels in the comparatively poorly ventilated enclosed spaces of the indoor environment are also a concern. Three billion people in the world still burn solid fuels for heating and cooking. In the United States, indoor carbon monoxide poisoning is also common problem. Nitrogen and sulfur oxides emitted from combustion processes are transported to local water sources and may make up a significant amount of nitrogen loading and acidification, respectively, to surface waters like the Chesapeake Bay and Tampa Bay, and lakes in the Adirondack Mountains of upstate New York, while emissions of air pollutants have led to melting of polar ice and exposure of Artic Beluga whales to toxic chemicals.

In the United States, federal regulation of air pollution largely began with the Air Pollution Control Act of 1955. This was followed by the Clean Air Act of 1963 and the Air Quality Act of 1967. The **Clean Air Act** (of 1970) was a major milestone in defining the federal government's responsibilities for protecting and improving the nation's air quality. The Clean Air Act (CAA) was signed into law in 1970 and was amended in 1977 and 1990. It provided legislative authority for the government to develop federal and state regulations to limit emissions from stationary sources (for example, industrial stacks) and mobile source (for example, automobile tailpipes). Table 11.1 provides some of the major provisions of the Clean Air Act and the **Clean Air Act Amendments**. One of the major regulatory programs of the Clean Air Act was the creation of the National Ambient Air Quality Standards (NAAQS), which will be discussed later in this chapter.

Air Pollutants
http://www.epa.gov/air/airpollutants.html

Air Quality Where You Live
http://www.epa.gov/air/where.html

History of Clean Air Act
http://epa.gov/oar/caa/caa_history.html

Table / 11.1

Major Provisions of the Clean Air Act and Amendments   (text obtained from http://www.epa.gov/air/).

Clean Air Act of 1970

- Enactment of the Clean Air Act of 1970 resulted in a major shift in the federal government's role in air pollution control. This legislation authorized the development of comprehensive federal and state regulations to limit emissions from both stationary sources and mobile sources.
- Enforcement authority was substantially expanded. The U.S. Environmental Protection Agency (EPA) was created on December 2,1970, in order to implement the various requirements included in these Acts.
- Authorized the establishment of National Ambient Air Quality Standards (NAAQS).
- Established requirements for State Implementation Plans (SIPs) to achieve the National Ambient Air Quality Standards.
- Authorized the establishment of New Source Performance Standards (NSPS) for new and modified stationary sources.
- Authorized the establishment of National Emission Standards for Hazardous Air Pollutants (NESHAPs).

Table / 11.1

(continued)

Clean Air Act Amendments of 1977

- Authorized provisions for the Prevention of Significant Deterioration (PSD) to protect clean (attainment) areas, particularly areas of special value, such as large national parks.
- Specified requirements for sources in areas that do not meet the National Ambient Air Quality Standard for one or more pollutants (called nonattainment areas). A **nonattainment area** is a geographic area that does not meet one or more of the federal air quality standards.

Clean Air Act Amendments of 1990

- Substantially increased the enforcement authority and responsibility of the federal government.
- Authorized new regulatory programs for control of acid deposition (that is, acid rain).
- The National Emission Standards for Hazardous Air Pollutants (NESHAPs) were incorporated into a greatly expanded program to control 189 toxic pollutants.
- Expanded and modified provisions concerning the attainment of National Ambient Air Quality Standards (NAAQS).
- Established a program to phase out the use of chemicals that deplete the stratospheric ozone layer.
- Established requirements for operating permits for all major air pollutant sources (Title V permits).

Application / 11.1   The EPA's Air Pollution Training Institute

The EPA's Air Pollution Training Institute (APTI) provides training for air pollution professionals. The goal is to facilitate professional development by enhancing the skills needed to understand and implement environmental programs and policies. The curriculum is divided by job function (for example, air toxics, permit writing, ambient monitoring) and includes classroom, self-instructional, and web-based training. Special courses and workshops are also offered. Learn more at: http://www.apti-learn.net

---

11.2   Scale and Cycles of Air Pollution

11.2.1   SCALE OF AIR POLLUTION ISSUES

Figure 11.2 shows that air pollution problems are associated with a wide variety of **spatial scales**, from urban, to regional, to global. Table 11.2 gives that the spatial scale of air pollution phenomenon can occur over distances that range from < 1 km to thousands of km. Emissions of air pollutants at ground level and near human activities (for example, traffic emissions and burning biomass in household cookstoves) lead to micro-environmental and local scale air pollution issues. These include direct personal exposures to combustion emissions, indoor air pollution, and inequality in air quality between neighborhoods.

In the atmosphere the emissions are transported, where they then mix and may react with emissions from industry and centralized energy generation facilities, leading to urban scale smog. Urban smog is characterized by broad areas with high levels of ozone

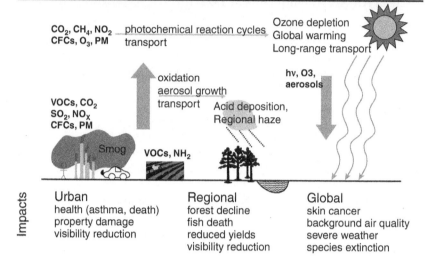

Air Pollution Issues and Scales

CO₂, CH₄, NO₂ CFCs, O₃, PM — photochemical reaction cycles / transport

CO_2, CH_4, NO_2, CFCs, O_3, PM — photochemical reaction cycles / transport

Ozone depletion
Global warming
Long-range transport

hv, O3, aerosols

oxidation
aerosol growth
transport

VOCs, CO₂
SO₂, NOₓ
CFCs, PM

$VOCs$, CO_2, SO_2, NO_x, CFCs, PM

Smog

VOCs, NH₂

Acid deposition,
Regional haze

Impacts

Urban
health (asthma, death)
property damage
visibility reduction

Regional
forest decline
fish death
reduced yields
visibility reduction

Global
skin cancer
background air quality
severe weather
species extinction

Figure 11.2 The Spatial Scale of Air Pollution Occurs over Urban, Regional, and Global Scales, Where It Results in Different Societal, Economic, and Environmental Impacts.

Table / 11.2

Spatial Scales of Air Pollution Phenomenon (based on Seinfeld and Pandis, 2006).

Phenomenon	Length Scale (km)
Indoor/micro-environmental pollution	0.001–0.1
Neighborhood air quality	0.1–1
Urban air pollution	1–100
Regional air pollution	10–1,000
Acid rain/deposition	100–2,000
Stratospheric ozone depletion	1,000–40,000
Greenhouse gas emissions	1,000–40,000

and fine particulate matter in a metropolitan region. Urban pollutants are further exported downwind, mixing and reacting with naturally occurring biogenic emissions and pollution from nearby urban regions, resulting in regional air pollution issues including acid deposition and regional haze. Finally, over longer timescales, many pollutants are transported to higher altitudes in the atmosphere. Once there, they can be transported over long distances, impacting air quality across continents. In this case, air pollutants can even interact with important natural atmospheric chemistry cycles and energy fluxes, leading to stratospheric ozone depletion and global climate change.

11.2.2 THE AIR POLLUTION SYSTEM

To consider the effectiveness of air pollution control measures, it is useful to conceptualize the air resource management system as a cycle that encompasses demand to effect (as shown in Figure 11.3). The origin of anthropogenic pollution, including air pollution, is caused from human demand for goods, services, and other activities (like travel), generated due to real and perceived needs for the betterment of economic and social well-being. This demand leads to emissions of air pollutants over the life cycle of the product or service or directly from the activity pursued. Emissions of air pollutants, which are transported and transformed in the atmosphere through a variety of multiphase processes, then lead to concentrations of pollutants in environmental media (that is, air, water, soils).

As explained in Chapter 6 during the discussion of environmental risk, when concentrations are high enough and lead to contact with a vulnerable receptor (human or environmental system), they may cause a variety of deleterious outcomes, including human health effects ranging from eye irritation to death, acid deposition and forest die off, visibility impairment in national parks, and global climate change. These effects, when they have had an adverse impact on human welfare, have led to attempts to regulate and control emissions through technological and management requirements. To some degree, there is also a feedback on the original demand, for example, educational

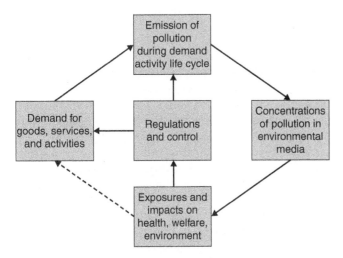

Figure / 11.3 Management of Air Pollution Follows a Cycle That Encompasses Demand to Effects.

efforts to reduce wasteful use of electricity and the associated air pollutants from production of electricity (for example, particulate matter, nitrogen oxides, sulfur oxides, CO_2, Hg).

The effectiveness of this cycle for reducing pollution and its effects is complicated by the other factors that impact the amount of pollution emitted by society. Remember in Chapter 5 we introduced the **IPAT Equation** (Equation 5.18) as a way to relate the environmental impact (I) to the influences of human population (P), affluence (A), and technology (T). The IPAT equation is rewritten here.

$$I = P \times A \times T \tag{11.1}$$

The IPAT equation can be written to understand air pollution generation (as the impact, I).

$$I = P \times D \times TF \tag{11.2}$$

In Equation 11.2, the amount of air pollution generated can be simplified to be the product of human population (P), per capita demand for, let's say, electricity or fossil fuel to power a vehicle (D, also often termed affluence or consumption), and a technology factor (TF, in this case, the ratio of air emissions per unit demand). In the case of a technology factor for vehicle emissions, TF would be reduced when a vehicle's fuel economy increased.

Equation 11.2 shows that as human populations increase, air emissions (that is, impact) will grow unless technology improves or demand per capita decreases. Fortunately, technologies and management structures that control emissions (or decrease concentrations at receptors) have improved over time. Hence, emissions (and effects) per unit demand of many traditional air pollutants have decreased in many areas of the world over the past several decades, despite substantial increases in population.

Unfortunately, the flip side of this coin is that demand per capita has also increased, in part due to new technologies that improve real or perceived quality of life (for example, more and larger televisions, cell phones, or even electric clothes dryers versus use of sun's energy for drying). As economies grow and societies amass wealth, demand per capita has usually increased. Further, there can be substantial impacts on social equity; some of the observed decreases in emissions per unit demand are local, with processes and their emissions often exported from rich to poor neighborhoods and countries. To create an air management system that is effective at multiple spatial scales, addressing the demand part of the air pollution system cycle cannot be overlooked. It is important for a reader to recognize that some of the demand leading to air emissions meets real needs and cannot be eliminated, but also to understand that technological improvements alone are unlikely to effectively address air pollution fully or equally.

Fuel Economy of Automobiles
http://www.fueleconomy.gov/

example/11.1 IPAT Equation and Improved Fuel Economy in Vehicles

The purpose of **Corporate Average Fuel Economy (CAFE) standards** is to reduce energy consumption by increasing the fuel economy of passenger cars and light trucks. They were first enacted by Congress in 1975. In December 2011, the National Highway Traffic Safety Administration (NHTSA) and the U.S. Environmental Protection Agency (EPA) issued joint final rules to further improve fuel economy (and reduce air emissions) for model years 2017 through 2025. The rules call for large improvements in fuel economy as shown in Figure 11.4c.

How might the IPAT equation help us understand the impact this improvement in fuel economy will have on environmental impact of air pollutants from driving for the time period 2010–2025?

solution

The IPAT equation can be written as

$$I = P \times D \times TF$$

In this particular situation of investigating how improvements in fuel economy might lessen the environmental impact from vehicle air pollutant emissions, P would equal the population of licensed drivers, D can be thought as the demand for using a vehicle (which can be measured in the number of vehicle miles traveled per licensed driver), and the technology term, TF (emissions per vehicle mile traveled), would decrease as fuel economy increases. Figure 11.4a–c shows these trends over the past few decades in the United States.

First, Figure 11.4a shows a steady increase of 128 million drivers on U.S. highways from 1950 to 2000. Notice the large increase in the P term over this 50-year time span. Currently approximately 88 percent of the driving age population is licensed to drive a motor vehicle and the number of licensed drivers is increasing at a similar rate to overall population. In 2010, the U.S. population was estimated to be 309 million, with expected increase of 21 percent to 375 million in 2025 (and 389 million in 2035).

Figure 11.4b shows that personal vehicle travel demand, measured as vehicle miles traveled (VMT) per licensed driver, grew at an average annual rate of 1.1 percent from 1970 to 2010, from about 8,700 miles per driver in 1970 compared to 12,700 miles per driver in 2010. Increased travel is caused by several factors that include greater income, declining costs associated with driving (determined by fuel economy and fuel price), and demographic changes (such as suburban sprawl). The Federal Highway Administration assumes that vehicle miles traveled per licensed driver will increase by an average of 0.2 percent per year, to 13,350 miles per driver in 2013. If we apply this annual rate, we will see an increase of 3 percent in 2025 over 2010.

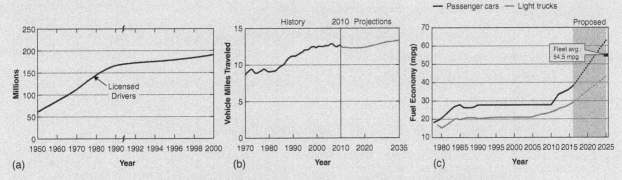

Figure / 11.4 (a) Licensed Drivers in United States from 1950 to 2000. (b) Vehicle Miles Traveled per Licensed Driver (1970–2035). (c) Fuel Economy (mpg) for New Passenger Cars and Light Trucks (Model Years 1980–2025).

(Figures adapted from the U.S. Federal Highway Administration and U.S. Energy Information Administration).

Figure 11.4c shows the improvements from vehicle and engine design expected if the new CAFE fuel economy standards are implemented by 2025. Because the technology factor is approximately inversely proportional to the fuel economy, this represents a decrease by 50% in 2025 over 2010.

Now going back to our IPAT equation adapted for our scenario, we can place the percent change in the particular terms that result in environmental impact in the year 2025 (I_{2025}):

$$I_{2025} = P_{2010}(1 + 21\,\text{percent}) \times D_{2010}(1 + 3\,\text{percent}) \times TF_{2010}(1 - 50\,\text{percent}) \tag{11.3}$$

The overall change in impact is this thus estimated as $I_{2025}/I_{2010} = (1.21) \times (1.03) \times (0.5) = 62\,\text{percent}$

This indicates that the expected improvements in fuel economy have the potential to reduce air emissions from use of vehicles by up to about 38 percent. However, a significant part of the technology-related reduction is offset by the large increases in the number of licensed drivers (P) and the expected increases in miles driven per vehicle (D). Other strategies to reduce the impact term, I, could include those related to demand management (reducing the D term) by providing mechanisms for drivers to access public or shared transport, incorporation of high occupancy vehicle lanes, improvements in bicycle and pedestrian paths, or changes in community design allowing closer access between daily activity centers.

Application / 11.2   Smart Growth

Smart growth is a term used to describe development of a community that protects natural resources, open space, and historical or cultural attributes of a community while also reusing land that is already developed. Table 11.3 lists the 10 principles of smart growth. Note how the principles are not based on design of vehicle-dominated communities and how they promote diversity in terms of housing, land types, and commuter accessibility. Table 11.4 identifies some environmental benefits of smart growth in terms of air quality, water quality, open-space preservation, and brownfields redevelopment.

Table / 11.3

Principles of Smart Growth

1. Mix land uses.

2. Take advantage of compact building design.

3. Create a range of housing opportunities and choices.

4. Create walkable neighborhoods.

5. Foster distinctive, attractive communities with a strong sense of place.

6. Preserve open space, farmland, natural beauty, and critical environmental areas.

7. Strengthen and direct development toward existing communities.

8. Provide a variety of transportation choices.

9. Make development decisions predictable, fair, and cost-effective.

10. Encourage community and stakeholder collaboration in development decisions.

SOURCE: From Smart Growth Online, available at www.smartgrowth.org.

Table / 11.4

Environmental Benefits of Smart Growth

Benefit	Comments
Improved air quality	• Siting new development in an existing neighborhood, instead of on open space at the suburban fringe, can reduce miles driven by as much as 58%. • Communities that make it easy for people to choose to walk, bicycle, or take public transit can also reduce air pollution by reducing automobile mileage and smog-forming emissions.
Improved water quality	• Compact development and open-space preservation can help protect water quality by reducing the amount of paved surfaces and by allowing natural lands to filter rainwater and runoff before it reaches drinking-water supplies. • Runoff from developed areas often contains toxic chemicals, phosphorus, and nitrogen; nationwide, it is the second most common source of water pollution for estuaries, the third most common for lakes, and the fourth most common for rivers.
Open-space preservation	• Preserving natural lands and encouraging growth in existing communities protects farmland, wildlife habitat, biodiversity, and outdoor recreation, and promotes natural water filtration. • A recent study in New Jersey found that, compared with less compact growth patterns, planned growth could reduce the conversion of farmland by 28%, open space by 43%, and environmentally fragile lands by 80%.
Brownfield redevelopment	• Cleaning up and redeveloping a brownfield can remove blight and environmental contamination, catalyze neighborhood revitalization, lessen development pressure at the urban edge, and use existing infrastructure.

SOURCE: From Environmental Protection Agency, www.epa.gov/smartgrowth.

11.3 Atmospheric Structure

For purposes of this book, we will just be concerned with the two lowest layers of the atmosphere: the troposphere and stratosphere. The air layer that extends from the ground surface 10–15 km in altitude is referred to as the **troposphere**. The layer that sits above the troposphere and extends from about 15 to 50 km in altitude is referred to as the **stratosphere**. The protective ozone layer resides in the stratosphere while urban ozone (also referred to as smog) occurs in the troposphere. It is the troposphere where pollution from anthropogenic and natural sources is emitted. The troposphere is an especially turbulent layer because of daily changes in the solar energy that causes daytime surface heating and nighttime cooling. Figure 11.5 shows this vertical structure of Earth's atmosphere that is defined based on the temperature profiles as a function of altitude (or pressure).

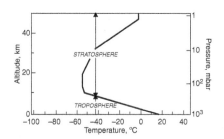

Figure / 11.5 Temperature Changes with Altitude and Pressure in the Atmosphere.

11.3.1 ATMOSPHERIC TEMPERATURE STRUCTURE

The troposphere is the most important layer relevant to air pollution issues. It is characterized by an average profile of decreasing temperature with altitude above the surface, with highly variable local profiles that change in time and space. Due to heating by the sun, the Earth's surface heats daily and radiates that heat, increasing the temperature of the air near the surface. Radiative heating drops off with distance from its source; therefore, the influence of the surfaces decreases rapidly with altitude and, on average, the temperature decreases with altitude. A warm troposphere is also maintained due to the presence of greenhouse gases (including water vapor and carbon dioxide) that preferentially absorb and locally reradiate long-wave radiation as was discussed in Chapter 4.

This structure of decreasing temperature with altitude as well as local variation in surface heating, drive substantial vertical and horizontal air movement and mixing, or weather, in the troposphere. However, localized vertical layers with distinct temperature profiles such as *inversions* (defined as increasing temperature with altitude) are also common; they can be formed due to several interacting meteorological processes, including cooling of the surface at night, elevated high-pressure systems, or interacting weather fronts. These stable layers are discussed later in this chapter and are often quite important to urban and regional air pollution events as they trap air and prevent pollutants from mixing with cleaner air. Therefore, high concentrations of air pollution often occur during times and areas influence by a stable air layer.

The stratosphere is also important to air pollution. It is characterized by a profile of increasing temperature with altitude. This structure is caused by heating due to absorption of shortwave ultraviolet (UV) radiation from the sun by the natural stratospheric ozone layer. Due to this very stable temperature structure, the stratosphere does not experience weather; rather it provides the ultimate inversion cap on tropospheric air movement and dilution of surface pollutants. However, pollutants can diffuse slowly into the stratosphere. One class of pollutants for which this is important is ozone depleting compounds (such as chlorofluorocarbons). These compounds do not react substantially in the troposphere, but they do react in the stratosphere where shorter wavelength (more energetic) radiation is present. In the stratosphere, chlorofluorocarbons are involved in catalytic reactions cycles that destroy the natural ozone layer, allowing more dangerous radiation to reach the surface of the earth.

11.3.2 ATMOSPHERIC PRESSURE AND DENSITY STRUCTURE

In addition to characteristic temperature structures, the atmosphere also has a characteristic density and pressure structure. Density decreases exponentially as the gravitational force also drops off with the square of distance. Therefore, the massive Earth holds gases comprising its atmosphere more tightly near the surface, and the majority of the earth's atmosphere lies in a very thin layer. As density decreases, so does pressure (the gravitational force per horizontal area due to the air above), and pressure follows a very similar profile as density, on average. Local variations in pressure (and density) also occur and are important in driving weather patterns.

11.3.3 COMPOSITION OF THE ATMOSPHERE

In order to understand air quality, we must understand the natural composition of the atmosphere and how that composition is changed by the presence of air pollution. As we mentioned at the beginning of this chapter, the natural atmosphere is largely made up of a few gases whose fractional composition does not vary much, except under very highly perturbed conditions. These gases include nitrogen (N_2, present at 78.1 percent by dry volume of the air mixture), oxygen (O_2, 20.9 percent), argon (Ar, 0.934 percent), and a few others (neon, helium, krypton, and xenon) present in trace quantities (with volume mixing ratios at the ppm_v level or below). Together these well-mixed gases make up about 99.9 percent of the natural atmosphere, but it is the composition of the gases and particulate matter that vary substantially in time and space, and are typically present in trace quantities, that are most important to air pollution issues.

Table 11.5 provides estimates of the composition of several common and varying components of the atmosphere. Water vapor is the most important of these. It is instrumental to the hydrological cycle and is also a potent greenhouse gas. Concentrations of water vapor vary substantially in time in space over about four orders of magnitude, with an average of about 1 percent. This variability is one reason why the composition of air is measured by dry volume (that is, not including water vapor volume in the denominator).

Carbon dioxide was historically considered a well-mixed gas, with a mixing ratio of 280 ppm_v at the turn of the 19th century. However, its emissions, due largely to burning of fossil fuels, have increased rapidly over the last 200 years. Average concentrations are now greater than 399 ppm_v, and higher concentration areas in urban regions are common. Because of its abundance, CO_2 is a very important greenhouse gas. Methane is another important greenhouse gas, with 25 times more absorption of Earth's infrared radiation per molecule than CO_2, but it is far less abundant. However, methane concentrations have also increased with the development of human civilization; concentrations in the mid 1800s were approximately 0.8 ppm_v.

The remaining compounds given in Table 11.5 are all commonly present pollutants with both natural and anthropogenic sources. Ozone is an important harmful component of urban smog in the troposphere but also serves a protective role in the stratosphere due to its absorption of UV radiation. Ozone is the most important example of a **secondary pollutant**. It has negligible emissions sources but forms in the atmosphere due to reactions of nitrogen oxides (NO_x; that is, NO plus NO_2) and reactive hydrocarbons in the presence of sunlight. Reactive hydrocarbons include numerous individual compounds; a few are ethane, ethene, propane, terpenes, benzene, and formaldehyde. There are a few different terms used to represent categories of hydrocarbon gases; they include reactive organic gases (ROGs) and volatile organic compounds

Table / 11.5

Approximate Concentrations of Common Varying Pollutants in Clean and Polluted Environments (based on Jacobson, 2012).

Gas	Clean	Polluted
Water vapor (H_2O)	3 ppm$_v$	<5,000 ppm$_v$
Carbon dioxide (CO_2)	399.5 ppm$_v$	<1,000 ppm$_v$
Methane (CH_4)	1.0–1.9 ppm$_v$	<2.5 ppm$_v$
Carbon monoxide (CO)	40–200 ppb$_v$	<10,000 ppb$_v$
Ozone (O_3)	10–100 ppb$_v$	<350 ppb$_v$
Sulfur dioxide (SO_2)	0.2–1 ppb$_v$	<30 ppb$_v$
Ammonia (NH_3)	1 ppb$_v$	<25 ppb$_v$
Formaldehyde (HCHO)	0.1–1 ppb$_v$	<200 ppb$_v$
Nitric acid (NO_3)	0.02–0.3 ppb$_v$	<50 ppb$_v$
Nitrogen dioxide (NO_2)	0.01–0.3 ppb$_v$	<200 ppb$_v$
Nitric oxide (NO)	0.005–0.1 ppb$_v$	<300 ppb$_v$
Particulate matter (PM)		
Coarse (size > 2.5 μm)	10 μg/m^3	<500 μg/m^3
Fine (PM$_{2.5}$) (size < 2.5 μm)	5 μg/m^3	<250 μg/m^3

(VOCs). These terms are used interchangeably for the purposes of this chapter.

Sulfur dioxide, emitted largely from burning fossil fuels, is the most important source of acid deposition and secondary particle formation. Ammonia, whose dominant source is livestock operations, also plays an important role in particle formation and acid neutralization. Particulate matter (PM), particularly small particles that can penetrate deep into the lungs, are the category of common pollutants that have been most strongly associated with human mortality and morbidity.

In terms of reporting concentration of air pollutants, remember from Chapter 2 we learned that *for ideal gases, volume ratios and mole ratios are equivalent*. This is clear from the ideal gas law, because at constant temperature and pressure, the volume occupied by a gas is proportional to the number of moles. Accordingly,

$$\text{ppm}_v = \frac{\text{moles } i}{\text{moles total}} \times 10^6 \qquad \textbf{(11.4)}$$

Also, the ideal gas law (Equation 2.6) can be used to convert gaseous concentrations between mass concentration (mass/volume) and volume mixing ratio (volume/volume) units as shown in the following example.

example/11.2 Conversion of Gas Concentration Between ppb$_v$ and $\mu g/m^3$

The concentration of O_3 is measured in air to be 75 ppb$_v$. What is this concentration in units of $\mu g/m^3$? Assume the temperature is 28°C and pressure is 1 atm. Remember that T expressed in K is equal to T expressed in °C plus 273.15.

solution

To accomplish this conversion, use the ideal gas law to convert the volume of O_3 to moles of O_3, resulting in units of moles/L. This can be converted to $\mu g/m^3$ using the molecular weight of O_3 (which equals 48).

First, use the definition of ppb$_v$ to obtain a unitless volume ratio for O_3:

$$75 \text{ ppb}_v = \frac{75 \text{ m}^3 O_3}{10^9 \text{ m}^3 \text{ air solution}}$$

Now convert the volume of O_3 in the numerator to units of mass. This is done in two steps. First, convert the volume to a number of moles, using a rearranged form of the ideal gas law (refer back to Equation 2.6), $n/V = P/RT$, with the given temperature and pressure:

$$\frac{75 \text{ m}^3 O_3}{10^9 \text{ m}^3 \text{ air solution}} \times \frac{P}{RT} = \frac{75 \text{ m}^3 O_3}{10^9 \text{ m}^3 \text{ air solution}} \times \frac{1 \text{ atm}}{8.205 \times 10^{-5} \dfrac{\text{m}^3\text{-atm}}{\text{mole-K}}(301 \text{ K})} = \frac{3.04 \times 10^{-6} \text{ mole} O_3}{\text{m}^3 \text{ air}}$$

In the second step, convert the moles of O_3 to mass of O_3 by using the molecular weight of O_3:

$$\frac{3.04 \times 10^{-6} \text{ mole} O_3}{\text{m}^3 \text{ air}} \times \frac{48 \text{ g} O_3}{\text{mole} O_3} \times \frac{10^6 \text{ } \mu g}{g} = \frac{146 \text{ } \mu g}{\text{m}^3}$$

These steps can be inverted to convert from a mass concentration to a volume mixing ratio.

11.4 Characteristics of Polluted Air

11.4.1 CRITERIA AIR POLLUTANTS

The Clean Air Act requires EPA to set national air quality standards for specific pollutants to safeguard human health and the environment. These air quality standards define the allowable concentrations of air pollutants that EPA has determined humans and the environment can be exposed to without significant adverse effects. EPA has established enforceable standards for six common air pollutants that may impact large populations or the environment. These six air pollutants are termed **criteria air pollutants**: (1) particle matter (PM), (2) carbon monoxide (CO), (3) nitrogen dioxide (NO_2), (4) sulfur dioxide (SO_2), (5) ozone (O_3), and (6) lead (Pb).

Air quality standards have been set at the national level for ambient air (that is, outside the home) are referred to as **National Ambient Air Quality Standards (NAAQS)**. Two types of standards are set, **primary standards** are designed to be protective of human health, while

secondary standards are designed to be protective of public welfare, including protection against environmental, cultural, and property damage. Table 11.6 provides the primary NAAQS for the six criteria air pollutants. Under the Clean Air Act, states are allowed to set stricter standards if they wish.

Table 11.7 lists six different ways that air pollution adversely impacts human health. Sensitive individuals are at greater risk and include older adults, children, diabetics, and people with preexisting heart and lung diseases (for example, heart failure/ischemic heart disease, asthma, emphysema, and chronic bronchitis). Table 11.8 summarizes the major sources of the criteria air pollutants and their specific health effects. Also included in this table is other important information about the criteria pollutants.

As with water quality analysis, particulate matter in air is a nonuniform combination of different compounds. Collectively, all particles with aerodynamic diameters $\leq 10\ \mu m$ are referred to as PM_{10}. All particles $\leq 2.5\ \mu m$ are referred to as $PM_{2.5}$. As discussed in Table 11.8, *primary particles* are generated directly from a source (for example, construction sites, agricultural fields, unpaved roads, fires, and smokestacks). *Secondary particles* are formed from reactions in the atmosphere (for example, important precursors include sulfur dioxide, ammonia, nitrogen oxides, and semi-volatile organics; they originate from power plants, industry, and vehicles).

Table / 11.6

National Air Quality Standards for the Six Criteria Air Pollutants (primary standards to protect human health).

Criteria Pollutant	National Ambient Air Quality Standards (NAAQS)
Particulate matter (PM_{10})	$150\ \mu g/m^3$ (24 h average)
Fine particulate matter ($PM_{2.5}$)	$35\ \mu g/m^3$ (24 h average) $12\ \mu g/m^3$ (annual average)
Carbon monoxide (CO)	$9\ ppm_v$ (8 h average) $35\ ppm_v$ (1 h average)
Nitrogen dioxide (NO_2)	$100\ ppb_v$ (1 h average) $53\ ppb_v$ (annual average)
Sulfur dioxide (SO_2)	$75\ ppb_v$ (1 h average)
Ozone (O_3)	$0.075\ ppm_v$ (8 h average)
Lead (Pb)	$0.15\ \mu g/m^3$ (quarterly average)

Table / 11.7

Six Ways Air Pollution Adversely Effects Human Health

Premature death

Decreased lung function

Increased susceptibility to respiratory infections

Aggravation of respiratory and cardiovascular disease

Increased frequency and severity of respiratory symptoms such as difficulty breathing and coughing

Effects on the nervous system, including the brain, such as IQ loss and impacts on learning, memory, and behavior

Table / 11.8

Sources, Health Effects, and Other Information Associated with Criteria Air Pollutants (adapted from EPA, 2012a).

Pollutant	Sources	Health Effects	Other
Particulate matter (PM)	Fuel combustion (for example, burning coal, wood, diesel); industrial processes; agriculture (plowing and burning fields); and emissions from unpaved roads.	Short-term exposures can aggravate heart or lung disease leading to respiratory symptoms, increased use of medication, hospital admissions, emergency room visits, and premature mortality. Long-term exposures can lead to the development of heart or lung disease and premature mortality.	Particulate matter is categorized by particle size; one grouping is into fine particles, or particles smaller than 2.5 μm in diameter ($PM_{2.5}$) and coarse particles, particles larger than that cutoff. Another commonly used category is PM_{10}, or particles smaller than 10 μm in diameter. *Primary particles* are generated directly from a source (for example, construction sites, agricultural fields, unpaved roads, fires, and smokestacks). *Secondary particles* are formed from reactions in the atmosphere (for example, precursors originate from power plants, industry, and vehicles).
Carbon monoxide (CO)	Fuel combustion (especially from vehicles).	Reduces the amount of oxygen reaching the body's organs and tissues. Aggravates heart disease, resulting in chest pain and other symptoms leading to hospital visits and emergency room visits.	Produced from incomplete combustion of fuels, usually arising from an insufficient amount of air for the amount of fuel. Inadequate air-to-fuel ratio can be due to poorly operated or maintained equipment, airflow limitations, or low temperatures. While high levels are rarely encountered in the ambient atmosphere, asphyxiation can occur in indoor environments, often through a combination of poorly functioning heating systems and inadequate ventilation.
Nitrogen oxides (NO_x) = (NO + NO_2)	Fuel combustion (for example, electric utilities, industrial boilers, vehicles) and wood burning. Results from the fact that air is rich in nitrogen.	Aggravate lung diseases leading to respiratory symptoms, hospital admissions, and emergency department visits. Increased susceptibility to respiratory infection.	Large amounts of air (that contains > 78% N_2) are used during the combustion of fossil fuels. The high temperature (and sometimes pressure) that can exist during fuel combustion produces NO, which is then quickly transformed to NO_2. Precursor to formation of ground-level ozone (O_3) (an important component of urban smog). NO_2 reacts to form nitric acid (HNO_3) in the atmosphere and important contributor to acid rain. Harms historical buildings and structures made of limestone or marble. *(continued)*

(continued)

Sulfur dioxide (SO$_2$)	Fuel combustion (especially high sulfur coal); electric utilities and industrial processes. Natural sources such as volcanoes.	Aggravates asthma and increased respiratory symptoms. Contributions to particle formation with associated health effects.	Sulfur is present in many raw materials, including coal, oil, iron, aluminum, and copper. SO$_2$ reacts with water vapor to form sulfuric acid (H$_2$SO$_4$), the most important contributor to acid rain and to secondary particle formation. Harms historical buildings and structures made of limestone or marble.
Ozone (O$_3$)	Only criteria pollutant that has no direct sources. This secondary pollutant is typically formed by chemical reaction of VOCs and nitrogen oxides (NO$_x$) in the presence of sunlight.	Decreases lung function and causes respiratory symptoms, such as coughing and shortness of breath. Aggravates asthma and other lung diseases leading to increased medication use, hospital admissions, emergency room visits, and premature mortality.	Can damage sensitive plants, reducing crop yields and forest productivity. Reduces visibility, causing haze in the atmosphere.
Lead	Smelters (metal refineries) and other metal industries. Combustion of leaded gasoline in piston engineer aircraft. Waste incinerators Battery manufacturing	Damages the developing nervous system, resulting in IQ loss and impacts on learning, memory, and behavior in children. Cardiovascular and renal effects in adults and early effects related to anemia.	Leaded gasoline and paints are still used in many parts of the world, especially sub-Saharan Africa and parts of Asia. Contributes 11% of the global environmental risk.

Application / 11.3   Can Carbon Dioxide (CO$_2$) Be Regulated as an Air Pollutant?

The answer to this question is yes. The Supreme Court ruled in 2007 that carbon dioxide and other greenhouse gases are covered by the Clean Air Act's broad definition of air pollutants. The Court said that EPA must decide whether greenhouse gases endanger public health or welfare, and whether emissions from new motor vehicles contribute to this air pollution. After considering the extensive scientific evidence, EPA issued endangerment and contribution findings in December 2009.

On June 26, 2012, the U.S. Court of Appeals for the D.C. Circuit upheld EPA's endangerment finding, its greenhouse gas emission standards for light-duty vehicles and its Tailoring Rule. This rule establishes a phased approach for applying certain Clean Air Act permitting requirements to stationary sources based on greenhouse gas emissions, focusing on large sources. The Court confirmed that EPA followed the science and the law in both these actions. In upholding the endangerment finding, the Court stated: "The body of scientific evidence marshaled by EPA in support of the Endangerment Finding is substantial." The court also confirmed that the Clean Air Act required EPA to regulate greenhouse gas emissions from cars and light trucks, and the court ruled that the litigants in the case were not harmed by EPA's Tailoring Rule.

Adapted from opening Statement of Regina McCarthy (Assistant Administrator for Air and Radiation, U.S. Environmental Protection Agency) at Congressional Hearing on EPA Regulation of Greenhouse Gases, June 29, 2012

11.4.2 HUMAN HEALTH IMPACTS AND DEFENSES TO PARTICULATE MATTER

Figure 11.6 shows how the human respiratory systems has several mechanisms to combat air pollution, especially that for larger particles. The downward turned nostril and tortuous nasal passageways separate large particles from the airstream; the largest cannot travel into the nose and others deposit due to interception by nose hairs and mucus. However, smaller "fine" particles and some particles with large aspect ratios can travel deep into the lungs and reach the respiratory zone. There these fine particles can foul the alveolar sacs of the respiratory system (see Figure 11.6) that provide an important interface for exchange of oxygen and carbon dioxide by the body. Ultrafine particles ($< 0.1~\mu$m) are believed to move through body membranes, including entering the bloodstream where they may impact the function of internal organs. Exposure to fine particles is also believed to cause cardiovascular problems such as heart attacks. The scientific attention paid in recent years to fine particles is one reason an air quality standard for particulate matter was developed for $PM_{2.5}$.

Globally, exposures to particulate matter concentrations that exceed or are close to the U.S. NAAQS values are quite large, especially in areas of the Americas, sub-Saharan Africa, and Asia (see Figure 11.7). Also note the many countries in sub-Saharan Africa are not listed in this figure because they do not have reported PM_{10} (or $PM_{2.5}$) data, yet they probably have high ambient concentrations of respirable particulate matter in urban areas.

Class Discussion
How do you think urbanization, population growth, and increased affluence impact the concentration of fine particulate matter in the developing world in the short and long term?

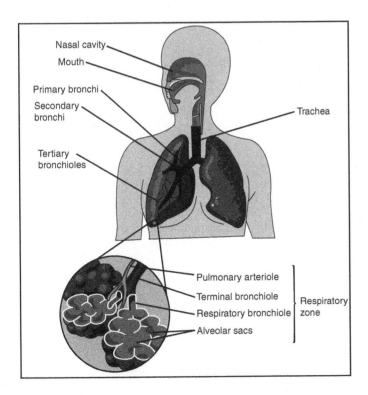

Figure / 11.6 **Major Parts of the Human Respiratory System** Fine particles can reach and be deposited in the respiratory zone.

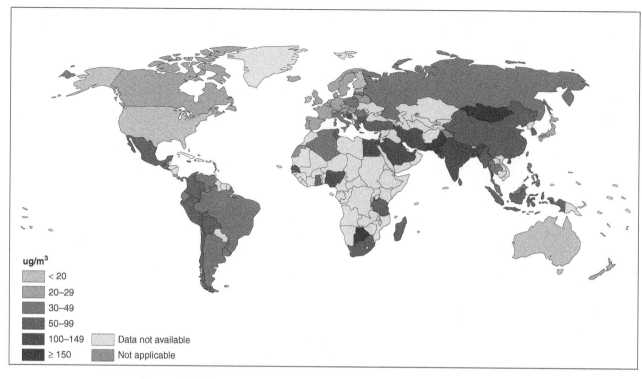

Figure / 11.7 **Global Exposure to Particles (PM$_{10}$) in Urban Areas for 2003–2010** (with permission of World Health Organization Map Production, Public Health Information and Geographic Information Systems (GIS), World Health Organization, 2012).

ug/m^3	
< 20	
20–29	
30–49	
50–99	
100–149	Data not available
≥ 150	Not applicable

Application / 11.4   Solid Fuels, Household Energy, Health, and Climate

Much of the world's poor, especially those living in developing counties, use solid biofuels (such as wood, animal dung, crop wastes, straw) as their primary household energy source for cooking, heating, and water purification (that is, boiling) (Bruce et al., 2000) (see Figure 11.8). Traditional fuel use and resulting exposure to indoor combustion emissions (including particulate matter, carbon monoxide, and air toxics) long have been recognized as a major cause of mortality and morbidity in the developing world, with disproportionate effects on women. Furthermore, recent carbon emissions inventories suggest that household emissions from cookstoves of black carbon (with short-lived climate forcing) are estimated to rank second behind CO_2 in regards to climate change, being responsible for 18 percent of the planet's warming (compared with 40 percent for CO_2).

The **energy ladder** (Figure 11.9) shows how fuel type may change as a household's social and economic status increases (Smith et al., 1993). As a household advances up the energy letter, they increase their expenditures for fuel but also decrease their emissions of particulate matter (in this case, measured as PM$_{10}$). Not every household follows the exact pattern depicted in the energy ladder. Households often continue to use solid fuels even when they have the economic ability to move up the energy ladder, either because the solid fuels are more accessible or because their newly acquired resources are used for other household needs.

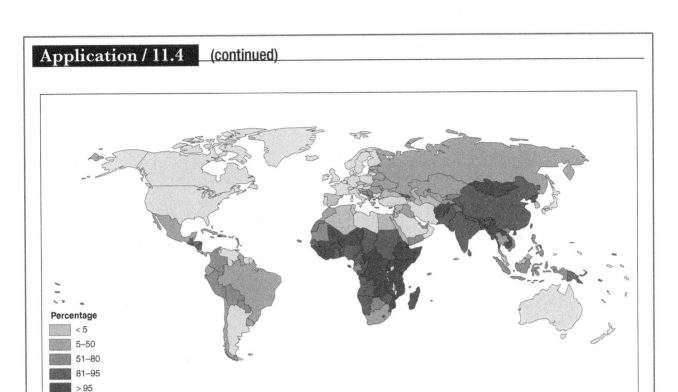

Figure / 11.8   **Global Population Living in Rural Areas That Use Solid Fuels**   (with permission of World Health Organization Map Production, Public Health Information and Geographic Information Systems (GIS), World Health Organization, 2012).

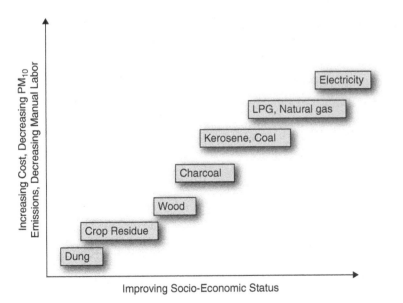

Figure / 11.9   Energy Ladder Showing How Changes in Fuel Type and PM_{10} Emissions Are Impacted by a Household's Social and Economic Status.

Table / 11.9

Major Sources of Air Pollutants

Source of Emissions	Specific Types
Stationary fuel combustion	Electric utilities, industrial boilers, industrial processes
Stationary industrial processes and other processes	Metal smelters, petroleum refineries, cement kilns, manufacturing facilities, solvent utilization
Mobile sources	Highway vehicles, nonroad sources such as recreational vehicles, construction equipment, marine vessels, aircraft, locomotives

Webinar: Test Results of Cook Stove Performance

http://www.pciaonline.org/ proceedings

Webinar: Health Effects of Particulate Matter from Wood Smoke

http://epa.gov/air/oaqps/eog/ broadcast.html#WoodSmoke072811

Partnership for Clean Indoor Air and Improved Cookstove Technology

www.PCIAonline.org

11.4.3   MAJOR SOURCES OF AIR POLLUTANTS

Table 11.9 lists the major sources of air pollutants and provides examples of the types of these sources. Note that air pollutants are emitted from **stationary sources** and **mobile sources**. As discussed in Table 11.9, pollutants emitted directly into the air are called *primary pollutants*, while pollutants like ozone and some particles are formed in the air and are called *secondary pollutants*. This formation is a function of weather (temperature, moisture, and winds) and also geographical features (for example, presence of sunlight is required for ozone formation). Some sulfate particles are formed by reactions after emissions of gaseous SO_2 into the atmosphere from coal-fired power plants and some industrial facilities.

Figure 11.10 provides information about the percent of emissions of criteria and other air pollutants that are emitted by a particular source

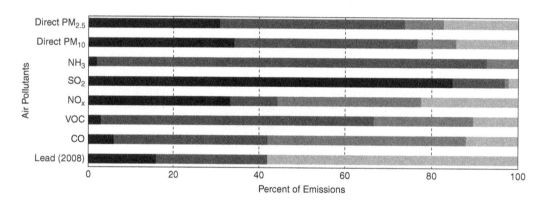

Figure / 11.10   Percent of Total 2010 Air Emissions by Source Category   (lead values are for 2008).

(Redrawn from EPA, 2012a).

category. Electric utilities account for > 60 percent of SO_2 emissions while agricultural processes contribute > 80 percent of ammonia (NH_3) emissions (accounted for in the other "processes" category in the figure). VOC emissions originate from solvent use (which is also accounted for in the "other processes" category of the figure). For CO emissions, approximately 60 percent of emissions are from mobile sources associated with highway vehicles and nonroad mobile sources. A significant fraction of NO_x emissions are associated with mobile sources of highway vehicles and nonroad mobile sources, along with stationary fuel combustion.

Lead Air Pollution

http://epa.gov/air/lead/

Application / 11.5   Air Pollution and Environmental Justice

Previously in this chapter, Table 11.7 listed six different ways that air pollution adversely impacts human health. Many times the burden of the environmental risk associated with air pollution is unfairly assigned to particular segments of the population (for example, age, gender, race, economic status).

Asthma is one health risk associated with air pollution, particularly ozone and SO_2, which are known to aggravate it. It is a chronic respiratory disease characterized by inflammation of the lung and airways that has symptoms ranging in severity from mild to life-threatening. It persists into adulthood, and annual health care expenses associated with asthma are estimated to be $50 billion (EPA, 2012b). Approximately 7 million U.S. children aged 0–17 suffer from asthma. Poor and minority children are assigned a greater burden of the disease. Figure 11.11 shows that the prevalence of asthma in U.S. children is greater in Puerto Rican children (16.5 percent), black children (16 percent), and American Indian/and Alaska Native

children (10.7 percent) than white children (8.3 percent). Though not shown in this figure, poorer children also have higher levels of asthma than children from households with higher income.

The disparities in the prevalence of asthma show in Figure 11.11 are also associated with outcomes of this illness. For example, black children with asthma are more likely to require hospitalization, have to visit an emergency room, and die from their asthma than white children (EPA, 2012b). Asthma also results in lost school days, and there is evidence that children with asthma do poorer academically. Underlying factors for having a higher prevalence of asthma could also be associated with exposure to air pollutants such as tobacco smoke or criteria air pollutants. Remember that exposure to ozone and SO_2 can aggravate asthma. Other underlying factors could be from genetics or exposure environmental allergens (for example, house dust mites, cockroach particles, cat and dog dander, and mold) (CDC, 2011).

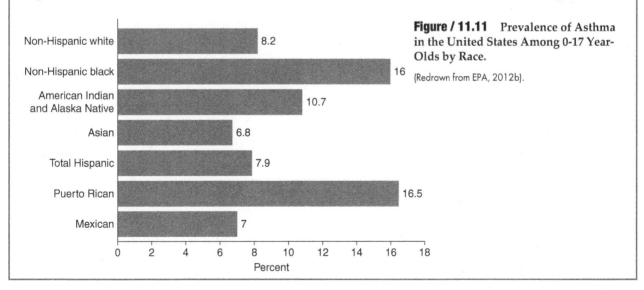

Figure / 11.11   Prevalence of Asthma in the United States Among 0-17 Year-Olds by Race.

(Redrawn from EPA, 2012b).

11.4.4 RECENT TRENDS IN CONCENTRATIONS OF AIR POLLUTANTS

Most air pollutants have shown a steady decline in air concentrations since the 1970s and especially since 1990 (EPA, 2012a). Areas that do not achieve ambient air quality concentrations below the required NAAQS are referred to as a **nonattainment area**. Table 11.10 gives recent trend in air pollutants in these nonattainment areas.

Identify Nonattainment Areas
www.epa.gov/airquality/greenbook

11.4.5 AIR QUALITY INDEX

The **air quality index (AQI)** provides an easy-to-understand way to relate the daily air pollution concentrations of criteria air pollutants measured in a particular geographic area to concentrations of air

Table / 11.10

Recent Trends in NAAQS Air Quality Criteria Air Pollutants in Nonattainment Areas

Pollutant	Sources
Particulate matter ($PM_{2.5}$)	Between 2001 and 2010 in the United States, the annual concentration of $PM_{2.5}$ declined by 24 percent and the 24 h concentrations for $PM_{2.5}$ declined by 28%.
	In 2010, 3 areas in the United States exceed the annual standard and 13 areas exceed the 24 h standard.
	In 2010, the highest annual average concentrations for $PM_{2.5}$ were reported in California, Indiana, Pennsylvania, and Hawaii. The highest 2 h concentrations were reported in California and Alaska.
	The winter use of woodstoves that occurs with cold temperature inversions can lead to seasonal concentrations that exceed the NAAQS.
Carbon monoxide (CO)	Concentrations of 8 h standard for CO decreased 52% between 2001 and 2010. No violations of the annual NAAQS standards were reported in 2010.
Nitrogen dioxide (NO_2)	The annual mean concentrations of NO_2 decreased 33% between 2001 and 2010. No violations of the annual standards for NO_2 were observed for the 8 and 1 h standards in 2010.
Sulfur dioxide (SO_2)	The annual mean concentrations of SO_2 decreased 50% between 2001 and 2010. The only sites in violation of annual standards in 2010 were located in Hawaii; these violations were believed to be because of volcanic eruptions.
Ozone (O_3)	Between 2001 and 2010, national average ground-level ozone concentrations were 13% lower. In addition, ozone nonattainment areas showed a 9% improvement in ozone concentration levels.
	Eight areas exceed the 8 h standard and 16 areas exceed the 24 h standard.
	The highest O_3 concentrations occur in California. The metropolitan monitoring sites showing the greatest improvement from 2001 to 2010 were South Bend (IN), Buffalo (NY), Chicago (IL), Milwaukee (WI), and Cleveland (OH).
Lead (Pb)	Concentrations of lead decreased 71% from 2001 to 2010. A typical average Pb concentration near a stationary source is about 8 times higher than other sites.
	In 2010, 34 sites around the United States exceeded the lead NAAQS.

SOURCE: EPA, 2012a.

Table / 11.11

Interpretation of the Air Quality Index

Air Quality Index (AQI) Numerical Value	Level of Health Concern	Meaning for General and Sensitive Populations
0–50	Good	Air quality is considered satisfactory, and air pollution poses little or no risk.
51–100	Moderate	Air quality is acceptable; however, for some pollutants there may be a moderate health concern for a very small number of people who are unusually sensitive to air.
101–150	Unhealthy for sensitive groups	Members of sensitive groups may experience health effects. The general public is not likely to be affected.
151–200	Unhealthy	Everyone may begin to experience health effects; members of sensitive groups may experience more serious health effects.
201–300	Very unlikely	Health warnings of emergency conditions. The entire population is more likely to be affected.
301–500	Hazardous	Health alert; everyone may experience more serious health effects.

pollutants. This information is then interpreted for the general public and sensitive populations. Five of the criteria air pollutants are tracked for the determination of the AQI in a particular location (that is, particulate matter, CO, NO_2, SO_2, ozone).

AQI values range from 0 (best air quality) to 500 (worst air quality). They are calculated for each pollutant by linear interpolation between threshold concentration cutoff values for each health concern range. An AQI value that equals 100 for an individual criteria pollutant corresponds to the national air quality standard for that pollutant. Values > 100 are considered unhealthy, values ≤ 100 are considered satisfactory. Table 11.11 provides the health concern meaning for AQI value ranges. The overall combined AQI value is set at that of the highest individual AQI value.

Figure 11.12 shows the number of days the AQI was exceeded in selected cities around the United States. Most of these exceedance days are now due to high ozone or particulate matter concentrations. These happen to be two criteria air pollutants for which weather conditions play an important role in formation. Note that all the areas reported in Figure 11.12 experienced a large decrease in the number of days on which the AQI was exceeded for the 2002–2010 period.

What's the Air Quality Index Where You Live?
http://www.airnow.gov/

Compare Your City's Air Quality to Other Cities
http://www.epa.gov/aircompare/index.htm

Air Quality in Your Neighborhood
http://www.scorecard.org

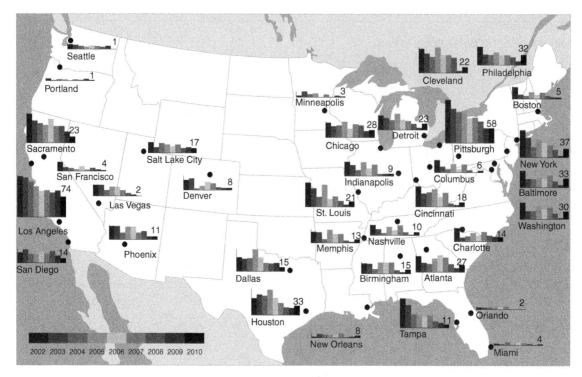

Figure / 11.12 Number of Days on Which Air Quality Index (AQI) Values Were Greater Than 100 during 2002–2010 in Selected Cities.

(Redrawn from EPA, 2012a).

example/11.3   Identifying the Air Quality Index (AQI)

Find the AQI in the nearest city to where you are taking this course. What is the level of health concern, and what does this AQI mean meaning for general and sensitive populations?

solution

On December 30, 2012, Dr. Mihelcic planned a trip canoeing near his home in Tampa, Florida. He went to the AirNow web site to determine the AQI for that day before going out. The AirNow web site (http://airnow.gov/) was developed by EPA, the National Oceanic and Atmospheric Administration (NOAA), the National Park Service, and tribal, state, and local agencies to provide the public with easy access to national air quality information. The AirNow web site provides daily AQI real-time AQI conditions and forecasts for upcoming days for over 300 cities across the United States.

The AQI index on this date was reported by AirNow to be 28. According to Table 11.11, the air quality is considered "good" and air pollution poses little or no risk to the health of general and sensitive populations.

Application / 11.6   How High Can the AQI Get?

On January 14, 2013, the AQI in Beijing was 341 and the $PM_{2.5}$ was measured as 291 $\mu g/m^3$. Two days earlier at the peak of this particular air pollution crisis, the AQI was determined to be 775 and the $PM_{2.5}$ was measured to be 886 $\mu g/m^3$. In contrast, on January 11, the five worst places for air quality in the United States were all in Utah (Logan, Ogden, Provo, Salt Lake City, and the Washakie Indian Reservation near the Utah-Idaho border). Salt Lake City (Utah) was suffering poor air quality from emissions that were compounded by the presence of a seasonal inversion. Its AQI was 142 compared to a value that day of 67 in San Francisco and a value of 23 in Las Vegas.

As described in Table 11.11, an AQI greater than 300 is considered hazardous to all humans, not just those from sensitive groups, who may have heart or lung ailments. The AQI of 142 suggests that members of sensitive groups may experience health effects. Residents of Beijing and many other cities in China were warned to stay inside their homes in the middle of January as the nation faced one of the worst periods of air quality in recent history. The Chinese government ordered factories to scale back emissions, while hospitals saw a 20–30 percent increase in patients complaining of respiratory issues.

How could the AQI get so high? Most fine particulate matter that is measured as $PM_{2.5}$ originates from the burning of fossil fuels and biomass that can include wood fires and agricultural burning. China has several factors that can cause poor air quality; a large population dependent on fossil fuels, especially coal, for heating, electricity, and transportation; economic activity that has poor regulation of industrial air emissions; changes in lifestyles that have led to increased dependence on the automobile and electricity; and local/regional metrological conditions all contribute to the major health problem (text on China adapted from NASA's Image of the Day Gallery, January 15, 2013).

Application / 11.7   Climate Change Impacts on Air Quality

EPA reports that climate change is expected to impact air quality by

- Producing 2–8 ppb_v increases in summertime average ground-level ozone concentrations in many regions of the country
- Further exacerbating ozone concentrations on days when weather is already conducive to high ozone concentrations
- Lengthening the ozone season
- Producing increases and decreases in particle matter pollution over different regions

The emissions reductions that have occurred with regulations, policy incentives, changes in personal behavior, and introduction of new technology are occurring faster than climate-driven impacts. Therefore, there may be ways to mitigate the impacts of climate change on air quality through further improvements in air quality.

11.4.6 HAZARDOUS AIR POLLUTANTS

Hazardous air pollutants are also known as toxic air pollutants or **air toxics**. They may cause cancer or noncancer human health impacts, as well as negative impacts on ecosystems. Remember from Chapter 6 that noncancer effects could include reproductive or birth defects. The EPA is required to control 187 **toxic air pollutants** under the Clean Air Act. Figure 11.13 shows the decrease in emission of most toxic air pollutants since 2003.

Figure 11.14 depicts the lifetime cancer risk for a person living in the United States that is associated with exposure to air toxics in a particular geographic region. The national average for cancer risk level from exposure to air toxics is currently 50 in a million. This cancer risk typically increases in urban areas and also for those who live in transportation corridors. In addition, almost 60 percent of the cancer risk is associated with two air toxics, formaldehyde and benzene. Exposure to diesel exhaust (a likely human carcinogen) is also large in the United States and is not included in this figure. It is also associated with living in urban areas and along transportation corridors.

Mercury

http://www.epa.gov/mercury/

Figure / 11.13 Change in Ambient Concentrations Reported at U.S. Toxic Air Monitoring Sites (2003–2010) Report is percent change in annual average concentration.

(Redrawn from EPA, 2012a).

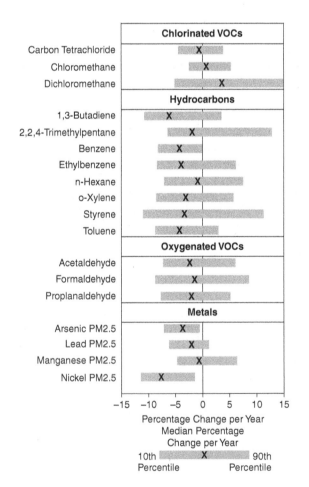

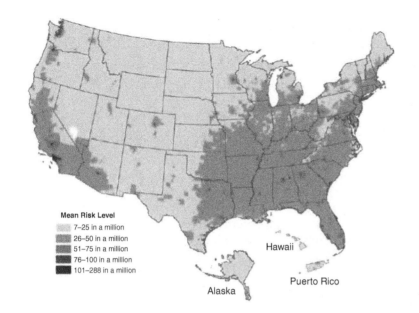

Figure / 11.14 Estimated Cancer Risk Associated with Exposure to Air Toxics Darker colors show greater cancer risk associated with exposure to toxic air pollutants.

(Figure redrawn from EPA, 2012a).

Mean Risk Level
7–25 in a million
26–50 in a million
51–75 in a million
76–100 in a million
101–288 in a million

Hawaii

Puerto Rico

Alaska

11.4.7   GROUND-LEVEL AND STRATOSPHERIC OZONE

Ozone exists in both the troposphere (that is, the ground level) and the stratosphere. Ozone is good when present in the stratosphere but bad when present in the troposphere where it is an important component of urban smog. Previously in this chapter, we have discussed the human health and environmental concerns associated with exposure to ground-level ozone. In this section, we will discuss the formation of ground-level ozone and reactions that result in depletion of the stratospheric ozone layer. This requires an understanding of photochemical reactions.

Photochemical reactions can occur through direct or indirect reactions of molecules with light, and they may be catalyzed by chemicals occurring naturally in the environment or by chemicals emitted through human activities. An example of photochemistry in our daily lives is the fading of fabric dyes exposed to sunlight. Another example is perhaps the most important photochemical reaction in the world, photosynthesis.

Light is differentiated according to its wavelength. Table 11.12 gives the entire *electromagnetic spectrum*. Light can be envisioned to consist of

Table / 11.12

Electromagnetic Spectrum   Photochemical reactions involving UV light are important in the creation of the ozone hole and in the formation of urban smog. Infrared light is important in understanding the greenhouse effect.

Wavelength (nm)	Range
<50	X-rays
50–400	UV
400–750	Visible (400–450 = violet and 620–750 = red)
>750	Infrared

small bundles of energy called *photons*, which can be absorbed or emitted by matter. The energy of a photon, E (units of joules), is computed as follows:

$$E = \frac{hc}{\lambda} \tag{11.5}$$

where h equals Planck's constant (6.626×10^{-34} J-s), c is the speed of light (3×10^8 m/s), and λ is the light's specific wavelength. Equation 11.5 shows that greater energy is contained in photons with a shorter wavelength.

This light energy can be absorbed by a molecule. A molecule that absorbs light energy has its energy increased by rotational, vibrational, and electronic excitation. This molecule typically has a very short time (a fraction of a second) to either use the energy in a photochemical reaction or lose it, most likely as heat.

All atoms and molecules have favored wavelengths at which they absorb light. That is, an atom or molecule will absorb light within a specific range of wavelengths. Greenhouse gases such as water vapor, CO_2, N_2O, and CH_4 absorb energy emitted by Earth as infrared light, while the major components of the atmosphere (N_2, O_2, Ar) are incapable of absorbing infrared light. It is this capture of energy released by Earth that partially contributes to the warming of the planet's surface. We discussed in Chapter 4 how anthropogenic emissions of greenhouse gases such as CO_2, CH_4, and CFCs have increased the amount of this "captured" energy.

GROUND-LEVEL OZONE Unlike the other criteria pollutants, ozone is not emitted directly from a stationary or mobile source. Instead, ozone is indirectly created by the emission of other air pollutants that undergo a complex sequence of chemical reactions that is driven by sunlight. The overall process of this photochemical system of reactions is simplified in one equation as follows:

$$RH + NO_x \xrightarrow{\text{sunlight}} O_3 + \text{other} \tag{11.6}$$

In Equation 11.6, RH is an abbreviation for reactive hydrocarbons that we earlier said for purposes of this book could be referred to as VOCs. Reactive hydrocarbons are a class of compounds including many commercial, industrial, and personal products (for example, benzene, propane, components of gasoline, and others). Some of these materials escape to the atmosphere as fugitive emissions or through daily activities (like filling up your vehicle's gas tank), and some may be released in small or large quantities through permitted discharges. The nitrogen oxide term (NO_x) is the sum of the two nitrogen species, NO and NO_2. They are produced during the combustion of fossil fuels with nitrogen-rich air. When RH and NO_x are present in the air and the sun rises, the formation of ozone commences.

The ozone concentration increases over the course of the day as the intensity of the sun increases and emissions of RH and NO_x (especially from vehicle exhaust) increase over the early part of the day. Ozone

Figure / 11.15   Nonattainment and Maintenance Areas for Ozone.

(From EPA 2012a).

concentrations typically peak in early afternoon at a similar time when the light energy from the sun is the greatest and the air temperature is highest (thus increasing the reaction rate). After the sun sets, the photochemical reaction summarized in Equation 11.6 stops. This allows reactions that destroy ozone as well as physical processes that remove ozone to dominate and reduce the ozone concentrations through the evening.

STRATOSPHERIC OZONE DEPLETION   Another example of molecules absorbing light energy is in the filtering of the UV light that enters Earth's atmosphere. O_2 molecules located above the stratosphere filter out (or absorb) most of the incoming UV light in the range of 120–220 nm, and other gases such as N_2 filter out the UV light with wavelengths smaller than 120 nm. This means no UV light with a wavelength below 220 nm reaches Earth's surface. All of the UV light in the range of 220–290 nm is filtered out by ozone (O_3) molecules in the stratosphere with a little help from O_2 molecules. However, O_3 alone filters a fraction of UV light in the range of 290–320 nm, and the remainder makes it to our planet's surface. Overexposure to this portion of the light spectrum can result in malignant and nonmalignant skin cancer, damage the human immune system, and inhibit plant and animal growth. Most of the UV light in the range of 320–400 nm reaches Earth's surface, but fortunately this type of UV light is the least harmful to the planet's biological systems.

Class Discussion

*Stratospheric ozone
depletion is a major public
health issue in New Zealand.
Why not in the United States? In
your discussion, consider differences
in geography, demographics, culture,
and governance.*

Stratospheric O_3 is formed by the reaction of atomic oxygen ($O\cdot$) with molecular oxygen (O_2) according to the following reaction:

$$O\cdot + O_2 \rightarrow O_3 \tag{11.7}$$

The ($O\cdot$) required in Equation 11.7 is derived from the reaction of O_2 with UV photons ($\lambda < 245$ nm) according to the following reaction:

$$O_2 + \text{UV photon} \rightarrow 2O\cdot \tag{11.8}$$

However, in the stratosphere, the majority of oxygen exists as O_2, so only a little $O\cdot$ is available. Therefore, even though there is little $O\cdot$ in the stratosphere relative to O_2, the small amounts of $O\cdot$ created here will react with the abundant O_2 to form ozone (O_3), according to Equation 11.7.

Equations 11.7 and 11.8 explain the natural formation of ozone in the stratosphere. They also provide insight as to why the concentration of ozone is much higher in the stratosphere (ppm_v levels) versus the troposphere (ppb_v levels). This is because the stratosphere contains much more $O\cdot$ than the troposphere, where $O\cdot$ is not produced by natural mechanisms in large amounts except under human-induced conditions of smog formation (discussed later in this section).

Ozone is destroyed naturally in the stratosphere by the reaction of ozone with UV photons or with atomic oxygen:

$$O_3 + \text{UV photon} \rightarrow O_2 + O\cdot \quad \text{or} \quad O_3 + O\cdot \rightarrow 2\,O_2 \tag{11.9}$$

Atomic oxygen can thus either react with O_2 to form more O_3 or else destroy O_3 to create O_2. Fortunately, the destruction reaction has a relatively high activation energy, so this natural destruction reaction occurs at a slow rate.

Ozone is destroyed continuously in the stratosphere, and the amount of species that destroy ozone moving into the stratosphere has been augmented in recent years by human activities. Although the chemical cycles that destroy ozone are complex, a illustrative shared catalytic mechanism is as follows:

$$X + O_3 \rightarrow XO + O_2 \tag{11.10}$$

$$XO + O\cdot \rightarrow X + O_2 \tag{11.11}$$

Overall reaction:

$$O\cdot + O_3 \rightarrow 2O_2 \tag{11.12}$$

The species X in these equations act as catalysts, speeding up the reaction between O_3 and O. The more common species of X have been identified in three categories: HO_x radicals ($OH\cdot, OOH\cdot$), NO_x radicals ($NO\cdot, NO_2\cdot$), and ClO radicals ($Cl\cdot, ClO\cdot$). In the stratosphere, HO radicals account for as much as 70 percent of the total ozone destruction. Although they are less abundant, BrO radicals can also efficiently catalyze ozone loss.

Free-radical chlorine atoms are very efficient catalysts in the destruction of ozone. Thus, the greatest threat to stratospheric O_3 is from chlorine-containing chemicals. Fortunately, 99 percent of stratospheric Cl is stored in nonreactive forms such as HCl and chlorine nitrate ($ClONO_2$). The amount of stratospheric chlorine has increased in recent decades, however, and during the Antarctic spring, a lot of this stored chlorine is released into the active catalytic forms, Cl· and ClO·. A naturally occurring chlorine-containing chemical is chloromethane (CH_3Cl), which is formed over the world's oceans and may be transported up into the stratosphere. CH_3Cl molecules can react with UV photons (wavelength of 200–280 nm) to produce chlorine-free radicals, Cl·.

The major anthropogenic source of chlorine is from the movement of chlorofluorocarbons (CFCs) into the stratosphere and subsequent release of chlorine-free radicals. CFCs (known commercially as Freon) were widely used in the northern hemisphere beginning in the 1930s. The three most commonly used CFCs were CFC-12 (CF_2Cl_2, used extensively as a coolant and refrigerant, and embedded in rigid plastic foam), CFC-11 ($CFCl_3$, used to blow holes in soft plastic such as cushions, carpet padding, and car seats), and CFC-13 ($CF_2Cl–CFCl_2$, used to clean circuit boards). CFCs are relatively stable in the troposphere, but after being transported up into the stratosphere, they can undergo photochemical reactions that release the catalytic chlorine-free radical, Cl·. For example, the breakdown of CFC-12 occurs as follows:

$$CF_2Cl_2 + UV \text{ photon } (200 - 280 \text{ nm}) \rightarrow CF_2Cl\cdot + Cl\cdot \qquad \textbf{(11.13)}$$

Another Cl· can subsequently be released from the $CF_2Cl\cdot$.

11.4.8 ODOROUS AIR

Odorous air is emitted from natural and engineered systems. Table 11.13 lists several odorous compounds associated with untreated municipal wastewater or agricultural feedlots and their associated waste management facilities. Also included in this table is the **odor threshold**, that is, the lowest concentration of the odor that can be detected by humans.

Table / 11.13

Odor Thresholds for Some Compounds Associated with Untreated Wastewater

Odorous Compound	Chemical Formula	Molecular Weight	Odor Threshold ppm$_v$	Characteristic Odor
Ammonia	NH_3	17	46.8	Ammoniacal
Dimethyl sulfide	CH_3-S-CH_3	62	0.0001	Decayed vegetable
Hydrogen sulfide	H_2S	34	0.00047	Rotten eggs
Methyl mercaptan	CH_3SH	48	0.0021	Decayed cabbage

SOURCE: Crites and Tchobanoglous, 1998.

The detection of odor is not a good metric for determining the health effect. This is because thresholds are not related to health effects. Thus, if an individual does not smell an odor, it does not mean that exposure to the odor is safe. Furthermore, while it may be initially easy to detect some odorous compounds at a low concentration, at higher concentrations and over time, the odors paralyze the ability of the body to sense the odor. Regulatory limits for odor emissions are also becoming increasingly stringent. For example, many municipal wastewater treatment plants must now control odorous emissions to nondetectable levels at the plant boundary.

11.4.9 INDOOR AIR POLLUTANTS

Chapter 6 stated that the indoor environment has become an important place for exposure to chemicals, one reason being that U.S. citizens now spend 85 percent of their time indoors. Of particular concern are indoor environments that are poorly ventilated, by natural or mechanical methods, and those that contain chemical-emitting building materials and carpeting, coatings, and adhesives, cigarette smoke, synthetic cleaning and fragrance products, and exposure to natural pollutants like radon. There are currently no enforceable standards for residential and commercial **indoor air quality** like there are for ambient air. However, limits and guidelines have been developed for industrial settings by the Occupational Safety and Health Administration (OSHA) and American Industrial Hygiene Association.

Some major sources of indoor air pollutants are provided in Table 11.14. The negative health effects associated with these pollutants

Table / 11.14

Sources of Indoor Air Pollution

Sources	Example Sources and Specific Pollutants
Combustion sources	Burning oil, gas, kerosene, coal, wood, and tobacco products can release fine particulate matter, CO, and hazardous air pollutants into the indoor environment.
Outdoor sources	Pesticides and outdoor air pollutants can enter the indoor environment through open windows, cracks, the ventilation intake, and from dust carried into the building on shoes. Radon can enter through the building's foundation.
Building materials and household furnishings	Asbestos and lead are found in building insulation and paint in older homes. Formaldehyde and other VOCs are emitted from pressed wood products found in the building structure and furniture. VOCs are also emitted from sealants, adhesives, and paints. Flame retardants are emitted from furniture, electronics, mattresses, and even baby clothes.
Household products	VOCs, hazardous air pollutants, and fragrances are emitted from common household cleaners, maintenance products, personal care and hobby products, and printer ink.
Dampness and water leaks	Biological pollutants such as mold.
Household inhabitants, pests, and their activities	Pets, people, rodents, insects, and plants all can shed or off-gas biological particles, allergens, and some gases. Inhabitant activities such as vacuuming with a poor quality filter also resuspend deposited particles into the air.

can be immediate or long term. Immediate effects span the range from irritation of eyes, throat, and nose to death by asphyxiation. Subjects can also suffer from headaches and fatigue and asthma. Long-term effects can result in respiratory disease, heart disease, and cancer.

Table 11.15 provides a list of ways to improve indoor air quality. Note that the main way to improve indoor air quality is through **source reduction** through careful selection of materials and consumable products that are used to construct, furnish, clean, and maintain the home or building. Ventilation is the second most important way to improve the indoor air environment. In fact, a large part of the LEED green building certification process is related to improving indoor air quality through source reduction and proper ventilation. Some materials used to promote energy efficiency in a building can emit VOCs. Accordingly, EPA developed a document titled *Healthy Indoor Environment Protocols for Home Energy Upgrades* that provides a set of best practices for improving indoor air quality in conjunction with home energy efficiency projects.

Table / 11.15

Methods to Control Indoor Air Pollution

Control Method	Description
Source control	Preferred and most effective because it eliminates the source of the pollutant. Strategically placed carpets as you enter the house can capture dust particles.
	Purchase and use vacuum cleaners with high capture and filtration efficiencies for small particles.
	Eliminate synthetic carpets and their adhesives that emit VOC and store pollutants. Use natural floor coverings (with zero or low VOC adhesives) such as wood, tile, and organic wool.
	Eliminate use of synthetic fragrances found in perfumes, candles, odor control devices, and cleaning products that contain synthetic chemicals.
	Purchase adhesives, paints, coverings, and furniture that have zero or very low VOC emission rates. Use of natural and organic products will minimize emissions of VOCs and other hazardous air pollutants such as flame retardants. Ensure product bottles remain sealed when not in use.
	Eliminate water leaks that release moisture into building structure.
	Purchase and use vacuum cleaners with high capture and filtration efficiencies for small particles. Do not let moisture or standing water accumulate.
	Clean up garbage and spills quickly. Clean air ducts, filters, and soiled/dusty materials regularly.
Ventilation improvements	Mechanical ventilation systems draw outside air into the building and can employ energy efficiency through use of air–air heat exchangers.
	Operating ceiling and attic fans can improve a home's ventilation rate.
	Strategic placement and use of windows that open and take advantage of natural ventilation.
	Use of well-designed chimneys and appliance ventilation are important to removing gaseous emissions from combustion and other human activities.
Air cleaners	This is a least preferred method of the pollution prevention hierarchy, because it typically is less effective than source control or improved ventilation.
	Cleaners are generally not designed to remove gaseous pollutants and fine particulate matter. Further, their effectiveness varies, and high throughput rates may be necessary for cleaning a large room.

11.5 Ambient Emissions and Emissions Control

11.5.1 TYPES OF EMISSIONS AND SOURCES

Table 11.16 gives how sources of emissions of air pollutants can be broken down into several categories. Emissions are usually reported in units of mass emitted per unit time (for example, g/s, lb/day, kg/day, metric tons/year). Stationary sources are facilities whose emissions are fixed in space. Types of stationary sources include **point sources**, whose emissions are released through stacks (called **stack emissions**). Stationary sources may also have **fugitive emissions**, which result from the unintended release of an air pollutant or lack of capture efficiency in a collection hood. **Area sources** are sources whose emissions are not associated with a stack or are individually too small to be treated as point sources. They can originate from many activities, for example, the refueling of your automobile and the use of solvents during painting and degreasing operations. **Mobile sources** are associated with transportation, both on roads and off-road, including boats, airplanes, and lawn equipment. They also include the release of air pollutants from construction and agriculture activities. Mobile emissions can be quite large for certain pollutants. For example, up to 33 percent of VOC emissions and 40 percent of NO_x emissions in the United States originate from highways.

Air Pollution from Burning Municipal Solid Waste (Santa Cruz, Bolivia)

(Courtesy of Heather Wendel Wright).

Table / 11.16

Definition of Some Major Anthropogenic Sources of Air Emissions

Emission Source	Major Sources
Point sources are emitted from a stack.	Power plants, industrial boilers, petroleum refineries, industrial surface coatings, and chemical manufacturing industries.
Area sources are those emissions that are not associated with a stack and are individually too small to be treated as point sources.	Solvents used for surface coating operations, degreasing, graphic arts, dry cleaning and gasoline stations from tank truck unloading, and vehicle refueling.
Mobile sources are categorized for highway and off-highway sources.	Highway sources include automobile, buses, trucks, and other vehicles traveling on local and highway roads. Off-highway sources are any mobile combustion sources such as railroads, marine vessel, off-road motorcycle, snowmobiles, farm, construction, industrial and lawn/garden equipment.
Fugitive emissions are releases of gases and particulates that are not through a collection system.	Usually associated with industrial activities. Emissions that escape capture by hoods are emitted during material transfer, are emitted to the atmosphere from a source area, or are emitted from process equipment. They can include those from pressurized equipment, leaks from valves and pipe connections, emissions from wastewater treatment ponds and waste storage tanks.

SOURCE: EPA.

11.5.2 EMISSIONS TRENDS

Since 1990, there has been a substantial reduction (59 percent) in the total U.S. emissions for the particulate matter (measured as $PM_{2.5}$ and PM_{10}), nitrogen oxides, VOCs, carbon monoxide (CO), lead, and sulfur dioxide (SO_2) (Figure 11.16). In terms of specific pollutant levels, direct $PM_{2.5}$ emissions have declined by more than half, SO_2 emissions by more than 60%, NO_x and VOC emissions by more than 40 percent. These large decreases in emissions occurred even as the U.S. population increased by 24 percent, gross domestic product increased by 65 percent, energy consumption increased by 15 percent, and vehicle miles traveled increased by 40 percent. Figure 11.16 also demonstrates one of the great positive outcomes of the regulatory approach implemented by the Clean Air Act and its amendments. Other actions besides regulatory approaches that reduced air emissions over that past 40 years included voluntary measures taken by industry and partnerships between local, state, tribal, and federal governments and environmental organizations. Personal choices by individuals and homeowners also resulted in the decrease in overall emissions (EPA, 2012a).

One important thing to observe from Figure 11.16 is the impact the economy has on emissions of air pollutants. During the economic downturn of 2007–2009, there is a noticeable drop in gross domestic product, vehicle miles traveled, overall energy consumption, and CO_2 emissions. This relates to not only changes in industrial output but also changes in personal and household income and increases in gasoline prices. Also seen from this figure is the decline in overall air emissions of the six criteria air pollutants even though the economy

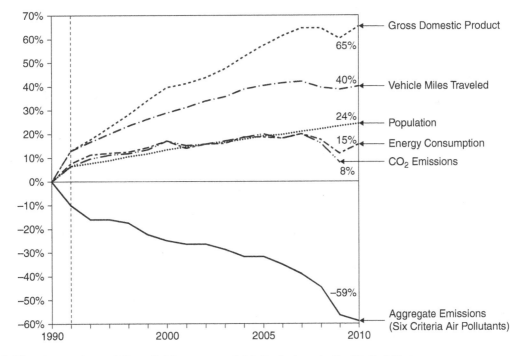

Figure / 11.16 **Comparison of Growth Measures and Air Emissions in the United States, 1990–2010** CO_2 emissions estimates are from 1990 to 2009.

(Redrawn from EPA, 2012a).

grew (as measured by gross domestic product) and the U.S. population and number of vehicle miles traveled also increased. In contrast, overall CO_2 emissions have increased by approximately 8 percent from 1990 to 2009.

11.5.3 EMISSIONS CONTROL

There are many ways to control the emissions of air pollutants. **Source reduction** and **demand management** strategies can be used to eliminate or reduce air emissions by changing activities that produce them. Application of green chemistry and engineering can also be used to eliminate the hazard and amount of air emissions. The successes of **regulatory approaches** (some that were discussed previously) included provisions of the Clean Air Act and its amendments that required major emitters of air pollution to obtain a permit from a state regulatory agency. This permit then specifies the types and quantities of air pollutants that may be released, along with legislated control strategies, such as types of control equipment required.

Market-based approaches have been considered as a control strategies since passage of the Clean Air Act Amendments in 1990. Market-based approaches attempt to create financial incentives to reduce emissions. Many such approaches allow for trading of emissions between sources at different scales (for example, from within a facility to across a state, region, or nation), so that sources that are less expensive to control will value implementing control measures. One example of this approach is to set an upper level of total allowable emissions for a particular pollutant. This *cap* is then theoretically lowered over time and individual businesses can determine if they wish to purchase emissions credits from the market, or invest in other emissions control strategies that can include source reduction, demand management, or control technology. A nationwide cap-and-trade market has been developed for sulfur dioxide emissions and the eastern states of the United States developed a similar market for NO_x emissions. California is currently implementing such a market for CO_2 emissions. Although cap-and-trade is an important strategy for emission reduction, it can be controversial because trading of emissions can lead to trading of effects for some pollutants, with potential environmental equity implications.

Voluntary approaches use education campaigns to promote source reduction or provide incentives to individuals or organizations that provide recognition or a financial incentive to reduce emissions. For example, a business could provide financial incentive to employees who participate in ride-sharing programs through recognition, bonuses, free passes to use mass transportation, or use of a shared vehicle supported by the company. In addition, homeowners can be educated on how their daily activities and purchasing behaviors impact release of air pollutants.

The last strategy to control air emissions is through the use of **control technologies**. Obviously treatment is a less preferred strategy for managing air emissions than source reduction strategies. One reason is based on conservation of mass principles. Treatments often convert pollutants of one type to another, or remove pollutants from the air into another media (that is, solid or liquid waste). Treatment

Learn more about Cap and Trade
http://www.epa.gov/airmarkt/trading/

Market Programs for SO₂ Reductions
http://www.epa.gov/airmarkets/
progsregs/arp/so2.html

processes attempt to reduce the hazard associated with air releases, but the hazard associated with the new waste stream should not be ignored.

ENGINE FUNDAMENTALS AND CONTROL OF AUTO EMISSIONS

As reported previously, air consists of 78.1 percent N_2 and 20.9 percent O_2. The production of NO_x is primarily from man-made combustion though natural production can occur from forest fires, lightning, bacterial action in soils, and oxidation of NH_3 in the atmosphere. In an engine or other process that combusts fossil fuels (for example, burning coal or natural gas to produce electricity), air is drawn into a combustion chamber and mixed with fuel. At high temperatures, the nitrogen and oxygen that occur naturally in the air react to form NO_x in the following general equation:

$$N_2 + O_2 \rightarrow NO_x \qquad \textbf{(11.14)}$$

Remember that NO_x is a term used to describe the combination of NO and NO_2. Approximately 80–90 percent of the NO_x that is formed in your car engine is NO, with the remainder being NO_2 according to the following more specific equations:

$$\boxed{N_2 + O_2 \rightarrow NO} \qquad \textbf{(11.15)}$$

$$\boxed{NO + \tfrac{1}{2}O_2 \rightarrow NO_2} \qquad \textbf{(11.16)}$$

In actuality, these two reactions are much more complex and depend on conditions in the engine and the fuel type. Temperature influences the formation of NO_x and higher temperatures tend to favor production of NO over NO_2. NO_x formed through the process described in the above two equations is referred to as thermal NO_x. Depending on the fuel type, nitrogen found in the fuel can also be oxidized to NO_x during combustion; this is called fuel NO_x and can account for up to about 50 percent of the total NO_x generated.

Attempts to minimize NO_x emissions from a vehicle can occur by several methods that include reducing peak temperatures and oxygen concentrations by recirculating flue gases, reducing the air to fuel ratio and staged combustions, and optimizing the valve or ignition timing in vehicles. Manufacturers also spend significant efforts to optimize the design and operation of their engines to reduce emissions of NO_x, CO, hydrocarbons, and particulates (especially important for diesel engines). Air emissions can also be treated after exiting the engine. For example, three-way catalytic converters are installed on modern vehicles to reduce emissions of three air pollutants: CO, hydrocarbons (that is, reactive hydrocarbons (VOCs)), and NO_x. The catalytic converter incorporates approximately 4 g of a precious metal such as platinum, palladium, or rhodium onto the enhanced surface area of the catalyst to catalyze the chemical reactions shown in Equations 11.17–11.19 (Hillier, 2001). Certain compounds present in the fuel can poison the catalyst, for example, lead that would be found in leaded gasoline:

$$2CO + O_2 \rightarrow 2CO_2 \qquad \textbf{(11.17)}$$

$$C_aH_{2a+2} + [(3a + 1)/2]O_2 \rightarrow aCO_2 + (a + 1)H_2O \qquad \textbf{(11.18)}$$

$$2NO_x \rightarrow xO_2 + N_2 \qquad \textbf{(11.19)}$$

Equation 11.18 is written specifically for the conversion of one type of hydrocarbon (an alkane) to CO_2 and water, where a is an integer. For example, for the alkane, hexane (C_6H_{14}), a would equal 6. Equation 11.19 represents the reduction of NO_x by a reducing agent present in the exhaust stream; reducing agents include CO, H_2, and unburned hydrocarbons. The use of the catalytic converter therefore results in lower emissions of CO, hydrocarbons, and NO_x, but note how the end products still result in the production of CO_2.

TRANSPORTATION DEMAND MANAGEMENT The EPA and Federal Highway Administration encourage **transportation demand management** as a way to reduce vehicle congestion and air emissions. Demand management strategies in the transportation sector focus on changing travel behavior of users of automobiles and light-duty trucks and also changing policies related to land use. For example, air pollutant emissions can be reduced by reducing the number of trips, the length of a trip, the mode of travel, and the time of day an individual or business conducts their travel. Methods of demand management are also often focused on reducing traffic during congested peak periods and have been shown to reduce emissions of all criteria pollutants.

Demand management strategies can also include trip reduction ordinances made by local, regional, or state government to encourage transportation alternatives and even encourage telecommunication substitutes for travel to a centralized workplace. These ordinances are typically aimed at employers and land developers. EPA (2007) also reports how local governments use methods of land usage, such as *in-fill development*, to encourage development of former industrial sites (that is, brownfields), declining suburban malls, vacant properties, and other underutilized land for redevelopment (EPA, 2007). In-fill development can thus reduce sprawl and the resulting air emissions associated with longer commutes from single vehicle modes of transportation. Table 11.17 provides an overview of 16 transportation demand management strategies.

© Steve Lovegrove/iStockphoto.

Traffic Congestion in the United States
http://mobility.tamu.edu/ums/

Table / 11.17

Examples of Transportation Demand Management Strategies That Reduce Emissions of Air Pollutants (adapted from U.S. Department of Transportation, 2013).

Park-and-ride facilities include the construction or expansion of parking lots in suburban or rural areas where people can park their vehicles and then join a carpool, vanpool, or transit service.	**High occupancy vehicle (HOV) lanes** use two important incentives (reduced travel time and improved trip time reliability) to maximize the person-carrying capacity of a roadway. The design and/or operation of the roadway shifts usage from single vehicle occupancy to priority treatment for HOVs, such as carpools, buses, and vans.
	(continued)

(continued)

Regional rideshare programs provide ride-matching services, employer outreach, and incentives to commute by carpool or vanpool (such as free gas cards, drawings, award programs, subsidies). Ridematching may be traditional (that is, people establish regular carpool routines) or dynamic (real-time matching of individuals who want to travel to/from similar locations).	**Vanpool programs** are particularly well suited for longer commutes. They use vans that typically carry 7–15 passengers and operate weekdays, traveling between one or two common pickup locations (typically a park-and-ride lot or a transit station) and the place of work. Vanpool programs will provide vehicles owned by an organization to commuters who share an employment destination. The vans are operated by a driver or by the commuters.
Bicycle and pedestrian projects/programs include a wide range of investments and strategies to facilitate and encourage nonmotorized travel. Examples of these strategies include safe bicycle paths and lanes, sidewalks, bicycle racks or lockers, pedestrian urban design enhancements, bicycle share programs, bicycle incentives, and incorporation of showering facilities into buildings.	**New or expanded bus or rail services** include additions to the provision of services through the establishment of new routes, increased frequency, hours of operation, or coverage of routes.
Improved transit service involves increasing the frequency or hours of service on existing transit routes. Increased frequency of service results in increased ridership, because transit becomes a more convenient option and increasing hours of service allows people to use the route at hours that were not previously available.	**Increased transit marketing, provision of more widely accessible transit information, and additional customer service** increase the number of people using public transportation. Provision of transit shelters, benches, maps, and visually pleasing aesthetics, or improving the comfort of buses and trains may be a supporting strategy to increase ridership. Service enhancements can also include improved transfer facilities and timing of transit services to reduce wait times.
Transit pricing strategies reduce the costs associated with using transit, thereby creating incentives for people to shift from other traveling modes. Fare reductions can be implemented systemwide, in specific fare-free or reduced-fare zones, or offered through employer-based benefits programs that are fully or partially paid by the employer.	**Parking pricing/management strategies** change the cost and/or convenience associated with driving a private vehicle, through pricing and management of parking on either end of the trip. While some policies increase the cost of parking through taxes or implementation of parking fees, some strategies reduce the supply of spaces through the creation of parking maximums for new development, regional parking caps, peak-hour parking bans, or curb-parking restrictions.
Road pricing change the costs to consumers operating private vehicles. Examples include new or increased tolls on roads and high occupancy toll lanes.	**Vehicle miles traveled-based pricing** imposes a fee based on miles driven. The fees are collected annually through the vehicle registration process or auto insurance program. Insurance premiums could be charged with a per-mile component and could be levied on a monthly or semiannual basis.
Fuel pricing strategy increases the tax rates applied to retail sales of motor fuels. Fuel pricing also creates an incentive for purchasing more fuel-efficient vehicles; overall vehicle stock changes may further affect emissions over the long term.	**Employer-based travel demand management programs** are designed to encourage employers to offer a range of worksite programs to reduce the number of vehicles using the road system during peak travel hours while providing a wide variety of mobility options including telecommuting.
Nonemployer-based travel demand management programs reduce noncommute trips to address the growth in nonwork trips. Examples of noncommute travel include special event travel (to sporting events and entertainment venues), tourism travel, and school-based travel.	**Land use strategies** include transit-oriented development and clustered activity centers. Integrating land use and transportation planning makes common destinations accessible by alternative modes of transportation, including transit, walking, and biking.

Application / 11.8 Traffic Congestion

The Texas Transportation Institute reported that in 2010 **traffic congestion** in the United States continued to increase, wasting 1.9 billion gallons of fuel and increasing emissions of air pollutants. The total cost of this congestion is estimated to exceed $100 billion, with a $750 cost to each commuter. In addition, studies now show that traffic congestion is increasing outside of the rush hour with 40 percent of traffic delays now reported during midday and overnight.

In the United States, traffic congestion was the norm by 1912. As early as 1907, there were reports that road-widening projects that had been expected to relieve congestion appeared to be doing the opposite. Woodrow Wilson commented in 1916 that motorists were using up roads almost as fast as they were made. By the 1920s, congestion had decreased vehicle speed to 4 mph on New York's Fifth Avenue.

History shows that building more roads and widening existing roads will never solve the problem of traffic congestion (see Table 11.18). Providing several options for travel is a key component of a sustainable accessibility plan. This not only relieves congestion but also provides substantial savings to the public. As one example, without rail transit to and from Manhattan, New York would require 120 new highway lanes and 20 new Brooklyn Bridges. Also U.S. taxpayers recover their $15 billion in investment in public transit with congestion cost savings alone.

*Last two paragraphs based on Alvord (2000).

Table / 11.18

What Actually Happens When Additional Vehicle Lanes Are Added to Reduce Congestion

Vehicular speeds increase, increasing the safety hazard to pedestrians and bicyclists
Pedestrian crossings increase in length and time, which makes walking less desirable
Lane crossing distances increase for bicyclists, which makes biking less desirable
Congestion relief is usually temporary; lanes eventually fill up
New roads and lanes cause additional loss of open space, because development takes place along the road corridor

example/11.4 Estimating Impact of Transportation Travel Demand Management

Park-and-ride facilities include the construction or expansion of parking lots where people can park their vehicles and then join a carpool, vanpool, or transit service. Typically, park-and-ride facilities are used in suburban areas. This strategy reduces air emissions by decreasing the number of single-occupancy vehicles on the road. A travel demand management action will add parking spaces to an existing park-and-ride facility that is not served by transit. The plan will add 60 parking spaces. Assume the new spaces will have a 70 percent estimated utilization rate and these individuals will join an existing carpool or vanpool. Also assume that 80 percent of users will have previously driven alone, the average commute that will be eliminated is 50 miles roundtrip (distance from lot to destination and return), and there are 250 operating days per year. What is the annual reduction in vehicle miles traveled from implementation of the park-and-ride facility?

solution

First we estimate the expected use of the new parking lot as follows:

$$= (\text{spaces added to lot}) \times (\text{estimated utilization rate}) = 60 \text{ spaces} \times 0.70 = 42 \text{ spaces}$$

Next we determine the expected number of people that will reduce their individual driving.

$$= (\text{spaces used}) \times (\text{percent of users who previously drove alone}) = 42 \text{ spaces} \times 0.80$$
$$= 33.6 \text{ fewer drivers per day}$$

The next step is to determine the annual reduction in vehicle miles traveled:

$$= (\text{number of fewer drivers per day}) \times (\text{estimated round trip}) \times (\text{total operating days})$$
$$= (33.6 \text{ fewer drivers per day}) \times (50 \text{ miles per driver}) \times (250 \text{ days})$$
$$= 420,000 \text{ annual reduction in vehicle miles traveled}$$

This reduction in vehicle miles driven per day can then be converted to reductions in air emissions by multiplying the reduction in vehicle miles driven by the amount of a particular criteria pollutant that is emitted per mile driven. This *emission factor* will be covered later in this chapter and is specific to an operating vehicle and the pollutant.

Example adapted from Department of Transportation, 2013.

EMISSIONS CONTROL TECHNOLOGIES Figure 11.17 shows the complexity of a typical air pollution treatment process using technological controls to remove particulate matter from an airstream originating from a two chambered incinerator. In this case, the primary air pollution control technology is a venturi scrubber that is followed by a mist eliminator. The presence of a stack indicates this is a stationary source; however, the incineration process may also produce fugitive emissions as well as the collection and treatment steps related to controlling the emissions.

Because of the great expense and regulatory requirements associated with treating air pollution emissions, pollution prevention activities can be substituted. This is similar to the demand management strategies discussed previously for controlling air emissions associated with the transportation. Table 11.19 provides a comparison of several control and pollutant prevention activities to manage some common air pollutants. For example, many manufacturers have replaced solvent-based coatings with water-based coatings. This eliminates VOC emissions and the resultant requirement for use of thermal oxidation to treat the effluent stream. To control mercury emissions from coal burning for electricity, one treatment method is to inject activated carbon into the flue gas from the combustion process. The activated carbon adsorbs mercury and is removed with other particulate matter by the effluent particle control technologies (an example of converting the pollutant from air to solid phase). Switching to

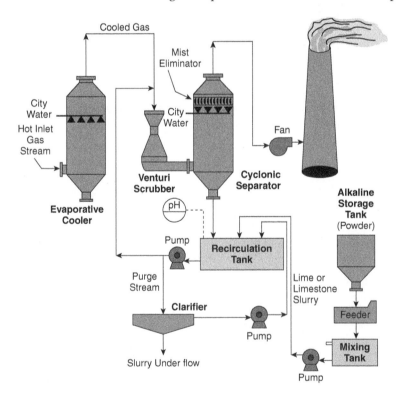

Figure 11.17 **Several Components of a Particulate Wet Scrubbing Air Pollution Control System** Not shown are the large number of fans, hoods, and ductwork that connect the process equipment and the air pollution control equipment together.

(Redrawn from EPA APTI).

Table / 11.19

Control Technologies and Pollution Prevention Activities to Control Emissions of Some Specific Air Pollutants

Pollutant(s)	Treatment Control Technology	Pollution Prevention Activity
Particulate matter	Electrostatic precipitator, bag house filters, particle scrubbers, and cyclones.	Reduce demand for electricity by providing incentives to purchase energy efficiency lighting and appliances. Energy Star programs provide consumers with information on energy efficiency of appliance purchases.
Sulfur dioxide	Flue gas desulfurization or cleaning sulfur from coal for combustion.	Reduce demand for coal generated electricity through demand strategies such as described above for particulate matter. Fuel switching to coal with lower sulfur content, a different type of fuel, or renewable energy source.
VOCs	Thermal or catalytic oxidation, adsorption systems (for example, activated carbon), and biofilters. Hoods and other technologies for capture of fugitive emissions.	Substitute solvent-based paints with water-based paints for the coatings used in the automobile and appliance industries. Improve fuel efficiency of highway and off-road vehicles.
NO_x	Catalytic converters for reduction in NO_x emissions from vehicular sources. Selective reduction, scrubbers, and adsorption systems for point sources.	Use demand management strategies that promote and provide incentives for the public to use mass transportation, walk, or cycle.
Ozone	Control of NO_x and VOCs as discussed above.	Reduce vehicle miles traveled and demand for electricity.
Mercury	Inject activated carbon into the flue gas followed by a particle collection technology.	Reduce demand for electricity. Reduce use of mercury in consumer, medical, and scientific products. Separate mercury-containing products from medical and municipal waste incineration streams.

renewable energy and reducing demand for electricity are alternative methods that would reduce the need for effluent treatment, as they reduce the mercury produced.

A control technology is associated with removal of specific air pollutants. Several examples that are commonly used to treat air emissions that contain particulate matter, VOCs, CO, SO_2, and odorous compounds are reviewed in Figure 11.18. Several of these technologies will be discussed in great detail in the section that follows. Readers should make note that it is always important to pair the proper materials and coatings of collection and treatment technologies to ensure potential corrosion does not impact system performance or worker safety.

Control Technology and Operating Principles	Illustration
Adsorption processes allow for gaseous pollutants like VOCs and SO$_2$ to be transferred from the air to a solid adsorbent. The use of adsorbants such as activated carbon was discussed in Chapter 8 for the treatment of water. The horizontal bed absorber shown on the right is commonly used for larger flow rates.	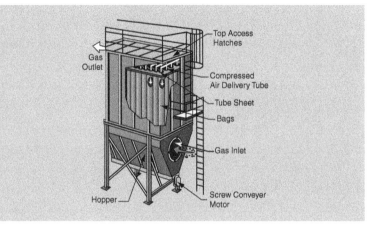
A **baghouse** allows for particulate matter to be filtered by a set of fabric filter bags. The bags are periodically shaken to remove the particulate matter into an underlying hopper from where it can be collected.	
A **biofilter** degrades gaseous pollutants (VOCs and odors) by use of microorganisms that reside on the filter media. The filter medium could be something such as lava rock or a mixture of compost and woodchips. Air is typically passed up through the bed. Periodically, moisture may be added to the filter.	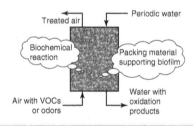
In a **cyclone**, particulate pollutants enter with the gas and are removed by centrifugal forces because the particles have more momentum and cannot turn with the gas. The air is moving in a helical pattern. Particles that impact the cyclone outer wall then fall by gravity into a hopper where they can be collected.	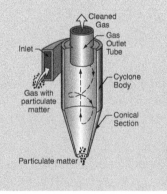

Figure / 11.18 Common Air Pollutant Emission Control Technologies (additional details on design and operation can be found in EPA (2012c,d), and Theodore (2008).

Control Technology and Operating Principles	Illustration
In an **electrostatic precipitator**, particulate matter is first charged by applying a high voltage, which produces ions that attach to the particles. The charged particles are then forced to collection plates. Rappers use impulse or vibrating methods to remove the particles from the collection plate so they fall into the hopper, where they can be removed.	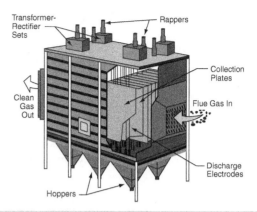
In a **packed bed absorption**, scrubber pollutants (such as SO_2, NH_3, HCl) are transferred from the air to liquid. A key performance indicator is how well equilibrium between the gaseous and aqueous phase (as determined by a Henry's constant) favors the liquid phase. Also important is the rate at which the pollutant transfers from the air to liquid phase.	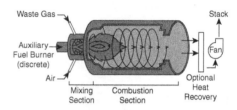
In a **thermal oxidizer**, the pollutant (VOC, CO, odor) is oxidized through high-temperature combustion. The system consists of a burner and reaction chamber. Removal is better with greater temperatures, turbulence, and gas residence times.	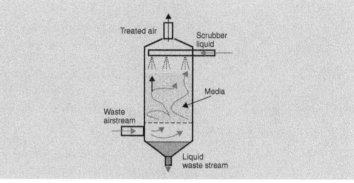
In a **venturi scrubber**, particulate matter and other pollutants such as SO_2 and HCl can be removed together by impaction of airborne pollutants on water droplets. A water spray can be added by several methods, including injection into flowing gas.	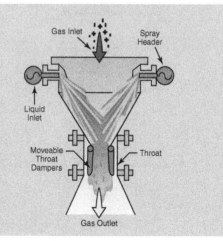

Figure / 11.18 (*continued*)

PARTICULATE MATTER   Particulate matter is typically controlled by the use of filtration, wet scrubbing, or mechanical collectors. The type of technology that is appropriate can be screened by using Figure 11.19. As shown in this figure, if the particulate matter contains sufficient moisture so it is wet or sticky, web scrubbers or a wet electrostatic precipitator would be the most appropriate control technology. If the airstream contains dry particulate matter (and no explosive gases), particles can be removed by a wider selection of control technologies. These include fabric filters (for example, a baghouse), electrostatic precipitators (ESPs), and wet scrubbers. Wet scrubbers are usually more expensive to construct and operate than fabric filtration systems, cyclones, and other mechanical systems.

ELECTROSTATIC PRECIPITATOR   An **electrostatic precipitator** is shown in Figures 11.18 and 11.20. They are commonly used to treat fly ash that originates from coal combustion during the generation of electricity. They are not used in situations where an explosive gas or vapor is present because of the sparking that occurs with the removal chamber.

Figure 11.19 had previously shown how the particle size range of the polluted airstream and regulatory requirements influence the selection of a control technology. For example, fabric filters like a baghouse are used to treat polluted air where there are a significant number of particles less than 0.5 μm in diameter and a high degree of removal is required. In contrast, while electrostatic precipitators can handle high volumes of air and operate at a low pressure drop, they are recommended for treatment of a larger size range of particles. Observe in

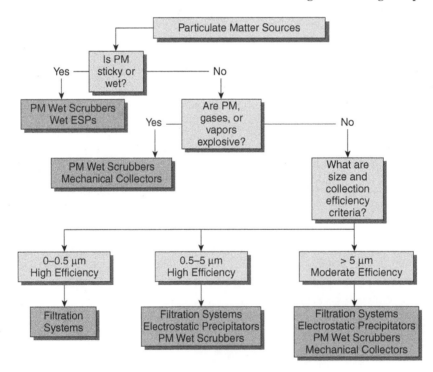

Figure / 11.19   General Applicability of Control Technologies for Particulate Matter.

(Redrawn from EPA, 2012d).

Figure / 11.20 Electrostatic Precipitator (ESP) The largest application is seen in removal of fly ash generated during combustion of coal to generate electricity. Gas flow rates associated with a 1,000 MW power plant may reach millions of cfm. The pressure drop may be only 1 inch versus 10–100 inches across a wet scrubber or filter baghouse. Low sulfur coals produce a fly ash with lower resistivity, which makes the particles more difficult to accept a charge. They are thus more difficult to collect with this technology and chemical addition into the gaseous stream may be required (Theodore, 2008).

(© Charles E. Rotkin/Corbis).

Figure 11.21 how the electrostatic precipitator has less removal efficiency for particles in the 0.1–0.5 μm size range. This is because the ability of the electrostatic precipitator to charge particles is poorer in this size range (EPA, 2012d).

BAGHOUSE **Bag filters** consist of a set of fabric filters that allow polluted air to pass through the fabric filter of the bag and flow out of the bag. In the process, particulate matter is collected on the outside of the bag. When several bag filters are housed together, the overall system is referred to as a **baghouse** (refer back to Figure 11.18). For a shaker

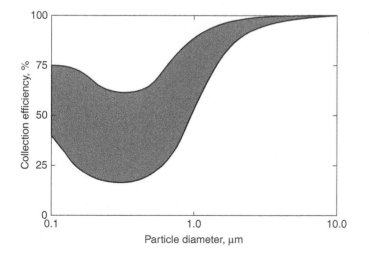

Figure / 11.21 Particle Size Efficiency Removal for an Electrostatic Precipitator.

(Redrawn from EPA, 2012d).

baghouse like the one depicted in Figure 11.18, the bag size diameter might be 8–20 ft in length with bag diameters of 5–12 in.

The surface of the filter consists of a fabric filter and some supporting mechanism to hold the fabric in place. During operation, particulate matter accumulates as a filter cake on the outside of the baghouse's many fabric filters, increasing both collection efficiencies and the pressure drop across the filters. To reduce the pressure drop, bags are cleaned periodically by mechanically shaking the accumulated particles off the fabric and into a hopper. The particulate matter is removed from the hopper and typically disposed of in a landfill. Bag filters are a commonly employed control technology, because collection efficiencies of particles can reach 99–99.5 percent particles in the size range of around 0.5–1 μm and larger particles are effectively removed (EPA, 2012e).

The failure of the bags in the baghouse can be observed and corrected quickly by daily inspection and monitoring. Table 11.20 lists some common causes for fabric failure. Baghouses may also fail because the shaker that regularly cleans the bag encountered a problem that results in excessive buildup of particles on the surface of the bag. This results in an increased pressure drop across the fabric filter. Other problems not related to fabric failures include plugging of the collection

Table / 11.20

Common Causes for Fabric Failures in a Baghouse (adapted from EPA, 2012e).

Cause	Results in	Reason
Improper bag installation	Holes or tears in bags Reduced bag strength	Lack of proper vendor instructions Poor access to bags by installer Improper tensioning or rough handling such as bending or stepping on the bags Bags to tight for cages Sharp edges on cages
High temperatures (fabric specific and maximum temperatures range from 180°F to 550°F)	Loss of fabric strength Attack on the bag's finish that causes abrasion	Improper fabric for service No high temperature alarm Continual operation at close to fabric temperature limits
Condensation	Alters adhesion characteristics of the particles to the bag materials resulting in mudding or blinding Chemical attack on the fabric	Unit not preheated or purged properly Air leakage or inadequate insulation
Chemical degradation	Attacks fibers of the fabric resulting in loss of strength	Change in manufacturing process
High-temperature drop	Increases bag abrasion that results in tears in the fabric	Poor cleaning, blind bags, increase in gas velocity
Bag abrasion	Worn or torn bags	Contact between bag and another surface High particle loadings Large particle inspection on bag

hopper because of failure to remove the particles that accumulate in the hopper (EPA, 2012e).

Baghouses are designed by considering variables that include characterization of the polluted airstream and selection of a gas-to-cloth ratio. The **gas-to-cloth ratio** is sometime referred to as the GC. It is defined as

$$\text{gas-to-cloth ratio} = \frac{\text{gas flow rate (ft}^3/\text{min)}}{\text{cloth area (ft}^2)} \qquad \textbf{(11.20)}$$

The gas-to-cloth ratio is determined based on the particular material being filtered (for example, coal, cement, limestone, sawdust), whether the fabric type is woven or felt, and the method of mechanical cleaning (that is, shaken, pulse jet, or reverse air). The selection of fabric type is influenced by the fabric's temperature limitations and its resistance to chemical attack from other compounds found in the polluted airstream. Table 11.21 provides examples of gas-to-cloth ratios for different fabrics and methods to remove the particles from the bag. Typically as the gas-to-cloth ratio increases, the pressure drop also increases (Theodore, 2008).

After the gas-to-cloth ratio is determined, the baghouse is sized by determining the total filter cloth area required for collection. Additional space is provided for a number of walkways and is influenced by the

Table / 11.21

Gas-to-Cloth Ratios (ft/min) for Designing a Baghouse

Material Being Removed in Filters	Fabric Type and Method of Particle Removal from Bags	
	Shaker/Woven Fabric; Reverse Air/Woven Fabric	Pulse Jet/Felt Fabric; Reverse Air/Felt Fabric
Cement	2.0	8
Coal	2.5	8
Fly ash	2.5	5
Gypsum	2.0	10
Iron ore	3.0	11
Lead oxide	2.0	6
Lime	2.5	10
Limestone	2.7	8
Rock dust	3.0	9
Sawdust	3.5	12
Spices	2.7	10

SOURCE: Theodore, 2008.

example/11.5 Baghouse Design

A baghouse that employs a shaker collection method is being designed to remove 99.75 percent of an incoming stream of particles originating from a cement plant. What fabric area is required for the baghouse if it treats 21,000 ft³/min of polluted air? How many bags are required if the bags are cylindrical and are 6 in. in diameter and 20 ft long? Assume the filter manufacturer has specified a woven fabric.

solution

Table 11.21 suggests that the appropriate air-to-cloth ratio for cement particles that are collected by shaking is 2.0 ft/min. The total fabric area required for the filter cloth (m² or ft²) required for collection is determined as

$$\frac{\text{inlet flow rate}}{\text{gas-to-cloth-ratio}} = \frac{21,000 \text{ ft}^3/\text{min}}{2.0 \text{ ft/min}} = 10,500 \text{ ft}^2$$

The number of cylindrical bags required is determined from the total area requirement and the surface area of each of the bags. The surface area of one bag (neglecting the top and bottom area of the cylinder) equals $\pi \times$ diameter $\times$ height:

$$\frac{\text{total filter area required}}{\pi \times \text{diameter} \times \text{height}} = \frac{10,500 \text{ ft}^2}{\pi \times 0.5 \text{ ft} \times 20 \text{ ft}} = 334 \text{ bags}$$

slope of the hopper. The total filter cloth area (m² or ft²) required for collection is determined as

$$\text{filter cloth area} = \frac{\text{inlet flow rate}}{\text{gas-to-cloth-ratio}} \qquad (11.21)$$

THERMAL OXIDIZER   As was described in Figure 11.18, a **thermal oxidizer** removes air pollutants such as VOCs through high-temperature combustion. They are also referred to as incinerators. Figure 11.18 showed that the system consisted of a burner and reaction chamber. Removal of organic compounds occurs at temperatures of 590–650°C (1,100–1,200°F) and most thermal oxidizers are operated at temperatures much higher. For example, hazardous waste incinerators are operated at temperatures as high at 1,800–2,200°F in an attempt to achieve near complete oxidation. Catalysts that contain platinum, copper, chromium, vanadium, nickel, and cobalt are now being used to decrease the operating temperature of catalytic oxidizers (typically for smaller systems that treat VOCs into the 650–880°F range). Unfortunately, heavy metals are not removed in a thermal oxidizer and some hazardous by-products can also be produced.

The conversion of a VOC to carbon monoxide and water vapor in a thermal oxidizer is shown as follows:

$$C_xH_y + \left(\frac{x}{2} + \frac{y}{4}\right)O_2 \rightarrow (x)CO + \left(\frac{y}{2}\right)H_2O \qquad \textbf{(11.22)}$$

The carbon monoxide is then converted to carbon dioxide as follows:

$$(x)CO + \left(\frac{x}{2}\right)O_2 \rightarrow (x)CO_2 \qquad \textbf{(11.23)}$$

In Equations 11.22 and 11.23, x and y are stoichiometric coefficients used to balance the chemical reactions. For example, for the chemical

---

example/11.6 Determining Stoichiometric Oxygen Requirements for a Thermal Oxidizer

A thermal oxidizer is used to treat 1,000 ft³/min of air that contains 1 lb of the VOC benzene (C_6H_6). Is there sufficient O_2 in the incoming air to achieve complete combustion to carbon dioxide and water?

solution

The correct stoichiometric equation for complete oxidation of benzene to the end products of water and carbon dioxide is written below. You can obtain this in one of two ways; you could refer back to the discussion on theoretical oxygen demand from Chapter 5 or write Equations 11.21 and 11.22 for benzene, and then add the two equations to obtain an overall reaction.

$$C_6H_6 + 7.5O_2 \rightarrow 6CO_2 + 3H_2O$$

The amount of oxygen that is required to oxidize the benzene is determined from the reaction stoichiometry as follows:

$$1\,lb\,C_6H_6 \times kg/2.205\,lb \times mole\,C_6H_6/78\,g\,C_6H_6 \times 7.5\,mole\,O_2/mole\,C_6H_6 \times 32\,g\,O_2/mole\,O_2 = 1.4\,kg\,O_2$$

This is the mass of oxygen required to oxidize the 1 lb (that is, 0.45 kg) of C_6H_6 that is found in every 1,000 ft³ of air that is processed every minute in the thermal oxidizer.

Next we must determine how much oxygen is present in the 1,000 ft³ of air. Because air at atmospheric pressure consists of 21 percent oxygen, there is 210 ft³ (equal to 5.95 m³) of oxygen in the 1,000 ft³ of air treated per minute. Assuming that the pressure is 1 atm and the temperature is 25°C, the number of moles/5.95 m³ of treated air is found from the ideal gas law.

$$n/V = n/5.95\,m^3 = P/RT = 1\,atm/8.2015 \times 10^{-5}\,m^3\text{-atm}/mole\text{-}K \times 298\,K$$

Solving this expression for n, which is the number of moles of O_2 in the 1,000 ft³ (or 5.95 m³) of air. This equals 243 moles of O_2.

This value can be converted to the mass of oxygen using the molecular weight of oxygen.

$$243\,moles\,O_2 \times 32\,g\,O_2/mole \times kg/1,000\,g = 7.8\,kg\,of\,O_2\,in\,the\,airstream$$

Because this value (7.8 kg O_2) is larger than the mass of oxygen required for complete oxidation (1.4 kg O_2), we have sufficient oxygen to perform the oxidation. Note that our solution did not account for the oxidation of nitrogen gas (which is 78 percent of the airstream) and would produce NO_x in the process.

Figure / 11.22 Coupled Effects of Temperature and Gaseous Residence Time on Pollutant Destruction in a Thermal Oxidizer.

(Redrawn from Theodore (2008) with permission of John Wiley & Sons, Inc.).

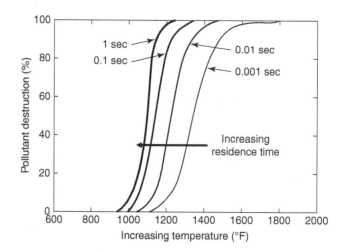

benzene (C_6H_6), x and y are both equal to 6. Both reactions consume oxygen, which is why performance of the reactor can be measured by the lack (or very low) amount of oxygen in the effluent gas (typically only 1–2 percent O_2 versus 21 percent in the input ambient air). Equation 11.23 also shows that complete oxidation of the VOC eventually results in production of the greenhouse gas CO_2. Pollution prevention activities like those described previously in Table 11.19 thus have added benefits besides just reducing emission of a VOC or other hazardous air pollutant; they also reduce CO_2 emissions directly through reduction of the energy needed for thermal oxidation.

Incomplete combustion in a thermal oxidizer can result in formation of aldehydes and organic acids which themselves can act as air pollutants. In addition, if the organic matter being oxidized contains sulfur or halogens, other unwanted by-products can be formed that can include SO_2, HCl, hydrofluoric acid, and phosgene. This may then require some type of web scrubber or absorption technology to be placed after the thermal oxidizer to remove these hazardous products.

The efficiency of pollutant removal is a function of the *temperature*, the gas residence *time* of the reactor, and degree of *turbulence* in the reactor. This is shown in Figure 11.22 where the percent of pollutant destruction is shown to increase with higher gas residence time and also higher temperature. Thus, an increase in temperature can accommodate a decrease in residence time (and vice versa). Because of the energy required to raise the temperature of the combustion chamber, it may be cost effective to increase the volume of the combustion chamber (which increases the gas residence time).

Recommended gas residences times are 0.2–2 s with a length-to-diameter ratio for the reactor of 2–3. Average gas velocities typically range from 10 to 50 ft/s and are used to discourage setting of particles and minimize dangers of fire hazard. The heat that is released during the oxidation can be recovered by direct heat recovery or through use of an external heat exchanger (Theodore, 2008).

example/11.7 Sizing a Thermal Oxidizer

A bench scale plug flow thermal oxidizer incinerator is being evaluated to operate at a temperature of 225°C and have a gas residence time of 0.3 s. If the pollutant (diethyl peroxide) enters the incinerator at a flow rate of 12.1 L/s and it is desired to remove 99.995 percent of the pollutant, what should the length of the thermal oxidizer be? The first-order rate constant for diethyl peroxide removal is 38.3/s at 225°C and the inside diameter of the thermal oxidizer (which is shaped like a cylinder) is 8 cm.

Remember from Chapter 4 that we developed an expression to describe pollutant removal in a plug flow reactor.

$$\frac{C_{out}}{C_{in}} = \exp\left(-\frac{kV}{Q}\right)$$

In this case, the gas residence time, t, equals V/Q. Therefore,

$$\frac{C_{out}}{C_{in}} = \exp(-38.3/\sec \times t)$$

Take the natural log of both sides of the expression

$$\ln[(1 - 0.99995)/1] = -38.3/\sec \times t$$

Solve for time = 0.259 s. The thermal oxidizer volume is determined as

$$T = 0.259\,s = \frac{V}{Q} = \frac{V}{12.1\,L/s}$$

which results in a volume of 3.13 L = 3.13×10^3 mL (or 3,130 cm).

Because the reactor is a cylinder, the length equals

$$\frac{V}{\pi D^2/4} = \frac{3,130\,cm^3}{\pi(8\,cm)^2/4} = 62.3\,cm$$

BIOFILTER   Emission and control of odorous gases is a growing concern in municipal wastewater collection and treatment and some industrial processes. The odor is primarily caused by hydrogen sulfide (H_2S) and other reduced sulfur compounds such as methyl mercaptan and dimethylsulfide. The presence of these chemicals can result in complaints from community members who live near a treatment plant (or pump station). These chemicals may also damage human health and corrode infrastructure and equipment. Odor problems can be addressed with control technologies such as scrubbing, adsorption, thermal oxidation, odor masking, and biofiltration.

A **biofilter** control technology (refer back to Figure 11.18) can treat odors and some VOCs. They are classified into two categories: biotrickling filters and biofilters. Biofiltration utilizes microorganisms that are attached to a packing material (synthetic or natural) to break down pollutants in a contaminated airstream that is passed through the packing material. If water is added consistently, the system is called a biotrickling filter. If the amount of water applied is minimal to just maintain sufficient moisture levels for microbial degradation, the system is called a biofilter.

Biofiltration units can also be constructed above or below the ground surface. They can be constructed as enclosed or open to the atmosphere.

Table / 11.22

Examples of Working Biofilters for Control of Odorous Air

Municipality	Description of Biofilter(s)	Major Odor Causing Compounds and Concentrations (ppb$_v$)	Type of Medium
Cedar Rapids (IA) Water Pollution Control Facility	Two full-scale biofilters, 120,000 cfm	H_2S (100,000 ppb$_v$) with smaller concentrations of organic sulfur compounds	Lava rock (6 ft depth)
Western Lake Superior Sanitary District (Duluth, MN)	Two-stage biofilter located at the Scanlon Pump Station 2,500 cfm Biofilter located at main treatment plant, 40,000 cfm	Methyl mercaptan (940 ppb$_v$) Dimethyl sulfide (21,200 ppb$_v$) Dimethyl disulfide (1,480 ppb$_v$) H_2S (93 ppb$_v$)	Compost/ wood chip (3.5 ft depth)
Metropolitan Council Environmental Services (Twin Cities, MN)	Twin Cities Metro Plant, 127,000 cfm Several smaller biofilters of different media (sand, woodchip/compost)	Primarily H_2S (concentration unknown) Organic sulfur compounds at lower levels	Compost/ wood chip (3 ft depth)

Open systems generally have a lower capital cost and are ideal when there are no space constraints. However, they may be impacted by heavy rain which can saturate the packing medium. Packing media typically consist of a combination of wood chips and compost, lava rock, or synthetic packing material. The media provides a high surface area/volume ratio where a biofilm of organisms can grow. Table 11.22 provides further information on several working biofilters used to control odorous air emissions from wastewater collection and treatment systems.

The primary organisms responsible for H_2S oxidation is believed to be members of the genus *Thiobacillus*. They are obligate chemo-autotrophic organisms that derive their energy from the oxidation of reduced sulfur compounds and obtain their carbon for growth from the fixation of atmospheric carbon dioxide. Although the optimum pH range for the growth of these organisms varies somewhat, they are predominantly acidophiles with optimum growth occurring under acidic conditions.

There are different ways to size a biofilter and many times a pilot scale study is performed before full scale design takes place. One method to size the volume of a biofilter is based on the empty bed residence time (EBRT):

Table / 11.23

Empty bed residence time to Achieve Removal of Many Odors

Empty Bed Residence Time (EBRT)	Application
>25 s	Industrial
30–60 s	Composting
5 s	Agricultural

SOURCE: Mann et al., 2002.

$$\text{EBRT} = \frac{V}{Q} \qquad (11.24)$$

where V is the volume of the filter (ft^3 or m^3) without the presence of the packing material and Q is the air flow rate (cfm or m^3/s). In this case, V is the volume of the filter without the presence of packing material.

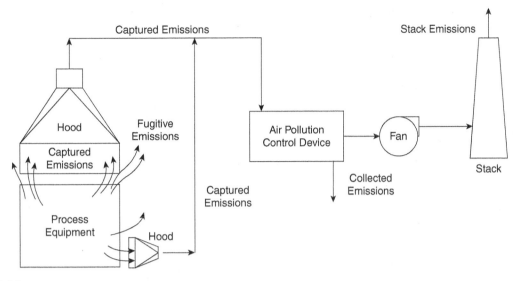

Figure / 11.23 Use of Hoods in an Industrial Process.

(Redrawn from EPA, 2012f).

HOODS **Hoods** are very common items that are paired with process equipment. The relationship of a hood with the process equipment and air pollution control technology is shown in Figure 11.23. Hoods are typically designed to operate under negative pressure, drawing air into the hood because the static pressure is lower inside the hood than in the surrounding air. The purpose of a hood is to capture a pollutant to prevent exposure of a worker to the pollutant and to minimize or eliminate fugitive emissions. These fugitive emissions can pass into the workspace and then leave the manufacturing facility through doors, window, and roof vents. Pollutants not captured by the hood are considered fugitive emissions and can be determined as:

$$\text{Fugitive emissions} = (\text{total emissions}) - (\text{emissions captured by a hood})$$

(11.25)

The emissions that exit the stack that follows the hood are then equal to:

$$\text{Stack emissions} = (\text{emissions captured by the hood}) \times \frac{100\% - \eta}{100\%}$$

(11.26)

where η is the collection efficiency of the hood in units of percent.

example/11.8 Determining Fugitive and Stack Emissions in an Industrial Setting

Determine the fugitive emissions and stack emissions from process equipment that generates 100 kg/h of VOCs. Assume the hood captures 95 percent of the VOCs and the collection efficiency of the air pollution control device is 95 percent.

solution

Equations 11.25 and 11.26 can be used to determine the quantity of fugitive and stack emissions.

$$\text{Fugitive emissions} = \text{total emissions} - \text{emissions captured by the hood}$$
$$= 100\,\text{kg/h} - 95\,\text{kg/h} = 5\,\text{kg/h}$$

$$\text{Stack emissions} = \text{Emissions captured by the hood} \times \frac{100\% - \eta}{100\%}$$
$$= 95\,\text{kg/h} \times \frac{100\% - 95\%}{100\%}$$
$$= 4.75\,\text{kg/h}$$

The total emissions is equal to the sum of the hood emissions and the stack emissions

$$= 5\,\text{kg/h} + 4.75\,\text{kg/h} = 9.75\,\text{kg/h}$$

Note the contribution to fugitive emissions for those not captured by the hood. You might want to just change the collection efficiency of the hood from 95 percent to 90 percent in this example to see the hood's importance in minimizing fugitive emissions. Of course, a more favorable action would be to use principles of green engineering to minimize the hazard of the air emissions so capture and treatment of air emissions is no longer required.

Problem adapted from EPA, 2012f.

11.6   Assessment of Emissions

Assessing the emissions associated with a particular source or industrial activity has become increasingly important for management of air quality. Figure 11.24 depicts five approaches to estimate air emissions and their hierarchy in terms of cost and reliability. More costly and reliable estimation methods (such as specific measurements) would be recommended when the resulting risks to human health or the environment are large or the adverse regulatory outcomes associated with making an error are high. In contrast, situations where the risk is low and adverse regulatory outcomes are also low, less expensive and less reliable methods may be employed. Table 11.24 provides an overview of some methods used to estimate air emissions.

As discussed in Table 11.24, an **emission factor** is a value that represents the ratio of the quantity of air pollutant emissions produced to the amount of pollutant-producing activity performed. The AP-42 is a series of documents EPA has used to document its emission factors. Thus, emission factors are sometimes referred to as AP-42 emission factors. Emissions can be estimated by multiplying the amount of a specific activity by a source specific emissions factor:

$$E = A \times \text{EF} \tag{11.27}$$

where E is the emissions in mass or mass per time (for example, kg or kg/day), A is a measure of the specific source activity (for example, liters of fuel used or chemical manufactured per unit time), and EF is the

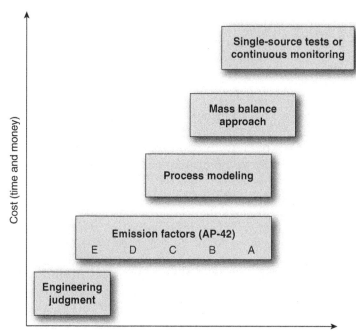

Figure / 11.24 Approaches to Estimate Air Emissions Compared with Their Reliability and Cost.

(Adapted from EPA Technology Transfer Network Clearinghouse for Inventories & Emissions Factors).

Table / 11.24

Overview of Methods to Estimate Air Emissions   (adapted from EPA Technology Transfer Network Clearinghouse for Inventories & Emissions Factors)

During *single-source testing or continuous monitoring*, a sample(s) is collected from the discharge point of an industrial airstream. This typically occurs at the stack discharge, during exhaust ventilation testing, or by sensors that provide continuous emissions monitoring. Difficulties can arise from the presence of extreme temperature, moisture, and discharge velocities experienced at the discharge point. Though the most reliable method, the results are applicable only to the conditions existing at the time of the testing or monitoring.

During a *mass balance approach*, a material balance is developed that is based on knowledge of the manufacturing process. Material balances should be performed only on components of pollutants that are conserved in the process, and not on materials that are consumed or chemically combined in the manufacturing process. They are most appropriate in situations where a high percent of the material is lost to the atmosphere. Appropriate examples include sulfur in fuel, or a solvent that is released during an uncontrolled coating process. The mass of a conserved pollutant component emitted can be estimated as follows:

$$M = M_{\text{in raw}} - M_{\text{product}} - M_{\text{accumulated}} - M_{\text{captured}} \qquad (11.28)$$

where $M_{\text{in raw}}$ is the mass of the pollutant component in the raw material feed, M_{product} is the mass of the component in the finished product, $M_{\text{accumulated}}$ is the mass of the component accumulated in the system, and M_{captured} is the mass of the component captured for recovery or disposed. M can then be converted to the mass of pollutant using the ratio of the molecular weight of the pollutant to the component. For example, a mass balance on sulfur in fuel may be performed, with M multiplied by 64/32 to estimated sulfur dioxide emissions.

Process modeling uses material and energy balance approaches to describe the flow of materials (and energy) through the processes of the overall manufacturing system. Most models are proprietary and specific to a manufacturing process.

Emission factors provide a representative value of emissions amounts associated with a specific amount of a pollutant producing activity. Activity amounts are then multiplied by an emission factor to estimate emissions.

emissions factor that is the amount of pollutant emitted per unit activity (for example, kg/L of fuel used, or kg/kg of product manufactured).

Emissions factors can be reported for both controlled and uncontrolled emissions. If an EF is reported for an uncontrolled emission, you can account for control of the emissions as follows:

$$E = A \times EF \times (1 - ER/100) \qquad \textbf{(11.29)}$$

where ER is expressed as a percent and is related to the control technology and the capture efficiency of the control system.

Emission factors have been developed for a large number of source activities from source sampling and mass balance calculations. They are tabulated in several locations, including (1) EPA's Clearinghouse for Inventories and Emissions (CHIEF) (www.epa.gov/ttn/chief/index.html), (2) AP-42 (Volume 1) by source type, (3) the searchable WebFIRE database (http://cfpub.epa.gov/webfire/), and (4) for mobile sources, EPA emission models such as MOVES and NONROAD. Emission factors are providing a rating from A to E (see Figure 11.24) depending on the accuracy and precision of the value. This is based on several factors that include the type and number of observations used to develop an emission factor. A is the best, E is the worst. This rating is an indicator of the accuracy and precision of a given emission factor

---

example/11.9 Using an Emission Factor Model to Estimate Carbon Monoxide Emissions from Oil Combustion

Consider an industrial boiler that burns 180,000 L of distillate oil per day. According to AP-42, the CO emission factor for industrial boilers that burn oil distillate is 0.6 kg CO per m³ of oil that is burned.

solution

The activity in this case is burning distillate oil in an industrial boiler. The CO emissions can be estimated using Equation 11.27.

$$E = A \times EF = 180{,}000 \text{ L oil/day} \times 0.6 \text{ kg CO/m}^3 \times \text{m}^3/1{,}000 \text{ L} = 108 \text{ kg CO/day}$$

---

example/11.10 Using an Emission Factor Model to Estimate Sulfur Dioxide Emissions from a Manufacturing Process

A manufacturing plant converts sulfur trioxide (SO_3 to sulfuric acid, H_2SO_4) at 97.5% efficiency. The plant produces 200 metric tons (that is, 200×10^6 g) of pure sulfuric acid every day. What are the daily emissions of SO_2 from the manufacturing facility?

solution

The emission factor for SO_2 air emissions from this process is related to the efficiency of the conversion process. AP-42 states the emission factor (in units of kg SO_2/metric ton processed) is determined as

$$EF = 682 - [6.82 \times \% \text{ conversion efficiency of } SO_3 \text{ to } H_2SO_4]$$

In our case, EF = $682 - [6.82 \times 97.5] = 682 - 665 = 17$ kg $SO_2/10^6$ g processed.

The emissions then can be estimated as

$$E = A \times F = 200 \times 10^6 \text{ g/day} \times 17 \text{ kg } SO_2/10^6 \text{ g} = 3{,}400 \text{ kg } SO_2/\text{day}$$

---

example/11.11 Mass Balance Approach Applied to Indirect Measurement of Air Emissions

A company purchases 80,000 L of a VOC-based coating. They apply 75,600 L of a coating onto their product and the remainder ends up on materials that are used for cleaning, become solid waste, and are subsequently disposed in a landfill. None accumulates in the manufacturing space, and there is no on-site treatment of the VOC emissions. The coating contains several VOCs that are emitted to the shop floor and enter the atmosphere through windows and vents. Determine the mass of VOC emitted last year to the atmosphere (kg/year). Part of the Material Safety Data Sheet (MSDS) is shown below:

Chemical Composition/Information on Ingredients

Chemical Name	CAS #	Percent by Weight
Ethyl benzene	100-41-4	20%
Xylene	1330-20-7	60%
Carbon black	1333-86-4	<10%
Water	7732-18-5	7%

Physical and Chemical Properties. Density 8.10 lb/gallon

solution

The mass of VOC per volume of coating (lb/gal or kg/L) is calculated using the VOC's density from the MSDS as:

$$\% \text{ by weight VOC}/100 \times 8.10 \text{ lb/gallon} = \text{lb VOC/gallon of coating}$$
$$= 80/100 \times 8.10 \text{ lb/gallon} \times \text{kg}/2.205 \text{ lb} \times \text{gallon}/3.78 \text{ L} = 0.78 \text{ kg/L}$$

The kg of VOC emitted in the year is

$$\text{mass VOC/gallon of coating} \times \text{gallons of coating used per year}$$
$$= 0.78 \text{ kg/L} \times 75,600 \text{ L/year} = 59,000 \text{ kg/year}$$

Note in our problem that we assumed none of the VOC is accumulated in the product. However, 4,400 L of the coating accumulates onto solid waste; therefore 3,432 kg of VOC (4,400 L × 0.78 kg/L) are transported to a landfill where they might be released to the atmosphere during transport or disposal.

11.7 Meteorology and Transport

11.7.1 FLOW FUNDAMENTALS

Air masses are a relatively homogenous and macroscale phenomena, extending upward for thousands of meters and covering hundreds of thousands of km². Figure 11.25 shows typical trajectories of air masses found in North America. Pollutants that are released into an air mass travel and are dispersed within the air mass. Air masses usually originate in specific regions that directly impact their temperature and humidity as well as the air pollutants that are emitted to them. For example, they can originate from ocean or continental locations or can be tropical or arctic depending on their latitude of origin (EPA, 2012g).

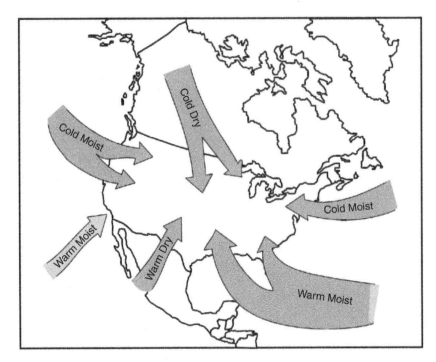

Figure / 11.25   Trajectories of Air Masses in North America.

(Redrawn from EPA, 2012g).

11.7.2   WINDS: DIRECTION, SPEED, AND TURBULENCE

Wind direction and speed has a great influence on the horizontal directional flow of air masses. Winds are named by the direction in which they originate. Thus, a west wind travels from west to east and northwest wind travels from the northwest toward the southeast. **Wind speed** increases with height and is also influenced by different surface elements that vary in roughness. Figure 11.26 shows some differences in wind speed as a function of height in urban, suburban, and level rural areas. As shown in this figure, the dense construction typical of a city provides a strong frictional force on the wind. This causes the wind to slow down, change direction, and be more turbulent. In contrast a level countryside has less frictional force and thus has stronger wind speeds near the surface (from EPA, 2012g).

11.7.3   ATMOSPHERIC STABILITY

The stability and mixing height of the atmosphere are important to understand how air pollution emissions are transported. While horizontal surface winds can disperse pollutants, it is the upward vertical mixing that greatly influences the concentration of a pollutant in ambient air. The potential for vertical mixing of a released air pollutant is primarily controlled by the degree of **atmospheric stability**. It is important that a reader carefully read and understand the terms listed in

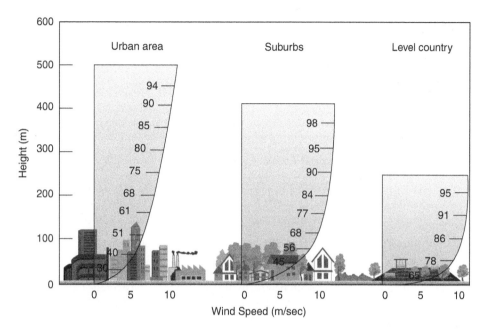

Figure / 11.26   Wind Speed Is Influenced by Height and Also by Different Surface Elements That Have Differing Roughness.

(Redrawn from EPA, 2012g).

Table 11.25 before moving on in this chapter. These items will ultimately assist you in understanding the vertical mixing of air pollutants and in our later efforts to estimate downwind concentrations of an air pollutant.

Table 11.25 describes how if the air temperature within an air parcel is warmer than surrounding air (which is usually true for air pollutants emitted during combustion and other industrial processes), it will be less dense than the cooler surrounding air, become buoyant, and rise. An air parcel that cools has the opposite effect; it will become denser and thus less buoyant, so it will descend.

An air parcel that begins to rise will cool at the dry adiabatic lapse rate (9.8°C/km) until it reaches the dew point. At this point, the air parcel will cool at the wet adiabatic lapse rate (6–7°C/km). Figure 11.27 shows the difference between these two lapse rates. Here you can visually see how the slope of the dry adiabatic lapse rate is greater than for the wet adiabatic lapse rate. If the lapse rate of the air parcel is less than the surrounding atmosphere (which is cooling down at a rate greater than 9.8°C/km), the air parcel will remain buoyant and continue to rise. This vertical mixing of an air parcel is very important in how the air pollutant is dispersed and carried away from (or returned to) a ground-level receptor.

Atmospheric stability is an important concept to understand in terms of the ability of the atmosphere to disperse and carry away air pollution emissions. Figure 11.28 provides descriptions and visual representations of different stability conditions that occur in the atmosphere. During **unstable conditions**, vertical movement of an air parcel in the atmosphere is encouraged upward or downward. As shown in Figure 11.28, the temperature difference between the environmental lapse rate (of the surrounding air) and the dry adiabatic lapse rate increases with height. This temperature difference enhances buoyancy

Important Terms to Understand the Vertical Mixing and Stability of Air   (adapted from EPA, 2012h).

Term	Description
Air parcel	A relatively well-defined body of air that acts as a whole and has a constant number of molecules. **Air parcels** can be thought of as the air inside a balloon. We assume the air parcel does not mix with surrounding air and the temperature within the parcel is uniform.
Buoyancy	An air parcel will expand and cool. If the air temperature within an air parcel is warmer than surrounding air (which is usually true for air pollutants emitted during combustion and other industrial processes), it will be less dense than the cooler surrounding air, become **buoyant**, and therefore rise. An air parcel that cools has the opposite effect; it will become denser and thus less buoyant, so it will descend.
Lapse rates	By definition, the **lapse rate** (Γ) is the ratio of the decrease in air temperature with increase in height ($\Gamma = -\Delta T / \Delta z$). It describes the lapse in temperature with altitude. A *positive lapse rate* is one where the temperature decreases with height. A *negative lapse rate* is one where the temperature increases with height. In the troposphere, the average environmental lapse rate is 6–7°C/km increase in altitude but can vary widely locally. *Remember, lapse rates are positive when the temperature decreases with altitude.*
Stable atmosphere	A **stable atmosphere** resists vertical motion and thus will have a low ability to disperse air pollutants that are emitted to it.
Dry adiabatic lapse rate	Adiabatic processes are ones where no transfer of heat or mass occurs across the boundaries of the air parcel. A dry air parcel rising in the atmosphere cools at a **dry adiabatic lapse rate** of 9.8°C/km. A dry parcel sinking in the atmosphere heats at a rate of 9.8°C/km.
Wet adiabatic lapse rate	A rising parcel of dry water that contains water vapor will cool at the dry adiabatic lapse rate until it reaches its dew point temperature (when the water vapor pressure equals the saturation vapor pressure at that temperature). Condensation releases latent heat into the air parcel; thus, the cooling rate of the parcel decreases. In the middle of the troposphere, the **wet adiabatic lapse rate** is about 6–7°C/km.
Environmental lapse rate	The **environmental lapse rate** is the actual temperature profile of the atmosphere as a function of altitude. It is also referred to as the prevailing or atmosphere lapse rate. Temperature usually decreases with height except in the case of a temperature inversion where the temperature of the atmosphere increases with height, thus preventing vertical mixing.
Mixing height	The **mixing height** is the maximum height an air parcel can ascend. It is usually the height at which a rising air parcel that is cooling at the dry adiabatic lapse rate intersects the ambient temperature profile. At this point the air parcel loses its buoyancy because it is no longer warmer than the surrounding air (it is the same temperature).
Mixing layer	The **mixing layer** is the air below the mixing height to the point of the air emission release. The larger the mixing layer, the greater volume of air into which air pollutants can be dispersed (and thus diluted).

of a rising air parcel that is cooling at the dry adiabatic lapse rate. During **stable conditions**, vertical movement of an air parcel is discouraged. Under very stable conditions, a cooler layer of air near the land surface is capped by an upper warmer air layer. This condition is referred to as inversion and prevents vertical motion of an air parcel. **Neutral stability** occurs when the environmental lapse rate is the same as the dry adiabatic lapse rate. The vertical movement of air is neither encouraged nor supported under these conditions (EPA, 2012h).

Figure / 11.27 Difference between Dry and Wet Adiabatic Lapse Rates.

Inversions occur when the temperature of the atmosphere increases with height (see Figure 11.28). In Figure 11.28, observe that the inversion is occurring at 1 km above the ground surface. Inversions are important to understanding air quality engineering because they result in very stable conditions that limit the vertical movement of an air parcel. The most common type of a surface inversion occurs in the late evening through the early morning, when the land surface cools rapidly after sunset. As the land surface cools, so does the layer of air close to the ground surface. If this air that resides close to the ground surface cools to a temperature lower than the air immediately above it, the air becomes very stable because the cap of warmer air prevents vertical motion. After the sun rises, the sun may be strong enough to disrupt the inversion layer and the inversion subsides as the day moves forward. However, in areas prone to heavy fog, the sun's energy may not be able to reach and erode the inversion layer.

Geography plays an important role in the occurrence of an inversion. For example, for locations in a valley, the valley floor can collect cooler air that moves down hillsides and mountains over the night. Inversions can occur along locations alongside a mountain as well. For example, in U.S. cities like Denver, inversions can occur when warm air originating in the west is forced over the top of a mountain creating the capping layer above a cooler air layer originating on the eastern slope of the Rocky Mountains. In Los Angeles, cooler air can be carried in off the ocean and stagnate near the valley's floor, while warmer air can be transported above this cooler air layer from the surrounding mountains and desert areas, creating a capping layer. In Los Angeles, the situation can be compounded because horizontal winds cannot carry the accumulating pollution to the east because of mountain barriers, which can be above the inversion layer. Inversions can also be compounded in the winter months by the presence of snow, which reflects the sun's energy and keeps the lower air mass cool (EPA, 2012h).

Knowledge of plume behavior and its relationship to atmospheric stability is critical for air quality management. Figure 11.29 provides a

Stability Condition	Figure showing temperature changes of a hypothetical air parcel (hashed line) and the surrounding air (environmental lapse rate) with height.
During **unstable conditions**, vertical movement of an air parcel in the atmosphere is encouraged upward or downward. Unstable conditions most commonly develop on sunny days with low wind speeds. The land surface quickly absorbs heat and transfers some heat to the surface air layer. This air warms, becomes less dense (and thus more buoyant) than the surrounding air so it rises vertically.	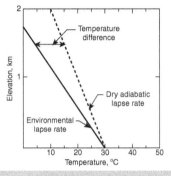
During **stable conditions**, vertical movement of an air parcel is discouraged. Under very stable conditions, a cooler layer of air near the land surface is capped by an upper warmer air layer. This condition is called an **inversion** and prevents vertical motion of an air parcel.	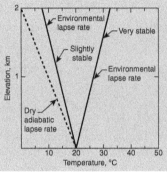
Neutral stability occurs when the environmental lapse rate is the same as the dry adiabatic lapse rate. The vertical movement of air is neither encouraged nor supported under these conditions. Neutral stability typically occurs on a windy day when cloud cover prevents strong heating or cooling of the land.	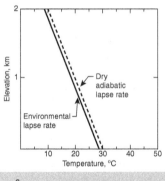
A temperature **inversion** occurs when a warmer layer of air resides above a cooler surface layer (from EPA, 2012h). Areas that are prone to inversions occur where large populations of humans reside. These areas include coastal zones, valleys, and locations near mountains.	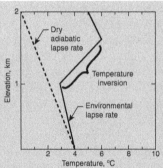

Figure / 11.28   Definitions of Atmospheric Stability Conditions.

SOURCE: EPA, 2012h

Looping—Occurs in highly unstable conditions. A rapid turnover of air causes turbulence. Looping plumes are usually favorable for dispersion of air pollutants and usually result in low exposure to low pollutant concentrations. However, there may be short episodes of exposure to higher concentrations of air pollutants where the plume loops downward to ground level.

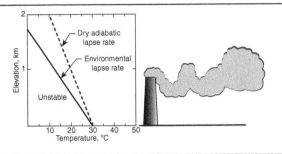

Fanning—Occurs in very stable conditions. An inversion prevents vertical motion of the plume, but horizontal motion of the plume is not prevented downwind.

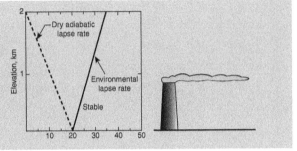

Coning—Occurs under neutral conditions where atmospheric conditions are slightly stable.

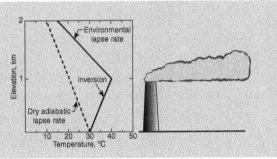

Lofting—Release of air pollutants occurs just above the inversion. The air above the inversion is unstable, which encourages vertical mixing above the inversion layer. In this case ground-level receptors are fortunate because stack height is above the elevation of the inversion.

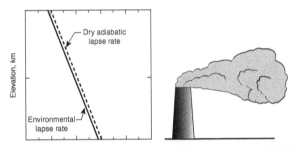

Fumigation—Air pollutants are released just below an inversion layer. The air below the plume in this case is unstable. Ground-level receptors can be exposed to high levels of air pollutants.

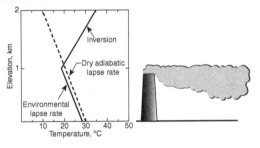

Figure / 11.29   Plume Type and the Influence of Atmospheric Stability.

SOURCE: EPA, 2012h

description of five types of plumes that one would commonly observe originating from a stack. In this case, the stack could be envisioned as a tall industrial stack or a shorter household chimney. The type of plume is controlled by the stability of the atmosphere so that information is provided as well. Note also how the different types of plume impact the potential for exposure of ground-level receptors to air pollutants, whether the exposure will occur to a more diluted form in the air parcel, or whether the plume is conducive to transporting the air pollutants away from the local source and to some downwind receptor.

11.7.4 TERRAIN EFFECTS ON ATMOSPHERIC STABILITY

We mentioned previously that land features impact turbulence and the horizontal speed of an air parcel. The vertical motion of an air mass is also impacted by lifting of air that occurs over terrain. For example, the horizontal and vertical direction of an air parcel is influenced by topographical features that can include hills, mountains, and valleys. In addition, other land features like the presence of water or trees can impact the movement of an air parcel. For example, water heats at a much slower rate during the day than concrete infrastructure because the water has a lower heat capacity (this was discussed in Chapter 4). The concrete can then release its heat back to the air at night while the water will not release heat to same degree. Also, tree-covered areas adsorb heat less than rocky slopes or bare ground do.

Land and the features on the ground heat up and cool down relatively quickly. In contrast, water heats and cools relatively slowly. In addition, water temperature does not vary greatly from day to day or week to week because water temperatures tend to follow a seasonal pattern. As shown in Figure 11.30, when the land surface heats up, the air adjacent to the land warms up and

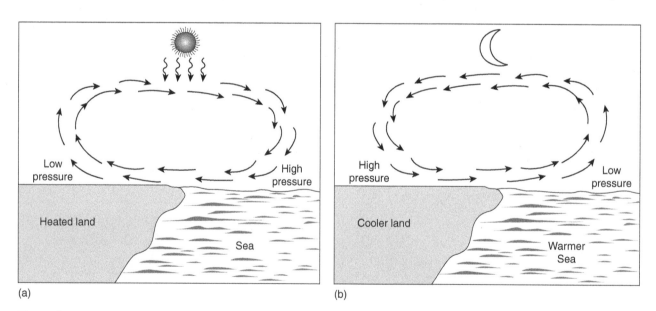

Figure / 11.30 Differential Heating Between Water and Land Leads to (a) Sea Breeze and (b) Land Breeze.

(Redrawn from EPA, 2012g).

becomes less dense and rise. The cooler air over the water is then drawn inland (known as a sea breeze). At night, the air over the land cools rapidly, which creates a return flow from land to the water that is called a land breeze. This movement of air can have a large impact on movement of air parcels when air emissions occur near large bodies of water.

11.8 Atmospheric Dispersion and the Gaussian Plume Dispersion Modeling

Estimation of concentrations downwind of a source is useful for air pollution management. Modeling is particularly necessary for predicting impacts of choices that have yet to be made in terms of permitting emissions and application of demand management and control technologies on particular air emissions. Gaussian dispersion estimation is a common approach that underpins a suite of models for estimating time-averaged air pollution concentrations downwind of an emission source.

11.8.1 FUNDAMENTALS OF DISPERSION MODELING

Dispersion models are fundamentally based on the observation that air pollution emissions are carried horizontally with the average wind flow, but also spread and dilute in the vertical and horizontal directions due to turbulent eddies, buoyancy in the vertical direction, and other fluctuations in wind direction. The equations used to approximate this behavior are grounded in physical laws, including Fick's law and the material balance (see Chapter 4), as well as statistical theory representing the approximately random effects of turbulence and wind fluctuations.

For **Gaussian models**, the population of pollutant mass is represented using Gaussian (also called normal) probability distribution functions. The basic steady-state **Gaussian plume equation** is provided in Equation 11.30, with the corresponding coordinate system depicted in Figure 11.31:

$$C(x,y,z) = \frac{S}{2\pi u \sigma_y \sigma_z} \exp\left(-\frac{y^2}{2\sigma_y^2}\right)\left\{\exp\left(-\frac{(z-H)^2}{2\sigma_z^2}\right)\right\} \qquad \textbf{(11.30)}$$

This model equation provided in Equation 11.30 estimates the time-averaged concentration of an air pollutant (C, mass/volume, for example, $\mu g/m^3$) at any spatial location (x, y, z) downwind of a source. In Figure 11.31 and Equation 11.30, x is the downwind distance, y is the crosswind coordinate, and z is the vertical coordinate (that is, the height). The first term to the right of the equal sign in Equation 11.30 provides the concentration directly at the centerline of the plume (coordinates of $x, 0, 0$). The second term adjusts the concentration as you move in the sideways (y) direction, and the third term adjusts the concentration in the vertical (z) direction. Here the source (or stack) is located at x, y, z

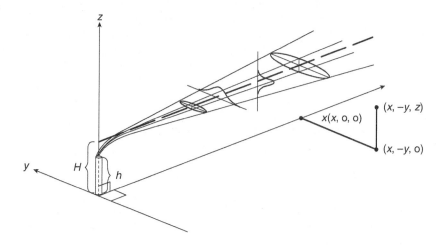

coordinates $0, 0, 0$ and at a height, h, above the ground surface. S (mass/time, for example, kg/day lb/h) is the continuous pollutant emissions rate from the source, and H is the **effective release height** of the plume (shown in Figure 11.32).

H represents the height at which the plume loses its vertical mean trajectory. Due to the momentum and buoyancy of the emissions, H is usually higher than the physical stack height (h), by a distance $\Delta h = H - h$, called the **plume rise**. Several approaches have been used to calculate plume rise. They generally combine empirical data fits with variable groups that estimate the momentum and buoyancy of the air emissions. A demonstrative historical plume rise equation is the Holland formula for neutral stability conditions:

$$\Delta h = \frac{v\,d}{u}\left(1.5 + 2.68 \times 10^{-3}P\left(\frac{T_s - T_s}{T_a}\right)d\right) \qquad \textbf{(11.31)}$$

Figure / 11.32   **Difference between Stack Height (h) and Effective Release Height (H).**

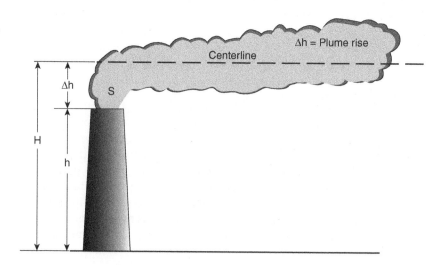

where v is the exit velocity of the stack gas (m/s), d is the inner diameter of the stack (m), T_s is the exit temperature of the stack gas (K), T_a is the ambient air temperature (K), and P is the ambient pressure (mb). Note in Equation 11.31 that the first constant is unitless and the second has units of $mb^{-1} m^{-1}$).

Once released, the center of the plume travels with the speed (u) and direction of the mean wind. Dispersion in the y and z directions of the total plume mass at any x location are each represented by a normalized Gaussian probability density function that has a mean at the plume centerline ($y = 0$, $z = H$) and standard deviations (σ_y and σ_z, respectively).

σ_y and σ_z are referred to as **dispersion coefficients** (also referred to as dispersion lengths). Dispersion coefficients increase with downwind distance (x). This means that the plume spreads farther away from the centerline and effective stack height. Accordingly, the maximum concentration of the air pollutant will occur along the plume's centerline. Also note how Equation 11.30 does not have an explicit term for x in the equation, but instead σ_y and σ_z indirectly account for the distance directly downwind from the emission because they are functions of x.

11.8.2 MODEL PARAMETERS

In order to apply Gaussian dispersion equations, a few parameters are required. These include dispersion coefficients, effective release height of the plume, and the mean wind speed at the stack height. As discussed in Section 11.6.2, mean wind speed increases with altitude from the ground surface and the gradient of the increase is dependent on the roughness of the surface. Wind speed (u) at the effective stack height (H) can be estimated from measured wind speed (u_0) measured at a reference height (z_0) using the following power law relationship:

$$u = u_0 (H/z_0)^p \qquad \textbf{(11.32)}$$

In Equation 11.32, z is the height and the exponent p is a function of the stability class of the air and surface roughness. Values range from 0.07 for smooth (rural) terrain under the most unstable atmospheric conditions (referred to as class A stability) to 0.6 for rough (urban) terrain for the most stable atmospheric stability (class F). Table 11.26 provides descriptions of the six classes (A–F) used to describe the stability of the atmosphere. You can see from Table 11.26 that the stability class is based on prevailing conditions of wind speed and solar insolation (or nighttime cloud cover).

Dispersion coefficients in the horizontal direction (σ_y) and vertical direction (σ_z) are also required in Equation 11.30. Values for them are largely based on empirical tracer study data that were obtained for a specific averaging time and are calculated using curve fitting equations that are categorized by atmospheric stability class. A commonly used set of curve fitting equations used to determine the dispersion coefficients in the x and y directions are the Briggs equations, provided in Table 11.27. In these equations, the value x is the

Table / 11.26

Description of the Classes of Stability of Air   (from Hanna et al., 1982).

Surface Wind (measured at 10 m)	Day-Time Isolation			Night-Time Cloudiness	
m/s	Strong	Moderate	Slight	Thinly Overcast or $\geq 4/8$ Cloudiness	$\leq 3/8$ Cloudiness
<1	A	A-B	B	—	—
2-3	A-B	B	C	E	F
3-5	B	B-C	C	D	E
5-6	C	C-D	D	D	D
>6	C	D	D	D	D

A-extremely unstable, B-moderately unstable, C-slightly unstable, D-neutral, E-slightly stable,
F-moderately stable

Table / 11.27

Equations to Estimate Horizontal (σ_y) and Vertical (σ_z) Dispersion Coefficients for Gaussian Plume Modeling. $10^2 < x$ (in m) $< 10^4$)

Class of Atmospheric Stability	σ_y, m	σ_z, m
Open-Country Conditions		
A	$0.22x(1 + 0.0001x)^{-\frac{1}{2}}$	$0.20x$
B	$0.16x(1 + 0.0001x)^{-\frac{1}{2}}$	$0.12x$
C	$0.11x(1 + 0.0001x)^{-\frac{1}{2}}$	$0.08x(1 + 0.0002x)^{-\frac{1}{2}}$
D	$0.08x(1 + 0.0001x)^{-\frac{1}{2}}$	$0.06x(1 + 0.0015x)^{-\frac{1}{2}}$
E	$0.06x(1 + 0.0001x)^{-\frac{1}{2}}$	$0.03x(1 + 0.0003x)^{-1}$
F	$0.04x(1 + 0.0001x)^{-\frac{1}{2}}$	$0.016x(1 + 0.0003x)^{-1}$
Urban Conditions		
A–B	$0.32x(1 + 0.0004x)^{-\frac{1}{2}}$	$0.24x(1 + 0.001x)^{-\frac{1}{2}}$
C	$0.22x(1 + 0.0004x)^{-\frac{1}{2}}$	$0.20x$
D	$0.16x(1 + 0.0004x)^{-\frac{1}{2}}$	$0.14x(1 + 0.0003x)^{-\frac{1}{2}}$
E–F	$0.11x(1 + 0.0004x)^{-\frac{1}{2}}$	$0.08x(1 + 0.0015x)^{-\frac{1}{2}}$

SOURCE: Hanna et al, 1982.

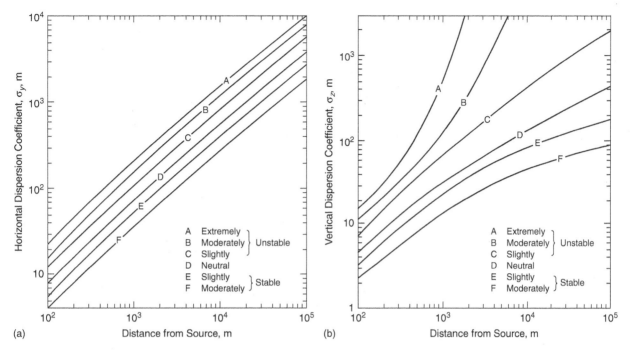

Figure / 11.33 Gaussian Dispersion Coefficients (that is dispersion lengths) for Different Conditions of Air Stability in the (a) Horizontal Direction (*y*) and (b) Vertical Direction (*z*).

(Redrawn from Hanna et al., 1982).

distance in the x direction downwind of the emission source. Graphs can also be used to estimate the dispersion coefficients as shown in Figure 11.33.

Looking at these equations and Figure 11.33, horizontal dispersion coefficients (σ_y) increase as the atmosphere becomes more unstable (change from F to A stability class) because under unstable conditions, vertical movement of an air parcel and resultant turbulence is encouraged. Likewise, you can see that atmospheric stability condition has a greater influence on vertical dispersion (σ_z) of the pollutant.

11.8.3 FORMS OF THE GAUSSIAN DISPERSION EQUATION

The basic plume equation presented in Equation 11.30 provides the most important concepts in atmospheric dispersion estimation but is rarely used in practice. Some important limiting assumptions of the basic equation are that the pollutant emission rate (S) is constant over the time period modeled, the wind speed is constant in time and with elevation, the pollutant mass is not lost or gained (for example, due to reaction or deposition), and there are no barriers to fluid flow or dispersion. Many forms of Gaussian dispersion equations have therefore been derived to model specific situations and to relax some of the basic plume assumptions.

The simplest form of the Gaussian dispersion equation that is used for estimation includes the barrier to vertical dispersion represented by the ground surface. To account for the surface barrier, a fictitious source at a height $-H$ (that is, below ground level) is used to add back the mass of pollutant that would have been mathematically dispersed into the ground. This results in an extra exponential term added to Equation 11.30 as shown in the following equation:

$$C(x,y,z) = \frac{S}{2\pi u \sigma_y \sigma_z} \exp\left(-\frac{y^2}{2\sigma_y^2}\right) \left\{ \exp\left(-\frac{(z-H)^2}{2\sigma_z^2}\right) + \exp\left(-\frac{(z-H)^2}{2\sigma_z^2}\right) \right\}$$

$$(11.33)$$

Equation 11.33 is referred to as the *ground reflection equation*. Absorption, rather than reflection, can be represented instead by subtracting (rather than adding) the final exponential term.

We are also most times concerned with the concentration of an air pollutant at ground level because this is where exposure of humans and crops would take place. In this case, the term z in Equation 11.33 can be set to zero ($z = 0$) so the reflection equation (Equation 11.33) is written as:

$$C(x,y,0) = \frac{S}{\pi u \sigma_y \sigma_z} \exp\left(-\frac{y^2}{2\sigma_y^2}\right) \exp\left(-\frac{H^2}{2\sigma_z^2}\right) \qquad (11.34)$$

Equation 11.34 will be used in our example below to estimate the downwind concentration of an air pollutant for a particular situation. In Equation 11.34, the maximum concentration of the air pollutant that occurs at ground level represents a balance between the distance required for the plume to reach the ground and the dilution of plume that occurs from dispersion. The downwind distance x at which the highest ground-level plume concentration occurs can be estimated as the point when the vertical dispersion coefficient equals:

$$\sigma_z = H/\sqrt{2} \qquad (11.35)$$

Thus, the effective release height of the plume (H) can be plugged into Equation 11.35 to first determine the vertical dispersion coefficient, σ_z. After obtaining σ_z, you can use the equations in Table 11.27 or curves in Figure 11.33 to back out the downwind distance, x, where the maximum ground-level concentration occurs. Strictly speaking, this approximation is valid only under moderately unstable (class C) to neutral conditions (class D) stability conditions.

Use of the Gaussian Plume Dispersion Model

A manufacturing process emits 2.4 g of the pollutant SO_2 every minute. The stack height is 15 m and there is zero plume rise. Assume the wind speed is 3 m/s and the horizontal dispersion coefficient (σ_y) is 25 m and the vertical dispersion coefficient (σ_z) is 15 m for this situation. What is the concentration of the air pollutant 0.5 km downwind of the release along the centerline? What is the concentration at this same distance downwind but at a location 100 m to the side at ground level?

solution

The problem is asking us to estimate the ground-level concentration of the pollutant in two locations. Both estimations are at ground level so we will use Equation 11.34. For the first question, we are estimating the concentration of the pollutant along the plume's centerline so the x, y, z coordinates are 500 m, 0, 0.

Equation 11.34 can be written as:

$$C(500,0,0) = \frac{2.4 \, \text{g/min} \times \dfrac{\text{min}}{60 \, \text{s}}}{\pi \times 3 \, \text{m/s} \times 25 \, \text{m} \times 15 \, \text{m}} \exp\left(-\frac{(0 \, \text{m})^2}{2 \times (25 \, \text{m})^2}\right) \exp\left(-\frac{(15 \, \text{m})^2}{2 \times (15 \, \text{m})^2}\right)$$

$$C = 6.9 \times 10^{-6} \, \text{g/m}^3 \times 10^6 \, \mu\text{g/g} = 6.9 \, \mu\text{g/m}^3$$

What happens to the ground-level concentration at this location if the stack height (h) is increased and we get some plume rise (Δh) so the value of H is increased to 50 m? Do the calculation yourself and note how the receptor located 0.5 km downwind along the center would now be exposed to a concentration of $0.044 \, \mu\text{g/m}^3$. What happened? Remember also that we can use Equation 11.35 to determine the x distance where the maximum concentration of the pollutant occurs.

For the second question we are asked to estimate the SO_2 concentration at x, y, z coordinates of 500, 100, 0.

$$C(500,100,0) = \frac{2.4 \, \text{g/min} \times \dfrac{\text{min}}{60 \, \text{s}}}{\pi \times 3 \, \text{m/s} \times 25 \, \text{m} \times 15 \, \text{m}} \exp\left(-\frac{(100 \, \text{m})^2}{2 \times (25 \, \text{m})^2}\right) \exp\left(-\frac{(15 \, \text{m})^2}{2 \times (15 \, \text{m})^2}\right)$$

$$C = 2.3 \times 10^{-9} \, \text{g/m}^3 \times 10^6 \, \mu\text{g/g} = 0.0023 \, \mu\text{g/m}^3$$

Note how the SO_2 concentration in the y direction is much lower than the concentration along the centerline of plume because of dispersion.

Fictitious sources that are formulated similarly can also be used to account for reflection due to a capping inversion layer, and multiple reflections between the ground and the inversion. Sources in a line, such as a roadway, are also often represented using an infinite sum of point sources. For instantaneous emissions, non-steady-state Gaussian puff dispersion equations are used. These equations follow discrete emissions puffs through changes in mean wind direction and speed. Many other adaptions have also been applied to account for other aspects that occur in practice, including elevated terrain, stack tip downwash, reaction or deposition of pollutant mass, and alternative averaging times.

In practice, most modeling of air pollution dispersion now uses computer modeling systems based on one or more of the analytical equations described above. The systems customize the equation form relevant to the conditions to be modeled and also estimate many of the necessary parameters. The EPA provides a clearinghouse of dispersion modeling programs that are appropriate for different applications at http://www.epa.gov/ttn/scram/dispersionindex.htm, along with guidance on their use for regulatory applications. The currently preferred model for steady-state (plume) estimation is the AERMOD system, while CALPUFF is preferred for non-steady-state (puff) modeling.

Key Terms

- adsorption
- air quality index (AQI)
- air parcel
- air toxics
- atmospheric stability
- area sources
- baghouse
- bag filters
- Clean Air Act
- biofilter
- buoyancy
- Clean Air Act Amendments
- control technologies
- Corporate Average Fuel Economy (CAFE) Standards
- criteria air pollutants
- cyclone
- demand management
- dispersion coefficients (σ_y and σ_z)
- dispersion models
- dry adiabatic lapse rate
- effective release height
- electrostatic precipitator (ESP)
- emission factor
- energy ladder

- environmental lapse rate
- fugitive emissions
- gas-to-cloth ratio
- Gaussian models
- Gaussian plume equation
- hazardous air pollutants
- hoods
- indoor air quality
- inversion
- IPAT Equation
- lapse rate (Γ)
- market-based approaches
- mixing height
- mixing layer
- mobile sources
- National Ambient Air Quality Standards (NAAQS)
- neutral stability
- nonattainment area
- odor threshold
- packed absorption bed
- photochemical reactions
- plume
- plume rise
- point sources
- primary standards

- regulatory approaches
- secondary pollutant
- secondary standards
- smart growth
- source reduction
- spatial scales
- stable atmosphere
- stable conditions
- stack emissions
- stationary sources
- stratosphere
- stratospheric ozone depletion
- thermal oxidizer
- toxic air pollutants
- traffic congestion
- Tragedy of the Commons
- transportation demand management
- troposphere
- unstable conditions
- venturi scrubber
- voluntary approaches
- wet adiabatic lapse rate
- wind direction
- wind speed

11.1 Carbon monoxide (CO) is measured to have a concentration of $103\,\mu g/m^3$. What is the concentration in (a) ppm_v, (b) ppb_v, and (c) percent by volume? Assume a temperature of 25°C and pressure of 1 atm.

11.2 If the atmospheric mass concentrations of nitrogen monoxide (NO) and nitrogen dioxide (NO_2) are 90 and $120\,\mu g/m^3$, respectively, what is the NO_x concentration in ppb_v? Assume a temperature of 30°C and pressure of 1 atm.

11.3 If the mass concentration of particulate matter is $12,500\,\mu g/m^3$, report this concentration as the number concentration (# particles/cm^3). Assume spherical particles of $0.5\,\mu m$ diameter with density of liquid water ($1\,g/cm^3$).

11.4 Formaldehyde is commonly found in the indoor air of improperly designed and constructed buildings. If the concentration of formaldehyde in a home is $1.2\,ppm_v$ and the inside volume is $600\,m^3$, what mass (in grams) of formaldehyde vapor is inside the home? Assume $T = 298\,K$ and $P = 1\,atm$. The molecular weight of formaldehyde is 30.

11.5 The National Ambient Air Quality Standard (NAAQS) for sulfur dioxide (SO_2) is $0.14\,ppm_v$ (24 h average). (a) What is the concentration in $\mu g/m^3$ assuming an air temperature of 25°C? (b) What is the concentration in moles SO_2 per 10^6 moles of air?

11.6 Table 11.5 provided information that suggested "clean" air might have a sulfur dioxide (SO_2) concentration of $<30\,ppb_v$, while polluted air might have a concentration of $1\,ppm_v$. Convert these two concentrations to $\mu g/m^3$. Assume a temperature of 298 K (note the difference units of concentrations, ppm_v versus ppb_v).

11.7 Carbon monoxide (CO) affects the oxygen-carrying capacity of your lungs. Exposure to $50\,ppm_v$ CO for 90 min has been found to impair one's ability to discriminate stopping distance; therefore, motorists in heavily polluted areas may be more prone to accidents. Are motorists at a greater risk of accidents if the CO concentration is $65\,mg/m^3$? Assume a temperature of 298 K.

11.8 Diesel engines emit very fine soot particles. In the atmosphere, these soot particles often agglomerate with other particles as they "age." Assume the agglomerated soot particles are initially suspended at a height of 22 m. They are spherical (diameter $= 0.5\,\mu m$) and have a density of $1.1\,g/cm^3$. (a) Calculate the terminal settling velocity of the soot particles. (b) How many hours will the soot particles remain suspended before settling by gravity to the ground? The density of air is $1.2\,kg/m^3$ and its fluid viscosity is $1.72 \times 10^{-4}\,g/cm\text{-}s$.

11.9 This problem allows you to think about how exposure impacts the concentration of air pollutants you are exposed to. (a) Maintain a diary for 1 full day and record all the locations you visit. Include the times of entry and exit for each location. Also record any interesting air quality information for each location. Calculate the percentage of time spent in each type of location. Summarize the data in a table. (b) In which location did you spend the most amount of time? The least? (c) Calculate the 24 h time integrated average *exposure concentration* (units of $\mu g/m^3$) to airborne particles based on your recorded activity patterns, using the average airborne PM_{10} concentration for different locations provided below.

Location	Average Airborne PM_{10} Concentration ($\mu g/m^3$)
Home	90
Office-factory	40
Bar-restaurant	200
Other indoor	20
In a vehicle	45
Outdoors	35

(d) Now add a $35\,\mu g/m^3$ "proximity effect" to one of your locations above that we will assume is from exposure to cigarette smoking at that location. How does this change your 24 h time integrated average *exposure concentration* (units of $\mu g/m^3$) to airborne particles?

11.10 (a) Define NAAQS and (b) identify the NAAQS pollutants.

11.11 On January 12, 2013, the AQI in Beijing (China) the AQI was determined to be 775. In contrast, on

January 11, 2013, Salt Lake City (Utah) has a reported AQI of 142 compared to a value of 67 in San Francisco's and a value of 23 in Las Vegas. Develop a table with three columns. List the four cities, the AQI reported above, and the third column should provide the level of health concern related to an AQI for general and sensitive populations.

11.12 The following data on U.S. population, number of licenses drivers, and number of vehicles was obtained from the U.S. Department of Transportation, Federal Highway Administration (http://www.fhwa.dot.gov/policyinformation/statistics/2010/dv1c.cfm). (a) Determine the rate of growth (on a per year basis) for the population, number of licensed drivers, and number of vehicles. (b) Are the rates of growth similar or different?

Year	Population	Drivers	Vehicles
1960	180	87	74
1961	183	89	76
1962	186	91	79
1963	188	94	83
1964	191	95	86
1965	194	99	90
1966	196	101	94
1967	197	103	97
1968	199	105	101
1969	201	108	105
1970	204	112	108
1971	207	114	113
1972	209	118	119
1973	211	122	126
1974	213	125	130
1975	215	130	133
1976	218	134	139
1977	220	138	142
1978	222	141	148
1979	225	143	152
1980	227	145	156
1981	230	147	158
1982	232	150	160
1983	234	154	164
1984	236	155	166
1985	239	157	172
1986	241	159	176
1987	243	161	179
1988	246	163	184
1989	248	166	187
1990	248	167	189
1991	252	169	188
1992	255	173	190
1993	258	173	194
1994	260	175	198
1995	263	177	202
1996	265	180	206
1997	268	183	208
1998	270	185	211
1999	273	187	216
2000	281	191	221
2001	285	191	230
2002	288	195	230
2003	291	196	231
2004	293	199	237
2005	296	201	241
2006	299	203	244
2007	301	205	247
2008	304	208	248
2009	307	210	246
2010	309	210	242

11.13 A travel demand management action is planned that will add parking spaces to an existing park-and-ride facility that is served by transit. The plan will add 120 parking spaces. Assume the new spaces will have a 95 percent estimated utilization rate and these individuals will use the available light rail and bus service. Also assume that the average commute that will be eliminated is 42 miles roundtrip (distance from lot to destination and return), and there are 250 operating days per year. (a) What is the annual reduction in vehicle miles traveled from implementation of the park and transit ride facility? (b) If the emission factor for reactive hydrocarbons is 0.23 g/mile driven and for NO_x is 0.40 g per mile driven, what is the estimated reduction in air emissions for both of these air pollutants over the year?

11.14 True or false? Compared to a baghouse with a high-pressure drop, a baghouse with a low-pressure drop would need a large fan and require more energy to move the gas through the baghouse.

11.15 A baghouse that employs a shaker collection method is being designed to remove 99.75 percent of an incoming stream of sawdust particles originating from a sawmill plant. What fabric area is required for the baghouse if it treats 15,000 ft^3/min of polluted air? How many bags are required if the bags are cylindrical and are 6 in. in diameter and 15 ft long? Assume the filter manufacturer has specified a woven fabric.

11.16 You are assigned to determine the number of filter bags required for an eight-compartment baghouse that uses pulse-jetting as the method to remove particulate matter from the bags. The following information is known: process gas exhaust rate is 100,000 ft^3/min and the recommended air-to-cloth ratio is 4 ft/min. The bags specified by the manufacturer have a diameter of 6 in and height of 10 ft. (a) what is the total required fabric area? (ft^2)? (b) What number of bags is required? (c) How many bags are in each compartment?

11.17 A thermal oxidizer incinerator operates as a plug flow reactor at a temperature of 250°C and has a gas residence time of 0.3 s. (a) If the pollutant (vinyl chloride) enters the incinerator at a flow rate of 3,000 m^3/min and it is desired to remove 99.99 percent of the pollutant, what should the length of the incinerator be? (b) What is the length of the incinerator if the desired removal is increased to 99.99995? The first-order rate constant for vinyl chloride removal is 45/s at 250°C and the inside diameter of the incinerator (which is shaped like a cylinder) is 1 m.

11.18 A tubular thermal oxidizer is operated at 225°C to remove toluene (C_6H_7) from a polluted airstream. The residence time for the gas is 1 s. (a) Write the balanced reaction for theoretical oxidation of toluene to carbon dioxide and water. (b) If the toluene

enters the incinerator at a flow rate of 2,500 ft^3/min and the reaction rate constant is 7.2/s at this temperature, what percentage of the toluene is removed? (c) What is the length required for the oxidizer if the inside diameter is 4 ft?

11.19 A biofilter design uses a media that consists of a mixture of wood chips and compost at a ratio of 1 to 3 parts wood chips to municipal compost. The biofilter has the dimensions $L = 8$ m, $W = 4.8$ m, depth = 0.35 m, and the air flow rate is 2.6 m^3/s. (a) What is the empty bed residence time (s)? (b) Using the table provided in the chapter that relates empty bed residence time to application, what application(s) might this biofilter be appropriate for?

11.20 Calculate the (a) stack emissions and (b) fugitive emissions if some process equipment that is associated with a hood system that generates 220 kg/h of VOCs, the hood capture efficiency is 87 percent, and the collection efficiency of the air pollution control technology used to treat the captured air emissions is 95 percent.

11.21 A manufacturing plant converts sulfur trioxide (SO_3) to sulfuric acid (H_2SO_4) at 97.5 percent efficiency. The plant produces 350 metric tons of sulfuric acid at 100 percent purity every day. A wet scrubbing system is installed to reduce the SO_2 emissions and has a removal efficiency of 97 percent. The emission factor for this process equals 17 kg SO_2/ metric ton of raw material processed. (a) What are the daily emissions of SO_2 from the manufacturing facility prior to treatment? (b) What are the daily emissions of SO_2 from the manufacturing facility after treatment?

11.22 Assume the emissions factor for release of mercury from a solid-waste incinerator is 0.107 lb per ton incinerated and for dioxin is 2.13×10^{-5} lb per ton solid waste incinerated. Estimate the daily emission rates (lb pollutant/day) and annual emissions rates (lb/year) of each of these hazardous air pollutants. Assume the incinerator processes 2.88×10^6 lb of solid waste per year.

11.23 Acetone is released from combustion of 100,000 metric tons of wood waste per year. The wood waste has an average heat content of 0.05 MM BTU/lb. The reported emission factor for acetone emitted from a waste bark-fired boiler with no emissions control is 9.5×10^{-3} kg/metric ton of wood burned. What are the annual emissions of acetone from this facility (metric tons per year)?

11.24 A company annually applies 25,000 L of a surface coating to their product. The surface coating contains the VOC acetone. The emissions are collected and an emissions control technology is installed that reduces the VOC emissions by 80 percent. The company is required to report their emissions to the atmosphere if they exceed 3,500 kg/year. For this coating process operation, there is no loss of the coating compound to the coating equipment and no loss to the system's liquid or solid waste streams. All the VOC applied is captured by the collection system prior to treatment. The Material Safety Data Sheet reports that the surface coating product contains 25 percent by weight acetone and the specific gravity of the material is 1.35 kg/L. Use a mass balance approach to estimate if the company must report their air emissions to the state regulatory agency.

11.25 A company applies 21,000 L of a coating every year that contains three VOCs; 0.26 kg xylene per liter of coating, 0.040 kg n-butyl alcohol per liter of coating, and 0.13 kg ethyl benzene per liter of coating. (a) What is the total mass of VOC in the coating (kg VOC/liter of coating)? (b) Estimate the total mass of VOCs released to the atmosphere every year (kg/year) assuming no collection or treatment of the VOCs takes place. (c) What is the total mass of VOCs released to the atmosphere assuming that 8 percent of the applied VOC is retained in a liquid waste stream and is discharged to the wastewater treatment plant and the company installs some control technology that captures and degrades 60 percent of the collected air emissions?

11.26 A company performs measurements that show the VOC xylene is present in the stack gas and the gas flow rate is 2,200 m³/min. The concentration of xylene was measured exiting the stack on a weekly basis and averaged 0.62 kg/h. The manufacturing process runs 6,700 h every year. If the reporting threshold for the xylene is 3,200 kg/year, use the source measurements to determine if the company exceeds the annual reporting threshold.

11.27 Between 1980 and 2000, the average CO emission factor of the vehicle fleet in Hillsborough County, Florida, dropped by almost half, from about 65 to 34 g of CO per vehicle-mile driven. However, the total miles driven in the county by all vehicles increased by 60 percent during this same time period. (a) Did countywide emissions of CO go up or down, and by how much between 1980 and 2000? (b) Vehicle exhaust is getting cleaner through a combination of engine improvements, emission control technologies, auto redesign, and fuel reformation. What are five transportation demand management strategies you can use to reduce emissions of air pollutants in an urban area?

11.28 Investigate sources of hazardous air pollutants emitted near your community. Go to Scorecard (www.scorecard.org) to gather data about air pollutant emissions. The Scorecard site makes the Toxics Release Inventory easily searchable; by entering your ZIP code, you can find a list of major air polluters in your area. (a) For your area, identify the top three five polluters and their total emissions. (b) Plot total environmental release for data from the top emitting company over the years of available data. (c) Describe the overall trend of emissions over time.

11.29 At the wet adiabatic lapse rate, the cooling rate of the air parcel is usually: (a) slower than the dry adiabatic lapse rate, (b) the same as the dry adiabatic lapse rate, or (c) faster than the dry adiabatic lapse rate?

11.30 In the following figure, an air parcel is displaced and becomes saturated at an elevation of 2 km. Which of the following stability conditions does the diagram depict? (a) stable only below 1 km, (b) stable only above 1 km, (c) unstable above 2 km? (problem from EPA, 2012h)

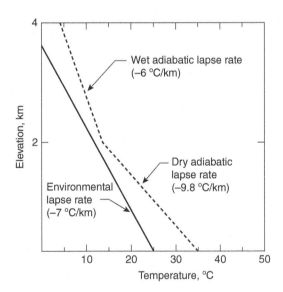

11.31 (a) Name the following three plume types. (b) Sketch a graph of elevation on the y-axis and air temperature on the x-axis that would describe the environmental lapse rate and the lapse rate of an air parcel being emitted from the stack for each of these three plume types.

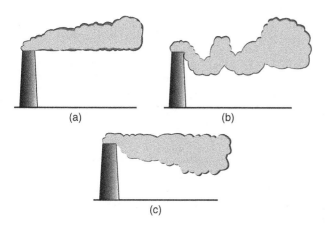

(a)

(b)

(c)

11.32 A fanning plume will occur when atmospheric conditions are generally (a) stable, (b) highly unstable, or (c) neutral?

11.33 For extremely unstable atmospheric stability conditions, what are values for the dispersion coefficient in the y and z directions 1,000 m downstream in an urban area from the point of the pollutant release? Estimate your values using two methods: (a) the correct Briggs equation, and (b) a graphical method that allows you to estimate the dispersion coefficients from established figures.

11.34 For neutral atmospheric stability conditions, what are values for the dispersion coefficient in the y and z directions 5 km downstream in a rural area from the point of the pollutant release? Estimate your values using two methods: (a) the correct Briggs equation and (b) a graphical method that allows you to estimate the dispersion coefficients from established figures.

11.35 For slightly unstable atmospheric conditions, what are the values the dispersion coefficients, σ_y and σ_z, 3 km downstream in a rural area? Provide your values for (a) the correct Briggs formula and (b) a graphical method that allows you to estimate the dispersion coefficients from established figures. (c) How do these values change if the atmosphere is moderately stable?

11.36 Wind is measured at 10 m above the ground surface at 2 m/s. Estimate the wind speed at an effective stack height of 40 m (a) in smooth rural terrain for unstable atmospheric conditions and (b) for rough urban terrain for the most stable atmospheric conditions.

11.37 What is the estimated wind speed at a 45 m effective stack height located in the country side? Assume there are very unstable atmospheric conditions near the stack and wind measurements made close to the stack location showed the wind speed was 3 m/s at a height of 5 m.

11.38 What is the ground-level concentration of a pollutant ($\mu g/m^3$) 250, 500, 750, and 1,000 m downwind for a stack release along the centerline of the plume? The pollutant is released from a 45 m tall stack at a rate of 8.5 g/s in an urban area. The plume rises an additional 10 m. The wind speed is 3 m/s, and there are slightly unstable atmospheric stability conditions.

11.39 NO_x is emitted from a 75 m high stack at a rate of 65 g/s. Calculate the ground-level concentration of NO_x 90 m from the centerline. The plume rises 20 m and the wind speed is 5 m/s. Assume $\sigma_y = 120$ m and $\sigma_z = 47$ m and there is reflection from the ground.

11.40 Estimate the downwind distance (x) at which the maximum pollutant concentration would occur at the ground surface for an emission that occurs from a stack that is 60 m high? Assume the pollutant is SO_2, it is emitted at a rate of 3,000 g/s, the wind speed is 4 m/s, the plume rises an additional 14 m after being emitted, and there are slightly unstable atmospheric conditions in this urban area. Calculate your value using two methods, equations and curves, provided in the chapter.

11.41 An air pollutant is released at a rate of 2 g/s from the top of a tall stack that is 110 m high. The plume initially rises upward an additional 10 m above the stack exit, after which it travels with a wind speed of 5 m/s. The atmosphere has moderately stable conditions in the open country. (a) What is the concentration of the pollutant ($\mu g/m^3$) in the very center of the plume 750 downwind of the stack and at the effective stack height? (b) What is the concentration of the pollutant ($\mu g/m^3$) at ground level 750 downwind of the stack? (c) At what distance downwind does the maximum ground-level concentration occur? (d) What is the concentration of the pollutant ($\mu g/m^3$) at this location you identify in part (c)?

11.42 An air pollutant is released at a rate of 0.72 g/s 4 m above ground level. The wind speed is 2 m/s. What is the maximum ground-level concentration 1 km downwind? Assume that $\sigma_y = 45$ m and $\sigma_z = 26$ m.

References

Alvord, K., 2000. *Divorce Your Car!* Gabriola Island: New Society.

Bruce, N., R. Perez-Padilla, and R. Albalak, 2000. Indoor air pollution in developing countries: a major environmental health challenge. *Bulletin of the World Health Organization, 78*: 1080–1092.

Centers for Disease Control and Prevention (CDC), 2011. CDC health disparities and inequalities report—United States, 2011. *MMWR 2011, 60*(Suppl): 84–86.

Crites, R., and G. Tchobanoglous, 1998. *Small and Decentralized Wastewater Management Systems*. Boston: McGraw-Hill.

Department of Transportation, Federal Highway Administration, 2013. Multi-Pollutant Emissions Benefits of Transportation Strategies-FHWA. http://www.fhwa.dot.gov/environment/air_quality/conformity/research/mpe_benefits/mpe03.cfm, accessed January 25, 2013.

Environmental Protection Agency (EPA), 2007. *Measuring the Air Quality and Transportation Impacts of Infill Development*, EPA 231-R-07-001.

Environmental Protection Agency (EPA), 2012a. *Office of Air Quality Planning and Standards, "Our Nation's Air: Status and Trends through 2010*, EPA-454/R-12-001, 32 pp.

Environmental Protection Agency (EPA), 2012b. President's Task Force on Environmental Health Risks and Safety Risks to Children: Coordinated Federal Action Plan to Reduce Racial and Ethnic Asthma Disparities, www.epa.gov/childrenstaskforce, May 2012. http://www.epa.gov/childrenstaskforce/federal_asthma_disparities_action_plan.pdf, accessed December 15, 2012.

Environmental Protection Agency (EPA), 2012c. Control Device Technology: A Quick Summary of Various Control Measures and Important Monitoring Characteristics. Presentation by Peter Westlin, EPA OAQPS. http://www.marama.org/calendar/events/presentations/2010_03Permit/Westlin_ControlDev_Mar10.pdf, accessed July 8, 2013.

Environmental Protection Agency (EPA), 2012d. APTI 413 Control of Particulate Matter Emissions. http://www.apti-learn.net, accessed July 8, 2013.

Environmental Protection Agency (EPA), 2012e. APTI 413 Control of Particulate Matter Emissions, Student Manual Chapter 7). EPA Air Pollution Training Institute (APTI). http://www.apti-learn.net,, accessed July 8, 2013.

Environmental Protection Agency (EPA), 2012f. APTI 415: Control of Gaseous Emissions, Chapter 3: Air Pollution Control Systems, 341 pp. EPA Air Pollution Training Institute (APTI). http://www.4cleanair.org/apti/415combined.pdf., accessed July 8, 2013.

Environmental Protection Agency (EPA), 2012g. APTI Virtual Classroom, Lesson 3: The Dynamic Structure of the Atmosphere, 34 pp. EPA Air Pollution Training Institute (APTI). http://yosemite.epa.gov/oaqps/EOGtrain.nsf/fabbfcfe2fc93dac85256afe00483cc4/59a3adbe2b6fb90885256b6d0064a13b/$FILE/Lesson%203.pdf, accessed November 26, 2012.

Environmental Protection Agency (EPA), 2012h. APTI Virtual Classroom, Lesson 4: Vertical Motion and Atmospheric Stability. 30 pp. EPA Air Pollution Training Institute (APTI). http://yosemite.epa.gov/oaqps/EOGtrain.nsf/fabbfcfe2fc93dac85256afe00483cc4/1c9d492b7ccef4fe85256b6d0064b4ee/$FILE/Lesson%204.pdf, accessed November 25, 2012.

Hanna, S. R., G. A. Brigss, and R. P. Hosker Jr., 1982. *Handbook of Atmospheric Diffusion, Office of Health and Environmental Research*. Washington, D.C.: Office of Energy Research, U.S. Department of Energy, 102 pp.

Hillier, V.A.W., 2001. *Fundamentals of Automotive Electronics*, 2nd ed. United Kingdom: Nelson Thornes org.

Jacobson, M. Z., 2012. *Air Pollution and Global Warming*, 2nd ed. Cambridge University Press, Cambridge, UK.

Mann, D. D., J. C. DeBruyn, and Q. Zhang, 2002. Design and evaluation of an open biofilter for treatment of odour from swine barns during sub-zero ambient temperatures. *Canadian Biosystems Engineering, 44*: 6.21–6.26.

Seinfeld, J. H., and S. N. Pandis, 2006. *Atmospheric Chemistry and Physics: From Air Pollution to Climate Change*. Hoboken: John Wiley & Sons, Inc.

Smith, K. R., 1993. Fuel combustion, air pollution exposure, and health: the situation in developing countries. *Annual Review of Energy and the Environment, 18*: 529–566.

Theodore, L., 2008. *Air Pollution Control Equipment*. Hoboken: John Wiley & Sons, Inc.

Turner, D. B., 1970. *Workbook of Atmospheric Dispersion Estimates*. Research Triangle Park: Environmental Protection Agency.

Answers to Selected Problems

Chapter 1. Sustainable Design, Engineering, and Innovation

1.7   a. You would choose an answer depending on your house, your lighting habits, your type of car and how you use it. You could ask questions like how many light bulbs you use in your home and for how long each day and how much you drive each year;
b. 99,000 lbs CO_2, 4,950 gallons;
c. 200 gallons, 400 lbs CO_2;
d. 76 gallons, 152 lbs CO_2

1.10   a. Plastic; b. No;
c. Plastic bags take less energy to produce

1.17 $E_{without\ solvent} = 1.4$ kg waste produced/kg of product;
$E_{with\ solvent} = 23.2$ kg waste produced/kg of product. These chemicals should be included if they are not recovered and recycled because they also contribute to the total waste of the process.

1.22   a. PV = $2,501; b. PV = $2,399

Chapter 2. Environmental Measurements

2.1   a. 0.41 mg/L; b. 0.21 mg/L

2.3   2.8 ppm_m

2.5   10.9 mg NH_3/L; 1.6 mg NO_2/L

2.7   5.28 mg S/L

2.9   490 mg Cd

2.11   a. 100 M Na^+, 100 M OH^-; b. 100 N OH^-, 100 N Na^+

2.13   a. i. 0.002 ppb_m, ii. 2 ppt_m, iii. 3.7×10^{-6} μM;
b. i. 0.002 ppm, ii. 2 ppb

2.15   a. $23 \frac{mgC_6H_6}{L}$; b. 2.3×10^4 ppb_m C_6H_6;
c. $3.3 \times 10^{-2} \frac{mole}{L}$ C_6H_6

2.17   0.74 μg/L

2.19   a. 0.5 mgO_2/L = 0.5 ppm_m, 8 mg O_2/L = 8 ppm_m;
b. 0.5 mg O_2/L = 1.6×10^{-5} moles O_2/L,
8 mg O_2/L = 2.5×10^{-4} moles O_2/L;

2.21 **a.** 8.9×10^{-2} ppm; **b.** $[CO] = 0.0000089\%$

2.23 0.7g

2.25 **a.** 384 $\mu g/m^3$; **b.** $6.0 \times 10^{-6} \frac{mole\ SO_2}{10^6\ m^3\ air}$

2.27 15 ppm_v

2.29 Yes (because 57 ppm_v > 50 ppm_v)

2.31 $4.1 \frac{mg\ P}{L}$

2.33 $2.2 \frac{mol\ Cl^-}{L}$

2.35 **a.** 100 ppb_m. 100 ppb > 60 ppb; therefore, the sample may pose a threat

2.37 5 ppm_m

2.39 **a.** Moderately hard; **b.** $34 \frac{mg\ Ca^{2+}}{L}$; **c.** $20.4 \frac{mg\ Mg^{2+}}{L}$

2.41 **a.** 3.10×10^9 metric tons, 6.84×10^{12} lbs; **b.** % of methane emissions = 18.6%; % of GHG emissions = 45.5%

2.43 FL brackish water = $1.32\ lb \frac{CO_{2e}}{m^3}$; CA brackish water = $0.73\ lb \frac{CO_{2e}}{m^3}$; FL sea water = $5.28\ lb \frac{CO_{2e}}{m^3}$; CA sea water = $2.92\ lb \frac{CO_{2e}}{m^3}$

2.45 $2.24 \times 10^5 \frac{lb\ CO_2}{yr}$

2.47 **a.** TSS = 170 mg/L; **b.** Because volatile solids consist primarily of organic matter, it can be concluded that approximately 70% (140/200) of the solids are organic.

Chapter 3. Chemistry

3.1 5.8 g

3.3 $\mu = 0.12$ mol/L; $\gamma(Mg^{2+}) = 0.35$; $\gamma(Fe^{3+}) = 0.096$; $\gamma(OH^-) = 0.77$; $\gamma(Cl^-) = \gamma(H^+) = 0.77$

3.5 **a.** Mn^{2+} is still being oxidzed and precipitate still forms; **b.** $[Mn^{2+}] = 2.5 \times 10^{-17}$ moles/L; **c.** $[Mn^{2+}] = 3.3 \times 10^{-4}$ moles/L

3.7 **a.** 3.2 g/m^3; **b.** 1,4-DCB, because a higher boiling point means a lower vapor pressure.

3.9 15.3 mg/L

3.11 $C_{eq-H20} = 96$ $\mu g/L$

3.13 **a.** pH = 3.8; **b.** pH = 12; **c.** pH = 4.4

3.15 **a.** 97%; **b.** 76%

3.17 An increase in temperature will increase the equilibrium constant.

3.19 **a.** $s = 3.0 \times 10^{-6}$ M; **b.** $s = 1.54 \times 10^{-6}$; **c.** Because activity coefficients for electrolytes are < 1, the K_{so} will increase, thus the solubility will increase.

3.21 **a.** $\Delta G^\circ = -78.4$ kJ/mole; **b.** No

3.23 94%

3.25 $\dfrac{d[C]}{dt} = \dfrac{d[A]}{3dt} = \left(\dfrac{2}{3}\right)\dfrac{d[B]}{dt} = -\dfrac{d[P]}{dt} = -\dfrac{d[Q]}{4dt}$

3.27 **a.** $\dfrac{d[S_2O_3^{2-}]}{dt} = \left(\dfrac{1}{2}\right)\dfrac{d[S_2O_3^{2-}]}{3dt} = -\left(\dfrac{1}{2}\right)\dfrac{d[SO_4^{2-}]}{dt} = -\dfrac{d[S_4O_6^{2-}]}{4dt}$;
b. 3rd order

3.29 999 min or 0.7 days

3.31 14 g N, 2g P, and 4g K are excreted every day in a Swedish person's urine.

3.33 **a.** First; **b.** Second; **c.** $t = 1.4 \times 10^{-3}$ s

3.36 20 years

3.37 3,780 Bq/L

3.39 $K_{25} = 0.37/day$; $K_5 = 0.055/day$

3.41 i. 37%; ii. 36%; iii. 16%; iv. 12%

Chapter 4. Physical Processes

4.1 $C_{out} = 11$ mg/L

4.3 **a.** 0.9 h; **b.** 1.25 pCi/L

4.5 **a.** 38 mg/s; **b.** 13 mg/m^3

4.7 1,691 lb/hr

4.9 $Q_5 = 3.2$ gpm

4.11 $\dot{m}_{outlet\ gas\ stream} = 243\frac{lb_m}{min}$

4.13 2 hours

4.15 $C_t = 64$ mg/L

4.17 $V = 720\ m^3$

4.19 **a.** $T_{out} = 47^\circ C$; **b.** $T_{out} = 53^\circ C$

4.20 **a.** 68.1 btu/$^\circ$F-day, 68.1 Btu/degree day;
b. 171 Btu/$^\circ$F-day, 171 Btu/degree day

4.25 6°C

4.27 a. $x = 1.5$ cm and 2.5 cm, $J = 10^{-8}$ mg/cm^2-s;
b. $m = 7.1 \times 10^{-8}$ mg/s at $x = 1.5$ and 2.5 cm;
d. The concentration profile in part (c) is changing due to the random motion of the molecules. The chemical is attempting to reach equilibrium through high concentration areas moving to areas with less concentration.

4.29 $J = 0.850 \frac{g}{m^2-s}$

4.31 $v_s = 2.4 \times 10^{-4}$ cm/s

Chapter 5. Biology

5.1 Viruses, Bacteria, macroinvertebrates, and protozoa.

5.3 a. Algae; b. Bacteria, c. Protozoa; d. Rotifers and microcrustaceans; e. Bacteria

5.5 $\mu_{max} = 1.0$/day

5.7 a. $X_{5 \text{ days}} = 213$ mgVSS/L;
b. $X_{20 \text{ days}} = 1.6 \times 10^7$ mgVSS/L

5.9 a. Exponential growth; b. 80 years for Boston to double its population from 12,000 to 24,000, doubles to 4 million in 50 years, the population of Boston will double to 8.8 million shortly (less than 30 years); c. The rate constant is 0.023/year.

5.11 a. 46 years

5.13 a. $X_t = 91,680$ mg/L; b. 8%

5.15 a. $P_{15 \text{ years at 1\% growth rate}} = 1,725$ people,
$P_{15 \text{ years at 2.5\% growth rate}} = 2,063$ people,
$P_{15 \text{ years at 5\% growth rate}} = 2,625$ people.
b. $P_{15 \text{ years at 1\% growth rate}} = 1,500$ people,
$P_{15 \text{ years at 2.5\% growth rate}} = 1,838$ people,
$P_{15 \text{ years at 5\% growth rate}} = 2,400$ people.

5.17 $Y = 0.2$ mg biomass/mg substrate

5.19 5.8 days

5.21 4.6 days

5.24 $\mu_{\text{toluene at 1 ppb}} = 0.0039$/day, $\mu_{\text{toluene at 1 ppm}} = 0.92$/day

5.25 a. $\mu_{\text{low strength WW}} = 0.65$/day;
b. $\mu_{\text{high strength WW}} = 0.81$/day; c. $\mu_{1000 \text{ mg VSS/L}} = 0.91$/day

5.27 Nitrogenous oxygen demand = 229 mg/L;
Theoretical carbonaceous oxygen demand = 129 mg/L;
Total theoretical oxygen demand = 358 mg/L

5.29 Nitrogenous oxygen demand = 228 mg/L;
Theoretical carbonaceous oxygen demand = 107 mg/L;
Total theoretical oxygen demand = 335 mg/L

5.31 a. 10 kg/day; b. 0.02 kgO$_2$/day

5.33 a. L_o = 38 mgO$_2$/L; b. k_{30} = 0.30/day, L_o = 26 mgO$_2$/L

5.35 L_o = 269 mgO$_2$/L

5.37 k_L = 0.20/day, L_o = 144 mgO$_2$/L

5.39 a. 0.30L_o; b. 0.16L_o; c. $y_{6\ days@T=20°C}$ = 0.51L_o,
$y_{6\ days@T=20°C}$ = 0.30L_o

5.40 minimum sample (mL) = 4 mL,
maximum sample (mL) = 15 mL

5.43 a. P_{max} = 1,369 mg/L; b. 0.04 molP/L; c. 3,838 mg/L

5.45 $[PCB116]_{large\ mouth\ bass}$ = 1.7 × 10^6 ng/kg;
$[PCB116]_{white\ perch}$ = 6.7 × 10^5 ng/kg

5.46 a. BAF = 8.9 × 10^6 L/kg; b. BAF = 4.5 × 10^6 L/kg;
c. BAF = 2.4 × 10^7 L/kg

Chapter 6 Environmental Risk

6.3 1. e; 2. and 3. a and d; 4. c; 5. b

6.5 Physical characteristics, toxicity, quantity generated,
history of the chemical

6.10 1. a.; 2. d.; 3. and 4. c and b

6.11 c. source reduction, d. source reduction, e. source
reduction, b. recycling, f. recycling, a. disposal,
g. disposal and source reduction

6.16 Hazard assessment, dose-response assessment,
exposure assessment, and risk characterization.

6.19 a. SF = 0.023 (mg/kg-day)$^{-1}$
b. The linear model of data is accurate

6.21 a. NOAEL is 70 ppm (3.5 mg/kg-day) and the RfD is
$3.5 × 10^{-2}$ mg/kg-day which is based off a 2-year rat
feeding study; c. 3.2 % effect of response/mg/kg/day;
d. 350 grams of grass; e. Spraying your lawn with
atrazine presents a serious risk to babies who eat grass.

6.23 $17.72 \frac{mg}{kg-day}$

6.25 $1.7 \times 10^{-3} \frac{mg}{kg-day}$

6.27 0.93 ppb

6.29 Exposure to arsenic in this case poses a noncarinogenic health risk.

6.31 There is an unsafe risk.

6.33 **a.** $[toxaphene_{fish}] = 0.32$ ppm (or 0.32 mg of toxaphene/kg of fish); **b.** Note that because the toxaphene bioconcentrates strongly in the food chain, an individual is exposed to a much greater mass of chemical by ingesting contaminated food versus drinking contaminated water. In this situation, the dose (and risk) increases with an increased in fish consumption. This amount could be much greater for segments of our population which consume more fish than the average person as well as for wildlife which depend upon fish for food. Also note that for chemicals that persist (do not degrade by natural mechanisms) and also bioconcentrate, what seems like a low water concentration may actually turn out to have a great environmental significance. This effect can be even more greatly magnified if the BCF is higher

6.35 **a.** CFR 40 Protection of Environment; **b.** CFR 49 Transportation; **c.** CFR 18 Conservation of Power and Water Resources; **d.** CFR 42 Public Health; **e.** CFR 23 Highways

6.37 **a.** Atrazine is *not* bioaccumulative but it is persistent in the environment.; **b.** Children are a high risk population since they have lower body mass than adults. A persistent and bioaccumulative chemical applied to grass where children play and may even ingest the grass, would be dangerous to human health.

Chapter 7 Water: Quantity and Quality

7.1 18 cm

7.3 **a.** $L_{SS} = 7980 \frac{lbs\ SS}{yr}$, $L_p = 12 \frac{lbs\ P}{yr}$, $L_N = 129 \frac{lbs\ P}{yr}$; **b.** Same as part a.; **c.** $L_{SS} = 67,800 \frac{lbs\ SS}{yr}$, $L_p = 100 \frac{lbs\ P}{yr}$, $L_N = 717 \frac{lbs\ P}{yr}$, **d.** % change$_{LSS}$ = 750% increase; % change$_{LP}$ = 733% increase; % change$_{LN}$ = 456% increase

7.5 $Q = 310 \frac{ft^3}{min} = 8.8 \times 10^6 \, cm^3/min$

7.6 $Q_{well} = 11,350 \frac{m^3}{day}$

7.12 Daily water demand = 13,200 gal/day,
Daily wastewater generated = $9,600 \frac{gal}{day}$

7.13 Daily water demand = 21,640 gal/day,
Daily wastewater generated = $16,860 \frac{gal}{day}$

7.15 Daily water demand = 9,000 gal/day;
Yearly water demand = 3,900,000 gallons/year

7.16 **a.** 2,743,200 gpd; **b.** Yes

7.19 Average flow rate = 1193 gpm

7.21 Storage = 1,257,000 gallons

7.22 **a.** Vmin = 4,500 gallons; **b.** D = 8 inch pipe

7.27 **a.** $DO_{sat} = 2.85 \times 10^{-4} \frac{mol \, O_2}{L}$; **b.** $9.1 \frac{mg \, O_2}{L}$; **c.** $9,100 \frac{\mu g}{L}$;
d. $9.1 \, ppm_m$ **e.** The answer to part b does not change.

7.29 $D = 6 \frac{mgO_2}{L}$

7.31 $DO_{act} = 6.2 \frac{mgO_2}{L}$

7.33 D = 4 mg/L

7.35 **a.** $DO_o = 4 \, mg/L$; **b.** $t_{crit} = 2.6$ days, $x_{crit} = 26$ km;
c. $D_t = 7.4 \, mg/L$

7.37 $CBOD_{5 \, after \, mixing} = 102 \, mg/L$

7.39 Maximum 5-day CBOD = 7.1 mg/L

7.41 **a.** critical time = 0.91 days; **b.** critical distance = 9.1 km

7.42 CBOD downstream = 49 mg/L

7.45 $4.1 \frac{mg \, P}{L}$

7.50 Fertilizers, sewage, atmospheric nitrogen, freshwater discharge, soil erosion.

7.51 **a.** 2,889 square miles; **b.** Smaller; **c.** Summer drought conditions

7.53 **a.** ~2,800 lb N/sq mi total nitrogen pollutant load at 5 percent of its wetlands; **b.** ~600 lb N/sq mi total nitrogen pollutant load at 15 percent of its wetlands

7.55 $v_a = 1.6 \frac{m}{day}$; t= 63 days

7.57 **a.** $R_{f\ trichoroethylene} = 30$, $R_{f\ hexachlorobenzene} = 3300$, $R_{f\ dichloromethane} = 3.9$; **b.** Dicholoromethane would be transported the furthest followed by trichloroethylene. Hexachlorobenzene would travel the least far.

Chapter 8: Water Treatment

8.2 The well with the nitrate concentration of 50 mg NO_3^-/L exceeds the 10 ppm standard.

8.4 9 mg/L alkalinity consumed

8.5 862,300 kg alum/yr

8.7 **a.** 5.3 kg/day; **b.** 235 $\frac{mgCaCO_3}{L}$ must be added

8.9 P = 5.7 kW

8.11 **a.** total hardness = 201 $\frac{mg\ CaCO_3}{L}$; **b.** TDS = 40 mg/L; **c.** x = 2 mg

8.13 Dosage of lime required for selective calcium removal = 197.5 mg/L as $CaCO_3$
Finished water hardness = 61.5 mg/L as $CaCO_3$

8.15 For CMFR: V = 4,286 m³ and t = 2.6 hours.
For PRF: t = 19 min and V = 528 m³

8.17 Vs = 8.7×10^{-3} m/h

8.19 Fraction of particles removed = 0.71

8.20 100% of the particles are removed

8.23 ES = 0.43 mm, UC = 2.74

8.25 Chick's law rate constant = −0.61, t = 15s

8.27 **a.** Ct product is a combined effect of disinfectant concentration and time of contact to achieve certain level of inactivation for a given microorganism.;
b. pH and temperature; **c.** Adenovirus and calicivirus;
d. C.parvum

8.28 **a.** [Cl] = 0.41 mg/L; **b.** chlorine demand = 0.21 mg/L

8.34 **a.** total surface area = 67,500 m²;
b. flux rate = 50 L/m² − h;
c. total number of fibers = 8,957,006, number of fibers per module = 3,980

8.35 **a.** total surface are = 19,440 m²;
b. flux rate = 75 $\frac{L}{m^2}$ − hr = 0.031 gpm/ft²

8.36 $q_{atrazine} = 23\,mg/g$; $q_{trichoroethylene} = 0.22\,mg/g$

8.39 FL brackish water $= 1.32\,lb\,\frac{CO_{2e}}{m^3}$;
CA brackish water $= 0.73\,lb\,\frac{CO_{2e}}{m^3}$;
FL sea water $= 5.28\,lb\,\frac{CO_{2e}}{m^3}$;
CA sea water $= 2.92\,lb\,\frac{CO_{2e}}{m^3}$

Chapter 9. Wastewater and Stormwater: Collection, Treatment, Resource Recovery

9.3 **a.** TSS $= 185\,mg/L$; **b.** Yes, the sample has appreciable organic matter at over half, 70%.

9.5 14 g N, 2g P, and 4g K are excreted every day in a Swedish person's urine.

9.7 $5.4\,\frac{mg\,P}{L}$

9.9 **a.** $144\,m^3$; **b.** w $= 4.5$ m, l $=10.7$ m; **c.** t $= 7.4$ min;
d. total air requirement $= 7.5\,m^3/min$;
e. grit volume $= 2.1\,\frac{m^3}{day}$

9.11 $A_{top} = 583\,m^2$; D $= 27.2$m; V$=1{,}749\,m^3$; $\theta = 1.2$ h

9.13 $Q_w X_w = 656\,kg/day$

9.15 $SRT_{min} = 20$ days

9.17 **a.** high; **b.** less; **c.** saturated; **d.** low; **e.** low; **f.** increased; **g.** increased

9.19 SVI $= 150$ mL/g

9.20 The sludge sample has poor settling characteristics.

9.24 V$=1{,}640\,m^3$; $\theta = 1.1$ h

9.25 **a.** F/M $= 0.33\,\frac{lbs\,BOD_s}{lbs\,MLVSS-days}$; **b.** Increase the solids retention time (SRT)

9.26 Pond area $= 14$ ha

9.29 **a.** 1,000 gallon tank, **b.** 1,200 gallon tank

9.30 The total depth to be dug is 6.8m

9.33 1 pig

9.35 **a.** 2,743,200 gpd; **b.** Yes

9.36 Future average flow rate $= 2.3$ MGD;
future maximum flow rate $= 5.3$ MGD

9.39 0.041 ft

9.41 Each will have an area of 150 ft^2 with a depth of 6 in.

9.42 V = 6,050 ft^3

9.45 38%–92%

Chapter 10. Solid Waste Management

10.1 **a.** Below; **b.** V = 1370 m^3; **c.** V = 869 m^3

10.2 V = 1430 m^3

10.6 Low moisture content = 17.3%;
typical moisture content = 43.5%

10.7 **a.** Dry paper mass for city 1 = 0.54 kg/person-day; dry
paper mass for city 2 = 0.69 kg/person-day;
b. 32.8%; fraction of dry weight = 31.1%

10.9 4,295 L air

10.11 Time to 90% decay for k for slowly biodegrading = 33
years; time to 90% for k for rapidly biodegrading = 10
years

10.13 t = 15 years

10.15 **a.** $V_{CH_4\ 1970} = 7.67 \times 10^9\ \frac{m^3 CH_4}{yr}$,
$V_{CH_4\ 2010} = 1.11 \times 10^{10}\ \frac{m^3 CH_4}{yr}$;
b. E_{1970} = 3,243 MW, E_{2010} = 4,693 MW

10.16 64%

10.18 Option 1, because the same 0.2 layer will be used to
cover 4 m instead of 3 m and that gives 25% more
refuse to cover ratio.

10.20 Cost of the smaller station = $740,740

10.25 0.30 kg

10.26 25%

10.28 **a.** total cost = 13,030,000 euros;
b. total cost = 79,880,000 euros

Chapter 11. Air Resources Engineering

11.1 **a.** 0.0899 ppm$_v$; **b.** 89.9 ppb$_v$; **c.** 8.99%

11.3 190,000 particles/cm^3

11.5   **a.** $367 \frac{\mu g}{m^3}$; **b.** $\frac{0.14 \text{ moles } SO_2}{10^6 \text{ moles air}}$

11.7   No, motorists are not at greater risk.

11.8   **a.** $v_s = 0.00087 \frac{cm}{s}$; **b.** $t = 70$ hr

11.12   **a.** $R_{pop} = \frac{2.54}{year}$; $R_{vehicle} = \frac{3.63}{year}$; $R_{drivers} = \frac{2.53}{year}$; **b.** Similar

11.13   **a.** $1.20 \times 10^6 \frac{miles}{yr}$; **b.** 275,000 grams hydrocarbon reduced, 479,000 grams NO_x reduced

11.15   total fabric area $= 4,290$ ft^2; $n_{\text{cylindrical bags}} = 182$ bags

11.17   **a.** $L = 13$ m; **b.** $L = 20$ m

11.19   **a.** EBRT $= 5.2$ s; **b.** Agriculture

11.21   **a.** $E = 5,950$ kg SO_2/day; **b.** 179 kg $\frac{SO_2}{day}$

11.23 $E_{acetone} = 1.05$ *tons*

11.25   **a.** $m_{\text{total VOC}} = 0.43 \frac{kg}{L} VOC$;
b. total mass of VOCs $= 9,030 \frac{kg}{yr}$;
c. total mass of VOCs released to the atmostphere $= 3,323 \frac{kg}{yr}$

11.27   **a.** Emissions would still go up; **b.** Examples include park-and-ride facilities, regional rideshare programs, new/expanded bus/rail service, bicycle and pedestrian projects/programs, vanpool programs

11.29   (a) Slower than the dry adiabatic lapse rate

11.32   (a) Stable

11.33   **a.** $\sigma_y = 270$ m, $\sigma_z = 170$ m; **b.** $\sigma_y \approx 220$ m, $\sigma_z \approx 450$ m

11.35   **a.** $\sigma_y = 289$ m, $\sigma_z = 190$ m; **b.** $\sigma_y \approx 220$ m, $\sigma_z \approx 200$ m

11.37 $u = 3.5$ m/s

11.39 $C(x,90,0) = 72$ μg/m^3 (when $x = 500$)

11.41   **a.** $C(750,0,120) = 1,120$ μg/m^3;
b. $C(750,0,0) \sim 0$ μg/m^3;
c. $x = 632$ m;
d. $C(632,0,0) = 23.0$ μg/m^3

Index

Water (*Continued*)
  reuse (*See* Water, reclamation)
  reuse guidelines, 499
  root mean square (RMS) velocity gradient and, 396
  satellite reclamation, 315, 482
  scarcity, 20, 247, 314, 316–317, 377, 484
  small systems, 433
  solids measurement, 58–60
  stormwater, 323, 442, 445, 500–514
  stress, 150, 316–317
  system layout, 331–333
  taste, 377, 379–383, 383, 422, 432
  total coliforms, 385–386
  treatment of wastewater (and resource recovery), 440–500
    aeration, 480–482
    attached growth processes, 472
    membrane bioreactors, 469–471
    natural treatment systems, 489–496
  turbidity, 175, 313, 337, 352, 377, 379–380, 392, 395, 410, 499
  unmetered flow, 326–327
  variation in flows, 321, 323
  wash out, 464, 468
  waste activated sludge, 457–458, 470, 483
Water Pollution Control Act (Clean Water Act), 2, 444
Watershed, 301–307, 337, 350, 352
Weight of evidence (risk), 269–270
Wet adiabatic lapse rate, 637–639

Wetlands, 159, 185, 208, 237, 299, 302, 305, 349–354, 494–496
  constructed, 185, 494–496
  free water surface (FWS) wetland, 494–496
  impact on nitrogen loading, 352
  restoration, 352
  subsurface flow wetland (SSF)
  types, 352
Wet weather flows, 500–505
Wet well, 323, 336–337
Wilderness, 2, 229, 237, 318
Wildlife corridor, 354
Wind direction, 636, 643, 649
Windrow (composting), 544–546
Wind speed, 636–637, 640, 645, 647
Winter stratification, 345–346
World Health Organization, 266, 378–380, 383, 387, 590, 594–595
World Summit on Sustainable Development, 7, 8

Y
Yield coefficient (in biology), 196–201, 461
Yield coefficient (in watershed runoff), 306

Z
Zero-order decay, 103–104, 121
Zero-order kinetics, 101–104, 106, 195
Zero-order reactions, 101–104, 106, 121, 195
Zero-waste, 29, 264, 525, 570